ENGINE MANAGEMENT SYSTEMS Manual

Martynn Randall

The Haynes Manual for maintenance, fault finding and repair

Haynes Publishing
Sparkford, Nr Yeovil, Somerset BA22 7JJ England

Haynes North America, Inc
861 Lawrence Drive, Newbury Park, California 91320 USA

Haynes Publishing Nordiska AB
Box 1504, 751 45 Uppsala, Sweden

Acknowledgements
We would like to thank all of those at Sparkford and elsewhere who helped us in the production of this manual, including Robert Bosch Ltd for permission to reproduce technical information.

We take great pride in the accuracy of information given in this manual, but vehicle manufacturers make alterations and design changes during the production run of a particular vehicle of which they do not inform us. No liability can be accepted by the authors or publishers for loss, damage or injury caused by any errors in, or omissions from, the information given.

A book in the Haynes Service and Repair Manual Series

Printed in the USA

ISBN 978 0 85733 746 7

British Library Cataloguing in Publication Data
A catalogue record for this book is available from the British Library

(4199-384)

Printed using 33-lb Resolute Book 65 4.0 from Resolute Forest Products Calhoun, TN mill. Resolute is a member of World Wildlife Fund's Climate Savers programme committed to significantly reducing GHG emissions. This paper uses 50% less wood fibre than traditional offset. The Calhoun Mill is certified to the following sustainable forest management and chain of custody standards: SFI, PEFC and FSC Controlled Wood.

GENERAL INFORMATION

MODEL SPECIFIC COMPONENT RENEWAL PROCEDURES

Introduction

This book is devoted to unravelling the mysteries of the Engine Management System (EMS) on the modern automobile. The book first describes the history and function of fuel injection and the EMS and then goes on to give a technical overview on how modern systems work. Other chapters describe general test routines and system descriptions, and component renewal procedures for 23 of the most popular petrol and diesel models. Even if the reader has no intention of actually attempting to investigate faults on his or her own vehicle, the book provides valuable insight into the electronic management of the modern car engine.

On the other hand, if you relish the task of electronic fault diagnosis, then this book will provide you with much of the background knowledge necessary to test the circuits and components on your engine. Generally we describe how to work using simple tools and equipment available from most good auto retailers. We will also mention where the use of specialised equipment is really necessary, and describe some of the common routines used by the professional garage trade.

This book does not go too deeply into electrical or electronic theory, but there are many other excellent publications available for that purpose. See 'Automotive Electrical & Electronic Systems' by Tony Tranter and Martynn Randall.

Many of our tests and routines may not be as specifically recommended by the Vehicle Manufacturer (VM). One of the reasons is because the VM may describe procedures for use with his own special test equipment that is not readily available to the aftermarket user. Another reason is that all too often the VM places too much reliance upon the ohmmeter as the primary testing tool.

The accepted test method used by all good electronic vehicle specialists is to place more weight on voltage readings, which tend to give more reliable results. We will discuss this issue at greater length in the appropriate sections later on. In almost all instances, our own tests follow well-defined test methods taught in independent training schools and used by many modern vehicle technical specialists.

The routines and test methods that we describe are perfectly safe to carry out on electronic systems, so long as certain simple rules are observed. These rules are actually the observation of good electrical practice. Be aware that damage resulting in the replacement of a very expensive Electronic Control Module (ECM or ECU) may be the result of not following the rules. Please refer to the Warnings section in Reference; these warnings will be repeated in the various chapters where necessary.

We have described the operation and test procedures of most of the modern EMS's fitted to a whole host of popular vehicles. The book then provides detailed descriptions of the renewal procedures for engine management system components, fitted to the 23 most popular petrol and diesel models sold over the last ten years..

Although we have tried to provide generic test procedures for many of the components fitted to modern EMSs, it is essential that a specific wiring diagram for the vehicle under test is obtained where serious tracing of wiring faults is required.

Throughout Europe, the USA and the Far East the various VMs tend to use their own particular terms to describe a particular component. Of course, all of these terms tend to be different, and the problem is exacerbated by translation from the various languages. This often leads to confusion when several terms are used to describe the same component. There have been attempts to bring all the VMs into line with a common standard for all, and one does now exist (J1930). It seems unlikely, however, that all VMs will adopt this particular standard, and we are unsure that the terms used are that meaningful anyway. So, the terms used in this book will tend to follow terms that are commonly used in the UK. To avoid confusion, we will tend to use similar terms over the whole range of EMSs, and alternatives will be listed in the Glossary.

Precautions

Petrol models

Warning: Many of the procedures in this Book require the disconnection of fuel lines and connections, which may result in some fuel spillage. Before carrying out any operation on the fuel system, refer to the precautions given in 'Safety first!', and follow then implicitly. Petrol is a highly dangerous and volatile liquid, and the precautions necessary when handling it cannot be overstressed.

Warning: Residual pressure will remain in the fuel lines long after the vehicle was last used. When disconnecting any fuel line, first depressurise the fuel system as described within the model specific Chapters.

Diesel models

Warning: It is necessary to take certain precautions when working on diesel fuel system components, particularly the fuel injectors. Before carrying out any operations on the fuel system, refer to the precautions given in 'Safety first!' at the beginning of this manual, and to any additional warning notes at the start of the relevant Sections. In particular, note that the injectors on direct injection diesel engines operate at extremely high pressures (approximately 1100 bar/ 16 000 psi), making the injector spray extremely hazardous.

Warning: Due to the high pressures generated, the manufacturing tolerances within the diesel fuel system components are so fine that any dirt or foreign body entering the system could cause catastrophic damage. Before opening any part of the fuel supply, injection or return system ensure the area around the component is scrupulously screen. Use a vacuum cleaner to remove any dirt or dust from the area surrounding the component(s) to be worked on.

Wait at least a minute after the engine stops before starting any work on the diesel injection system. Try to ensure the engine and fuel system has cooled sufficiently before commencing any work.

All models

Electronic control units are very sensitive components, and certain precautions must be taken to avoid damage to these units as follows.

When carrying out welding operations on the vehicle using electric welding equipment, the battery and alternator should be disconnected.

Although the underbonnet-mounted modules will tolerate normal underbonnet conditions, they can be adversely affected by excess heat or moisture. If using welding equipment or pressure-washing equipment in the vicinity of an electronic module, take care not to direct heat, or jets of water or steam, at the module. If this cannot be avoided, remove the module from the vehicle, and protect its wiring plug with a plastic bag.

Before disconnecting any wiring, or removing components, always ensure that the ignition is switched off.

After working on fuel injection/engine management system components, ensure that all wiring is correctly reconnected before reconnecting the battery or switching on the ignition.

Working on your car can be dangerous. This page shows just some of the potential risks and hazards, with the aim of creating a safety-conscious attitude.

General hazards

Scalding

• Don't remove the radiator or expansion tank cap while the engine is hot.

• Engine oil, automatic transmission fluid or power steering fluid may also be dangerously hot if the engine has recently been running.

Burning

• Beware of burns from the exhaust system and from any part of the engine. Brake discs and drums can also be extremely hot immediately after use.

Crushing

• When working under or near a raised vehicle, always supplement the jack with axle stands, or use drive-on ramps. ***Never venture under a car which is only supported by a jack.***

• Take care if loosening or tightening high-torque nuts when the vehicle is on stands. Initial loosening and final tightening should be done with the wheels on the ground.

Fire

• Fuel is highly flammable; fuel vapour is explosive.

• Don't let fuel spill onto a hot engine.

• Do not smoke or allow naked lights (including pilot lights) anywhere near a vehicle being worked on. Also beware of creating sparks (electrically or by use of tools).

• Fuel vapour is heavier than air, so don't work on the fuel system with the vehicle over an inspection pit.

• Another cause of fire is an electrical overload or short-circuit. Take care when repairing or modifying the vehicle wiring.

• Keep a fire extinguisher handy, of a type suitable for use on fuel and electrical fires.

Electric shock

• Ignition HT voltage can be dangerous, especially to people with heart problems or a pacemaker. Don't work on or near the ignition system with the engine running or the ignition switched on.

• Mains voltage is also dangerous. Make sure that any mains-operated equipment is correctly earthed. Mains power points should be protected by a residual current device (RCD) circuit breaker.

Fume or gas intoxication

• Exhaust fumes are poisonous; they often contain carbon monoxide, which is rapidly fatal if inhaled. Never run the engine in a confined space such as a garage with the doors shut.

• Fuel vapour is also poisonous, as are the vapours from some cleaning solvents and paint thinners.

Poisonous or irritant substances

• Avoid skin contact with battery acid and with any fuel, fluid or lubricant, especially antifreeze, brake hydraulic fluid and Diesel fuel. Don't syphon them by mouth. If such a substance is swallowed or gets into the eyes, seek medical advice.

• Prolonged contact with used engine oil can cause skin cancer. Wear gloves or use a barrier cream if necessary. Change out of oil-soaked clothes and do not keep oily rags in your pocket.

• Air conditioning refrigerant forms a poisonous gas if exposed to a naked flame (including a cigarette). It can also cause skin burns on contact.

Asbestos

• Asbestos dust can cause cancer if inhaled or swallowed. Asbestos may be found in gaskets and in brake and clutch linings. When dealing with such components it is safest to assume that they contain asbestos.

Special hazards

Hydrofluoric acid

• This extremely corrosive acid is formed when certain types of synthetic rubber, found in some O-rings, oil seals, fuel hoses etc, are exposed to temperatures above 400°C. The rubber changes into a charred or sticky substance containing the acid. *Once formed, the acid remains dangerous for years. If it gets onto the skin, it may be necessary to amputate the limb concerned.*

• When dealing with a vehicle which has suffered a fire, or with components salvaged from such a vehicle, wear protective gloves and discard them after use.

The battery

• Batteries contain sulphuric acid, which attacks clothing, eyes and skin. Take care when topping-up or carrying the battery.

• The hydrogen gas given off by the battery is highly explosive. Never cause a spark or allow a naked light nearby. Be careful when connecting and disconnecting battery chargers or jump leads.

Air bags

• Air bags can cause injury if they go off accidentally. Take care when removing the steering wheel and/or facia. Special storage instructions may apply.

Diesel injection equipment

• Diesel injection pumps supply fuel at very high pressure. Take care when working on the fuel injectors and fuel pipes.

Warning: Never expose the hands, face or any other part of the body to injector spray; the fuel can penetrate the skin with potentially fatal results.

Remember...

DO

• Do use eye protection when using power tools, and when working under the vehicle.

• Do wear gloves or use barrier cream to protect your hands when necessary.

• Do get someone to check periodically that all is well when working alone on the vehicle.

• Do keep loose clothing and long hair well out of the way of moving mechanical parts.

• Do remove rings, wristwatch etc, before working on the vehicle – especially the electrical system.

• Do ensure that any lifting or jacking equipment has a safe working load rating adequate for the job.

DON'T

• Don't attempt to lift a heavy component which may be beyond your capability – get assistance.

• Don't rush to finish a job, or take unverified short cuts.

• Don't use ill-fitting tools which may slip and cause injury.

• Don't leave tools or parts lying around where someone can trip over them. Mop up oil and fuel spills at once.

• Don't allow children or pets to play in or near a vehicle being worked on.

Vehicles covered

Index of vehicles covered

Model	Engines	System
BMW		
3-Series E46 1998 to 2003		
Petrol	1796 cc N42 (4 cylinder Valvetronic)	DME ME9
	1895 cc M43TU	DME BMS46
	1995 cc N42 (4 cylinder Valvetronic)	DME ME9
	2171 cc M52TU	DME MS42
	2494 cc M52TU	DME MS42
	2494 cc M54	DME MS43
	2793 cc M52TU	DME MS42
	2979 cc M54	DME MS43
Citroën		
Saxo 1996 to 2001		
Petrol	954 cc TU9 CDZ	Bosch MA3.1
	1124 cc TU1M+ HDZ	Bosch MA3.1
	1360 cc TU3JP KFX	M Marelli 1AP
	1587 cc TU5JP NFZ	Bosch MP5.1/5.2
	1587 cc TU5JP4 NFX	M Marelli 1AP41
Xsara Picasso 2000 to 2002		
Petrol	1587 cc TU5JP NFZ	Bosch MP7.2
	1587 cc TU5JP NFV	Bosch ME7.4.4
Diesel	1997 cc DW10TD RHY	Bosch EDC15
Fiat		
Punto 1999 to 2003		
Petrol	1242 cc 188 A4 000	Weber Marelli 59F
	1242 cc 16v 188 A5 000	Bosch ME7.3H4
Ford		
Fiesta 1995 to 2001		
Petrol	1242 cc Zetec-SE	Ford EEC-V
	1298 cc Endura-E	Ford EEC-V
	1388 cc Zetec-SE	Ford EEC-V
	1596 cc Zetec-SE	Ford EEC-V
Diesel	1753 cc Endura-DI	Ford EEC-V/PCU
Focus 1998 to 2001		
Petrol	1388 cc Zetec-SE	Ford EEC-V
	1596 cc Zetec-SE	Ford EEC-V
	1796 cc Zetec-E	Ford EEC-V
	1989 cc Zetec-E	Ford EEC-V
Diesel	1753 cc Endura-DI	Ford EEC-V/PCU
Focus 2001 to 2004		
Petrol	1388 cc Zetec-SE	Ford EEC-V
	1596 cc Zetec-SE	Ford EEC-V
	1796 cc Zetec-E	Ford EEC-V
	1989 cc Zetec-E	Ford EEC-V
Diesel	1753 cc Endura-TDDI	Ford EEC-V/PCU
	1753 cc DuraTorq-TDCi	Ford EEC-V/IDM
Galaxy 1995 to 2000		
Petrol	1998 cc NSE DOHC 8V	Ford EEC-V
	2295 cc Y5B DOHC 16V	Ford EEC-V
Diesel	1896 cc 1Z, AHU, AUG, AFN	EDC

Model	Engines	System
Ford (continued)		
Ka 1996 to 2002		
Petrol	1298 cc Endura-E	Ford EEC-V
Mondeo 1993 to 2000		
Petrol	1597 cc Zeta/Zetec/Zetec-E	Ford EEC-V
	1796 cc Zeta/Zetec/Zetec-E	Ford EEC-IV/V
	1988 cc Zeta/Zetec/Zetec-E	Ford EEC-IV/V
	2544 cc Duratec V6	Ford EEC-V
Mondeo 2000 to 2003		
Petrol	1798 cc Duratec-HE	Black Oak
	1999 cc Duratec-HE	Black Oak
	2544 cc Duratec V6	Black Oak
Diesel	1998 cc DuraTorq-Di	Ford EEC-V
	1998 cc DuraTorq-DCi	Black Oak EDC
Land Rover		
Freelander 1997 to 2002		
Petrol	1796 cc K16	MEMS 1.9 / 3
Diesel	1951 cc TD4	DDE 4.0
	1994 cc L-Series	EDC
Peugeot		
206 1998 to 2001		
Petrol	1124 cc TU1JP HFZ/HFY	M Marelli 1AP
	1360 cc TU3JP KFX	Bosch MP7.3 or M Marelli 1AP
	1587 cc TU5JP NFZ	Bosch MP7.2
Diesel	1868 cc DW8 WJZ/Y	Lucas
	1997 cc DW10TD RHY	EDC 15C2
307 2001 to 2004		
Petrol	1360 cc TU3JP KFW	Sagem S2000
	1587 cc TU5JP4 NFW	Bosch ME7.4.4
	1997 cc EW10J4 RFN	M Marelli 4.8P
Diesel	1398 cc DV4TD 8HZ	EDC 16
	1997 cc DW10ATED/TD RHZ/RHY	EDC 15C2
406 1999 to 2002		
Petrol	1761 cc XU7JP4	Bosch MP5.1.1, Bosch MP7.3 or Sagem S2000
	1749 cc EW7J4	Sagem S2000
	1997 cc EW10J4	M Marelli 4.8P
Diesel	1997 cc DW10TD/ATED	EDC 15C2
	2197 cc DW12TED4	EDC 15C2
Renault		
Clio 1998 to 2001		
Petrol	1149 cc D7F	Sagem SEFi, M Marelli SEFi or Siemens-Sirius SEFi
	1390 cc 8V E7J 764	Siemens-Fenix 5
	1390 cc 16V K4M 712, K4J 713	Siemens-Fenix 5
	1598 cc 8V K7M 744	Siemens-Fenix 5
	1598 cc 16V K4M 748	Siemens-Sirius
Diesel	1870 cc F8Q 630/632/662	EPIC
Mégane & Scénic 1999 to 2002		
Petrol	1390 cc E7J 764	Siemens-Fenix 3
	1390 cc 16V K4J 750	Siemens-Sirius 32
	1598 cc K4M 700/701	Siemens-Sirius 32
	1998 cc F4R 740/741	Siemens-Sirius 32

Vehicles covered

Model	Engines	System
Rover/MG		
25 & MGZR		
Petrol	1120 cc K16	MEMS 3
	1396 cc K16	MEMS 3
	1589 cc K16	MEMS 3
	1796 cc K16	MEMS 3
Diesel	1994 cc 20T2N	EDC
Vauxhall/Opel		
Corsa 2000 to 2003		
Petrol	973 cc Z10XE	Bosch ME1.5.5
	1199 cc Z12XE	Bosch ME1.5.5
	1389 cc Z14XE	Multec (S)
Diesel	1686 cc Y17DT/DTL	ECDV5
Astra & Zafira 1998 to 2004		
Petrol	1598 cc 8V X16SZR	Multec
	1389 cc X14XE, Z14XE	Multec (S)
	1598 cc 16V X16XEL, Z16XE	Multec (S)
	1796 cc X18XE1, Z18XE, Z18XEL	Simtec 70/71
	1998 cc X20XEV	Simtec 70
	2198 cc Z22SE	GMPT-E15
Diesel	1686 cc Y17DT	ECDV5
	1700 cc X17DTL	EDC MSA15M
	1994 cc X20DTL, Y20DTL/DTH	EDC 15M
Vectra 1999 to 2002		
Petrol	1598 cc 8V X16SZR	Multec
	1598 cc 16V X16XEL, Z16XE	Multec(S)
	1796 cc X18XE1, Z18XE	Simtec 70/71
	1799 cc X18XE	Simtec 56.5
	1998 cc X20XEV	Simtec 56.5
	2198 cc Z22SE	GMPT-E15
Diesel	1994 cc X20DTL/DTH, Y20DTH	EDC 15M
	2171 cc Y22DTR	EDC 15M
VW		
Golf & Bora 1998 to 2000		
Petrol	1390 cc AHW, AKQ	M Marelli 4AV
	1390 cc APE, AXP	Bosch ME7.5.10
	1595 cc 8V AEH, AKL, APF	Simos 2 or 3.3
	1597 cc 16V ATN, AUS	M Marelli 4LV
	1781 cc AGN	Bosch M3.8.5
	1781 cc AQA, ARZ	Bosch ME7.5
	1781 cc AGU	Bosch M3.8.3
	1984 cc APK	Bosch M5.8.2
	1984 cc AQY	Bosch M5.9.2
Diesel	1896 cc	EDC
Polo 1994 to 1999		
Petrol	999 cc AER, ALL	Bosch Motronic 9.0
	1043 cc AEV	MonoMotronic 1.2.3
	1296 cc ADX	MonoMotronic 1.3
	1390 cc AEX, AKV, ANX, APQ	Bosch Motronic 9.0
	1598 cc AEA	MonoMotronic
	1598 cc AEE	M Marelli 1AV
Diesel	1716 cc AHG, AKU	EDC
	1896 cc AGD	EDC

A brief history of Fuel Injection and the Engine Management System

Mechanical fuel injection

Fuel injection is almost as old as the oldest motor car. As long ago as 1881, when most of the pioneer automotive engineers were engaged in carburettor design, a Frenchman named Eteve was issued with a patent for a compressed air metering device. A German patent was issued to J. Spiel in 1883 for devising a method of injecting fuel into a flame-filled chamber linked to the engine cylinders. About the same time in Kent, England, Edward Butler devised an engine equipped with an injection system that forced fuel under pressure through a hollow-stem inlet valve.

The first production engine equipped with fuel injection was manufactured by the Charter Gas Engine Company of Sterling, Illinois, USA in 1887 *(see illustration 1.1)*. This engine was actually a stationary engine, and fuel was gravity-fed from a tank into the injector body via the throttle valve. Deutz in Europe developed its own stationary engine with low-pressure-fed kerosene into the inlet port. Three hundred models of the single-cylinder engine were built between 1898 and 1901.

Orville Wright was not blind to the attractions of fuel injection, and his 1903 Flier was equipped with fuel injection *(see illustration 1.2)*. Fuel-injected aircraft suffered less from carburettor icing and carburettor fires, and this meant less chance of them falling out of the sky. For this reason, fuel injection evolved more quickly in aircraft engines, although progress was still painfully slow. A high-pressure plunger pump and the calibrated injector principle was introduced by Leon Levavassuer in 1906, while 1912 saw the first involvement of Bosch, who adapted a two-stroke outboard engine utilising a lubrication pump as the injector pressure pump.

The First World War and the years leading up to 1920 saw considerable evolution of the carburettor, and reliability was achieved in aircraft engines to the extent that fuel injection was placed on the back burner, where it was to languish for a good number of years. During the 1930's, the German DVL (Deutsche

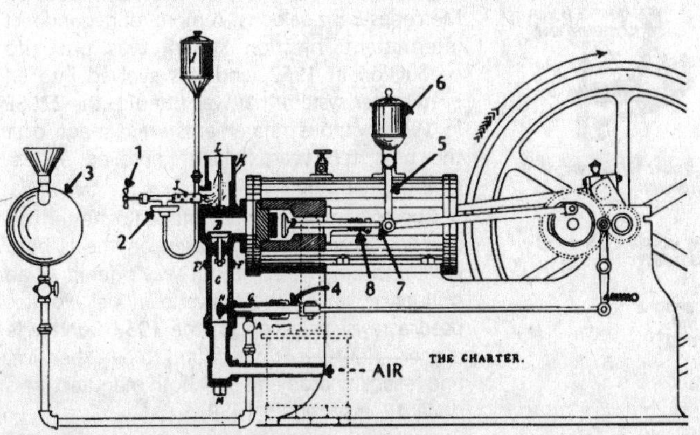

1.1 *As early as 1887, the Charter engine combined the basic elements needed for low-pressure port-type fuel injection*

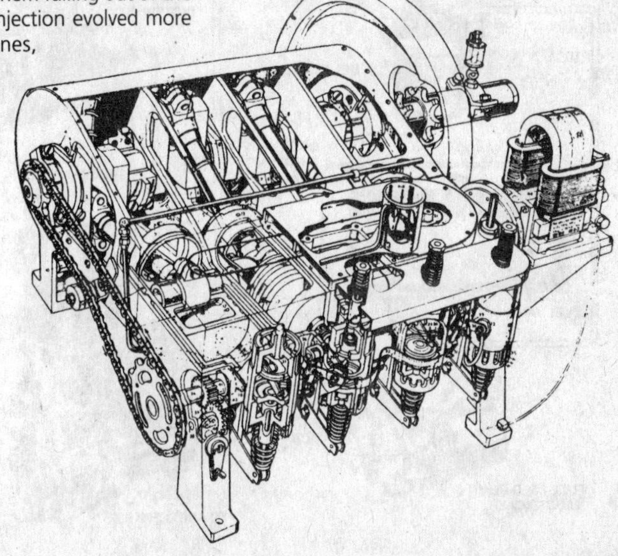

1.2 *The Wright brothers used a gear-type pump to force fuel into the intake manifold of their 28 hp four-cylinder engine in 1903*

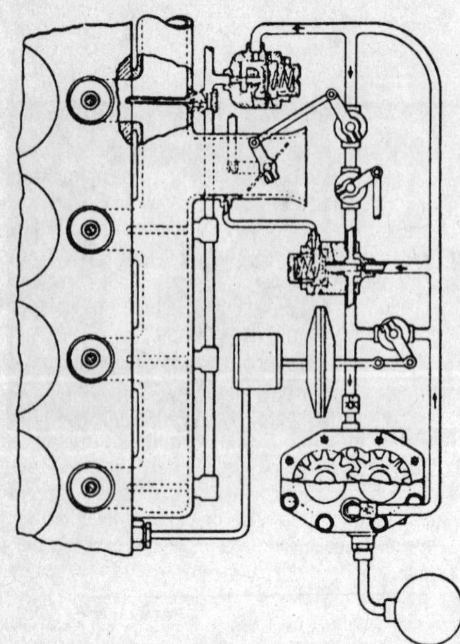

1.3 ***Constant-flow port injection was patented by Ed Winfield in 1934. A gear-type pressure pump sent fuel into a rail, where a pressure regulator controlled fuel flow in accordance with throttle position and manifold vacuum***

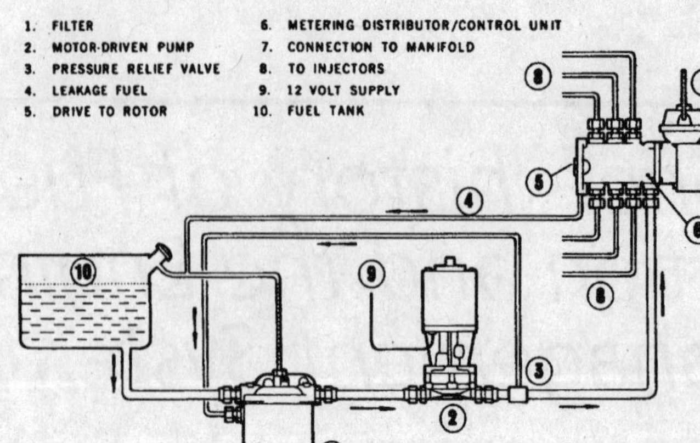

1.4 Lucas fuel injection used a motor-driven fuel pump to draw fuel from the tank. The metering distributor was driven at half engine speed, and delivered metered and timed fuel charges to each injector in sequence

Versuchsanstalt für Luftfahrt – otherwise known as the German Aviation Test Centre), Bosch, BMW and Mercedes-Benz all began development programmes of aviation-based high-pressure fuel injection systems. When Mercedes-Benz introduced the 1200 HP DB-601 V-12 engine to the world in 1937, fuel injection became the premier fuel system in aircraft engines. However, the design was inevitably akin to the diesel system, with direct injection into the combustion chamber. SU carburettors of Birmingham, England developed a fuel injection system for Rolls-Royce Merlin aircraft engines during the latter stages of World War II. This system was later used on the American Patton tank during the Korean war of 1950 to 1953.

Fuel injection in automobiles between the 1930's and 50's mainly appeared on racing engines; one early pioneer was Ed Winfield, who obtained a patent for a port-type injection system *(see illustration 1.3)*. A similar system appeared on the splendid Indy Offenhausers of the 1950's. The fuel system fitted to the Offenhauser engines was designed by Stuart Hilborn, and this system was also based on the low-pressure port principle. Fuel was continuously injected at constant pressure through a throttle body at each inlet port, via a spray nozzle inside the port area. The Hilborn system was used into the late 1960's, and European racing car manufacturers soon began to follow suit.

Lucas developed a system for the 1956 D-Type Jaguar, which then won the Le-Mans 24 hour race *(see illustration 1.4)*. The production version that followed was a commercial disaster, however, due to the very high cost. The only manufacturer ever to fit it in production was Maserati, who fitted it to the 3500 Gti in 1961. Further development of the Lucas system saw it introduced to Formula one racing engines, where it was used alongside Kugelfischer fuel injection in the 1960's and 1970's. A production version was fitted to some Triumph sports cars, where it gained a reputation for unreliability.

Meanwhile, Mercedes-Benz used their vast experience in fuel-injected aircraft engines to develop a fuel-injected version of the 300SL sports car that went into production in 1954 *(see illustration 1.5)*. This system was further refined when fitted to the W-196 Formula one racing car and racing versions of the 300SL. However, these direct injection systems owed much to diesel technology, and were considered too noisy in operation for the sophisticated Mercedes-Benz saloons. A more refined indirect intermittent injection system was put into production in 1957, and this evolved into an even better system that was fitted to the 220SE in 1958. Various refinements were made over the next ten years which improved system operation and horsepower.

Chevrolet experimented with a Rochester fuel injection system, overtly based on the Hilborn system, during 1957, and it was offered as an optional extra on the Corvette model. Pontiac used a revised version on the 1957 Bonneville models. However, it was found to be expensive, and reliability problems saw both manufacturers discard the system in 1959.

In the late 1960's Bendix developed a system that was soon challenging the domination of the Hilborn system for Indy racing cars. The

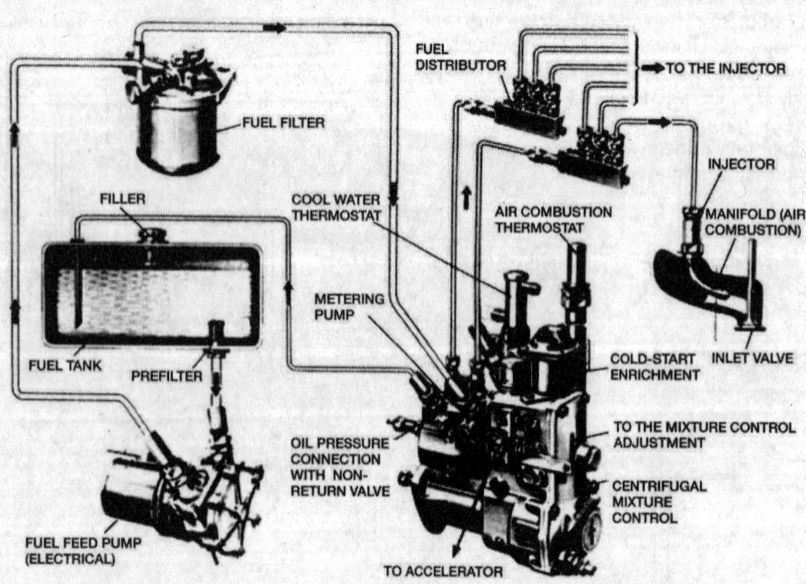

1.5 Schematic of the intermittent port injection system introduced on the Mercedes 220 SE in 1958

Bendix system was adopted by 32 out of the 33 runners for the 1971 Indy 500, and was unchallenged during the remainder of the decade.

A number of fuel injection systems manufactured by smaller companies appeared and disappeared over the period from 1950 to about 1986. In modern times, fuel injection has become the province of the bigger players, and there is now no place in the market for the smaller operation.

Kugelfischer were undoubtedly the most successful of the smaller companies, and their injection systems were fitted to a number of production European vehicles during the 1960's until the company was absorbed by Bosch in 1974 *(see illustration 1.6)*.

In 1973 Bosch introduced the K-Jetronic mechanical injection system *(see illustration 1.7)*. The 'K' system was destined to become the most successful mechanical injection system ever, due to its reliable effective operation and low emissions. With the introduction of electronic controls in 1984, the 'KE' Jetronic was born, with improved starting and pollution control. This hybrid part mechanical and part-electronic system is still in service well into the 1990's. But times they were a-changing and from slow beginnings, Electronic Fuel Injection systems gradually evolved until all mechanical systems, other than K-Jetronic, have now given way to the onward march of electronics.

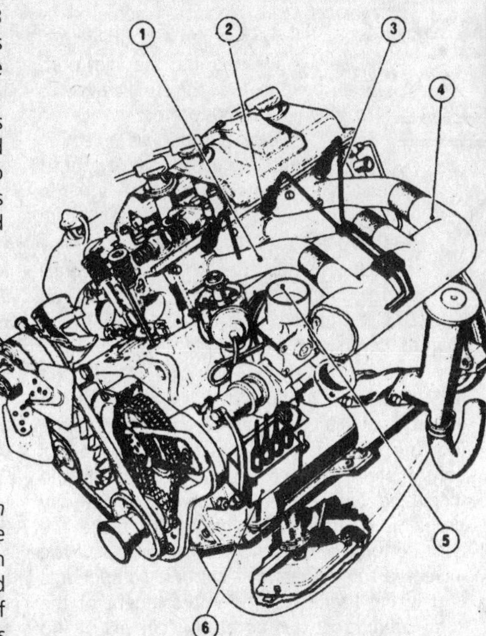

1.6 Kugelfischer system on a Peugeot 404

1 Intake runner
2 Injector
3 High-pressure fuel line
4 Manifold
5 Throttle body
6 Injection pump

Electronic petrol injection

Many features of the modern automobile were actually tried and discarded by pioneers many years before the technology became available that made their employment possible. The use of automotive electronic controls is a prime example; although attempts were made in the 1930's and the 1950's, a viable electronic system was not put into production until 1966.

The very first recorded use of an electronic valve to spray fuel occurred in 1932, when an engineer named Kennedy – an employee of the Atlas Imperial Diesel Engine Company – adapted a six-cylinder, spark-ignition, oil-burning marine engine. The system was devoid of transistors (the transistor was not invented for another 16 years) and any of the components that make up the modern system. In 1934, Kennedy installed a six-cylinder engine equipped with the same system into a truck, He then successfully drove the truck from Los Angeles to New York and back. When Atlas Imperial went bust, the Kennedy system disappeared into obscurity.

On February 4, 1957, the Bendix Corporation filed for a patent on automotive Electronic Fuel Injection systems. Thirty-nine claims were cited, and this foresight gave Bendix effective world-wide patent rights to all forms of Electronic Fuel Injection. The patent was granted on April 18, 1961. Bendix originally began development of the

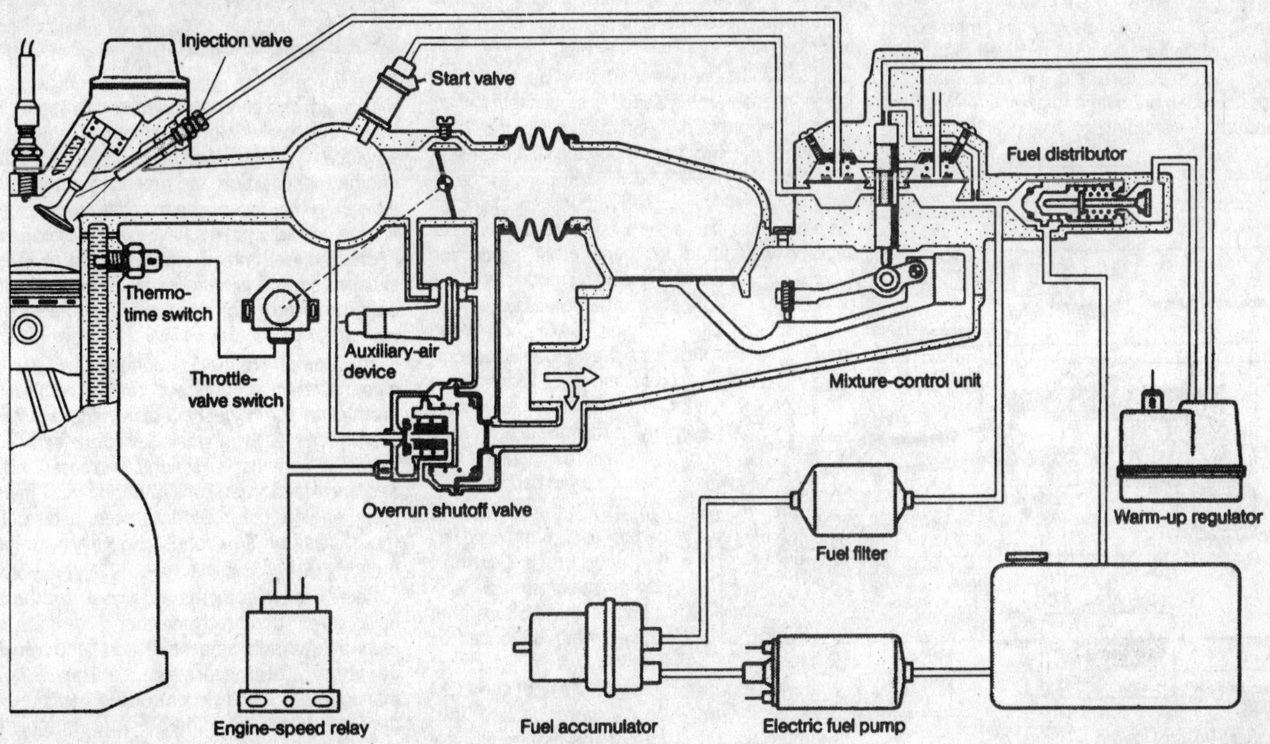

1.7 The Bosch K-Jetronic system, with fuel cut-off on the overrun, permitted fuel savings as well as a cleaner exhaust

© Robert Bosch Limited

Chapter 1

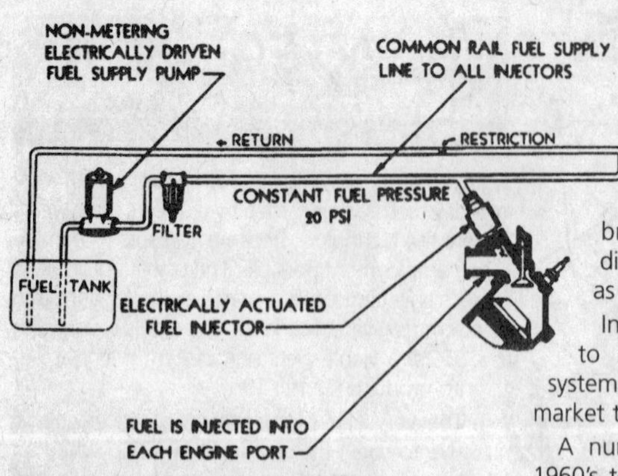

1.8 The Bendix Electrojector system featured timed injection into the intake ports, with a 20 psi common-rail fuel line

Electrojector system in 1952, under the supervision of Robert W. Sutton *(see illustration 1.8)*. Although many years and $1,000,000 were spent in development, the system never reached production. The Electrojector system was made, tested and developed into a working system with great potential; but the high cost of components, harsh underbonnet environment, and the primitive technology available during the 1950's proved a bridge too far. However, much of the initial work was to prove fruitful some years later.

During early development, whenever the vehicle was driven under power cables during a road test, the induced current would trigger the modulator, and the injectors would open to fill the cylinders with neat fuel! Fortunately, this problem was eventually overcome.

The basic Electrojector system comprised an Electronic Control Unit that received data from sensors on inlet manifold pressure, ambient temperature, air pressure and engine rpm. After a computation according to speed and load, the injectors were electrically actuated by solenoid valves, so that a timed fuel pulse was injected into the inlet port. Fuel pressure in the fuel rail was maintained at 20 psi by use of an electrical fuel pump. The injection signal was triggered by a set of contact breakers mounted in the ignition distributor, and actuated by the same cam as the ignition contact breakers.

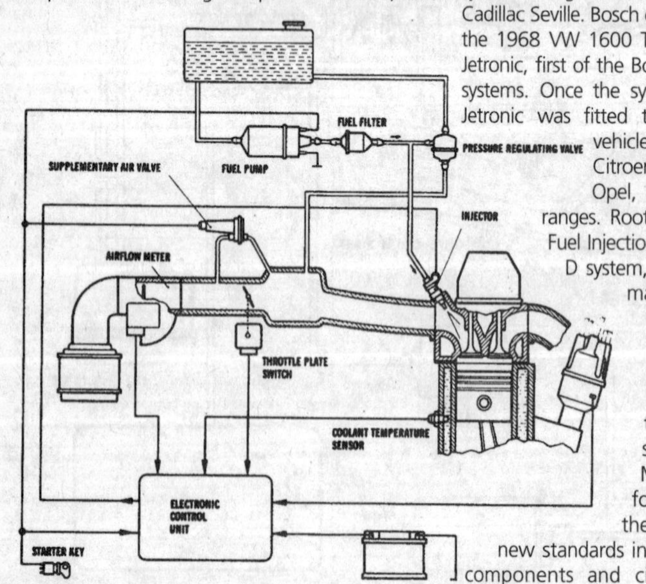

© Robert Bosch Limited

1.9 Schematic view of the Bosch L-Jetronic ignition/injection system

In 1966, Bendix granted Bosch a licence to manufacture electronic fuel injection systems in Germany and Brazil, and rights to market the systems world-wide.

A number of factors emerged in the late 1960's to bring together the last pieces of the electronic injection jigsaw. These included the looming issue of pollution control and petrol consumption, the ever-increasing complexity and increasing cost of the carburettor, driveability problems created by attempting to make the carburettor cope with a lean mixture, the increased ruggedness of electronic components and wiring harnesses, the improvement of the semi-conductor, integrated circuits and transistors. A viable injection system was now possible, and because the carburettor was failing to meet pollution and driveabilty expectations, designers began turning to fuel injection as the long-term solution to current problems.

By 1966, Associated Engineering (a British company) had developed the AE-Brico Electronic Fuel Injection system, which was fitted in 1969 to the Aston Martin DB6 as an alternative to the standard Weber carburettors. However, its production life was short-lived, and the AE-Brico system was never fitted to any other production vehicle.

In 1967 Bendix re-started their electronic fuel injection programme, and this resulted in the system being fitted in production to the 1976 Cadillac Seville. Bosch equipped the USA version of the 1968 VW 1600 Type 3 flat-four with the D-Jetronic, first of the Bosch Electronic Fuel Injection systems. Once the system had been proven, D-Jetronic was fitted to many other production vehicles, including models from the Citroen, Lancia , Mercedes-Benz, Opel, Renault, SAAB and Volvo ranges. Roots of the modern Electronic Fuel Injection system were laid with the D system, and modern systems have many similarities with the system that was first manufactured back in 1967.

In 1973 the D-Jetronic began to be replaced by the new Bosch L-Jetronic system *(see illustration 1.9)*. Many of the systems that followed were influenced by the basic L system, which set new standards in design. Many of the ECM components and circuits were packed into integrated circuits. Further development saw EGR, Lambda operation and closed-loop control.

In 1978 Bendix formed a partnership with Renault, and the Renix injection system, first marketed in 1984 on the Renault 25, was the result. But Renault subsequently sold its share in Renix to Bendix-France, which then continued to develop and improve the Renix Electronic Fuel Injection system. Previously, in 1966, Renault itself developed an Automatic Transmission unit with electronic controls which went into production on the Renault 16 in 1969.

By 1978 the first Engine Management System had appeared, in the form of the Bosch Motronic, and this was first fitted to the BMW 732i. The fuel system was basically L-Jetronic, but the ECM now contained the circuits to control the primary ignition function. Later on, Motronic would also be adapted to automatically control the idle speed, and handle self-diagnostics. Meantime, the first self-diagnostic capability was produced on a Cadillac equipped with a Bendix digital system in 1981.

As the 1980's gave way to the 1990's, Electronic Fuel Injection development gained speed, and a wide variety of systems have now been developed. In some systems the EMS was connected to the automatic transmission, traction control and/or other electronic controls. Single point injection, sequential injection, and distributorless ignition are just some of the later developments. Self-diagnostics have also moved on, and where the early systems were limited in the number of fault codes, later systems are capable of logging perhaps one hundred or more.

Electronic diesel injection

Development of the diesel engine, and particularly the engine management system, has been relatively slow compared with the advances which have been made in petrol engine fuel injection and management systems. However, in recent years, electronic diesel engine control systems have been developed to improve diesel engine efficiency and to reduce exhaust emissions. Almost all modern engines use some form of electronic engine control system.

For a diesel engine to operate efficiently, it is essential that the correct amount of fuel is injected at the correct pressure, and at exactly the right time. Even small deviations can cause increased exhaust emissions, increased noise, and increased fuel consumption. In a typical diesel engine, the injection process takes only a thousandth of a second, and only a minute quantity of fuel is injected.

Recent developments in diesel technology have given us direct injection, common rail technology, multi-stage injection, and variable geometry turbochargers. The engine management systems control the fuel pressure generated by the high-pressure pump, the precise timing of the electronically controlled fuel injectors, along with the operation of the EGR system, and the output of the turbocharger.

Technical overview of Fuel Injection and the Engine Management System

Definition of the engine management system

Petrol engines

An engine management system (EMS) is essentially an electronic control system whereby the engine ignition, fuelling and idle speed functions are controlled by an electronic control module (ECM). The distributor, when used, is provided purely to distribute the HT spark to the correct cylinder in firing order. It is important to understand that the EMS does not control these functions separately. During the different engine control operations, the ECM synchronises timing and injection so that the various inputs and outputs work hand in hand, and not as separate entities *(see illustration 2.1)*.

The modern EMS employs digital technology and has a high degree of self-diagnostic capability. In addition, an EMS may also control other vehicle functions such as the air conditioning and communicate with other vehicle ECMs such as those controlling the ABS, automatic transmission or traction control when fitted.

However, in early systems the ECM did not always control the idle speed and the self-diagnostic function was often of limited performance.

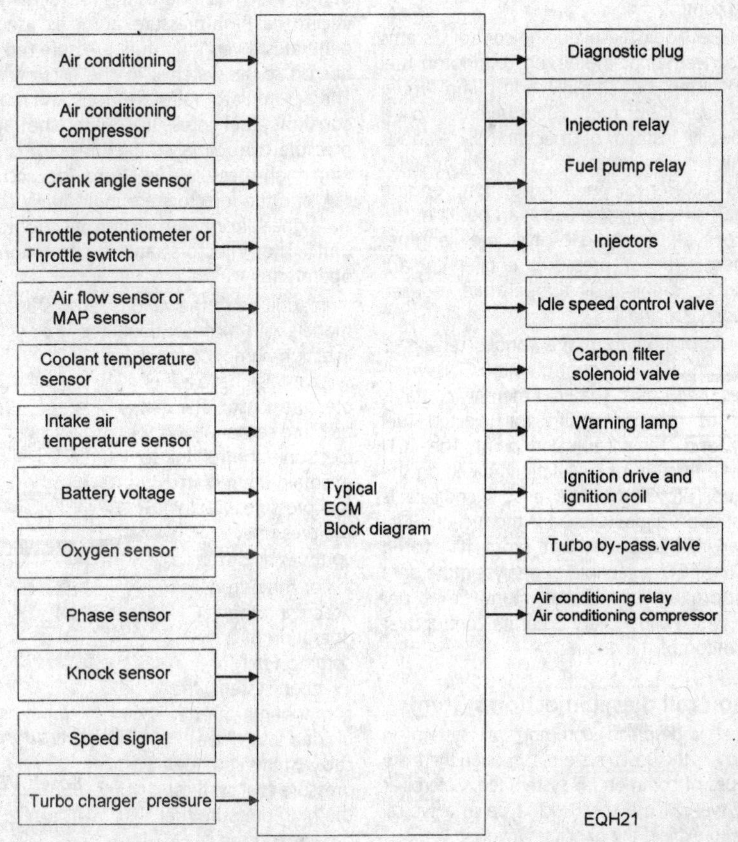

2.1 Typical petrol engine block diagram

Chapter 2

Diesel engines

For many years diesel engines were loved and loathed for being essentially crude mechanical devices, relying on solid engineering to function in a workmanlike manner, with the minimum of electronic/electric components. However, environmental pressures and spiralling fuel costs have forced the development of the diesel engine to the point where now, it can be argued that now they are more complex than their petrol equivalents.

Electronic control using a conventional diesel injection pump

The function of the fuel injection pump is to supply fuel to the injectors at the correct pressure, at the correct moment in the combustion cycle, and for the length of time necessary to ensure efficient combustion. A conventional (mechanically-controlled) fuel injection pump uses an accelerator cable (connected to the driver's accelerator pedal), and various mechanical add-on devices (such as cold start injection advance, fast idle units, turbo boost pressure sensors, etc) to provide control of the fuel injection timing and the quantity of fuel injected. Even with these add-on devices, it has become increasingly difficult for a mechanical diesel control system to keep pace with modern demands on engine refinement and exhaust emission control.

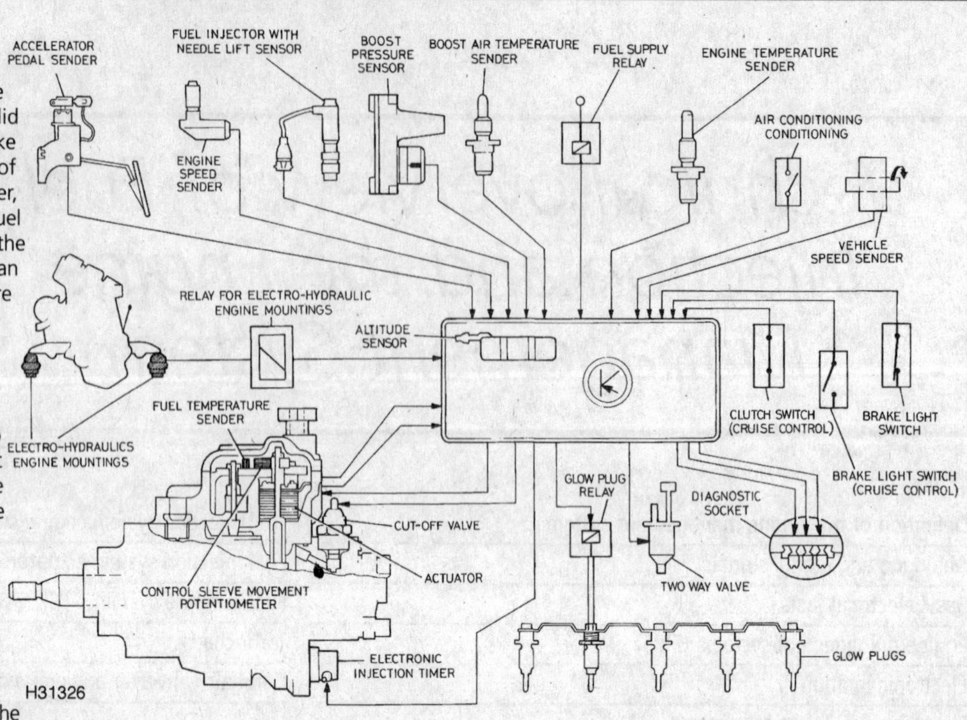

2.2 Electronic diesel control system components fitted to an Audi 2.5 litre engine

Many electronic diesel engine control systems use a conventional in-line or distributor fuel injection pump, but the injection pump timing and the quantity of fuel injected are controlled electronically instead of mechanically. Various electronic sensors are used to measure variables such as accelerator pedal position, engine crankshaft speed, engine camshaft position, the mass of air passing into the engine, turbocharger boost pressure, engine coolant temperature, ambient air temperature, etc *(see illustrations 2.2 and 2.3)*.

The information from the various sensors is passed to an electronic control unit (ECU), which evaluates the signals. The ECU memory contains a series of mapped values for injected fuel quantity, and start-of-injection point. The ECU performs a number of calculations based on the information provided by the sensors, and selects the most appropriate values for the fuel quantity and start-of-injection point from its stored values. The ECU is capable of analysing the data and performing calculations many times per second, which allows very accurate control over the operation of the engine.

Common rail diesel injection systems

The most widespread common rail system in current use is the Bosch system. Although there are other types of common rail system (eg, Caterpillar system), we will use the Bosch type as a typical example to explain the principles involved.

The common rail system derives its name from the fact that a common rail, or fuel reservoir, is used to supply fuel to all the fuel injectors. Instead of an in-line or distributor fuel pump, which distributes the fuel directly to each injector, a high-pressure pump is used, which generates a very high fuel pressure (up to 1350 bar on some systems) in the accumulator rail. The accumulator rail stores fuel, and maintains a constant fuel pressure, with the aid of a pressure control valve. Each injector is supplied with high-pressure fuel from the accumulator rail, and the injectors are individually controlled via signals from the system electronic control unit. The injectors are electromagnetically-operated.

In addition to the various sensors used on models with a conventional fuel injection pump, common rail systems also have a fuel pressure sensor. The fuel pressure sensor allows the electronic control unit to maintain the required fuel pressure, via the pressure control valve.

For the purposes of describing the operation of a common rail injection system, the components can be divided into three sub-systems; the low-pressure fuel system, the high-pressure fuel system and the electronic control system.

Low-pressure fuel system

The low-pressure fuel system may consist of the following components:

Fuel tank.
Fuel lift pump.
Fuel filter/water trap.
Low-pressure fuel lines.
Fuel cooler

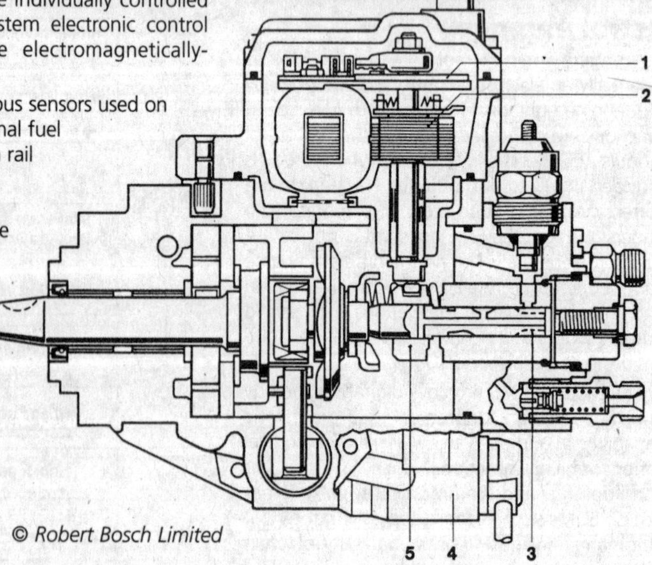

2.3 Bosch VE injection pump with electronic diesel control

1 Control spool position sensor
2 Electrical actuator
3 Pump plunger
4 Cavity pressure control solenoid
5 Control spool

The low-pressure system (fuel supply system) is responsible for supplying clean fuel to the high-pressure fuel circuit.

High-pressure fuel system

The high-pressure fuel system consists of the following components *(see illustration 2.4)*:

High-pressure fuel pump with pressure control valve.
High-pressure accumulator rail with fuel pressure sensor.
Fuel injectors.
High-pressure fuel lines.

After passing through the fuel filter, the fuel reaches the high-pressure pump, which forces it into the accumulator rail, generating pressures of up to 1350 bar. As diesel fuel has a certain elasticity, the pressure in the accumulator rail remains constant, even though fuel leaves the rail each time one of the injectors operates: additionally, a pressure control valve mounted on the high-pressure pump ensures that the fuel pressure is maintained within pre-set limits.

The *pressure control valve* is operated by the ECU. When the valve is opened, fuel is returned from the high-pressure pump to the tank, via the fuel return lines, and the pressure in the accumulator rail falls. To enable the ECU to trigger the pressure control valve correctly, the pressure in the accumulator rail is measured by a *fuel pressure sensor*.

The electromagnetically-controlled fuel injectors are operated individually, via signals from the ECU, and each injector injects fuel directly into the relevant combustion chamber. The fact that high fuel pressure is always available allows very precise and highly flexible injection in comparison to a conventional injection pump: for example combustion during the main injection process can be improved considerably by the pre-injection of a very small quantity of fuel.

Electronic control system

The electronic control system consists typically of the following components:

Electronic control unit (ECU).
Fuel lift pump.
Crankshaft speed/position sensor.
Camshaft position sensor.
Accelerator pedal position sensor.
Turbocharger boost pressure sensor.
Air temperature sensor.
Coolant temperature sensor.
Air mass meter.
Fuel pressure sensor.
Fuel injectors.
Fuel pressure control valve.
Preheating control circuit.
EGR valve actuator.

The information from the various sensors is passed to the ECU, which evaluates the signals. The ECU contains electronic 'maps' which enable it to calculate the optimum quantity of fuel to inject, the appropriate start of injection, and even pre- and post injection fuel quantities, for each individual engine cylinder under any given condition of engine operation.

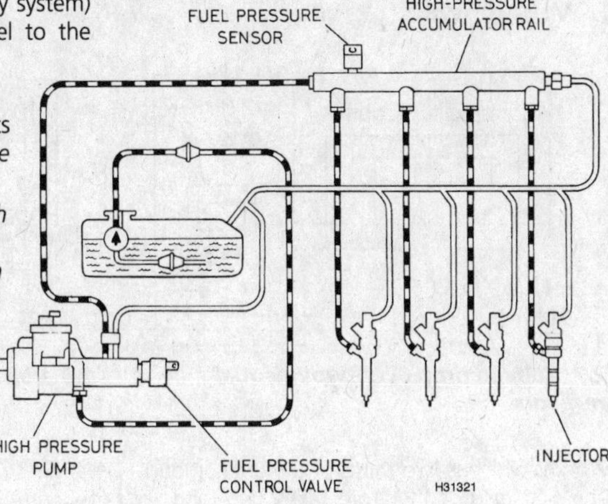

2.4 Schematic view of a high-pressure fuel system – Bosch common rail

Additionally, the ECU carries out monitoring and self-diagnostic functions. Any faults in the system are stored in the ECU memory, which enables quick and accurate fault diagnosis using appropriate diagnostic equipment (such as a suitable fault code reader).

Analogue and digital signals

An analogue signal is any continuous signal that can change by an infinitely small amount. Typically, an analogue signal is measured by an instrument that uses a needle to progressively sweep across a fixed scale. Any change in the signal will cause the needle to move by a similar amount. One example would be the throttle potentiometer/accelerator pedal position sensor. As the throttle is opened and closed, the voltage output signal from the throttle pot increases and decreases and an analogue signal is passed to the ECM.

On the other hand, a digital signal is represented by a code that has two states, on and off. In simple terms, the signal consists of a series of digital pulses where the pulse width or frequency (number of pulses) is used to indicate a specific value *(see illustration 2.5)*.

Many of the signals passed to the ECM from the engine sensors are analogue signals. Because the ECM works in a digital fashion, all analogue signals must pass through an analogue to digital converter (ADC) when the signal will be stored by the ECM in digital format. If the sensor is able to output data in digital form, so much the better. A digital signal from a digital sensor does not need converting, and processing by the ECM is therefore much faster.

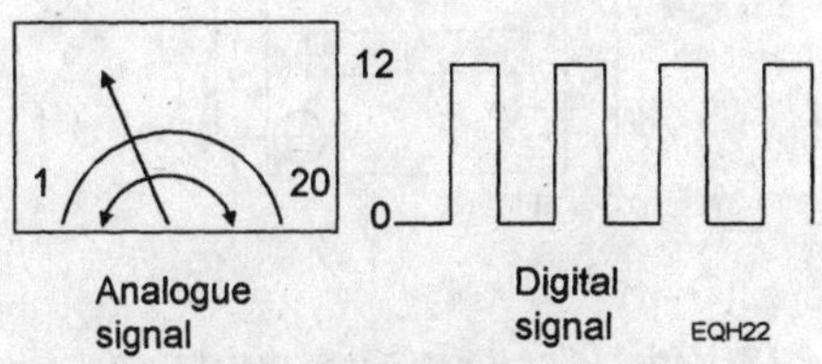

2.5 Analogue and digital signal

Basic electrical facts

General

The reader needs to be conversant with a basic understanding of electrical measurements and terms if he is to obtain the most benefit from this section. If you do not understand the meaning of voltage, voltage drop, resistance, amperes, and series and parallel circuits, then you will need to study a basic automotive electrical textbook before you go any further. One such textbook is "Automobile Electrical & Electronic Systems" by Tony Tranter (available from Haynes Publishing).

Moving on, in order to gain a better comprehension of the EMS it is important to grasp the following principles. We will start off by looking at Ohm's Law.

Ohm's Law

Ohm's Law: I = V/R when V = Volts, R = Resistance (ohms) and I = current (amps).

The Law can be expressed as V = IR or R = V/I, but the expression I = V/R is the one we shall use most in the following descriptions.

Nominal battery voltage (nbv)

In any automotive circuit, the voltage will vary at any moment according to the vehicle operating conditions. With the engine stopped, voltage at the battery will be somewhere between 12.0 and 12.6 volts. During engine start-up when the engine is cranked upon the starter motor, the voltage will reduce to between 11.0 and perhaps 9.5 volts. Once the engine is running, the alternator will regulate the voltage to between 13.5 and 15.0 volts.

Because the operating conditions vary so much it is obvious that asking the engineer to check for 12 volts during a test procedure will be incorrect in most cases. For this reason we have coined the term 'nbv' which stands for nominal battery voltage. Where nbv is mentioned in the text, the voltage to be measured will depend on whether the engine is at rest, cranking or running.

Note: *nbv will vary in any particular vehicle and manufacturer's specifications should be obtained for accuracy of testing. However, the following figures are a good 'rule of thumb' guide.*

Engine at rest	*12.0 to 12.6 volts*
Engine cranking	*9.5 to 11.0 volts*
Engine running	*13.5 to 15.0 volts**

** Voltage may be slightly lower at idle*

We will continue to use 12.0 volts during the examples set in this Chapter, since this value is a known constant and it also makes the calculations easier. In a real-life situation you would use the exact voltage value.

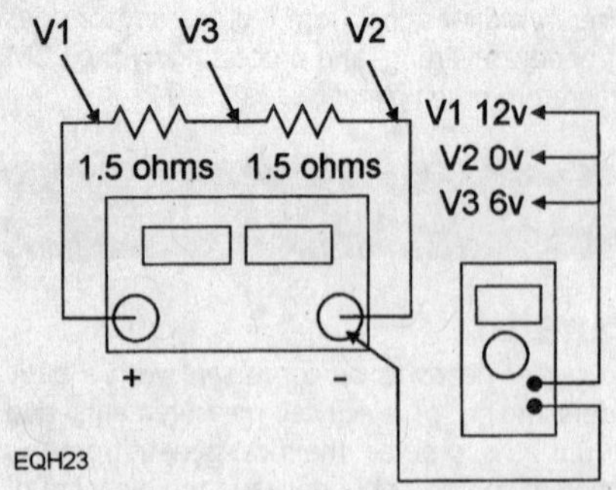

2.6 Voltage drop over two equal resistors

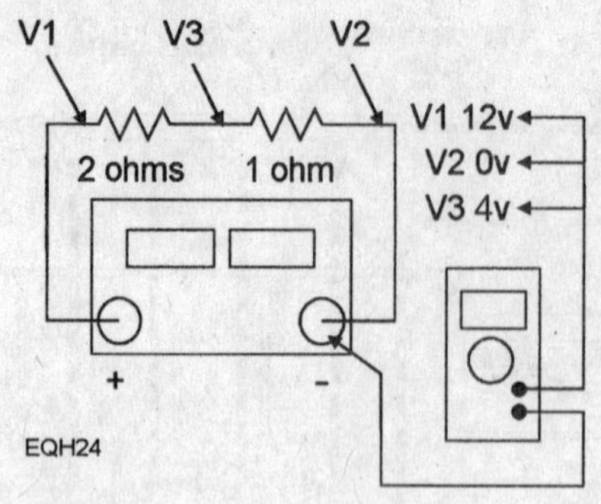

2.7 Voltage drop over two unequal resistors

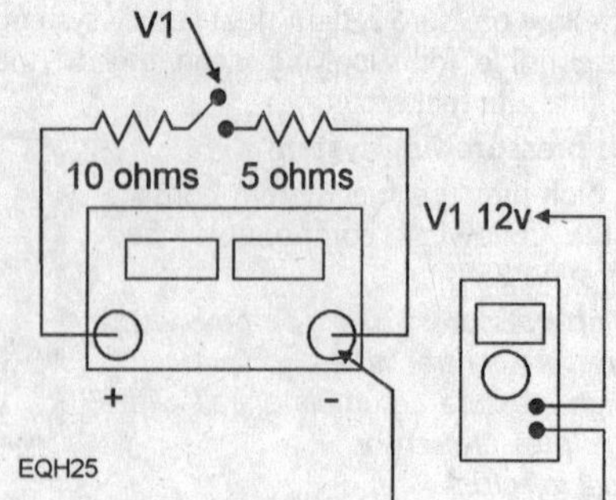

2.8 Voltage drop over an unequal circuit

Circuit fundamentals

In the first example, a 12-volt circuit consists of a wire connecting two resistors of equal value to a voltage supply and earth. We will use two resistors of (say) 1.5 ohms.

A voltmeter may be used to measure the voltage at the various points in the circuit. 12 volts is measured in the supply to the first resistor, and zero volts is measured in the earth part of the circuit. The voltage drop measured over the whole of any circuit must equal 12 volts. The voltage measured between the two resistors is 6 volts. This is because each resistor will take an equal amount of the available voltage. By applying Ohm's Law (12 volts divided by 3.0 ohms total resistance) the current flowing in the circuit is calculated at 4 amps. This could be represented in a real automotive situation by a coil and ballast resistor. The current flowing in the circuit will vary according to changes in voltage and/or resistance *(see illustration 2.6)*.

For example: 15.0 volts divided by 3.0 ohms total resistance gives a current flow of 5 amps.

For the purist, Ohm's Law may be used to calculate the voltage drop in the circuit so: V = IR or 4 (amps) x 1.5 (ohms) = 6 volts. Since this value is the voltage drop, we need to deduct the voltage drop as follows. 12.0 – 6.0 = 6.0. 6.0 volts is therefore the voltage value measured between the resistors in the circuit.

If the two resistors are now substituted for unequal values ie. 2 ohms and 1 ohms respectively, then the equation changes. The total resistance is 3 ohms and the first resistor is 2 ohms which is two-thirds of the total amount. The first resistor therefore takes two-thirds of the voltage (8 volts) and the second takes 4 volts. The voltages measured at the three points of the circuit will now be 12, 4, and 0 *(see illustration 2.7)*. **Note:** *It is not the actual resistor values that are important but the comparative values.*

If the supply wire to the first resistor is disconnected, the voltage measured at this point will be 12 volts. Similarly, on disconnecting the cable between the two resistors; then the voltage measured at this point will also be 12 volts. This is because resistors only function in a live circuit when current is flowing (see illustration 2.8).

If the voltage at the earth side of a resistance is greater than zero, then there must be a further resistance, such as a bad connection in the earth circuit *(see illustration 2.9)*.

If the resistor at B is made to vary, then the voltage between the resistors will also vary. This is the principle by which a variable analogue signal is sent from a sensor to the ECM. For example, as the CTS thermistor varies in resistance according to temperature, a variable voltage signal from is sent to the ECM for evaluation. The ATS, AFS, CTS, TPS and many other sensors operate on the same principle. In practice, we do not know the value of the first resistor located in the ECM. However, we do know the value of the variable resistor under various conditions, and we can then ascertain whether the sensor is functioning within its operating parameters. When the resistance varies, so does the voltage and although the voltages measured in this circuit will remain constant irrespective of resistance, measuring the voltage signal is much more useful than measuring the resistance *(see illustration 2.10)*.

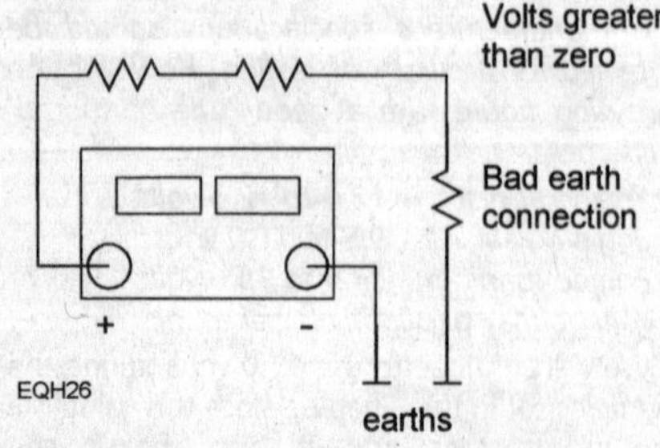

2.9 Voltage drop with bad earth connection

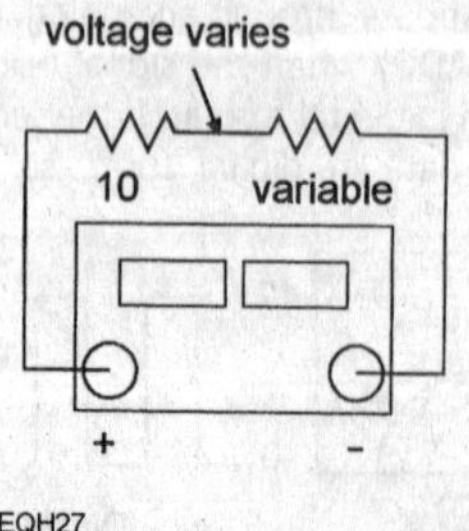

2.10 Variable voltage when resistance varies

Parallel circuits

Many automotive wiring circuits can be essentially identified as series circuits. However, there are also a number of parallel circuits and the most common form exists in the fuel injection wiring circuit. The injection circuit may be wired in banks of two, three or four injectors.

To find the current flow through each injector we could apply Ohm's Law to each injector branch. In our example, the injector resistance is 4 ohms – again a common value.

Thus: 12 volts divided by 4 ohms = 3 amps. The current flow through the whole injector circuit would be 3 amps x 4 cylinders = 12 amps.

We might want to know the parallel (equivalent) resistance of the injector circuit. This resistance could be measured from the supply terminal at the relay (typical) to the ECM pin. The resistance is calculated as follows.

1/R = 1/R1 + 1/R2 + 1/R3 + 1/R4.
Thus, 1/R = 1/16 + 1/16 + 1/16 + 1/16
which becomes 1/R = 4/16 or 1/R = 1/4
and inverted R/1 = 4/1 ohms.

Thus the total parallel resistance of the circuit will be measured as 4 ohms. This assumes of course that the circuit is fault-free. Let us take a look at what could happen to a circuit with one or more faults that will increase the circuit resistance. Let us say that one injector circuit has a high resistance of 160,000 ohms. If we use the above formula the values are as follows:-

1/R = 1/16 + 1/16 + 1/16 + 1/16000,
which becomes 1/R = 30001/160000
or R = 5.33 ohms.

Such a small difference in resistance between a good and a bad value is very easy to overlook and great care must therefore be taken when evaluating the injector circuit by measuring the resistance of the parallel circuit.

Assuming that the resistance of a single injector is 16 ohms, the values that are likely to be obtained with various configurations of the injector circuit are as follows:

Four injectors in bank

Res. value (ohms)	Condition
4 to 5	*all injectors ok*
5 to 6	*one injector suspect*
8 to 9	*two injectors suspect*
16 to 17	*three injectors suspect*

Three injectors in bank

Res. value (ohms)	Condition
5 to 6	*all injectors ok*
8 to 9	*one injector suspect*
16 to 17	*two injectors suspect*

Two injectors in bank

Res. value (ohms)	Condition
8 to 9	*all injectors ok*
16 to 17	*one injector suspect*

Note: *When the injector resistance is other than 16 ohms, the values will be different to those in the above table.*

Dwell and duty cycle

Traditionally, a dwell angle is defined as being the number of degrees through which a distributor rotates when the CB points are closed. However, in modern times we should consider the wider context of the meaning of 'dwell'. A good definition of a dwell angle would be the rotational angle during which a device is active or switched 'on' *(see illustration 2.11)*.

A more meaningful term might be 'duty cycle', and we could expand duty cycle to include any switchable device. Such devices need not be confined to the rotating type. Duty cycle could thus be measured in terms of degrees of rotation, percentage (%) of time 'on' – (or even off), or time on or off in milliseconds (ms). All we need is the appropriate meter. Usually, dwell is measured in degrees, but if we use either (%) or (ms), it is more appropriate to refer to duty cycle.

Devices capable of providing a dwell or duty cycle are usually actuators connected to the ECM or some other type of electronic module, and are normally switched quite rapidly. Examples would include the ignition coil, Hall-effect pulse generator, fuel injector, ISCV, VSS, CFSV etc – note that this list is by no means exhaustive.

The modern automotive multimeter usually contains functions to measure both dwell angle and duty cycle %. Some can even measure duty cycle in ms. By connecting the meter between the pulse terminal and earth on actuators such as the coil, ignition module, hall-effect switch, injector, ISCV or in fact any other device that is rapidly switched, a duty cycle may be obtained. By comparing this value with known operating parameters, correct operation of the device can be determined.

To convert dwell degrees to dwell percent and vice versa, use the following formulae:

Dwell° x cyls/360 x 100 = Dwell%
eg 45° x 4/360 x 100 = 50%
Dwell% / 100 x 360/cyls = Dwell°
eg 50% / 100 x 360 / 6 = 30°

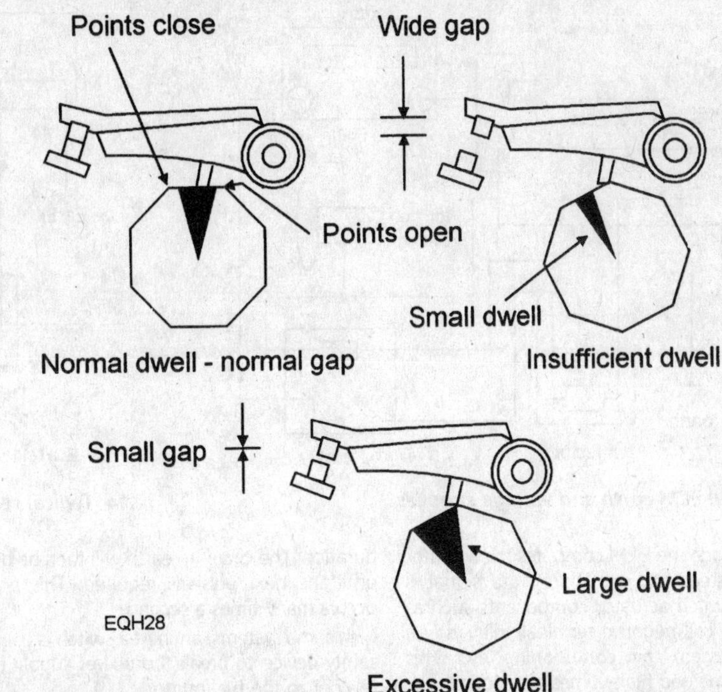

2.11 Conventional dwell angle

Most automotive actuated devices are switched to earth, although the occasional device that is switched to live may be discovered. For example, if battery voltage is provided to an actuator such as an ISCV and the ECM switches the ISCV 'on'. A voltmeter would measure the voltage at a level of zero – ie earth. As the ISCV is switched off, the voltage returns to battery voltage level. This occurs many time a second and a digital voltmeter could also be used to measure the average switching voltage, which for an ISCV would usually be 6 to 7 volts *(see illustration 2.12)*.

In addition, the frequency meter facility provided on many digital multi-meters could also measure the frequency of pulse, and thus we have several methods of measuring the pulsing of a device. Since the pulse wire is the earth wire, measurement must always be made at the earth connection. All the devices mentioned above can be measured in a similar fashion.

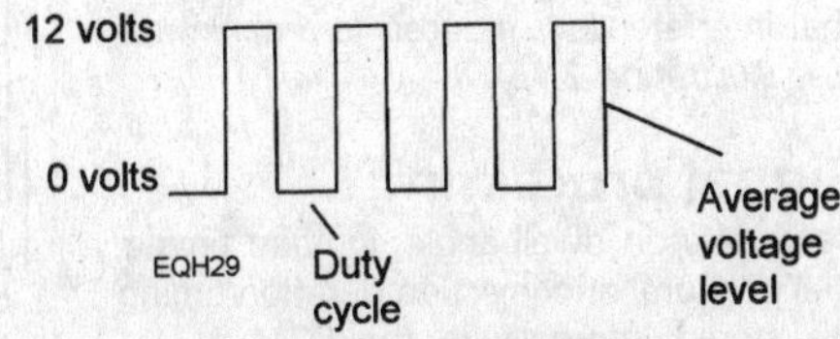

2.12 Typical ISCV waveform (graphical example of voltage switching at the ISCV as seen on an oscilloscope)

Engine management systems (EMS)

Basic electronic control module (ECM) operation

This section describes the operation of a typical EMS. Although the components of each specific system may vary, the principles involved are fairly constant and this description is relevant to many of the modern systems. You will need to refer to a specific system for a more detailed description of how that particular system works.

Sensors and actuators

A sensor is an input device that provides variable information on an engine function. Examples include the airflow sensor (AFS), crank angle sensor (CAS), coolant temperature sensor (CTS), throttle potentiometer sensor (TPS) etc, and these provide data on load, rpm, temperature, throttle opening etc. This data is signalled to the ECM, which then analyses the results and computes an output signal. The output signal is used to actuate an output device. An actuator is a device that is switched or actuated by the ECM for a period of time depending on the input signal. Examples include the fuel injector, idle speed control valve (ISCV), EGR valve, fuel pressure regulating valve, carbon filter solenoid valve (CFSV), relay driver etc.

Put in simple computer terms, the process is summed up as input, compute, and output. Input is determined by the signals from the

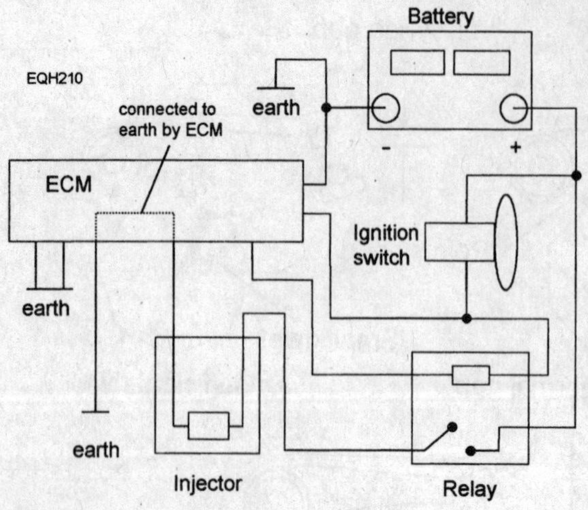

2.13 Typical ECM earth and voltage supplies

2.14 Typical sensor wiring

engine sensors, the ECM computes the actuator values required, and an appropriate signal is output to switch actuator components such as the injector, coil negative terminal, idle system, emission system, air conditioning etc. This process is repeated many times per second.

ECM power supplies and earths

A permanent voltage supply is usually applied from the vehicle battery to one of the ECM pins. This allows the self-diagnostic function to retain data of an intermittent nature. Once the ignition is switched on, a voltage supply is made to the ECM, ignition coil(s), injectors, ISCV and other devices. The voltage supplies may be made direct from the ignition switch or come from the system relay/fusebox. The ignition 'on' supply to the ECM effectively powers up the ECM so that it is ready to go about its business.

When a petrol engine is cranked or run, a signal from the ignition causes the ECM to earth the fuel pump relay so that the fuel pump will run. Ignition and injection functions are also activated. Actuators such as the injectors and ISCV are supplied with nbv from the main relay or ignition switch and the ECM completes the circuit by pulsing the relevant actuator wire to earth *(see illustration 2.13)*. The same process occurs with a modern diesel engine, except that EGR is activated (temperature dependant), and most common-rail systems do not have a lift pump between the fuel tank and high-pressure pump.

There are normally several earth connections to the ECM. However, only one or two will normally be used as the actual ECM system earth. The other earth connections are provided to allow the ECM to complete various sensor and actuator circuits. In effect, these earth paths are secondary paths and only used when the ECM switches an actuator.

For example, to create an injection pulse, the ECM will use an earth to complete the circuit for the instant required to give correct pulse duration. The circuit to earth will then be broken until the next pulse is required. This practice occurs many times a second.

In some systems, an inertia switch is used as a safety device to break the power supply to the relay or to the fuel pump.

Reference voltage

During normal engine operation, battery voltage could vary between 9.5 (cranking) and 14.5 (running). To minimise the effect on engine sensors (for which the ECM would need to compensate), the voltage supplies from the ECM to the engine sensors are made at a constant voltage (known as a reference voltage) of 5.0 volts *(see illustration 2.14)*.

Once the ECM has powered up then, the majority of sensors (other than those that generate a voltage such the CAS and OS), are supplied with the 5.0 volt reference.

In most instances the engine sensors are not directly connected to earth. Rather, the sensor earth path is connected to one or two common earth return connections that connect directly to the ECM. This ECM pin is not usually connected directly to earth. The ECM internally connects that pin to earth via one of the ECM pins that are directly connected to earth.

Signal shielding

To reduce radio frequency interference or RFI, some signals are protected with a shielded cable. The shielded cable is often connected to the main ECM earth pin, so that interference is reduced to a minimum *(see illustration 2.15)*.

Signal processing

Basic data on dwell angle, ignition timing, fuel pressure, and injection duration/timing are stored internally in the ECM in a number of two- or three-dimensional maps. These maps or look-up tables allow the ECM to look-up the correct ignition timing and injection duration/timing according to speed, temperature and load.

Accessing data from a table is a far quicker method of selecting data than by computing it. If the system is well designed, the map will contain settings for almost every possible engine load and speed. Even so, the map would be enormous if every single rpm or load setting was listed, and rpm may be mapped every 5 rpm and the load logged in a similar fashion. The ECM will interpolate in-between values so that most operating conditions are extensively covered. Some early systems were not always well mapped, and this created a 'black hole' for some operating conditions with consequent deterioration of performance.

In addition, on petrol engines, the ignition point and injection duration are jointly processed by the ECM so that the best moment for ignition and fuelling are determined for every operating condition.

The main engine load sensor is either the AFS or MAP sensor and engine speed is determined from the Crankshaft Position Sensor or CAS. Correction factors are then applied for starting,

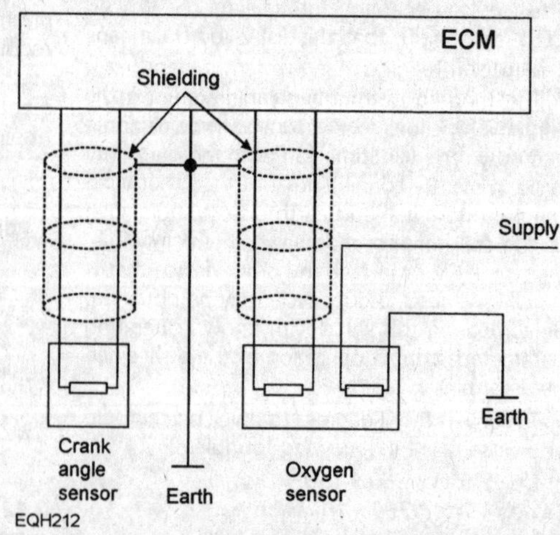

2.15 Sensor shielding to reduce RFI

idle, deceleration and part and full-load operation. The main correction factor is engine temperature from the CTS. On petrol engines, minor corrections to timing and air-fuel ratio (AFR) are made with reference to the battery voltage, ATS and TPS signals. On diesel engines, alterations to injection timing and duration are made with reference to the coolant temperature, intake air temperature, fuel temperature, and accelerator pedal position sensor.

In petrol models with an ISCV or a stepper motor, the ECM accesses a different map for idle running conditions, and this map is implemented whenever the engine speed is at idle. Idle speed during warm-up and normal hot running conditions are maintained automatically by the idle control function. However, in most systems the ECM makes small adjustments to the idle speed by advancing or retarding the timing, and this results in an ignition timing that is forever changing during engine idle.

In petrol models with an catalytic converter and OS, the ECM monitors the OS signal and controls the injection pulse so that the AFR is always around Lambda = 1.0 (0.97 to 1.03 is a typical range for Lambda control). Lambda is represented by the Greek symbol 'λ'. This mode of operation is known as 'closed-loop'. If the AFR was placed into closed-loop under all operating conditions, the vehicle would be pretty much undriveable during certain driving situations. For this reason, during cold start and warm-up, sharp acceleration and wide-open throttle conditions, the ECM slips into open-loop operation when the AFR is allowed to move outside of the Lambda operating range.

If the engine speed exceeds a pre-designated engine rpm, the ECM cuts off injector operation as a safety precaution. Fuel is also cut during deceleration to maintain smooth running and as an economy measure. Fuel is re-introduced when the deceleration speed falls below a certain rpm. After a period of closed throttle running; when the throttle is opened again, the ECM re-introduces fuel injection in a controlled manner to prevent a loss of smoothness.

Self-diagnostic function

The modern engine management system has a self-test capability that regularly examines the signals from the engine sensors and in some instances the actuators. In the event of a fault being present the ECM internally logs a code.

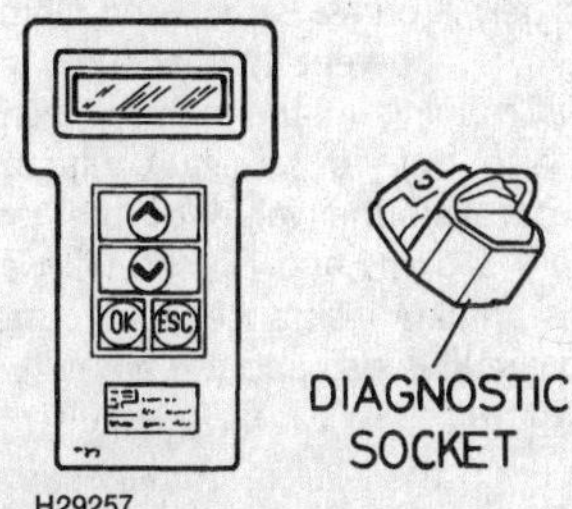

2.16 FCR and typical serial port/diagnostic socket

This code can be extracted from an output terminal, known as a diagnostic plug/socket or serial port, by a suitable fault code reader *(see illustration 2.16)*.

To a large degree, the format and type of data to be output used to be determined by the vehicle manufacturer (VM). The function of the fault code reader (FCR) is to initiate the VM's program and to make the best of what is actually available. In other words, if the VM does not make certain information available, then it is not possible to access such information through the serial port. For example, apart from reading fault codes and clearing fault codes, it may be possible to obtain datastream information on the signal output from the sensors, or make adjustments to CO or timing, or fire the injector, ISCV, relays and other actuators or log sensor data during a road test. But these functions will only be available if the VM has made them available. However, it is usually still possible to obtain such information in other more traditional ways, and we will explore these methods in Chapter 3.

Some vehicles are provided with a dash-mounted warning lamp. When the ECM detects that a major fault is present, it earths a dedicated ECM pin and the warning lamp on the dash will light whilst the engine is running. The lamp remain lit until the fault is no longer present. If the fault clears, the code will remain logged until wiped clean with a suitable FCR, or when the battery is disconnected. However, not all vehicles utilise a warning lamp; those without one will require interrogation by a FCR to determine whether a fault is logged or not. Some ECMs retain codes for faults of an intermittent nature, and this is a valuable aid to fault diagnosis.

Codes emitted by an ECM may be designated as 'slow codes' or 'fast codes'. Slow codes (sometimes termed flash codes) are fault codes emitted by an EMS that are slow enough to be displayed on an LED lamp or on a dash-mounted warning lamp. Fast codes are digital fault codes emitted by an EMS that are too fast to be displayed on an LED lamp or on a dash-mounted warning lamp. A digital FCR instrument is required for capturing fast codes.

European On-Board Diagnosis (EOBD)

All new petrol engined models sold after 01/01/2000, all existing models with petrol engines from 01/01/2001, and all diesel engined models sold after 2005 in Europe, must be equipped with EOBD. This is a higher standard of system monitoring, where any emission related faults must be notified to the driver by illuminating the MIL (Malfunction Indicator Light) in the instrument cluster, the fault (and the distance travelled since the fault occurred) must be recorded for later retrieval.

EOBD also attempts to standardise the type of diagnostic plug (16-pin J1962), the plug position, and the configuration of the plug

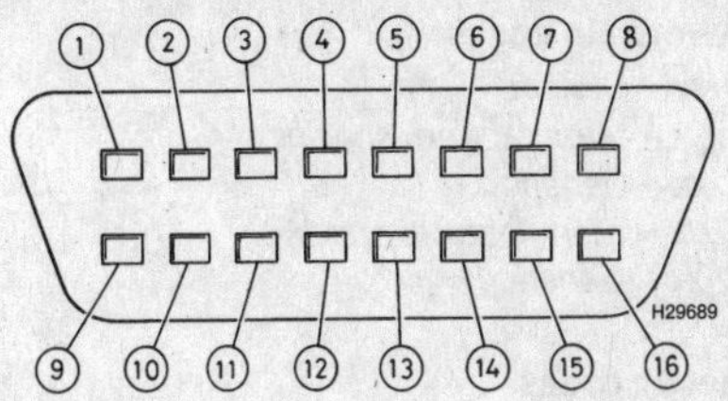

2.17 EOBD 16-pin diagnostic connector details

terminals. The diagnostic plug must be within reach of the driver's seat, tools must not be required to reveal the plug, although it may have a removable cover. Details of the locations of the diagnostic connector fitted to most vehicles equipped with EOBD are given in Chapter 5.

The different pins of the plug *(see illustration 2.17)* are used by different manufacturers in different ways, depending on which communication protocol is used. There are four protocols available:

Protocol	**Connector pins used**
J1850 VPW	*2, 4, 5 and 16*
ISO 9141 – 2	*4, 5, 7, 15 and 16*
J1850 PWM	*2, 4, 5, 10 and 16*
KWP2000 (ISO14230)	*4, 5, 7, 15 and 16*

Generally speaking, European and Asian manufacturers use ISO / KWP protocols, whilst General Motors (Vauxhall/Opel) use J1850 VPW, and Ford use J1850 PWM.

The plug's pins are configured as follows:

Pin	
2	*J1850 BUS+*
4	*Chassis ground*
5	*Signal ground*
6	*CAN High (J-2284)*
7	*K-line*
10	*J1850 BUS*
14	*CAN Low (J-2284)*
15	*L-line*
16	*Battery power*

Manufacturers may use additional pins for other purposes.

To a certain extent, EOBD standardises the fault codes generated by the self-diagnosis system. These five digit alphanumeric codes are made up as follows:

Letter	**System**
B	*Body*
C	*Chassis*
P	*Powertrain*
U	*Network*

The second digit determines whether the code is a generic EOBD code (where the definition of the code is stated in the EOBD regulations, and will be the same for all manufacturers), or a Manufacturer specific code (where definition of the code is determined by the vehicle manufacturer).

Powertrain codes
P0xxx – Generic
P1xxx – Manufacturer specific
P2xxx – Generic
P30xx – Manufacturer specific
P34xx – Generic

Chassis codes
C0xxx – Generic
C1xxx – Manufacturer specific
C2xxx – Manufacturer specific
C3xxx – Generic

Body codes
B0xxx – Generic
B1xxx – Manufacturer specific
B2xxx – Manufacturer specific
B3xxx – Generic

Network communication codes
U0xxx – Generic
U1xxx – Manufacturer specific
U2xxx – Manufacturer specific
U3xxx – Generic

The third digit determines the specific system/sub-system within the vehicle that is associated with the recorded fault code:

Digit	System/sub-system
1	*Fuel and Air metering*
2	*Fuel and Air metering (injector circuit only)*
3	*Ignition system or misfire*
4	*Auxiliary emission control system*
5	*Vehicle speed control and idle control system*
6	*Computer output circuits*
7	*Transmission*
8	*Transmission*

Note: *Manufacturers are not obliged to adhere to the above table for their manufacturer specific codes.*

The fourth and fifth digits of the code identify the section related to the code generation.

Limited operating strategy (LOS)

In addition to the self-test capability, the modern EMS normally contains a 'limp-home' facility that is usually termed LOS or Limited Operating Strategy. This means that in the event of a serious fault in one or more of the sensor circuits, and the system perceives that a sensor is operating outside of its design parameters, the EMS will substitute a fixed default value in place of the defective sensor. A fault that puts the circuit outside of its design parameters would normally be a short circuit or when the circuit is virtually open. Other sensor faults may not necessarily cause a code to be logged.

The LOS is a safety system which allows the engine to operate at a reduced efficiency level. and allows the vehicle to be driven to a service area. Some LOS systems are so smart that the driver may be unaware that a fault has occurred during most of the vehicle's operating conditions.

However, since the substituted values are usually those of a hot (sometimes warm) engine, cold starting and running during the warm-up period may be less than satisfactory. Also, failure of a major sensor, ie the AFS or the MAP sensor, will tend to make driving conditions less easy.

The instrument panel warning light (where fitted) is switched on to indicate that a fault has occurred. Some systems (for example Ford) may also lock the timing to a set figure (with no timing advance) and allow the fuel pump to run continuously.

Adaptive systems

The ECM is adaptive to changing engine operating characteristics, and constantly monitors the data from the various sensors (ie AFS or MAP, ATS, CTS. TPS etc). As the engine or its components wear, the ECM reacts to new circumstances by adopting the changed values as a correction to the basic map.

When the adaptive map is used in conjunction with the oxygen sensor (OS) on petrol engines, the ECM is able to respond much more quickly and retain tighter control over the changing gases in the exhaust system. During closed-loop operation, the basic injection value is determined by the values stored in the map for a specific rpm and load. If the basic injection value causes exhaust emissions outside of the Lambda value (ie 0.97 to 1.03 AFR) the mixture would be too rich or too lean and the OS would signal the ECM, which in turn will correct the mixture. However, this response takes a little time and so the ECM learns a correction value and adds this 'adaptive' values to the basic map. From now on, under most operating conditions, the emissions will be very close to Lambda and so, after reference to the OS signal, the ECM will only need to make small corrections to keep it that way.

Adaptation and correction of the map occurs during the following engine operations.
a) CFSV operation
b) ISCV operation
c) Idle speed and mixture adjustment
d) Part-load mixture adjustment

Operation of the CFSV introduces a combustible mixture to the engine that is compensated for by the fuel evaporation adaptive correction values after detection by the OS.

At idle speed the system will settle down to idle at the best speed for each individual application. Most adaptive systems will lose their settings if the battery is disconnected. Once the battery is reconnected and the engine is restarted, the system will need to go through a relearning curve. This usually occurs fairly quickly, although idle quality may be poor until the adaptive process is completed.

Not all systems are affected by battery disconnection, and the Rover MEMS is an example of a system that uses non-volatile memory to retain adaptive settings when the battery is disconnected.

When one or more system components have been renewed, the ECM will need to relearn the new values, and this can sometimes create operating problems until the ECM has completed the process.

Rogue adaptive function

The danger with an adaptive function is that sometimes an erroneous signal may be adopted as a valid measurement, and this may create an operating problem. If the erroneous signal is not serious enough to generate a fault code, the fault may remain undetected.

In some instances the ECM can become confused, and the adaptive values could become corrupted. This may cause operational problems and a system check will reveal 'no fault found'. Disconnecting the vehicle battery may effect a cure, since the re-calibration will reset the ECM default base values.

Electronic ignition

General description

Reduced exhaust emissions figures highly in automotive design these days, and much of the rapid spread of electronics in engine controls is driven by the need to reduce pollution to acceptable levels. Where the ignition system is inherently more reliable, and the need for regular tune-ups are reduced, exhaust emissions will also be reduced.

A very rich mixture is easy to burn because the molecules of fuel are packed closely together. Once fired, the spark propagates easily and the flame front spreads easily and quickly. As the mixture becomes weaker, a more powerful spark is required.

But actually a powerful spark is of little use in burning a lean mixture. What is more important is the energy to prolong the spark. Mainly this is because there are less molecules of fuel in the mixture, and the spark will easily 'blow out'. Thus, the spark must be prolonged for about 1.3 to 1.5 ms, and adequate energy is required in the primary circuit to accomplish this task.

In a standard CB points ignition system with a 12 volt battery, the primary resistance of the coil is approximately 3 ohms. By applying Ohm's Law : I = V/R, it is apparent that about 4 amps of current is running in the primary circuit. The same is also true for a circuit with a ballast resistor, because the value of the resistor is (typically) 1.5 ohms and is used with a coil of 1.5 ohms primary resistance. Total current in a ballasted circuit is also about 4 amps.

Because the level of secondary HT voltage depends upon the level of primary current, it is obvious that for an increase in secondary output, there must be first an increase in primary current.

In older technology systems the maximum current that a set of CB points will handle is about 4 to 5 amps and this limits the amount of energy developed by the ignition coil. Furthermore, as the CB wears it causes misfires, retarded timing, higher emissions and a need for regular 'tune-ups'. In addition, because the dwell angle is fixed, coil performance is unsatisfactory at both low and high engine speeds. We will examine the reasons for this in the next paragraph. The modern electronic ignition system is capable of high output at all engine speeds, and has now become the definitive ignition system so replacing the once-ubiquitous CB.

Constant-energy ignition

Electronic ignition, used with a coil of lower primary resistance, produces increased primary current resulting in increased coil output. Coupled with higher coil energy, this will produce an arc of longer duration at the spark plug, and enable a weaker mixture to be ignited with a corresponding improvement in economy and exhaust emissions. Improved reliability, better control of timing and longer periods between 'tune-ups' are other benefits over the conventional CB system.

Virtually all modern types of electronic ignition use the variable-dwell current-limiting constant-energy system.

If Ohm's Law: I = V/R is applied to a 12-volt system with a coil of 0.75 ohms primary resistance, it is apparent that the potential current will be about 16 amps. Because this current flow would be far too high for most conditions, the electronic ignition module utilises an in-built current limiter device that limits the energy to a pre-determined figure of about 8 amps to avoid circuit overload and coil overheating.

As rpm and voltage rise and dwell increases with a equivalent rise in current flow, current limitation occurs earlier. As rpm, voltage and dwell levels decrease, with an equivalent drop in current flow, current limitation occurs later. Current is thus always limited to around 8 amps, irrespective of engine operating conditions. The advantage of having a high current reserve ensures a rapid build-up of coil energy. This results in a faster spark rise time, and sufficient energy is always available to maintain current flow across the spark plug gap for the required spark duration of about 1.3 to 1.5 ms.

Dependence by the ignition system on the battery voltage, temperature and engine speed is therefore much reduced.

At slow engine speeds, when the ignition cycle is relatively long, the dwell angle remains small. Also, because the current flows for shorter periods of time, the coil remains cool. As the engine speed increases, the dwell angle progressively increases to provide the necessary high speed coil saturation.

Let us now look at how this works in real terms. First a four-cylinder engine with a CB system. Refer to chart 'A'.

Chart A – CB points system

RPM	Dwell °	Dwell (time)
800	*45° (50%)*	*16 ms*
1600	*45°*	*8 ms*
3200	*45°*	*4 ms*
6400	*45°*	*2 ms*

A dwell of between 3 and 6 ms is required for good coil saturation. Referring to the CB chart, it is apparent that as rpm rises the points cannot 'dwell' closed for a long enough period to ensure a good spark. Good coil saturation is essential if the spark is to be adequate. It is obvious that this system will run out of spark at high engine rpm. The quality of the spark varies considerably, and due to the (comparatively) prolonged duration of points closure at low speeds, the primary circuit becomes overheated thus reducing the longevity of the CB and coil.

Let us now apply the same logic to an electronic system on a four-cylinder engine. Refer to chart 'B'.

Chart B – Electronic system

RPM	Dwell °	Dwell (time)
800	*10°*	*6 ms*
1600	*20°*	*6 ms*
3200	*40°*	*6 ms*
6400	*80°*	*6 ms*

The variable factor is now the dwell angle. Please note that not all vehicle manufacturers use a module 'on' time of 6 ms, but most will fall within the range of 3-6 ms. It is apparent that if the dwell time remains constant (ie constant energy) the spark would also be generally constant over most of the engine operating range. A degradation of spark would, however, be inevitable at high engine rpm because the dwell angle for our four-cylinder example could not exceed 90°. This concept then, is the basis for all modern electric ignition systems, irrespective of whether they are part of an EMS or a simple stand-alone electronic ignition. The short dwell time at low speeds has the further advantage of allowing the coil to remain cool under these conditions.

Early electronic ignition systems used a fixed dwell with some of the limitations of the conventional system, ie secondary output was restricted at higher engine speeds. In addition, these systems normally used a ballast resistor and by-pass to regulate current and voltage in a similar fashion to breaker points operation.

In the early days of electronic ignition, some manufacturers turned to capacitor discharge systems to provide a powerful spark. However, this spark only burned for about 0.1 ms, and has proved inadequate to the task (with one notable exception – the SAAB multiple spark system).

Ignition

Data on load (AFS or MAP), engine speed (CAS), engine temperature (CTS) and throttle position (TPS) are collected by the ECM, which then refers to the digital map stored within its microprocessor. This map contains an advance angle for each operating condition, and thus the best ignition advance angle for a particular operating condition can be determined. The ECM looks-up the correct dwell duration and timing point, and signals the amplifier – which in turn switches the coil negative terminal to achieve correctly-timed ignition *(see illustration 2.18)*.

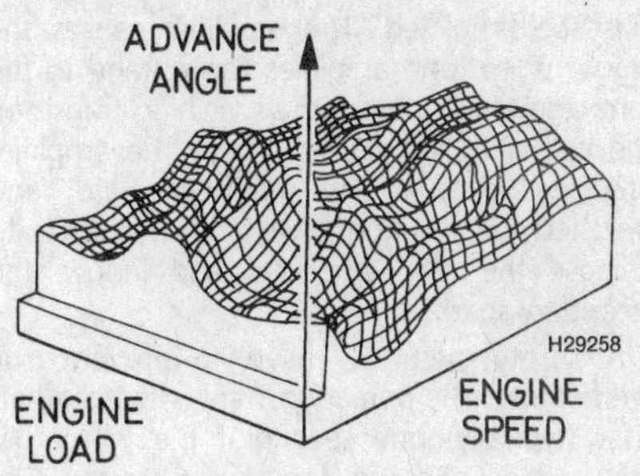

2.18 Typical timing map

Air and coolant temperatures are also monitored by the ECM, and the timing is automatically retarded by a knock sensor (where so equipped) if engine knock occurs.

Ignition amplifier operation

In a conventional ignition, the CB switches the negative side of the coil on and off to produce a spark. The electronic system is very similar in operation but uses (typically) a pulse generator and amplifier to achieve the same result.

A pulse generator provides correctly-timed electrical pulses for the amplifier to trigger the ignition by switching off the coil negative (-) terminal. The pulse generator can be an inductive magnet located inside the distributor or adjacent to the flywheel. Or it may use a Hall-effect generator to produce the pulse *(see illustration 2.19)*.

The voltage of the generated pulse is too weak to operate the switching transistor and

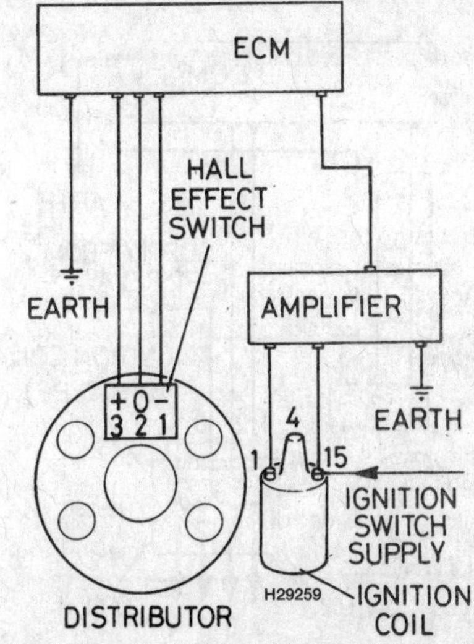

2.19 Typical ignition with Hall-effect trigger

must be amplified. The amplifier senses the trigger pulse, and amplifies the voltage to the correct level to operate the switching transistor. The coil is thus switched on by the amplifier circuitry to build the magnetic field, and switched off by the switching transistor to collapse the magnetic field and induce the secondary spark.

A safety circuit is used to prevent coil overheating. The primary current is switched off after (typically) one second if the ignition is switched on and the engine not started. The amplifier also contains the constant energy limiting circuitry.

EMS ignition operation

The EMS works very much along the principles of the above systems, and is usually triggered by a CAS or Hall-effect switch. The amplifier may be an integral part of the ECM, or may be a separate amplifier that is switched by the ECM. One advantage of a separate amplifier, is that if the amplifier fails, it is less costly to renew than a new ECM.

A typical sequence of events is as follows. The CAS or Hall-effect switch signals the ECM which 'looks up' the correct ignition dwell time and timing advance from data received from its sensors. The ECM then sends a signal to the amplifier (internal or external) which in turn switches the coil negative terminal. Where the ignition amplifier is combined with the ECM, the current-limiting function is often controlled by the ECM.

Distributor

The early distributor used in an EI system functioned much the same as its CB counterpart with mechanical and vacuum advance units. As the ECM developed, the mechanical and vacuum advance functions were removed from the distributor, and are now embodied in the ECM map. The modern distributor (where fitted) contains secondary HT components (distributor cap, rotor and HT leads) and serves to distribute the HT current from the coil secondary terminal to each spark plug in firing order.

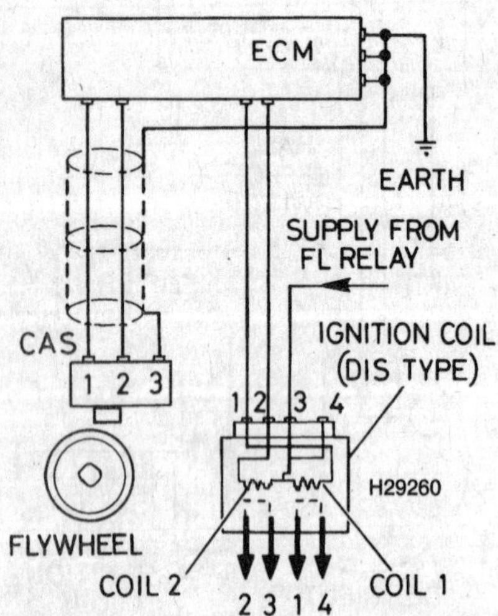

2.20 Typical direct ignition system (DIS) with CAS trigger

DIS (direct ignition system – sometimes termed 'distributorless')

Two reasons for discarding the distributor:

1) *An ignition coil should be capable of producing enough voltage to bridge the rotor gap and spark plug electrodes, with enough in reserve to provide a fairly good spark burn time. Even so, as much as 10 kV (10,000 volts) can be required to bridge the rotor gap. If the energy required to produce this voltage can be saved, the energy could be used more productively to prolong the burn time* (see illustration 2.20).
2) *The distributor cap and rotor account for a high proportion of electrical breakdowns and ignition problems on cars with EI. Further, RFI from a defective ignition system can disrupt ECM operation.*

By discarding the distributor and firing both spark plugs together, the coil energy saved can be used to maintain the burn time across the plug gap. About 3 kV is still required to fire the 'wasted spark' plug, but this is far less than that required to bridge the rotor gap.

Wasted spark DIS operation

To understand what is actually happening, we first need to look at some scientific theory. Whatever the polarity of the engine earth, the plug body and earth electrode will always have a positive polarity. Also, the spark plug central electrode will always be negative polarity. Polarity in this context has nothing to do with whether the vehicle is positive- or negative-earth.

The negative-earth EI system provides a spark of negative polarity from the ignition coil, and the spark will jump from the central electrode to the side (earth) electrode. The negative spark is provided by the coil high tension output when the coil negative terminal is connected to earth through a CB or other switching circuit.

In a positive-earth system, the spark will jump from the earth electrode (still positive polarity) to the central electrode (still negative polarity). Thus the system provides a positive spark. Electrons move more readily from negative to positive, and more readily from a hot object (plug central electrode) to a cold object (plug earth electrode).

Therefore, less HT voltage is required to fire the plug when a negative-polarity spark is used. This means that with conventional ignition, there is less wear in a negative spark system, and until recently that was the norm. When the coil low tension terminals are incorrectly connected to give reverse polarity, 20 to 40% more voltage is required to bridge the spark plug gap.

However, in our distributorless system the spark travels through the central electrode of one plug to earth (negative spark). The spark then travels through earth and the earth electrode to the central electrode of the companion cylinder (positive spark) and back to the coil. The circuit is thus completed. In our four-cylinder engine, we therefore have two negative and two positive sparks.

This principle was used with some success in Citroën 2CV and Dyane models, and was designed for simplicity without the expense of a distributor. Coil output was very low, and the Citroën would tend to foul plugs on the positive spark plug. Many owners swapped the plugs every 3000 miles to even up wear.

In the modern DIS system, an EMS is used in conjunction with a low primary resistance coil, CAS and other engine sensors to control the coil, spark energy and ignition timing. Because of the very high level of coil energy, a positive spark is no longer considered a problem.

Although the ignition system may be termed DIS, the basic operation is similar to models with conventional ignition. In a DIS or so called 'wasted spark' system, a double-ended coil is used to fire two plugs at the same time. This means that the system can only be used where two cylinders rise and fall together. Two pairs of coils will therefore be required for a four-cylinder engine, and three pairs of coils for a six-cylinder engine.

One cylinder will fire on the compression stroke and the companion cylinder will fire on the exhaust stroke where the spark is 'wasted'. Each ignition coil receives a voltage supply from the ignition switch, and a separate dwell connection to the amplifier (or ECM). In addition, separate connections for each coil are made between the ECM and the amplifier. In effect, the ECM and amplifier contains two separate circuits so that each coil can be switched individually and alternately.

A CAS is used to trigger the timing moment for cylinders 1 and 4, and the engine ECM calculates the correct timing for cylinders 2 and 3.

Ignition timing adjustment

Few modern systems allow adjustments to the ignition timing, irrespective of whether the model is equipped with DIS ignition or a distributor. Where adjustment is possible, some systems utilise an octane adjuster to enable small adjustments to be made. Alternatively, the ECM may be placed into a base timing mode with the aid of a handheld FCR, and the base timing may then be adjusted by turning the distributor.

Octane coding

An octane coding connector or a jumper cable may be provided, to enable the ECM to adopt certain timing characteristics that suit the various octane grades of fuel. In some instances, when the ECM recognises the installation of a certain octane coding connector, the ECM may modify

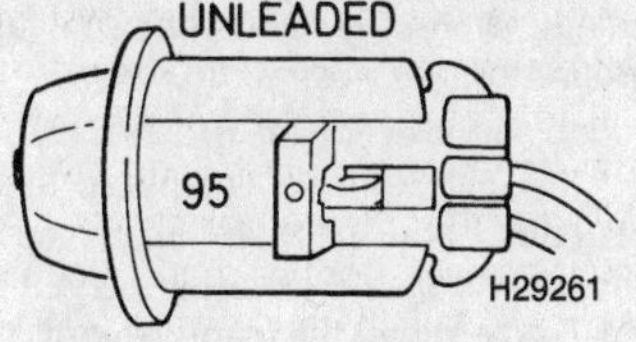

2.21 *Typical octane coding plug*

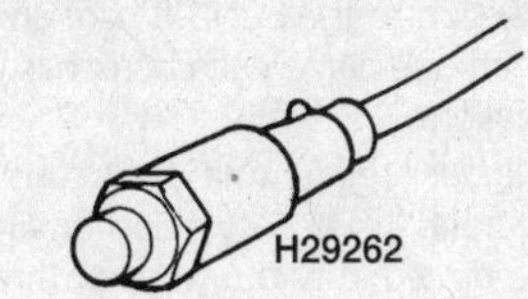

2.22 *Typical knock sensor (KS)*

both the basic timing and fuel injection maps *(see illustration 2.21)*.

The ECM provides a 5.0 volt reference feed to the octane connector multi-plug. When this cable is connected to earth, via a resistor (octane plug), a voltage of lower value than 5.0 volt will be returned to the ECM. Once the ECM sees the lower voltage value, appropriate adjustments are made to its internal program.

Knock sensor

The optimal ignition timing (at engine speeds greater than idle) for a given high-compression engine is quite close to the point where engine knock occurs. However, running so close to the point of knock occurrence, means that knock will certainly occur on one or more cylinders at certain times during the engine operating cycle.

Since knock may occur at a different moment in each individual cylinder, the ECM employs a knock control processor to pinpoint the actual cylinder or cylinders that are knocking. The knock sensor is mounted on the engine block, and consists of a piezo-ceramic measuring element that responds to engine noise oscillations. This signal is converted to a voltage signal that is proportional to the level of knock, and returned to the ECM for evaluation and action *(see illustration 2.22)*.

The knocking usual frequency is in the 6 to 15kHz frequency band.

The ECM will analyse the noise from each individual cylinder and set a reference noise level for that cylinder based upon the average of the noise over a pre-determined period. If the noise level exceeds the reference level by a certain amount, the ECM identifies the presence of engine knock.

Initially, timing will occur at its optimal ignition point. Once knock is identified, the knock control microprocessor retards the ignition timing for that cylinder or cylinders by degrees. After knocking ceases, the timing is advanced until the reference timing value is achieved or knock occurs once more when the timing is once more retarded. This process continually occurs so that all cylinders will consistently run at their optimum timing.

Inductive magnet signal generator

The inductive trigger is mounted in the distributor, and is essentially a permanent magnet and pole piece attached to a stator. The two most common types in current service are the pick-up limb or the annular coil *(see illustration 2.23)*.

If a reluctor containing one arm for every cylinder is connected to the distributor cam and rotated within the magnetic field, an AC voltage will be generated.

The AC voltage will fluctuate between positive and negative, and when the arm on the reluctor and stator are directly opposite, the voltage will change from positive to negative. This is the moment of trigger. Two wires connect the trigger unit to the module, and the module will amplify this signal to switch off the coil negative terminal. Without amplification, the signal would be too weak to accomplish this aim.

From the switch-off point, the module can determine the switch-on point and therefore calculate the correct dwell duration.

Crank angle sensor (CAS)

The CAS works on the same principle as the inductive signal generator. A number of steel pegs or pins are set at regular intervals around the circumference of the flywheel or crankshaft. Typically, a pin may be set at every 10° – thirty-six in all – but the number does vary between different VM's. The flywheel thus becomes a reluctor.

A permanent magnet inductive signal generator is mounted in close proximity to the flywheel, where it radiates a magnetic field. As the flywheel spins and the pins are rotated in the magnetic field, an alternating (AC) waveform is delivered to the ECM to indicate speed of rotation.

If a pin is intentionally omitted at two points on the flywheel, or by contrast a double pin is used, the signal will vary at these points, and a reference to TDC will be returned to the ECM. The location of the positional signal is not at TDC, but may be at 90° BTDC or some other point fixed by the VM.

In addition, as the flywheel spins, the missing pins or double pins cause a variance of the signal which is returned to the ECM as reference to the TDC position.

Although most modern systems utilise a single CAS, some of the older systems use two CAS – one for RPM and one for position. The waveform produced by each type of CAS will be slightly different.

The peak-to-peak voltage of the speed signal varies according to engine speed, and can vary from 5 volts at idle to over 100 volts at 6000 rpm. Because computers prefer their data as on/off signals, an analogue-to-digital converter (ADC) transforms the AC pulse into a digital signal *(see illustration 2.24)*.

When used, the CAS provides the primary signal to initiate both ignition and fuelling.

Hall-effect trigger operation

Hall-effect ignition is usually fitted to vehicles equipped with a distributor, and this is where the Hall switch is located. The module supplies a voltage slightly under nbv to the Hall-effect switch in the distributor. An earth return wire completes the circuit back to the module. Opposite the Hall switch is a magnet whose field causes the switch to return a small voltage back to the module. Attached to the distributor shaft is a rotor with the same number of cut-outs as cylinders. Passing the rotor between the switch and the magnet will cause the switch to be turned off and on. As the cut-out space proceeds past the switch, a voltage is returned to the amplifier via a third wire termed the output wire. When the solid portion comes between the switch and magnet, the voltage is turned off as the magnetic field is diverted. The number of voltages returned per four-stroke engine cycle will equal the number of cut-outs. Essentially then, the voltage signal returned is either voltage or no voltage, and the waveform produced is that of a square wave.

The main advantage of the Hall-effect signal is that output is in digital form, and this means faster processing by the ECM. However, Hall-effect ignition is usually fitted to vehicles equipped with a distributor, and since the modern trend is to remove the distributor, Hall-effect ignition is tending to be less used in modern vehicles.

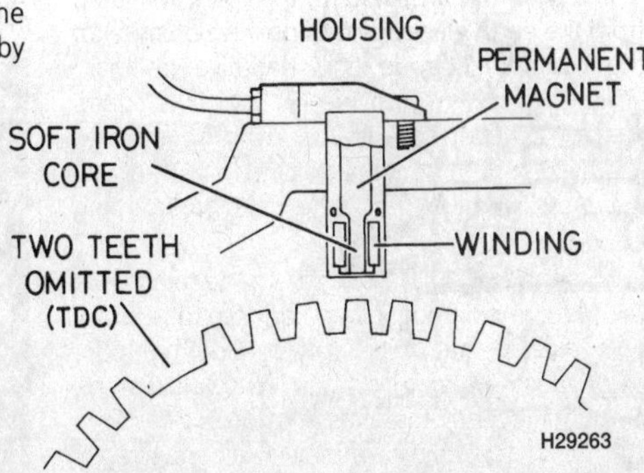

2.23 *Bosch distributor with inductive trigger*

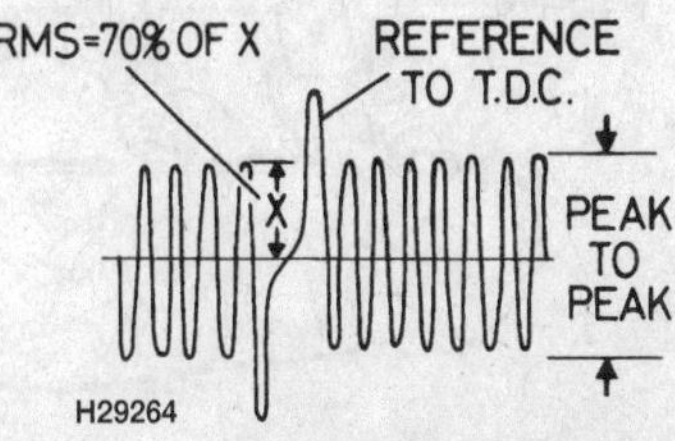

2.24 *AC voltage waveform from inductive trigger*

When used, the Hall-effect sensor provides the primary signal to initiate both ignition and fuelling.

Electronic fuel injection systems

General

The AFR in carburettor systems varies according to intake air temperature and density. This makes accurate control of the correct mixture strength virtually impossible. In contrast, the electronic fuel injection system uses a number of sensors to monitor all conditions that will affect the AFR, and responds very quickly to the changing operating conditions. This means that very accurate control of the mixture is possible.

The ECM is programmed to calculate a basic fuel injection duration. This is the time that the ECM will actually hold the injector open, and is usually measured in milliseconds (ms). Data from the engine sensors, which include the AFS or MAP sensor, the TPS or throttle pot, and the CTS and ATS is collected and used by the ECM to modify this figure so that the quantity of injected fuel is computed exactly for all conditions of temperature, speed and load. The ECM will also actuate the ISCV to automatically control idle speed depending on coolant temperature and engine load at idle.

There are several different kinds of injection system in current use, but the three main types are:

Simultaneous multi-point fuel injection (MPi)
Single point injection (SPi)
Sequential fuel injection

Simultaneous multi-point injection (MPi)

This is the most common type of injection system in current use. A number of injectors are looped together in a parallel 'bank', with a single connection to the ECM. Where an engine has more than one bank, each bank has its own ECM connection.

In a 4-cylinder engine, one bank connects all of the injectors. In a 6-cylinder engine, the injectors are placed in two groups of three, and an 8-cylinder engine has two groups of four; V6 and V8 engine injector groups are termed left or right 'bank'. In 12-cylinder engines, the injectors are placed into four groups of three cylinders. Two power resistors control two groups each.

The injectors are triggered from a reference signal, which may originate from the ignition system or from a timing pulse at the CAS. Normally, the injectors operate twice per complete engine cycle. Half of the fuel required is injected onto the back of a closed inlet valve, waiting for it to open, and the other half is injected as the valve opens for the intake stroke. Once the valve has opened, the fuel enters the cylinder in the normal way *(see illustration 2.25)*.

This system is fairly effective and usually works quite well. It is also cheaper to develop than a full-blown sequential system, which makes it very popular amongst vehicle manufacturers.

However, a particular type of fault can sometimes afflict vehicles with a high mileage (or sometimes even a low mileage). What happens is that a build-up of carbon on the back of the inlet valve allows the injected fuel to 'soak' into the carbon whilst awaiting its turn to be injected into the cylinder. This can cause a slight hesitation on acceleration, as the air fuel ratio becomes too lean. A decoke is the only real cure.

Single point injection (SPi)

Sometimes known as 'throttle body injection', the petrol SPi system is no longer used in recent years. Although less costly than the MPi equivalent, a single injection point is not able to control the fuel injection process with the accuracy and flexiblilty demanded by the later emission standards. SPi uses the same sensors as the MPi systems. A single injector (normally of the current-controlled type) injects fuel into a distributing manifold in much the same fashion as a carburettor *(see illustration 2.26)*.

Although the injection of fuel is much more precise, the problems of manifold heating becomes critical and the warm-up period must be carefully controlled if driveability is not to be impaired. Furthermore, the manifold is said to be of the 'wet' type. This term means that fuel is present in the manifold. In contrast, a multi-point injection system is said to be of the 'dry' type because fuel is injected into the inlet stub to the valve, and thus only air is present in the manifold.

Current-controlled/pulse-modulated, and peak-and-hold injector

Some systems rely on the principle that more current is required to open the injector than to actually keep it open. The injector solenoid circuit is also earthed – but for only about one millisecond (ms: one thousandth of a second) – which is just long enough to open the injector. The opening circuit is then switched off, and another circuit rapidly closed and opened, to apply a small holding current to the injector so that it remains open for the correct time duration. The pulsing is so blindingly fast that the injector does not have time to close, and current flow is therefore much reduced. This type of system is known as 'current-controlled' or 'pulse-modulated'. A variation of this method is termed 'peak-and-hold'. After the initial pulse, the second circuit holds the injector open without the rapid pulse modulation.

Sequential fuel injection

Both simultaneous multi-point and single point injection systems have succumbed to the sequential multi-point system, where the injectors open in cylinder sequence. Emissions can be significantly reduced with this type – particularly if the engine suffers a mechanical or ignition problem. The sequential system uses the same sensors as the multi-point systems. However, an additional sensor pinpoints the correct cylinder for the sequential system. The most common form of sensor used to indicate a specific cylinder is one that generates an AC signal similar to the signal generated by the CAS. This sensor is often attached to the camshaft. A Hall-effect device located in the distributor is another sensor sometimes employed.

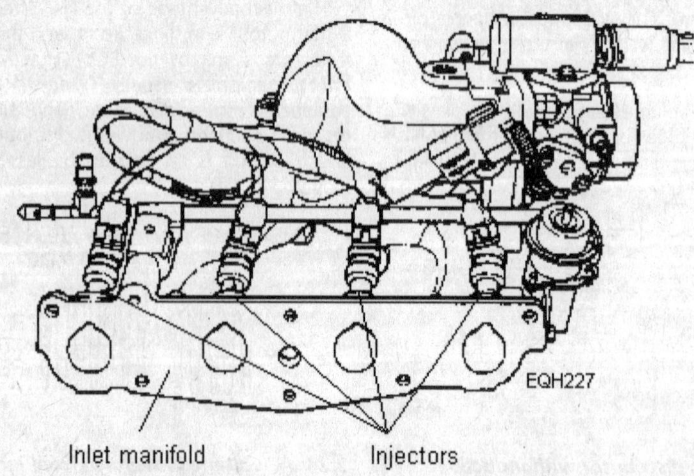

2.25 Multi-point injection system

2.26 Nissan single point injection system and hot film AFS

Fuel injection system engine sensors

The electronic fuel injection system comprises a number of components. The components can be divided into two groups – sensors and actuators. Sensors send data to the ECM so that it can actuate the injector and idle control correctly. The electronic fuel injection system therefore consists of the following components.

All engine sensors have a resistance that responds to either temperature or load. As the temperature or load varies, the resistance value of the sensor also varies. In turn, a resistance variation will cause a variation in the voltage being returned to the ECM. By measuring this voltage, according to certain parameters, we can ascertain correct operation of any particular sensor at a range of engine temperatures or loads.

Injection system sensors can generally be divided into several groups. The first and second group contain sensors that are termed variable signal output types. We could define the sensors in these groups as being either a two-wire or a three-wire type. The third group may have a combination of two- and three-wire sensors incorporated into its body. An example of this type is the AFS fitted to many Bosch systems. This AFS combines the AFS, ATS and CO pot into the AFS unit – all with a common earth. The final group contains sensors that are basically switches (ie Lambda and TPS).

The two-wire sensors utilise an earth wire and a 5-volt supply wire in a circuit that begins and ends at the ECM. The supply wire also doubles as the output wire in the following manner. Once the supply and earth wires are connected to the sensor, the resistance value of the sensor causes the voltage value of the supply to vary. Thus, if we take an example of a two-wire CTS, the supply value of 5 volts will reduce (typically) to between two and three volts if the engine is cold (20°C), and to between 0.6 and 0.8 volts once the engine has become warm (80°C). Examples would include the ATS, CTS and CO pot.

The three-wire sensor has a voltage supply of 5 volts, an earth connection (often made through the ECM) and an output (signal) wire. The output wire returns a variable voltage signal to the ECM. The two most common forms of output are by resistance track and wiper arm, or via a transducer. Examples would include the AFS and TP (wiper arm) and MAP (transducer).

Load sensor

The sensor which detects the load placed upon the engine may be the AFS, MAP sensor or even the TPS. The ECM will detect the load from the signal that is returned from whichever of the three components is used. The AFS signals the volume of air that passes into the engine, the MAP sensor signals the absolute pressure from the depression in the inlet manifold with respect to atmospheric pressure, and the TPS signals the throttle position. However, the first two are more commonly used to signal the main load, and although the TPS is sometimes used to signal load without aid of another sensor, in practice the signal from the TPS is more usually used to trim the signal that is returned from one of the other two sensors.

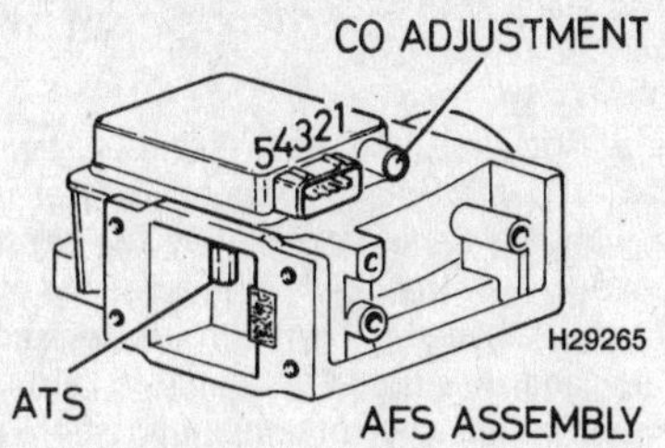

2.27 Motronic type airflow sensor (AFS)

Vane-type airflow sensor (AFS)

The AFS is normally located between the air filter and the throttle body. As air flows through the sensor, it deflects a vane (flap). The greater the volume of air, the more will the flap be deflected. The vane is connected to a wiper arm which wipes a potentiometer resistance track and so varies the resistance of the track. This allows a variable voltage signal to be returned to the ECM *(see illustration 2.27)*.

Three wires are used by the circuitry of this sensor, and it is often referred to as a three-wire sensor. A 5-volt reference voltage is applied to the resistance track with the other end connected to the sensor earth return. The third wire is connected to the wiper arm.

From the voltage returned, the ECM is able to calculate the volume of air (load) entering the engine and this is used to calculate the main fuel injection duration. To smooth out inlet pulses, a damper is connected to the AFS vane. The AFS exerts a major influence on the volume of fuel injected.

Hot-wire airflow sensor (AFS) – often termed 'mass airflow meter'

Because the volume, temperature and density of air, at all altitudes is accurately measured – the hot-wire or hot-film AFS is becoming increasingly popular as a technically superior alternative to the vane and MAP sensor types *(refer to illustration 2.26)*.

The AFS may be mounted in the airflow trunking between the air filter and the engine, or on the throttle body. A voltage of either 5 or 12 volts, according to system, is applied to the AFS unit.

Air passes through the AFS body into the engine. A small quantity of air is drawn into a by-pass channel containing two wires. These wires are known as the sensing wire and the compensating wire.

A small current is applied to the compensating wire, which remains unheated. As air passes over the wire, its resistance and current change and the AFS is able to determine the temperature of the incoming air. The sensing wire is heated to a temperature of 100°C above that of the compensating wire. Air passing over the sensing wire causes it to become cooler, and its current and resistance values change. More current is passed through the sensing wire so that it remains 100°C above that of the compensating wire. An output voltage, proportional to the current applied to the sensing wire, is returned to the ECM.

The value of this voltage is directly related to the volume, temperature and density of air introduced into the engine. The beauty of the hot-wire system allows automatic compensation for altitudes from sea-level to mountain top, and the ECM will correctly calculate the AFR under virtually all conditions.

Manifold absolute pressure (MAP) sensor

The MAP sensor is often used as an inexpensive alternative to the airflow meter. It is less accurate than the various forms of AFS but is often chosen on grounds of cost rather than its technical attributes. The MAP sensor measures the manifold vacuum or pressure, and uses a transducer to convert the signal to an electrical signal which is returned to the ECM. The unit may be designed as an independent sensor that is located in the engine compartment or integral with the ECM *(see illustration 2.28)*.

Used in both MPi and SPi systems, the MAP sensor is particularly popular in SPi systems.

A vacuum hose connects the MAP sensor and the inlet manifold. Manifold vacuum/pressure acts upon the MAP sensor diaphragm, and the ECM internally converts the pressure into an electrical signal. MAP is calculated from the

2.28 Manifold absolute pressure (MAP) sensor assembly

1 MAP sensor
2 Mounting screws (some units are simply clipped into place)
3 MAP sensor vacuum line (goes to the intake manifold)
4 MAP sensor electrical connector (usually goes to the main harness)

formula: Atmospheric Pressure less Manifold Pressure = Manifold Absolute Pressure.

Using the speed/density method, the ECM calculates the AFR from the MAP signal and the speed of the engine. The speed/density method is based on the theory that the engine will draw in a fixed volume of air per revolution. This method can never be as accurate as using an AFS to accurately measure airflow and then calculating the AFR from the volume or mass of air that is entering the engine.

When manifold vacuum is high (ie idle condition), MAP is moderately low and the ECM provides less fuel. When manifold vacuum is low (ie wide-open throttle), MAP is high and the ECM provides more fuel.

MAP sensors may take one of two forms. The most common form is the analogue sensor where the voltage output is proportional to the load. The other form is that of a digital sensor which is mainly used by the Ford EEC IV/V system.

A digital MAP sensor sends a square waveform in the form of a frequency. As the load increases, the frequency also increases and the time in ms between pulses becomes shorter. An ECM will respond much faster to a digital signal, because an analogue-to-digital converter is no longer necessary.

Where the manifold is of the 'wet' type (ie SPi), the changing pressures in the manifold will cause fuel to enter the vacuum hose, where it may eventually reach the MAP sensor. Installation of a fuel trap and careful routing of the vacuum hose may slow down the ingress of fuel. However, once fuel reaches the MAP sensor, its diaphragm may be adversely affected. If the MAP sensor is an independent unit, renewal is comparatively inexpensive. A new ECM may be necessary for the integral types.

The inlet manifold on MPi models is a 'dry' manifold. Since fuel does not enter the manifold – due to injection onto the back of the inlet valve – there is no risk of fuel being drawn into the MAP sensor to contaminate the diaphragm, and a fuel trap is not used.

Cylinder identification sensor (sequential injection engines only)

In non-sequential systems, the ECM does not recognise number one cylinder, or indeed even the firing order. This is because it is deemed unnecessary. When the crankshaft or distributor provides a timing signal, the correct cylinder is identified by the mechanical position of the crankshaft, camshaft, valves and ignition rotor.

On models fitted with sequential injection, the ECM must determine which cylinder is on its firing stroke, and the CID sensor provides the appropriate signal. The CID sensor operates on the inductive principle, and is a permanent magnet device mounted adjacent to the camshaft. The operation of this sensor differs according to VM, and a more detailed description will be described in the appropriate Chapter.

Air and coolant temperature sensors

These sensors are thermistors that usually work on the NTC principle. A number of vehicles fitted with the Renix system may use sensors that work on the PTC principle. As the temperature changes, the thermistor resistance changes, and thus the ECM is able to calculate the air and engine coolant temperature from the level of voltage (or current) that is registered on the sensor signal wire *(see illustration 2.29)*.

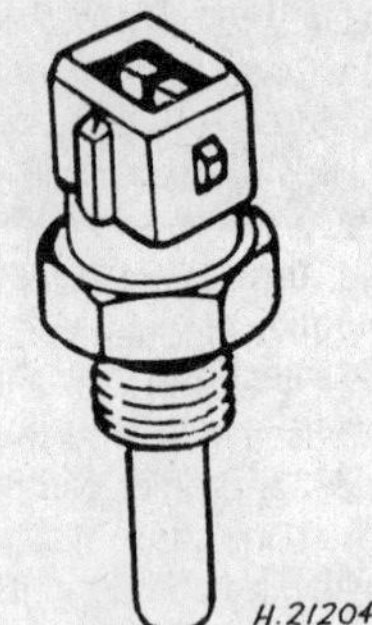

2.29 Typical air or coolant temperature sensor (ATS or CTS)

NTC & PTC (negative & positive temperature co-efficient)

An NTC resistor decreases (negatively) in resistance as the temperature rise, and a PTC type increases (positively) on temperature rise. Both types are example of the two-wire resistor, but in one case the voltage increases and in the other case the voltage decreases.

Air temperature sensor (ATS)

The ATS is a two-wire thermistor that measures the air temperature in the manifold. Because the density of air varies in inverse proportion to the temperature, the ATS signal allows more accurate assessment of the volume of air entering the engine.

The open circuit supply to the sensor is applied at 5.0 volt reference level, and the earth path is usually made through the sensor return. A variable voltage signal is returned to the ECM based upon the air temperature.

The ATS may be mounted in various locations according to the design features of a particular vehicle or model. Locations may include the inlet manifold, air filter or AFS intake. The location has an important bearing on the air temperature range that is sensed in any particular engine.

The range of the air temperature signal will vary considerably between a cold and a hot engine according to the location of the ATS. For example, when the ATS is located in the air filter inlet or in the AFS intake, the air temperature will vary according to engine compartment heat, and the range is likely to be in the range of 20°C to 40°C. However, when the ATS is mounted in the inlet manifold or in the throttle body (some SPi systems), the temperature range will be much greater because the air temperature may reach 70°C when the engine is hot.

Some vehicles may utilise two sensors, one to measure ambient air temperature in the engine compartment, and the other to measure the intake air temperature. This may be particularly important in turbocharged applications.

Although the majority of vehicles are equipped with an ATS that conforms to the NTC principle, a number of vehicles that are fitted with the Renix system may use sensors that work on the PTC principle.

Coolant temperature sensor (CTS)

The CTS is a two-wire thermistor that measures the coolant temperature. The CTS is immersed in the engine coolant, and contains a variable resistor that usually operates on the NTC principle.

When the engine is cold, the resistance is quite high. Once the engine is started and begins to warm-up, the coolant becomes hotter, and this causes a change in the CTS resistance. As the CTS becomes hotter, the resistance of the CTS reduces (NTC principle) and this returns a variable voltage signal to the ECM based upon the coolant temperature.

The open circuit supply to the sensor is at a 5.0 volt reference level, and this voltage reduces to a value that depends upon the resistance of the CTS resistor. Normal operating temperature is usually from 80°C to 100°C. The ECM uses the CTS signal as a main correction factor when calculating ignition timing and injection duration.

A number of vehicles fitted with the Renix system may use sensors that work on the PTC principle. In this case, the resistance and voltage will rise as the temperature rises.

Throttle signals

Throttle position may be signalled by a simple switch (TS), a potentiometer (TPS), or even a combination switch that contains both a TS and TPS. Sometimes, both switches are present in a system as separate components.

Throttle potentiometer/position sensor (TPS)

A TPS is provided to inform the ECM of idle position, deceleration, rate of acceleration and wide-open throttle (WOT) conditions. The TPS is a potentiometer with three wires. A 5-volt reference voltage is supplied to a resistance track with the other end connected to the sensor earth return. The third wire is connected to an arm which wipes along the resistance track and so varies the resistance and voltage of the signal returned to the ECM *(see illustration 2.30)*.

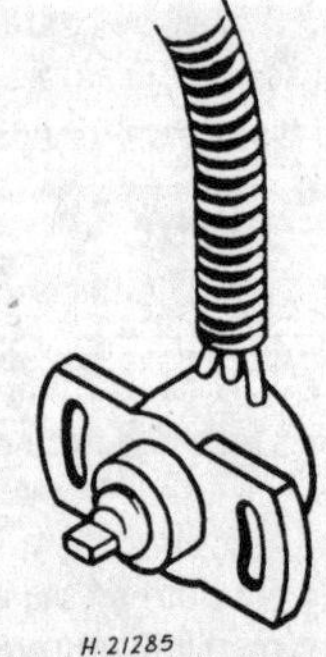

2.30 Typical throttle potentiometer sensor (TPS)

From the voltage returned, the ECM is able to calculate idle position (usually less than 0.7 volts), full-load (approximately 4.5 volts) and also how quickly the throttle is opened. During full-load operation, the ECM provides additional enrichment. During closed throttle operation above a certain rpm (deceleration), the ECM will cut off fuel injection. Injection will be reintroduced once the rpm returns to idle or the throttle is opened. It is adjustable on some models.

Throttle switch (TS)

The throttle switch informs the ECM when the engine is in idle mode. An additional contact is also usually provided to indicate wide-open throttle (WOT). Additional enrichment is often provided at idle and during full-throttle running. Each TS contact has two conditions – that of closed or open and the ECM is therefore able to recognise three different engine situations:

1) Throttle closed (idle contact closed).
2) Throttle open (idle contact open and WOT contact open).
3) Throttle wide open (idle contact open and WOT contact closed).

The TS is adjustable on some models.

Mixture adjustment

Catalyst-equipped petrol vehicles often have no form of CO adjustment. Where adjustment is possible, turning the screw only effects small changes in CO – and then only at the idle position. Once the throttle is opened from its idle stop, the volume of fuel injected depends entirely upon the pulse duration. There are two different types of CO adjustment in current use:

1) An air screw that varies air volume through an idle passage. As the screw position is varied, airflow acting upon the metering flap varies, causing the flap to change its idle position. The changed position results in an altered signal to the ECM, and the idle mixture will either increase or decrease. Usually fitted to older type vehicles.
2) A potentiometer ('pot') that varies the voltage signal being returned to the ECM. This sensor may be mounted upon the ECM, on the airflow meter, or even upon the inner engine compartment wing. It may be of the two- or three-wire type.

CO pot (non-cat petrol models only)

The CO pot mixture adjuster is a potentiometer that allows small changes to be made to the idle CO. The CO pot may be mounted directly on the ECM, or may be a separate three-wire sensor that is mounted in the engine compartment or on the AFS. In the case of the three-wire sensor, a 5.0 volt reference voltage is applied to the sensor and connected to the sensor earth return circuit. The third wire is the variable CO pot signal *(see illustration 2.31)*.

As the CO pot adjustment screw is turned, the change in resistance returns a voltage signal to the ECM that will result in a change in CO.

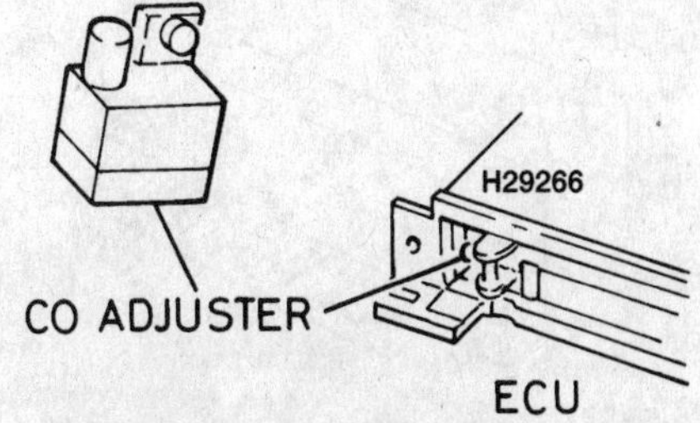

2.31 CO pot, external and ECM-located

Fuel injection system actuators

Fuel injector

The injector is a solenoid-operated valve that delivers an exact amount of fuel according to an opening duration signal from the ECM. Voltage to the injector is applied from the main relay or ignition switch, and the earth path is completed by the ECM for a period of time (called pulse duration). The pulse duration is very much dependent upon engine temperature, load, speed and operating conditions. When the magnetic solenoid closes, a back EMF voltage of up to 60 volts is induced *(see illustrations 2.32a and 2.32b)*.

Essentially, the injector is simply earthed by the ECM for a calculated period of time, During this time, the injector solenoid operates and fuel is injected. In modern diesel engines, the injector performs what are known as pre- or pilot injections. Several short injections are made immediately prior to the main injection period to 'soften' the combustion process. This results in lower exhaust emissions, and enhanced refinement (reduced diesel 'knock').

A fine filter is used to prevent debris damaging the precision action. However, gums and lacquers can build-up on this filter and on the injection pintle, eventually reducing fuel flow. Injection fouling is a serious problem on many injection systems.

In petrol engines, the fuel injectors are mounted in the inlet stubs to the engine inlet valves, so that a finely-atomised fuel spray is directed onto the back of each valve. When injectors are pulsed simultaneously, fuel will briefly rest upon the back of a valve before being drawn into a cylinder. On modern direct injection diesel engines, the injectors are mounted in the roof of the combustion chamber so that the fuel can be injected directly into the recess in the crown of the piston.

Idle speed control

Most modern vehicles have fully-automatic idle control, with no means of manual adjustment. Where adjustment is possible, this is usually effected by placing the idle system outside of the ECM's jurisdiction, and setting a base idle speed with a by-pass idle speed air screw.

Most modern petrol systems utilise an ECM-controlled ISCV or stepper motor to maintain idle speed under engine load. This prevents poor idle and stalling with heavy electrical loads and a lean mixture; the automatic idle system is also used to provide a higher idle during engine warm-up.

A small volume of air is allowed to by-pass the throttle plate. This air may pass through a hose, or through a port in the inlet manifold. The ISCV is mounted *in situ*, thus allowing the by-pass air to pass through the body of the valve. If the amount of air in the by-pass is varied, the idle speed can also be varied. On modern diesel engines, the idle speed is monitored by the ECM via the crankshaft position sensor, and controlled by means of the injection timing and quantity.

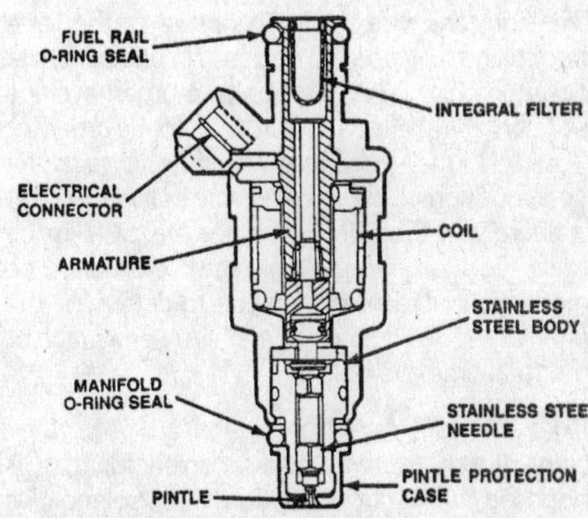

2.32a Petrol fuel injector

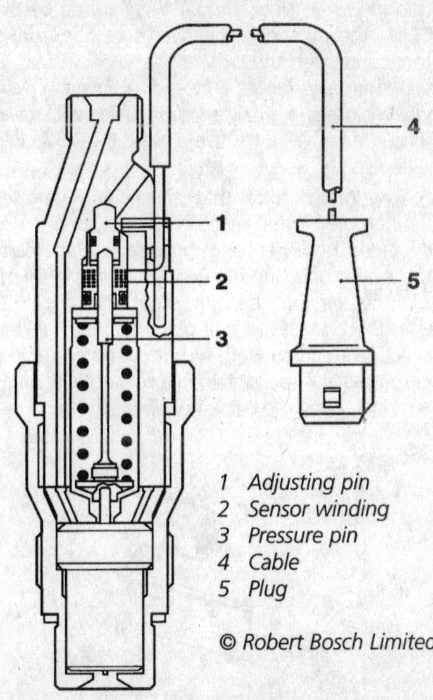

1 Adjusting pin
2 Sensor winding
3 Pressure pin
4 Cable
5 Plug

© Robert Bosch Limited

2.32b Diesel fuel injector with needle lift sensor

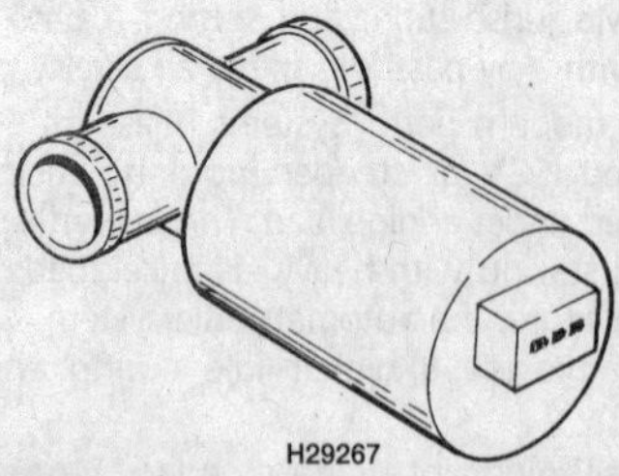

2.33 Idle speed control valve (ISCV), three-wire

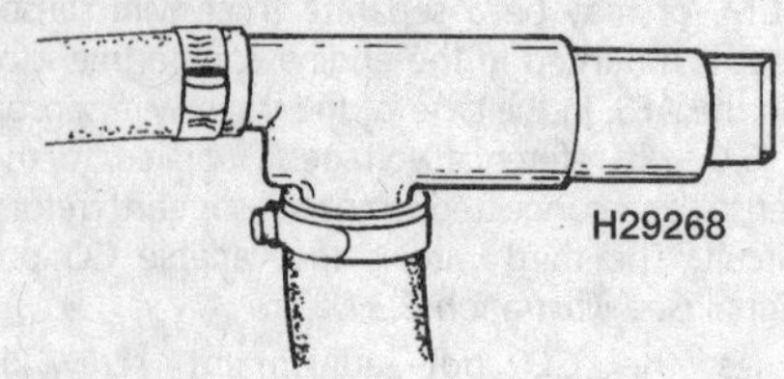

2.34 Idle speed control valve (ISCV), two-wire

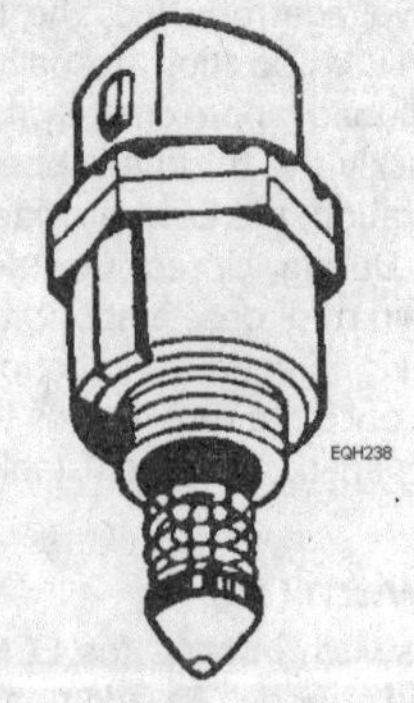

2.35 Idle control, stepper motor

ISCV (Bosch 3-wire)

Early Bosch systems use an ISCV attached to an DC electric motor that can be rotated clockwise or anti-clockwise by virtue of two opposing circuits *(see illustration 2.33)*. The motor is supplied with a voltage supply and two earth paths that are made through the ECM. When one path is earthed, the motor will attempt to rotate in one direction, and vice-versa. This prevents the valve from being fully opened or closed in one particular direction. The valve will thus take up an average position that reflects circuit bias to be open or closed. Normally, this bias would be towards the open position. By varying the time that each circuit is energised, the ECM will place the ISCV in the exact position required. A duty cycle can be measured on each earth circuit to determine the opening or closing time period as a percentage of the total time available.

ISCV (Bosch 2-wire)

Later Bosch systems use a solenoid that is opposed by a strong spring. The solenoid is supplied with a voltage supply and one earth path, made through the ECM *(see illustration 2.34)*. When the ISCV is earthed by the ECM, the solenoid will overcome the spring force and open the ISCV. If the solenoid fails, it will normally fail-safe closed. However, even when closed, a small amount of air will travel through the valve to give a basic (but low) idle speed.

The longer time that the ECM energises (holds open) the ISCV, the further open it will become. However, this process occurs many times a second (the frequency is about 110 Hz), and by varying the time that the circuit is energised, the ECM will place the ISCV in the exact position required. A duty cycle for the time energised can be obtained on the ISCV earth terminal, or at the corresponding ECM pin.

Ford ISCV

Ford use an ISCV which is very similar in operation to the later Bosch two-wire type. However, due to the use of a diode, the ISCV waveform is distorted to that of a sawtooth.

Stepper motor

The stepper motor may take several forms, and here are described the two most common systems:

1) *A motor is used to drive a valve which opens or closes an air by-pass passage in the inlet manifold (see illustration 2.35).*
2) *A motor is used to increment the throttle plate by so many steps, thereby allowing more or less air through the opening.*

Usually the motor is fed with a battery voltage supply connected to four earth paths. By pulsing the motor using a combination of the earth paths, the motor can be stepped to the correct position.

Auxiliary air valve (AAV)

The AAV is a control valve that is used to increase the idle speed during cold engine operation. It is mainly used in fairly early systems, and is not controlled by the ECM. The AAV is mounted in a hose that by-passes the throttle plate. A circular gate valve responds to temperature and allows extra air to by-pass the throttle when the engine is cold. During cold engine operation, the valve is open and so engine idle speed is increased. As the engine warms-up, the gate valve gradually closes until it is fully closed at normal operating temperature. In fact, the AAV is usually fully closed within 2 to 4 minutes of the engine starting from cold *(see illustration 2.36)*.

The AAV is usually connected to the fuel pump relay output terminal, so that it will only operate during the time that the engine is actually running. Radiated heat from the engine will affect valve operation, and allow the valve to remain closed when the engine is hot and is not being operated.

Vehicle speed sensor (VSS)

The VSS is used to advise the ECM of vehicle speed. Typically. The VSS operates upon the Hall-effect principle, and is often mounted on the gearbox or speedometer.

A voltage of approximately 12 volts is applied to the VSS from the ignition switch. As the speedometer cable turns, the Hall switch is alternately turned on and off, to return a square wave signal to the ECM. The signal frequency usually denotes the vehicle speed.

Fuel injection system relays

Due to the variations between the different systems, it is impossible to mention all the various relay combinations. The following is generally correct, although there are a number of systems that may use other methods.

One system relay may be used to control the whole fuel injection system. In that instance, the relay will have double contacts. Alternatively, two relays may be used to control the system. In that instance, the output from the main system relay will be used to supply the energising voltage to the second (fuel pump) relay.

In this book, the relay terminal numbers are usually annotated to the DIN standard, to which most (but not all) European VM's subscribe. In some instances, however, the VM will use his own method of terminal marking *(see illustrations 2.37 and 2.38)*.

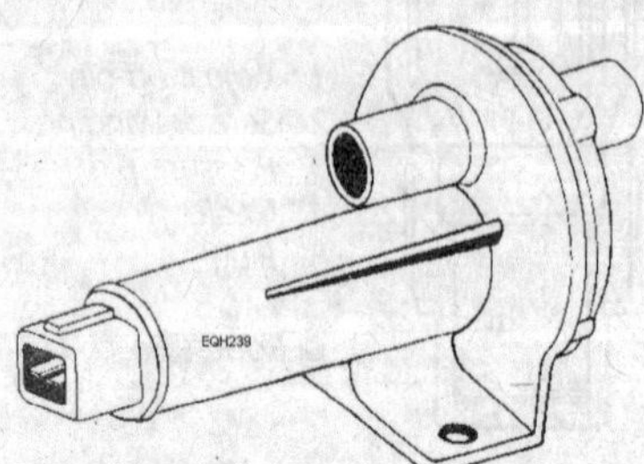

2.36 Auxiliary air valve (AAV)

2.37 Standard four-pin relay

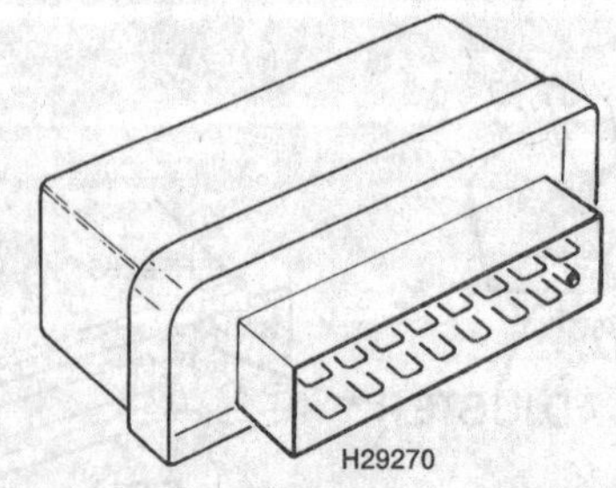

2.38 Fifteen-pin relay

DIN method of terminal annotation

30 Supply voltage direct from the battery positive terminal.
31 Earth return direct to the battery (little used in modern systems, may be found in older fuel injection systems).
85 Relay earth for energising system. May be connected direct to earth, or connected to earth through the ECM.
85b Relay earth for output. May be connected direct to earth, or connected to earth through the ECM.
86 Energising system supply. May be applied from the battery positive or through the ignition switch.
87 Output from the first relay or first relay winding. This terminal will often provide power to the second relay terminal 86 and provide voltage to the ECM, injectors, ISCV etc.
87b Output from the second relay or second relay winding. Often provides power to the fuel pump and OS etc.

Note: *In some relays, the terminals 30 & 87/87b and/or terminal 85 & 86 may be reversed. This is so that the internal relay diode protection path is reversed, and prevents voltage spikes during relay actuation from reversing and creating a short circuit.*

Main and fuel pump relays (separate relays)

Where this set-up is used, the engine electrical system is controlled by a main fuel injection relay and a fuel pump relay. A permanent voltage supply is made to the main relay terminal 30 and fuel pump relay terminal 30 from the battery positive terminal. When the ignition is switched on, a voltage supply is made to fuel injection terminal 86, and this energises the relay winding which is connected to earth. This causes the relay contacts to close and terminal 30 is connected to the output circuit at terminal 87 or 87b. A voltage supply is thus output at this terminal, which normally supplies voltage to the injectors, the ECM and the ISCV. In addition, voltage is supplied to the fuel pump relay terminal 86.

When the ignition is switched on, the ECM briefly earths the fuel pump relay contact 87 or 87b at the appropriate ECM pin. This energises the relay winding, which closes the relay contact and connects voltage from terminal 30 to the output terminal 87 or 87b, thereby providing voltage to the fuel pump circuit. After approximately one second, the ECM opens the circuit and the pump stops. This brief running of the fuel pump allows pressure to build within the fuel pressure lines, and provides for an easier start.

The fuel pump relay circuit will then remain open until the engine is cranked or run. Once the ECM receives a speed signal, the relay winding will again be energised by the ECM, and the fuel pump will run until the engine is stopped.

2.39 Inertia switch

Main and fuel pump relays (single 'dual' relay)

Where this set-up is used, the electrical system is protected by a single relay with dual contacts. There are a number of different kinds of dual relay, and each will be described more specifically under the appropriate Chapter heading. Some relays (eg GM dual type) use the DIN annotation method, other types (eg Magneti Marelli dual type) use other methods of annotation. However, the following statements are generally true.

A permanent voltage supply is made to the relay terminals from the battery positive terminal. When the ignition is switched on, a voltage supply is connected to earth which energises the first relay winding. When the relay winding is energised, this causes the relay contacts to close and a voltage supply is output to the appropriate ECM pin and to the injectors, ISCV and a further terminal of the relay.

When the ignition is switched on. the ECM briefly earths the fuel pump driver at the appropriate ECM pin. This energises the second relay winding, which closes the second relay contact and connects voltage to the fuel pump circuit. After approximately one second, the ECM opens the circuit and the pump stops. This brief running of the fuel pump allows pressure to build within the fuel pressure lines, and provides for an easier start.

The second circuit will then remain open until the engine is cranked or run. Once the ECM receives a speed signal from the CAS, the second winding will again be energised by the ECM, and the fuel pump, ignition and injection will run until the engine is stopped.

Inertia switch

The inertia switch is a safety cut-out switch used to isolate the fuel pump in the event of a very sharp deceleration – ie a collision. Once the switch has been activated, the electrical supply to the fuel pump remains open circuit until the inertia switch has been reset *(see illustration 2.39)*.

Fuel pressure system

Fuel pump – petrol models

The VM uses a number of different methods to provide fuel to the injection rail. A system may be designed with a single external pump or a single internal pump, or both internal and external.

Generally, both internal and external pumps are of the 'wet' variety, in that fuel actually flows through the pump and the electric motor. There is no actual fire risk, because the fuel drawn through the pump is not in a combustible condition. The systems for MPi and SPi are fairly similar, although the SPi operates at a much lower pressure.

The fuel pump normally provides much more fuel than is required, and surplus fuel is thus returned to the fuel tank via a return pipe. In fact, a maximum fuel pressure in excess of 5 bar is possible in MPi systems. To prevent pressure loss in the supply system, a non-return valve is provided in the fuel pump outlet. When the ignition is switched off, and the fuel pump ceases operation, pressure is thus maintained for some time. A fuel pulsation damper may also be fitted in the fuel line – close to the fuel filter – to reduce noise from the injectors.

External fuel pump

A roller-type fuel pump, driven by a permanent magnet electric motor mounted close to the fuel tank, draws fuel from the tank and pumps it to the fuel rail via a fuel filter *(see illustration 2.40)*.

Mounted on the armature shaft is an eccentric rotor, holding a number of pockets arranged around the circumference – each pocket containing a metal roller. As the pump is actuated, the rollers are flung outward by centrifugal force to act as seals. The fuel between the rollers is forced to the pump pressure outlet. A fuel pressure damper is

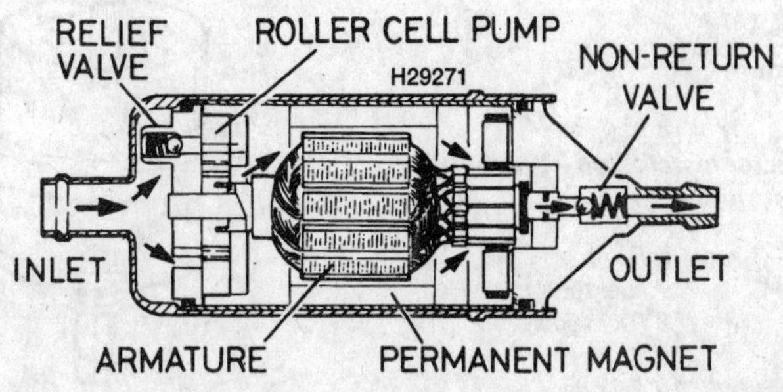

2.40 Roller-type external fuel pump

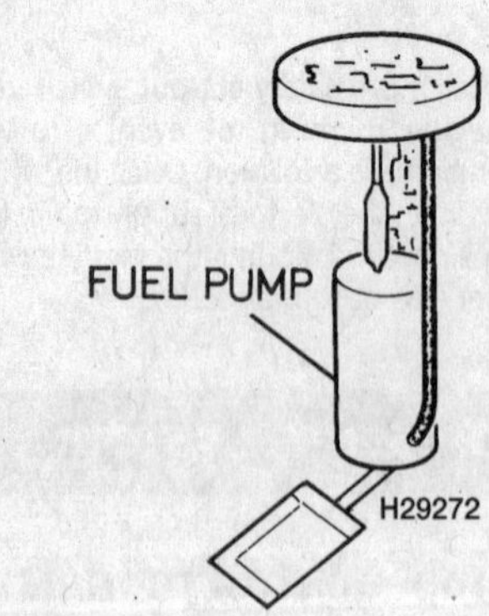

2.41 Internal fuel pump

mounted before the fuel rail, to reduce the pulsating effects of pump operation. The appearance of the damper is similar to the pressure regulator, but without a vacuum pipe.

Internal fuel pump

The fuel pump is mounted vertically in the fuel tank, and comprises an outer and inner gear assembly, termed a 'gerotor'. Once the pump motor becomes energised, the gerotor rotates, and as the fuel passes through the individual teeth of the gerotor, a pressure differential is created. Fuel is drawn through the pump inlet, to be pressurised between the rotating gerotor teeth and discharged from the pump outlet into the fuel supply line *(see illustration 2.41)*.

Fuel pump – diesel models

Traditionally, diesel systems have been equipped with two pumps. First, the lift pump moves the fuel at low pressure from the fuel tank to the injection pump, then from the injection pump to

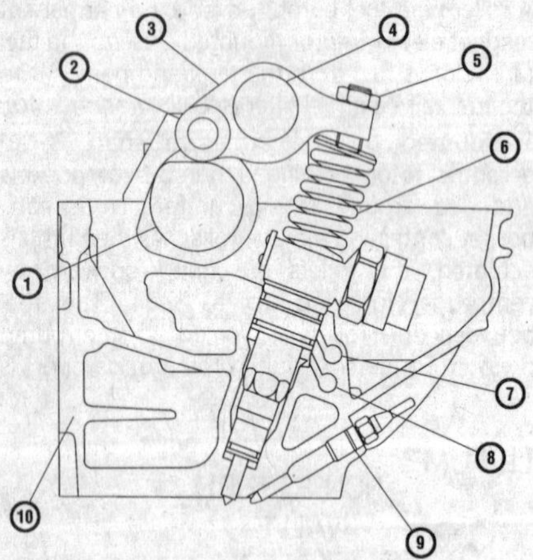

2.42 Pump injector installation – Bosch/VW pump injector system

1 Camshaft injection lobe
2 Rocker roller
3 Rocker
4 Rocker shaft
5 Ball-pin adjuster
6 Pump injector
7 Fuel return line
8 Fuel supply line
9 Glow plug
10 Cylinder head

the injectors at high pressure. The lift pump may be mounted on the engine and driven by a dedicated lobe on the camshaft/auxiliary shaft, or electrically driven and mounted in the fuel – often combined as a module with the fuel level sensor. On some modern common-rail systems, no lift pump is fitted. The high-pressure pump draws the fuel from the tank, pressurises it and delivers it to the common-rail (often referred to as an accumulator).

The exception to this is the 'pump-injector' system pioneered by VAG. Here, fuel is drawn from the tank by a camshaft driven pump, and delivered at low pressure to each injector. At the top of each injector is a piston and chamber. A rocker arm operated by a lobe on the camshaft forces the piston down to pressurise the fuel in the chamber. With the fuel pressurised, the ECM controls the exact injection quantity and duration via electronic solenoids in each injector *(see illustration 2.42)*.

Fuel pressure regulator – petrol engines

The fuel pump supplies fuel to the fuel rail at a pressure that exceeds the required system pressure. The function of the fuel pressure regulator is to maintain a constant pressure, and allow excess fuel to flow back to the tank via a fuel return line *(see illustration 2.43)*.

The pressure regulator is fitted on the outlet side of the fuel rail, and consists of two chambers separated by a diaphragm. The upper chamber contains a spring which exerts pressure upon the lower chamber and closes off the outlet diaphragm. Pressurised fuel flows into the lower chamber, and this exerts pressure upon the diaphragm. Once pressure exceeds a pre-designated value, the outlet diaphragm opens, and excess fuel flows back to the fuel tank via the fuel return line.

It is important that fuel pressure in the fuel rail is maintained at a constant pre-set pressure. This is usually about 2.5 to 3.0 bar in MPi systems, and 1.0 bar in SPi systems. When constant pressure is achieved, the opening time of the fuel injector will be the prime factor that influences the quantity of injected fuel delivered to the engine, as determined by the ECM, and not on a variable fuel pressure. If the fuel pressure were to vary, so too would the volume of fuel delivered through the injector for any given period of injection open time.

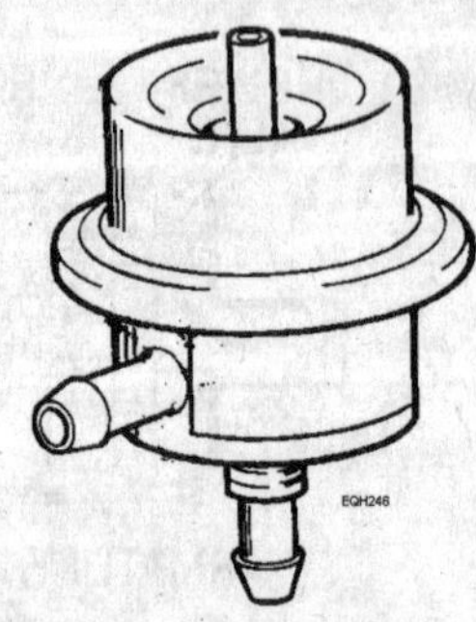

2.43 Fuel pressure regulator

In MPi systems, there is another factor that creates a variation in fuel pressure, and this is the negative pressure in the inlet manifold. As the depression in the inlet manifold varies according to throttle opening, so too will the depression acting upon the injector needle. This will affect the volume of fuel that is injected. By connecting the pressure regulator to the inlet manifold via a short length of vacuum hose, the fuel pressure will increase on acceleration to give a richer mixture and reduce during cruising to give a leaner mixture. However, whatever the throttle opening, the pressure drop across the injector is held at a constant pressure above the pressure in the inlet manifold.

At idle speed with the vacuum hose disconnected, or with the engine stopped and the pump running, or at full-throttle, the system fuel pressure for a typical MPi system will be approximately 3.0 bar. At idle speed (vacuum hose connected), the fuel pressure will be approximately 0.5 bar under the system pressure.

Fuel pressure regulator – diesel engines

On diesel engines, the fuel pressure is regulated by the design of the injection pump (traditional non-common rail engines) or by a pressure regulating valve incorporated into the high-pressure pump (common-rail engines). Surplus pressure is allowed to 'leak-back' into the fuel return system.

Turbocharger

General

The power output for any given engine is limited by the volume of air that the engine can 'inhale'. Simply put, the more air that an engine can cram into a cylinder, the greater is the volumetric efficiency, and the more power will be produced. Obviously, as the air intake is increased, so too must the fuel intake so that the AFR is maintained.

Turbocharging is a method of compressing the inlet air so that the maximum charge can be forced into a cylinder under high pressure or boost.

Exhaust gases are used to drive a turbine and impeller to compress the intake air. The turbine is mounted in the exhaust system, fairly close to the exhaust manifold. Compressing the air generates heat, so the air tends to expand and lose some of its efficiency. A method of overcoming this problem is to use an intercooler, which is a kind of 'air radiator', so that the air is cooled on its way to the inlet manifold, ensuring

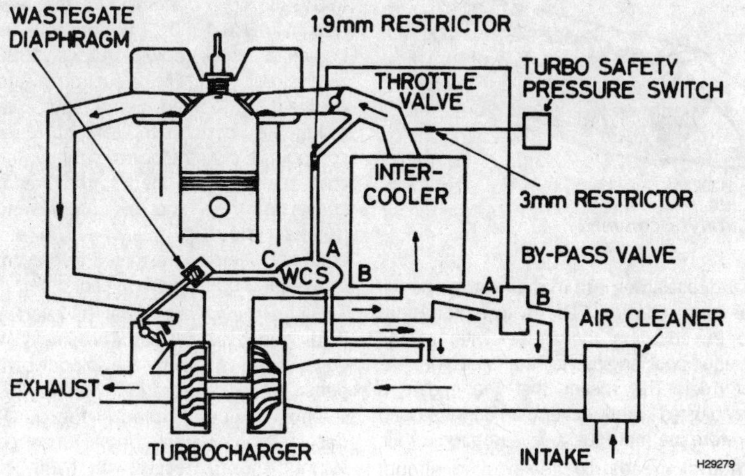

2.44 Typical turbocharged circuits

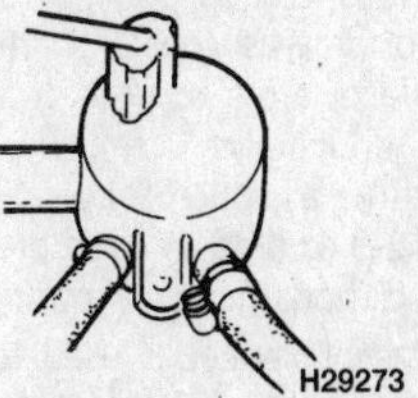

2.45 Wastegate control solenoid (WCS)

maximum compression of the air charge *(see illustration 2.44)*.

Boost control

As the volume of exhaust gases increase, the turbine is driven ever faster; at high engine speeds, there is a danger of excessive turbine speeds and over-pressurisation causing engine damage. Traditionally, this was overcome by using a wastegate control valve to actuate a wastegate flap in the exhaust system – upstream of the turbo *(see illustration 2.45)*.

When the boost control valve receives a signal from the ECM, a solenoid valve is actuated to open a channel. This channel allows compressed air to flow from a point before the throttle plate, through the boost valve to the wastegate actuator. The compressed air acts upon the wastegate diaphragm, which mechanically opens a flap inside the exhaust pipe. Some of the exhaust gases flow through the flap and by-pass the turbine so that the boost pressure is reduced.

A hose connects the turbo compressor to the diaphragm of the wastegate actuating valve. As

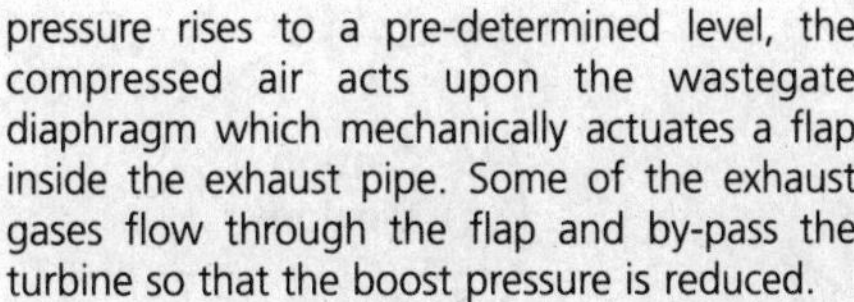

pressure rises to a pre-determined level, the compressed air acts upon the wastegate diaphragm which mechanically actuates a flap inside the exhaust pipe. Some of the exhaust gases flow through the flap and by-pass the turbine so that the boost pressure is reduced.

In its simplest form, this provides adequate protection from over-pressurisation. However, by utilising a wastegate control solenoid (WCS) under control of the ECM, maximum turbo boost can be varied to improve power under acceleration and at different engine speeds. The WCS is located in the hose from the turbo to the wastegate, and acts to allow all or part of the boost to be applied to the wastegate. When the valve is open, air is vented back to the low-pressure side of the induction system.

At low engine speeds (typically less than 2500 rpm), boost is negligible and the valve is not actuated until rpm rises above that level. The ECM tends to pulse the valve with a fixed frequency, and the duration of the pulse is varied so that the valve is open for longer or shorter time periods as desired.

On modern diesel engines, boost pressure is controlled by means of a variable boost turbocharger *(see illustration 2.46)*. With this design, the exhaust gases entering the turbocharger pass through a variable size venture. When the engine speed is low and the gas speed is slow, the venture diameter is reduced. This has the effect of speeding up the gasses just before they meet the turbocharger wheel. This maintains a high turbine speed, which improves turbocharger performance at low speed. At higher engine speeds when the gasses are moving much faster, the venturi is enlarged. Using this system allows the turbocharger to operate at close to maximum efficiency at a greater range of engine speeds.

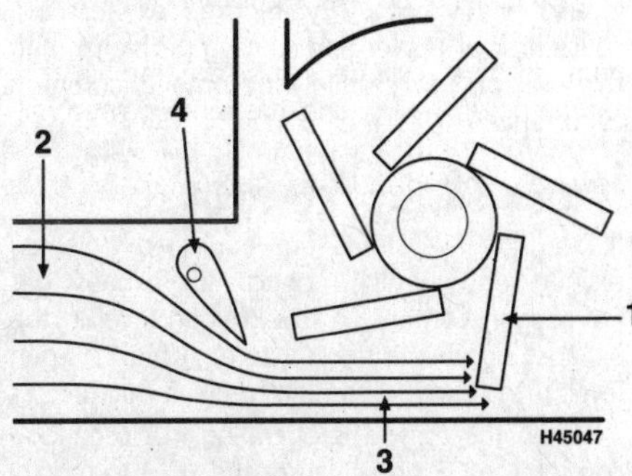

2.46 Variable geometry turbocharger at low engine speed

1 Turbine wheel
2 Low speed exhaust gases
3 High speed exhaust gases
4 Adjustable vane

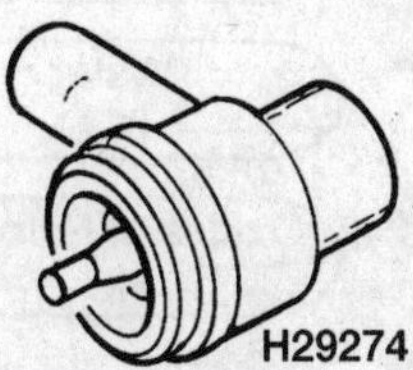

2.47 Turbo by-pass valve

Air by-pass

A phenomenon known as 'turbo-lag' can cause hesitation during initial acceleration as the turbo takes a moment to come to speed. This can be very noticeable on acceleration after a period of deceleration when the turbine has been slowed by back pressure in the turbocharger housing. Turbo-lag can be reduced by use of an air by-pass valve *(see illustration 2.47)*.

A sensing pipe connects the by-pass valve with the inlet manifold. When the turbine supplies compressed air to the manifold, the compressed air pushes upon the air by-pass valve and it remains shut. During deceleration or light load when the turbo is inactive, the manifold contains depressed air (a vacuum) and the depression will open the air by-pass valve. Air pressure from the impeller wheel is circulated throughout the turbocharger housing, and prevents a back pressure forming. The turbine slows very little, and turbo-lag is much reduced when the accelerator is re-applied.

Catalytic converter and emission control

General

All new petrol vehicles sold in the UK from January 1st 1993 onwards are equipped with a catalytic converter, and implement a closed-loop control system so that exhaust emissions may be reduced. Many diesel vehicles are fitted with a catalytic converter, but a closed loop system similar to that fitted to petrol engines cannot be used, because a diesel engine always operates with excess air, and hence oxygen, in the exhaust gas.

Catalytic converter

A catalyst is something which promotes a chemical reaction, but itself remains unaffected by the reaction. The catalytic converter consists of a stainless steel housing containing a ceramic monolith element with a honeycomb of passages called cells. There are 400 cells per square inch, giving an internal surface area of 3.55 metres.

The element is coated with a rough-surfaced aluminium oxide washcoat and fired in a kiln to give a surface area of 1-2 football pitches (depending on catalyst size). The washcoat is coated with a microscopically-thin layer containing 2-3 grams of the precious metals platinum and rhodium *(see illustration 2.48)*.

A steel mesh blanket is used to protect the monolith from heat and road vibrations. Some form of heat shielding is placed between the vehicle underbody and exhaust, and thus the passenger compartment is heat-insulated from the high operating temperature of the converter.

The catalyst is like a secondary combustion chamber, and CO and HC are oxidised into H_2O and CO_2. NOx is broken down by a process known as reduction, where oxygen and nitrogen are forced apart. The oxygen combines with CO to produce CO_2 and N_2.

A weak mixture with a high level of O_2 is good for the efficient oxidation of CO and HC. On the other hand, a relatively rich mixture with some CO aids the reduction of NOx. A compromise is reached by adjusting the air-fuel ratio of the catalyst-equipped engine to the stoichiometric ratio of 14:1. This means that the engine is perhaps adjusted slightly richer than desirable, and will therefore use more fuel *(see illustration 2.49)*.

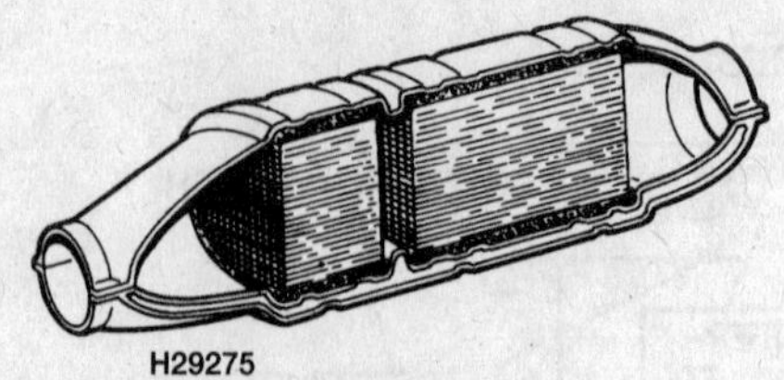

2.48 Catalytic converter

A catalyst needs to reach a minimum temperature of 300°C before it begins to work efficiently, and a working temperature of 400-800°C is more desirable. As the temperature rises over 800-1000°C, the precious metals will begin to break down. Above 1000°C, the catalyst will melt. Excess fuel or misfires causes overheating. Leaded petrol and excessive oil residue also destroys the catalyst. Here the lead compounds clog the pores of the washcoat, and coat the precious metals, thus reducing the conversion rate and eventually rendering it useless. The fuel filler pipe in a catalyst-equipped vehicle is restricted to prevent the use of anything other than unleaded petrol.

When new, the catalyst may emit H_2S (hydrogen sulphide) gas. This smells like rotten eggs, and is caused by the sulphur contained in petrol. Under deceleration when the AFR is lean, sulphur trioxide is stored in the catalyst. After deceleration, when the AFR enriches, this sulphur trioxide reacts with hydrogen in the exhaust to be emitted as H_2S. Although H_2S is toxic, the emission is considered to be quite safe. Generally, the smell becomes less pronounced after a few thousand miles.

An engine with a catalytic converter, but without an ECM and an oxygen/Lambda sensor and therefore operating in 'open-loop' control, will convert approximately 50% of emissions. However, in an engine with a catalytic converter, and with an ECM and oxygen/Lambda sensor ('closed-loop' control), the emission conversion rate is likely to be more than 90%.

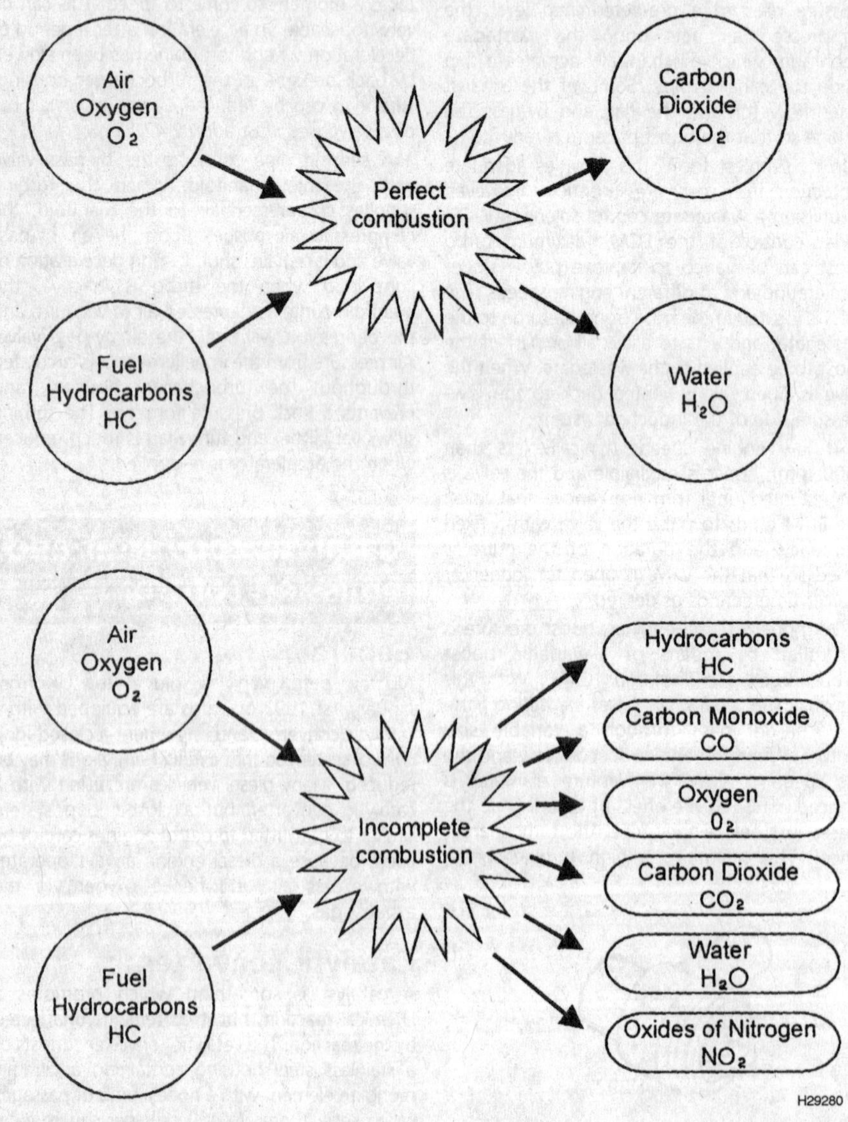

2.49 Combustion chart

Precautions

DO NOT

a) Turn the engine off at engine speeds above idle rpm.
b) Start the vehicle by towing.
c) Use fuel or oil additives.
d) Drive when the engine burns oil.
e) Park over dry leaves or long grass.

It is essential that an engine with a catalyst is operating correctly at all times. Any engine problem or misfire that results in unburnt fuel passing into the catalyst will destroy it in very quick order. The unburnt fuel causes the catalyst to overheat. Once temperatures over 900°C are attained, the catalyst substrate will melt. This will destroy the catalyst, and a blocked exhaust is also highly likely. The blocked exhaust will cause lack of power and starting problems, and this will almost certainly become a common occurrence.

'Closed-loop' control

'Closed-loop' systems are equipped with an oxygen sensor which monitors the exhaust gas for oxygen content. A low oxygen level in the exhaust signifies a rich mixture. A high oxygen level in the exhaust signifies a weak mixture.

When the engine is operating in closed-loop control, the OS signal causes the ECM to modify the injector pulse so that the AFR is maintained close to the stoichiometric ratio. By controlling the injection pulse, during most operating conditions, so that the air/ fuel ratio is always in a small 'window' around the Lambda point (ie Lambda = 0.97 to 1.03), almost perfect combustion could be achieved. Thus the catalyst has less work

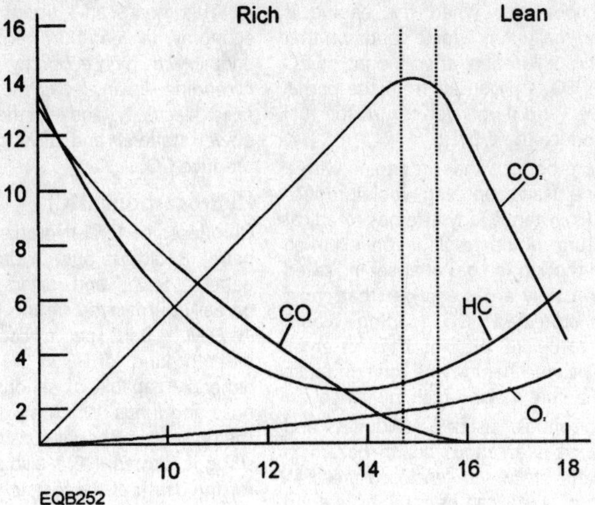

2.50 AFR chart

to do, and it will last longer with fewer emissions at the tail pipe *(see illustration 2.50)*.

The closed-loop control is implemented during engine operation at coolant temperatures near normal operating temperature. When the coolant temperature is low, or the engine is at full-load or on the overrun, the ECM will operate in 'open-loop'. When operating in open-loop, the ECM allows a richer or leaner AFR than the stoichiometric ratio. This prevents engine hesitation, for example, during acceleration with a wide-open throttle.

Oxygen sensor (OS)

An oxygen sensor is a ceramic device placed in the exhaust manifold on the engine side of the catalytic converter on petrol vehicles. Various names have been given to this sensor, and it could equally be called a Lambda sensor, oxygen sensor or even an exhaust gas oxygen sensor (EGOS) *(see illustration 2.51)*.

The quantity of oxygen remaining after combustion is an excellent indicator of a deficit or surplus of air (rich or weak mixture). The oxygen sensor returns a signal to the ECM, which can almost instantaneously (within 50 ms) adjust the injection duration. By controlling the engine electronically so that the AFR is always at the Lambda point, no matter the load or speed, almost perfect combustion could be achieved.

Essentially, the OS contains two porous platinum electrodes. The outer surface electrode is exposed to exhaust air and coated in porous ceramic. The inner surface electrode is exposed to ambient atmospheric air.

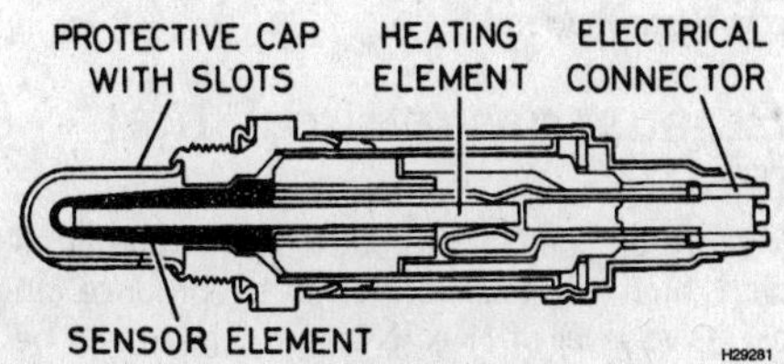

2.51 Oxygen sensor (OS)

There are now two different types of OS in use. The first and most commonly used OS utilises a zirconia element. A voltage signal is generated by the difference in oxygen levels at the two electrodes, and this is transmitted to the ECM. This voltage is inversely proportional to the level of oxygen in the exhaust manifold. The ECM uses this signal to adjust the injector opening time to maintain Lambda = 1.0 ± 0.02.

The OS closed-loop voltage is quite low, and switches between 100 millivolts (weak) to 1.0 volt (rich). The signal actually takes the form of a switch, and switches very quickly from weak to rich.

The second type of OS is made from titania, and functions by a change in resistance which alters a digital signal that originates from the ECM. Response of the titania OS is much faster than those made from zirconia, and the signal is much more stable over a range of exhaust temperatures.

Unfortunately, driveability would be impaired if the engine was set at Lambda for its entire driving cycle. Controlling the engine to the Lambda ideal would cause hesitation, flat spots and a lack of smoothness that would make a car pretty near undriveable. Therefore, when the engine is under Lambda control (ie cruising) it is in 'closed-loop' operation. Under acceleration or during the warm-up period, the system goes into 'open-loop' operation, and a richer mixture is allowed to prevent hesitation.

Heated OS

The OS only produces a signal when the exhaust gas, has reached a minimum temperature of approximately 300°C. In order that the OS will reach optimum operating temperature as quickly as possible after the engine has started, the OS contains a heating element. The OS heater supply is usually applied from the fuel pump relay output terminal. This ensures that the heater will only operate whilst the engine is running.

Engine faults

Although the oxygen sensor method of regulating the AFR works perfectly well with a properly operating engine, things go awry once faults develop. Almost any problem that affects combustion will see an increase of O_2 in the exhaust. For example, a misfire due to an ignition or mechanical problem means that combustion is incomplete, and the level of O_2 in the exhaust will rise.

The ECM will interpret this as a lean mixture and increase the fuel injection duration. When the fuel injection method is Multi-point simultaneous injection or single point injection, all cylinders will be enriched. This is the main reason why use of the sequential injection method will almost certainly increase. With sequential injection, only the afflicted cylinder will receive more fuel.

Evaporation control (activated carbon filter and control valve)

An activated carbon canister is employed in catalyst equipped vehicles to aid evaporative emission control on petrol vehicles. The carbon canister stores fuel vapours and a control valve (CFSV) or (CFCV) is used to control the vapours stored in the canister. Depending upon the system, the control valve may be actuated by the ECM, or mechanically actuated according to temperature. Once the control valve is actuated, fuel vapours are drawn into the inlet manifold to be burnt by the engine during normal combustion. Alternatively, a very simple evaporative purge system may simply be operated by the position of the throttle *(see illustration 2.52)*.

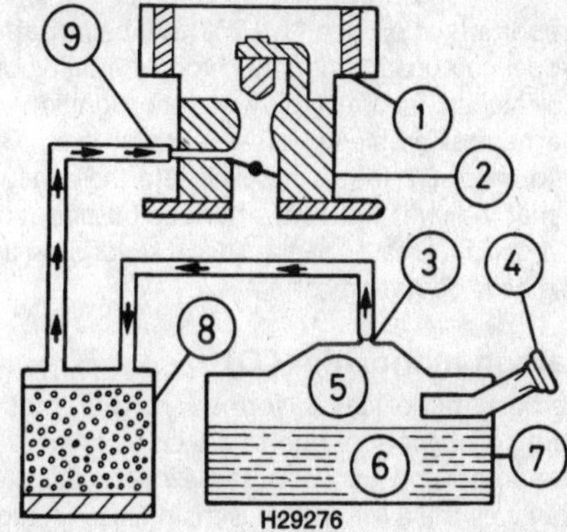

2.52 Mechanically-controlled evaporation circuit

1 Air filter
2 Throttle body
3 Restrictor
4 Sealed fuel cap
5 Fuel vapour
6 Fuel
7 Fuel tank
8 Charcoal canister
9 Purge hose

ECM-actuated

When the engine is stopped, the carbon filter solenoid valve (CFSV) is open. As soon as the ignition is switched 'on' the CFSV closes. So that engine performance will not be affected, the CFSV remains closed during cold engine operation and also during engine idle. Once the engine coolant temperature reaches normal operating temperature and the throttle is partially open (normal cruise conditions with a hot engine) the CFSV will be modulated (pulsed) on and off by the ECM. Once the CFSV is actuated by the ECM, fuel vapours are drawn into the inlet manifold to be burnt by the engine during normal combustion.

Mechanically-actuated

Vacuum is usually applied to the carbon filter control valve (CFCV) through a thermistor. The CFCV remains closed during cold engine operation, so that engine performance will not be affected, and the vacuum supply is arranged so that vacuum to the CFCV is not available during engine idle. Once the engine coolant temperature reaches normal operating temperature and the throttle is partially open (normal cruise conditions with a hot engine) vacuum is applied to the CFCV so that fuel vapours are drawn into the inlet manifold to be burnt by the engine during normal combustion.

Throttle-actuated

When the throttle is closed, the purge line to the canister is closed. After the engine is started and the throttle is opened, vacuum acts upon the canister to draw the vapours into the inlet manifold, to be burnt by the engine during normal combustion.

Analysis of exhaust gases

Oxygen (O_2)

Oxygen is a harmless gas that is present in about 21% of air, and is necessary for proper combustion. O2 consists of two oxygen atoms and is measured in % volume. A small proportion of oxygen (1 – 2%) will be left after proper combustion. Too much or too little would indicate an incorrect air/fuel ratio, ignition or mechanical problems, or an exhaust leak. The amount of O2 that is expelled into the exhaust is that which is left over after combustion, and is a good indicator of the AFR – so long as the engine is operating correctly.

Carbon monoxide (CO)

Carbon monoxide is formed by the partial burning of the fuel due to a lack of oxygen. A low proportion of CO in the exhaust indicates how well the air/fuel ratio is maintained. A high proportion of CO in the exhaust indicates a rich fuel mixture, choked air filter, choked PCV valve or low idle speed. Low CO would indicate a lean fuel mixture or a vacuum leak, or even an exhaust leak. CO (and HC) emissions decrease as load (heat and temperature) rises to give a more efficient engine.

The CO content in the exhaust is an indicator of the AFR, but only when the engine is functioning normally. Any engine condition that causes a misfire will reduce the amount of CO that is burnt. CO is produced from the partial burning of fuel, and if you don't burn the fuel you won't produce the CO.

It is therefore obvious that an engine with a burnt valve or a dead spark plug would produce less CO. In this instance, any attempt to adjust the fuel mixture would result in an over-rich mixture, even though the gas analyser indicated a lean mixture. Only a gas analyser that could calculate a 'corrected' CO reading would provide the complete picture. It is essential therefore that all mechanical and ignition faults are corrected before making fuel adjustments.

CO is a poisonous, tasteless, colourless and odourless gas. It is a serious health hazard in dense city traffic or in semi-enclosed areas. A concentration of 0.3% can be fatal if breathed in continually for 30 minutes. CO combines with red blood cells at the expense of oxygen, and causes suffocation. By weight, CO accounts for about 47% of air pollution, but is thought to have little effect on the environment. One molecule of CO contains one atom of carbon and one of oxygen, and it is measured in % volume. CO is inversely proportional to the AFR – the less fuel, the lower the CO.

Carbon dioxide (CO_2)

CO_2 is the product of an efficient engine. With low CO and HC levels, the percentage of CO_2 in the exhaust is likely to be 13 to 15%. Less than 8% CO_2 indicates an incorrect AFR, or a misfire or leaky exhaust. CO_2 is directly proportional to the AFR, but inversely proportional to CO. The less fuel, the higher the CO_2. At speeds over 2000 rpm, the level will be 1-2% higher than at idle, due to an increase in engine efficiency.

One molecule of CO_2 contains one atom of carbon and two of oxygen. CO_2 is chemically stable, and does not easily react with other substances. Not poisonous, it is produced by all breathing animals, including fish. Oxygen is inhaled and CO_2 exhaled at a concentration of about 5%. CO_2 is absorbed by all green plants by a process called 'photo-synthesis', which only happens in daylight, and which also releases O_2 into the atmosphere.

Any burning process produces CO_2, and the contribution from automotive sources is said to be less than half that of industrial and domestic sources. The contribution from people and animals is insignificant. A heavy concentration of CO_2 is like placing a blanket over the atmosphere, and this prevents heat loss by radiation. At present, more CO_2 is being produced than is being consumed, and the disappearance of the rain forests is another significant factor. As the forests fade away, less CO_2 is absorbed; the increase in atmospheric concentration is said to contribute towards 'global warming' and the so-called 'greenhouse effect'.

In the automobile, the only way to produce less CO_2 is to burn less petrol, or even none at all. This means an efficient engine with good economy (a lean-burn engine), or a diesel engine with high economy, or even a car with no engine – ie an electric motor. But electric cars need electricity, and electricity is produced by power stations, and most power stations also produce CO_2.

Hydrocarbons (HC)

Composed of 15% hydrogen and 85% carbon, petrol is almost pure hydrocarbons. HC is a generic term, and refers to unburnt and partially-burnt hydrocarbon fuel. It is measured in PPM – parts per million. There are many different kinds of HC in the exhaust and HC is generally capable of serious damage to eyes, nose and lungs. When mixed with NOx and in the presence of bright sunshine, photochemical smog is formed. HC is also said to be a reason for the death of the forests.

During combustion, the hydrogen atoms combine with the O_2 atoms to produce H_2O. The carbon atoms combine with O_2 atoms to produce CO_2. High levels of HC in the exhaust signifies ignition problems such as defective plugs or HT leads, incorrect timing, vacuum leaks, incorrect air/fuel ratio or engine mechanical faults. In fact, anything that causes inefficient engine operation will increase the level of unburnt HC in the exhaust.

As the AFR weakens, the HC emissions increase due to a lean misfire. This is why a black exhaust is often the result of a too-lean idle mixture. Careful design of the combustion chamber can overcome this problem.

Oxides of nitrogen (NOx)

NOx is a term for a group of poisonous gases formed due to high temperatures (exceeding 1300°C). and high compression. There are many different kinds of NOx (ie NO, NO_2, NO_3 etc) and they are all lumped together under the term NOx, the N representing one nitrogen atom and Ox representing any number of oxygen atoms.

The nitrogen content of air passes through the combustion process unchanged until high temperature (more than 1370°C) and high pressures are reached. Under these conditions, nitrogen and oxygen react to form nitrogen monoxide – sometimes called Nitric oxide – (NO). The breeding conditions for NOx are wide-open throttle, acceleration and high-speed cruising. When NO is produced in the presence of strong sunshine, NO_2 (nitrogen dioxide), ozone (O_3) and NO_3 (nitrogen nitrate) are the result. NO_2 is a light brown gas commonly called 'smog'. Unfortunately, NOx emissions reach their peak at Lambda = 1, the so-called perfect combustion point.

Exhaust gas re-circulation (EGR)

Modern engines that run at a high temperature with high compression tend to produce an excessive level of NOx. NOx production can be reduced by recycling a small amount of exhaust gas into the combustion chamber. This

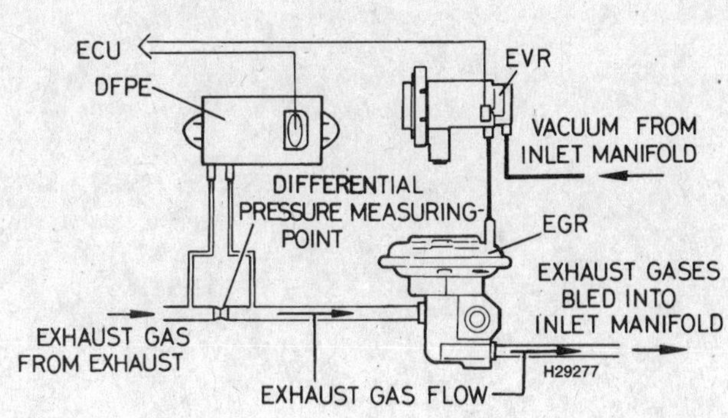

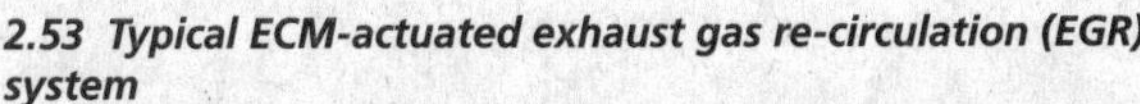
2.53 Typical ECM-actuated exhaust gas re-circulation (EGR) system

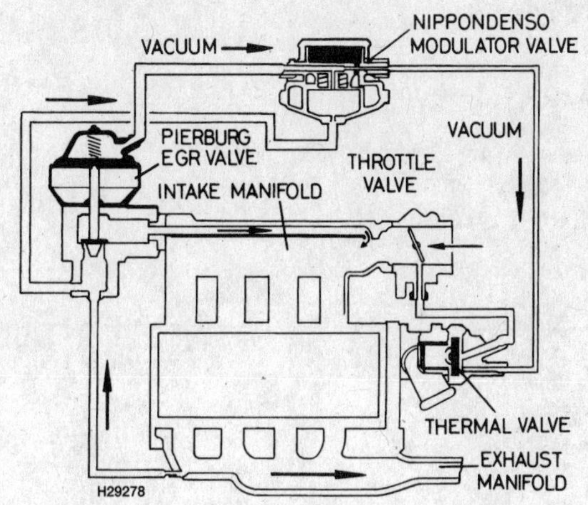

2.54 Typical mechanically-actuated exhaust gas re-circulation (EGR) system

procedure effectively lowers the combustion temperature, and reduces the incidence of NOx. So long as the recycling of the exhaust gas is properly controlled, engine operation will be little affected.

EGR operation in modern engines is controlled by the ECM, as distinct from earlier vehicles which used mechanical means. EGR control only occurs after the engine has attained normal operating temperature and the engine is operating under part-load conditions. The ECM monitors the voltage returned from the EGR sensor, and adjusts the signal to actuate the EGR vacuum control. The vacuum control device opens the EGR valve, which allows a finely-metered supply of exhaust gas to be introduced into the inlet manifold *(see illustration 2.53 and 2.54)*.

Particle filter systems – diesel vehicles

Particle filters and traps reduce the level of smoke particles released into the air by the diesel engine exhaust. This is an evolving technology; not all vehicle manufacturers use particle filters, but they still manage to satisfy the 'Euro 4' emission standards (mandatory in 2005).

The particle filter used by Peugeot and Citroën (PSA) is of the soot burn-off type. It works as follows *(see illustration 2.55)*.

The soot particles are trapped in a block of filter material. The filter is carefully designed to allow the exhaust gases to flow through it, whilst trapping the soot particles; however there obviously comes a point when the filter will become blocked due to the large number of particles trapped. As the trapped particles in the filter build up, there will be a resistance to the exhaust gas flow. Pressure sensors on either side of the filter detect this resistance; when it exceeds a certain level the ECM triggers a cleaning cycle.

Because the diesel engine always operates with excess air, the exhaust gas contains enough oxygen that at temperatures above 550°C, soot will burn off of its own accord. The exhaust gas temperature in a diesel engine is normally between 150°C and 200°C, which is not high enough to burn off the soot. When the cleaning cycle is triggered, an additive is injected into the fuel which increases the exhaust gas temperature and burns off the soot. The additive is stored in a container next to the fuel tank.

If the filter is still clogged following a cleaning cycle, the ECU will go into 'limp home' mode and engine performance will be severely impaired. The filter must then be cleaned by a dealer workshop or diesel specialist using a special diagnostic tool in conjunction with the system's ECM (this cleaning may also be specified as a routine maintenance operation, typically every 48 000 miles). The engine's high pressure injection system is utilized to inject fuel into the exhaust gases during the post injection period; this causes the filter temperature to increase sufficiently to oxidize the particulates, leaving an ash residue. The filter is then removed from the exhaust system, and the ash residue flushed away with water.

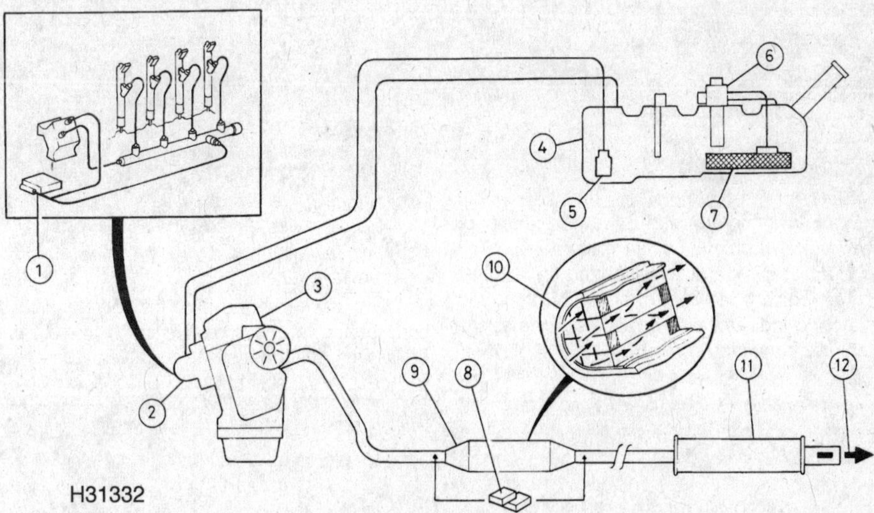

2.55 Schematic view of emission control system using soot burn-off filter

1 Common rail injection system
2 High pressure fuel pump
3 Engine
4 Fuel tank
5 Fuel lift pump
6 Additive injector and regulator
7 Additive tank
8 Exhaust gas pressure sensors
9 Catalytic converter
10 Burn-off filter
11 Silencer
12 Exhaust gases

Notes

Test equipment, training and technical data

Introduction

Testing the modern automobile engine is a serious business. To be good at it, you need to seriously invest in three areas. We can liken the three areas to the good old three-legged stool. In our automotive stool, the legs are equipment, training and information. Kick one leg away, and the others are left a little shaky. Those with serious diagnostic intentions will make appropriate investments in all three areas.

That is not to say that those without the best equipment, or the necessary know-how, or the information, are completely stuck. It will just require a little more time and patience, that's all.

However, we do have to say at the outset that although quite a lot can be accomplished with a digital multi-meter (DMM), a number of tests and checks do require specialist equipment. For the sake of completeness, we will detail test procedures that do require the use of sophisticated equipment.

Fault diagnosis then, and your method of diagnosis, will largely depend upon the equipment available. and your expertise. There is a definite trade-off in time against cost. The greater the level of investment in equipment and training, the speedier the diagnosis. The less investment, the longer it will take. Obvious, really!

3.1 Two typical high-impedance DMMs with similar performance but different sets of leads and probes. The left unit is equipped with alligator clips, and the right unit with spiked probes. Using the alligator clips frees your hands for other tasks, whilst the probes are useful for backprobing multi-plug connectors

Equipment

Within the motor trade, there are several different approaches to testing vehicles and diagnosing faults. Let us take a look at the various options.

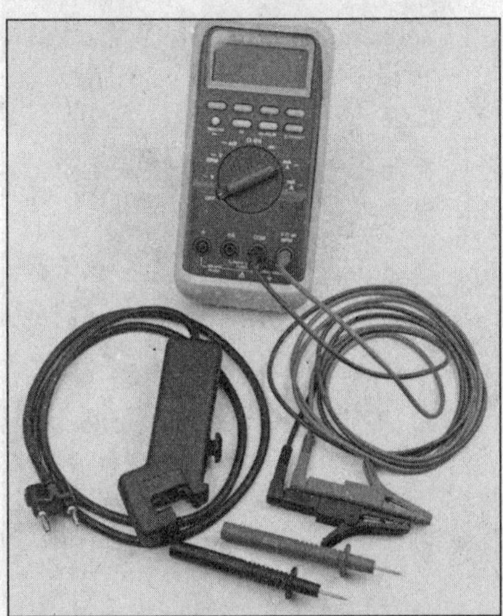

3.2 Top-of-the-range Fluke DMM with a multitude of features and attachments

Multi-meter

This is the equipment required for the most basic approach. These days, the meter will probably be digital, and must be designed for use with electronic circuits. An analogue meter or even a test lamp may be used, so long as it meets the same requirements as the digital meter. Depending on the sophistication of the meter, the DMM can be used to test for basic voltage (AC and DC), resistance, frequency, rpm, duty cycle, temperature etc. A selection of thin probes and banana plugs for connecting to a break-out box (BOB) will also be useful. Use a splitpin or hairgrip for probing into square or oblong multi-plugs. The round DMM probes can 'spread' the connection and turn a good connection into a bad one *(see illustrations 3.1 to 3.3)*.

If the fault is a straightforward electrical fault, the meter will often be adequate. However, the drawback is that it cannot analyse the complex electrical waveforms produced by many electronic sensors and actuators, and test results can sometimes be misleading.

3.3 A cooking thermometer like this can be useful as a last resort. Most good DMMs incorporate a thermometer function and a selection of probes that may be more effective

Programmed test equipment

This kind of proprietary equipment will interface between the ECM and the ECM multi-plug. This equipment checks the input and output signals moving between the ECM and its sensors and actuators. If one or more of the signals are outside of pre-programmed parameters, the equipment will display the erroneous signals as faults. Once again, other test equipment may be required to pinpoint the actual fault.

Oscilloscope (with or without DMM and engine analyser)

An oscilloscope is essentially a graphic voltmeter. Voltage is rarely still, and tends to rise and fall over a period of time. The oscilloscope (or 'scope) measures voltage against time, and displays it in the form of a waveform. Even when the voltage change is very rapid, the scope can usually capture the changes. Circuit faults can often be spotted much faster than when using other types of test instrument. Traditionally, the 'scope has been used for many years to diagnose faults in the primary and secondary ignition systems of conventional non-electronic vehicles. With the advent of electronics, the 'scope has become even more important, and when a labscope function is available, analysis of complex waveforms is possible. This equipment is often used in conjunction with other equipment, for speedy diagnosis of a wide range of problems. The large engine analyser and 'scope is now giving way to a plethora of smaller handheld 'scopes that pack great diagnostic power into portable form.

Exhaust gas analyser

These days the state-of-the-art gas analyser comes with the ability to measure all five of the gases present in the exhaust pipe, and it also calculates the Lambda ratio. Less-expensive gas analysers are available that will measure one, two, three or four gases. However, the better the gas analyser, the easier it gets. The gas analyser is now a recognised diagnostic tool. Faults in ignition, fuelling and various mechanical engine problems can be diagnosed from the state of the various gases present in the exhaust.

Battery saver

Actually, 'battery saver' is a misnomer, since the function of this device is to hold power to permanently live circuits whilst the battery is removed or changed. The live circuits may provide power to the radio security and station memory, and to the ECM adaptive memory, etc.

Jump leads with surge protection

It is possible to destroy an ECM if unprotected jump leads are used to provide emergency power to the battery. Rather than use jump leads, it is far safer to charge the battery before attempting to start the vehicle. A poor engine or chassis earth, flat battery or tired starter motor, and unprotected jump leads are a recipe for total disaster.

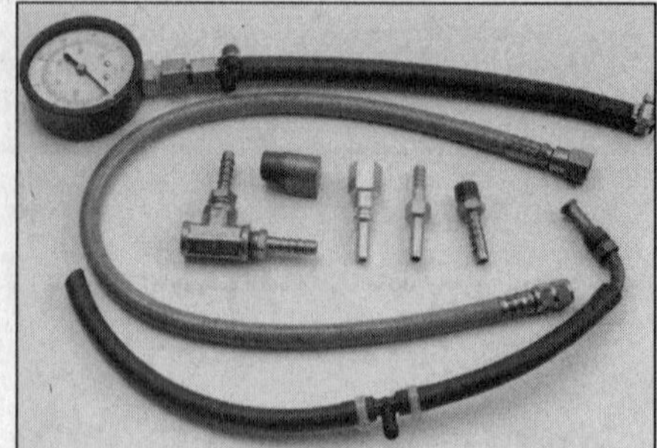

3.4 Fuel pressure gauge and adapter kit

Flight recorder

Useful for capturing a 'snapshot' of the signals produced by an EMS during a road test or some other operating condition for later analysis in the workshop. The recorder is usually set to record signals just before and after a particular incident. Particularly useful for diagnosis of intermittent faults, the flight recorder function may be built into the FCR or other portable equipment.

ECM testing equipment

Usually the province of those companies that specialise in the repair of the ECM, and not available for purchase by the garage or workshop. One company (ATP) offer an ECM test via a modem over the telephone network if the ECM is taken to one of their agents. Other ECM testing companies require that the ECM is sent to them by post for evaluation.

Fuel pressure test kit

Fuel pressure is vitally important to the well-being of the fuel-injected engine and a proper test gauge that will measure fuel pressures up to 7.0 bar is essential. The pressure gauge is normally supplied with a kit of adapters to connect it to a wide range of disparate fuel systems *(see illustration 3.4)*.

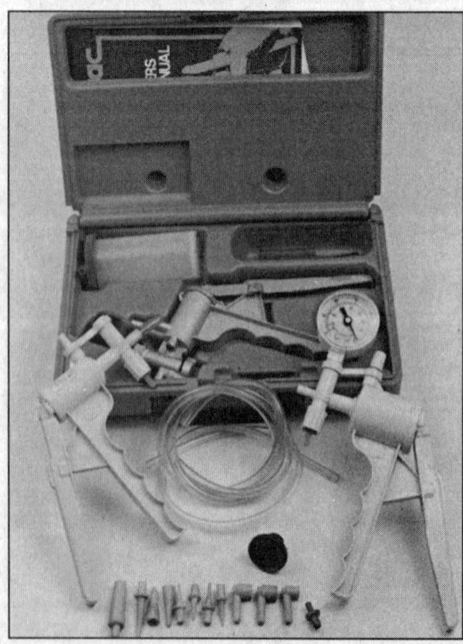

3.6 Vacuum pump kit

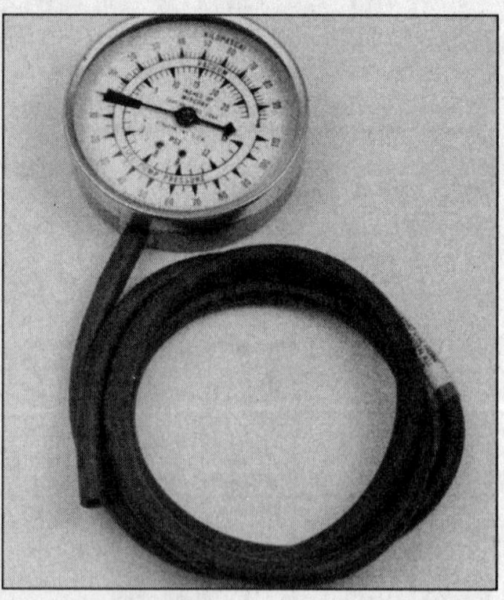

3.5 Vacuum gauge

Vacuum gauge

As useful as it always was. The vacuum gauge takes the pulse of the engine from a connection to the inlet manifold, and is useful for diagnosing a wide range of timing and mechanical faults, including a blocked exhaust system or vacuum leak *(see illustration 3.5)*.

Vacuum pump

The vacuum pump can be used to check the multitude of vacuum-operated devices that are fitted to many modern vehicles *(see illustration 3.6)*. A crude vacuum pump can be constructed from a bicycle pump. Reverse the washer in the pump, and the pump will then 'suck' instead of 'blow'.

Spark jumper

Useful for attaching to an HT lead to check for a spark. If you hold the lead from a modern high-output ignition system whilst cranking the engine, you may get quite a shock when testing for spark. Apart from curling your hair, the ignition system may also be damaged.

Feeler gauges

Still useful for measuring the various clearances at the CAS, TS, spark plug, valve clearances etc.

Hairdryer and cold spray

Useful for gently heating and cooling components during a test where heat may be contributing to failure.

HT lead puller

Ideal for safely breaking the HT lead-to-spark plug seal and then safely disconnecting the lead *(see illustration 3.7)*. How many times have you pulled at a lead to have it disintegrate into your hand?

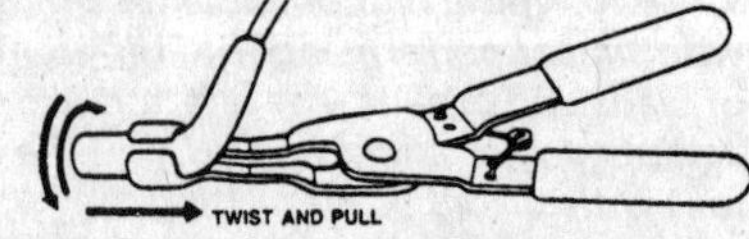

3.7 Ignition HT lead removal tool. Makes safe removal of HT leads a doddle

Exhaust back-pressure tester

Useful for checking for exhaust back-pressure. Screws into the OS hole. The presence of back-pressure indicates an exhaust blockage.

Variable potentiometer

Because of the widespread use of 'limp home' or LOS in the modern EMS, disconnecting a sensor such as the CTS may have little effect on the running of the engine. The ECM will assume a fault, and place a fixed value as replacement for that sensor. However, it is useful to be able to vary the resistance sent to the ECM and note the effect. One answer is to use a potentiometer with a variable resistance. If this is connected in place of the CTS resistor, then ECM response, injection duration and CO may be checked at the various resistance values that relate to certain temperature.

Noid light

A noid light is a small inexpensive light with inbuilt terminals for checking the signal to the injector. The injector harness is detached at the injector, and the noid light plugged into the injector harness. If the engine is then cranked, the light will flash if the injector is being pulsed by the ECM *(see illustration 3.8)*.

Jumper wires

Useful for checking out circuits, and bridging or 'by-passing' the relay *(see illustration 3.9)*.

Break-out box (BOB)

The BOB is a box containing a number of connectors that allows easy access to the ECM input and output signals, without directly probing the ECM pins. The BOB loom terminates in a universal connector. A multi-plug harness of similar construction to the ECM harness is interfaced between the ECM and its multi-plug, and the other end is connected to the BOB loom. The BOB will now intercept all signals that go to and from the ECM. If a DMM or an oscilloscope or any other suitable kind of test equipment is connected to the relevant BOB connectors, the ECM signals can be easily measured. The main drawback is the number of different ECM multi-plug connectors required for a good coverage of electronic systems. Small BOB's are also available for measuring values at components where it is difficult to connect the test equipment *(see illustration 3.10)*.

3.8 Injector noid light

There are three main reasons why use of a BOB is desirable in order to access the signals:

1) *Ideally, the connection point for measuring data values from sensors and actuators is at the ECM multi-plug (with the ECM multi-plug connected). The ECM multi-plug is the point through which all incoming and outgoing signals will pass, and dynamically testing at this point is considered to give more accurate results.*
2) *In modern vehicles, the multi-plug is becoming more heavily insulated, and removing the insulation or dismantling the ECM multi-plug so that back-probing is possible, is becoming more difficult, and in some instances, almost impossible. To a certain extent, the same is true of components, with certain components becoming increasingly difficult to backprobe or even probe at all.*
3) *ECM multi-plug terminals (pins) are at best fragile, and frequent probing or backprobing can cause damage. Some pins are gold-plated, and will lose their conductivity if the plating is scraped off. Using a BOB protects the pins from such damage.*

Fault code reader (FCR) - sometimes called a 'scanner'

A number of manufacturers market test equipment for connecting to the EMS serial port/diagnostic socket. These general-purpose FCR's allow data on a wide range of vehicles and systems to be obtained. The FCR could be used to obtain and clear fault codes, display datastream information on the state of the various sensors and actuators, and also "fire" the system actuators. The FCR is very useful for pointing the engineer in the direction of a specific fault. However, other test equipment may be required to pinpoint the actual fault, and the faults detected may be limited by the self-diagnostics designed into the vehicle ECM *(see illustration 3.11)*.

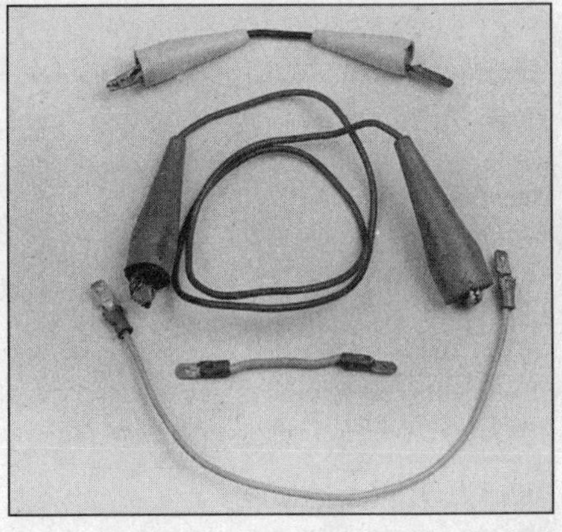

3.9 A selection of temporary jumper wires

Franchised vehicle dealer or authorised repairer

Will often use dedicated test equipment that relies on programmed test methods. The equipment will interface with the ECM, usually through the serial port/diagnostic socket, and lead the engineer through a programmed test procedure. Depending on its sophistication, the test equipment may be able to test most circuits, or may refer the engineer to test procedures using additional equipment. This equipment is dedicated to one vehicle manufacturer, and may not be available to other workshops outside of the franchised network *(see illustration 3.12)*.

3.10 Using a BOB to obtain voltage at the ECM pins

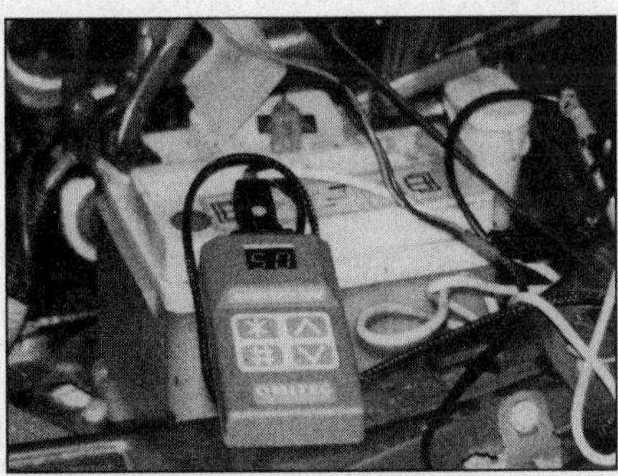

3.11 A typical FCR or 'scanner'

3.12 The Rover TESTBOOK – a laptop computer-based piece of test equipment that contains a very sophisticated and interactive test programme

Major suppliers of diagnostic equipment

Note: *The details below are correct at the time of writing (Autumn 2004).*

Alba Diagnostics Ltd
Bankhead Avenue
Bankhead Industrial Estate
Glenrothes, Fife
Scotland, KY7 6JG
Tel: 01592 774333
www.brakefluidtester.com

ASNU (UK) Ltd
14 King George Avenue
Bushey
Herts, WD23 4NT
Tel: 0208 4204494

ATP Electronic Developments Ltd
Unit 7
Hemlock Way
Cannock
Staffordshire, WS11 2GF
Tel: 01543 467466
www.atpelectronics.co.uk

AutoDiagnos Ltd
St. Christopher House
217 Wellington Road South
Stockport
SK2 6NG
Tel: 0161 2094470
www.autodiagnos.com

Crypton Ltd
Crypton Technology Business Park
Bristol Road
Bridgwater, Somerset, TA6 4BX
Tel: 01278 436200
www.cryptontechnology.com

Fluke (UK) Ltd
52 Hurricane Way
Norwich, Norfolk, NR6 6JB
Tel: 0207 9420700
www.fluke.co.uk

Gunson Ltd
Bristol Road
Bridgwater, Somerset, TA6 4BX
Tel: 01278 436240
www.gunson.co.uk

Intermotor
Occupation Road
Hucknall, Nottingham, NG15 6DZ
Tel: 0115 9528000
www.intermotor.co.uk

Lucas Aftermarket Operations
Stratford Road
Solihull, Birmingham, B90 4AX
Tel: 0121 5065999

Omitec Instrumentation Ltd
Hopton Industrial Estate
London Road
Devizes, Wiltshire, SN10 2EU
Tel: 0870 2400050
www.omitec.com

Robert Bosch Ltd
PO Box 98
Broadwater Park, Denham
Uxbridge, Middx, UB9 5HJ
Tel: 01895 834466
www.bosch.co.uk

SPX UK Ltd
Genoa House
Everdon Park
Heartlands Business Park
Daventry, Northants, NN11 5YJ
Tel: 01327 303400
www.spx.com

Sykes-Pickavant Ltd
Churchbridge Works,
Walsall Road
Cannock, Staffs, WS11 8JR
Tel: 01922 702000
www.sptools.co.uk

Training courses

There are a number of companies that specialise in training for the motor industry. The same training courses are usually available to the general public.

Note: *The details below are correct at the time of writing (Autumn 2004).*

AA External Training Courses
Contact Centre
Carr Ellison House
William Armstrong Drive
Newcastle-upon-Tyne, NE4 7YA
Tel: 0800 551188

Crypton Ltd
Bristol Road
Bridgwater, Somerset, TA6 4BX
Tel: 01278 436210

Delphi Diesel Aftermarket
Stratford Road, Shirley
West Midlands, B90 4DT
Tel: 0121 7466000

Lucas Test
International Training Centre
Unit 7, Mica Close
Tamworth
Staffs
B77 4QH
Tel: 0827 63503

Automotive Diagnostics Solutions
Preston Technology Management Centre
Marsh Lane
Preston
Lancs
PR1 8UD
Tel: 01772 201597

Autodiagnos Ltd.
St Christopher House
217 Wellington Road South
Stockport
Ches
SK2 6NG
Tel: 0870 2406331

Facom UK ltd.
Diagnostic research centre
Lancaster House
Bowerhill Industrial Estate
Melksham
Wilts
SN12 6TT
Tel: 01225 791883

General test procedures

General tests

Introduction

Generally speaking, test results obtained using a voltmeter or oscilloscope (particularly recommended) are more reliable and may reveal more faults than the ohmmeter. Voltage tests are much more dynamic and are obtained with voltage applied to the circuit, which is far more likely to reveal a problem than if the circuit is broken and the component measured for resistance. In some instances, disconnecting a multi-plug may break the actual connection that is at fault, and the circuit test may then reveal 'no fault found'.

In addition, the oscilloscope may reveal some faults that the voltmeter fails to find; the 'scope is particularly useful for analysing and displaying the complex signals and waveforms from some sensors and actuators. With the proliferation of small, portable handheld oscilloscopes at a cost of less than £2500, the 'scope is not quite in the province of the home mechanic, but every workshop that is serious about fault diagnosis should certainly have one.

For the purposes of this book, we will generally test the majority of components with reference to the voltmeter. Resistance or continuity tests using an ohmmeter will be mentioned where appropriate.

Ideally, the connection point for measuring data values from sensors and actuators is at the ECM multi-plug (with the ECM multi-plug connected). The ECM multi-plug is the point through which all incoming and outgoing signals will pass, and dynamically testing at this point is considered to give more accurate results. However, for a variety of reasons it is not always possible to test at the ECM multi-plug, and other points of testing may also give satisfactory results.

Chapter 4

Voltage tests

Connecting equipment probes

1 Connect the voltmeter negative probe to an engine earth.

2 Use the positive probe to backprobe for voltage at the actual terminals of the component under test *(see illustration 4.1)*. **Note:** *This procedure will give acceptable results in most instances, and is the one we would recommend to non-professionals.*

3 Alternatively, if possible, peel back the insulated boot to the ECM multi-plug, and backprobe the terminals using the equipment probes *(see illustration 4.2)*.

4 If the ECM terminals are not accessible, then ideally connect a BOB between the ECM and its multi-plug. This is the preferred method, and will avoid any possibility of damage to the ECM terminals. **Note:** *Refer to Warning No 3 (in Reference) before disconnecting the ECM multi-plug.*

5 Otherwise, the ECM multi-plug could be disconnected and the ECM multi-plug terminals probed for voltages. **Note:** *This procedure is mainly used for checking voltage supplies to the ECM and integrity of the earth connections.*

6 Unless otherwise stated, attach the voltmeter negative test lead to an earth on the engine, and probe or backprobe the component terminal under test with the voltmeter positive test lead. **Note:** *DO NOT push round tester probes into square or oblong terminal connectors. This leads to terminal deformation and poor connections. A splitpin is the correct shape for inserting into square or oblong terminals.*

7 In this book, the multi-plug diagram usually shows the terminals of the harness connector. When back-probing the multi-plug (or viewing the sensor connector terminals), the terminal positions will be reversed.

Probing for supply or reference voltage

8 Ignition key on, component multi-plug connected or disconnected as stated in the appropriate test. Probe or backprobe for nominal battery voltage or the reference voltage 5.0 volt supply.

Probing for signal voltage

9 Ignition key on, component multi-plug connected. Backprobe for nominal battery voltage or the reference voltage 5.0 volt supply.

Earth or return

10 Method one: Ignition key on, component multi-plug connected. Backprobe for 0.25 volts max.

11 Method two: Component multi-plug

4.1 The art of backprobing for DC voltage. Circuit multi-plugs connected and the ignition switched 'on': Attach the negative probe to earth and push the positive probe past the boot until it makes contact with the terminal connection. This is a dynamic test of a live circuit and makes for accurate diagnosis of voltage faults

connected or disconnected: Voltmeter positive test lead attached to the supply or reference terminal, and the voltmeter negative test lead attached to the earth or return terminal. The voltmeter should indicate supply voltage if the earth is satisfactory.

Resistance tests

1 Ensure that the ignition key is off, and that the circuit or component under test is isolated from a voltage supply.

2 DO NOT push round tester probes into square terminal connectors. This leads to deformation of the terminal and poor connections.

3 Circuits that begin and end at the ECM are best tested for resistance (and continuity) at the ECM multi-plug (after it has been disconnected) *(see illustration 4.3)*. **Note:** *Refer to Warning No 3 (in Reference) before disconnecting the ECM multi-plug.*

4 The use of a BOB is also recommended for resistance tests, but the BOB must be connected to the ECM multi-plug, and NOT to the ECM itself.

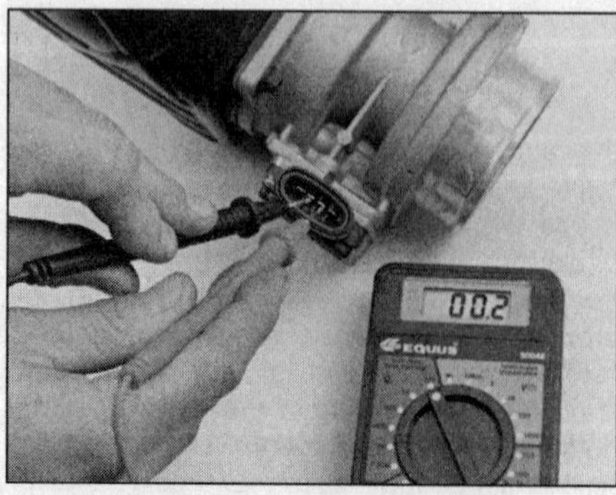

4.3 Measuring resistance: Detach the circuit multi-plug, select the appropriate resistance range and then touch the probes to the two terminal under test

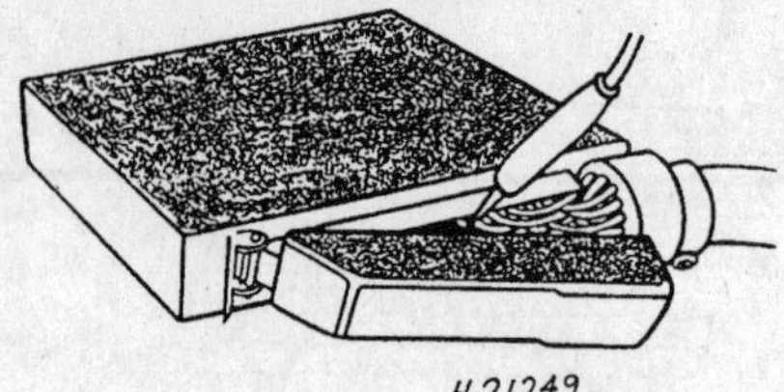

4.2 Backprobing at the ECM multi-plug

5 If the resistance test for a sensor circuit is made at the ECM multi-plug pins, and the sensor has a common connection to the ECM (either through a 5.0 volt reference supply and/or a sensor earth return), the multi-plug connectors for the remaining components must be disconnected. If this procedure is not followed, the results may be inaccurate.

6 When checking continuity of a circuit or continuity to earth, the maximum resistance should be less than 1.0 ohm.

7 When checking the resistance of a component against specifications, care should be taken in evaluating the condition of that component as the result of a good or bad test result *(see illustration 4.4)*. A component with a resistance outside of its operating parameters may not necessarily be faulty. Conversely, a circuit that measures within its operating parameters may still be faulty. However, an open-circuit or a very high resistance will almost certainly be indicative of a fault. The ohmmeter is more useful for checking circuit continuity than it is for indicating faulty components.

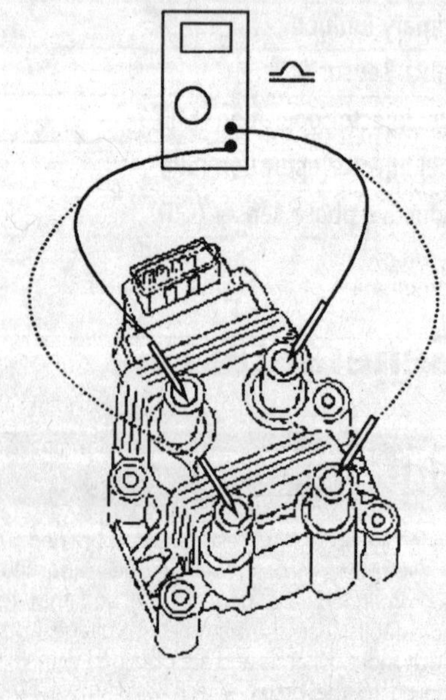

4.4 Checking secondary resistance on a typical DIS coil

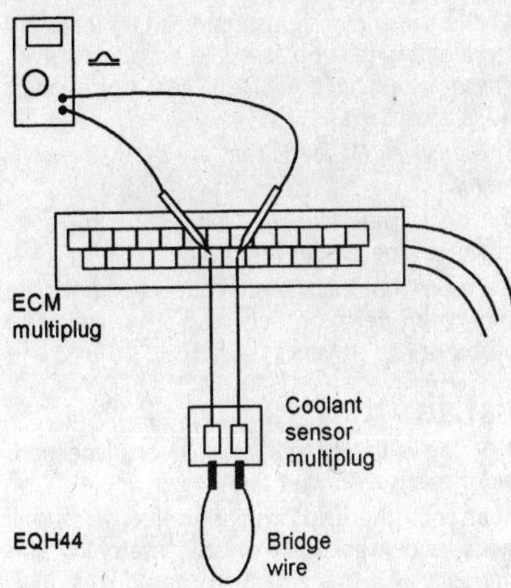

4.5 Check continuity of circuit between the ECM and the component multi-plug

Checking for continuity of circuit

Note: *These tests can be used to quickly check for continuity of a circuit between most components (sensors and actuators) and the ECM.*

8 Disconnect the ECM multi-plug. Refer to Warning No 3 in Reference.

9 Disconnect the component multi-plug.

10 Connect a temporary bridge wire between terminals 1 and 2 at the component multi-plug.

11 Identify the two ECM pins which are connected to the component under test.

12 Connect an ohmmeter between the two pins at the ECM multi-plug. The meter should display continuity of the circuit.

13 If no continuity, check for a break in the wiring or a bad connection between the ECM pin and its corresponding terminal at the multi-plug *(see illustration 4.5)*.

14 Move one of the ohmmeter probes and touch to earth. The ohmmeter should display an open-circuit.

15 If the component is connected to the ECM by more than two wires, repeat the test using a combination of two wires at one time.

4.6 Connect the positive dwell meter probe to coil negative terminal No 1, and measure the duty cycle under varying engine operating speeds

Duty cycle tests

Connecting equipment probes

1 Connect the negative dwell meter probe to an engine earth.

2 Use the positive probe to backprobe the signal terminal of the component under test.

3 Make the duty cycle tests at various engine temperatures, with the engine cranking or running and at different engine speeds *(see illustration 4.6)*.

Possible dwell meter problems

Use of dwell meter during primary cranking tests

4 Although meaningful readings can generally be obtained with most modern DMMs, it is true that some may not be totally accurate during a cranking test on the primary ignition. This occurs when the meter's own preset trigger level may not be suitable for capturing the true voltage level of the component being tested.

Use of dwell meter during injector tests

5 Where the injector is either the current-controlled type or the peak-and-hold kind, very few dwell meters may be capable of registering the rapid pulsing to earth or the current holding that occurs during the second stage of the pulse duration. The meter may only register the switch-on circuit of approximately 1.0 or 2.0%. This means that the injector duty cycle reading will be inaccurate, and not representative of the total pulse width seen in the circuit.

Variable potentiometer

It can be very useful to make certain tests on an engine at various operating temperatures. If the engineer has to wait for the engine to become cold, or reach its normal operating temperature, or indeed reach any other important temperature, the task of testing can be irksome and prolonged *(see illustration 4.7)*. Most fuel injection ECMs (and some ignition ECMs) recognise the engine temperature by monitoring the voltage signal returned from the CTS. **Note:** *In a very few instances, this signal may be returned from the oil temperature sensor (OTS) in addition to, or instead of, the CTS.*

If a variable resistor (or potentiometer) is connected between the CTS or OTS terminals, the engine temperature may be simulated over the entire engine operating temperature range. Obtain a variable resistor or potentiometer ('pot'). A simple pot can be obtained from an electrical/electronic component store. Although the simple pot type is adequate for most tests, we recommend the use of the best quality pot that you can obtain. A good quality pot will give more 'feel' and better control of the engine. The pot range should be from 1 ohm to 100 000 ohms.

4.7 Using a variable potentiometer to vary the CTS resistance. Voltage change can be measured and the engine can be fooled into thinking it is cold or hot. This means that simulated cold running tests can be accomplished with the engine hot and without waiting for it to cool

Variable potentiometer (test methods)

1 The following procedures should be followed when using the pot with either the OTS or CTS.

2 Disconnect the CTS multi-plug.

3 Connect the pot between the two multi-plug terminals.

4 Set the pot to the correct resistance for the temperature that you wish to simulate.

5 Vary the resistance and make the test procedures as required.

6 On some engines you will set fault codes during test procedures, and these codes must be erased after testing is completed.

7 Refer to the Fault Code section in the relevant system Chapter for instructions on how to clear fault codes.

Compression test

Petrol engines

Care must be taken when making a compression test on an petrol engine with electronic ignition, fuel injection and/or a catalytic converter. As a general rule, disconnect the coil negative terminal, fuel system fuse or relay and, where the VM recommends, the ECM.

Disabling the ignition and fuel systems

If the cranking test is to be accomplished, the ignition must be disabled to prevent the engine from starting. In an injection engine, the fuel system must also be disabled. Reference should be made to the manufacturer's recommended method of disabling the engine. However, the following methods may be suitable in some instances.

a) Use a spark gap jumper to connect the coil to distributor HT lead (king lead) to engine earth. Injection (where fitted) will still occur, and this method is unsuitable for wasted-spark ignition.

b) Disconnect the wire from the coil (-) terminal. Many injection systems are pulsed from the coil (-) terminal, and disabling the primary ignition will also disable the injection system.

c) *An engine with an EMS (one combined ECM to control both fuel and ignition) may still inject, even with the coil (-) terminal disconnected. Disconnecting the CAS will therefore disable both ignition and fuel. The coil (-) terminal should also be disconnected.*
d) *Disconnect the fuel pump, or the main relay, or the electrical connection to the injectors.*

Diesel engines

A compression tester specifically intended for diesel engines must be used, because of the higher pressures involved. The tester is connected to an adapter which screws into the glow plug hole or replaces the injector itself. It is unlikely to be worthwhile buying such a tester for occasional use, but it may be possible to borrow or hire one – if not, have the test performed by a garage.

Unless specific instructions to the contrary are supplied with the tester, observe the following points:

a) *The battery must be in a good state of charge, the air filter must be clean, and the engine should be at normal operating temperature.*
b) *All the glow plugs or injectors should be removed before starting the test.*
c) *The stop solenoid and fuel metering control wiring must be disconnected, to prevent the engine from running or fuel from being discharged. On common rail engines, disconnect the injector wiring connectors.* ***Note:*** *As a result of the wiring being disconnected, faults may be stored in the ECU memory. These must be erased after the compression test.*

Secondary (HT) ignition system testing – vehicles with distributor

Secondary circuit tuning

A 'scope is invaluable for analysis of the ignition secondary circuit. However, much of the secondary circuit can be easily tested with inexpensive equipment and a careful visual inspection.

Overview of secondary 'tuning'

The ignition system should provide sufficient voltage to bridge the rotor and spark plug gaps, with an adequate coil voltage in reserve. If the coil reserve is lower than the voltage required by the ignition system, a misfire will be the result.

Tuning, in the modern sense, is building an adequate coil reserve so that the engine will run powerfully and economically. Over a period of time (not necessarily mileage), it is possible for faults to develop and components to wear or become dirty so that the reserve of coil voltage becomes too low for reliable running. The engine may then become difficult to start, or develop a misfire or a hesitation, and engine operation will become less than smooth and uneconomical. A high percentage of engine faults can be attributed to a low coil reserve, and faults resulting in excessive HT could lead to disruptive ECM operation or premature ECM failures.

Pre-testing

1 Ensure that the battery, starter motor and alternator are operating properly. Faults in these areas can cause problems in other areas. For example, a defective alternator diode can cause RFI which can disrupt ECM operation and cause it to erratically switch the ignition and fuelling systems.

2 The engine should also be mechanically sound, with fresh engine oil of the correct grade.

3 Many engine running problems are caused by dirty or faulty secondary HT components. A careful check of the ignition secondary, and a renewal of all worn or faulty components, will reap dividends and help to build a healthy voltage reserve.

4 Where possible, the vehicle manufacturer's original equipment parts should be fitted as replacements in the HT circuit. Pattern parts rarely perform as well, or for as long, as the originals, and premature failure is often the norm.

Coil tower

5 Carefully check the coil, and in particular the coil tower, for a skeletal etching that denotes tracking.

6 Clean and polish the coil tower. A polished tower will reduce tracking in damp conditions. Any leakage from the coil will reduce the coil voltage reserve.

Coil secondary resistance

7 Disconnect the coil LT wires, and connect the ohmmeter probes between the coil tower (usually terminal 4) and one of the two LT terminals. Compare the measured value to the specification in the system Chapter. In most instances, the value will be within the range 5000 ohms to 15 000 ohms. If the probe is moved from one LT terminal to the other, the measured value should be very similar to the first reading obtained.

HT leads

8 Carefully remove the HT leads from the spark plugs. If you are less than careful, it is possible to damage the HT lead connection ends, resulting in an increase in HT lead resistance.

9 Check the HT leads for damage, poor insulation, pinhole leaks in the insulation, or poor connections at either end.

10 Measure the HT lead resistance with an ohmmeter.

11 Similarly check all HT suppression caps that may be fitted.

12 Remove and discard suppression devices that are not fitted as original equipment.

13 Do not use unsuppressed HT leads in any engine that has electronic components – a disrupted ECM is usually the result.

14 In some engines, unsuppressed HT leads are used along with suppressed HT caps (eg BMW). If fitted as original equipment, then this type of lead is acceptable.

15 Renew all HT leads that are excessively oily or dirty.

16 Check that the HT leads are correctly installed in the proper firing order.

17 Ensure that the HT lead connection ends are making proper contact with the coil and distributor cap towers.

Distributor cap

18 Remove the distributor cap for inspection, even if this is a difficult task.

19 Inspect the cap insulation for scratches, cracks and tracking – both internally and externally.

20 Check for oil or moisture on the internal surfaces. Signs of oil seepage could suggest a defective distributor seal; oil will reduce the quality of the spark by bleeding the HT current to earth. Water may be present in the engine oil (oil change overdue, head gasket fault, etc), and find its way into the cap along with the oil mist through a faulty distributor seal.

21 Inspect the HT towers for corrosion, and the internal terminals for pitting and erosion.

22 Clean off all the old dirt, grease, cold start, electrical spray etc from the outside of the distributor cap, and polish the surface. A polished cap will reduce tracking in damp conditions.

23 If there is any doubt about the condition of the distributor cap, it should be renewed along with the rotor arm *(see illustration 4.8)*.

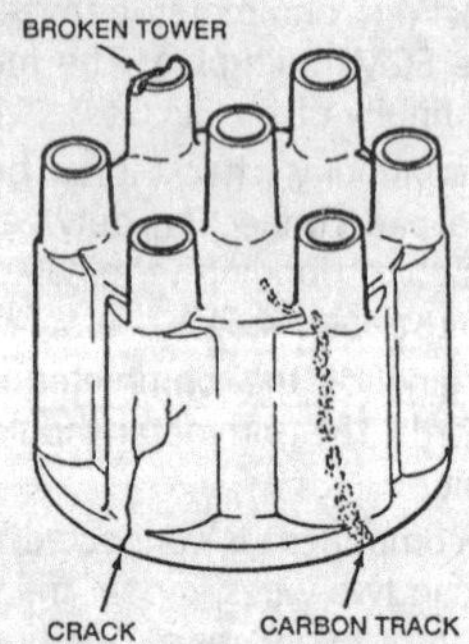

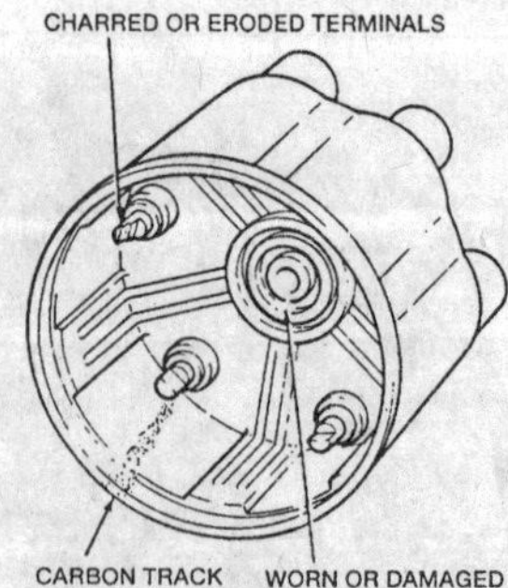

4.8 Carefully inspect the distributor cap for defects and renew it, along with the rotor arm, if there is any doubt about the condition

CARBON DEPOSITS

Symptoms: Dry sooty deposits indicate a rich mixture or weak ignition. Causes misfiring, hard starting and hesitation.

Recommendation: Check for a clogged air cleaner, high float level, sticky choke and worn ignition points. Use a spark plug with a longer core nose for greater anti-fouling protection.

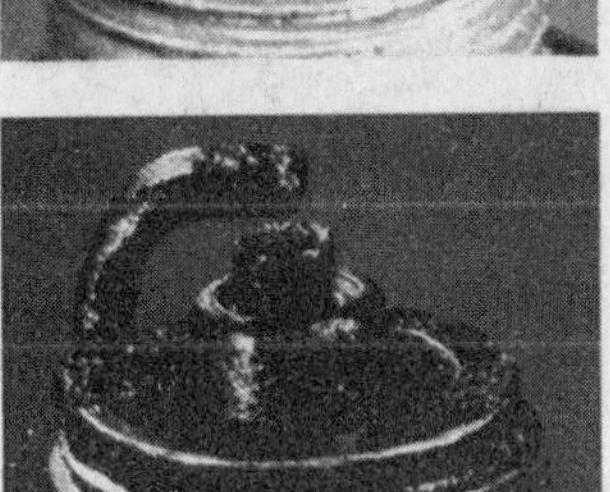

OIL DEPOSITS

Symptoms: Oily coating caused by poor oil control. Oil is leaking past worn valve guides or piston rings into the combustion chamber. Causes hard starting, misfiring and hesition.

Recommendation: Correct the mechanical condition with necessary repairs and install new plugs.

TOO HOT

Symptoms: Blistered, white insulator, eroded electrode and absence of deposits. Results in shortened plug life.

Recommendation: Check for the correct plug heat range, over-advanced ignition timing, lean fuel mixture, intake manifold vacuum leaks and sticking valves. Check the coolant level and make sure the radiator is not clogged.

PREIGNITION

Symptoms: Melted electrodes. Insulators are white, but may be dirty due to misfiring or flying debris in the combustion chamber. Can lead to engine damage.

Recommendation: Check for the correct plug heat range, over-advanced ignition timing, lean fuel mixture, clogged cooling system and lack of lubrication.

HIGH SPEED GLAZING

Symptoms: Insulator has yellowish, glazed appearance. Indicates that combustion chamber temperatures have risen suddenly during hard acceleration. Normal deposits melt to form a conductive coating. Causes misfiring at high speeds.

Recommendation: Install new plugs. Consider using a colder plug if driving habits warrant.

GAP BRIDGING

Symptoms: Combustion deposits lodge between the electrodes. Heavy deposits accumulate and bridge the electrode gap. The plug ceases to fire, resulting in a dead cylinder.

Recommendation: Locate the faulty plug and remove the deposits from between the electrodes.

NORMAL

Symptoms: Brown to grayish-tan color and slight electrode wear. Correct heat range for engine and operating conditions.

Recommendation: When new spark plugs are installed, replace with plugs of the same heat range.

ASH DEPOSITS

Symptoms: Light brown deposits encrusted on the side or center electrodes or both. Derived from oil and/or fuel additives. Excessive amounts may mask the spark, causing misfiring and hesitation during acceleration.

Recommendation: If excessive deposits accumulate over a short time or low mileage, install new valve guide seals to prevent seepage of oil into the combustion chambers. Also try changing gasoline brands.

WORN

Symptoms: Rounded electrodes with a small amount of deposits on the firing end. Normal color. Causes hard starting in damp or cold weather and poor fuel economy.

Recommendation: Replace with new plugs of the same heat range.

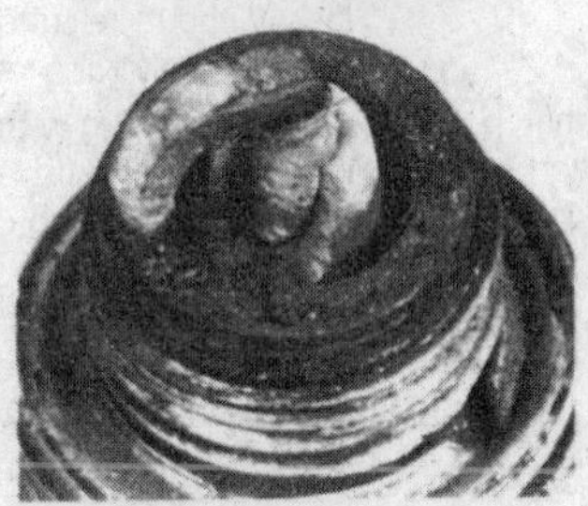

DETONATION

Symptoms: Insulators may be cracked or chipped. Improper gap setting techniques can also result in a fractured insulator tip. Can lead to piston damage.

Recommendation: Make sure the fuel anti-knock values meet engine requirements. Use care when setting the gaps on new plugs. Avoid lugging the engine.

SPLASHED DEPOSITS

Symptoms: After long periods of misfiring, deposits can loosen when normal combustion temperature is restored by an overdue tune-up. At high speeds, deposits flake off the piston and are thrown against the hot insulator, causing misfiring.

Recommendation: Replace the plugs with new ones or clean and reinstall the originals.

MECHANICAL DAMAGE

Symptoms: May be caused by a foreign object in the combustion chamber or the piston striking an incorrect reach (too long) plug. Causes a dead cylinder and could result in piston damage.

Recommendation: Remove the foreign object from the engine and/or install the correct reach plug.

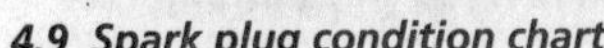

4.9 Spark plug condition chart

Rotor arm

24 Check the rotor arm for tracking.

25 If the rotor arm contains a resistor, use an ohmmeter to check the resistance value, and compare with the specified value. If no figure is quoted, the value will normally be approximately 1000 to 5000 ohms. Excessive rotor resistance will dramatically increase secondary output, which results in a reduced coil voltage reserve.

26 DO NOT polish the rotor tip, as this will increase the rotor air gap and result in a reduced coil voltage reserve.

Spark plugs

27 The spark plugs should generally be renewed according to the VM recommendation.

28 Where operating conditions are known to be 'severe service' including low-mileage and short-journey operation, plugs should be renewed more frequently.

29 Remove the spark-plugs and inspect them carefully. Spark plug condition provides many clues to the observant and it is possible to detect engine condition, mixture strength and many other faults from 'reading' the plugs *(see illustration 4.9 on the previous page)*.

30 Renew spark plugs with eroded electrodes, a cracked insulator (internal and external), heavy carbonisation or glazing.

31 A new set of plugs will often work wonders; remember to re-gap new plugs to the specified gap.

Spark plug removal

32 Before removing a plug, check that it has not been cross threaded by observation of the angle of installation. Cross threading usually occurs due to sloppy installation and aluminium cylinder heads are particularly at risk.

33 Extreme care must be observed during the offering up and entry of the plug into the spark plug aperture.

34 Removal is best done from a cold engine, particularly where the cylinder head is made from aluminium. Several VMs warn against removing plugs from a hot engine, in any case.

35 Slacken the plug two or three turns, and then use an air line to blast away debris and dirt from the plug wells, as it may enter the cylinders on removing a plug.

36 If the plug is overtight, squirt penetrating oil into the well, and gently move the plug backwards and forwards until it loosens.

37 A badly-worn or wrong-size spanner will round the plug hexagons, making removal difficult.

Spark plug type

38 The plug must be the correct size and heat range. Sizes differ in length from 3/8" to 3/4", and in diameter from 10 mm to 18 mm. Within these sizes, the seats may be tapered or use a sealing washer.

39 It is quite possible to fit a plug with an incorrect seat, length or heat range with sometimes disastrous results.

40 If a longer plug than standard is fitted (eg 3/4" instead of 1/2"), there would be a danger of the piston fouling the spark plug tip. Also, the threads projecting into the combustion chamber will carbon up, making removal very difficult.

41 When a shorter plug than standard is fitted (eg 1/2" instead of 3/4"), the first few threads of the spark plug hole will also tend to carbon up, making the later fitting of the correct plug a very difficult task without re-cutting the thread. Incomplete combustion will also result from the shrouded spark.

42 A spark plug with the wrong heat range can be a very expensive mistake. A hotter-than-standard spark plug can cause overheating, pre-ignition and piston failure. A colder plug is usually safer, but will tend to foul around town, as it does not get hot enough to burn off the low-speed deposits that form on the insulator.

43 When an engine is worn and tends to regularly oil foul one or more plugs, a temporary cure may be to fit a plug with a hotter heat range. The hotter plug will burn the oily deposits more readily and be less prone to fouling.

> *Warning: Fitting a hotter spark plug than standard is not recommended because of the danger of engine failure if the engine is driven at medium to high engine or vehicle speeds, or under heavy load. When high speeds are attempted with hotter plugs than normal, there is a strong danger that the hotter plug will cause pre-ignition and piston failure with very little warning*

Spark plug servicing

44 Spark plug cleaning with a wire brush is not recommended as this can cause plug damage and tracking.

45 Oiled plugs are caused by defective piston rings or valve guides. A compression test should identify the fault, which must be remedied if efficient engine operation is to be maintained. A dirty insulator can also lead to tracking of the spark to earth.

Spark plug gap

46 Always check the gap when inspecting or renewing spark plugs. A gap that is too narrow or too wide will cause poor starting, poor performance and poor economy.

Spark plug installation

47 Where the thread in the cylinder head is dirty, use a spark plug thread restorer, with plenty of grease, to clean up the threads before installation.

48 Be very careful that debris does NOT enter a cylinder.

49 Install the spark plug by hand until the faces are felt to make contact, then tighten (using a

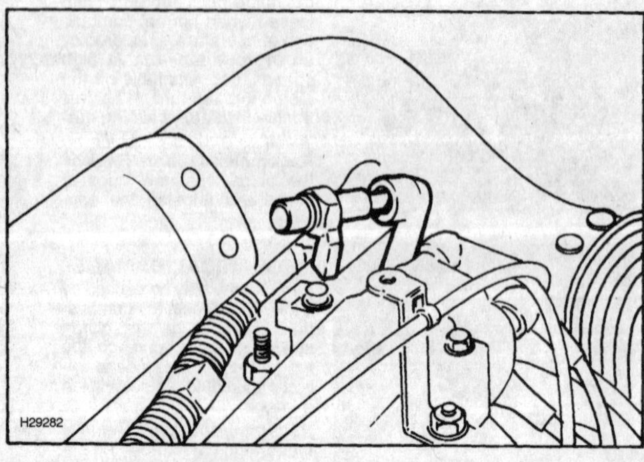

4.10 *Use a spark jumper to test for an HT spark from the coil 'king' lead*

4.11 *Pass an insulated probe over the HT leads, cap and coil. If an insulation leak is present, a spark will jump to earth through the tip of the probe*

plug tool) by a further 1/4 turn (flat seat plugs) or by a further 1/16 turn (taper seat plugs). DO NOT overtighten the plugs – you'll risk stripping the threads out of the cylinder head!

50 If using a torque wrench, the correct figure for taper-seat plugs is 8 to 15 lbf ft. If the spark plug is overtightened, subsequent removal can be extremely difficult.

51 Taper-seat plugs are easily overtightened on installation due to the wedging effect of the taper. The plug may break at the boss during a subsequent attempt at removal. In this instance, the plug may need to be extracted with a special tool. In extreme cases, the head will need to be removed to facilitate plug removal.

52 Taper seat plugs installed in certain engines may fit less tightly into the threads of the spark plug hole than is desirable. In the first few hundred miles, combustion fumes may escape via the loose-fitting plug, giving a very slight compression loss. Eventually, carbon will seal this leak, but this may make subsequent removal more difficult. Where this is noticed (the spark plug boss will be carbon-contaminated) the manufacturer's original-equipment spark plug recommendation should be used.

Checking for secondary HT voltage (non-runner test)

53 Check for an HT spark to the plugs. Use one of the following methods whilst cranking the engine on the starter, referring to Warning No 2 in Reference. A powerful spark should consistently bridge the gap:

a) Use a set of insulated pliers to hold one of the HT leads 6 mm from an engine earth.
b) Connect a spark jumper between the HT lead and cylinder head (see illustration 4.10).
c) An alternative method can be used with some diagnostic equipment, when the equipment can detect rpm from an HT lead via an inductive probe. Connect the equipment's inductive trigger probe to an HT lead. Crank the engine. If the equipment registers rpm, the ignition system must be providing a spark. Having tested one lead, test the other three in a similar fashion.

54 No spark: Check for primary ignition switching. If no switching, check the primary ignition system. Refer to the specific Chapters for the system under test.

55 Check for an HT spark from the coil tower. Use one of the three methods listed above whilst cranking the engine on the starter. Refer to Warning No 2 in Reference. A powerful spark should consistently bridge the gap.

56 No spark from the coil tower: If the coil primary is switching, yet no secondary output, the coil is suspect.

57 Before renewing the coil, check the main HT lead for a high resistance (greater than 30000 ohms) and that it is making a good contact with the inner coil tower.

58 Check the coil secondary resistance. Attach an ohmmeter between the coil tower and each primary coil terminal in turn. Compare with the specification in the system Chapter.

59 If there is output from the coil secondary, but no spark at the plugs, check the following components:

a) HT leads. Mark each HT lead so that it may be removed and returned to its correct location.
b) Ensure that the HT leads are pressed firmly into full contact with the coil and distributor cap towers. Loose or poorly fitting HT leads are a common HT fault.
c) Remove the leads (one by one) and measure the resistance with an ohmmeter. Maximum HT lead resistance should be no more than 30 000 ohms.
d) Visually inspect each lead. Replace a lead where the insulation is cracked or chafed or insulating boots & connectors are loose.
e) Check that the HT leads are connected to the correct coil tower in the correct firing order.
f) Remove the distributor cap and check the cap and rotor for defects.
g) Check for wet or damp HT components.
h) Check the spark plugs for condition, gap and correct type. Many faults in poor running electronic engines can actually be traced to worn out spark plugs, renew them if at all in doubt.

Engine running tests

60 Some of the above faults that might prevent the engine from starting may also be present when the engine is running badly:

a) Check the HT leads.
b) Check the coil secondary resistance.
c) Check the spark plugs for condition, gap and for correct type.

Secondary insulation

61 An insulation failure occurs when the spark is allowed to track to earth.

62 Secondary insulation can be tested as follows:

a) Run the engine at idle. A dark garage or workshop will help the result, by making any 'stray' HT voltage easier to spot.
b) Use an insulated probe with one end connected to earth.
c) Slowly pass the probe over the HT leads, distributor cap and ignition coil.
d) The probe should be held no more than 6 mm above the secondary component under test.
e) If a spark jumps from an HT lead or coil to earth through the tip of the probe, the component has inadequate insulation, and should be renewed (see illustration 4.11).

Faulty secondary HT circuit

63 The following list of faults are common reasons for HT failure:

a) Fouled spark plugs.
b) Spark plugs with eroded electrodes or wide gaps.
c) Reversed coil polarity.
d) High resistance HT leads.
e) Incorrect HT leads.
f) Corroded coil towers.
g) Poor primary connections.
h) Over suppression.
i) Faulty HT can cause either electronic component failure or a spurious RFI signal. RFI can disrupt ECM operation.

Secondary (HT) ignition system testing – vehicles without distributor (DIS)

Testing the DIS secondary circuit

1 Testing the DIS secondary HT circuit is similar in many respects to testing the secondary with a distributor. Check the following points, and also refer to the secondary tests with distributor.

2 Visually inspect and check the HT leads. The maximum resistance of a lead of any length should not exceed 30 000 ohms.

3 Visually inspect and check the spark plugs.

4 Check the spark plugs for condition, gap and correct type. Many faults in poor-running electronic engines can actually be traced to worn-out spark plugs; renew them if at all in doubt.

5 Because two plugs fire together in a wasted-spark DIS ignition system, a faulty HT lead will affect both cylinders.

6 Check that the HT leads are connected to the correct coil tower, in the correct firing order.

7 The coil towers are normally marked with the appropriate cylinder numbers. Connection of HT leads to the wrong coil tower can cause engine damage.

8 Visually inspect and check the DIS coil unit and towers.

9 Check the coil secondary resistance as follows:

a) Detach the HT leads from the coil unit.
b) Mark the leads so that they can be replaced in the correct position.
c) Measure the resistance between each pair of matched HT terminals and compare to the manufacturer's specifications (refer to illustration 4.4).

Analysis

10 A spark at only one plug of a set (ie 1 and 4 or 2 and 3) indicates an HT lead or plug fault.

11 Failure of both plugs to spark would indicate a coil secondary or primary fault.

12 Failure of all plugs to spark would indicate a coil secondary or primary fault, or possibly a primary trigger (ie CAS) fault.

System sensor and actuator tests

 *Important: The following test procedures are general in nature, and should be followed with reference to the Chapter which covers the system under test, and to the system wiring diagram.*

Primary trigger – crank angle sensor (CAS)

1 Inspect the CAS multi-plug for corrosion and damage.

2 Check that the terminal pins in the CAS multi-plug are fully pushed home and making good contact with the CAS.

3 Remove the CAS from the engine block. Inspect the end surface for corrosion and damage.

4 Measure the CAS resistance, and compare to the specifications in the system Chapter. Refer also to the resistance tests at the end of this section.

5 A fault in any of the above areas are common reasons for a poor or inaccurate signal from the CAS. **Note:** *These tests are also applicable to RPM and TDC sensors, and distributor-located inductive triggers.*

Checking the CAS output with an AC voltmeter (engine cranking)

Note: *This test is more likely to be made if the engine is a non-runner.*

6 Detach the CAS or ECM multi-plug *(refer to Warning No 3 in Reference)* and connect an AC voltmeter between the two terminals leading to the CAS. If a third wire is present, it will be a shield wire.

7 Crank the engine. A minimum AC RMS voltage of about 0.7 should be obtained, although most good sensors will provide an output of more than 1.4 AC RMS voltage *(see illustration 4.12)*. **Note:** *The AC voltmeter at least proves that a signal is being generated by the CAS. However, the AC voltage is an average voltage, and would not clearly indicate damage to the CAS lobes or that the sinewave is regular in formation.*

Checking the CAS resistance with an ohmmeter

8 Detach the CAS or ECM multi-plug *(refer to Warning No 3 in Reference)* and connect an ohmmeter between the two terminals leading to the sensor.

9 Record the resistance and compare it with the specified value in the system Chapter *(see illustration 4.13)*. **Note:** *Even if the resistance is within the quoted specifications in the system Chapter, this does not prove that the CAS can generate an acceptable signal.*

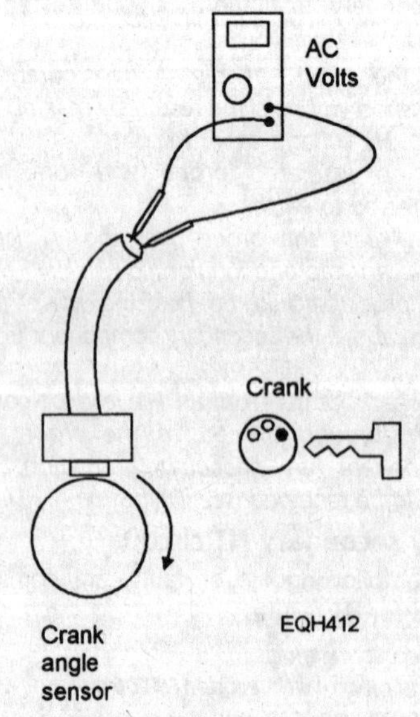

4.12 Check the CAS output with an AC voltmeter

Checking the CAS shield connection

10 The CAS may have a shield wire (not in all cases). Locate the wiring multi-plug connector or disconnect the ECM multi-plug *(refer to Warning No 3 in Reference).*

11 Attach an ohmmeter probe to one of the sensor terminals (1 or 2).

12 Attach the other ohmmeter probe to the shield wire terminal. A reading of infinity should be obtained.

13 Move the ohmmeter probe from the shield wire terminal, and connect it to earth. A reading of infinity should also be obtained. **Note:** *The shield wire on the CAS in some systems is connected to the CAS earth return wire. In such a case, continuity will be registered on the ohmmeter and this is normal for that vehicle. Refer to the wiring diagrams for the system under test to determine how the CAS is wired.*

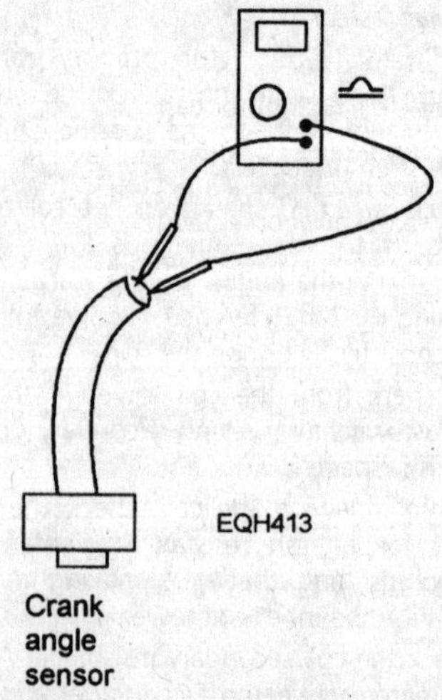

4.13 Measure the CAS resistance

Primary trigger – Hall-effect sensor (HES)

1 In most systems the HES is located in the distributor. A flywheel-mounted HES is found in some VAG systems.

2 Inspect the HES multi-plug for corrosion and damage.

3 Check that the terminal pins in the HES multi-plug are fully pushed home and making good contact with the HES.

Quick test (non-runner, no spark)

4 Remove the main HT lead from the distributor cap centre tower, and connect it to the cylinder head via a spark jumper.

5 Detach the HES multi-plug at the distributor.

6 Identify the supply, signal and earth terminals.

7 Briefly flash a small jumper lead between the (O) and (-) terminals on the HES harness multi-plug *(see illustration 4.14)*.

8 If a spark jumps across the spark jumper terminals to the cylinder head, the coil and amplifier are capable of producing a spark, and the Hall switch in the distributor is suspect.

HES test procedures

9 Roll back the rubber protection boot to the HES multi-plug.

10 Connect the voltmeter negative or dwell meter probe to an engine earth.

11 Identify the supply, signal and earth terminals.

12 Connect the voltmeter positive or dwell meter probe to the wire attached to the HES signal terminal.

13 Allow the engine to idle.

14 An average voltage of approximately 7 to 8 volts, or a duty cycle of approximately 35% should be obtained.

Signal voltage or duty cycle signal not available

15 Stop the engine.

16 Remove the distributor cap.

17 HES multi-plug connected, ignition on, voltmeter positive probe connected to the signal terminal *(see illustration 4.15)*.

18 Turn the engine over slowly. As the trigger vane cut-out space moves in and out of the air gap, the voltage should alternate between 10.0 to 12.0 volts and zero volts.

Signal voltage not available

19 Disconnect the HES multi-plug at the distributor.

20 Probe output terminal 2 (O) of the harness multi-plug with the voltmeter positive probe. A voltage of between 10 and 12 volts should be obtained.

21 No voltage from the ECM to terminal 2: Check for continuity of the signal wiring between the HES and the ECM. Recheck for voltage at the ECM terminal.

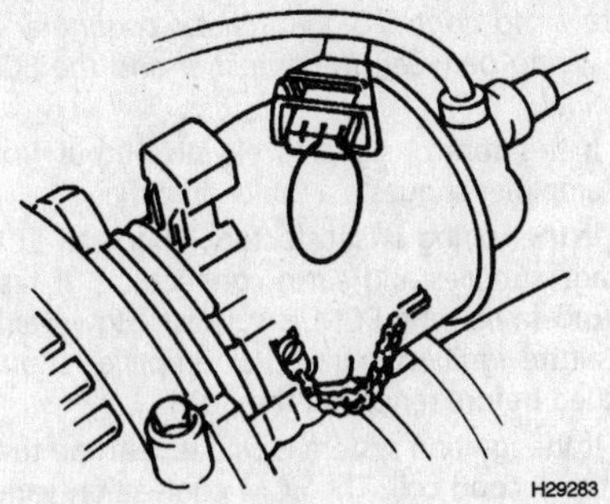

4.14 Very briefly flash the 'O' and (-) terminals at the HES multi-plug to check for a spark

22 No voltage available at the ECM: Check all voltage supplies and earth connections to the ECM. If the voltage supplies and earth connections are satisfactory, the ECM is suspect.

23 Check the voltage supply (10 to 12 volts) at HES terminal number 1 (+). If the supply is unsatisfactory, check for continuity of the wiring between the HES and the ECM.

24 Check the earth connection at HES terminal number 3 (-).

25 If the voltage supply and earth are ok, the HES in the distributor is suspect.

Primary ignition

1 Check the coil terminals for good clean connections.

2 Clean away accumulations of dirt, and the residue from a maintenance spray. The residue will attract dirt, and often leads to bleeding of the HT current to earth.

3 Inspect the ignition coil for signs of tracking, particularly around the coil tower area.

4 Although the following tests are accomplished with the aid of a basic dwell meter, an oscilloscope is a more suitable instrument for analysing the signals generated by the primary ignition.

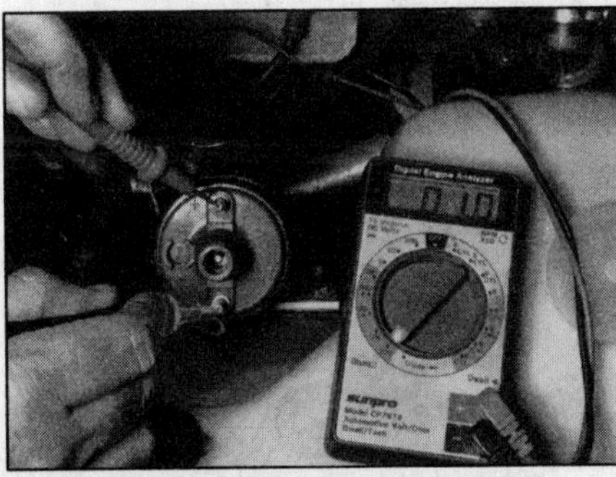

4.16 Check the primary resistance. Disconnect the low tension wires and connect the ohmmeter between the positive and negative terminals

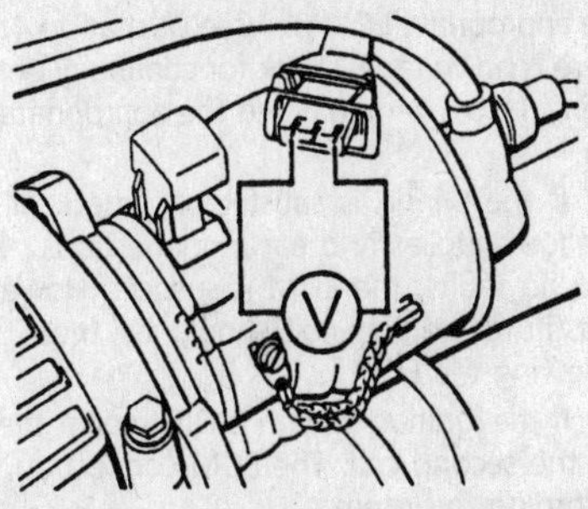

4.15 Connect the voltmeter between the HES (+) and (-) terminals. A voltage of between 10 and 12 volts should be obtained

Engine non-runner test procedures

5 Connect the negative dwell meter probe to an engine earth.

6 Connect the positive dwell meter probe to the coil negative (-) terminal (usually marked 1 in Bosch systems).

7 Crank the engine on the starter.

8 A duty cycle reading of approximately 5 to 20% should be obtained. If the instrument can measure the value in milliseconds, then this is even more useful.

9 Good primary signal: The primary ignition (including the CAS or HES) are providing an acceptable signal.

Primary signal not available (amplifier inside the ECM)

10 Check the primary trigger for a good signal (refer to CAS or HES test).

11 Switch the ignition on.

12 Check for a voltage supply to the coil positive (+) terminal (15). If there is no voltage, check the wiring back to the supply (usually the ignition switch, but could be one of the relays).

13 Check for voltage to the coil negative (-) terminal (1). If there is no voltage, remove the wire to the coil (-) terminal and recheck. If there is still no voltage, check the coil primary resistance *(see illustration 4.16)*. Refer to the specifications in the system Chapter.

14 With voltage at nbv level, check for a short to earth between the coil No 1 terminal and the appropriate ECM pin.

15 Detach the ECM multi-plug (refer to Warning No 3 in Reference) and check for nbv at

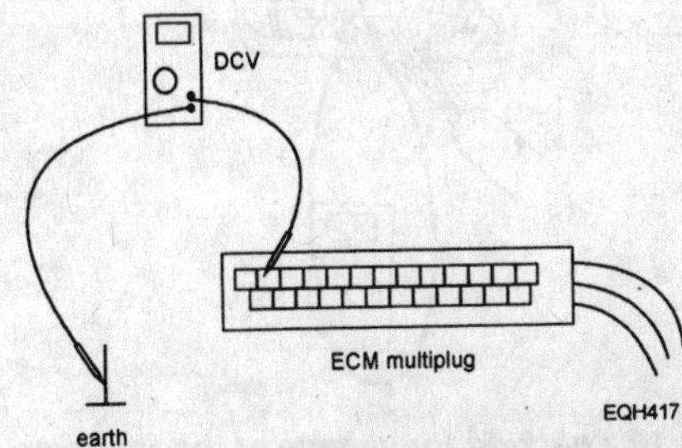

4.17 Detach the ECM multi-plug and check for nbv at the ECM primary ignition terminal

the appropriate ECM pin *(see illustration 4.17)*. If there is no voltage, check for continuity between the coil No 1 terminal and the appropriate ECM pin.

16 If the wiring is satisfactory, check all ECM voltage supplies and earth connections. If tests find no faults, the ECM is suspect. However, a substitute ignition coil should be tried before renewing the ECM.

17 If the ignition system is DIS, repeat the tests for the second coil. The ECM connection varies according to system.

Primary signal not available (separate external amplifier)

18 Check the primary trigger for a good signal (Refer to CAS or HES test).

19 Switch the ignition on.

20 Check for a voltage supply to the coil positive (+) terminal (15). If there is no voltage, check the wiring back to the supply (usually the ignition switch, or one of the system relays).

21 Check for voltage to the coil negative (-) terminal (1).

a) *No voltage: Remove the wire to the coil (-) terminal and recheck.*
b) *Still no voltage: Check the coil primary resistance, the coil is suspect.*
c) *Voltage at nbv level: Check for a short to earth between the coil No 1 terminal and the amplifier. If the wiring is ok, the amplifier is suspect.*

22 Disconnect the amplifier multi-plug *(refer to Warning No 3 in Reference)*. Check for voltage at the amplifier terminal that is connected to the ignition coil terminal 1 *(see illustration 4.18)*. If there is no voltage, check for continuity of wiring between the amplifier and ignition coil terminal No 1.

23 Check for voltage to the amplifier from the ignition switch.

24 Check the amplifier earth connection.

25 Crank the engine and check for a control signal from the ECM to the amplifier. **Note:** *Although it is possible to use a dwell meter to check for a duty cycle signal from the ECM to the amplifier, the integrity of the signal may be difficult to establish. Once again, an oscilloscope is more likely to make sense of this signal. If there is no control signal, check continuity of the wiring between the amplifier and the ECM terminal.*

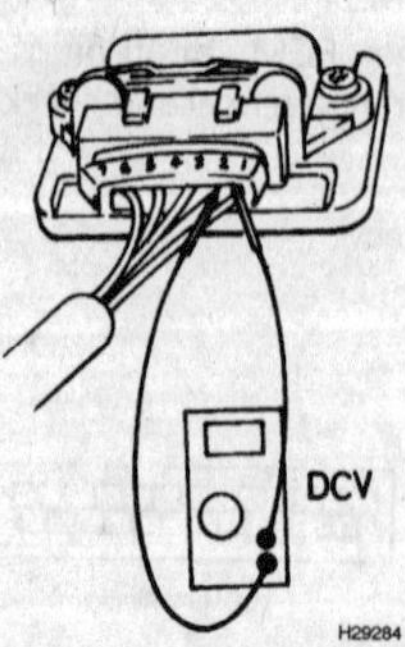

4.18 Checking for voltage at the amplifier terminal that is connected to the ignition coil terminal No 1

26 If the control signal is ok, no output from the amplifier suggests a faulty amplifier.

27 If the wiring is satisfactory, check all ECM voltage supplies and earth connections. If tests find no faults, the ECM is suspect. However, a substitute ignition coil and/or amplifier should be tried before renewing the ECM.

28 If the ignition system is DIS, repeat the tests for the second coil. The ECM connection varies according to system.

Engine running test procedures

29 Connect the negative dwell meter probe to an engine earth.

30 Connect the positive dwell meter probe to the coil negative (-) terminal (usually marked 1 in Bosch systems).

31 Run the engine at idle and various speeds. Record the duty cycle values – approximate values are as follows:

Idle speed: 5 to 20%
2000 rpm: 15 to 35%
3000 rpm: 25 to 45%

32 It is important that the duty cycle in % increases in value as the engine rpm is raised. If your DMM can measure the duty cycle in ms, the reading should not change much in value as the engine rpm is raised.

33 Check the amplifier earth.

34 Check that devices such as a radio suppresser or a burglar alarm have not been connected to the coil primary (-) terminal. *All other tests and any detailed primary analysis really requires an oscilloscope.*

Knock sensor (KS)

1 Inspect the KS multi-plug for corrosion and damage.

2 Check that the terminal pins in the KS multi-plug are fully pushed home and making good contact with the KS.

3 Attach the probe of an inductive timing light to the HT lead of No 1 cylinder.

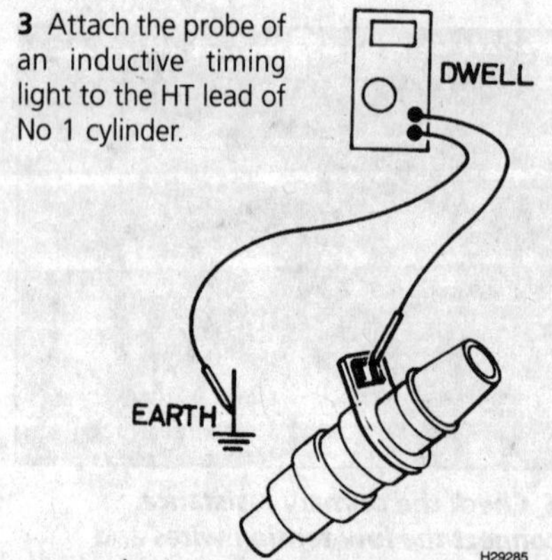

4.19 Checking the fuel injection pulse

4 Allow the engine to idle.

5 Gently tap the engine block close to No 1 cylinder.

6 The timing should be seen to retard.

Fuel injector operation (MPi)

General fuel injector test procedures

1 Inspect the injector multi-plugs for corrosion and damage.

2 Check that the terminal pins in the injector multi-plugs are fully pushed home and making good contact with the injector.

3 Check for corrosion in the connection plugs between the relay and the injector, and the ECM and the injector. Corrosion in connection plugs is a common reason for poor injector performance.

4 Roll back the rubber protection boot to the injector multi-plug *(see illustration 4.19)*.

5 Connect the negative dwell meter probe to an engine earth.

6 Identify the supply and signal terminals. **Note:** *An injector dwell reading will only be obtained upon the signal terminal, which is the wire connecting the injector to the ECM. If you cannot obtain a reading, reconnect the probe to the other terminal and retry.*

7 Connect the positive dwell meter probe to the wire attached to the Injector signal terminal.

8 Although the following tests are accomplished with the aid of a basic dwell meter, an oscilloscope is a more suitable instrument for analysing the signals generated by the electronic fuel injector circuits.

9 Initially, the probe can be connected to the signal terminal of any one of the injectors.

Current-controlled or peak-and-hold injection circuits (dwell meter testing)

10 If the injector is of the current-controlled kind, very few dwell meters are capable of registering the second stage of the pulse duration. The meter may only register the switch-on circuit of approximately 1.0 or 2.0%. This means that the injector duty cycle reading will be inaccurate, and not representative of the total pulse width seen in the circuit. A small number of DMMs are available that can actually measure this circuit satisfactorily.

Engine non-runner test procedures

11 Crank the engine.

12 A duty cycle reading (injector duty cycle) of approximately 5 to 10% should be obtained. If the dwell meter can measure the value in milliseconds, this could be even more useful.

Good injector signal

a) *Check for an injector pulse on the other injectors.*

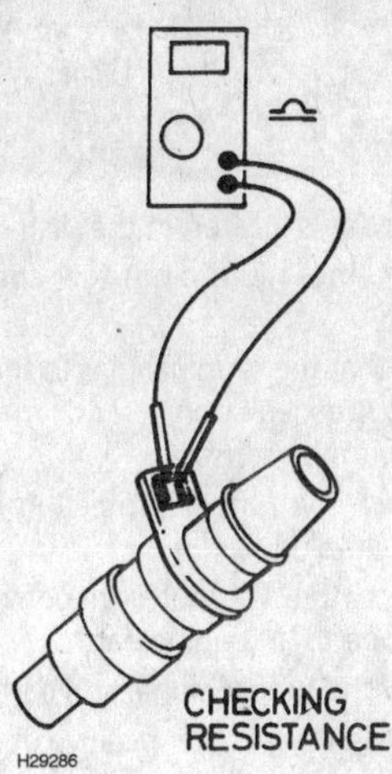

4.20 Checking the injector resistance

b) If the injector signal is satisfactory and if the primary ignition signal is also providing an acceptable signal, the fault is unlikely to be related to the ECM.

Poor or no injector signal on one or more injectors

Note: *In some Motronic systems, the frequency of injection increases for several seconds during initial cranking.*

13 Check the fuel pressure and fuel flow.

14 Check the CAS or HES for a good signal.

15 Check the voltage at the signal terminal of the injector multi-plug. Battery voltage should be obtained.

16 No voltage: Check the injector resistance and the injector voltage supply *(see illustration 4.20).*

17 Disconnect the ECM multi-plug (see Warning No 3 in Reference).

18 Switch on the ignition.

19 Use a jumper lead to very briefly touch each one of the injector actuator pins in the ECM multi-plug to earth *(see illustration 4.21).*

20 If the injector actuates, check the ECM main voltage supplies and earths. If the tests reveal no fault, the ECM is suspect.

21 If the injector does not actuate, check for nbv at the ECM pin.

a) Voltage present: the injector is suspect.

b) No voltage present: Check for continuity of wiring between the injector multi-plugs and the ECM multi-plug.

22 If the injector circuit is banked or sequential, individually check each connection to the ECM.

Duty cycle too long or too short

23 Check the CTS.

24 Check the AFS or MAP sensor. **Note:** *If the ECM has entered LOS due to a fault in one of the sensors, the engine may generally behave quite well whilst the engine is hot, but may be difficult to start when cold.*

Engine running tests

25 Run the engine at various speeds. Record the duty cycle and compare to the approximate values in the following table. When the engine is cold, the values will slightly increase.

Engine speed	Duty cycle
Idle speed	*3 to 6%*
2000 rpm	*7 to 14%*
3000 rpm	*11 to 16%*
Slow throttle increase	*11 to 16%*
Rapid throttle increase	*20+%*
Deceleration	*Zero (release throttle at 3000 rpm)*

26 Evaluate as follows:

a) The duty cycle in % should increase in value as the engine rpm is raised.

b) Under rapid acceleration, the duty cycle should show a great increase in value.

c) Under deceleration, when the engine is hot, the duty cycle should drop to zero (digital meter) and reappear as the engine speed sinks below approximately 1200 rpm.

d) Where the meter does not drop to zero, check the throttle valve for correct adjustment and the TPS or TS for correct operation.

e) Noise from the injectors should also temporarily disappear as the cut-off operates.

f) Please note that a slow-responding digital meter may not show the drop to zero and this should also be considered.

Duty cycle that is too long or too short

27 Check the CTS.

28 Check the AFS or MAP sensor.

Injector resistance tests

29 Remove each injector multi-plug, and measure the resistance of the injector between the two terminals.

a) Current-controlled injectors: 4 ohms.

b) Most other systems: 16 ohms.

30 Parallel injector circuits: Refer to Chapter 2 for a description of the merits of parallel circuit measurement.

Fuel injector operation (SPi)

1 Inspect the injector multi-plug for corrosion and damage.

2 Check that the terminal pins in the multi-plug are fully pushed home and making good contact with the injector.

3 Check for corrosion in the connection plug between the relay and the injector, and the ECM and the injector. Corrosion in connection plugs is a common reason for poor injector performance.

4 Roll back the rubber protection boot to the injector multi-plug.

5 Connect the negative dwell meter probe to an engine earth.

6 Identify the supply and signal terminals.

7 Connect the positive dwell meter probe to the wire attached to the injector signal terminal *(see illustration 4.22).* **Note:** *The majority of SPi systems utilise current control, and the average dwell meter will not accurately measure this kind of injection signal. An oscilloscope is therefore recommended for signal tests on the majority of SPi systems.*

Engine non-runner test procedures

8 Crank the engine.

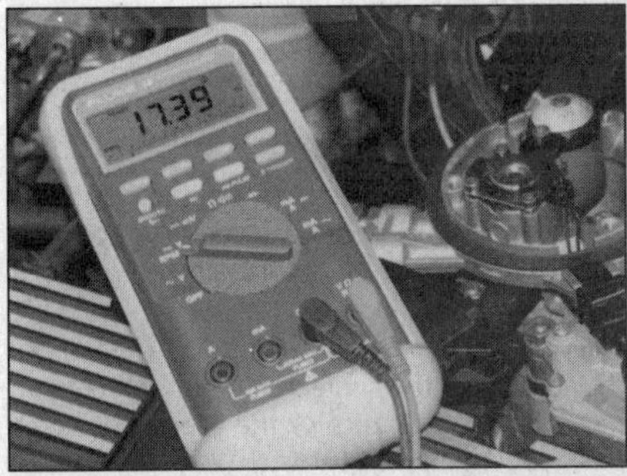

4.22 Backprobing the single point injector and measuring frequency. The frequency of injection may change during a cold start or during acceleration

Jumper lead

ECM multiplug

earth

EQH421

4.21 Use a jumper lead to very briefly touch an injector actuator pin in the ECM multi-plug to earth

9 A duty cycle reading (injector duty cycle) of some description should be obtained. If the dwell meter can measure the full pulse width value in milliseconds, this could be even more useful. If a signal is obtained, this at least indicates that the ECM is capable of switching the injection circuit. However, it does not prove that the signal is totally satisfactory.

Good injector signal

10 If the injector signal is satisfactory and if the primary ignition signal is also providing an acceptable signal, the fault is unlikely to be related to the ECM.

Poor or no injector signal

11 Check the fuel pressure and fuel flow.

12 Check the CAS, HES or other primary trigger for a good signal.

13 Check the voltage at the signal terminal of the injector multi-plug. Battery voltage should be obtained.

14 No voltage:

a) Check the injector resistance.
b) Check the ballast resistor resistance (where fitted).
c) Check for continuity of wiring between the injector multi-plug and the ECM multi-plug.
d) Check the voltage supply to the injector.

15 Disconnect the ECM multi-plug (see Warning No 3 in Reference).

16 Switch on the ignition.

17 Use a jumper lead to very briefly touch the injector actuator pin in the ECM multi-plug to earth.

18 If the injector actuates, check the ECM main voltage supplies and earths. If the tests reveal no fault, the ECM is suspect.

19 If the injector does not actuate: check for nbv at the ECM pin:

a) Voltage present: The injector is suspect.
b) No voltage present: Check for continuity of wiring between the injector multi-plug and the ECM multi-plug.

Incorrect pulse width (if an accurate measurement can be made)

20 Check the CTS.

21 Check the MAP sensor. **Note:** *If the ECM has entered LOS due to a fault in one of the sensors, the engine may generally behave quite well whilst the engine is hot, but may be difficult to start when cold.*

Engine running tests

22 Please refer to Section 13, which describes test procedures applicable to checking both MPi and SPi operation in a running engine.

Resistance tests

23 Remove the injector multi-plug and measure the resistance of the injector between the two terminals.

24 Where a ballast resistor is fitted, remove the resistor multi-plug and measure the resistance of the ballast resistor between the two terminals.

Inductive phase sensor (CID)

1 The inductive phase sensor which identifies the cylinders for sequential injection operation may be fitted inside the distributor or mounted upon the camshaft.

2 The timing of the phase sensor and the primary trigger is particularly important in sequential fuel injected vehicles. If the phasing is out of synchronisation, at best the engine may sink into LOS mode, with loss of power and increased emissions. At worst, the engine may fail to start.

3 Reasons for phasing errors:

a) Incorrectly-adjusted distributor (only if the distributor is adjustable).
b) Slack timing belt (very common fault).
c) Misaligned timing belt.

4 Inspect the CAS multi-plug for corrosion, and damage.

5 Inspect the CID multi-plug for corrosion, and damage.

6 Check that the connector terminal pins are fully pushed home and making good contact with the CID multi-plug.

7 Measure the CID resistance and compare it with the specified value in the system Chapter.

CID output

8 Detach the CID or ECM multi-plug (refer to Warning No 3 in Reference)

9 Connect an AC voltmeter between the two terminals at the CID or at the corresponding multi-plug terminals at the ECM. **Note:** *Better results are usually obtained by probing the + terminal, although the signal can often be obtained on the CID earth return.*

10 Crank the engine. A minimum AC RMS voltage of about 0.40 volts should be obtained.

11 Reconnect the CID or ECM multi-plug.

12 Backprobe the CID signal and earth terminals.

13 Start the engine and allow it to idle. A minimum AC RMS voltage of about 0.75 volts should be obtained.

14 No signal, or a very weak or intermittent signal:

a) Measure the CID resistance.
b) Check the sensor for damage, dirt or oil contamination.
c) Check the camshaft or distributor for damage

CID resistance

15 Detach the CID or ECM multi-plug, and connect an ohmmeter between the two terminals leading to the sensor.

16 Record the resistance and compare it with the specified values in the system Chapter.

Hall-effect phase sensor (CID)

1 The sensor is located in the distributor

2 Inspect the multi-plug for corrosion and damage.

3 Check that the terminal pins in the multi-plug are fully pushed home and making good contact with the sensor.

4 Roll back the rubber protection boot to the sensor multi-plug.

5 Connect the voltmeter negative or dwell meter probe to an engine earth.

6 Identify the supply, signal and earth terminals.

7 Connect the voltmeter positive or dwell meter probe to the wire attached to the sensor signal terminal.

8 Allow the engine to idle.

9 An average voltage of around 2.5 volts, or an approximate duty cycle of 50% should be obtained.

Signal voltage or duty cycle signal not available

10 Stop the engine.

11 Remove the distributor cap.

12 Sensor multi-plug connected, ignition on, voltmeter positive probe connected to the signal terminal.

13 Turn the engine slowly by hand. As the trigger vane cut-out space moves in and out of the air gap, the voltage should alternate between 5.0 volts and zero volts.

Signal voltage not available

14 Disconnect the sensor multi-plug at the distributor.

15 Probe output terminal 2 (O) of the harness multi-plug with the voltmeter positive probe:

a) No voltage from the ECM to terminal 2: Check for continuity of the signal wiring between the sensor and the ECM.
b) Recheck for voltage at the ECM terminal.
c) If no voltage is available at the ECM, check all voltage supplies and earth connections to the ECM. If the voltage supplies and earth connections are satisfactory, the ECM is suspect.

16 Check the voltage supply (5.0 volts) at sensor terminal number 1 (+). If the supply is unsatisfactory, check for continuity of the wiring between the sensor and the ECM.

17 Check the earth connection at sensor terminal number 3 (-).

18 If the voltage supply and earth are ok, the sensor in the distributor is suspect.

Airflow sensor (AFS)

1 Inspect the air trunking from the AFS and check for splits, poor fitting or damage. A large vacuum leak at this point will cause the engine to fire but fail to continue running, and a small vacuum leak will adversely affect the AFR.

2 Inspect the AFS multi-plug for corrosion, and damage.

3 Check that the terminal pins in the AFS multi-plug are fully pushed home and making good contact with the AFS.

4 The AFS may be of the vane, hot-wire or hot-film type, depending on system.

Vane-type AFS voltage tests

5 Roll back the rubber protection boot to the AFS multi-plug.

6 Connect the voltmeter negative probe to an engine earth.

7 Identify the supply, signal and earth terminals.

8 Connect the voltmeter positive probe to the wire attached to the AFS signal terminal *(see illustration 4.23)*.

9 Remove the air trunking.

10 Remove the air filter box so that the AFS flap can be easily opened and closed.

11 Open and close the AFS flap several times and check for smooth operation. Also check that the flap does not stick *(see illustration 4.24)*.

12 Switch on the ignition, engine stopped. A voltage of approximately 0.20 to 0.30 volts should be obtained.

13 Open and close the flap several times, and check for a smooth voltage increase to a maximum of 4.0 to 4.5 volts. **Note:** *If a digital voltmeter is used, then it is useful for it to have a bar graph facility. The smoothness of the voltage increase can then be more easily seen.*

14 Start the engine and allow it to idle. A voltage of approximately 0.5 to 1.5 volts should be obtained.

15 Open the throttle to no more than 3000 rpm. A voltage of approximately 2.0 to 2.5 volts should be obtained.

16 Snap open the throttle. A voltage greater than 3.0 volts should be obtained.

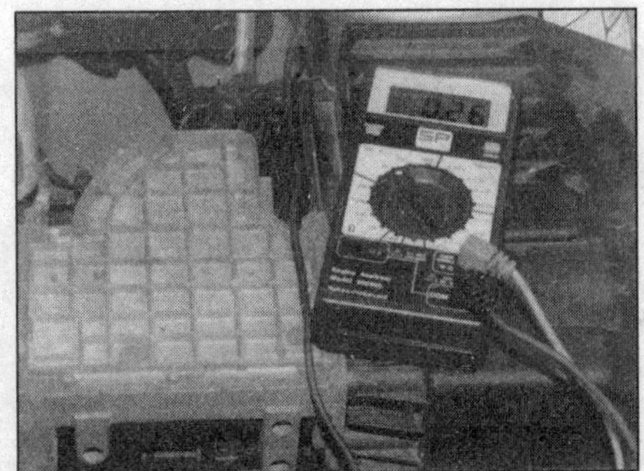

4.23 Backprobing the AFS for voltage

Erratic signal output

17 An erratic output occurs when the voltage output is stepped, or drops to zero or becomes open circuit.

18 When the AFS signal output is erratic, this usually suggests a faulty signal track or sticking flap. In this instance, a new or reconditioned AFS may be the only cure. If the carbon signal track is thought to be dirty, remove the AFS top cover and try cleaning the track carefully – do not use strong solvents, however.

19 Sometimes the wiper arm becomes disengaged from the signal track at certain points during its traverse. This can also give an erratic output.

a) Remove the AFS top cover, and check that the wiper arm touches the track during its swing from the open to the closed position.

b) Carefully bending the arm so that it touches the signal track can cure this fault.

Signal voltage not available

20 Check for the 5.0-volt reference voltage supply at the AFS supply terminal.

21 Check the earth return connection at the AFS earth terminal.

22 If the supply and earth are satisfactory, check for continuity of the signal wiring between the AFS and the ECM.

23 If the supply and/or earth are unsatisfactory, check for continuity of the wiring between the AFS and the ECM.

24 If the AFS wiring is satisfactory, check all voltage supplies and earth connections to the ECM. If the voltage supplies and earth connections are satisfactory, the ECM is suspect.

Signal or supply voltage at nbv level

25 Check for a short to a wire connected to the battery positive (+) terminal or a switched supply voltage.

Vane-type AFS resistance tests

26 Connect an ohmmeter between the AFS signal terminal and supply terminal, or between the AFS signal terminal and earth terminal.

27 Open and close the AFS flap several times, and check for a smooth resistance change. As the AFS flap is moved slowly from the closed to the fully-open position, the AFS resistance may increase and decrease in a series of steps. This is normal. If the AFS resistance becomes open- or short-circuit, a fault is revealed.

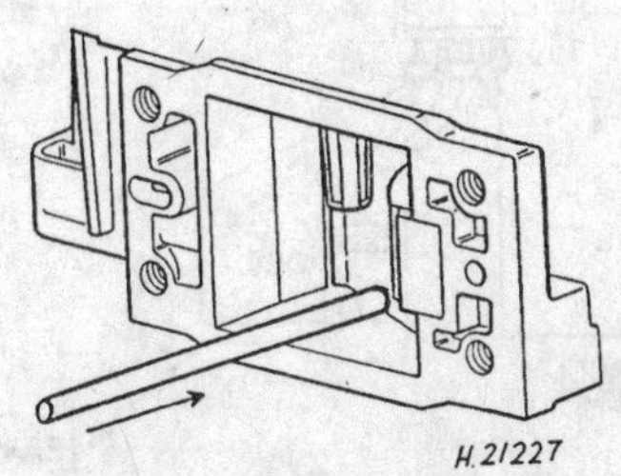

4.24 Check the vane AFS flap for a smooth opening and closing action

28 We are not providing resistance specifications for the AFS in this book. It is less important that the resistance of the AFS remains within arbitrary values, than that the operation is correct.

29 Connect an ohmmeter between the AFS earth terminal and supply terminal. A stable resistance should be obtained.

30 Renew the AFS if the resistance is open-circuit or shorted to earth.

31 Refer to the comments on resistance readings in Section 3.

Hot-wire/hot-film type AFS tests

Note: *The voltage measurements are based on the Vauxhall 16-valve engines with Motronic 2.5. The readings from other vehicles should be similar.*

32 Switch on the ignition. A voltage of approximately 1.4 volts should be obtained.

33 Start the engine and allow it to idle. A voltage of approximately 1.9 to 2.3 volts should be obtained.

34 Snap open the throttle several times. The voltage will not increase significantly over the idle value during this off-load test. **Note:** *If a digital voltmeter is used, then it is useful for it to have a bar graph facility. The smoothness of the voltage increase can then be more easily seen.*

35 It is less easy to test the AFS Hot Wire signal output, because It is impossible to simulate full-load conditions in the workshop without putting the vehicle on a chassis dynamometer (rolling road). However, the following test procedure will usually prove if the signal output is consistent.

36 Disconnect the air trunking so that the hot wire is exposed.

37 Switch on the ignition.

38 Use a length of plastic tubing to blow air over the hot wire.

39 It should be possible to plot a voltage curve, although the curve will be much steeper than that obtained with the engine running.

Erratic signal output

40 An erratic output occurs when the voltage output is stepped, or drops to zero or becomes open-circuit.

41 Check the AFS resistance by connecting an ohmmeter between AFS terminals 2 and 3. A resistance of approximately 2.5 to 3.1 ohms should be obtained.

42 When the AFS signal output is erratic, and all supply and earth voltages are satisfactory, this suggests a faulty AFS. In this case, a new or reconditioned AFS may be the only cure.

Signal voltage not available

43 Check for the battery voltage supply to AFS terminal No 5.

44 Check the earth return connection at AFS terminal No 2.

45 Check the earth connection at AFS terminal No 1.

46 If the supply and earths are satisfactory, check for continuity of the signal wiring between the AFS and the ECM.

47 If the supply and/or earths are unsatisfactory, check for continuity of the supply and / or earth wiring between the AFS and the ECM.

48 If the AFS wiring is satisfactory, check all voltage supplies and earth connections to the ECM. If the voltage supplies and earths are satisfactory, the ECM is suspect.

MAP sensor (analogue and digital types)

1 Inspect the MAP sensor multi-plug for corrosion and damage.

2 Check that the terminal pins in the MAP sensor multi-plug are fully pushed home and making good contact with the MAP sensor *(see illustration 4.25)*.

3 Where the MAP sensor is located internally in the ECM, voltage tests are not possible.

MAP sensor test procedures (Analogue: external and internal MAP sensor)

4 Use a 'T' connector to connect a vacuum gauge between the inlet manifold and the MAP sensor *(see illustration 4.26)*.

5 Allow the engine to idle. If the engine vacuum is low (less than 425 to 525 mm Hg), check for the following faults:

a) A vacuum leak.
b) A damaged or perished vacuum pipe.
c) A restricted vacuum connection.
d) An engine problem – eg a misaligned cambelt.
e) A leaky MAP diaphragm (inside the ECM if the MAP sensor is internal).

6 Disconnect the vacuum gauge, and connect a vacuum pump in its place.

7 Use the pump to apply vacuum to the MAP sensor until approximately 560 mm Hg is reached.

8 Cease pumping, and the MAP sensor diaphragm should hold pressure for a minimum of 30 seconds at this vacuum setting.

MAP sensor test procedures (Analogue: external MAP sensor)

9 Roll back the rubber protection boot to the MAP sensor multi-plug.

10 Connect the voltmeter negative probe to an engine earth.

11 Identify the supply, signal and earth terminals.

12 Connect the voltmeter positive probe to the wire attached to the MAP sensor signal terminal.

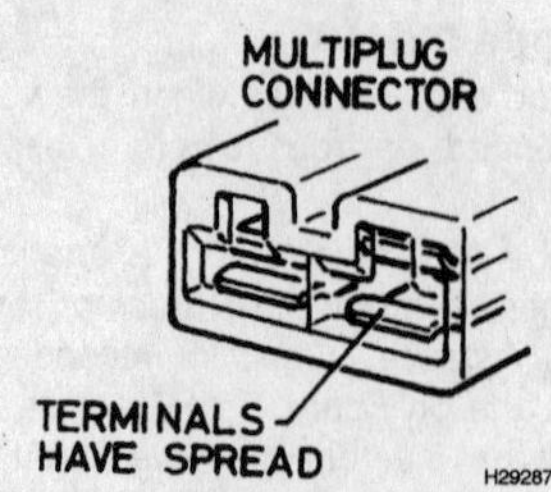

4.25 Check that the terminal pins in the multi-plug are fully pushed home and making good contact with the MAP sensor

13 Disconnect the vacuum pipe from the MAP sensor.

14 Connect a vacuum pump to the sensor.

15 Switch the ignition on.

16 Compare the ignition-on voltage to that specified in the system Chapter.

17 Apply vacuum as shown in the table, and check for a smooth voltage change.

Erratic signal output

18 An erratic output occurs when the voltage output is stepped, drops to zero, or becomes open-circuit. This usually suggests a faulty MAP sensor. In this instance, a new sensor is the only cure.

MAP sensor test procedures (Digital: external MAP sensor only)

19 Set the DMM to the volts scale.

20 Switch on the ignition.

21 Identify the supply, signal and earth terminals.

22 Connect the voltmeter positive probe to the wire attached to the MAP sensor signal terminal. An average voltage of approximately 2.5 volts should be obtained. If not, refer to the 'Signal voltage not available' tests below.

23 Set the meter to the tachometer 4-cylinder scale (all engines).

24 Disconnect the vacuum hose to the MAP sensor.

25 Connect the positive DMM probe to the signal terminal, and connect the negative probe to the earth terminal.

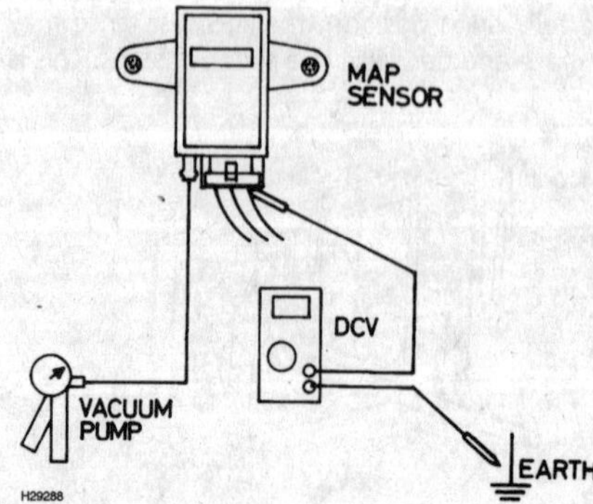

4.26 Using a vacuum gauge and a voltmeter to check the MAP sensor signal

26 An rpm reading of 4500 to 4900 should be obtained.

27 Attach a vacuum pump to the MAP sensor hose connection. During the following tests, the vacuum should hold steady at all of the pressure settings:

a) Apply 200 mb, the rpm should drop by 525 ± 120 rpm.
b) Apply 400 mb, the rpm should drop by 1008 ± 120 rpm.
c) Apply 600 mb, the rpm should drop by 1460 ± 120 rpm.
d) Apply 800 mb, the rpm should drop by 1880 ± 120 rpm.

28 Release the pressure when the measured value should return to the original setting of 4500 to 4900.

29 Renew the MAP sensor if it fails to behave as described.

Signal voltage not available (both Analogue and Digital external types)

30 Check the reference voltage supply (5.0 volts).

31 Check the earth return.

32 If the supply and earth are satisfactory, check for continuity of the signal wiring between the MAP sensor and the ECM.

33 If the supply and/or earth are unsatisfactory, check for continuity of the wiring between the MAP sensor and the ECM.

34 If the MAP sensor wiring is satisfactory, check all voltage supplies and earth connections to the ECM. If the voltage supplies and earth connections are satisfactory, the ECM is suspect.

Signal or supply voltage at nbv level

35 Check for a short to a wire connected to the battery positive (+) terminal or a switched supply voltage.

Other checks

36 Check for excessive fuel in the vacuum trap or hose.

37 Check for a faulty vacuum hose or a vacuum leak.

38 Check for mechanical, ignition or a fuel fault resulting in low engine vacuum.

Air temperature sensor (ATS)

1 The ATS has only a minor effect on engine operation. However, if the ATS is suspect, the following tests could be made.

2 The ATS may be located in the inlet tract of the AFS, or in the inlet manifold:

a) If located in the AFS, the ATS shares a common earth return.
b) Both types of ATS are examples of two-wire sensors, and test procedures are similar.

3 Inspect the ATS multi-plug for corrosion, and damage.

4 Check that the terminal pins in the ATS or AFS multi-plug are fully pushed home and making good contact with the ATS.

ATS test procedures

5 Roll back the rubber protection boot to the ATS (or AFS multi-plug if the ATS is located in the AFS).

6 Connect the voltmeter negative probe to an engine earth.

7 Identify the signal and earth terminals.

8 Connect the voltmeter positive probe to the wire attached to the ATS signal terminal *(see illustration 4.27)*.

9 With the engine off, and ignition on, a voltage of approximately 2.0 to 3.0 volts, depending upon air temperature, is likely to be obtained. Refer to the ATS chart for typical voltages at various temperatures.

10 The signal voltage will vary according to the temperature of the air in the AFS inlet tract or inlet manifold. As the engine compartment or inlet manifold air rises in temperature, the voltage signal passed to the ECM will reduce. When the engine is cold, the air temperature will match the ambient temperature. After the engine is started, the temperature of the air in the engine compartment and the inlet manifold will rise. The temperature of the air in the inlet manifold will rise to approximately 70 or 80°C. which is a much higher temperature than that of the air in the engine compartment.

11 When undergoing tests at various temperatures, the ATS can be warmed with a hairdryer or cooled with something like 'Freezit', which is an ice-cold aerosol spray, sold in electronic component shops. As the ATS is heated or cooled, the temperature will change and so too will the resistance and voltage.

ATS voltage and resistance table (typical)

Temp (°C)	Resistance	Volts
0	*4800 – 6600*	*4.00 – 4.50*
10	*4000*	*3.75 – 4.00*
20	*2200 – 2800*	*3.00 – 3.50*
30	*1300*	*3.25*
40	*1000 – 1200*	*2.50 – 3.00*
50	*1000*	*2.50*
60	*800*	*2.00 – 2.50*
80	*270 – 380*	*1.00 – 1.30*
110		*0.50*
Open-circuit		*5.0 ± 0.1*
Short to earth		*Zero*

12 Check that the ATS voltage corresponds to the temperature of the ATS. A temperature gauge is required here.

13 Start the engine and allow it to warm-up to normal operating temperature. As the engine warms up, the voltage should reduce in accordance with the ATS chart.

4.27 Backprobing for an ATS signal (ATS located in air filter box)

14 Make the following tests and checks if the ATS signal voltage is zero (supply is open-circuit or shorted to earth) or at 5.0 volt level (ATS is open-circuit).

Zero volts at the ATS signal terminal

15 Check that the ATS signal terminal is not shorted to earth.

16 Check for continuity of the signal wiring between the ATS and the ECM.

17 If the ATS wiring is satisfactory, yet no voltage is output from the ECM, check all voltage supplies and earth connections to the ECM. If the voltage supplies and earth connections are satisfactory, the ECM is suspect.

5.0 volts at the ATS signal terminal

18 This is the open circuit voltage, and will be obtained in the event of one or more of the following conditions:

a) The signal terminal in the ATS (or AFS) multi-plug is not making contact with the ATS.

b) The ATS is open circuit.

c) The ATS earth connection is open-circuit.

Signal or supply voltage at nbv level

19 Check for a short to a wire connected to the battery positive (+) terminal or a switched supply voltage.

Resistance tests

20 A resistance test may be made at various temperatures, and a comparison made with the temperature/resistance chart. Refer to paragraph 11 for a method of heating/cooling the ATS.

21 When the ATS resistance is within the stated parameters for a cold engine (20°C) the coolant temperature should also be within ± 5°C of that figure.

CO potentiometer ('pot')

1 The CO pot may be located in the AFS, or may be a separate sensor, located in the engine compartment or directly attached to the ECM.

a) If located in the AFS, the CO pot shares a common earth return.

b) Both the AFS located CO pot and the separate CO pot in the engine compartment are examples of three-wire sensors, and test procedures follow similar paths.

c) The CO pot attached to the ECM cannot be tested separately; a new ECM is required if the CO pot fails.

2 Inspect the CO pot or AFS multi-plug for corrosion and damage.

3 Check that the terminal pins in the multi-plug are fully pushed home and making good contact with the CO pot or AFS.

CO pot test procedures

4 Roll back the rubber protection boot to the CO pot multi-plug (or AFS multi-plug if located in the AFS).

5 Connect the voltmeter negative probe to an engine earth.

6 Identify the supply, signal and earth terminals.

7 Connect the voltmeter positive probe to the wire attached to the CO pot signal terminal.

8 A voltage of approximately 2.5 volts should be obtained in most systems.

9 Record the exact voltage so that the voltage can be reset to the exact value after tests are complete.

10 Remove the tamperproof cap from the adjustment screw.

11 Turn the adjustment screw one way and then the other. The voltage should vary smoothly.

CO pot voltage does not alter during adjustment

12 Check for the 5.0 volt reference voltage supply to the sensor.

13 Check the earth return connection to the sensor.

14 If the supply and earth are satisfactory, check for continuity of the signal wiring between the CO pot and the ECM.

15 If the supply and/or earth are unsatisfactory, check for continuity of the supply and/or earth wiring between the CO pot or AFS (as appropriate) and the ECM.

16 If the AFS wiring is satisfactory, check all voltage supplies and earth connections to the ECM. If the voltage supplies and earths are satisfactory, the ECM is suspect.

Coolant temperature sensor (CTS)

1 Inspect the CTS multi-plug for corrosion and damage.

2 Check that the terminal pins in the CTS multi-plug are fully pushed home and making good contact with the CTS. **Note:** *Poor contact and corrosion are common reasons for an inaccurate signal from the CTS.*

3 Roll back the rubber protection boot to the CTS multi-plug.

4 Connect the voltmeter negative probe to an engine earth.

5 Identify the signal and earth terminals.

6 Connect the voltmeter positive probe to the wire attached to the CTS signal terminal.

7 Allow the engine to become cold. With the engine off, and ignition on, a voltage of approximately 2.0 to 3.0 volts, depending upon temperature, is likely to be obtained. Refer to the CTS chart for typical voltages at various temperatures.

CTS voltage and resistance table (typical)

Temp (°C)	Resistance	Volts
0	*4800 – 6600*	*4.00 – 4.50*
10	*4000*	*3.75 – 4.00*
20	*2200 – 2800*	*3.00 – 3.50*
30	*1300*	*3.25*
40	*1000 – 1200*	*2.50 – 3.00*
50	*1000*	*2.50*
60	*800*	*2.00 – 2.50*
80	*270 – 380*	*1.00 – 1.30*
110		*0.50*
Open-circuit		*5.0 ± 0.1*
Short to earth		*Zero*

8 Check that the CTS voltage corresponds to the temperature of the CTS.

9 Start the engine, and allow it to warm up to normal operating temperature. As the engine warms up, the voltage should reduce in accordance with the CTS chart.

10 A common problem may occur where the CTS varies in resistance (and voltage) outside of its normal range. If, for example, the CTS voltage is normally 3 volts cold/0.5 volts hot, a faulty CTS may give a voltage of 1.5 volts cold/1.25 volts hot, resulting in the engine being difficult to start when cold and running richer than normal when hot. This will NOT result in the generation of a fault code, because the CTS is still operating within its design parameters. Renew the CTS if this fault occurs.

11 Make the following tests and checks if the CTS signal voltage is zero (supply is open-circuit or shorted to earth) or at 5.0 volt level (CTS is open-circuit).

Zero volts obtained at the CTS signal terminal

a) Check that the CTS signal terminal is not shorted to earth.

b) Check for continuity of the signal wiring between the CTS and the ECM.

c) If the CTS wiring is satisfactory, yet no voltage is output from the ECM, check all voltage supplies and earth connections to the ECM. If the voltage supplies and earth connections are satisfactory, the ECM is suspect.

5.0 volts obtained at the CTS signal terminal

12 This is the open-circuit voltage, and will be obtained in the event of one or more of the following conditions:

a) The signal terminal in the CTS multi-plug is not making contact with the CTS.

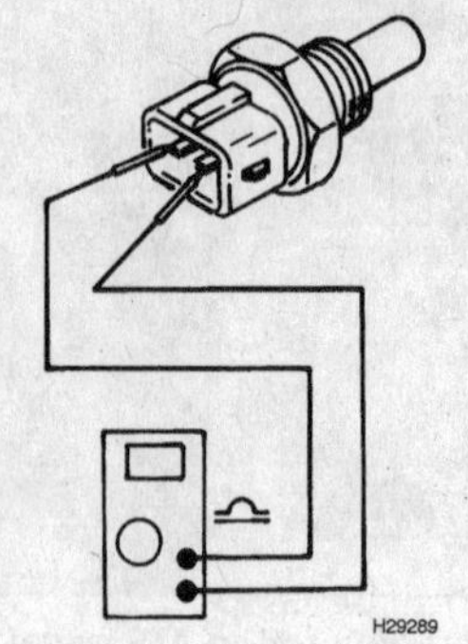

4.28 Checking the CTS resistance

b) The CTS is open-circuit.

c) The CTS earth connection is open-circuit.

Signal or supply voltage at nbv level

13 Check for a short to a wire connected to the battery positive (+) terminal or a switched supply voltage.

Resistance tests

CTS on vehicle

14 A resistance test may be made at various temperatures and a comparison made with the temperature/resistance chart *(see illustration 4.28)*. When the resistance is within the stated parameters for a cold engine (20°C) the coolant temperature should be within ± 5°C of that figure.

15 An allowance should be made for a temperature obtained by probing the outside of the CTS or coolant passage. This is because the actual temperature of the coolant may be hotter than the surface temperature of the CTS.

CTS off vehicle

16 The recommended method is described as follows.

17 Remove the CTS from the vehicle.

18 Place the CTS in a suitable container of water and measure the temperature of the water.

19 Measure the resistance of the CTS, and check the resistance against the temperature chart.

20 Heat the water, periodically measuring the water temperature and the CTS resistance, and comparing the resistance with the temperature chart.

Throttle switch (TS)

1 Inspect the TS multi-plug for corrosion and damage.

2 Check that the terminal pins in the TS multi-plug are fully pushed home and making good contact with the TS.

Poor contact and corrosion are common reasons for an inaccurate signal from the TS.

3 The three wires to the TS multi-plug connector are earth, idle signal and WOT signal.

4 Roll back the rubber protection boot to the TS multi-plug.

5 Connect the voltmeter negative probe to an engine earth.

6 Identify the idle signal, WOT signal and earth terminals.

7 Engine stopped, ignition on.

8 Connect the voltmeter positive probe to the wire attached to the TS idle signal terminal.

9 Zero volts should be obtained. If the meter indicates 5.0 volts, loosen the screws and adjust the TS so that zero volts is obtained.

Zero volts cannot be obtained (throttle closed)

a) Check the throttle valve position.

b) Check the TS earth connection.

c) Make the TS resistance tests (below).

10 If the voltage is satisfactory with the throttle closed, crack open the throttle; the switch should 'click' and the voltage rise to 5.0 volts.

Voltage low or non-existent (throttle open)

a) Check that the TS idle terminal is not shorted to earth.

b) Disconnect the TS multi-plug and check for 5.0 volts at the multi-plug idle terminal. If no voltage, make the following checks.

c) Check for continuity of the idle signal wiring between the TS and the ECM.

d) If the TS wiring is satisfactory, check all voltage supplies and earth connections to the ECM. If the voltage supplies and earth connections are satisfactory, the ECM is suspect.

Voltage satisfactory (throttle open)

11 Reconnect the voltmeter probe to the wire attached to the TS full-load signal terminal.

12 With the throttle in either the idle or just open positions, the meter should indicate 5.0 volts.

Voltage low or non-existent (throttle closed or just open)

a) Check the earth connection.

b) Check that the TS full-load terminal is not shorted to earth.

c) Disconnect the TS multi-plug and check for 5.0 volts at the full-load multi-plug terminal. If no voltage, make the following checks.

d) Check for continuity of the full-load signal wiring between the TS and the ECM.

e) If the TS wiring is satisfactory, check all voltage supplies and earth connections to the ECM. If the voltage supplies and earth connections are satisfactory, the ECM is suspect.

Voltage satisfactory (throttle closed or just open)

13 Fully open the throttle. As the throttle angle becomes greater than 72°, the voltage should drop to zero volts. If the voltage does not drop, the TS is suspect.

Resistance tests

14 Disconnect the TS multi-plug.

15 Connect an ohmmeter between the TS earth terminal (sometimes marked 18) and terminal 2 (idle contact).

16 With the TS closed, the ohmmeter should indicate very close to zero ohms.

17 Slowly open the throttle; as the TS cracks open, it should click and the resistance should become open-circuit and remain so – even as the throttle is opened fully.

18 Reconnect the ohmmeter between the earth terminal (sometimes marked 18) and terminal 3 (full-load contact). With the TS closed, the ohmmeter should indicate an open circuit.

19 Slowly open the throttle, as the TS cracks open, it should click and the resistance should remain open circuit until the throttle angle becomes greater than 72°, when the resistance should change to continuity of approximately zero ohms.

20 If the TS does not behave as described, and it is not prevented from opening or closing fully by the binding of the throttle linkage, the TS is suspect.

Throttle potentiometer sensor (TPS)

1 Inspect the TPS multi-plug for corrosion, and damage.

2 Check that the terminal pins in the TPS multi-plug are fully pushed home and making good contact with the TPS. **Note:** *Poor contact and corrosion are common reasons for an inaccurate signal from the TPS.*

3 Roll back the rubber protection boot to the TPS multi-plug.

4 Connect the voltmeter negative probe to an engine earth.

5 Identify the supply, signal and earth terminals.

6 Connect the voltmeter positive probe to the wire attached to the TPS signal terminal.

7 Switch on the ignition, engine stopped. In most systems, a voltage less than 0.70 volts should be obtained.

8 Open and close the throttle several times, and check for a smooth voltage increase to a maximum of 4.0 to 4.50 volts. **Note:** *If a digital voltmeter is used, then it is useful for it to have a bar graph facility. The smoothness of the voltage increase can then be more easily seen.*

Erratic signal output

9 An erratic output occurs when the voltage output is stepped, or drops to zero or becomes open-circuit.

10 When the TPS signal output is erratic, this usually suggests a faulty potentiometer. In this instance, a new or reconditioned TPS is the only cure.

Signal voltage not available

11 Check for the 5.0 volt reference voltage supply at the TPS supply terminal.

12 Check the earth return connection at the TPS earth terminal.

13 If the supply and earth are satisfactory, check for continuity of the signal wiring between the TPS and the ECM.

14 If the supply and/or earth are unsatisfactory, check for continuity of the wiring between the TPS and the ECM.

15 If the TPS wiring is satisfactory, check all voltage supplies and earth connections to the ECM. If the voltage supplies and earths are satisfactory, the ECM is suspect.

Signal or supply voltage at nbv level

16 Check for a short to a wire connected to the battery positive (+) terminal or a switched supply voltage.

Resistance tests

17 Connect an ohmmeter between the TPS signal terminal and supply terminal or the TPS signal terminal and earth terminal.

18 Open and close the throttle several times and check for a smooth resistance change. If the TPS resistance becomes open- or short-circuit, a fault is revealed.

19 We are not providing resistance specifications for the TPSs described in this book. For one thing, many vehicle manufacturers do not publish test values. Also, it is less important that the resistance of the TPS remains within arbitrary values, than the operation is correct.

20 Connect an ohmmeter between the TPS earth terminal and supply terminal. A stable resistance should be obtained.

21 Renew the TPS if the resistance is open-circuit or shorted to earth.

22 Refer to the comments on resistance readings in Section 3.

Vehicle speed sensor (VSS)

Note: *These test procedures describe testing of the most common type of VSS which operates upon the Hall-effect principle.*

1 The VSS is usually located on the gearbox.

2 Inspect the VSS multi-plug for corrosion, and damage.

3 Check that the terminal pins in the VSS multi-plug are fully pushed home and making good contact with the VSS.

4 Roll back the rubber protection boot to the VSS multi-plug.

5 Connect the voltmeter negative or dwell meter probe to an engine earth.

6 Identify the supply, signal and earth terminals.

7 Connect a voltmeter positive or dwell meter probe to the wire attached to the VSS signal terminal.

Checking for a VSS signal

8 The drive wheels must rotate for a signal to be generated. This may be accomplished by using one of the two following methods:

a) Push the vehicle forward.

b) Place the vehicle upon a ramp, or jack up the vehicle so that the drive wheels can freely turn.

9 Rotate the wheels by hand so that a duty cycle or voltage can be obtained.

No signal or an erratic duty cycle or voltage

10 VSS multi-plug disconnected, ignition on.

11 Check the voltage at the signal terminal. A voltage between 8.5 and 10.0 volts should be obtained.

12 Check the voltage supply at the VSS supply terminal. A voltage slightly less than nbv should be obtained.

13 Check the VSS earth connection.

Supply and earth voltages ok

14 The VSS is suspect or the VSS is not being rotated by the speedometer drive (ie broken cable or gearbox fault).

No signal voltage

15 Check the voltage at the ECM multi-plug terminal.

a) If voltage is satisfactory at the ECM, check the diode in the wire between the ECM and VSS and continuity of the signal wiring.

b) If no voltage is available at the ECM, check all voltage supplies and earth connections to the ECM. If the voltage supplies and earth connections are satisfactory, the ECM is suspect.

Idle speed control valve (ISCV)

1 Inspect the ISCV multi-plug for corrosion, and damage.

2 Check that the connector terminal pins are fully pushed home and making good contact with the ISCV multi-plug.

3 Allow the engine to idle.

4 Check that the idle speed is within its operating limits.

5 Load the system by switching on the headlamps, rear screen heater and heater motor onto high. The idle speed should barely change.

6 If possible, squeeze one of the air hoses. The

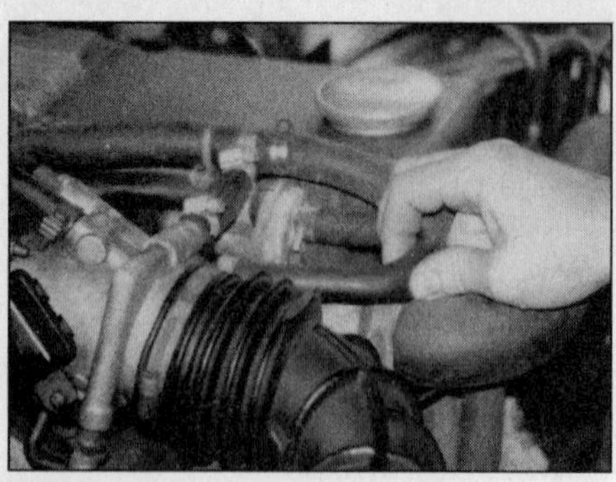

4.29 Squeeze an idle air hose to check ISCV response

idle speed should surge and then return to normal *(see illustration 4.29)*.

7 If the idle condition meets the above criteria, it is unlikely to be at fault.

8 The following faults will adversely affect idle integrity, and these components should be checked before attempting diagnosis of the ISCV.

a) Engine mechanical fault.
b) Incorrect ignition timing.
c) An induction vacuum leak.
d) Incorrect CO level.
e) Clogged air filter.
f) An incorrectly adjusted throttle valve.
g) Carbon-fouled throttle plate.
h) An incorrectly-adjusted TS or TPS.

Test procedures (two-wire)

9 Detach the ISCV multi-plug.

10 Identify the supply and signal terminals.

11 Measure the ISCV resistance. See below.

12 Use a jump lead to connect a battery voltage supply to the ISCV supply terminal.

13 Use a second jump lead to connect the ISCV earth terminal to earth. **Note:** *The connection should be made only briefly.*

14 The ISCV should actuate.

15 Roll back the rubber protection boot to the ISCV multi-plug.

16 Connect the negative instrument probe to an engine earth.

17 Connect the voltmeter positive or dwell meter probe to the wire attached to the ISCV signal terminal.

18 Start the engine and allow it to idle.

19 Hot engine: A varying voltage of average 7.0 to 9.0 volts or a duty cycle of 40 to 44% will be obtained. A frequency of 110 can also be measured.

20 When the engine is cold or placed under load, the voltage will decrease and the duty cycle will increase. **Note:** *The reading on a digital voltmeter will indicate the average voltage.*

21 Load the engine by switching on the headlamps, rear screen heater and heater motor onto high. The average voltage will decrease and the duty cycle will increase. The frequency of pulse should remain constant.

22 If an air leak or another fault is present resulting in more air by-passing the throttle, the ISCV duty cycle will be lower than normal as the ECM pulses the ISCV less open.

23 When more load is placed upon the engine, the ECM pulses the ISCV more open (larger duty cycle) to increase the idle speed.

24 In addition, if the engine is mechanically unsound or the throttle valve is dirty, the ECM may pulse the ISCV more open to increase the idle speed. This may result in an uneven idle, and a larger-than-normal duty cycle.

ISCV signal not available

25 Check the ISCV resistance. See below.

26 Ignition on. Check for nbv at the supply terminal.

27 No voltage, trace the wiring back to the main relay or ignition switch as appropriate.

28 Disconnect the ECM multi-plug (see Warning No 3 in Reference).

29 Switch on the ignition.

30 Use a jumper lead to very briefly touch the actuator pin in the ECM multi-plug to earth.

a) If the ISCV actuates, check the ECM main voltage supplies and earths. If tests reveal no fault, the ECM is suspect.
b) If the ISCV does not actuate: check for continuity of of wiring between the ISCV multi-plug and the ECM.

Resistance checks

31 Remove the ISCV multi-plug.

32 Connect an ohmmeter between the two terminals. A resistance of 8 to 16 ohms should be obtained.

Test procedures (three-wire)

33 Detach the ISCV multi-plug.

34 Check the ISCV resistance. See below

35 Identify the supply and signal terminals. The three wires to the ISCV multi-plug are supply and two signal wires.

36 Use a jump lead to connect a battery voltage supply to the ISCV middle terminal.

37 Use a second jump lead to connect one of the outer ISCV terminals to earth. The ISCV should actuate.

38 Remove the second jump lead and use it to connect the other outer ISCV terminals to earth. The ISCV should actuate in the reverse direction.

39 If the ISCV does not actuate, or operation is poor the ISCV is suspect.

40 Roll back the rubber protection boot to the ISCV multi-plug.

41 Connect the voltmeter negative or dwell meter probe to an engine earth.

42 Connect the voltmeter positive or dwell meter probe to the wire attached to one of the two ISCV signal terminals.

43 Start the engine and allow it to idle.

44 Hot engine: A varying voltage or a duty cycle of either approximately 31% or 69% will be obtained *(see illustration 4.30)*. The duty cycle obtained will depend upon which terminal the instrument is connected.

45 When the engine is cold or placed under load, the voltage will decrease and the duty cycle will increase. **Note:** *The reading on a digital voltmeter will indicate the average voltage.*

46 Load the engine by switching on the headlamps, rear screen heater and heater motor onto high. The average voltage will decrease and the duty cycle will increase.

47 If an air leak or another fault is present resulting in more air by-passing the throttle, the ISCV duty cycle will be lower than normal as the ECM pulses the ISCV less open.

48 When more load is placed upon the engine, the ECM pulses the ISCV more open (larger duty cycle) to increase the idle speed.

49 In addition, if the engine is mechanically unsound or the throttle valve is dirty, the ECM may pulse the ISCV more open to increase the idle speed. This may result in an uneven idle and a larger than normal duty cycle.

50 Switch the voltmeter positive or dwell meter probe to the wire attached to the other one of the two ISCV signal terminals.

51 Hot engine: A varying voltage or a duty cycle of either approximately 31% or 69% will be obtained. The duty cycle obtained will depend upon which terminal the instrument is connected.

ISCV signal not available

52 Check the ISCV resistance

53 Ignition on. Check for nbv at the supply terminal.

54 No voltage: Trace the wiring back to the main relay or ignition switch as appropriate.

55 Disconnect the ECM multi-plug see Warning No 3 in Reference.

56 Switch on the ignition.

57 Use a jumper lead to very briefly touch one of the two actuator pins in the ECM multi-plug to earth.

a) If the ISCV actuates, check the ECM main voltage supplies and earths. If the tests reveal no fault, the ECM is suspect.
b) If the ISCV does not actuate, check for continuity of of wiring between the ISCV multi-plug and the ECM.

58 Switch the jumper lead to very briefly touch the other ISCV actuator pin in the ECM multi-plug to earth. Evaluate the results as in paragraph 57.

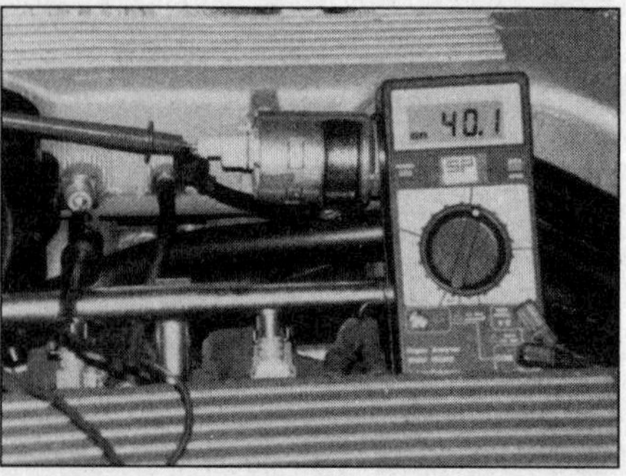

4.30 Backprobing for a typical dwell at the ISCV, engine at idle speed

Resistance checks

59 Remove the ISCV multi-plug.

60 Connect an ohmmeter between the centre terminal and one of the outer terminals. A resistance of 20 ohms should be obtained.

61 Reconnect the ohmmeter between the centre terminal and the other outer ISCV terminal. A resistance of 20 ohms should be obtained.

62 Reconnect the ohmmeter between the two outer ISCV terminals. A resistance of 20 ohms should be obtained.

Throttle body and manifold heater (TBH)

Quick check

1 Start the engine when cold and feel the area around the throttle body or inlet manifold (as appropriate), if the heater is working, this area should become hot quite quickly. Take care not to burn your fingers!

Throttle body heater tests

2 Allow the engine to idle.

3 Attach the voltmeter negative probe to an earth.

4 Attach the voltmeter positive probe to the heater supply connector. Battery voltage (nbv) should be obtained.

No voltage supply

a) Check the TBH supply.

b) Check continuity of the wiring between the relay and the heater.

Nbv present, but heater does not operate

a) Check the heater resistance.

b) Check the heater earth.

ECM voltage supplies and earths

Note: *Refer to the relevant Warnings in the Reference section at the end of this book before starting work.*

1 Inspect the ECM multi-plug for corrosion and damage.

2 Check that the terminals in the ECM multi-plug are fully pushed home and making good contact with the ECM pins. **Note:** *Poor contact and corrosion are common reasons for inaccurate signals from the ECM.*

3 Voltage supplies and earths are best measured at the ECM multi-plug. Use one of these test methods.

a) Peel back the ECM multi-plug insulation (not always possible) and backprobe the ECM multi-plug pins.

b) Attach a BOB between the ECM and its multi-plug, and probe the box for voltages.

c) Detach the ECM from its multi-plug, and probe for voltages at the multi-plug pins.

4 Attach the voltmeter negative probe to an engine earth for the ECM connected tests.

5 Identify the various types of connection and the relevant ECM pins from a wiring diagram for the vehicle in question.

Note: *Not all of the following connections will be available in any particular system.*

ECM battery supply pin test

6 This pin is directly connected to the battery (+) terminal and a constant voltage should be available at all times, even with the ignition key turned Off.

7 ECM multi-plug connected: Backprobe the relevant ECM pin – nominal battery voltage (nbv) should be obtained.

8 ECM multi-plug disconnected:

a) Attach the voltmeter negative probe to an ECM earth pin.

b) Attach the voltmeter positive probe to the relevant ECM pin: Nominal battery voltage (nbv) should be obtained.

9 No voltage: Check the supply back to the battery.

ECM cranking supply pin test

10 This pin is connected to the ignition switch starter terminal, and a battery voltage will only be available during engine cranking.

11 ECM multi-plug connected:

a) Backprobe the relevant ECM pin.

b) Crank the engine on the starter: Voltage should only be obtained during cranking.

12 ECM multi-plug disconnected:

a) Attach the voltmeter negative probe to an ECM earth pin.

b) Attach the voltmeter positive probe to the relevant ECM pin

c) Crank the engine on the starter: Voltage should only be obtained during cranking.

13 No voltage: Check the supply back to the ignition switch starter terminal.

ECM supply from the ignition switch test

14 This pin is connected to the ignition switch, and voltage should be available at all times whilst the ignition is on

or the engine is running.

15 ECM multi-plug connected:

a) Backprobe the relevant ECM pin.

b) Switch on the ignition: nbv should be obtained.

16 ECM multi-plug disconnected:

a) Attach the voltmeter negative probe to an ECM earth pin.

b) Attach the voltmeter positive probe to the relevant ECM pin.

c) Switch on the ignition: nbv should be obtained.

17 No voltage: Check the supply back to the ignition switch.

ECM supply from the main system relay test

18 This pin is connected to the main relay, and voltage should be available at all times whilst the ignition is on

or the engine is running. Supply may be made to more than one ECM pin.

19 ECM multi-plug connected:

a) Backprobe the relevant ECM pin.

b) Switch on the ignition: nbv should be obtained.

20 ECM multi-plug disconnected:

a) Attach the voltmeter negative probe to an ECM earth pin.

b) Attach the voltmeter positive probe to the relevant ECM pin.

c) Switch on the ignition: nbv should be obtained.

21 No voltage: check the supply back to the main system relay and make the relay checks.

ECM earth connections test

22 ECM multi-plug connected, ignition on:

a) Attach the voltmeter negative probe to an engine earth

b) Attach the voltmeter positive probe to the earth terminal under test: The voltmeter should indicate 0.25 volts maximum.

23 ECM multi-plug disconnected, ignition on or off:

a) Attach the voltmeter negative probe to the earth terminal under test.

b) Attach the voltmeter positive probe to the ECM battery supply or directly to the battery positive terminal: The voltmeter should indicate nbv if the earth is satisfactory.

ECM coding earth pins test

Note: *The coding pins are used to code the ECM for certain vehicle configurations (some systems only).*

24 ECM multi-plug connected, ignition on:

a) Attach the voltmeter negative probe to an engine earth.

b) Attach the voltmeter positive probe to the coding earth pin under test: The voltmeter should indicate 0.25 volt maximum if the coding earth is connected. 5.0 volts if the coding earth is not connected.

ECM relay driver pins test (relays and ECM connected)

Note: *Depending on system, the ECM may drive the main, fuel pump or OS relay winding to earth.*

Main relay driver

25 Identify the ECM relay driver pins.

26 With the ignition off, backprobe the ECM

main relay driver pin with the voltmeter positive probe: nbv should be obtained. If not, check the relay and the relay wiring.

27 Switch on the ignition: The voltage should drop to near zero.

a) If not, disconnect the ECM multi-plug (see Warning No 3 in Reference), and connect a temporary jumper lead from the driver pin to earth.

b) Relay operates: Check all voltage supplies and earth connections to the ECM. If the wiring is satisfactory, the ECM is suspect.

c) Relay does not operate: Check the relay and the relay wiring.

Note: *In some systems, the main relay winding is connected directly to earth.*

Pump relay driver

28 The main relay driver operation (previous test) must be satisfactory before commencing this test, including when the main relay winding is directly connected to earth.

29 With the ignition on, backprobe the pump relay driver with the voltmeter positive probe: nbv should be obtained. If not, check the relay and the relay wiring.

30 Crank or run the engine: The voltage should drop to near zero.

a) If not, disconnect the ECM multi-plug (see Warning No 3 in Reference), and connect a temporary jumper lead from pin 3 to earth.

b) Relay operates: Check all voltage supplies and earth connections to the ECM. If the wiring is satisfactory, the ECM is suspect.

c) Relay does not operate: Check the relay and the relay wiring.

31 Essentially, the tests for any additional relay drivers are similar to the pump driver tests.

Inertia switch

1 The inertia switch is a safety device, designed to isolate the fuel pump or cut the engine electrical system in the event of a crash. It can sometimes be affected by heavy deceleration or a thump close to its location.

2 Reset the inertia switch by pressing down the reset button.

3 If voltage is still not available at the fuel pump, continue with the tests.

4 Inspect the inertia switch terminal connections for corrosion and damage.

5 Check that the terminal connections are making good contact with the switch.

6 Study a specific wiring diagram to identify the circuit which the inertia switch protects. Typical circuits are:

a) Relay output to the fuel pump.

b) Relay supply.

c) Relay driver circuit to the ECM.

7 Check the supply voltage and earth connections to the inertia switch.

System relays

Quick relay tests

1 If the engine does not run, or a relay fed component does not function, the following method is a quick way of determining whether the relay is defective:

a) Check for a supply voltage at the component(s) supplied by the relay.

b) If voltage is not available, by-pass the relay (see below) and retest the component for voltage or attempt to run the engine.

c) If the engine runs or voltage is now available, test the relay (see below) or renew the relay.

d) If voltage is not available, check for supply, earth and output voltages at the relay terminals (see illustration 4.31). Trace supply faults back to the source. Check for a blown fuse or fusible link in the supply line.

Common relay terminal connections (standard relays)

2 Dual relays operate in a similar fashion, but may use different numbers. Some Citroën, Peugeot, Renault and Far Eastern systems (including Japanese manufacturers) may use a numbering system based on the numbers one to five or six.

Main relay terminal 30: Supply from the battery positive terminal. Constant voltage available.

Main relay terminal 86: Supply from the battery positive terminal or the ignition switch. Either constant or switched voltage available.

Main relay terminal 85: Relay winding, connected to earth or ECM driver terminal. Voltage almost zero when ignition switched on.

Main relay terminal 87: Output terminal supplies voltage to ECM, ISCV, injectors etc. Battery voltage available when ignition switched on.

Pump relay terminal 30: Supply from the battery positive terminal. Constant voltage available.

Pump relay terminal 86: Supply from the main relay terminal 87 or the ignition switch. Either constant or switched voltage available.

Pump relay terminal 85: Relay winding, ECM driver terminal. Voltage less than 1.25 volts when engine cranking or running

Pump relay terminal 87: Output terminal supplies voltage to fuel pump and sometimes OS heater. Battery voltage available when engine cranking or running

Terminal 85a and 85b similar to terminal 85 depending on use.

Terminal 87a and 87b similar to terminal 87 depending on use.

Citroën, Peugeot and Fiat 15-pin relay terminals (typical)

1 Relay output terminal: Usually connected to fuel pump circuit.

4.31 ***Test the relay, probe for voltages***

2 Battery supply to relay: Supply from the battery positive terminal. Constant voltage available.

3 Battery supply to relay: Supply from the battery positive terminal. Constant voltage available.

4 Relay output terminal: Components supplied vary depending on system.

5 Relay output terminal: Components supplied vary depending on system.

6 Relay output terminal: Components supplied vary depending on system.

7 Relay earth or driver terminal

8 Battery supply to relay: Supply from the battery positive terminal. Constant voltage available.

9 Relay output terminal: Usually connected to fuel pump circuit.

10 Relay earth or driver terminal.

11 Battery supply to relay: Supply from the battery positive terminal. Constant voltage available.

12 Unused

13 Relay output terminal: Components supplied vary depending on system.

14 Supply from the ignition switch: Switched voltage available.

15 Battery supply to relay: Supply from the battery positive terminal. Constant voltage available.

Note: *Although the function of the above terminal numbers are generally as stated, there are wide differences in how the relay is wired in any particular application.*

By-passing the relay

3 Remove the relay from the relay multi-plug.

4 Connect a fused (15 amp) jumper lead between the battery supply terminal (usually terminal 30) and the output terminal (usually terminal 87) on the terminal block where power to the fuel pump or other fuel injection components is required.

5 Do not run the fuel pump continually under this condition; disconnect the by-pass whenever a particular test is completed.

Testing the relay

6 Remove the relay from the terminal block.

7 Connect an ohmmeter across terminals 30 and 87.

8 Attach a wire between terminal 86 and a 12 volt supply.

9 Attach a wire between terminal 85 and earth.

10 The ohmmeter should indicate continuity.

Fuel pump and circuit

1 Locate the fuel pump. Typical fuel pump locations are:

a) Bolted to the chassis outside of the fuel tank.
b) Located inside the fuel tank. Access to the in-tank pump is often gained by burrowing under the rear passenger seat.

2 Connect the voltmeter negative probe to an earth.

3 Identify the supply and earth terminals.

4 Connect the voltmeter positive probe to the wire attached to the fuel pump supply terminal.

5 Crank the engine or by-pass the fuel pump relay: nbv should be obtained.

Voltage supply not available

a) Check the fuel pump fuse (where fitted).
b) Check the fuel pump relay.
c) Check the inertia switch (where fitted).
d) Check continuity of the wiring.

6 Attach the voltmeter positive probe to the fuel pump earth terminal.

7 Crank the engine or by-pass the relay. A voltage of 0.25 maximum should be obtained *(see illustration 4.32)*.

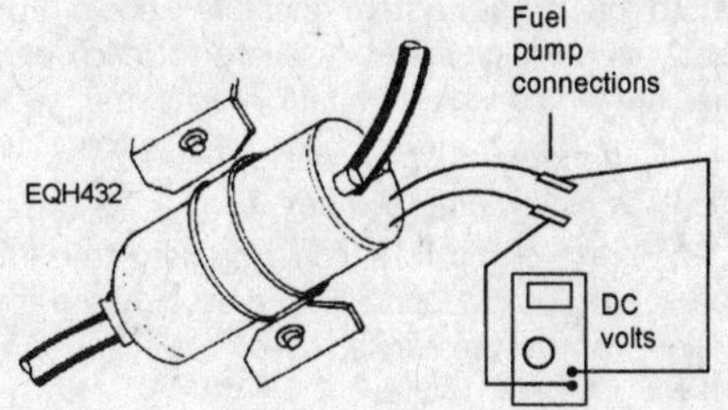

4.32 Check the voltage to the fuel pump

Fuel pressure system (MPi and SPi) – petrol models only

Note: *The fuel pressure system operates at high pressure. Refer to Warning No 12 in Reference, and take all possible care.*

1 The fuel system should be de-pressurised before the fuel filter is changed or work is carried out upon the fuel lines.

De-pressurise the fuel system (single 'dual' relay)

2 Remove the fuel pump fuse, or disconnect the earth wire on the fuel pump.

3 Crank or briefly run the engine until it stops.

Note: *System pressure cannot be relieved by removing a single fuel system relay. The single relay also supplies power to the injectors, which would be disabled by removal of the relay.*

De-pressurise the fuel system (separate fuel pump relay)

4 Remove the fuel pump relay.

5 Crank or briefly run the engine until it stops.

Check the fuel system volume

Note: *Fuel system volume is a good indicator of system integrity. If the pressure is satisfactory, yet the volume is low, the reason for low volume should definitely be investigated.*

6 Disconnect the fuel return line at a point after the pressure regulator, and place into a suitable graduated container *(see illustration 4.33)*.

7 Operate the fuel pump by allowing the engine to idle, or by-pass the relay.

8 Measure the volume of fuel flow.

a) A good MPi system will provide in excess of 2.0 litres per minute.
b) A good SPi system will provide in excess of 1.0 litres per minute.

9 If the fuel volume is low, check for the following faults:

a) Clogged fuel filter.
b) crushed or blocked fuel lines.
c) Blocked fuel tank outlet.

Check the fuel pressure (engine stopped)

Note: *A high-pressure fuel gauge and adapters for the fuel system under test is required for these tests.*

10 De-pressurise the fuel system. See above.

11 Connect the fuel pressure gauge into the fuel supply pipe *(see illustration 4.34)*:

a) MPi: before the fuel rail.
b) SPi: before the injector on the throttle body.

12 Reconnect the fuel lines.

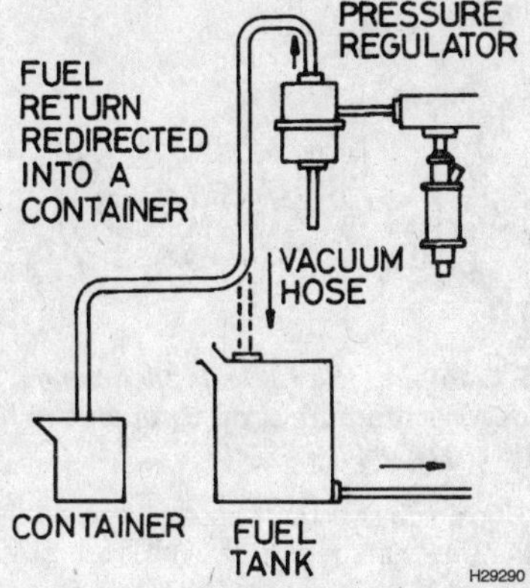

4.33 Fuel flow volume test. Disconnect the fuel return at a point after the pressure regulator, and direct the flow of fuel into a suitable container

13 By-pass the relay. An alternative method of pressurising the fuel system is to switch the ignition on and off several times.

14 Record the fuel pressure, and compare with the specified values in the system Chapter.

a) In most MPi systems, the system pressure will be either 2.5 or 3.0 bar.
b) In most SPi systems, the system pressure will be approximately 1.0 bar.

Check the fuel pressure (engine running, MPi only)

15 Reconnect the fuel pump relay.

16 Start the engine and allow it to idle.

17 Connect a gas analyser (optional).and record the CO value.

18 Record the fuel pressure which should be approximately 0.5 bar under the pressure obtained with the engine stopped.

19 Remove the vacuum hose to the pressure regulator, and plug it.

20 Record the fuel pressure which should now be comparable to the pressure obtained whilst the engine was stopped. The CO value should be slightly richer than before.

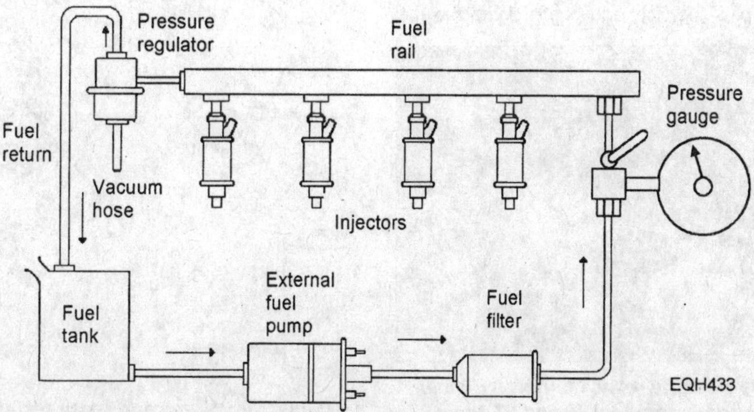

4.34 Connect a pressure gauge and check the fuel pressure in the fuel rail

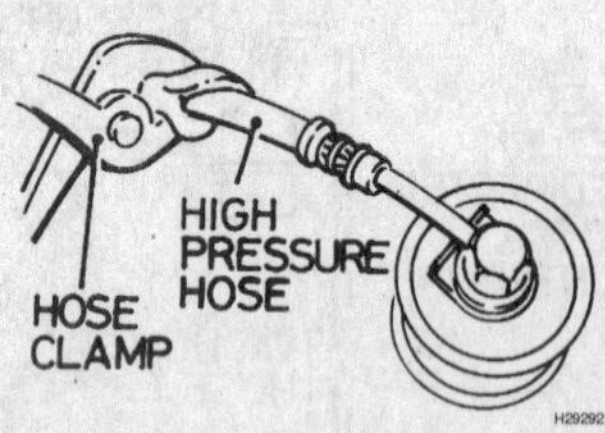

4.35 Clamping the fuel supply hose to check for a pressure drop through the fuel pump check valve

21 Unplug the vacuum hose and reconnect it to the pressure regulator. The fuel pressure should decrease by approximately 0.5 bar, and the CO level should also return to the normal idle value.

22 If the fuel pressure is low, carry out the 'Low fuel pressure' tests below.

23 If the fuel pressure is high, carry out the 'High fuel pressure' tests below.

24 If the fuel pressure is satisfactory, carry out the 'Maximum fuel system pressure' test.

Low fuel pressure test (MPi and SPi)

25 Briefly clamp the fuel return line from the pressure regulator back to the fuel tank *(see illustration 4.35)*.

26 If the pressure now rises, the pressure regulator is faulty.

27 A low pressure coupled with a much slower pressure rise could be due to a severe supply line or filter blockage. This would normally be discovered in a fuel volume test.

28 Check the injectors for leakage.

29 If the pressure is still low, and there are no other fuel leaks, suspect a faulty fuel pump.

High fuel pressure test (MPi and SPi)

30 Disconnect the fuel return line from the pressure regulator.

31 Attach a temporary pipe from the pressure regulator to a container, and run the engine.

32 If the fuel pressure remains high, renew the pressure regulator.

33 If the fuel pressure is now satisfactory, check the return line for a blockage.

Maximum fuel system pressure test

34 By-pass the relay and briefly clamp the fuel return pipe. The system pressure should now reach its peak pressure. A failure to reach peak pressure would indicate a faulty fuel pump.

a) MPi: between 4 and 6 bar.
b) SPi: approximately 3.0 bar.

35 A brake clamp hose could be used to clamp the return hose during this procedure on the majority of fuel pressure systems.

4.36 By-pass the relay. Connect a jumper lead between terminal 30 and 87 when power will be supplied to the relay output

Residual pressure drop test (MPi and SPi)

36 De-pressurise the fuel system.

37 Connect the fuel pressure gauge into the fuel supply pipe (before the fuel rail).

38 Reconnect the fuel lines.

39 Allow the engine to idle at normal operating temperature.

40 Record the pressure and switch off the ignition.

41 The pressure should not drop more than 0.5 bar in 60 seconds.

Note: *A drop in residual pressure will not usually affect running but may cause difficult starting, particularly when the engine is hot.*

High residual pressure drop (MPi and SPi)

42 Switch off the ignition and at the same moment clamp the fuel return pipe.

43 If the pressure drop is still higher than specified:

a) Test the fuel pump check valve.
b) Test the injectors.

44 If the pressure drop is now acceptable, the pressure regulator is suspect.

Fuel pump check valve test (MPi and SPi)

45 Switch off the ignition whilst clamping the fuel supply line before the fuel rail. The pressure should remain high to confirm that the fuel pump check valve is faulty. Check valves are sometimes supplied separately to the pump. Otherwise, renew the pump.

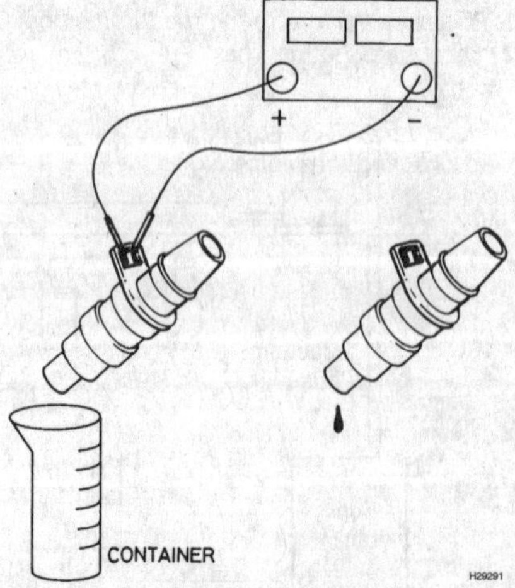

4.37 Injector volume and spray test. See checking the injectors for flow

Checking the injectors for flow (MPi and SPi)

46 Remove the injectors from the cylinder head, and use a dedicated fuel injection test rig to test for injector flow. This is the more satisfactory method of injector testing. Fuel Injection specialists who can carry out the work are available in most areas.

47 An alternative method (MPi only):

a) Remove the injectors from the cylinder head and place into separate graduated containers. The injectors must remain connected to the fuel rail and pressure regulator.
b) By-pass the relay (see illustration 4.36).
c) Remove the electrical connections from all of the injectors.
d) Use a jumper lead to connect the injector supply terminal to a 12-volt supply.
e) Use a second jumper lead to connect the injector earth terminal to an earth.
f) Apply power for 15 seconds exactly.
g) Similarly test each of the injectors in turn.
h) The difference in quantity sprayed from each injector should be within 5cc.
i) Each injector should emit a fine conical spray (see illustration 4.37).
j) Repeat the test to ensure consistent results.
k) On removal of the jumper leads, the injector should leak no more than one drop in 60 seconds.

Visual flow test (SPi)

48 Remove the air filter.

49 Allow the engine to idle and then stop the engine.

50 Visually check for leaking fuel from the injector. If the injector is leaking, it will need to be renewed or serviced.

51 Remove the injector from the throttle body, and use a dedicated fuel injection test rig to test for spray pattern and injector flow.

52 If a test rig is not available in-house, take the injector to a specialist who can carry out the work.

Oxygen sensor (OS)

1 Inspect the OS multi-plug for corrosion and damage.

2 Check that the terminal pins in the OS multi-plug are fully pushed home and making good contact with the OS.

3 Roll back the rubber protection boot to the OS multi-plug.

4 Connect the voltmeter negative probe to an engine earth.

5 Identify the terminals. Depending upon system there will be up to four terminals:

a) OS heater earth.
b) OS heater supply.
c) OS signal.
d) OS return or earth.

6 Connect the voltmeter positive probe to the wire attached to the OS signal terminal.

7 If an MOT-specification four-gas analyser with Lambda is attached to the exhaust system, the following values should be obtained.

a) CO: as vehicle specification.
b) HC: less than 50 ppm
c) CO_2: greater than 15.0
d) O_2: less than 2.0
e) Lambda: 1.0 ± 0.03

8 Run the engine to operating temperature.

9 Raise the engine speed to 3000 rpm for 30 seconds. This will raise the temperature of the OS so that switching should occur.

10 Hold the engine speed at a steady 2500 rpm. If the engine is allowed to idle for prolonged periods, the OS will become cool and switching may stop.

11 Check for OS switching. See below for full details and analysis.

OS heater tests

12 Check for nbv at the OS heater supply terminal.

a) No voltage: Trace the supply wiring back to the relay or ignition switch as appropriate.
b) Check the OS heater earth connection

OS signal output

Condition	Voltage
Engine running (hot, at 2500 rpm)	*0.2 to 1.0 volts*
Throttle fully-open	*1.0 volt constant*
Fuel cut-off	*0 volt constant*
Switching frequency	*1 sec intervals (approx)*

Check for switching

1 All closed-loop catalyst vehicles monitor the presence of oxygen in the exhaust system and adjust the injector output to keep the AFR within Lambda 1.0 ± 0.03. The switching of the OS is fundamental to the proper operation of the injection system. Checking that OS switching is correctly occurring is a vitally-important test routine.

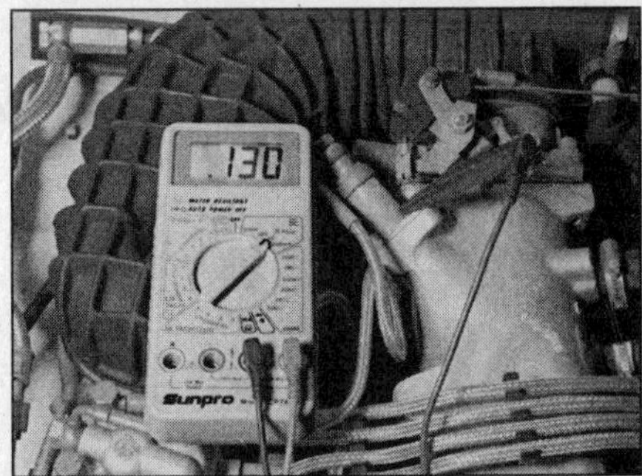

4.38 Oxygen sensor switching voltage low. 0.130 volts is equivalent to 130 millivolts. A low voltage signifies a weak mixture

2 Attach a suitable oscilloscope or voltmeter to the OS switching wire.

3 Increase the engine speed to between 2500 and 3000 rpm for a period of 3 minutes in order to heat the OS and light the catalyst.

4 Allow the engine to fast idle and check for OS switching.

5 The OS voltage should switch high and low from approximately 200 mv to 800 mv at a frequency of 8 to 10 times every 10 seconds (1 Hz) *(see illustration 4.38)*. **Note:** *A digital voltmeter will indicate an average voltage of approximately 450 mv. A sluggish OS may appear to be switching correctly, and may not reveal that the voltage is slightly high. An oscilloscope is the more accurate form of test equipment, and will reveal most faults. However, if the voltmeter has a max. and min. function, the range of average switching will be more easily spotted.*

No OS switching

a) Check the self-diagnostic system for fault codes. If the OS has failed, the ECM will either go into open-loop, or use a fixed voltage of approximately 0.45 to establish Lambda = 1.0.
b) Check the OS heater circuit (heated OS only, 2, 3 or 4 wire types). Refer to the 'Oxygen sensor' tests in the system specific Chapter.
c) If the OS heater circuit has failed, the OS may never (or only occasionally) reach operating temperature.
d) Snap-accelerate the engine: As the AFR goes rich, the OS should give a high voltage.
e) If the exhaust is equipped with a CO inspection port before the catalyst, measure the CO vol % and HC at the port. If the cat is operating efficiently, the following tests may not be so productive when the CO is measured at the tailpipe.
f) Increase the engine speed to between 2500 and 3000 rpm for 3 minutes to heat the OS and light the catalyst.
g) Allow the engine to fast idle.
h) Place the system in open-loop by disconnecting the multi-plug to the OS.
i) MPi engines: Remove the vacuum hose from the fuel pressure regulator, and seal the hose end.
j) SPi engines: Briefly clamp the fuel return line from the pressure regulator back to the fuel tank.
k) The CO should increase, and the OS voltage should switch high.
l) Return the system to closed-loop operation by reconnecting the multi-plug to the OS.
m) The CO should return to normal as the engine responds to the rich mixture. This proves that the OS and ECM can handle a rich mixture.
n) MPi engines: Refit the hose to the pressure regulator.
o) Place the system in open-loop by disconnecting the multi-plug to the OS.
p) Pull the dipstick halfway out or detach a vacuum hose to introduce a vacuum leak.
q) The CO should decrease, and the OS voltage should switch low.
r) Return the system to closed-loop operation by reconnecting the multi-plug to the OS.
s) The CO should return to normal as the engine responds to the lean mixture. This proves that the OS and ECM can handle a weak mixture.

Carbon filter solenoid valve (CFSV)

1 Inspect the CFSV multi-plug for corrosion and damage.

2 Check that the connector terminal pins are fully pushed home and making good contact with the CFSV multi-plug.

CFSV test procedures

3 Roll back the rubber protection boot (where possible) to the CFSV multi-plug.

4 Identify the supply and signal terminals.

5 Switch the ignition on.

6 Check for nbv at the CFSV supply terminal. If there is no voltage, trace the wiring back to the battery, ignition switch or relay output as appropriate.

7 Check the CFSV resistance (see below).

8 Disconnect the ECM multi-plug *(see Warning No 3 in Reference)* and use a jumper lead to very briefly touch the switching pin in the ECM multi-plug to earth.

a) If the CFSV actuates: Check the ECM main voltage supplies and earths. If tests reveal no fault, the ECM is suspect.
b) If the CFSV does not actuate: Check for continuity of wiring between the CFSV and the ECM switching pin.

Resistance check

9 Remove the multi-plug and measure the resistance of the CFSV between the two terminals. The resistance of the CFSV is typically around 40 ohms.

Notes

Fault diagnosis

The modern 'tune-up'

Due to improved reliability and developments in electronic engine and fuel management systems, tuning as a regular process is gradually changing to diagnostics – the science of fault diagnosis. Although the reason for a tune-up may be receding, with fewer engine adjustments being possible, regular diagnostic checks are desirable to identify possible problem areas and pinpoint faults. Quick and accurate diagnosis is particularly important where a catalyst is fitted, because certain catalyst destruction will follow when more than a small amount of unburnt fuel is allowed to pass through the exhaust system.

Radiated RFI can also be a problem, and this can emanate from items such as the secondary HT circuits or the alternator. Excess RFI can disrupt ECM operation, and can affect EMS operation – particularly where both ignition and fuelling functions are located in the same ECM.

The modern tune-up is concerned with ensuring that the engine is running at maximum efficiency, and this can be accomplished with the minimum of tools or equipment. Refer to Chapter 3 for details on the various kinds of equipment required.

Diagnosis of automotive faults is sometimes a time-consuming process. Unless you are very lucky and stumble across the fault immediately, the best and ultimately quickest method is to follow a logical test pattern that checks, tests and evaluates all possibilities.

1) Interrogate the customer (even if you are the customer)
Use an Incident check list to log details of the incident and record under what conditions the problem occurs. This is an important pre-requisite to determining test procedures, and also prevents misunderstandings between workshop and customer.

2) Basic inspection
Faults in engines with electronic systems are often similar to problems found in engines with non-electronic controls. Make a basic inspection, and follow a sequence of visual checks & adjustments and the problem area can often be quickly diagnosed.

3) Evaluation of the general engine functions using diagnostic equipment
Connect appropriate diagnostic equipment, and run through a series of mechanical and electrical checks. These checks are recommended for a basic analysis of the general engine and electrical system condition. Many of the checks can be completed using basic test meters.

4) Fault code reader (FCR) diagnosis – also refer to the appropriate Chapter for the system under test
If possible, connect a FCR to the serial port/diagnostic socket and interrogate the self-diagnostic function for fault codes. If a FCR is not available, flash codes can sometimes be obtained by following the procedures detailed for the particular system under test. If a fault code is present, test the relevant circuit by following the system sensor and actuator tests.

5) Symptom-related fault diagnosis – also refer to the appropriate Chapter for the system under test
If a fault code is not available, follow the symptom-related fault diagnosis charts, and systematically test the circuits and components that might be responsible for the fault condition.

Incident check list

Customer name: .... Date: ....

Vehicle make: .... Model and year : ....

Speedometer mileage reading: .... Registration number: ....

VIN: .... Engine number: ....

Starting ability

		No start O	cold O	hot O	
		Difficult to start O	cold O	hot O	after brief stop O
Starting affected by throttle position		easier O	harder O		

Idle speed

Fast idle when cold or during warm-up		yes O	no O		
Stalls when cold or during warm-up		yes O	no O		
Hot idle		unstable idle O	high idle O	low idle O	
		engine stalls O	occasionally O	frequently O	

Driving concerns

	Hesitation O	from low speed O	from high speed O		
	Inlet backfire O	cold O	hot O		
	Exhaust backfire O	cold O	hot O		
	Surge O	detonation O	lack of power O		

Time of incident

	Morning (first thing) O	during day O	night O		
Frequency	occasionally O	all of the time O	certain conditions (please state) O		

Weather conditions

	Not affected O	fine O	wet/damp O	snowing O	other O
Temperature	hot O	cold O	cool O	humid O	

Engine conditions

	Cold O	warm-up O	hot O		
	2,000 rpm O	4.000 rpm O	6000 + rpm O		
	Idle O	acceleration O	overrun O		
Check Engine warning light	illuminated O	not illuminated O			

Road conditions

	In town O	in suburbs O	motorway O	off-road O	

Driving conditions

	Not affected O	starting O	idling O	at high speeds O	
	Accelerating O	cruising O	decelerating O	turning (RH/LH) O	
Vehicle speed (mph)	0-30 O	30-50 O	50-70 O	70-90 O	90+ O

Other comments

....

....

Basic inspection

No matter what is the problem, the following checks are an essential pre-requisite to the use of diagnostic equipment. In many instances, the fault will be revealed during these procedures. Make a careful visual inspection of the following items. Not all checks will be appropriate for all engines. This basic inspection can save a great deal of valuable diagnostic time. Worn but electrically-sound components do not always fail tests.

- ☐ Check the engine oil level, oil condition and positive crankcase ventilation (PCV) circuit condition. Maintenance of the lubrication system is particularly important for good engine operation. In catalyst-equipped vehicles, contaminated oil, a poorly-maintained PCV system or an oil-burning engine will contaminate the catalyst in a very short period of time.
- ☐ Check the coolant level and coolant system condition. Maintenance of the cooling system is particularly important for good engine operation. An engine that is overcooled or running too hot will cause an incorrect CTS signal to be passed to the EMS, which may result in incorrect output signals. This will affect timing and fuelling actuation.
- ☐ Check the automatic transmission fluid level and condition.
- ☐ Check the battery condition.
- ☐ Check the battery for security.
- ☐ Check the battery electrolyte level.
- ☐ Check the battery cables and connections.
- ☐ Check the drivebelt(s) condition and tension.
- ☐ Remove the spark plugs and check the condition. Renew if necessary.
- ☐ Check that the spark plug electrode gap is correct.
- ☐ Check that the spark plug type is the correct type for the vehicle.
- ☐ Remove the distributor cap and check the condition, both external and internal. Look for cracks or signs of tracking.
- ☐ Look for oil or water that may have seeped into the cap through a defective seal.
- ☐ Check the rotor condition and measure the resistance where appropriate.
- ☐ Check the coil tower condition. Look for cracks or signs of tracking.
- ☐ Check for poor or corroded electrical connections
- ☐ Check for freedom from vacuum leaks from the vacuum hoses, inlet manifold, AFS trunking, oil dipstick seal and rocker cover seal.
- ☐ Check the breathing system condition. Clean away accumulated sludge, and ensure that the hoses are clear.
- ☐ Check the air filter condition. Renew if it is even slightly dirty.
- ☐ Check the exhaust system condition.
- ☐ Check the fuel system condition. Check for fuel leaks, and for worn or broken components. If available, the probe from a gas analyser with HC meter can be passed over the fuel and evaporation pipes and hoses. If the HC meter registers a measurement, that component may be leaking fuel or vapour.
- ☐ Visually inspect all connections, multi-plugs and terminals. Check for corrosion and loose or displaced terminals.
- ☐ Check the throttle body for a carbon build-up – usually as a result of fumes from the breathing system. The carbon can cause a sticking or jacked-open throttle, which can cause idle, cruising and other running problems. Carburettor cleaning fluid usually cleans away the carbon nicely.

Chapter 5

Evaluate main engine functions using diagnostic equipment

Generally, an engine analyser or various pieces of hand-held diagnostic equipment should be used to locate and diagnose faults in the following areas:

a) *Mechanical condition of engine.*
b) *Vacuum.*
c) *Compression.*
d) *Battery.*
e) *Charging system.*
f) *Ignition secondary circuit.*
g) *Fuel system.*
h) *Exhaust system.*

The list below is based on a professional check list, and enables analysis of the basic mechanical and electrical condition of the engine. Many of the checks require only a DMM or other low-cost gauges. However, some tests may require specialist equipment.

Customer name: .... Date: ....

Vehicle make: .... Model and year : ....

Speedometer mileage reading: .... Registration number: ....

Fault reported: ....

Tests | Results

1 Battery volts *(12.4+, nominal battery volts)* ______

2 Cranking volts *(9.6+, disable the engine and crank it for 15 seconds)* ______

3 Battery recover volts *(12.0+, battery voltage 15 seconds after cranking ceases)* ______

4 Cranking amps *(current drawn during cranking)* ______

5 Charging amps after cranking *(immediately after cranking, start engine and raise engine speed to 3000 rpm, the current drawn should be within 10% of maximum alternator output)* ______

6 Charging volts, off load *(engine speed 3000 rpm, all electrical loads switched off, voltage should be less than rated maximum charging voltage).* ______

7 Charging volts, on load *(engine speed 3000 rpm, heated rear window and heater blower switched on, voltage should be 12.0 to 13.0 volts)* ______

8 Volts at coil + *(11.0+, ignition on)* ______

9 Cranking volts at coil + *(9.0+, during engine cranking)* ______

10 Primary ignition dwell in % or ms at idle ______

11 Primary ignition dwell in % or ms at 2000 rpm ______

12 Check that the coil polarity is correct ______

13 Check for good secondary insulation ______

14 Check the basic Ignition timing and adjust (where possible) ______

15 Check ignition timing advance ______

16 Check the CAS adjustment (where possible) ______

17 Check the throttle valve position adjustment (where possible) ______

18 Check the throttle pot or switch adjustment (where possible) ______

19 Check the CO and base idle speed values and adjust (where possible) ______

20 Check the engine vacuum at idle speed *(580 to 750 mbar)* ______

21 Check the engine vacuum at 2500 rpm ______

22 Idle speed ______

23 Lambda switching *(rich/weak/OK)* ______

24 Gas analysis:

	Idle	***2000 rpm***	***3000 rpm***	***Acceleration***
O_2				
CO				
CO_2				
HC				

Fault code reader (FCR) diagnosis

Serial data testing – introduction

1 As a general rule, it is usually more beneficial to work through the checks listed in 'Basic inspection' and 'Engine evaluation' before connecting the FCR. The reason for this is clear – electrical and HT faults may adversely affect ECM operation, giving incorrect or spurious results, and causing much confusion. Only after electrical and HT problems have been resolved should the operation of the ECM and its sensors be evaluated.

Testing self-diagnostic systems

2 Is the warning lamp (if fitted) illuminated while the engine is running?. If so, this is indicative of a system fault. **Note:** *Be aware that some lamps do not illuminate for faults that are designated as minor faults.*

3 Connect a FCR to the serial port connector/diagnostic socket, and interrogate the ECM for fault codes, or trigger the flash codes if this is possible.

4 If faults are logged, use the appropriate system test procedures to check out the relevant circuits.

5 If fault codes are not logged, use the FCR to view the datastream (live data on system sensors and actuators, and not available for all systems) or follow the symptom-driven fault diagnosis charts.

Limitations of self-diagnostic systems

6 Some may see the FCR as a panacea for solving all electronic problems with the car, but reading the fault code is only the beginning. To a large degree, the information decoded by the FCR is provided by the software designed into the vehicle ECM. The FCR makes the most of this information, but if certain facilities or data are not designed to be output at the serial port, these facilities will not be available to the FCR.

7 In many instances, the FCR can provide the answer to a puzzling fault very quickly. However, it will not provide all the answers, because some faults (including actual ECM faults), may not even generate a fault code.

8 There are a number of distinct limitations to self-diagnostic systems:

a) The basic data extracted from the EMS by the FCR is laid down by the vehicle manufacturer, and the self-diagnostic system and FCR must work within those limitations.

b) A code will not be logged if the ECM is not programmed to recognise that a particular component is faulty.

c) Spurious codes can be triggered by electrical or secondary HT faults.

d) One or more spurious codes can be triggered by a faulty component that may or may not trigger a code by itself.

e) The fault code indicates a faulty circuit, and not necessarily a component. For example, a code indicating a CTS fault may be caused by a faulty sensor, wiring fault, or corroded connector. Always check the wiring and connectors, and apply proper tests to the component before judging it to be faulty.

f) Limited range or out of range sensors. If the sensor remains within its design parameter, even if the parameters are incorrect for certain operating conditions, a fault code will not be logged. For example, a faulty CTS will generate a fault code if it is open-circuit or shorted to earth. However, if the CTS is stuck at either the hot or cold resistance, a code may not be generated, although the engine will indeed run badly at some temperatures.

g) Some vehicle systems are capable of logging faults that occur intermittently, and others are not.

h) In some instances, a fault code may be lost the moment that the ignition is switched off, and due allowance should be made for this kind of system.

i) Older vehicles with basic electronic fuel injection systems may not support self-diagnosis.

Using a FCR

9 The FCR can be used for the following tasks. Some more sophisticated FCRs may interact with the ECM and allow a diagnostic trail to be followed.

a) Reading fault codes.

b) Clearing fault codes.

c) Datastream testing (not all systems, for example Ford EEC IV cannot provide datastream).

d) Actuator testing.

Dynamic test procedures

10 Use the FCR to interrogate the ECM via the serial port/diagnostic socket. See later in this Chapter for details of the locations of the diagnostic socket fitted to vehicles equipped with EOBD (European On-Board Diagnostics).

11 Once the FCR has diagnosed one or more faults, further tests are usually required, and the technician may use the FCR (where possible), or it may be necessary to use a DMM or an oscilloscope to complete diagnosis. Refer to the Sensor and Actuator tests in the specific Chapter for the system under test.

12 Once the FCR has logged a fault, a datastream enquiry (some systems only) is a quick method of determining where the fault might lie. This data may take various forms, but is essentially electrical data on voltage, frequency, dwell or pulse duration, temperature etc, provided by the various sensors and actuators. Unfortunately, such data is not available in all vehicle systems, and datastream is not an option if you are working with flash codes. Since the data is in real time, various tests can be made, and the response of the sensor or actuator evaluated.

13 Driving or actuating the system actuators such as the idle control valve, relays and injectors through the ECM is an excellent method of testing effectiveness of the actuator and associated wiring circuit. If the actuator operates when driven in this fashion, you have proved that there is little wrong with the circuit or component. This procedure is not possible for engine sensors.

14 Use an oscilloscope or DMM to check voltages at the faulty component. Compare with the vehicle specifications in the relevant system Chapter.

15 Use an ohmmeter to check the faulty circuit for continuity of the wiring and component resistance. Compare with the specifications in the relevant system Chapter.

16 A faulty circuit should be tested and any faults that are discovered should be repaired. The FCR should then be used to clear the errors and the ECM interrogated once again to see if other fault codes are still present.

17 An important point to bear in mind is that the ECM will only log faults in the electronic circuits; mechanical, ignition secondary or fuel problems will still require to be diagnosed using time-honoured methods.

18 Check that the faults remain cleared and do not return.

Intermittent faults

19 Wiggle the component wiring, apply heat from a hairdryer, or freeze with a cold spray.

20 Intermittent faults can be extremely difficult to find, and on-road testing is often desirable, with codes or datastream information being generated as the fault occurs. Take the vehicle for a road test with the FCR attached.

Fault diagnosis

Emission test failure – non-catalyst/open-loop catalyst petrol vehicles

Requirements for the (UK) MOT test

Vehicle registered	Registration letter	CO limit (% vol.)
01/08/1975 to 31/07/1986	'P' suffix to 'C' prefix	4.50
01/08/1986 onwards	'D' prefix onwards	3.50

Maximum HC limit for all vehicles: 1200 ppm. The test is manual in operation, and normally carried out at idle speed.

Emission test

1 Run the engine to normal operating temperature. Connect a gas analyser to the exhaust tailpipe. If an MOT emission test is being executed, the gas analyser must meet relevant legislation.

2 CO test at normal idle speed. Measure the CO vol. % and HC level.

3 If either the CO vol. %, or the HC level or both values are greater than the statutory specifications, the vehicle has failed the emission test. *Diagnosis is more straightforward with non-catalyst equipped vehicles because there is no OS to alter the mixture. The measurements are therefore more open to interpretation.*

Fault analysis

Note: *Because the vehicle's own CO limit will usually be lower than the statutory limit, the vehicle's CO value should always be used when attempting to diagnose faults as a result of the CO measurement. Similarly, because the statutory HC limit is set so high, an HC value of 300 ppm would be a more realistic ceiling when attempting to diagnose faults as a result of the HC measurement. An HC measurement serious enough to allow failure of the UK MOT test is indicative of a fairly serious fault or faults.*

4 The following analysis is made with comparison to the vehicle's normal CO level and an HC limit of 300 ppm.

High CO and higher-than-normal HC

- Rich mixture

Low CO and high HC

- Fouled spark plug
- Misfire
- Weak mixture

Low CO and low or normal HC

- Fouled injector
- Exhaust system leak

Emission test failure – catalyst-equipped petrol vehicles

Requirements for the (UK) MOT test

Vehicle registered	Registration letter	CO limit (% vol.)
01/08/1992 onwards	from 'K' prefix	VM specification

Fast idle speed:	CO, HC and Lambda test
Idle speed:	Idle speed band, CO test

The test is automatic, and values are compared to VM's values. Where VM values are not available, the following default values are used.

Fast idle speed	**CO**	**HC**	**Lambda**
2500 to 3000 rpm	0.3%	200 ppm	1.0 ± 0.03
Idle speed	**CO**		
500 to 1100 rpm	0.3%		

Emission test

1 Run the engine to normal operating temperature. Connect a gas analyser to the exhaust tailpipe. If an MOT emission test is being executed, the gas analyser must meet current legislation.

2 Increase the engine speed to between 2500 and 3000 rpm for a period of 3 minutes to heat the OS and light the catalyst.

3 CO test at either normal idle speed or fast idle. Measure the CO %.

4 If the CO vol. % is either greater than the VM specifications or higher than 0.3%, the vehicle has failed.

5 Increase the engine speed to between 2500 and 3000 rpm for a period of 3 minutes to heat the OS and light the catalyst.

6 Retest the engine (recheck the CO value).

7 If the CO vol. % is either greater than the VM specifications or higher than 0.3%, the vehicle has failed the CO vol % emission test. Refer to 'Fault analysis'.

8 HC test at fast idle speed. Measure the HC level.

9 If the HC level is either greater than the VM specifications or higher than 200 ppm, the vehicle has failed the HC emission test. Refer to 'Fault analysis'.

10 Lambda test at fast idle speed. Measure the Lambda ratio.

11 If the Lambda ratio is either outside of the VM specifications or higher than 1.03, the vehicle has failed the Lambda emission test. Refer to 'Fault analysis'.

Fault analysis

12 Diagnosis is less straightforward with closed-loop catalyst and OS-equipped vehicles as compared to non-catalyst vehicles, because of the tendency of the OS to alter the mixture. If the engine is operating properly without faults, the situation is quite clear. However, when a combination of fuel or sensor faults, exhaust and vacuum leaks and ignition or mechanical faults are present, the OS will attempt to compensate and the effect on the various gases may lead to some initial confusion.

Important note: *Exhaust leaks, and any mechanical/ignition faults, must be put right before attempts are made to correct the AFR.*

13 If the catalyst requires replacement:

a) Ensure that faults are not present in the mechanical, ignition or fuel systems that might lead to premature failure of the replacement cat.

b) Ensure that the engine is not being run on leaded fuel. All cat-equipped engines must run on unleaded fuel. Although the cat-equipped vehicle utilises a small filler cap to prevent filling with leaded fuel at the petrol filling station, the determined person may still manage to find a way to introduce leaded fuel to the petrol tank.

14 The reasons for an emission failure are many and diverse. The following reasons for failure are based on the data obtained during the test procedure. In addition, attach a suitable DMM or oscilloscope to the OS, and increase the engine speed to between 2500 and 3000 rpm for a period of 3 minutes to heat the OS and light the catalyst. Allow the engine to fast idle, and record the following data.

a) Lambda ratio
b) OS switching at the correct frequency
c) CO vol. %
d) HC value
e) O_2 content
f) CO_2 content

15 A faulty catalyst is indicated if the following were found:

a) The CO vol. % is either greater than the VM specifications, or higher than 0.3%.
b) The HC level is either greater than the VM specifications, or higher than 200 ppm.
c) The Lambda ratio is higher than the VM minimum specification, or greater than 0.97.
d) The O_2 content is between 0.5 and 1.5%.
e) The OS is switching high and low at the correct frequency.

16 If the catalytic converter was definitely lit during the test, the symptoms suggest that the cat is faulty. Confirm diagnosis in one of the following ways:

a) If the exhaust is equipped with an CO inspection port before the cat, measure the CO vol % and HC at the port. The readings at the tailpipe should be lower than the readings obtained at the inspection port. Similar readings indicate that the cat is inefficient and is not 'catalysing' the CO or HC gases.

b) Allow the cat to cool. Measure the CO vol % and HC at the exhaust tailpipe. The readings should be slightly higher than the readings obtained when the cat is lit. Similar readings indicate that the cat is inefficient and is not 'catalysing' the CO or HC gases.

c) Measure the exhaust temperature both before and after the cat, if the cat is operating efficiently, the temperature after the cat will be approximately 55°C higher than the temperature before. If the two temperatures are approximately equal, the cat is probably not functioning.

Rich mixture and exhaust leak

- The CO vol. % is either greater than the VM specifications or higher than 0.3%.
- The HC level is either greater than the VM specifications or higher than 200 ppm.
- The Lambda ratio is higher than the VM minimum specification or greater than 0.97.
- The O_2 content is between 0.5 and 1.5%.
- The OS is not switching and reading high.

17 The symptoms suggest that the mixture is probably rich together with a small hole in the exhaust which is deceiving the gas analyser into calculating an incorrect Lambda ratio.

18 Check the self-diagnostic system for fault codes.

19 Make the rich mixture checks Refer to 'Fault diagnosis'.

Faulty OS and exhaust leak

- The CO vol. % is either greater than the VM specifications or higher than 0.3%.
- The HC level is either greater than the VM specifications or higher than 200 ppm,
- The Lambda ratio is higher than the VM minimum specification or greater than 0.97.
- The O_2 is between 0.5 and 1.5%
- The OS is not switching and reading low.

20 The symptoms suggest that the OS is probably faulty together with a small hole in the exhaust which is deceiving the gas analyser into calculating an incorrect Lambda ratio.

21 Check the self-diagnostic system for fault codes.

Rich mixture

- The CO vol. % is either greater than the VM specifications or higher than 0.3%.
- The Lambda ratio is higher than the VM minimum specification or greater than 0.97.
- The O_2 content is lower than 0.5%.
- The OS is not switching and reading high.

22 The symptoms suggest that the mixture is too rich.

23 Check the self-diagnostic system for fault codes.

24 Make the rich mixture checks Refer to 'Fault diagnosis'.

Slightly rich mixture, with or without a faulty catalytic converter, and a hole in the exhaust system

- The CO vol. % is either greater than the VM specifications, or higher than 0.3%.
- The Lambda ratio is higher than the VM minimum specification, or greater than 0.97.
- The O_2 content is higher than 1.5 %.
- The CO_2 content is also likely to be low.
- The HC value is less than 250 ppm.

25 The symptoms suggest a slightly rich mixture, with or without a faulty catalytic converter, and a hole in the exhaust system.

26 A hole in the exhaust pipe, upstream of the OS, will cause the OS to return a lean signal to the EMS, which is deceived into forcing the mixture rich.

27 Check the self-diagnostic system for fault codes.

28 Make the rich mixture checks. Refer to 'Fault diagnosis'.

Engine misfire

- The CO vol. % is either greater than the VM specifications, or higher than 0.3%.
- The Lambda ratio is higher than the VM minimum specification, or greater than 0.97.
- The O_2 content is higher than 1.5 %.
- The HC is higher than 200 ppm (emission test HC failure).

29 The symptoms suggest a misfire. A high CO level is not usually associated with a misfire. However, the closed-loop catalyst vehicle may produce extra CO at the exhaust pipe due to one or more of the following reasons:

a) The misfire is caused by a very rich mixture and the high O_2 in the exhaust is deceiving the gas analyser into calculating an incorrect Lambda ratio.

b) The vehicle has been driven for some time with a misfire present and the efficiency of the catalytic converter has been so greatly impaired that a higher than normal CO level can be measured at the exhaust tailpipe. The Lambda ratio is correct.

c) The misfire has caused a high level of O_2 content to be exhausted. The OS returns a lean signal to the EMS which is deceived into forcing the mixture rich. The same high O_2 in the exhaust is deceiving the gas analyser into calculating an incorrect Lambda ratio.

30 Check the self-diagnostic system for fault codes.

31 Make the rich mixture checks. Refer to 'Fault diagnosis'.

32 Make the misfire checks. Refer to 'Fault diagnosis'.

Rich mixture and engine misfire

- The CO vol. % is either greater than the VM specifications, or higher than 0.3%.
- The Lambda ratio is lower than the VM minimum specification, or lower than 0.97.
- If the O_2 content is higher than 0.5%, the symptoms suggest a rich mixture with a misfire.

33 Check the self-diagnostic system for fault codes.

34 Make the rich mixture checks. Refer to 'Fault diagnosis'.

35 Make the misfire checks. Refer to 'Fault diagnosis'.

Essential pre-fault diagnosis routines

1 Firstly, no matter what is the fault, always make a basic inspection.

2 Next, use an engine analyser or DMM to make diagnostic checks.

3 Then use a FCR to check the self-diagnostic system for fault codes.

a) If one or more codes are present, a serious fault may have occurred in the circuit that has raised the code. In that case, the system may have entered LOS and this will turn the dash-mounted warning lamp on (some systems).

b) When the EMS enters LOS, it will usually substitute a fixed value for the faulty circuit and the engine may behave fairly normally. The EMS should operate in 'open-loop' control and may substitute a fixed value for the OS output of 0.45.

c) Check the relevant circuits as described under the system specific test routines, put right any faults and clear the fault codes. Refer to the general self-diagnostic routines.

4 Now continue with the following test procedures, with reference to the specific Chapter for the system under test.

Non-starter, cold engine

1 Check for a broken cambelt

- Engines with distributor: Remove the distributor cap. Carefully turn the engine over and check that the rotor also turns.
- DIS engines: Remove the rocker cover. Carefully turn the engine over and check that the valve gear also turns.
- Even if the cambelt (or timing chain) is not broken, check for misalignment.
- Remove the cambelt cover and inspect (diesel engines).

2 Sluggish rotation of the engine whilst the starter motor is engaged

- Check that the battery and starter motor are not defective.
- Incorrect ignition timing can give symptoms not unlike a defective battery or starter motor.
- Check that the engine is being lubricated by the correct grade of oil. An incorrect grade of oil, or oil that has seriously deteriorated, can cause sluggish rotation of the engine.
- If the engine has been recently rebuilt, engine tolerances that are on the tight side can lead to very sluggish rotation of the engine.

3 Check the fuel supply

- If a gas analyser with HC facility is available, crank the engine (do not disable the ignition); an HC reading of more than 4000 ppm should be displayed.
- An HC meter that cannot display measurements of 4000 ppm or more will give an indication of fuel passing through the engine up to the meter's maximum HC potential, which should be of some help.
- Low HC: Insufficient fuel is being supplied to the cylinders. Check for an engine compression or injection system failure.
- HC measurement greater than 4000 ppm: Check for ignition or mechanical faults. It is also possible that too much fuel is being injected.
- Check the condition of the air filter. A clogged air filter could cause a mixture that is too rich for good starting.
- Test for a cranking injection signal. Record whether the signal is too long or too short. Refer to the non-runner tests under 'Injection' tests in the appropriate system Chapter.
- If no injection signal is obtained during cranking, this indicates an electrical supply fault or a major component failure.

4 Check the ignition for a spark

- *Refer to the 'Secondary ignition' non-runner tests in Chapter 4 for methods on how to check for a spark.*
- If no spark, refer to the 'Secondary' non-runner test procedures in Chapter 4.
- If no ignition secondary signal, check the HT leads, distributor cap, rotor, coil tower and coil secondary resistance. Faulty HT can cause either electronic component failure or a spurious signal. The spurious signal can disrupt ECU operation (particularly in an EMS).
- If no ignition primary signal, make the 'primary trigger' and 'primary ignition' tests. Refer to the appropriate system Chapter.
- The fuel injection in many EFI engines is triggered from the ignition primary circuit. If there is no ignition, there will be no injection. The foregoing is not always true where an EMS controls both ignition and fuel from the same ECU.
- If the system contains a CID (phasing) signal (often used in sequential injection engines), check that the two signals are in phase. In some systems, out-of-phase signals can result in a non-start or poor running situation.
- Remove the spark plugs and check for engine compression. Before commencing the test, disable the engine to prevent it from starting
- Check the spark plugs.

5 No ignition or injection signal

- **Note:** *If primary ignition and injection are established, then the fault lies in the secondary HT, air intake or fuel pressure system, or is a mechanical or timing fault.*
- If the primary trigger is an CAS or Hall-effect trigger, check for proper operation of these devices. Refer to the non-runner tests under the 'CAS' or 'HES' tests in the appropriate system Chapter.

6 Primary trigger satisfactory, yet no output from ECM

- Check the inertia switch (if fitted).
- Check the voltage supplies and earth connections to the ECM.
- Check the fuel system pressure.
- Check the voltage supply to the fuel pump.

7 Check the CTS operation

- Refer to the 'CTS' tests in the system specific Chapter.
- Check for a voltage signal according to temperature.
- Check for a poor connection at the CTS. A slightly higher than normal resistance at the connection will add to the CTS resistance, and cause an over-rich mixture during cranking.
- A very high CTS resistance could place the EMS in LOS. The fixed value used by the EMS in LOS is normally that of a warm engine. Ease of cold starting will therefore deteriorate.
- Engine without LOS will fail to start if the CTS circuit is shorted or open-circuit. The ECM will recognise a very high or a very low resistance as a very cold or a very hot engine, and will apply the injection duration accordingly.

8 Check the AFS operation (if fitted)

- Refer to the 'AFS' tests in the system specific Chapter.
- Check for a voltage output signal.
- Check for poor or sticky mechanical operation.

9 Check the MAP sensor operation (if fitted)

- Refer to the 'MAP sensor' tests in the system specific Chapter.
- Check for a voltage output signal.
- Check for a vacuum leak or displaced hose, manifold to MAP sensor.

10 Check for a major inductive vacuum leak, displaced or incorrectly-routed vacuum hose

11 Check the evaporation system (if fitted)

- Refer to the 'CFSV' tests in the system specific Chapter.
- Check the carbon filter canister.
- Check the CFSV circuit. A fault could result in a vacuum leak.

12 Check the EGR system (if fitted)

- Refer to the 'EGR' tests in the system specific Chapter.
- A fault could result in a vacuum leak or permanent leakage of exhaust into the inductive system.

13 Check for failure of the ATS, TS or TPS

- A failure of one or more of these minor components may not be serious enough to cause total system failure. However, starting may become harder than normal. Refer to the system specific routines in the relevant Chapter under 'component' tests.

14 Check the idle control system

- The engine may fire but not run if the ISCV or circuit is faulty. Refer to the 'ISCV, stepper motor or AAV' tests in the system specific Chapter.

15 Check the CSV and TTS (only on some vehicles – prior to 1992)

- A failure of one or more of these two components may not be serious enough to cause total system failure. However, starting may become harder than normal. Refer to the system specific routines under 'CSV' and 'TTS' tests.

16 Check for injector mechanical faults

- If the vehicle has been laid up for some time, the injector(s) may be stuck (open or closed). In this case, either too much fuel will be injected or too little.
- Fouling of the injector pintle(s) may cause too little fuel or an inefficient spray pattern that may affect starting.

17 Check for a blocked catalyst

- Remove one of the spark plugs or injectors. Crank the engine. If the cat is blocked, the exhaust gases will escape to atmosphere through the open plug hole and the engine will probably start-up.

18 ECM suspect

- If all signals and voltages are present at the ECM multi-plug, yet an injection signal is not available, the ECM is suspect and could be tested by substitution.

Non-starter, hot engine

1 Cambelt (or timing chain) misalignment

2 Crank the engine and check for an ignition spark or secondary signal

3 Check the spark plugs

- Refer to spark plug tests under 'Secondary' ignition tests in Chapter 4.

4 Mixture too rich

- Check the CO % value. If the mixture is adjustable, set to VM specifications

5 Mixture too weak

- Check the CO % value. If the mixture is adjustable, set to VM specifications

6 Check the condition of the air filter

- A clogged air filter could cause a mixture that is too rich for good starting.

7 Check the fuel system pressure

- Refer to the 'fuel pressure' tests in the system specific Chapter.
- In particular, check for a high pressure loss after engine shutdown.

8 Check for fuel vaporisation

- Refer to the relevant component tests in the system specific Chapter.
- If a fuel rail temperature sensor or switch is fitted, check its operation.

9 Test for a cranking injection signal

- Refer to the 'Injection' tests in the system specific Chapter.
- Record whether the signal is too long or too short.

10 Check the CTS

- Refer to the 'CTS' tests in the system specific Chapter.
- Check for a voltage signal according to temperature.

11 Check for injector mechanical faults

- Fouling of the injector pintle(s) may cause too little fuel or an inefficient spray pattern that could affect starting.
- One or more leaking injectors may cause an over rich mixture that could affect starting.

12 Check the MAP sensor operation (if fitted)

- Refer to the 'MAP sensor' tests in the system specific Chapter.
- Check for a voltage output signal.
- Check for a vacuum leak or displaced hose, manifold to MAP sensor. A vacuum leak could cause an over rich mixture.

13 Check the evaporation system (if fitted)

- Refer to the 'CFSV' tests in the system specific Chapter.
- Check the carbon filter canister.
- Check the CFSV circuit. A fault could result in a vacuum leak.

14 Check the EGR system (if fitted)

- Refer to the 'EGR' tests in the system specific Chapter.
- A fault could result in a vacuum leak or permanent leakage of exhaust into the inductive system.

15 Check for failure of the ATS, TS or TPS

- A failure of one or more of these minor components may not be serious enough to cause total system failure. However, starting may become harder than normal. Refer to the component tests in the system specific Chapter.

16 Check the CSV and TTS (only on some vehicles – prior to 1992)

- A failure of one or more of these two components could cause fuel to be injected or dribbled when the engine is warm. Refer to the system specific routines under 'CSV' and 'TTS' tests.

Misfire

1 Check the primary ignition system

- Refer to the running tests under the system 'primary' tests. Pay particular attention to the following primary components:
- Dwell angle at idle, 2000 and 3000 rpm.

2 Check the secondary ignition system

- Refer to the running tests under the system 'secondary' tests. Pay particular attention to the following secondary components: Spark plugs, HT leads, coil secondary, distributor cap & rotor arm (if fitted).

3 Check the ignition timing

- If timing marks are fitted and specifications are available, compare to specs at idle, 2000 rpm and 3000 rpm.
- If timing marks are not fitted or specifications are not available, make your own marks on the pulley and timing case. Start the engine and check that timing at idle is around 0 to 15° BTDC. If the timing is under ECM control, the marks will be unsteady as the ECM alters the timing to control the idle speed. As the throttle is opened the timing should smoothly advance.

4 Check for a mixture that is too rich

5 Check for a mixture that is too lean

6 Check the AFS operation (if fitted)

- Refer to 'AFS' tests in the system specific Chapter.
- Pay particular attention to the smoothness of the voltage output as the engine rpm increases.

7 Check the TPS or TS operation

- Some engines may utilise a TS or even a combined TPS/ TS. Refer to the 'TPS' or 'TS' tests in the system specific Chapter.
- TPS operation (if fitted). Pay particular attention to the smoothness of the voltage output as the throttle is opened.
- TS operation (if fitted). Pay particular attention to the TS adjustment at idle and full-throttle (where appropriate).

8 Check for cambelt (or timing chain) misalignment

9 CID signal out of phase

- If the system contains a CID (phasing) signal (often used in sequential injection engines), check that the two signals are in phase. An out-of-phase signal can result in poor performance.
- Check that the distributor has not been moved from its datum position (where appropriate). A tiny movement out of position can result in an out-of-phase signal.
- An misaligned cambelt or even a stretched belt can result in an out-of-phase signal.

10 Poor connections

- Check for poor or corroded connections in the electrical harness, and at the sensor and actuator multi-plug connectors.

11 Check the compression pressures

- Remove the spark plugs or injectors/glow plugs.
- Before commencing the compression test, disable the engine to prevent it from starting.

Mixture too rich

1 Check the condition of the air filter

- A clogged air filter could result in a mixture that is too rich for good emissions or performance.

2 Check the condition of the engine oil and PCV system

- Contaminated oil and excessive fumes can result in high CO and HC levels.

3 Check the CO vol % at idle, 2000 and 3000 rpm

- Adjust the mixture (if possible), otherwise check for EMS or fuel injection system faults. *Be aware that attempting to adjust the CO value when system faults are present will result in an incorrect setting.*

4 A lean mixture setting at idle can result in excessive HC emissions

5 A high fuel pressure could result in a rich mixture

- Refer to the system specific routines under 'fuel pressure' tests.
- Check the fuel system pressure
- Check for a faulty pressure regulator
- Check for a clogged fuel return line.

6 Check for injector mechanical faults

- Fouling of the injector pintle(s) may cause too little fuel or an inefficient spray pattern that may affect emissions and engine performance.
- One or more leaking injectors may result in an over-rich mixture that may affect emissions or engine performance.

7 OS (where fitted)

- Check the OS for switching. Refer to the system specific routines under 'oxygen sensor' tests.
- A hole in the exhaust system, upstream of the OS, will cause the OS to return a lean signal to the EMS which is deceived into forcing the mixture rich.

8 Coolant temperature sensor (CTS)

- Check for a voltage signal according to temperature. Refer to the 'CTS' tests in the system specific Chapter.
- A high voltage or resistance or an out-of-range sensor could result in a rich mixture.

9 Check the MAP sensor operation (if fitted)

- Refer to the 'MAP sensor' tests in the system specific Chapter.
- Check for a voltage output signal.
- Check for a vacuum leak or displaced hose, manifold to MAP sensor. A vacuum leak could result in an over-rich mixture.

10 AFS operation (if fitted)

- Refer to the 'AFS' tests in the system specific Chapter.

11 Check the injection signal, and record the pulse duration

- Refer to the 'injection' tests in the system specific Chapter. Check the CTS, AFS or MAP sensor and other fuel system sensors.
- A pulse duration that is too long could result in a rich mixture.

12 Check the evaporation system (if fitted)

- Refer to the 'CFSV' tests in the system specific Chapter.
- Check the carbon filter canister.
- Check the CFSV circuit. A fault could result in a leakage of fuel vapours into the inductive system.
- In most systems, the evaporation purge system should not operate whilst the engine is cold or during the warm-up period or whilst the engine is idling.

13 Check the EGR system (if fitted)

- Refer to the 'EGR' tests in the system specific Chapter.
- A fault could result in a permanent leakage of exhaust fumes into the inductive system.
- In most systems, the EGR system should not operate when the engine is cold, during warm-up, nor when the engine is idling.

Mixture too lean

1 Check the CO vol % at idle, 2000 and 3000 rpm

- Adjust the mixture (if possible), otherwise check for EMS or fuel injection system faults. *Be aware that attempting to adjust the CO value when system faults are present will result in an incorrect setting.*

2 Lean misfire

- A lean mixture setting at idle can result in a lean misfire that will result in excessive HC emissions.

3 Low fuel pressure

- A low fuel pressure could result in a lean mixture. Refer to the system specific routines under 'fuel pressure' tests.
- Check the fuel system pressure.
- Check for a faulty pressure regulator.
- Check for a faulty fuel pump.

4 Check the injection signal and record the pulse duration

- Refer to the 'injection' tests in the system specific Chapter. Check the CTS, AFS or MAP sensor and other fuel system sensors.
- A pulse duration that is too short could result in a lean mixture.

5 Coolant temperature sensor (CTS)

- Refer to the 'CTS' tests in the system specific Chapter.
- Check for a voltage signal according to temperature.
- A low voltage or resistance or an out-of-range sensor could result in a lean mixture.

6 Check for injector mechanical faults

- Fouling of the injector pintle(s) may cause too little fuel or an inefficient spray pattern that may affect emissions and engine performance.

7 Check for an induction vacuum leak

- Manifold induction leak.
- Displaced or incorrectly routed vacuum hose.
- Other possible sources of vacuum leaks are at the rocker or cam box seal, or at the dipstick seal.

8 Check the evaporation system (if fitted)

- Refer to the 'CFSV' tests in the system specific Chapter.
- Check the carbon filter canister.
- Check the CFSV circuit. A fault could result in a vacuum leak into the inductive system.
- In most systems, the evaporation purge system should not operate whilst the engine is cold, during the warm-up period, or whilst the engine is idling.

9 Check the EGR system (if fitted)

- Refer to the 'EGR' tests in the system specific Chapter.
- A fault could result in a vacuum leak into the inductive system.
- In most systems, the EGR system should not operate when the engine is cold, during warm-up, or when the engine is idling.

10 Check the pulse air system (if fitted)

- Refer to the system specific routines under 'pulse air' tests.
- A fault could result in a vacuum leak into the exhaust system.
- In most systems, the pulse air system should only operate when the engine is cold, and then only for a few minutes after start-up.

Lack of power

1 Check the primary ignition system

- Refer to the running tests under the system 'primary' test.
- Check the dwell angle at idle, 2000 and 3000 rpm.

2 Check the secondary ignition system

- Refer to the running tests under the system 'secondary' tests. Pay particular attention to the following secondary components: Spark plugs, HT leads, coil secondary, distributor cap & rotor arm (if fitted)

3 Check the ignition timing

- If timing marks are fitted and specifications are available, compare to specifications at idle, 2000 rpm and 3000 rpm.
- If timing marks are not fitted or specifications are not available, make your own marks on the pulley and timing case. Start the engine and check that timing at idle is approximately 0 to 15° BTDC. If the timing is under ECM control, the marks will be unsteady as the ECM alters the timing to control the idle speed. As the throttle is opened, the timing should smoothly advance.

4 Check for a mixture that is too rich

5 Check for a mixture that is too lean

6 Check the condition of the air filter

- A clogged air filter could cause a mixture that is too rich for good performance.

7 Check for a blocked fuel filter

8 Check the AFS operation (if fitted)

- Refer to the 'AFS' tests in the system specific Chapter.
- Pay particular attention to the smoothness of the voltage output as the engine rpm increases.

9 Check the TPS or TS operation

- Some engines may utilise a TS or even a combined TPS/TS. Refer to the 'TPS' or 'TS' tests in the system specific Chapter.
- TPS operation (if fitted). Pay particular attention to the smoothness of the voltage output as the throttle is opened.
- TS operation (if fitted). Pay particular attention to the TS adjustment at idle and full-throttle (where appropriate).
- Check that full throttle is achieved when the accelerator pedal (in the car) is fully open.

10 Check for cambelt (or timing chain) misalignment

11 CID signal out of phase

- If the system contains a CID (phasing) signal (often used in sequential injection engines), check that the two signals are in phase. An out-of-phase signal can result in poor performance.
- Check that the distributor has not been moved from its datum position (where appropriate). A tiny movement out of position can result in an out-of-phase signal.
- An misaligned cambelt or even a stretched belt can result in an out-of-phase signal.

12 Coolant temperature sensor (CTS)

- Check for a voltage signal according to temperature. Refer to the 'CTS' tests in the system specific Chapter.
- A high voltage or resistance, or an out-of-range sensor, could result in a rich mixture.

13 Check the injection signal, and record the pulse duration

- Refer to the 'injection' tests in the system specific Chapter. Check the CTS, AFS or MAP sensor and other fuel system sensors.
- A pulse duration that is incorrect could result in lack of power.

14 Check for injector mechanical faults

- Fouling of the injector pintle(s) may cause too little fuel or an inefficient spray pattern that could affect starting.
- One or more leaking injectors may cause an over-rich mixture that may affect engine performance.

15 Check the CO vol % at idle, 2000 and 3000 rpm

- Adjust the mixture (if possible), otherwise check for EMS or fuel injection system faults. *Note that attempting to adjust the CO when system faults are present will result in an incorrect setting.*
- A lean mixture setting at idle can result in a lean misfire that will result in excessive HC emissions.

16 Low fuel pressure

- A low fuel pressure could result in a lean mixture. Refer to the 'fuel pressure' tests in the system specific Chapter.
- Check the fuel system pressure.
- Check for a faulty pressure regulator.
- Check for a faulty fuel pump.

17 Check the MAP sensor operation (if fitted)

- Refer to the 'MAP sensor' tests in the system specific Chapter.
- Check for a voltage output signal.
- Check for a vacuum leak or displaced hose, manifold to MAP sensor. A vacuum leak could result in an over-rich mixture.

18 Check for an induction vacuum leak

- Manifold induction leak.
- Displaced or incorrectly-routed vacuum hose.
- Other possible sources of vacuum leaks are at the rocker or cam box seal, or at the dipstick seal.

19 Poor connections

- Check for poor or corroded connections in the electrical harness, and at the sensor and actuator multi-plug connectors.

20 Check the engine oil grade

- An incorrect grade of oil, or oil that has seriously deteriorated, can cause hydraulic tappets (where fitted) to jack open, thereby affecting the valve timing to restrict engine speed.

21 Check that the engine is not running in LOS

22 Check for a blocked catalytic converter

- Normally, the cat should last 50 000 miles. Premature failure is almost always due to adverse operating conditions.
- In some cars, a blocked cat may be revealed by the distinctive sound of a 'hissing,' noise.
- Grounding the catalyst against speed humps or curbs can lead to damage that may lead to an exhaust blockage.
- An engine misfire or rich mixture will raise the cat temperature so high that the substrate will melt. The cat will be destroyed in a very short time and this often leads to a blocked exhaust.
- When a faulty cat is diagnosed, the reason for failure must be ascertained if the replacement is not to meet the same fate.

23 Check the compression pressures

- Remove the spark plugs and check the compression pressures. Before commencing the compression test, disable the engine to prevent it from starting.

Hesitation or uneven running

1 Check the primary ignition system

- Refer to the running tests under the system 'primary' tests.
- Check the dwell angle at idle, 2000 and 3000 rpm.

2 Check the secondary ignition system

- Refer to the running tests under the system 'secondary' tests. Pay particular attention to the following secondary components: Spark plugs, HT leads, coil secondary, distributor cap & rotor arm (if fitted).

3 Check the ignition timing

- If timing marks are fitted and specifications are available, compare to specifications at idle, 2000 rpm and 3000 rpm.
- If timing marks are not fitted or specifications are not available, make your own marks on the pulley and timing case. Start the engine and check that timing at idle is approximately 0 to 15° BTDC. If the timing is under ECM control, the marks will be unsteady as the ECM alters the timing to control the idle speed. As the throttle is opened the timing should smoothly advance.

4 Check for a mixture that is too rich

5 Check for a mixture that is too lean

6 Check the condition of the air filter

- A clogged air filter could cause a mixture that is too rich for good performance.

7 Check for a blocked fuel filter

8 Check the CO vol % at idle, 2000 and 3000 rpm

- Adjust the mixture (if possible), otherwise check for EMS or fuel injection system faults. *Note that attempting to adjust the CO when system faults are present will result in an incorrect setting.*
- A lean mixture setting at idle can result in a lean misfire that will result in excessive HC emissions.

9 Check the injection signal, and record the pulse duration

- Refer to the 'injection' tests in the system specific Chapter. Check the CTS, AFS or MAP sensor and other fuel system sensors.
- A pulse duration that is incorrect could result in lack of power.

10 Low fuel pressure

- A low fuel pressure could result in a lean mixture. Refer to the 'Fuel pressure' tests in the system specific Chapter.
- Check the fuel system pressure.
- Check for a faulty pressure regulator.
- Check for a faulty fuel pump.

11 Coolant temperature sensor (CTS)

- Check for a voltage signal according to temperature. Refer to the 'CTS' tests in the system specific Chapter.
- A high voltage or resistance, or an out-of-range sensor, could result in a rich mixture.

12 Check for injector mechanical faults

- Fouling of the injector pintle(s) may cause too little fuel or an inefficient spray pattern that could affect starting.
- One or more leaking injectors may cause an over-rich mixture that may affect engine performance.

13 Check the MAP sensor operation

- Refer to the 'MAP sensor' tests in the system specific Chapter.
- Check for a voltage output signal.
- Check for a vacuum leak or displaced hose, manifold to MAP sensor. A vacuum leak could result in an over-rich mixture.

14 Check the AFS operation

- Refer to the 'AFS' tests in the system specific Chapter.
- Pay particular attention to the smoothness of the voltage output as the engine rpm increases.

15 Check the TPS or TS operation

- Some engines may utilise a TS or even a combined TPS/ TS. Refer to the 'TPS' or 'TS' tests in the system specific Chapter.
- TPS operation (if fitted). Pay particular attention to the smoothness of the voltage output as the throttle is opened.
- TS operation (if fitted). Pay particular attention to the TS adjustment at idle and full-throttle (where appropriate).
- Check that full throttle is achieved when the accelerator pedal (in the cab) is fully open.

16 Check for an induction vacuum leak

- Manifold induction leak.
- Displaced or incorrectly routed vacuum hose.
- Other possible sources of vacuum leaks are at the rocker or cam box seal, or at the dipstick seal.

17 Check the evaporation system (if fitted)

- Refer to the 'CFSV' tests in the system specific Chapter.
- Check the carbon filter canister.
- Check the CFSV circuit. A fault could result in a vacuum leak into the inductive system.
- In most systems, the evaporation purge system should not operate whilst the engine is cold, during the warm-up period, nor whilst the engine is idling.
- If fuel vapours are introduced into the inlet system at the wrong moment, the engine will hesitate. Disconnect and plug the vapour hose to the throttle body. Road test the engine. If the engine runs satisfactorily, test the CFSV circuits.

18 Check for cambelt (or timing chain) misalignment

19 CID signal out of phase

- If the system contains a CID (phasing) signal (often used in sequential injection engines), check that the two signals are in phase. An out-of-phase signal can result in poor performance.
- Check that the distributor has not been moved from its datum position (where appropriate). A tiny movement out of position can result in an out-of-phase signal.
- An misaligned cambelt or even a stretched belt can result in an out-of-phase signal.

20 Poor connections

- Check for poor or corroded connections in the electrical harness, and at the sensor and actuator multi-plug connectors.

21 Carbon build-up

- A build-up of carbon on the back of the inlet valves can result in the injected fuel 'soaking' into the carbon.
- This causes a lean mixture and a 'flat spot' that may be worse when the engine is cold. Excessive deposits are usually caused by seriously-deteriorated engine oil and fumes being drawn through the breather system into the throttle body.

22 Lean-mixture hesitation

- The modern engine is designed to give good economy and low emissions. Mapped EMS settings tend to be on the lean side and, depending on engine tolerances, this means that some engines will run too lean with a hesitation or flat spot at certain engine speeds.
- A number of 'tweaks' have been employed over the years to combat flat spots in injection engines. These 'tweaks' can actually make the situation worse if they are used in an attempt to overcome a system fault. The 'tweaks' should only be used as a last resort on non-catalyst engines. The catalyst engine with closed-loop control will always attempt to control the AFR within the Lambda window.
- Soldering a resistor into the CTS supply wire. The resistor size is determined as follows. Attach a pot to the CTS circuit, and enrich the mixture by dialling in a slightly higher resistance than usual at normal operating temperature. If the engine responds more positively during a road test, a resistor value of the difference between the higher value and the normal resistance should be soldered into the supply wire to the CTS.
- Fitting injectors with slightly better flow rates or enlarged nozzles.
- Drilling a tiny hole in the MAP sensor vacuum hose, the slightly lower vacuum signal will cause the ECM to enrich the mixture.
- Varying the vane AFS spring tension so that less force is required to move the AFS to any particular open position. Less air will flow and the AFR is enriched.
- Catalyst engines: Placing a small air bleed in the exhaust system. Weld a boss to the exhaust pipe above the OS. Attach a small carburettor jet to the boss, so that a small air leak bleeds air into the exhaust pipe. The OS will pass a slightly lean signal to the ECM, which will increase the injection duration to enrich the mixture. The jet will clog with carbon during engine operation, and will need to be regularly checked and cleaned.

> *Caution: This procedure will almost certainly result in a statutory emission test failure.*

Engine stalling or cutting out at idle speed (also poor running on warm-up)

Note: *One or more of the following faults are likely to be present when the engine is prone to stalling. Only faults likely to affect the idle condition have been listed. The engine may also display other symptoms, in which case other fault charts may need to be consulted.*

1 Check the primary ignition system

- Refer to the running tests under the system 'primary' tests.
- Check the dwell angle at idle, 2000 and 3000 rpm.

2 Check the secondary ignition system

- Refer to the running tests under the system 'secondary' tests. Pay particular attention to the following secondary components: Spark plugs, HT leads, coil secondary, distributor cap & rotor arm (if fitted).

3 Check the ignition timing

- If timing marks are fitted and specifications are available, compare to specs at idle, 2000 rpm and 3000 rpm.
- If timing marks are not fitted or specifications are not available, make your own marks on the pulley and timing case. Start the engine, and check that timing at idle is around 0 to 15° BTDC. If the timing is under ECM control, the marks will be unsteady as the ECM alters the timing to control the idle speed. As the throttle is opened the timing should smoothly advance.

4 Check for a mixture that is too rich

5 Check for a mixture that is too lean

6 Check the condition of the air filter

- A clogged air filter could cause a mixture that is too rich for good performance.

7 Low base idle speed

- Refer to the 'adjustments' in the system specific Chapter.
- Adjust the base idle speed (if possible).

8 Check the CO vol % at idle

- Adjust the mixture (if possible), otherwise check for EMS or fuel injection system faults. *Note that attempting to adjust the CO when system faults are present will result in an incorrect setting.*
- A lean mixture setting at idle can result in a lean misfire that will result in excessive HC emissions.

9 Check the engine is not in LOS

- If in LOS, the idle speed may be too low during hot engine operation, and too high during cold engine operation.

10 Coolant temperature sensor (CTS)

- Check for a voltage signal according to temperature. Refer to the 'CTS' tests in the system specific Chapter.
- A high voltage or resistance, or an out-of-range sensor, could result in a rich mixture

11 Check for injector mechanical faults

- Fouling of the injector pintle(s) may cause too little fuel or an inefficient spray pattern that could affect starting.
- One or more leaking injectors may cause an over rich mixture that may affect engine performance.

12 Check for a vacuum leak at the injector seals

13 Check the throttle body for a carbon build-up

- This is usually as a result of fumes from the breathing system. The carbon can cause a sticking or jacked-open throttle, which can cause idle, cruising and other running problems.

14 Check the throttle body for wear, particularly around the spindle area

15 Check the TPS or TS operation

- Some engines may utilise a TS or even a combined TPS/TS. Refer to the 'TPS' or 'TS' tests in the system specific Chapter.
- TPS operation (if fitted). Pay particular attention to the smoothness of the voltage output as the throttle is opened.
- TS operation (if fitted). Pay particular attention to the TS adjustment at idle and full-throttle (where appropriate).
- Check that full-throttle is achieved when the accelerator pedal (in the cab) is fully open.

16 Check for an induction vacuum leak

- Manifold induction leak.
- Displaced or incorrectly-routed vacuum hose.
- Other possible sources of vacuum leaks are at the rocker or cam box seal, or at the dipstick seal.

17 Check the evaporation system (if fitted)

- Refer to the 'CFSV' tests in the system specific Chapter.
- Check the carbon filter canister.
- Check the CFSV circuit. A fault could result in a vacuum leak or permanent leakage of fuel vapours into the inductive system.
- In most systems, the evaporation purge system should not operate whilst the engine is cold, during the warm-up period, nor whilst the engine is idling.
- If fuel vapours are introduced into the inlet system at the wrong moment, the engine will hesitate. Disconnect and plug the vapour hose to the throttle body. Road test the engine. If the engine runs satisfactorily, test the CFSV circuits.

18 Check the EGR system (if fitted)

- Refer to the 'EGR' tests in the system specific Chapter.
- A fault could result in a vacuum leak into the inductive system.
- In most systems, the EGR system should not operate whilst the engine is cold, during the warm-up period, nor whilst the engine is idling.

19 Check the ISCV/stepper motor or AAV as appropriate (idle control system)

- Refer to the relevant 'component' tests in the system specific Chapter.

20 If the ECM permanent voltage supply lost or disrupted

- In this event, the adaptive function of the ECM will lose its idle values, and stalling will often occur.
- Check the in-line fuse (if fitted) or supply from the battery (+).
- Systems with non-volatile memory will retain idle values, even after battery disconnection.

21 Check the TBH (where fitted) operation

- Refer to the 'TBH' tests in the system specific Chapter.
- If the TBH is faulty, stalling may occur in low temperatures, or during damp humid conditions, due to the icing phenomenon.
- If the throttle body is coolant-heated, check that the throttle body is heated sufficiently after the engine has reached normal operating temperature. If throttle body heating is insufficient, stalling may occur in low temperatures, or during damp humid conditions, due to the icing phenomenon.

22 Check the valve clearances

- Tight valve clearances, in particular, may contribute to stalling.

23 Check the oxygen sensor (OS) operation

- A defective OS may cause poor idle and 'hunting'. Disconnect the OS multi-plug. If the idle quality improves, the OS is suspect. Refer to the 'oxygen sensor' tests in the system specific Chapter.

24 Stalling after engine deceleration

- Check the throttle valve position and the TS or TPS adjustment. The ECM must receive a throttle-closed signal to activate the engine deceleration fuel injection cut-off routine.

Poor running during the warm-up period

The faults that will affect the engine during the warm-up period are generally those that will also affect the engine idle quality. In addition, most running faults are accentuated during periods of cold running.

Engine knock

- Spark plugs worn or wrong heat range. Refer to spark plug test routines under the 'secondary ignition' tests in Chapter 4.
- Defective knock sensor. Refer to the 'knock sensor' tests in the system specific Chapter.
- Build-up of carbon deposits in cylinders.
- Cooling fault.
- Over-advanced ignition timing (generally in distributors with mechanical advance). Refer to 'adjustments' in the system specific Chapter.
- Fuel of low octane.

Engine backfires through the inlet valve

1 Check the primary ignition system

- Refer to the running tests under the system 'Primary' tests.
- Check the dwell angle at idle, 2000 and 3000 rpm.

2 Check the secondary ignition system

- Refer to the running tests under the system 'secondary' tests. Pay particular attention to the following secondary components: Spark plugs, HT leads, coil secondary, distributor cap & rotor arm (if fitted).

3 Check the ignition timing

- If timing marks are fitted and specifications are available, compare to specs at idle, 2000 rpm and 3000 rpm.
- If timing marks are not fitted or specifications are not available, make your own marks on the pulley and timing case. Start the engine, and check that timing at idle is approximately 0 to 15° BTDC. If the timing is under ECM control, the marks will be unsteady, as the ECM alters the timing to control the idle speed. As the throttle is opened, the timing should smoothly advance.

4 Check for a mixture that is too lean

5 Check for an induction vacuum leak

- Manifold induction leak.
- Displaced or incorrectly routed vacuum hose.
- Other possible sources of vacuum leaks are at the rocker or cam box seal, or at the dipstick seal.

6 Carbon build-up

- A build-up of carbon on the back of the inlet valves can result in the injected fuel 'soaking' into the carbon.
- This causes a lean mixture, and could cause a backfire that may be worse when the engine is cold. Excessive deposits are usually caused by seriously-deteriorated engine oil and fumes being drawn through the breather system into the throttle body.

7 Check for a leaking or burnt inlet valve

- Remove the spark plugs and check the engine compression pressures.
- Before commencing the test, disable the engine to prevent it from starting.

Engine backfires through the exhaust valve

1 Coolant temperature sensor (CTS)

- Check for a voltage signal according to temperature. Refer to the 'CTS' tests in the system specific Chapter.
- The ECM may be operating under LOS, or the CTS resistance may be too low when the engine is hot.

2 Check for injector mechanical faults

- Fouling of the injector pintle(s) may cause too little fuel or an inefficient spray pattern that could affect starting.
- One or more leaking injectors may cause an over-rich mixture that may affect engine performance.

3 Check for a leaking or burnt exhaust valve

- Remove the spark plugs and check the engine compression pressures.
- Before commencing the test, disable the engine to prevent it from starting

Chapter 5

Intermittent faults

1 Intermittent faults are by far the most difficult to deal with, and we can offer only a sketchy guide at best on how to deal with this kind of problem.

2 Carefully follow the procedures described in above. Although you may not necessarily find the problem in those system specific routines, you will have a pretty good idea of what it is not.

3 Most intermittent faults are caused by loose, corroded or poor wiring or connectors.

4 Check for wiring that is tight, as it may pull at the connector as the engine vibrates. The connection may be momentarily broken and then remade.

5 Check for wiring with poor or broken insulation that is shorted, or chafed wires that may occasionally touch to earth.

6 Drive the vehicle and attempt to reproduce the fault under the conditions in which it usually occurs.

7 If the tachometer needle drops to zero during the incident, the fault may be in the primary ignition.

8 If the instrument display switches off, the fault may be in the voltage supply or ignition switch.

9 Connect a dwell meter or voltmeter to suspect circuits, and run the engine or drive the system. Some meters display maximum and minimum values, and this could help to pinpoint a poor supply or earth.

10 Some kinds of test equipment have a 'flight recorder' facility that could attach to either the serial port or interfaces between the ECM and its multi-plug. The flight recorder records the data from virtually all of the sensors, actuators and control functions over a period of time. The engine may then be driven until the fault occurs. Pressing a button will cause approximately 30 seconds of data before and after the incident to be saved to the test equipment memory for later evaluation in the workshop.

11 Run the engine, wiggle wiring connections and gently tap relays and components. If the engine cuts out, the fault may have been revealed.

12 Use a FCR to interrogate the ECM for fault codes. If a code is present, clear the code and drive the vehicle to see if the code returns. One or two occurrences of a fault in a particular circuit may be enough to set a code.

13 If you cannot find an electrical fault, consider the possibility of poor lubrication or the wrong grade of oil. These problems can cause jacking of hydraulic tappets and loss of power (where applicable).

14 Check the valve clearances. Tight clearances can cause loss of power as the clearance closes when the engine becomes hot.

Plug locations

According to the EOBD regulations, all 16-pin diagnostic plugs should be located within reach of the driver's seat, normally under the facia. It may be necessary to remove a cover to expose the plug, but no tools should be needed to remove the cover.

Index of vehicles

Alfa Romeo

GTE	2001	156	2003
147	2002		

Audi

A2	2002	A4	2001
A3	2001	A6	2002

BMW

3 & 5 Series	2002	X5	2002
730d	2004	Z4	2003

Citroën

Berlingo	2004	C8	2004
C3	2004	Xsara	2003
C5	2004		

Daewoo

Kalos	2003	Nubira	2003
Matiz	2003	Tacuma	2003

Daihatsu

Charade	2004	Sirion	2002
Copen	2004	Terios	2003
Cuore	2002	YRV	2003

Fiat

Doblo	2004	Punto	2003
Idea	2004	Stilo	2004
Multipla	2004	Ulysse	2004
Panda	2004		

Ford

Fiesta	2003	Galaxy	2003
Focus	2004	Ka	2003
Focus C-Max	2004	Mondeo	2002

Honda

Accord	2004	HRV	2004
Civic	2004	S2000	2004
CRV	2001	Stream	2004
Jazz	2004		

Hyundai

Accent	2001	Matrix	2004
Amica	2004	Santa Fe	2002
Coupe	2004	Sonata	2004
Elantra	2004	Terracan	2004
Getz	2004	Trajet	2004

Isuzu

Rodeo	2004

Jaguar

X-Type	2002

Kia

Carens	2004	Sedona	2004
Magentis	2002	Sorento	2004
Rio	2004		

Land Rover

Freelander	2004	Defender	2001
Discovery	2004	Range Rover	2004

Lexus

IS200	2002

Mazda

2	2004	6	2004
3	2004	MX5	2004
323F	2001	RX8	2004

Mercedes

A-Class	2001	M-Class	2004
C-Class	2004	S-Class	2001
CLK	2001	SLK	2004

Mini

Cooper	2004

Mitsubishi

Carisma	2004	Grandis	2004
Evolution VII	2004	Shogun Warrior	2004

Nissan

Almera	2004	Terrano	2004
Micra	2004	X-Trail	2004
Primera	2002		

Peugeot

106	2002	406	2001
206	2000	607	2003
307	2003	Partner	2004

Proton

Impian	2004	Satira	2004
Jumbuck	2004	Wira	2004

Renault

Clio	2001	Megane	2004
Espace	2002	Scenic	2004
Kangoo	2004	Vel Satis	2004
Laguna	2004		

Rover/MG

City Rover	2004	75	2001
25 / ZR	2002	TF	2003
45	2003		

Saab

9-3	2003	9-5	2001

SEAT

Alhambra	2004	Ibiza	2004
Arosa	2004	Leon	2003

Skoda

Fabia	2004	Superb	2003
Octavia	2004		

Subaru

Impreza	2004	Legacy	2004
Forrester	2004		

Suzuki

Alto	2004	Swift	2001
Ignis 5-door	2004	Vitara	2004
Ignis 3-door	2004	Wagon-R	2004
Jimny	2004		

Toyota

Avensis	2001	Land Cruiser	2004
Avensis	2004	MR2	2004
Celica	2001	Rav 4	2004
Corolla	2004	Yaris	2004

Vauxhall

Agila	2004	Meriva	2004
Astra-G	2003	Omega	2001
Astra-H	2004	Vectra	2003
Corsa-C	2003	Zafira	2001

VW

Beetle	2003	Polo	2001
Golf	2001	Touran	2004
Passat	2002		

Alfa Romeo GTE 2001
Remove the lower facia panel from the driver's side to expose the 16-pin connector

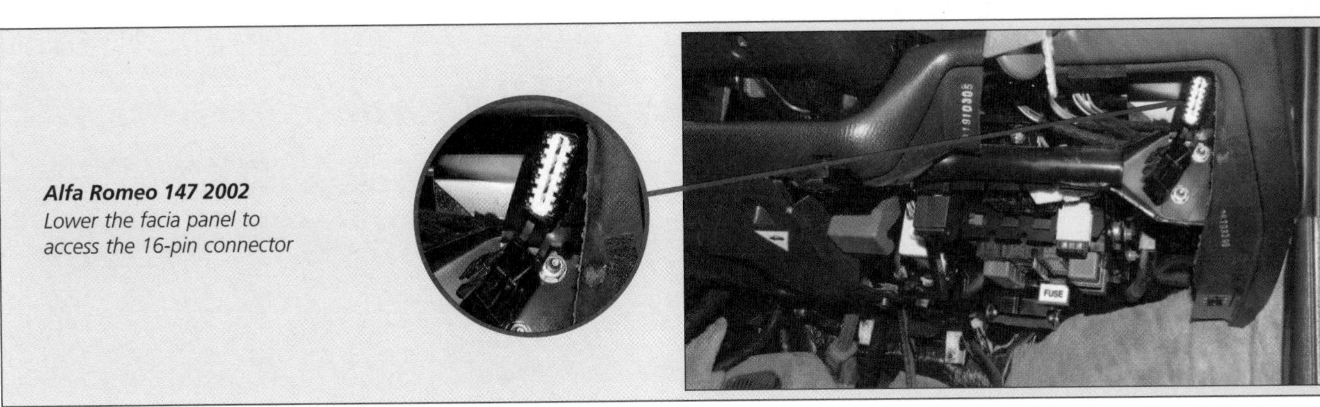

Alfa Romeo 147 2002
Lower the facia panel to access the 16-pin connector

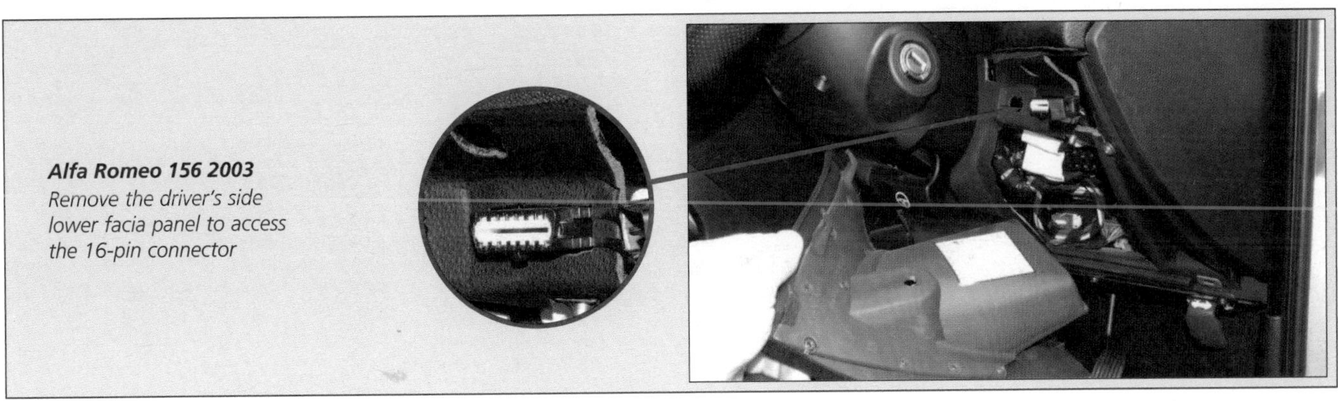

Alfa Romeo 156 2003
Remove the driver's side lower facia panel to access the 16-pin connector

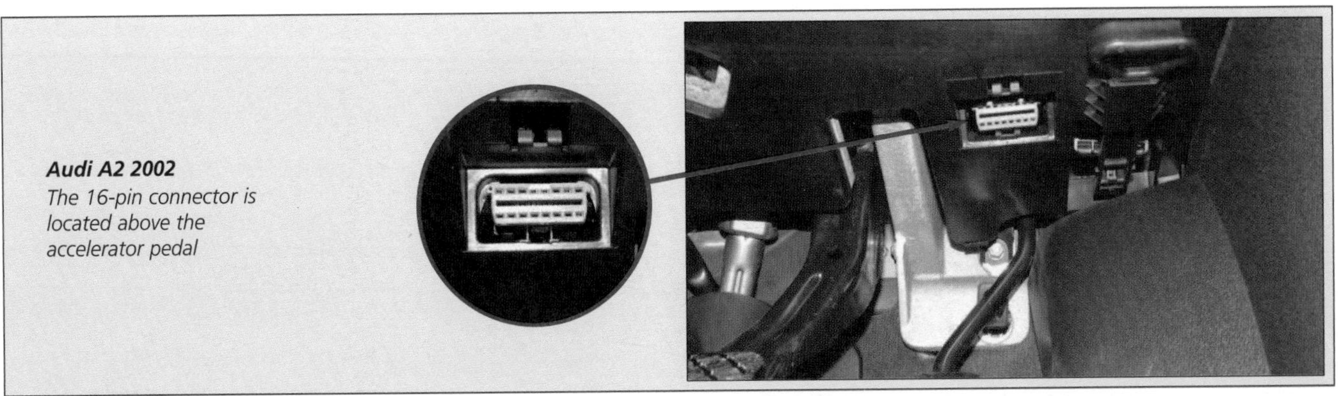

Audi A2 2002
The 16-pin connector is located above the accelerator pedal

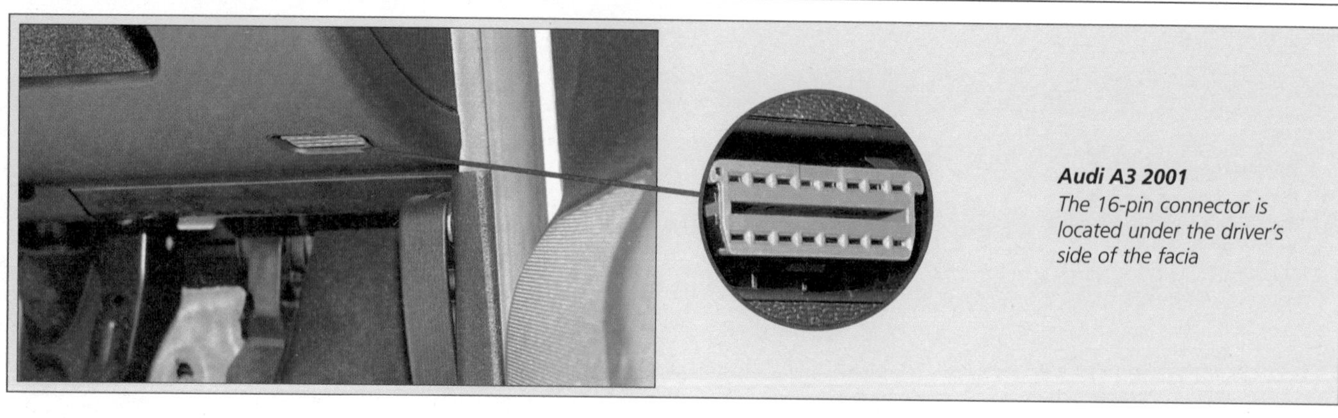

Audi A3 2001
The 16-pin connector is located under the driver's side of the facia

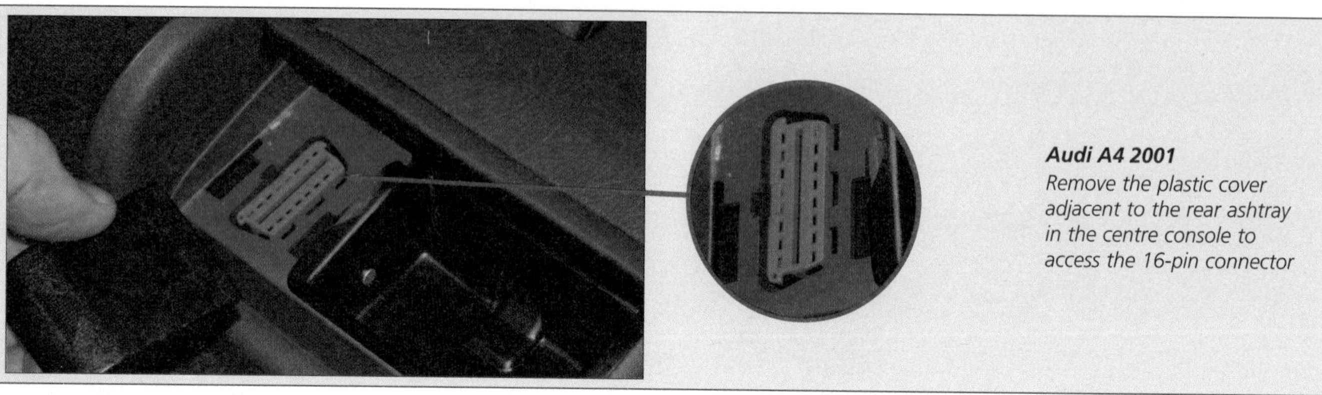

Audi A4 2001
Remove the plastic cover adjacent to the rear ashtray in the centre console to access the 16-pin connector

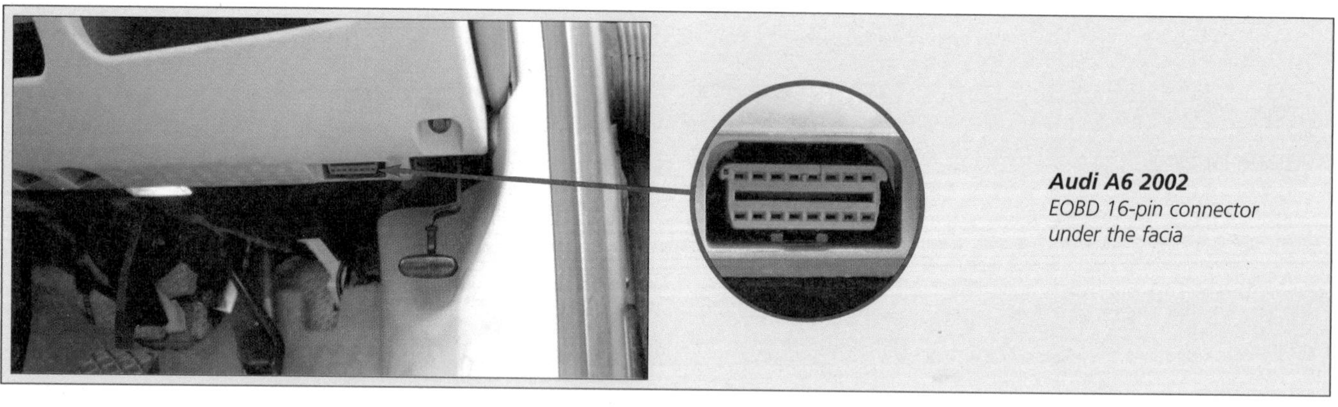

Audi A6 2002
EOBD 16-pin connector under the facia

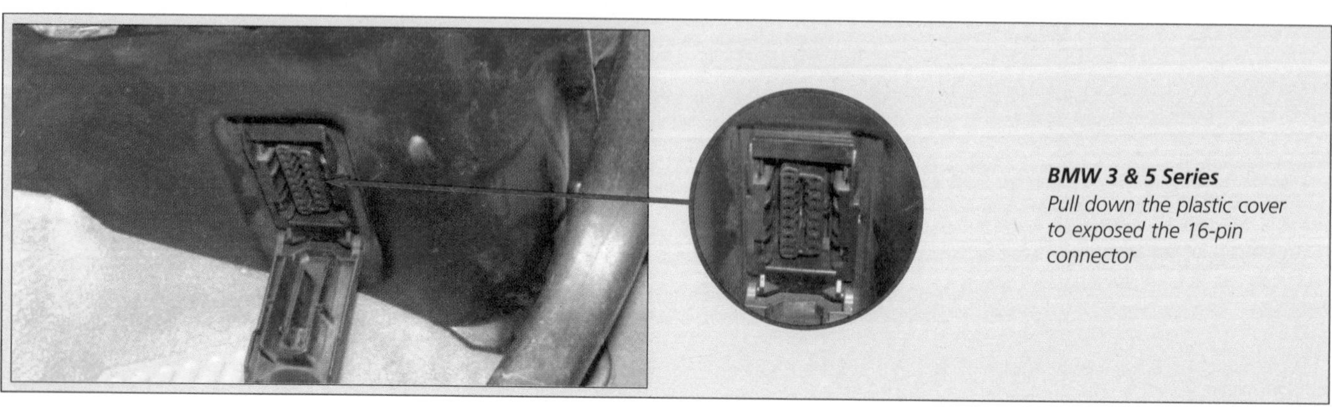

BMW 3 & 5 Series
Pull down the plastic cover to exposed the 16-pin connector

BMW 730d 2004
Prise of the plastic cover in the driver's side kick panel to access the 16-pin connector

BMW X5 2002
Prise open the plastic cover to access in the driver's side lower facia panel to access the 16-pin connector

BMW Z4 2003
Pull down the plastic cover in the lower facia panel to expose the 16-pin connector

Mini models
EOBD 16-pin connector (note the plastic cover)

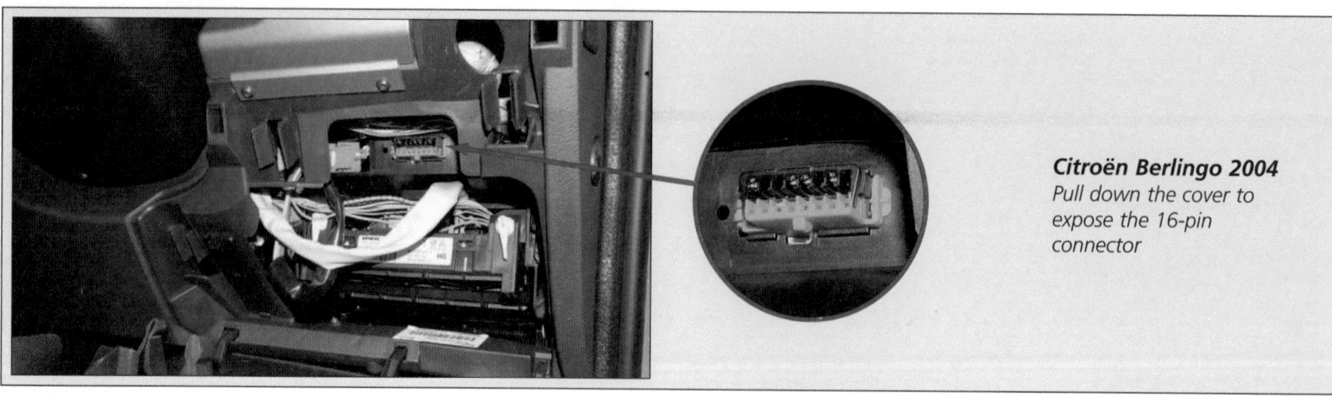

Citroën Berlingo 2004
Pull down the cover to expose the 16-pin connector

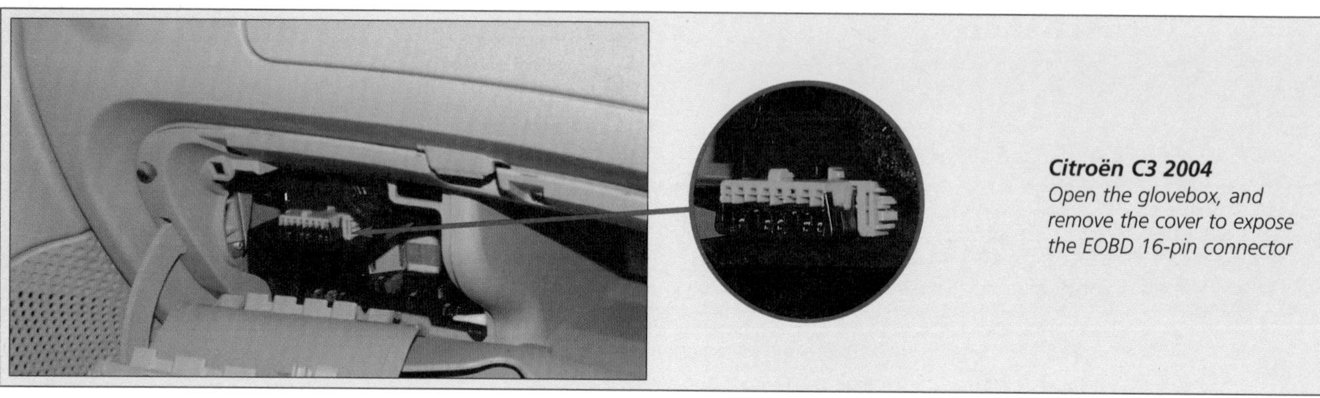

Citroën C3 2004
Open the glovebox, and remove the cover to expose the EOBD 16-pin connector

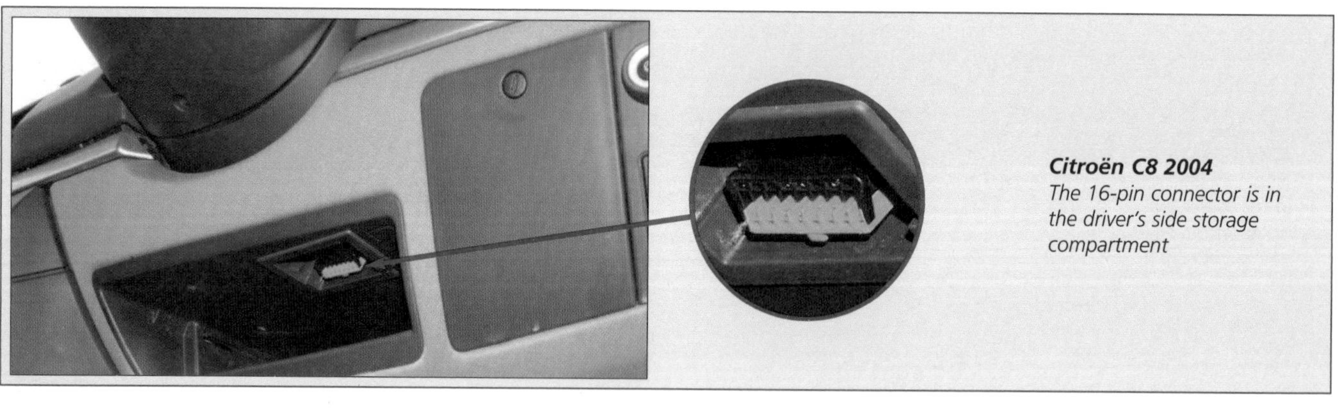

Citroën C8 2004
The 16-pin connector is in the driver's side storage compartment

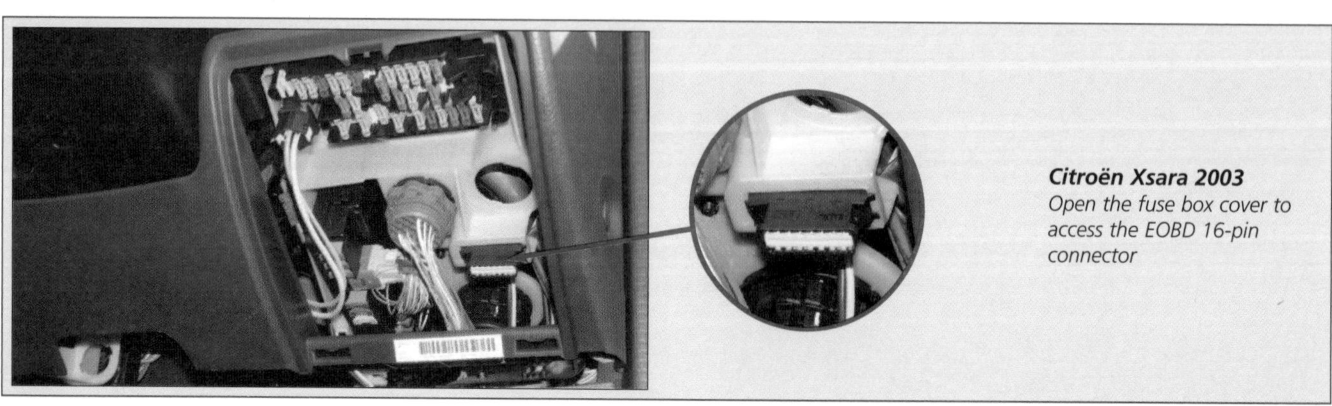

Citroën Xsara 2003
Open the fuse box cover to access the EOBD 16-pin connector

Citroën C5 2004
Pull out the trim panel from the passenger's side glovebox to expose the 16-pin connector

Daewoo Kalos 2003
The 16-pin connector is located under the facia

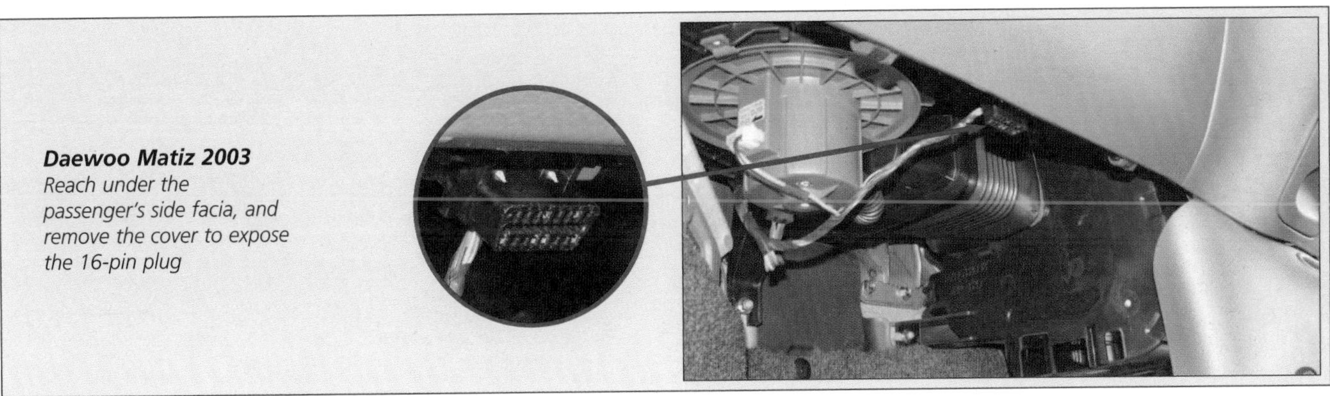

Daewoo Matiz 2003
Reach under the passenger's side facia, and remove the cover to expose the 16-pin plug

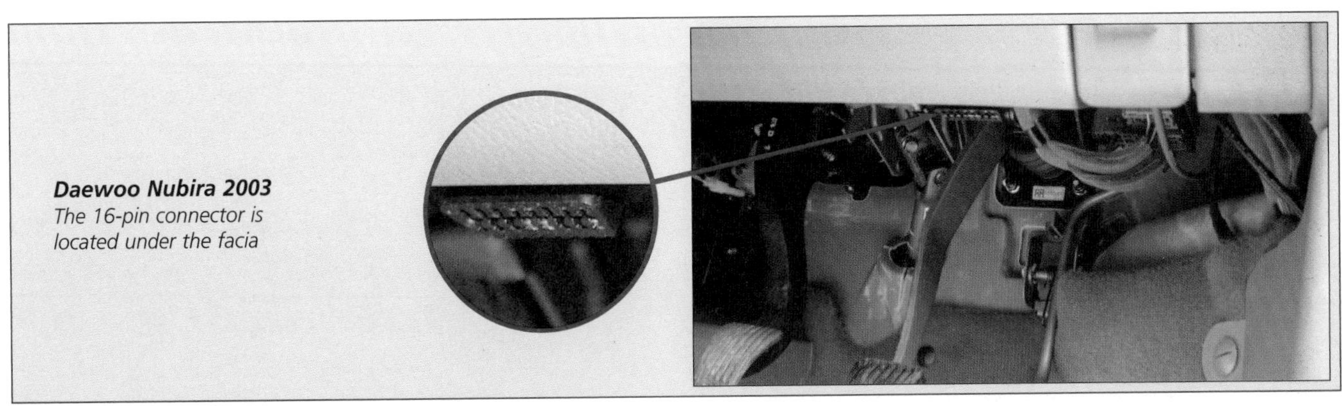

Daewoo Nubira 2003
The 16-pin connector is located under the facia

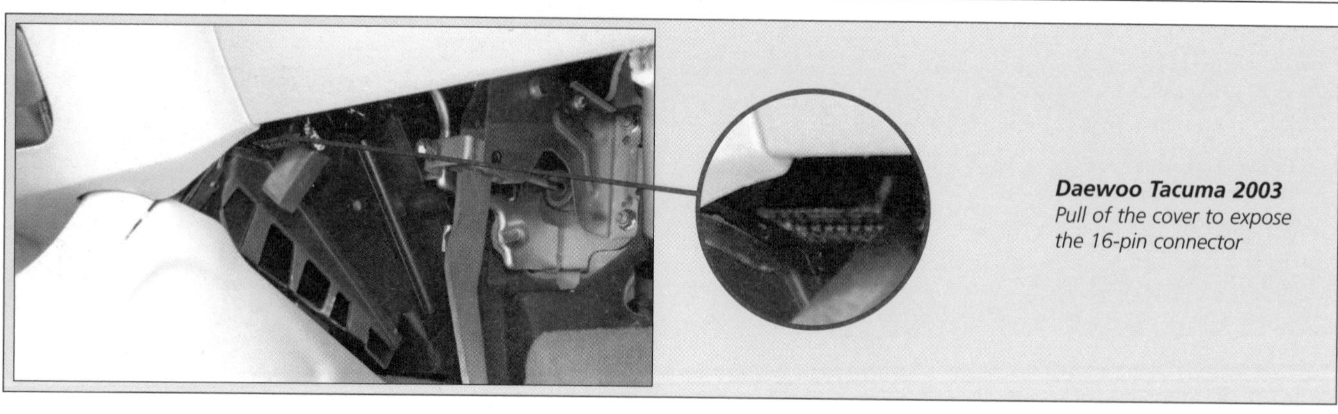

Daewoo Tacuma 2003
Pull of the cover to expose the 16-pin connector

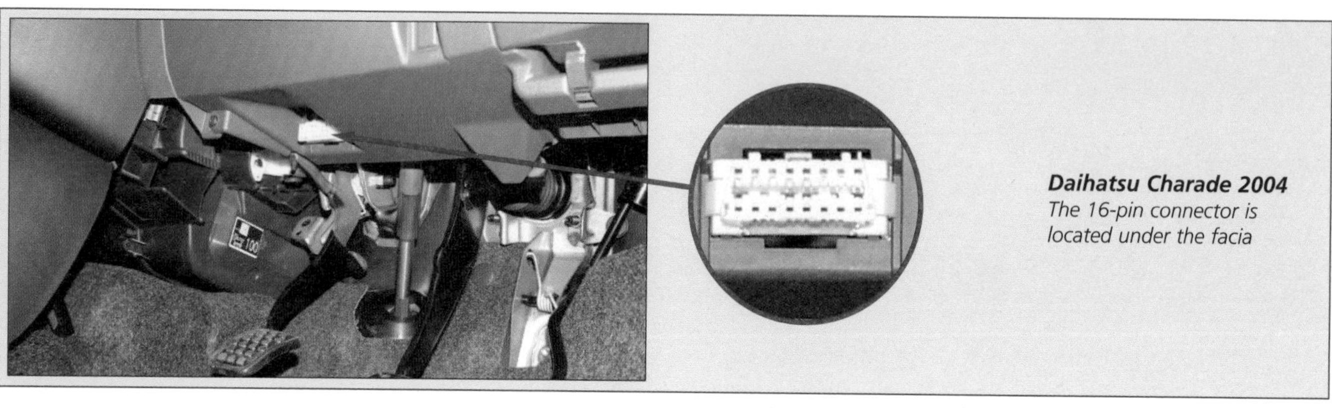

Daihatsu Charade 2004
The 16-pin connector is located under the facia

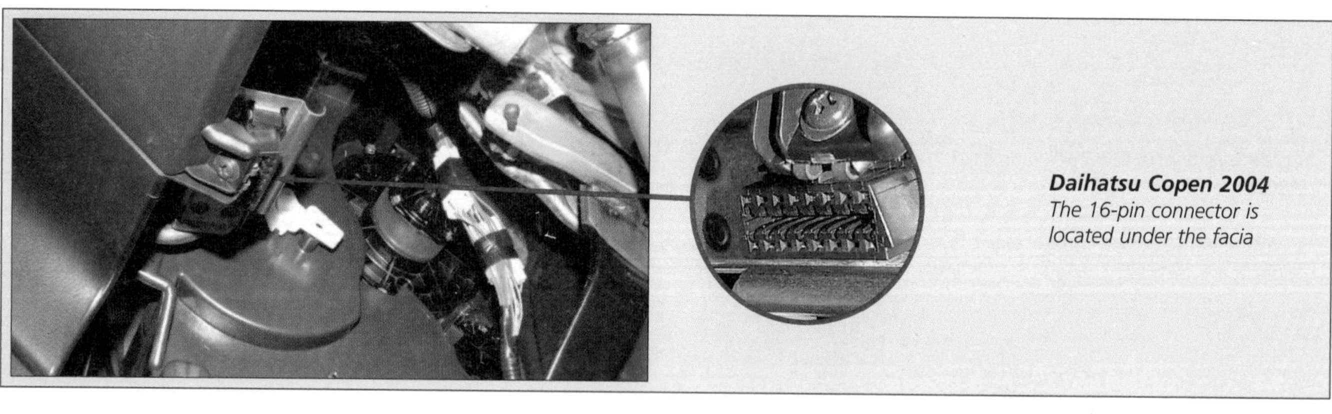

Daihatsu Copen 2004
The 16-pin connector is located under the facia

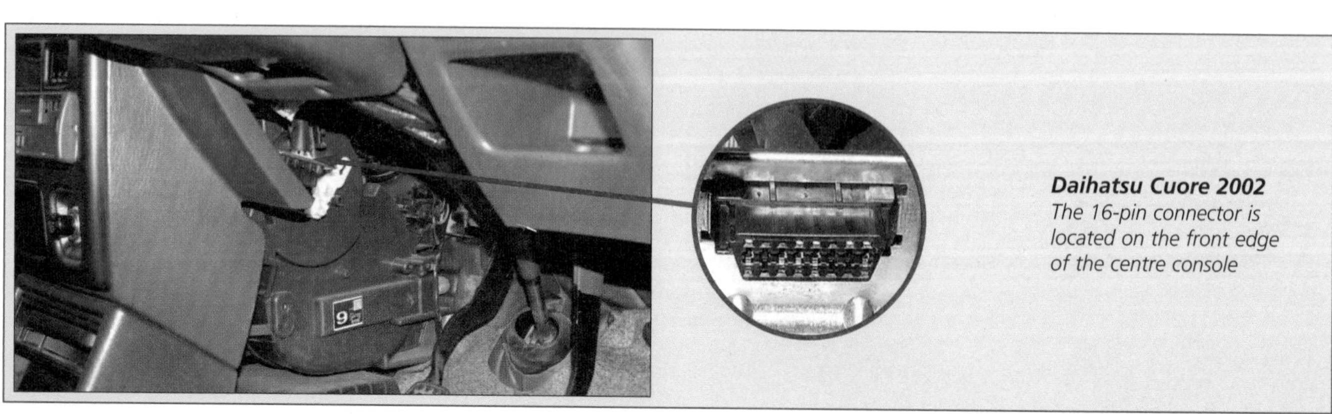

Daihatsu Cuore 2002
The 16-pin connector is located on the front edge of the centre console

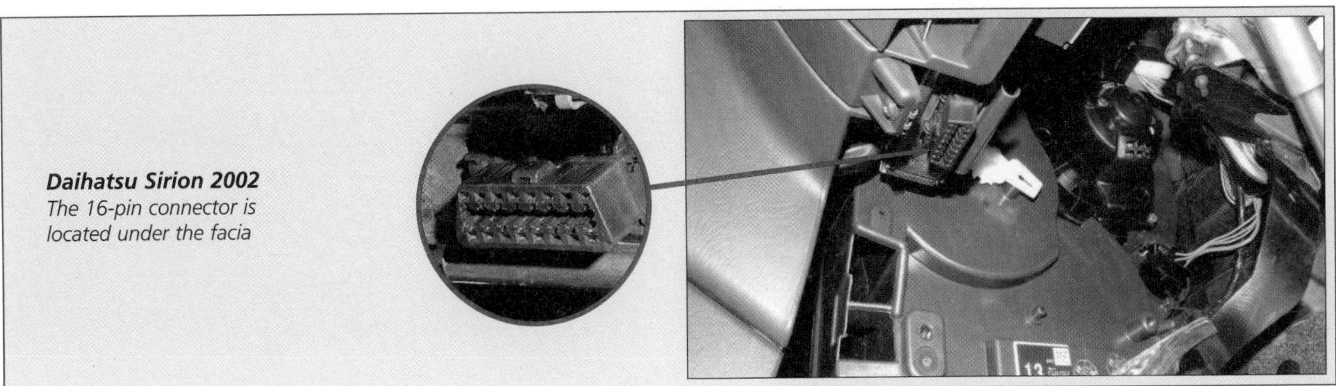

Daihatsu Sirion 2002
The 16-pin connector is located under the facia

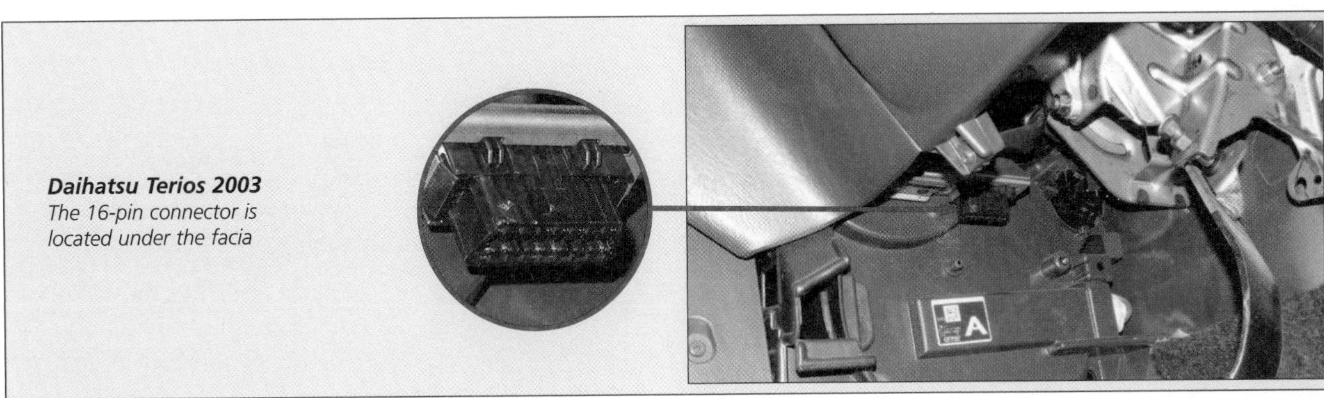

Daihatsu Terios 2003
The 16-pin connector is located under the facia

Daihatsu YRV 2003
The 16-pin connector is located under the facia

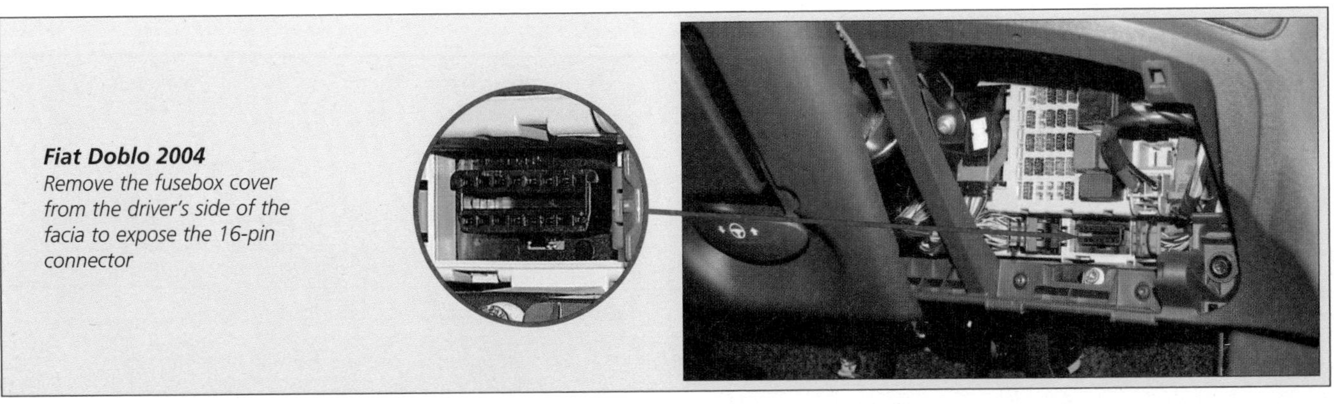

Fiat Doblo 2004
Remove the fusebox cover from the driver's side of the facia to expose the 16-pin connector

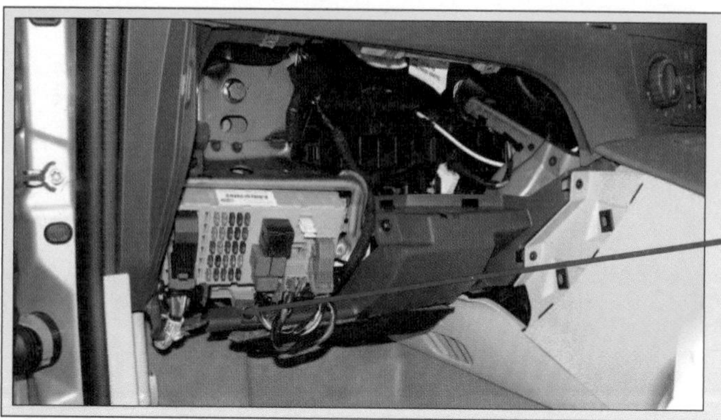

Fiat Idea 2004
Undo the screws, and remove the lower facia panel from the passenger side to access the 16-pin connector

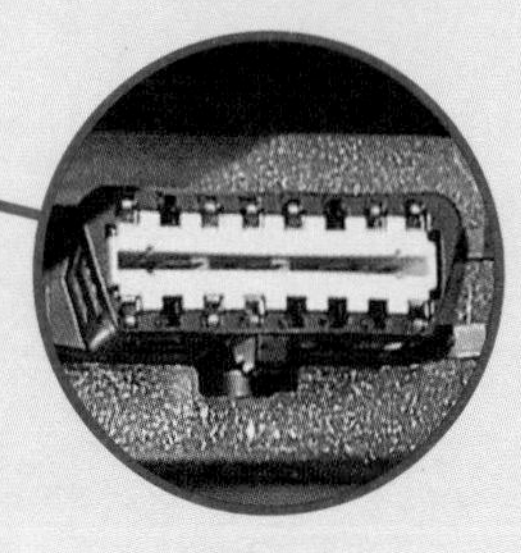

Fiat Multipla 2004
Remove the trim panel from the driver's side of the facia to access the 16-pin connector

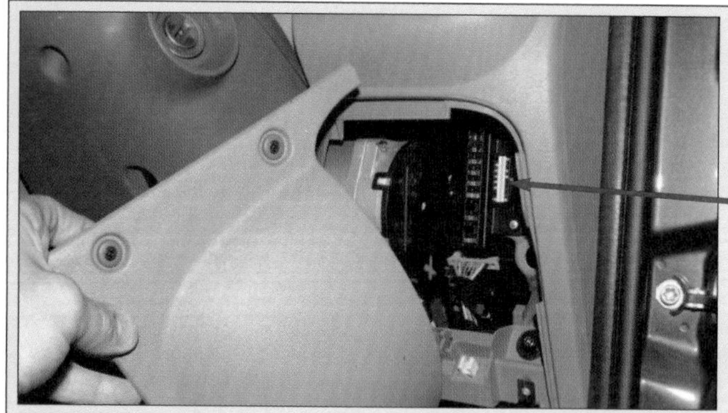

Fiat Panda 2004
Undo the fasteners, and remove the trim panel from the driver's side of the facia

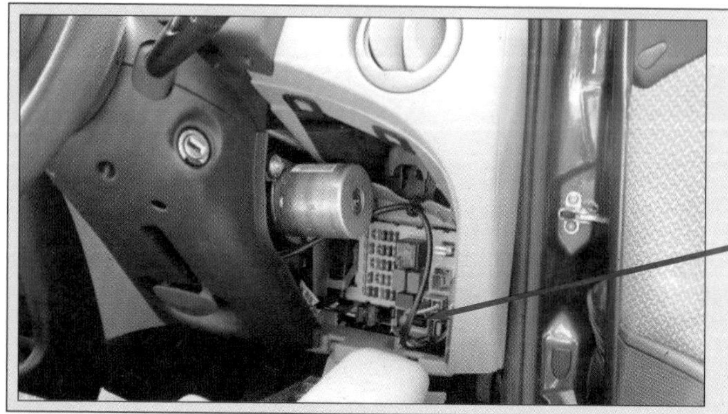

Fiat Punto 2003
Remove the lower facia panel to access the 16-pin connector

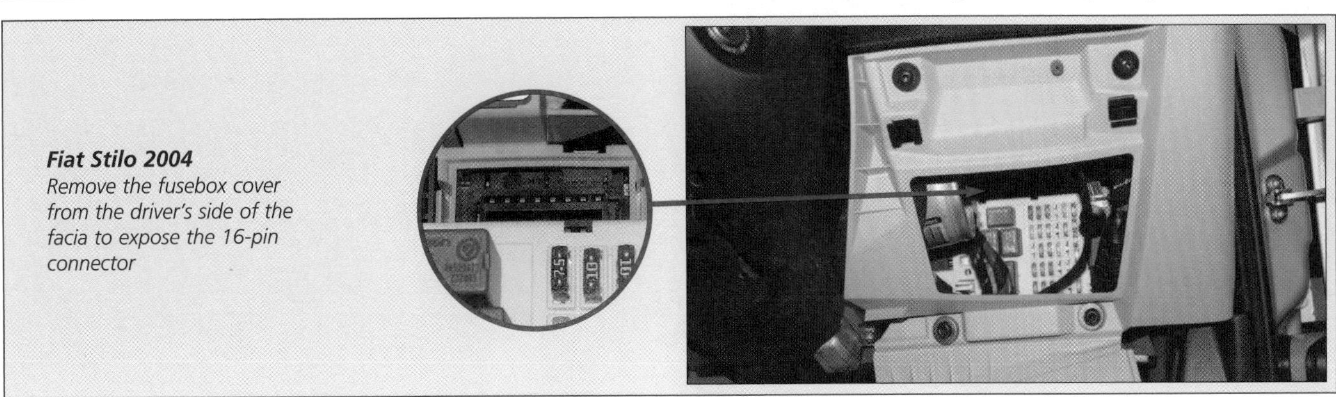

Fiat Stilo 2004
Remove the fusebox cover from the driver's side of the facia to expose the 16-pin connector

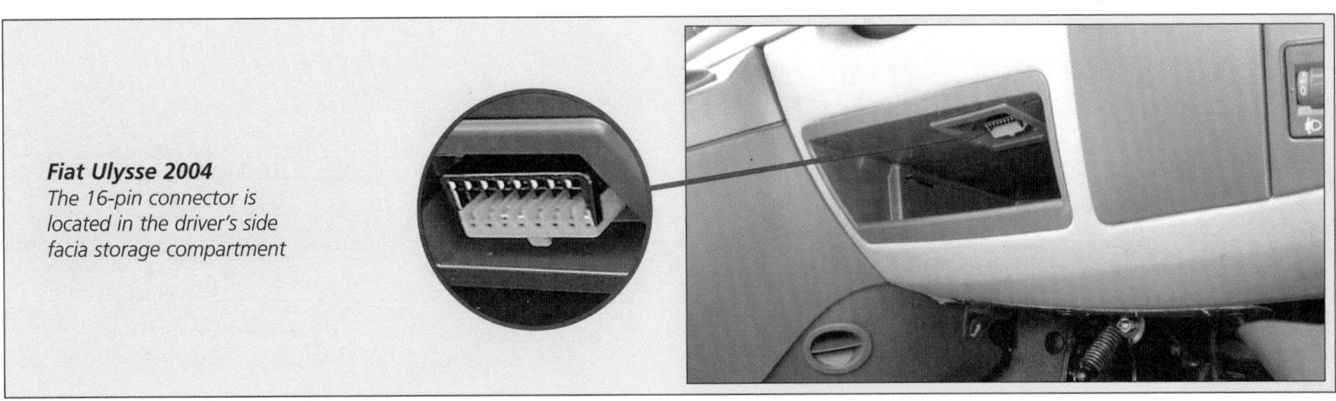

Fiat Ulysse 2004
The 16-pin connector is located in the driver's side facia storage compartment

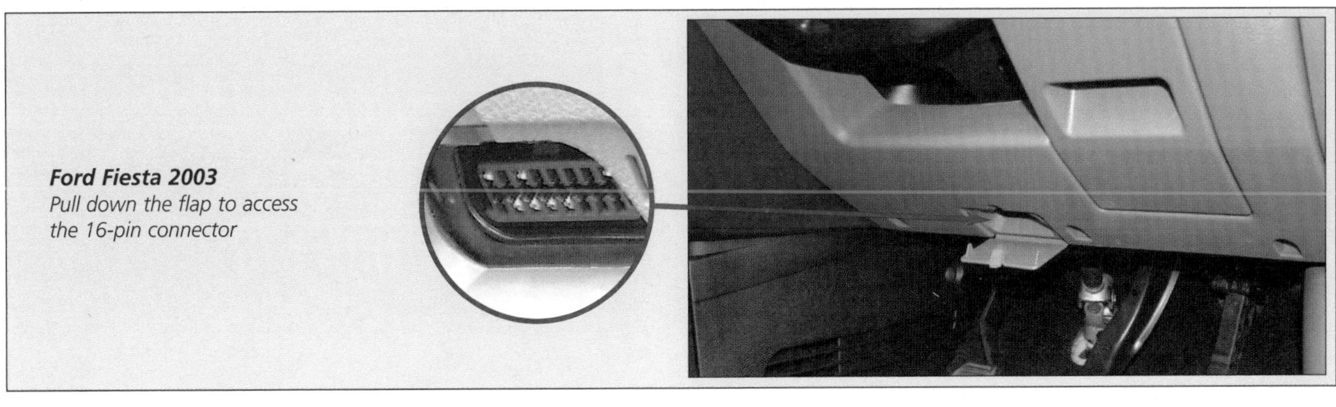

Ford Fiesta 2003
Pull down the flap to access the 16-pin connector

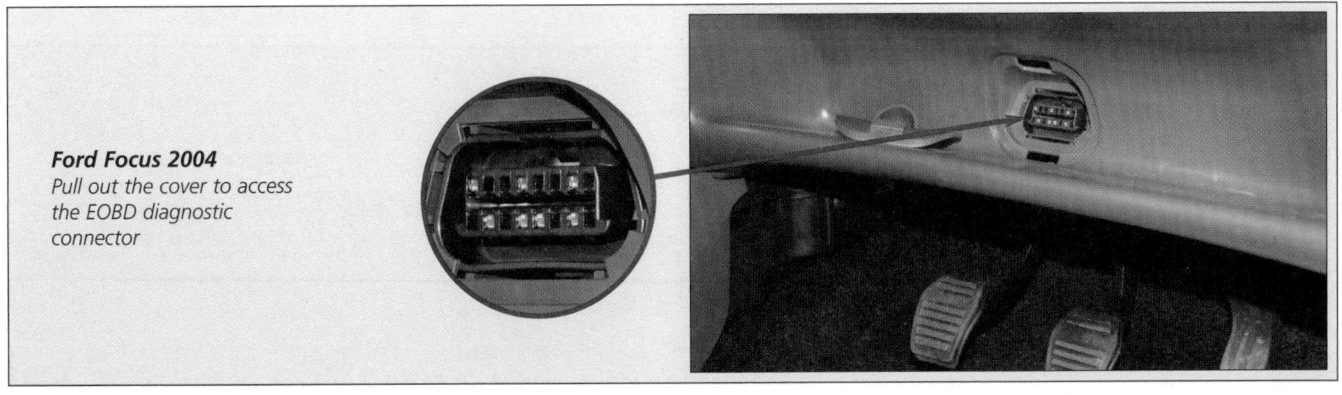

Ford Focus 2004
Pull out the cover to access the EOBD diagnostic connector

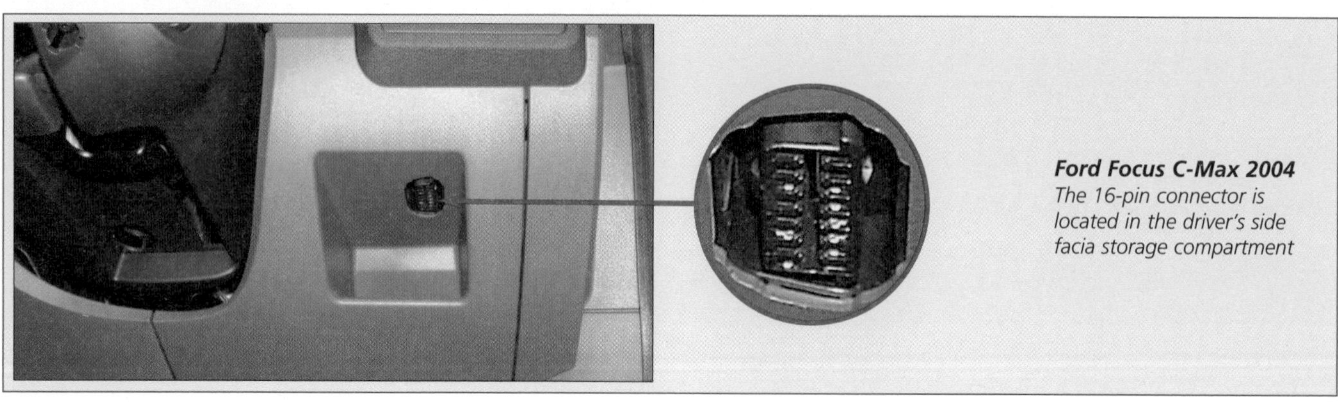

Ford Focus C-Max 2004
The 16-pin connector is located in the driver's side facia storage compartment

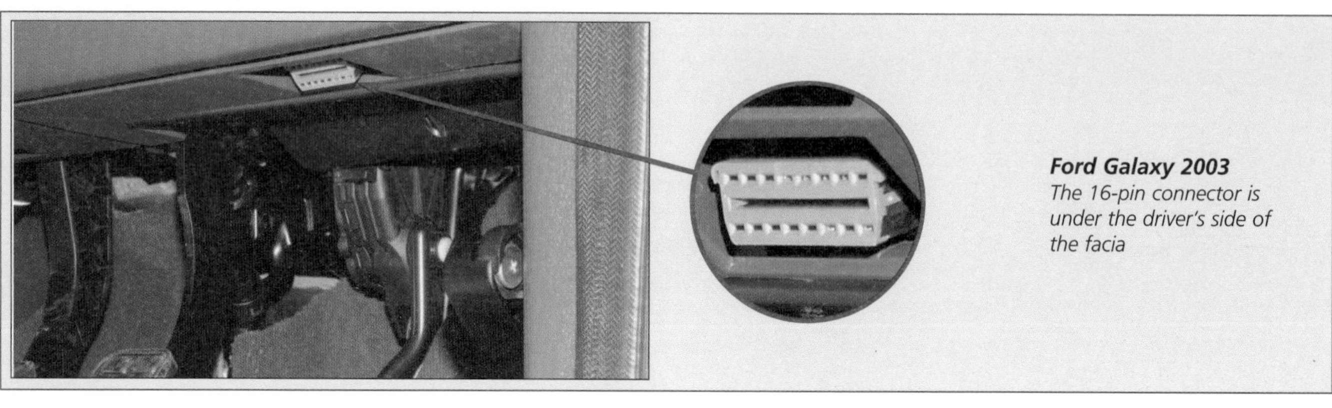

Ford Galaxy 2003
The 16-pin connector is under the driver's side of the facia

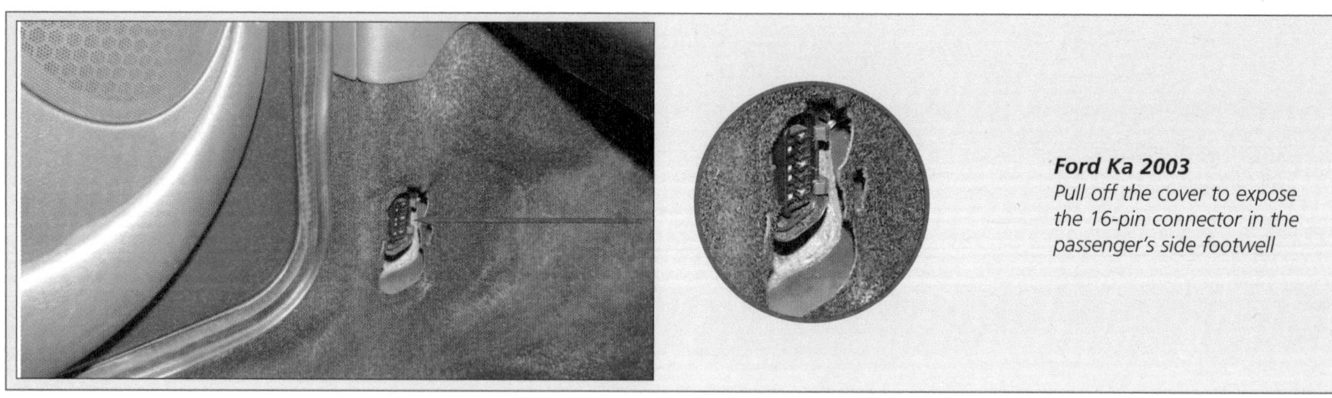

Ford Ka 2003
Pull off the cover to expose the 16-pin connector in the passenger's side footwell

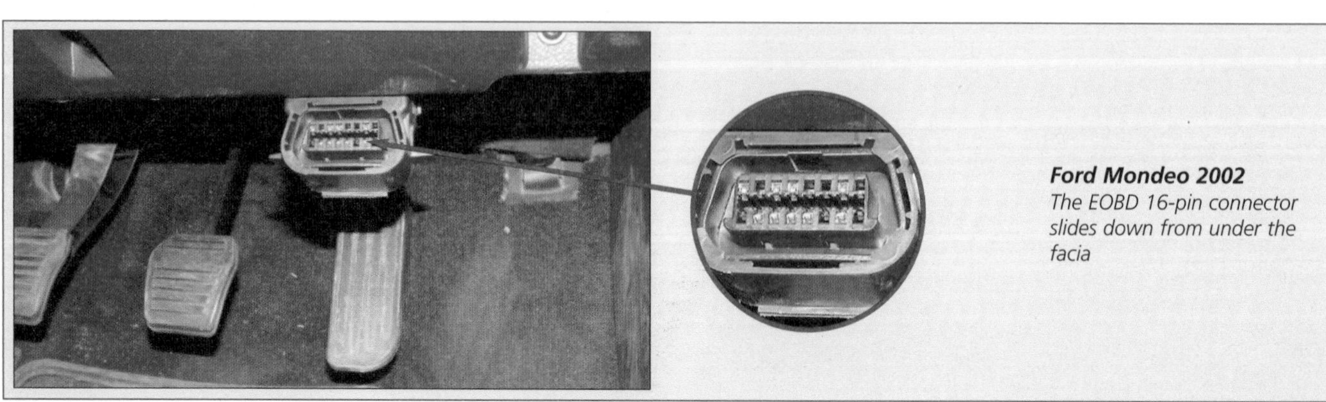

Ford Mondeo 2002
The EOBD 16-pin connector slides down from under the facia

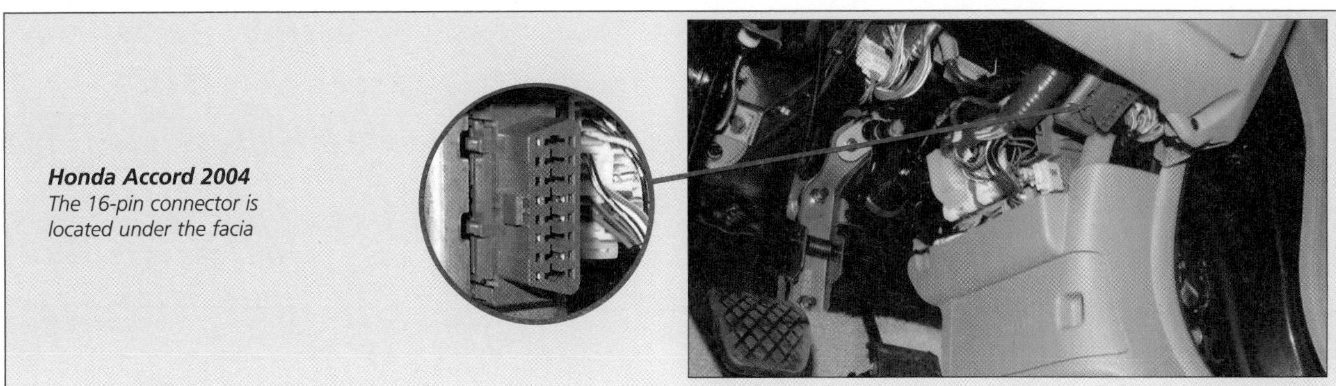

Honda Accord 2004
The 16-pin connector is located under the facia

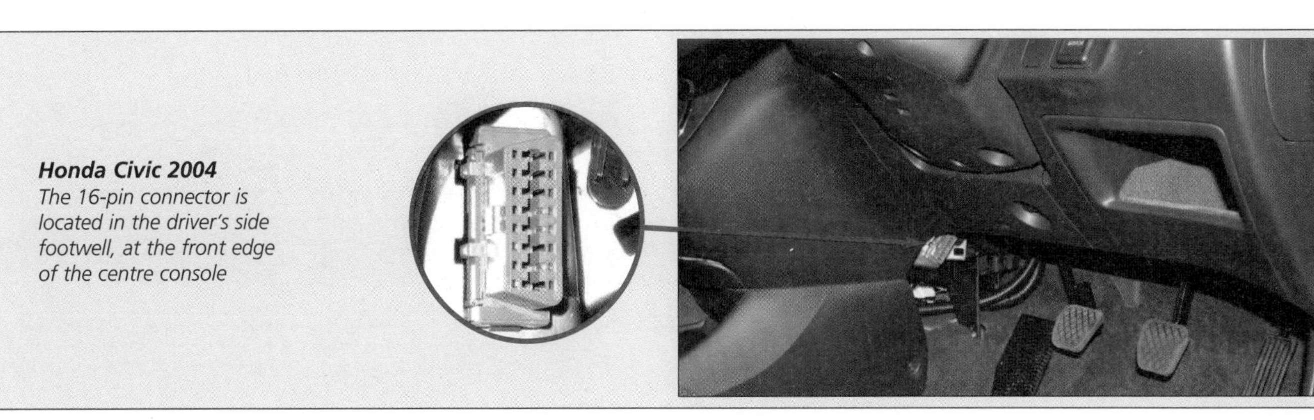

Honda Civic 2004
The 16-pin connector is located in the driver's side footwell, at the front edge of the centre console

Honda CRV 2001
EOBD 16-pin diagnostic connector in the passenger side footwell

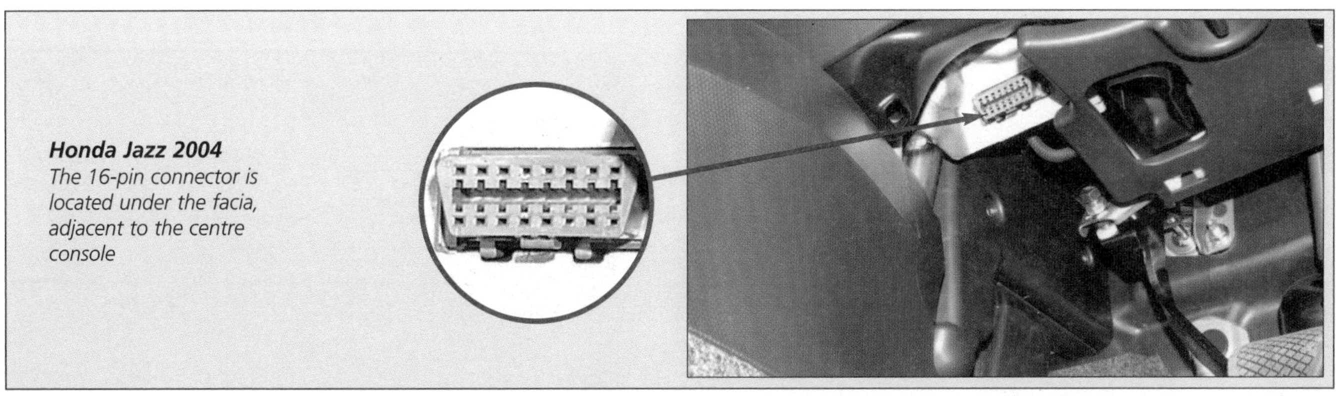

Honda Jazz 2004
The 16-pin connector is located under the facia, adjacent to the centre console

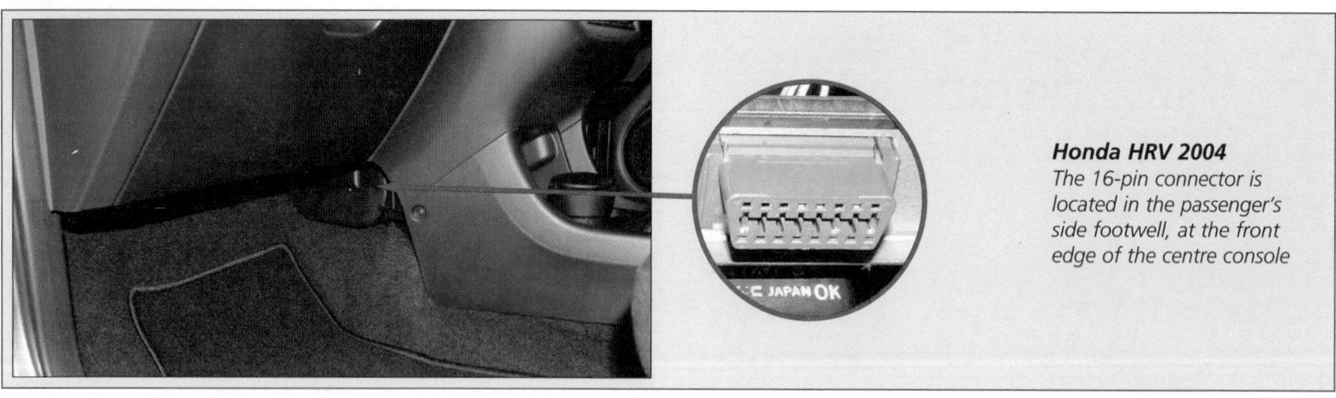

Honda HRV 2004
The 16-pin connector is located in the passenger's side footwell, at the front edge of the centre console

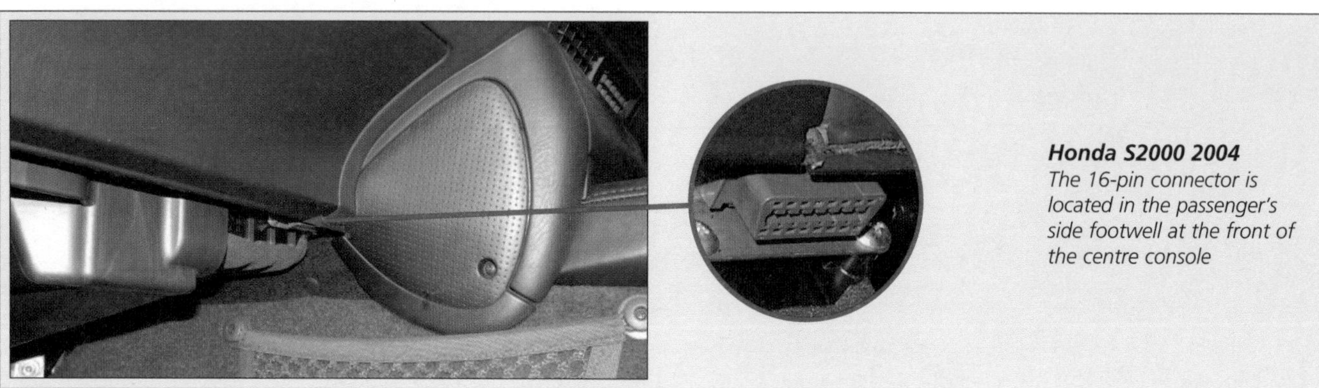

Honda S2000 2004
The 16-pin connector is located in the passenger's side footwell at the front of the centre console

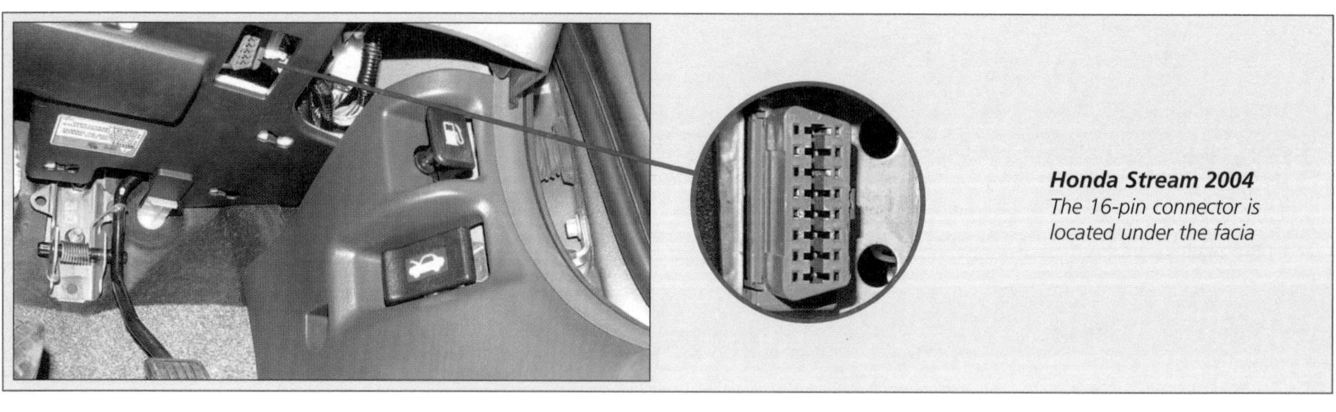

Honda Stream 2004
The 16-pin connector is located under the facia

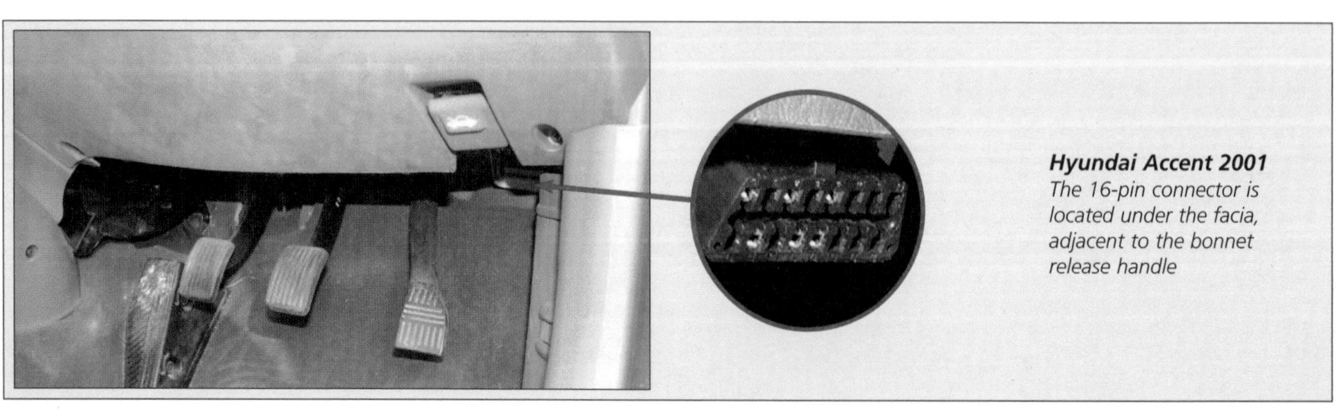

Hyundai Accent 2001
The 16-pin connector is located under the facia, adjacent to the bonnet release handle

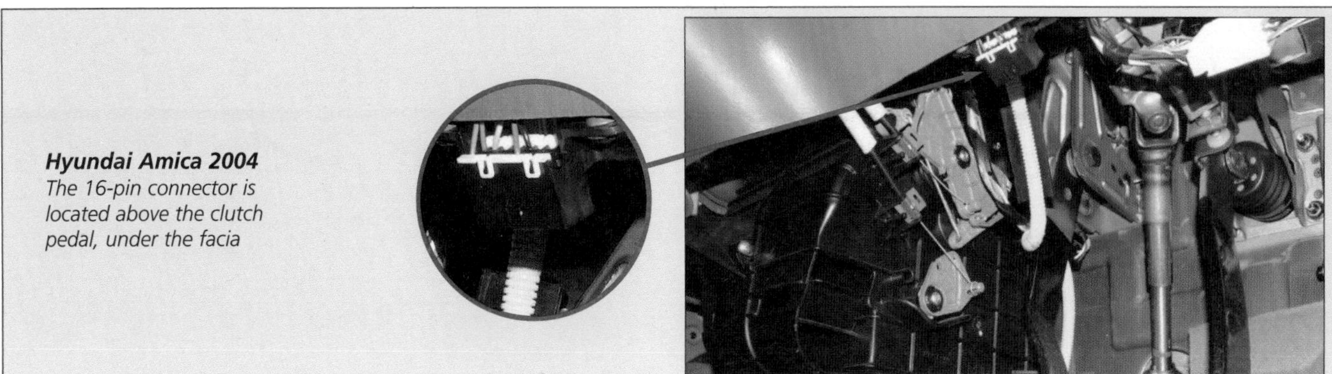

Hyundai Amica 2004
The 16-pin connector is located above the clutch pedal, under the facia

Hyundai Coupe 2004
Remove the fusebox cover from the facia to expose the 16-pin connector

Hyundai Elantra 2004
The 16-pin connector is located under the facia, adjacent to the bonnet release handle

Hyundai Getz 2004
The 16-pin connector is located under the facia, above the driver's pedals

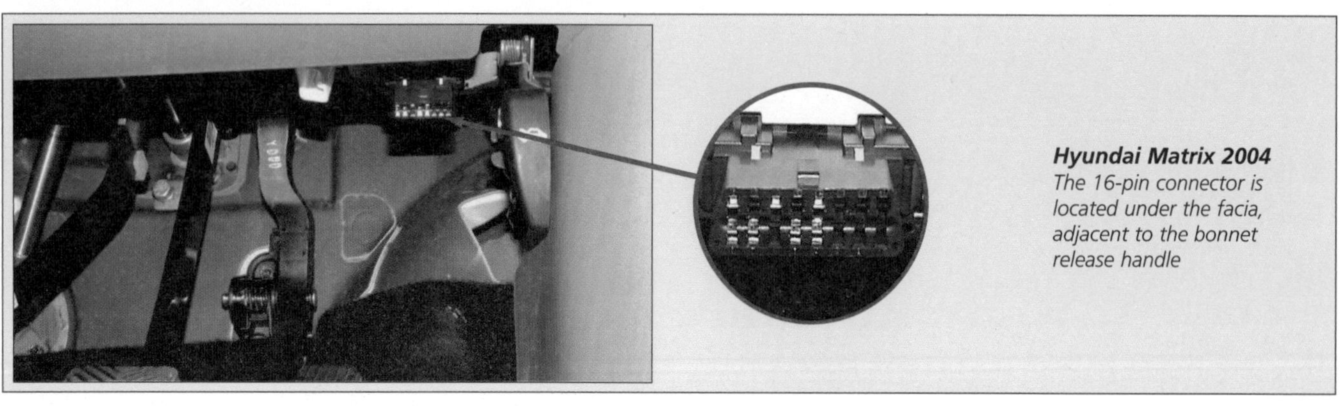

Hyundai Matrix 2004
The 16-pin connector is located under the facia, adjacent to the bonnet release handle

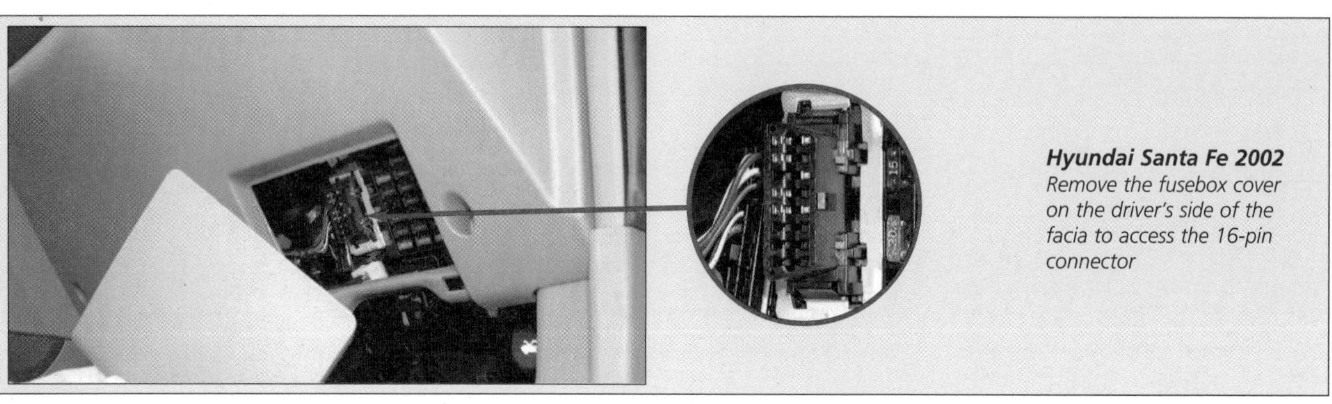

Hyundai Santa Fe 2002
Remove the fusebox cover on the driver's side of the facia to access the 16-pin connector

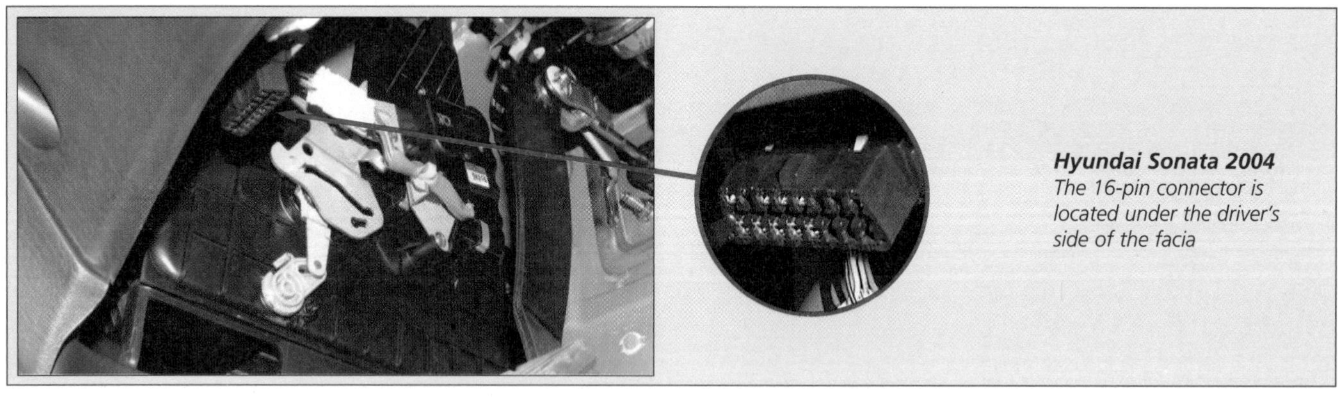

Hyundai Sonata 2004
The 16-pin connector is located under the driver's side of the facia

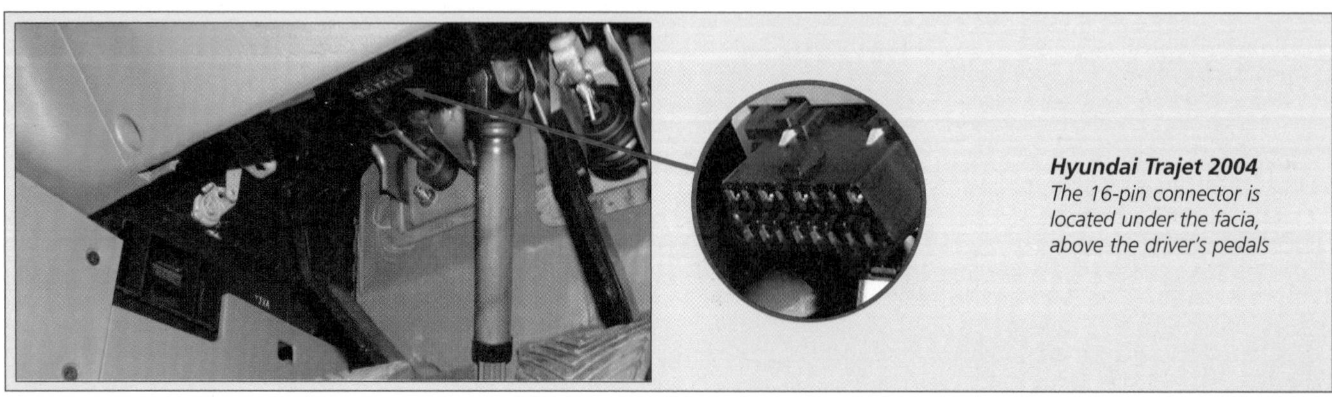

Hyundai Trajet 2004
The 16-pin connector is located under the facia, above the driver's pedals

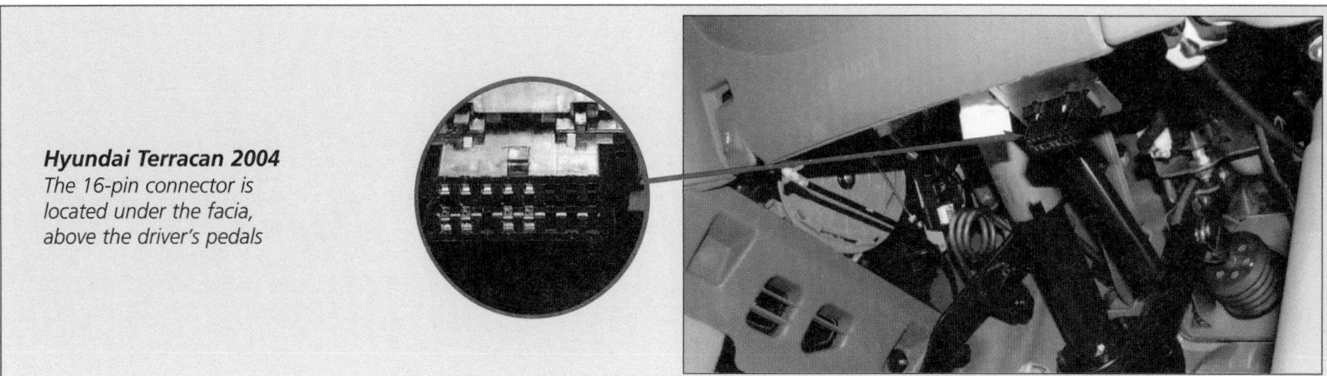

Hyundai Terracan 2004
The 16-pin connector is located under the facia, above the driver's pedals

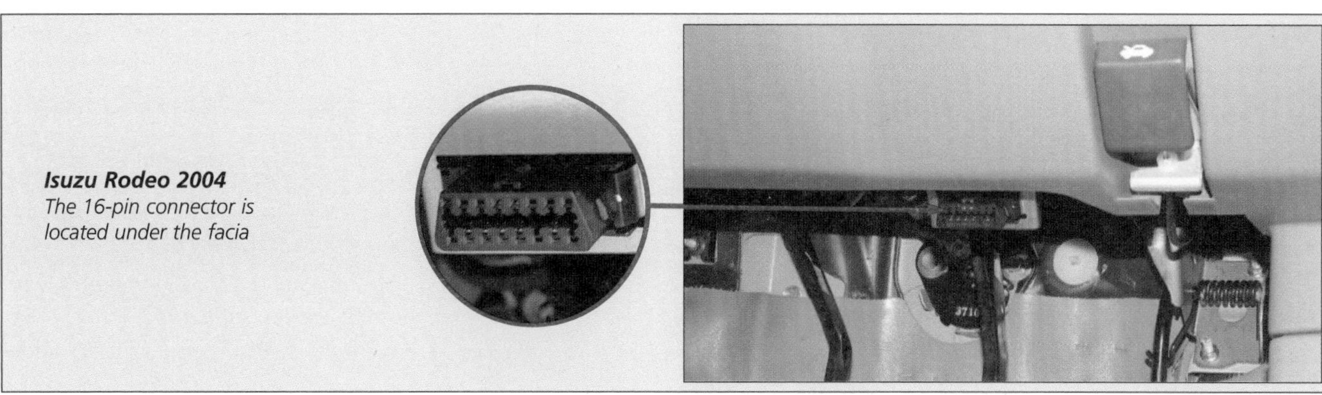

Isuzu Rodeo 2004
The 16-pin connector is located under the facia

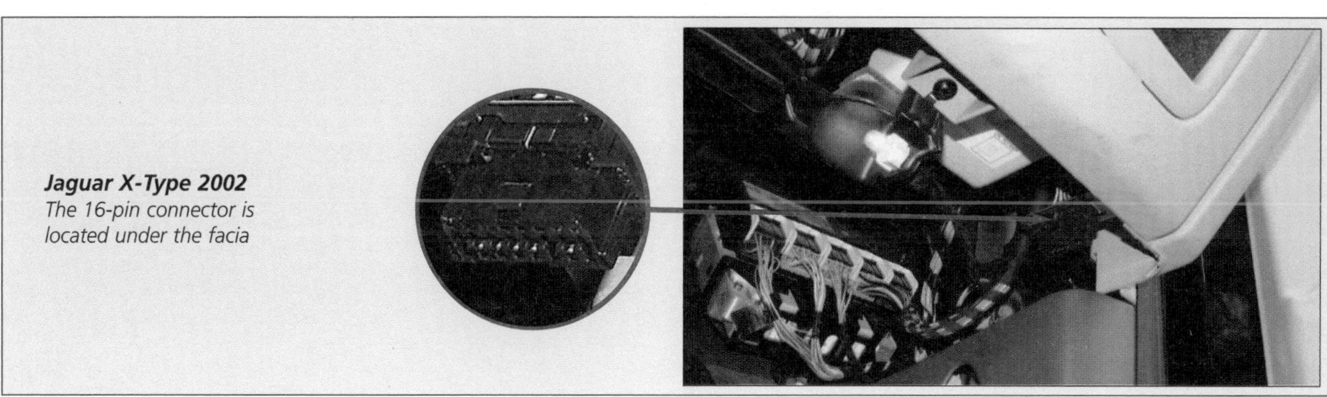

Jaguar X-Type 2002
The 16-pin connector is located under the facia

Kia Carens 2004
The 16-pin connector is located under the driver's side of the facia

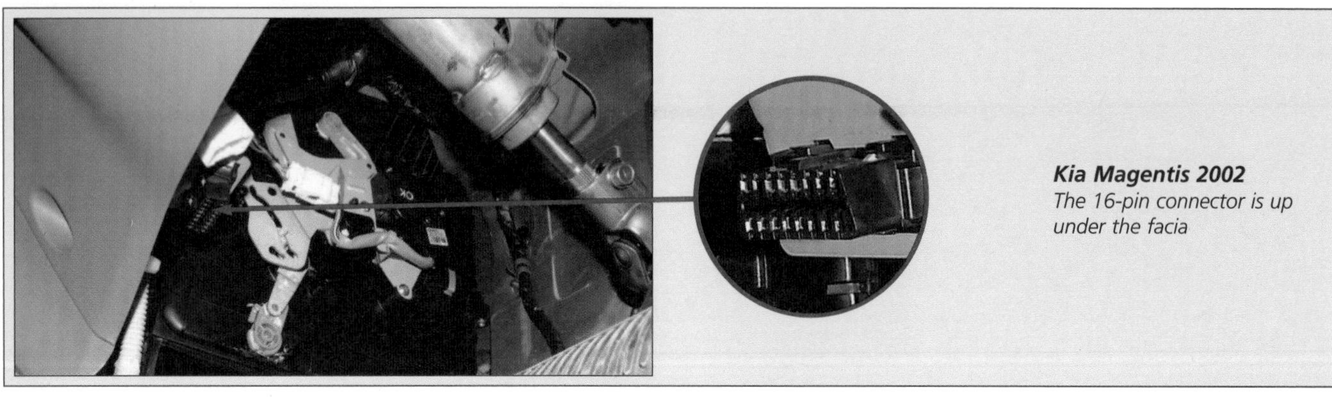
Kia Magentis 2002
The 16-pin connector is up under the facia

Kia Rio 2004
The 16-pin connector is located under the driver's side of the facia

Kia Sedona 2004
The 16-pin connector is located under the driver's side of the facia

Kia Sorento 2004
The 16-pin connector is located under the driver's side of the facia

Land Rover Freelander 2004
EOBD 16-pin connector in the passenger's footwell at the front of the centre console (note the plastic cover)

Land Rover Discovery 2004
EOBD 16-pin connector under the facia

Land Rover Defender 2001
EOBD 16-pin connector between the front seats

Range Rover 2004
EOBD 16-pin connector at the outer edge of the facia

Lexus IS200 2002
The 16-pin connector is located under the facia, adjacent to the bonnet release handle

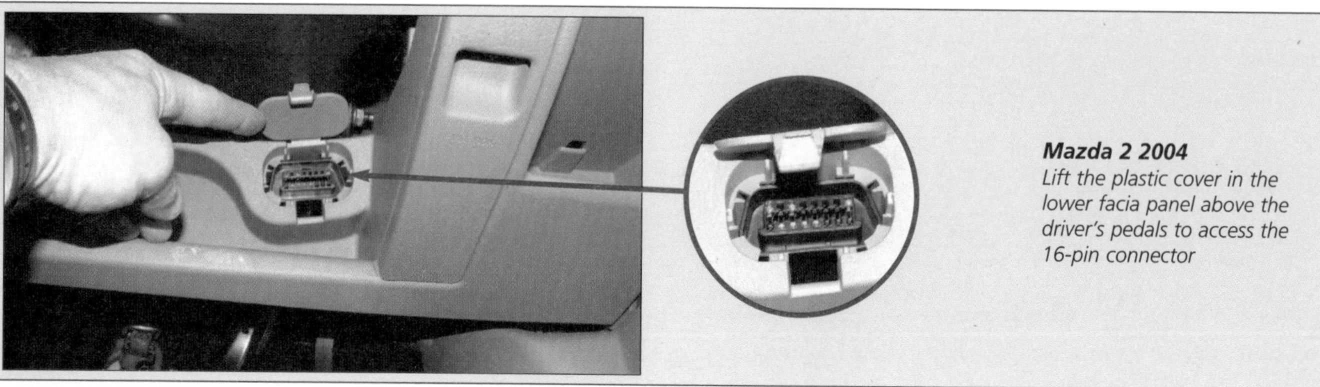

Mazda 2 2004
Lift the plastic cover in the lower facia panel above the driver's pedals to access the 16-pin connector

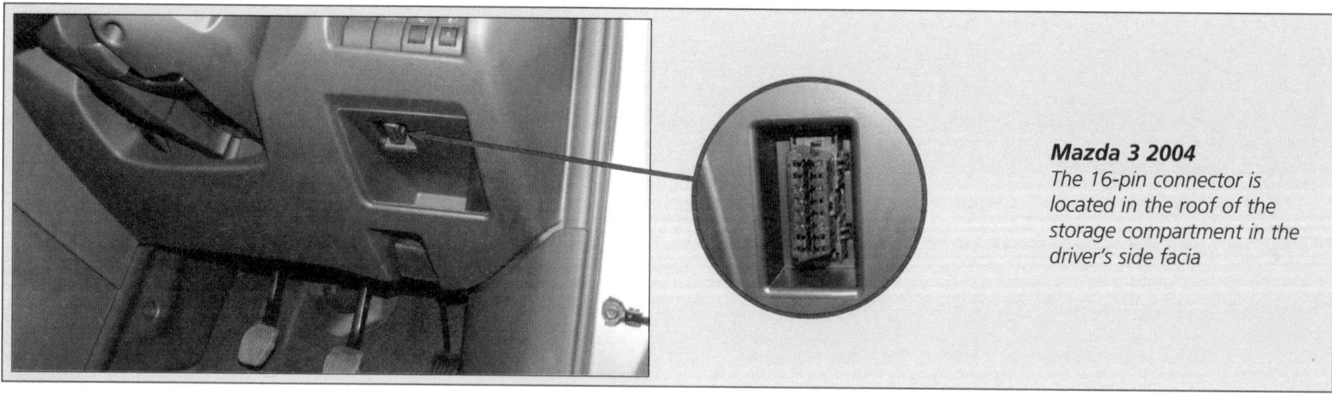

Mazda 3 2004
The 16-pin connector is located in the roof of the storage compartment in the driver's side facia

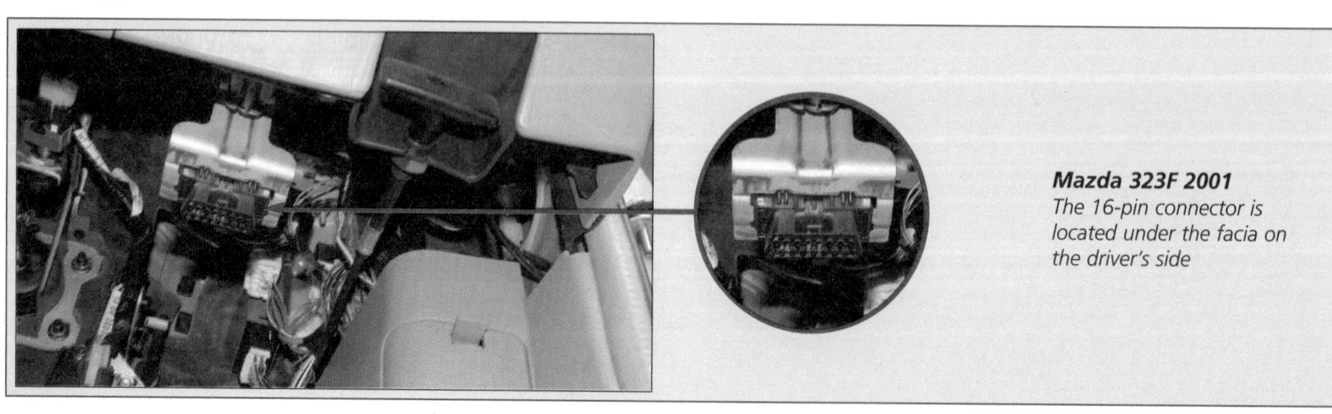

Mazda 323F 2001
The 16-pin connector is located under the facia on the driver's side

Mazda 6 2004
The 16-pin connector is located above the driver's pedals

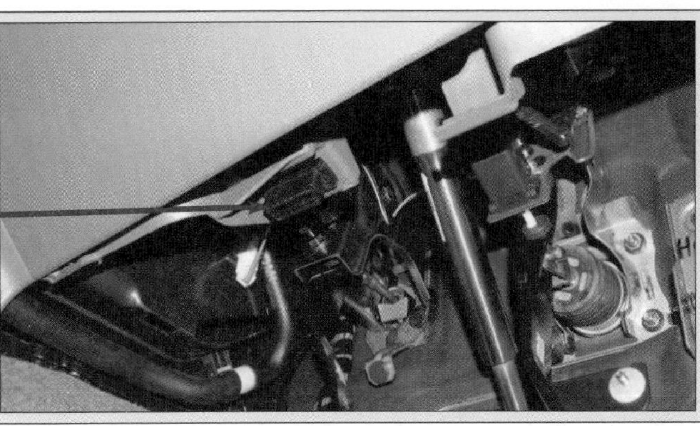

Mazda MX5 2004
Remove the fusebox cover above the bonnet release handle to access the 16-pin connector

Mazda RX8 2004
The 16-pin connector is located adjacent to the bonnet release handle

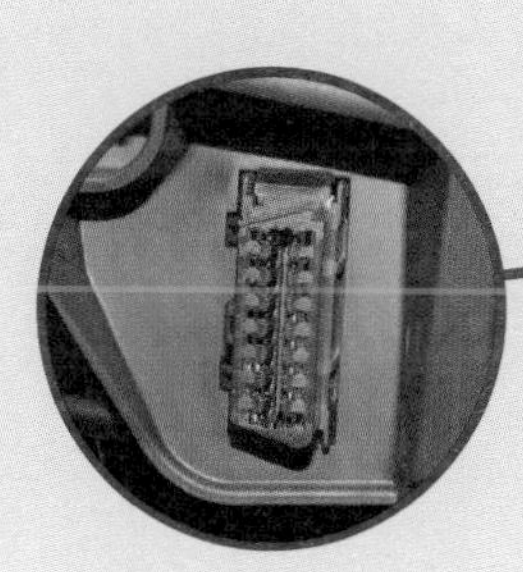

Mercedes A-Class 2001
Pull down the cover to expose the 16-pin connector under the driver's side of the facia

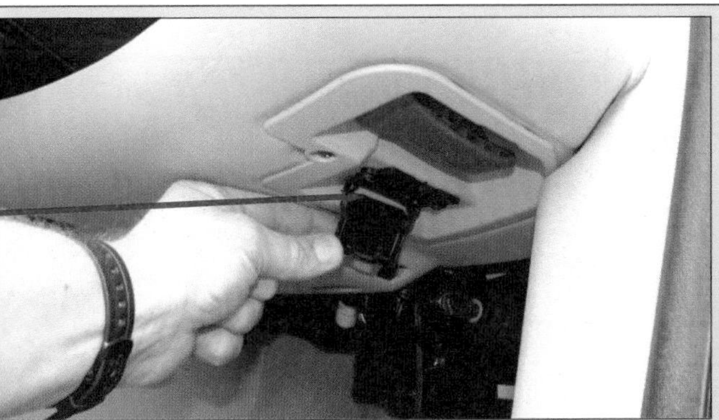

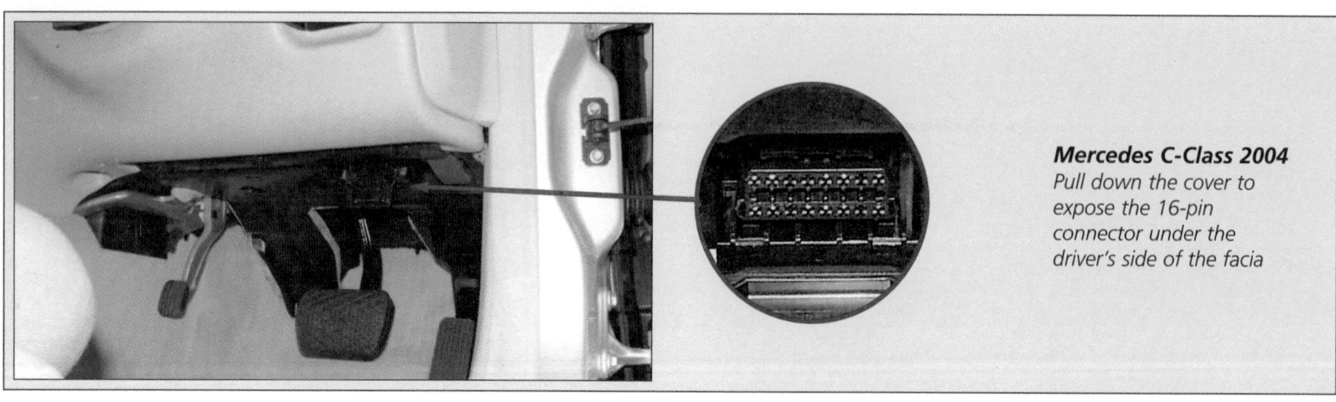

Mercedes C-Class 2004
Pull down the cover to expose the 16-pin connector under the driver's side of the facia

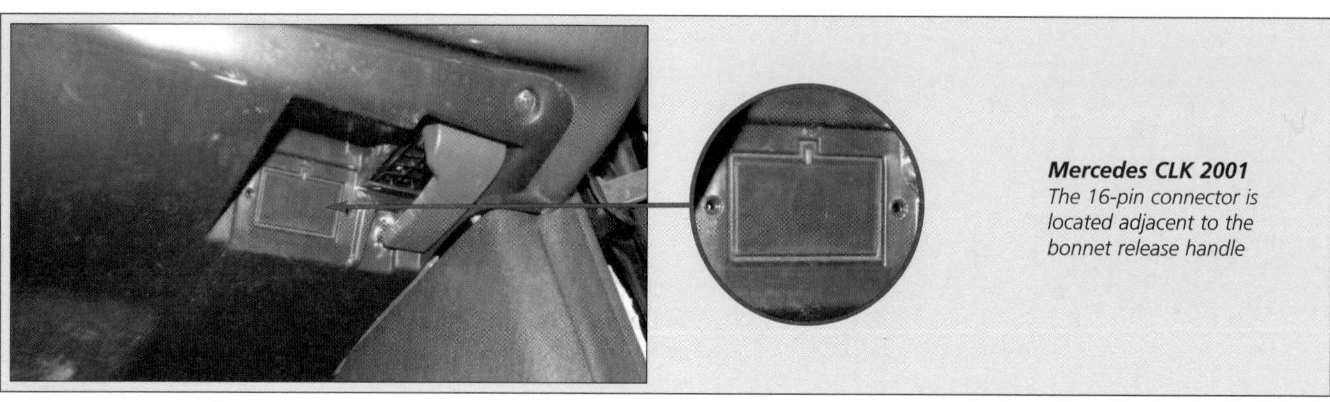

Mercedes CLK 2001
The 16-pin connector is located adjacent to the bonnet release handle

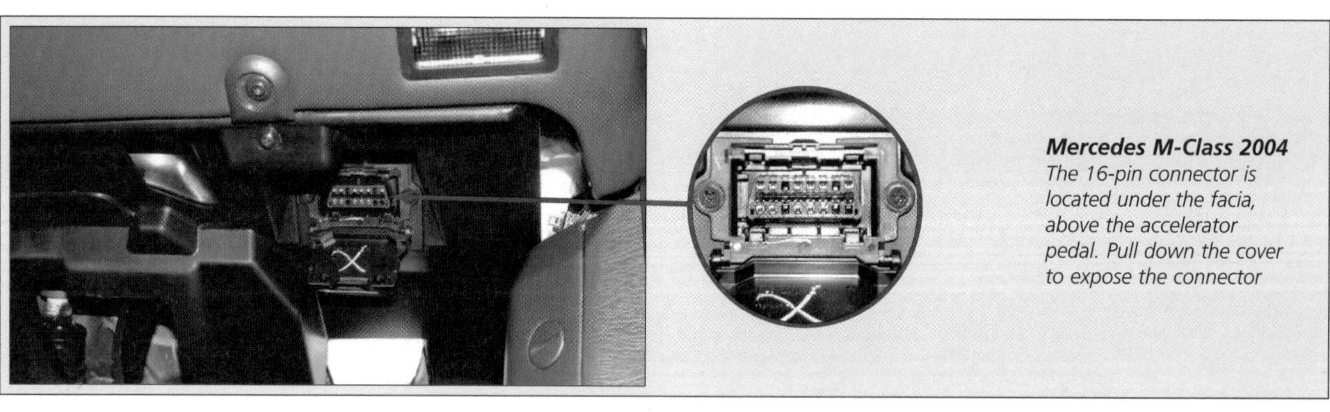

Mercedes M-Class 2004
The 16-pin connector is located under the facia, above the accelerator pedal. Pull down the cover to expose the connector

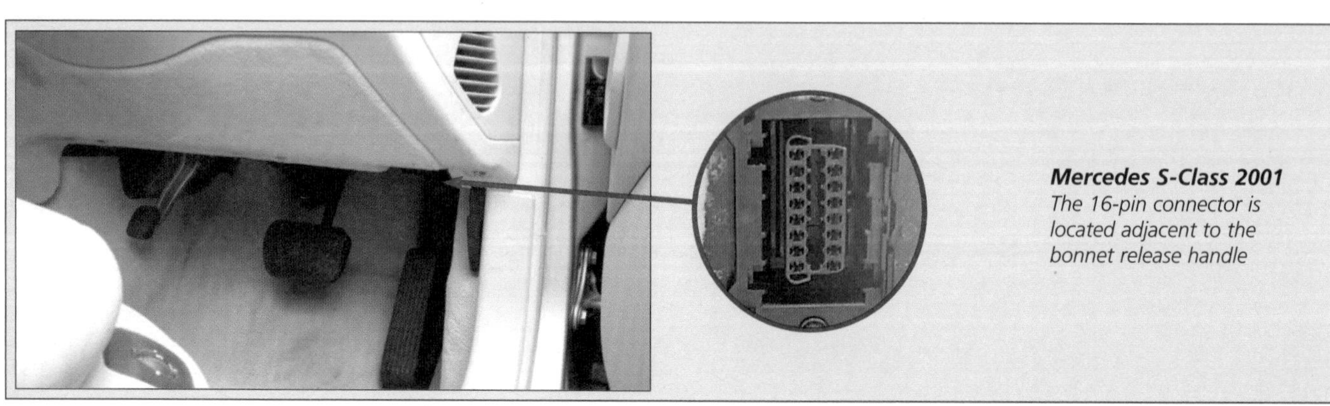

Mercedes S-Class 2001
The 16-pin connector is located adjacent to the bonnet release handle

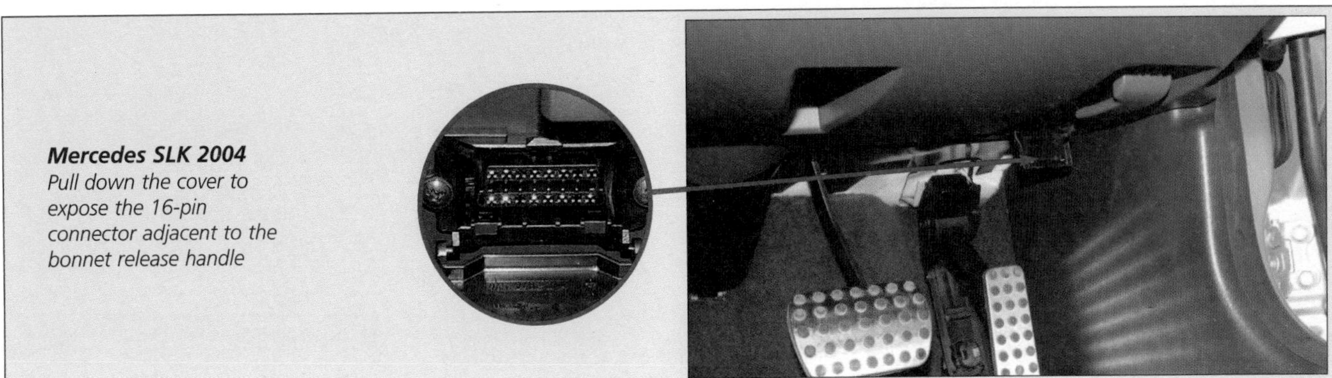

Mercedes SLK 2004
Pull down the cover to expose the 16-pin connector adjacent to the bonnet release handle

Mitsubishi Carisma 2004
The 16-pin connector is located in the passenger's side footwell, at the front edge of the centre console

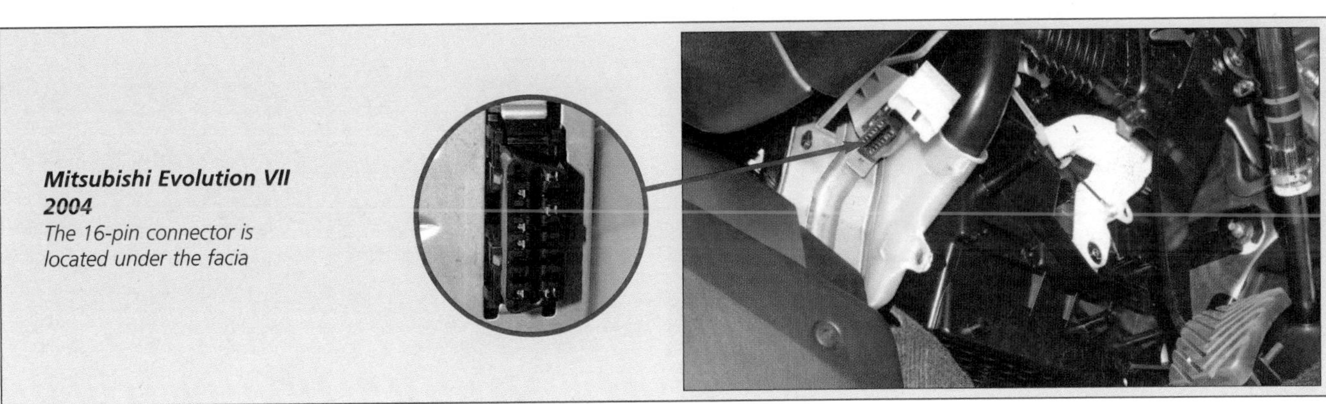

Mitsubishi Evolution VII 2004
The 16-pin connector is located under the facia

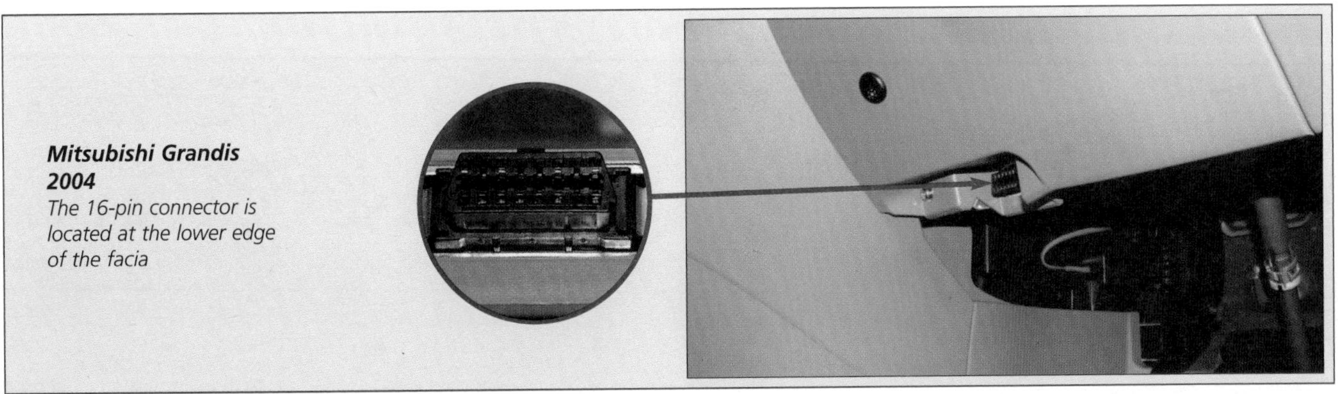

Mitsubishi Grandis 2004
The 16-pin connector is located at the lower edge of the facia

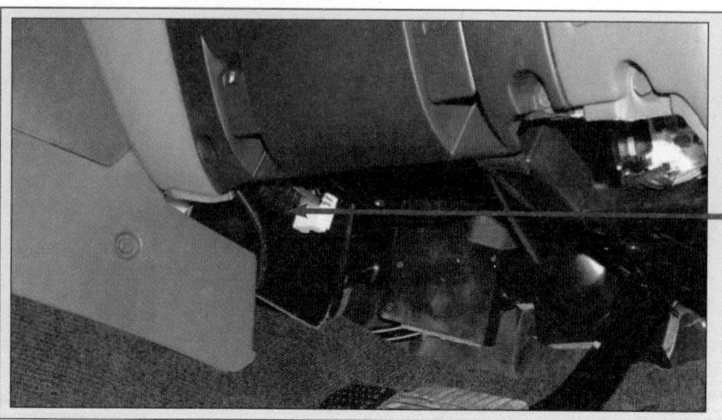

Mitsubishi Shogun Warrior 2004
The 16-pin connector is located under the facia

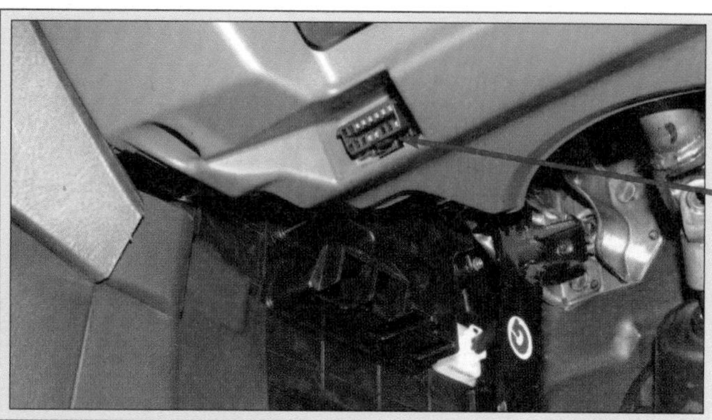

Nissan Almera 2004
The 16-pin connector is located under the facia

Nissan Micra 2004
Remove the panel on the driver's side of the facia to expose the 16-pin plug

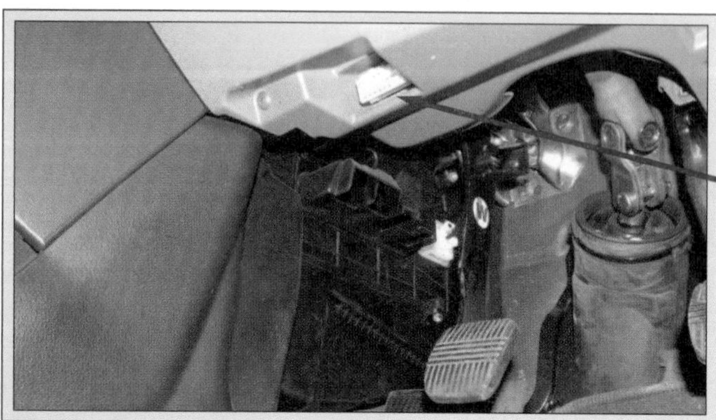
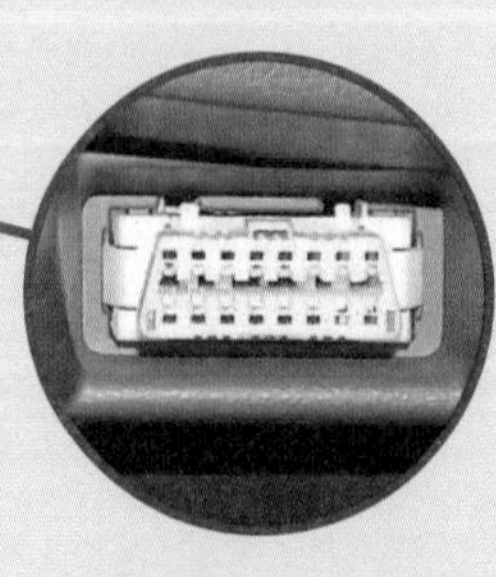

Nissan Primera 2002
The 16-pin connector is located at the lower edge of the facia

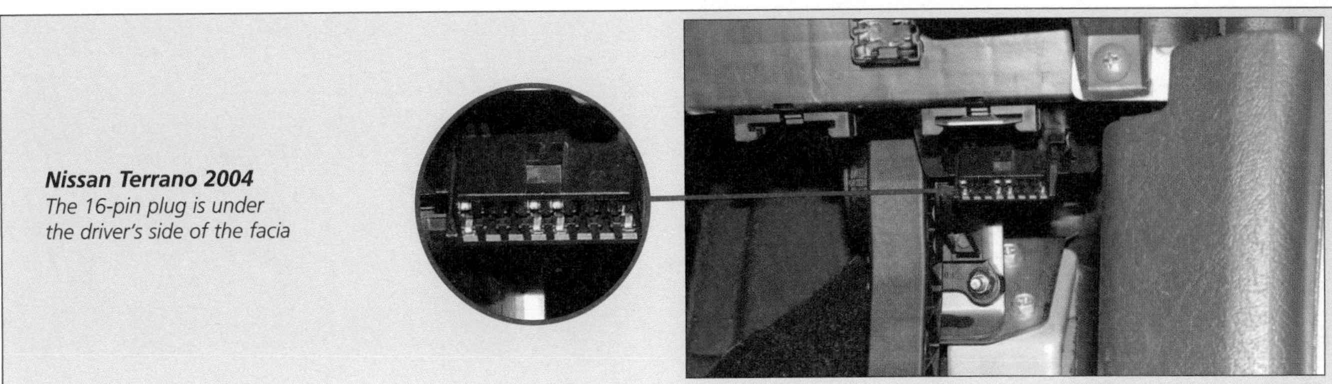

Nissan Terrano 2004
The 16-pin plug is under the driver's side of the facia

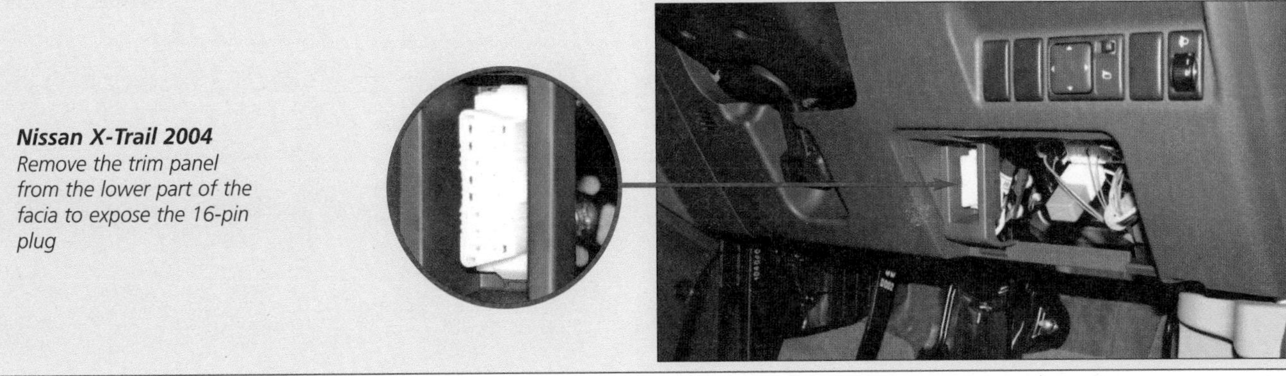

Nissan X-Trail 2004
Remove the trim panel from the lower part of the facia to expose the 16-pin plug

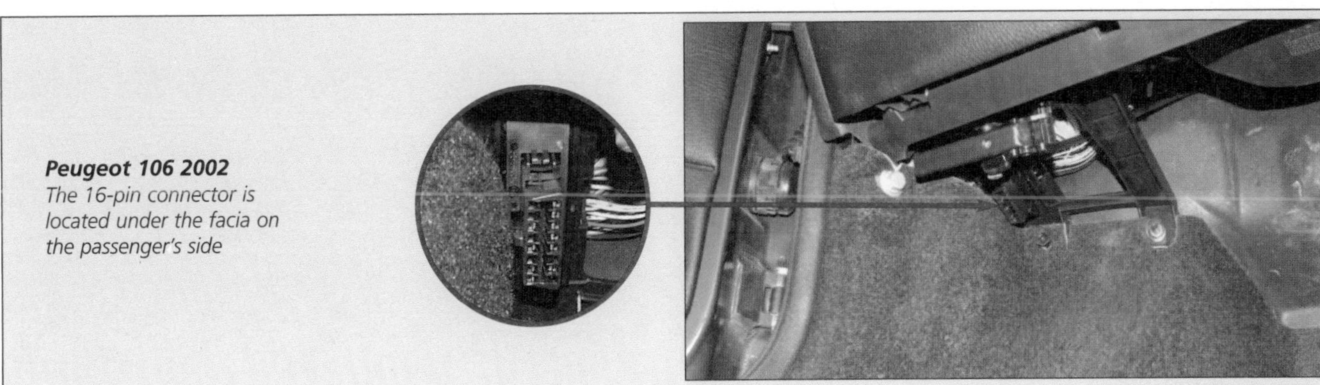

Peugeot 106 2002
The 16-pin connector is located under the facia on the passenger's side

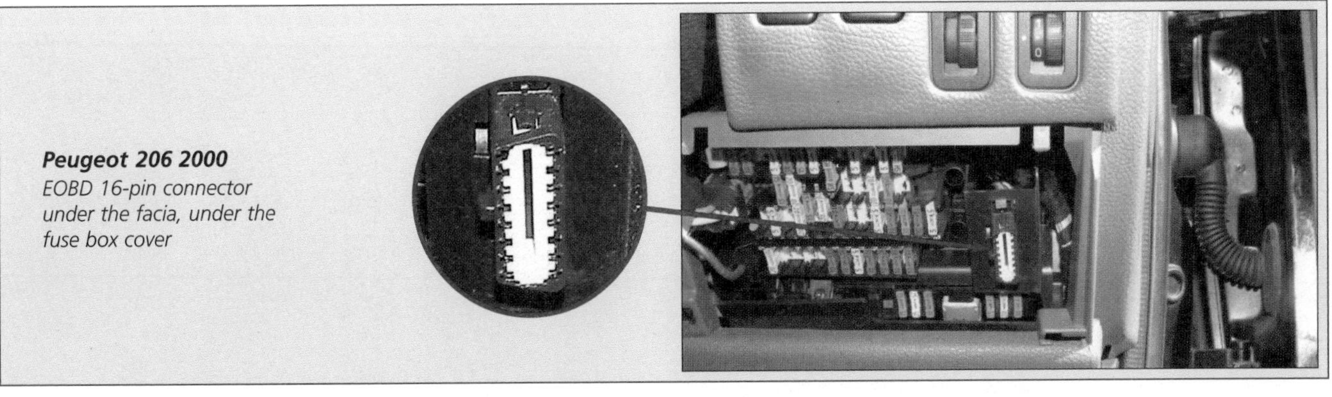

Peugeot 206 2000
EOBD 16-pin connector under the facia, under the fuse box cover

Peugeot 307 2003
Use a coin to undo the fastener, then remove the trim panel from the facia to expose the 16-pin plug

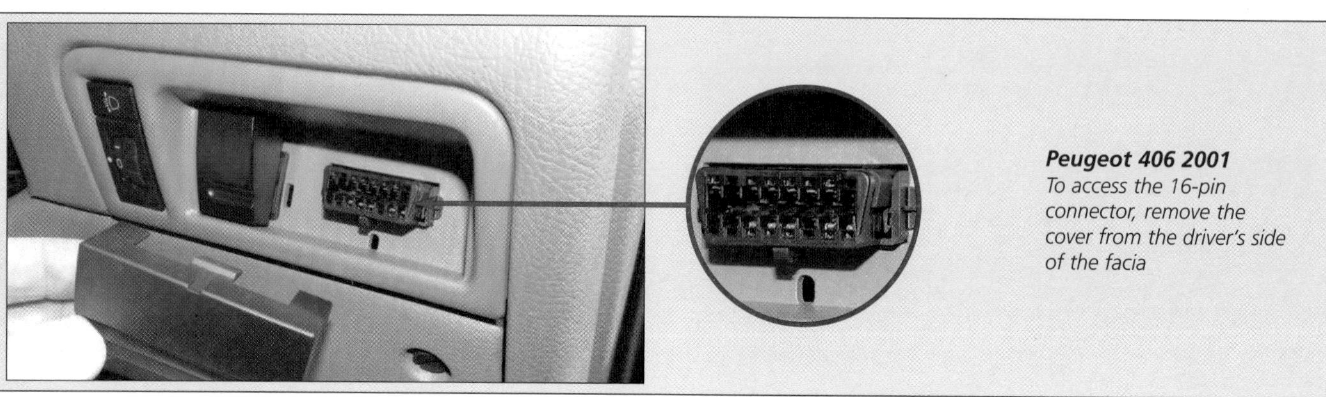

Peugeot 406 2001
To access the 16-pin connector, remove the cover from the driver's side of the facia

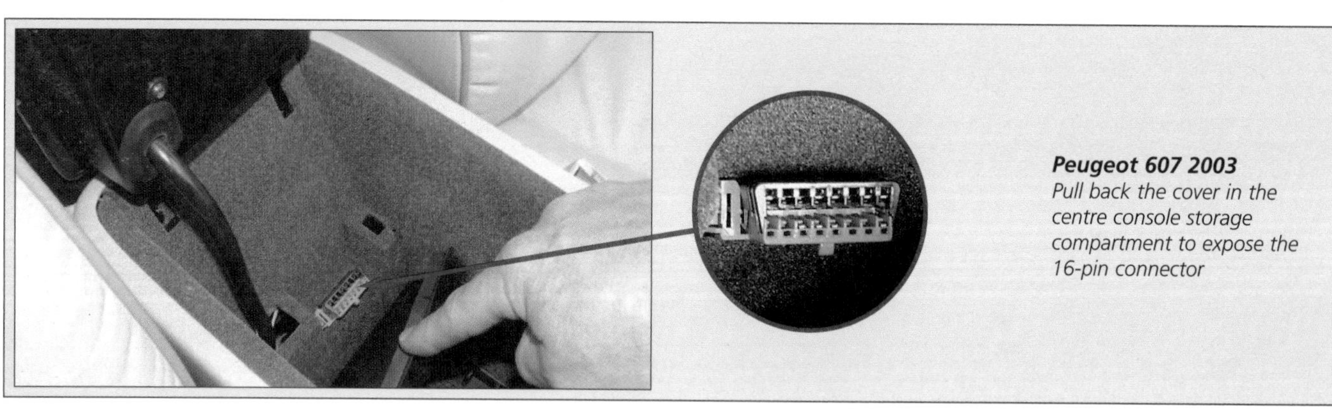

Peugeot 607 2003
Pull back the cover in the centre console storage compartment to expose the 16-pin connector

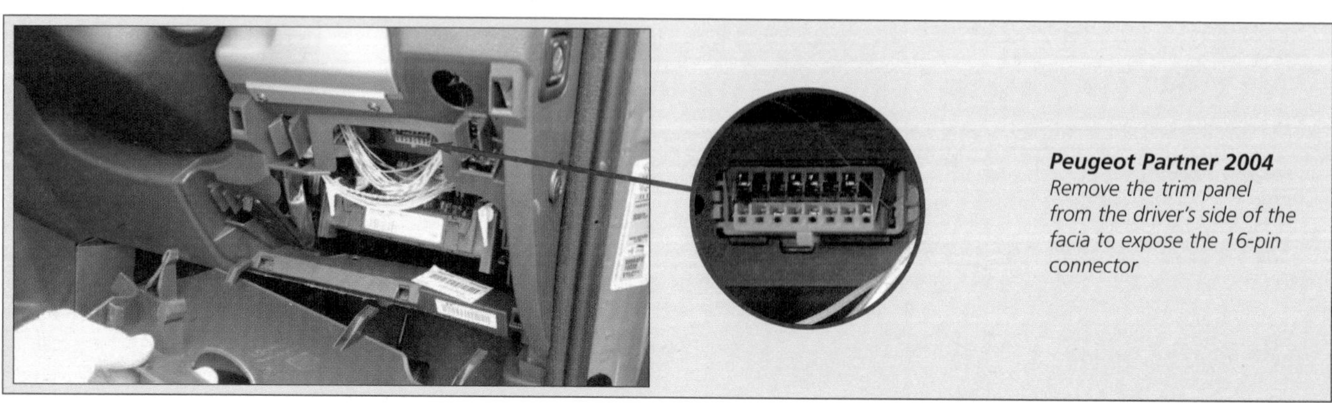

Peugeot Partner 2004
Remove the trim panel from the driver's side of the facia to expose the 16-pin connector

Proton Impian 2004
Pull down the carpet at the front of the centre console to access the 16-pin connector

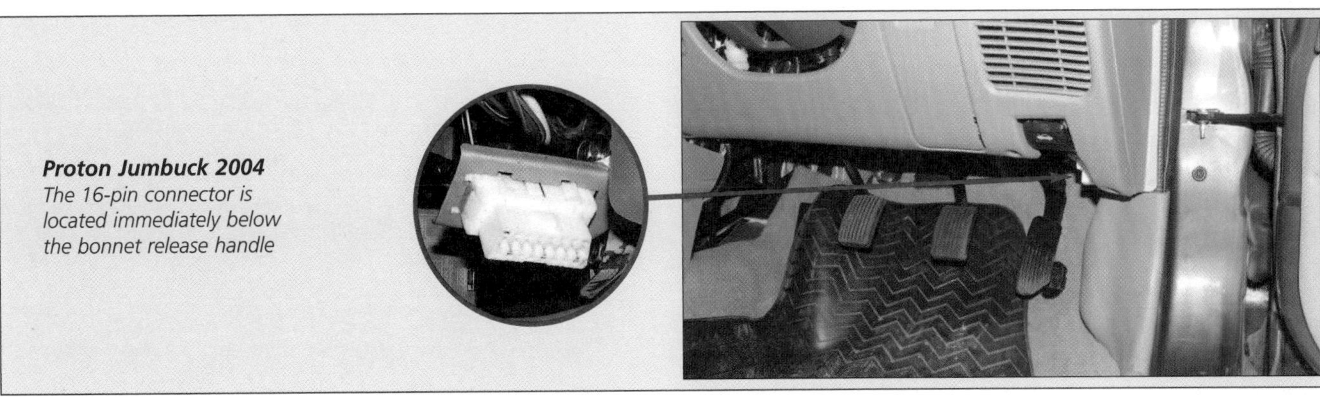

Proton Jumbuck 2004
The 16-pin connector is located immediately below the bonnet release handle

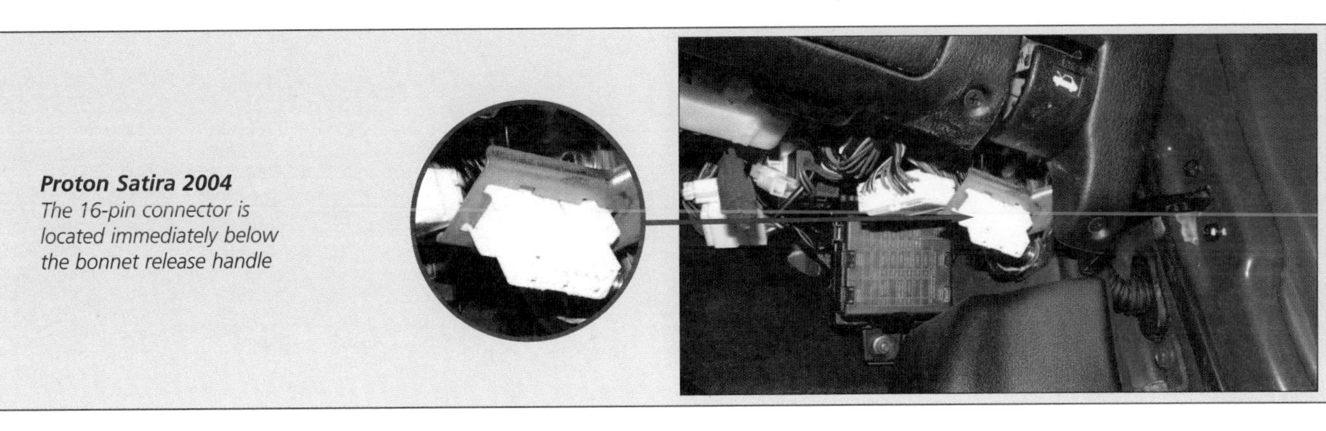

Proton Satira 2004
The 16-pin connector is located immediately below the bonnet release handle

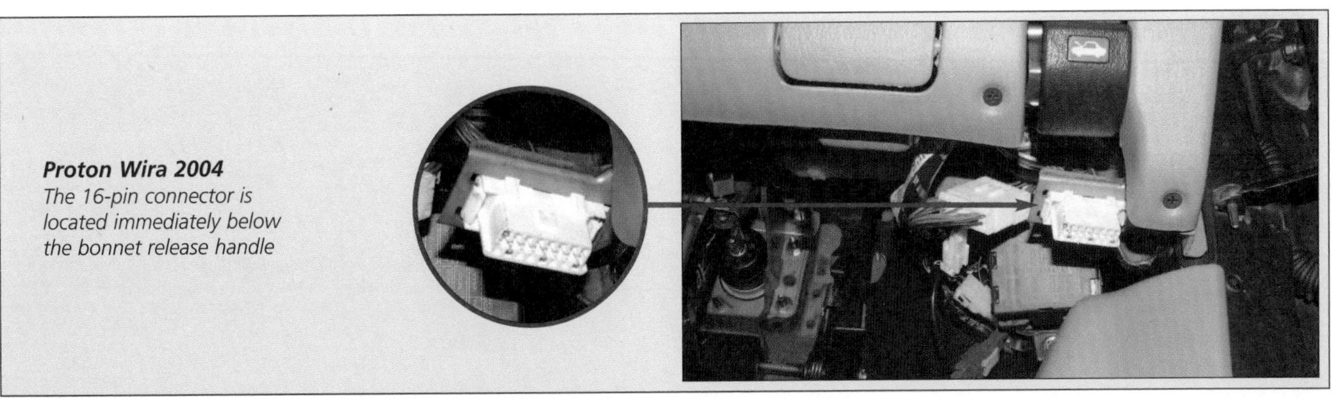

Proton Wira 2004
The 16-pin connector is located immediately below the bonnet release handle

Renault Clio 2001
Remove the ashtray in the centre console to access the 16-pin connector

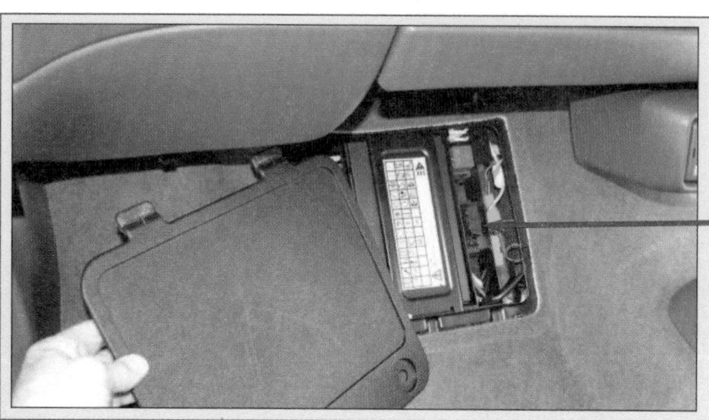

Renault Espace 2002
Remove the fuse box cover from the passenger's side footwell to expose the 16-pin connector

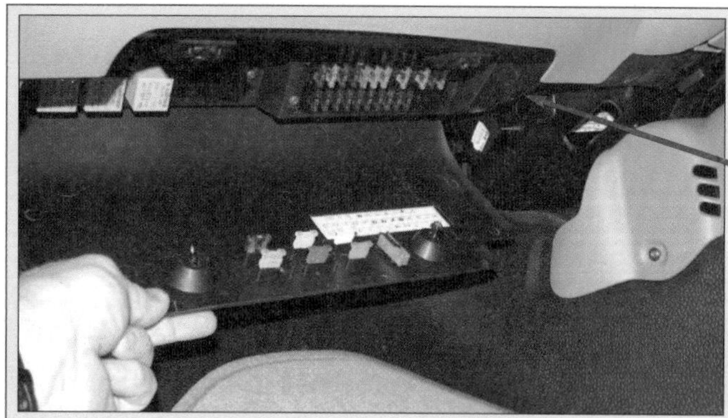

Renault Kangoo 2004
Open the fusebox cover on the passenger's side of the facia to expose the 16-pin connector

Renault Laguna 2004
Remove the ashtray/storage compartment from the centre console, the pull up the plastic cover to access the 16-pin connector

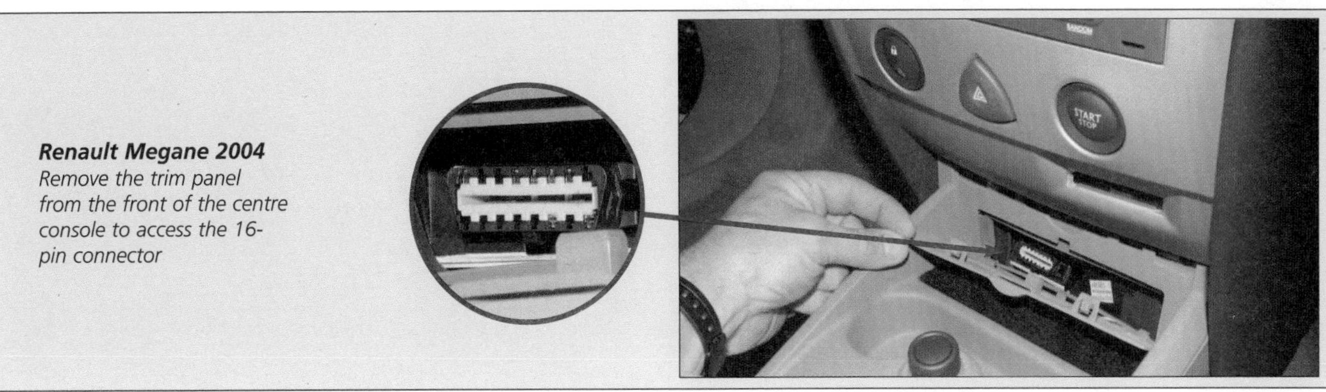

Renault Megane 2004
Remove the trim panel from the front of the centre console to access the 16-pin connector

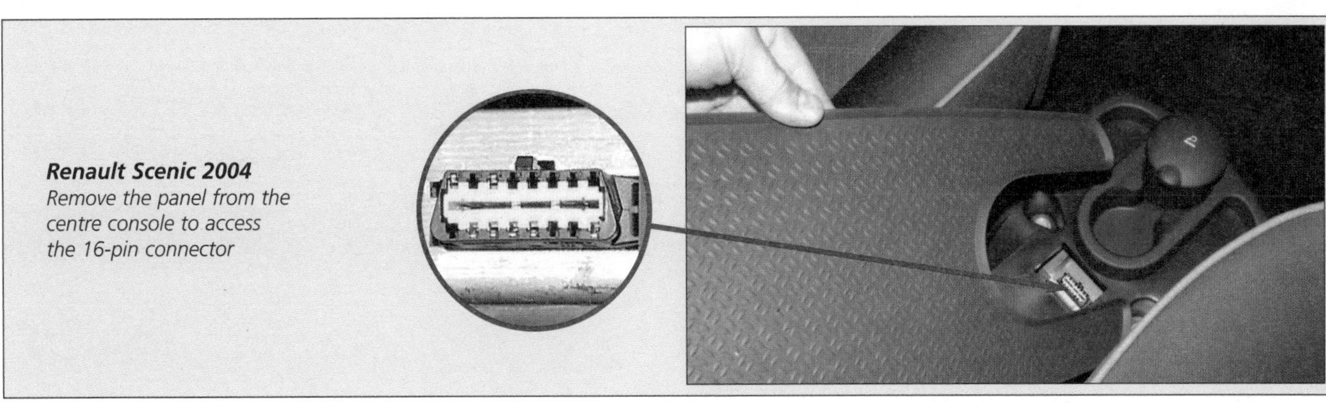

Renault Scenic 2004
Remove the panel from the centre console to access the 16-pin connector

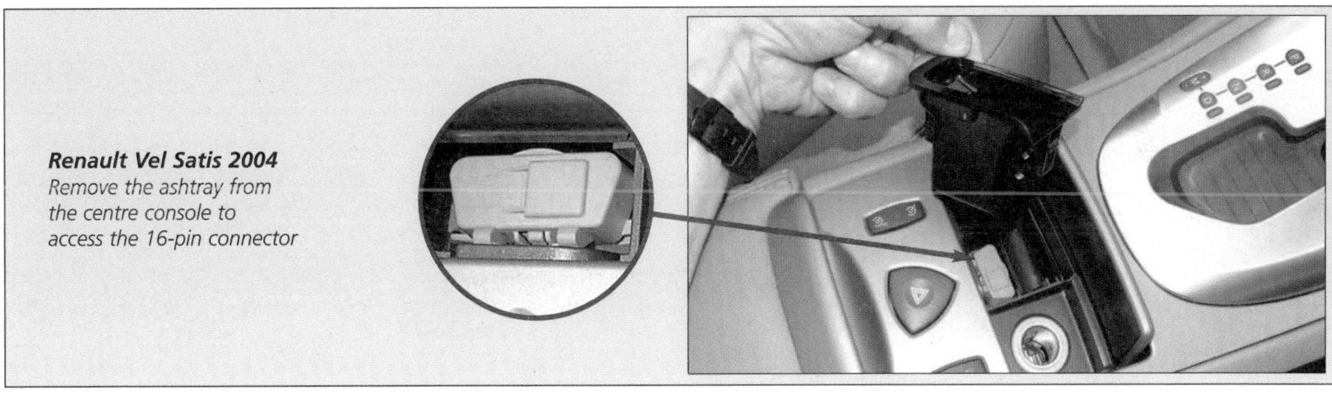

Renault Vel Satis 2004
Remove the ashtray from the centre console to access the 16-pin connector

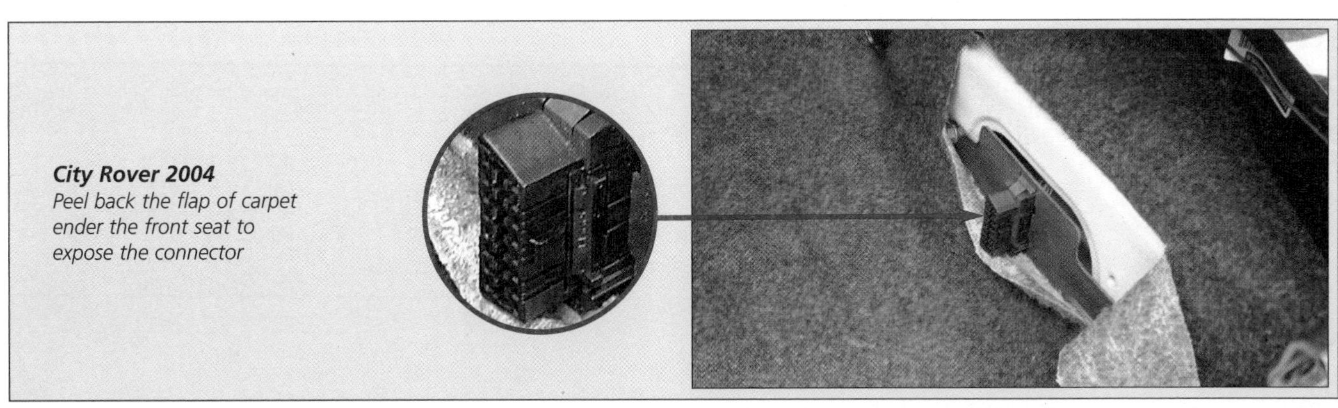

City Rover 2004
Peel back the flap of carpet ender the front seat to expose the connector

Rover 25 / MG ZR 2002
The 16-pin connector is located in the driver's footwell at the front edge of the centre console

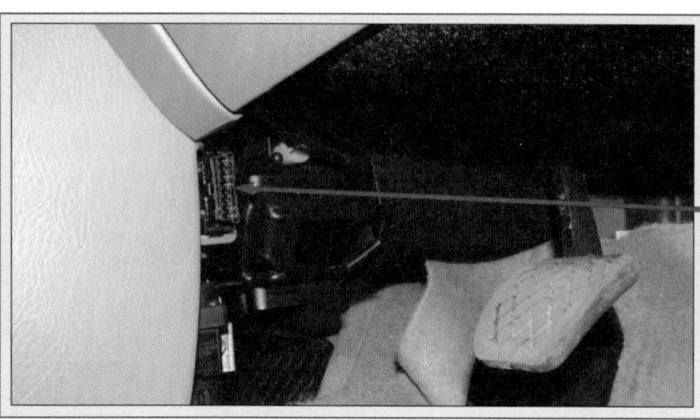

Rover 45 2003
The 16-pin connector is located in the driver's footwell at the front edge of the centre console

Rover 75 2001
EOBD 16-pin connector under the facia

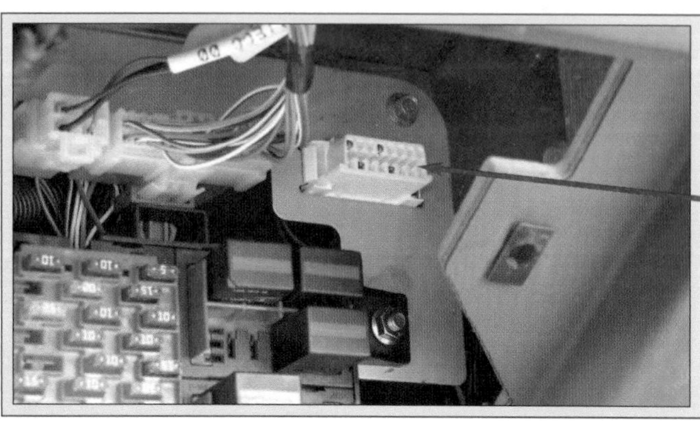

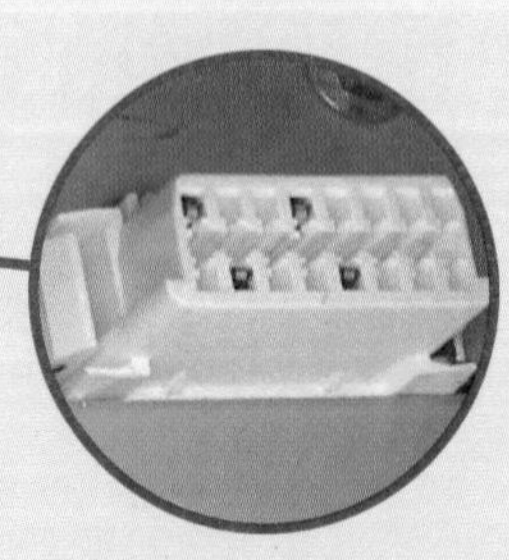

MG TF 2003
Open the fusebox cover under the driver's side of the facia to expose the 16-pin connector

Saab 9-3 2002
Remove the cover to access the 16-pin connector

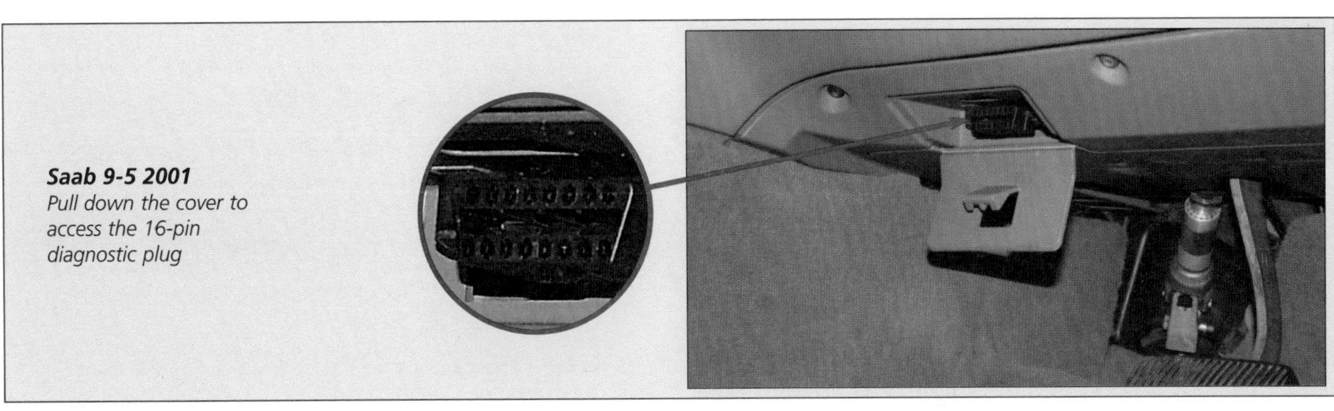

Saab 9-5 2001
Pull down the cover to access the 16-pin diagnostic plug

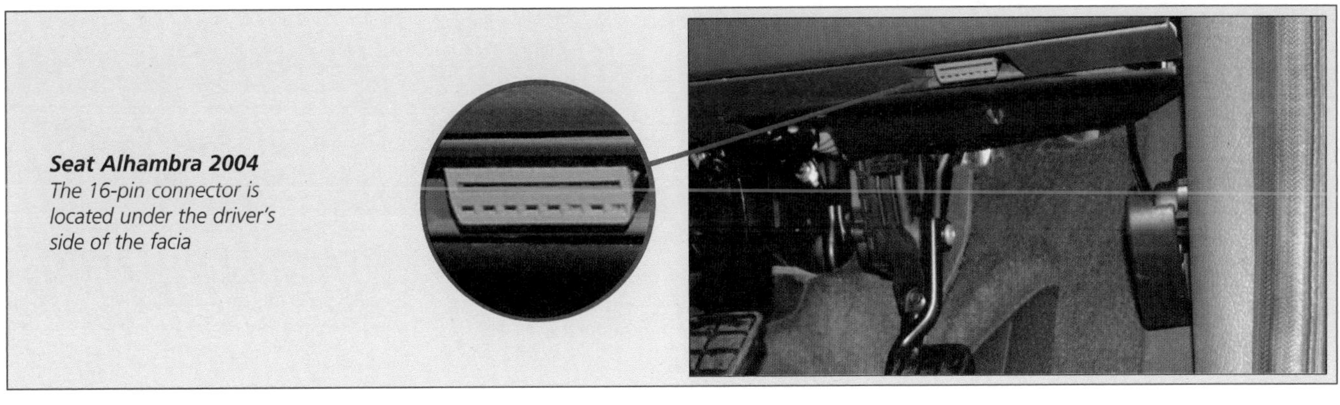

Seat Alhambra 2004
The 16-pin connector is located under the driver's side of the facia

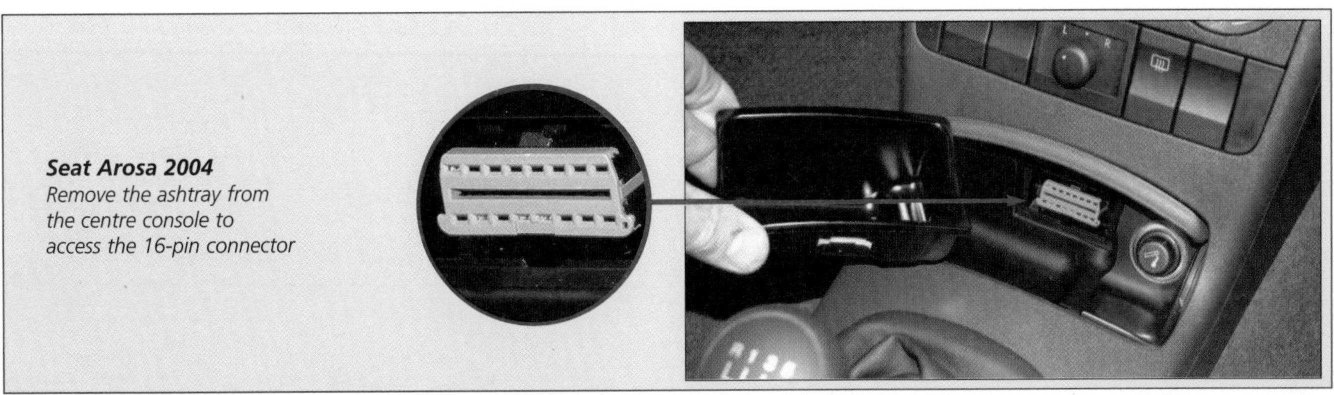

Seat Arosa 2004
Remove the ashtray from the centre console to access the 16-pin connector

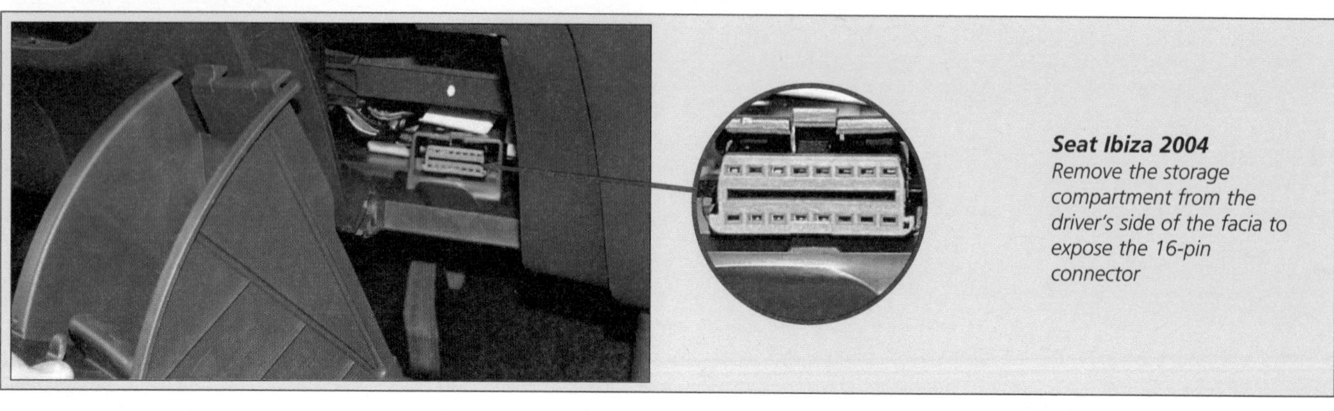

Seat Ibiza 2004
Remove the storage compartment from the driver's side of the facia to expose the 16-pin connector

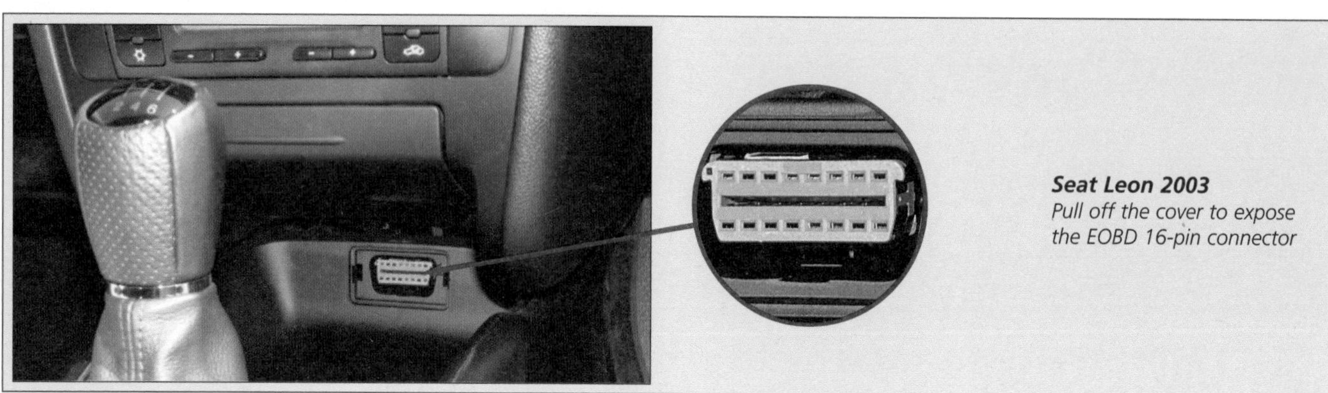

Seat Leon 2003
Pull off the cover to expose the EOBD 16-pin connector

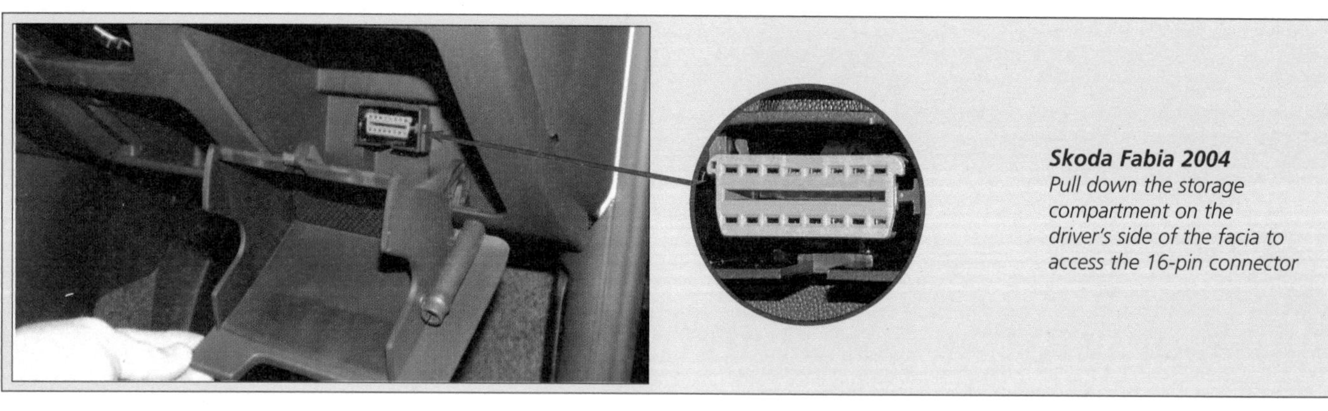

Skoda Fabia 2004
Pull down the storage compartment on the driver's side of the facia to access the 16-pin connector

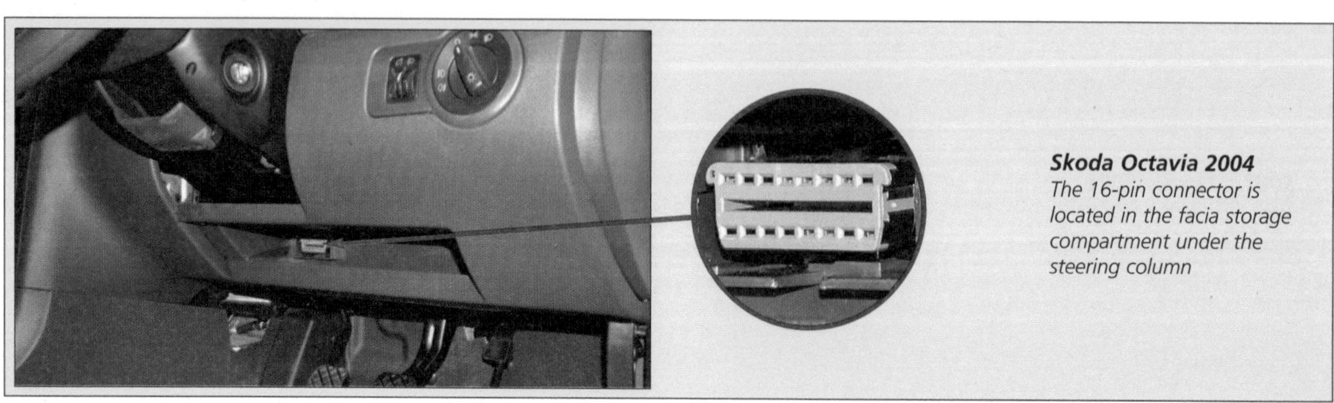

Skoda Octavia 2004
The 16-pin connector is located in the facia storage compartment under the steering column

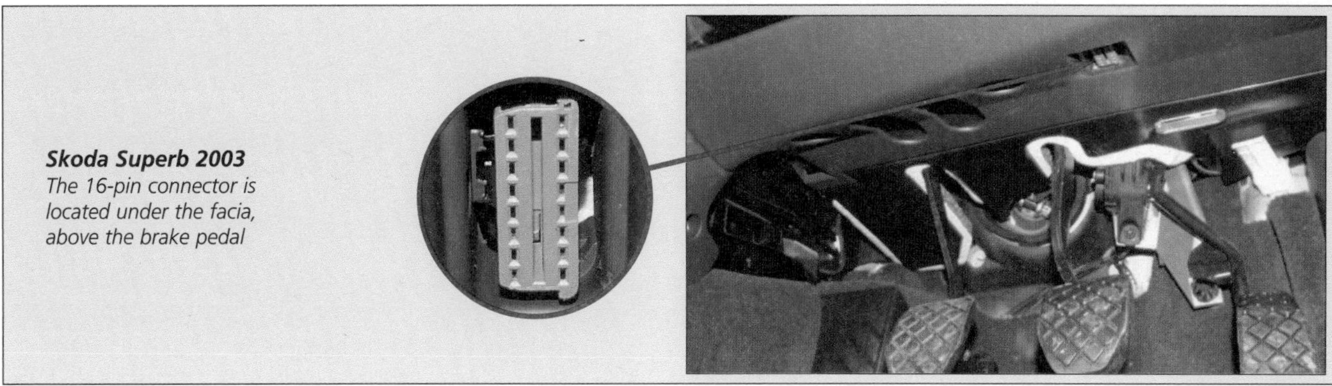

Skoda Superb 2003
The 16-pin connector is located under the facia, above the brake pedal

Subaru Impreza 2004
The 16-pin connector is located under the facia

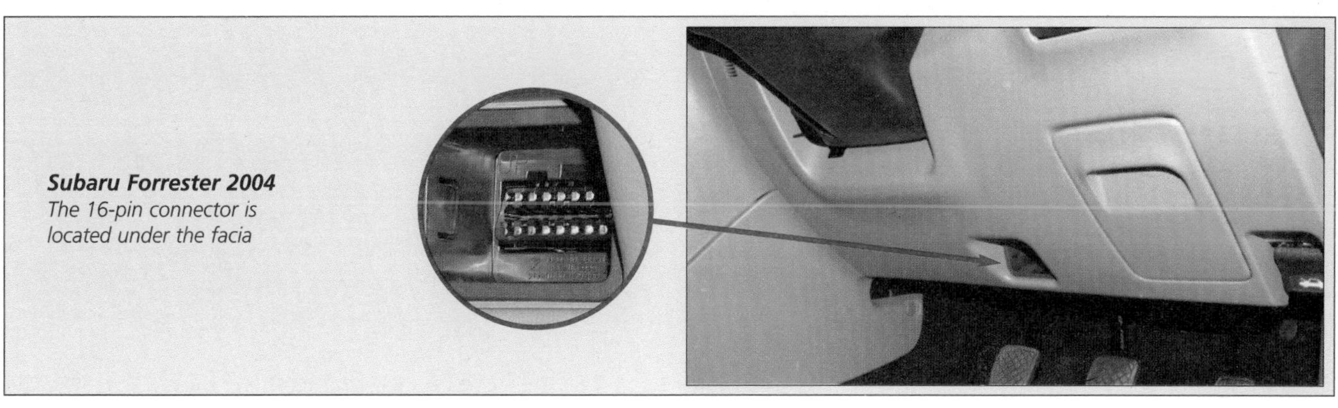

Subaru Forrester 2004
The 16-pin connector is located under the facia

Subaru Legacy 2004
The 16-pin connector is located under the facia

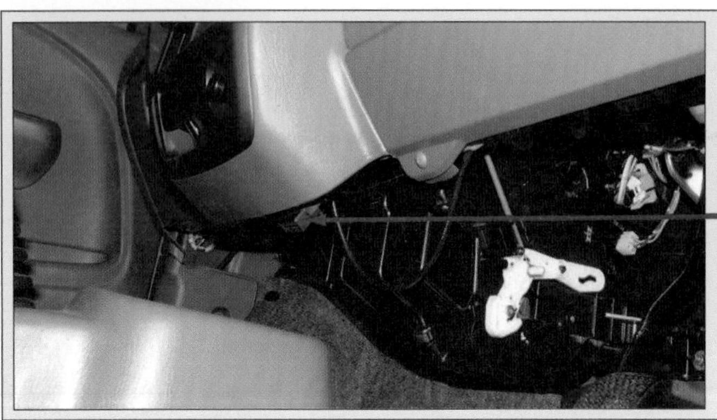

Suzuki Alto 2004
The 16-pin connector is located under the lower edge of the facia centre panel

Suzuki Ignis 5-door 2004
The 16-pin connector is located under the facia, adjacent to the bonnet release handle

Suzuki Ignis 3-door 2004
The 16-pin connector is located under the facia, adjacent to the heater housing

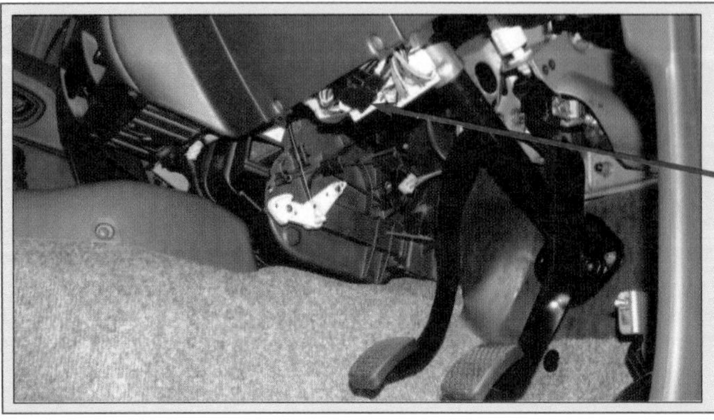

Suzuki Jimny 2004
The 16-pin connector is located under the facia, adjacent to the heater housing

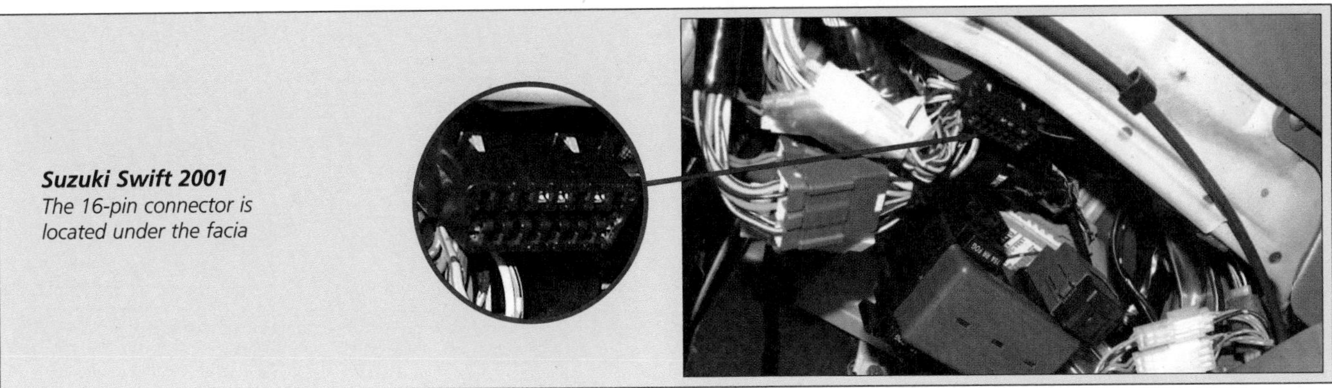

Suzuki Swift 2001
The 16-pin connector is located under the facia

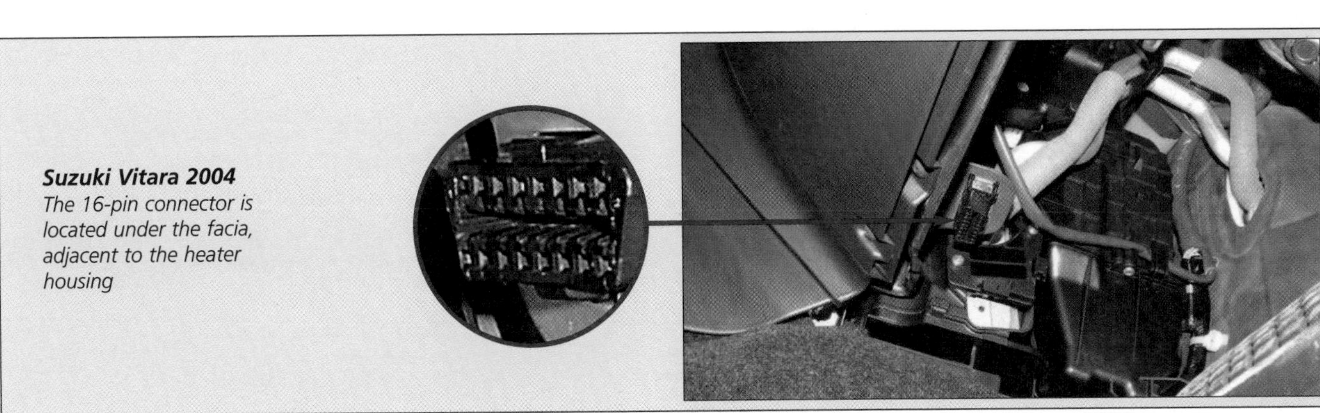

Suzuki Vitara 2004
The 16-pin connector is located under the facia, adjacent to the heater housing

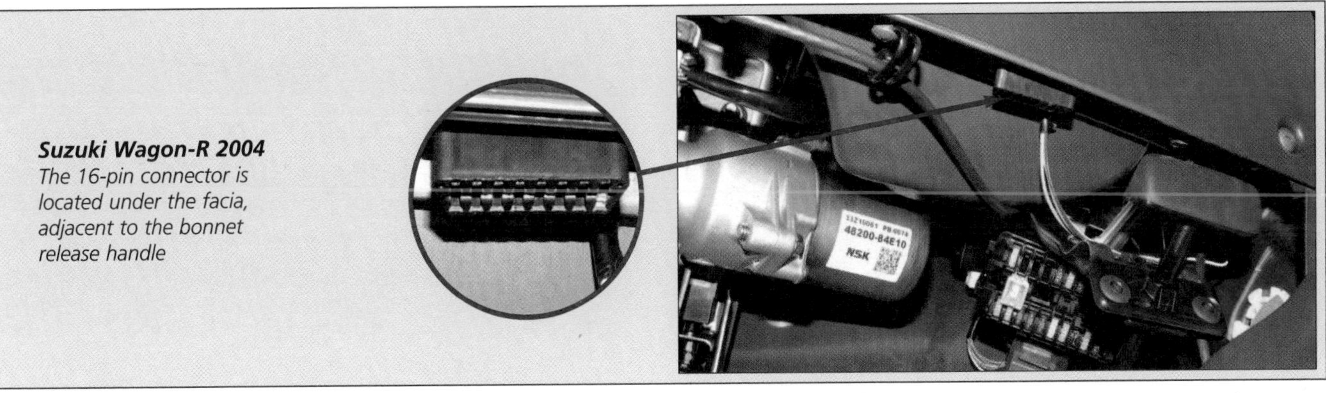

Suzuki Wagon-R 2004
The 16-pin connector is located under the facia, adjacent to the bonnet release handle

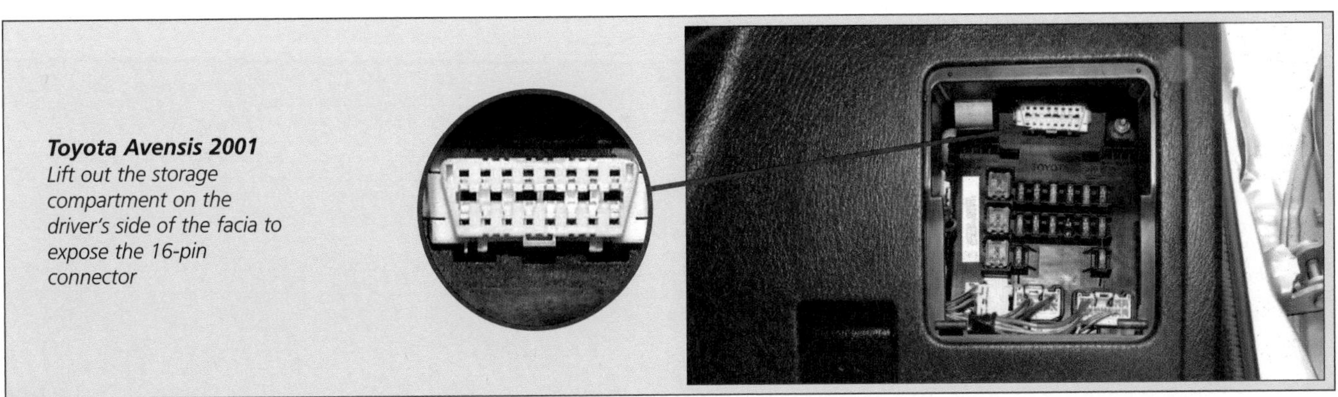

Toyota Avensis 2001
Lift out the storage compartment on the driver's side of the facia to expose the 16-pin connector

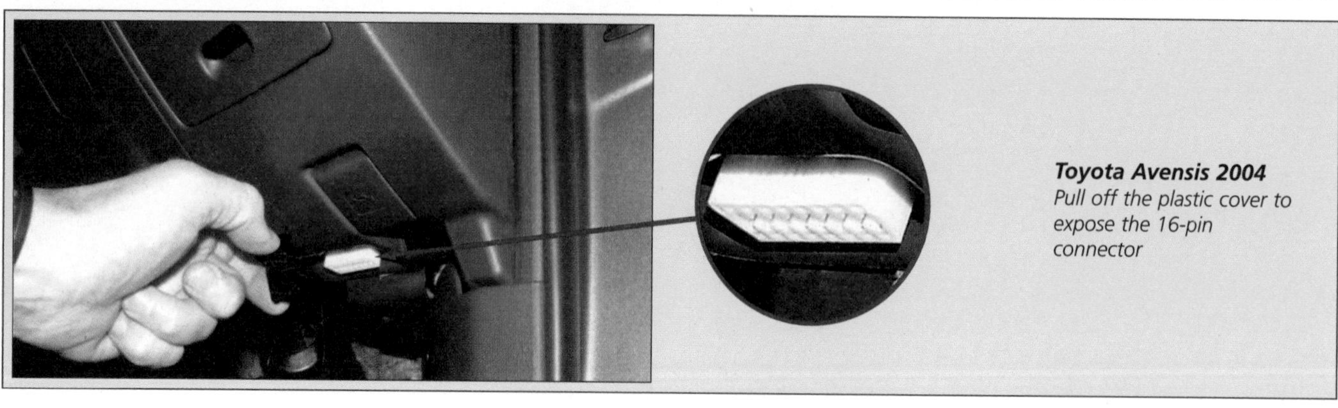

Toyota Avensis 2004
Pull off the plastic cover to expose the 16-pin connector

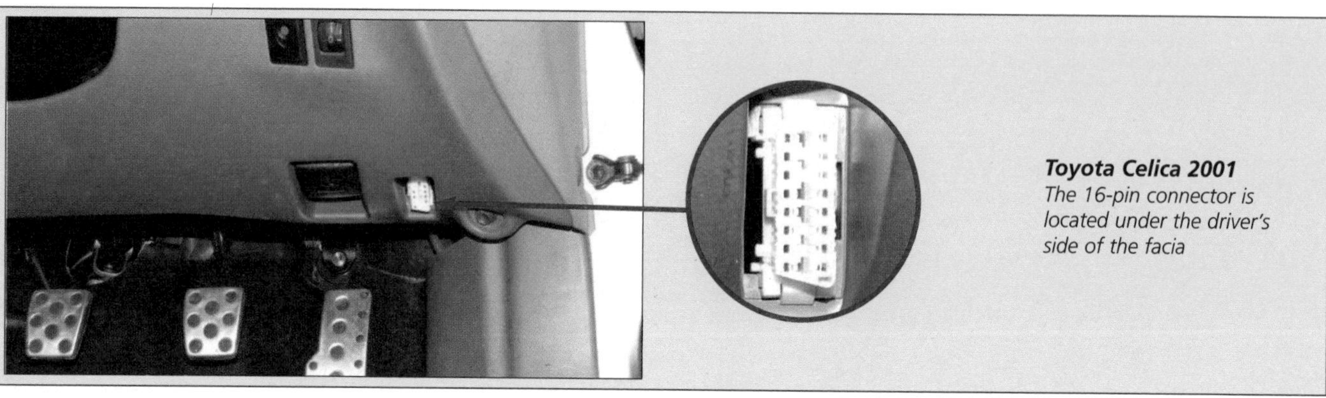

Toyota Celica 2001
The 16-pin connector is located under the driver's side of the facia

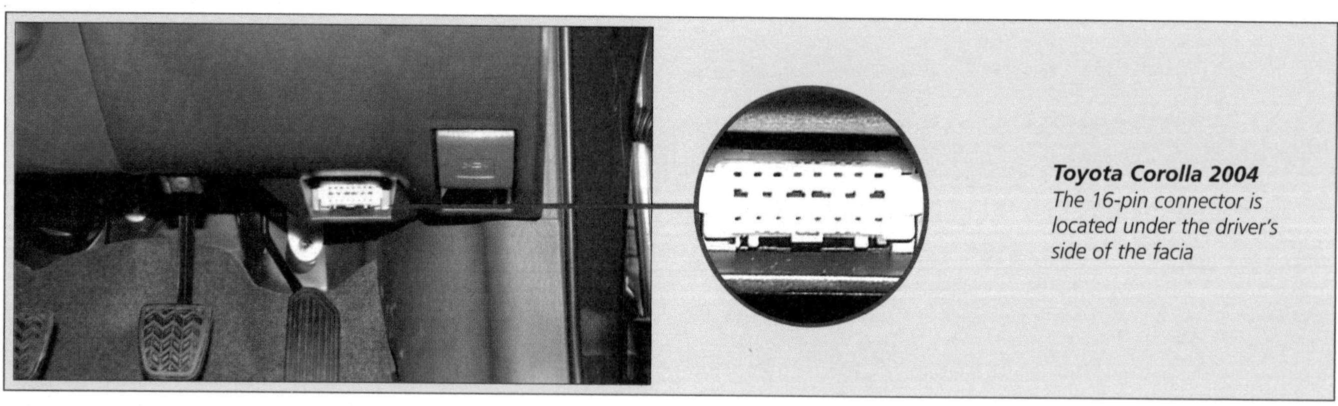

Toyota Corolla 2004
The 16-pin connector is located under the driver's side of the facia

Toyota Land Cruiser 2003
The 16-pin connector is located above the driver's pedals

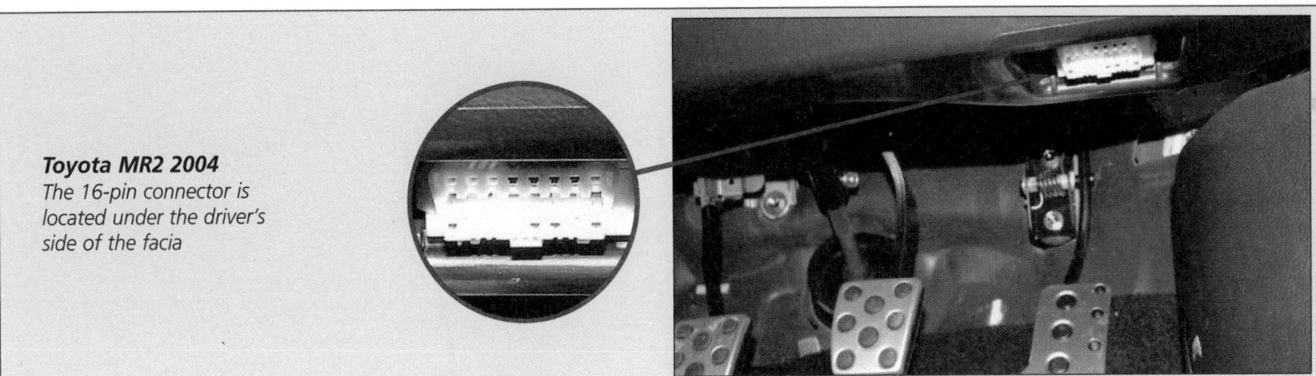

Toyota MR2 2004
The 16-pin connector is located under the driver's side of the facia

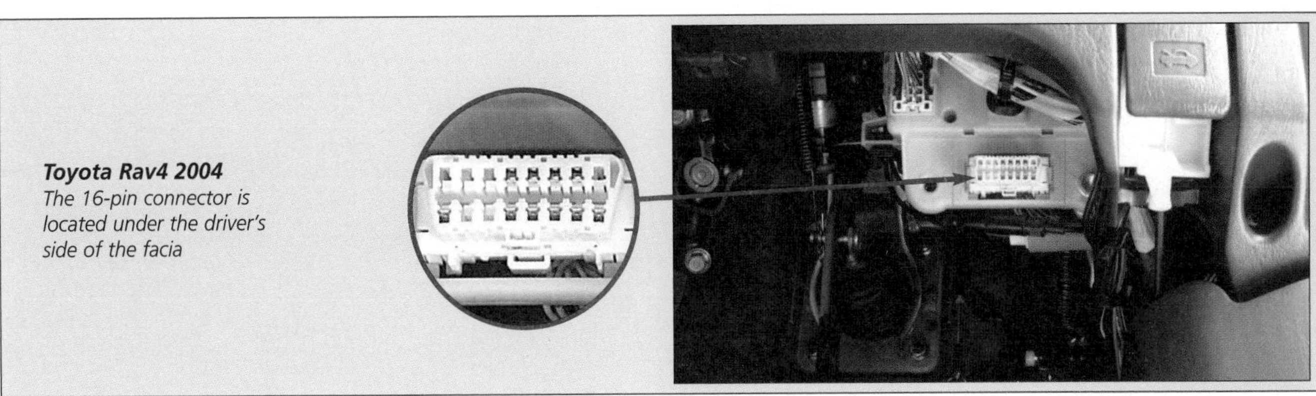

Toyota Rav4 2004
The 16-pin connector is located under the driver's side of the facia

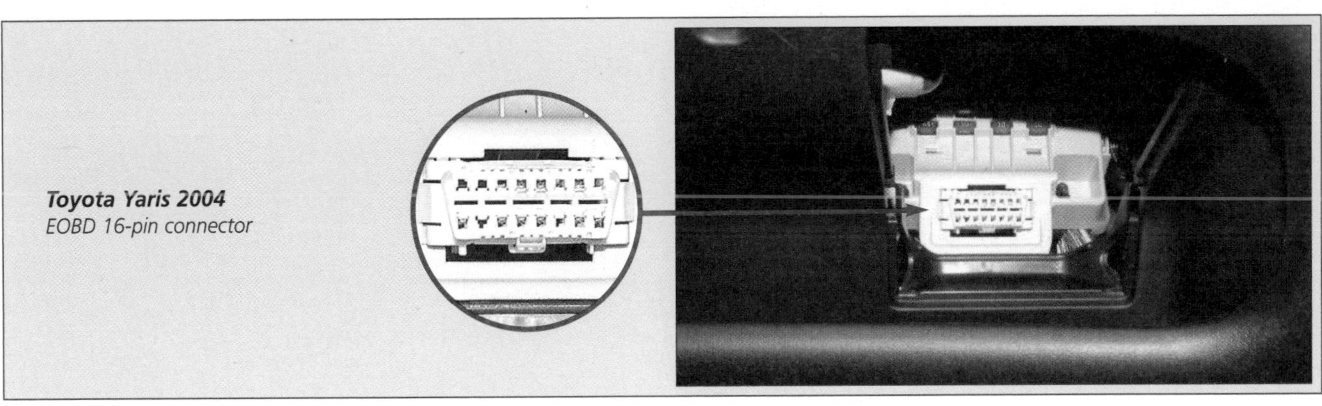

Toyota Yaris 2004
EOBD 16-pin connector

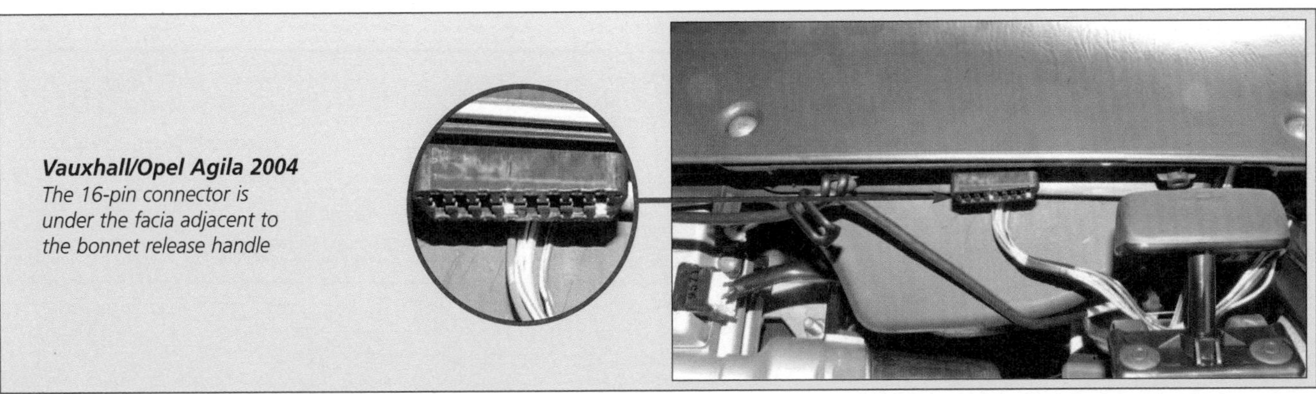

Vauxhall/Opel Agila 2004
The 16-pin connector is under the facia adjacent to the bonnet release handle

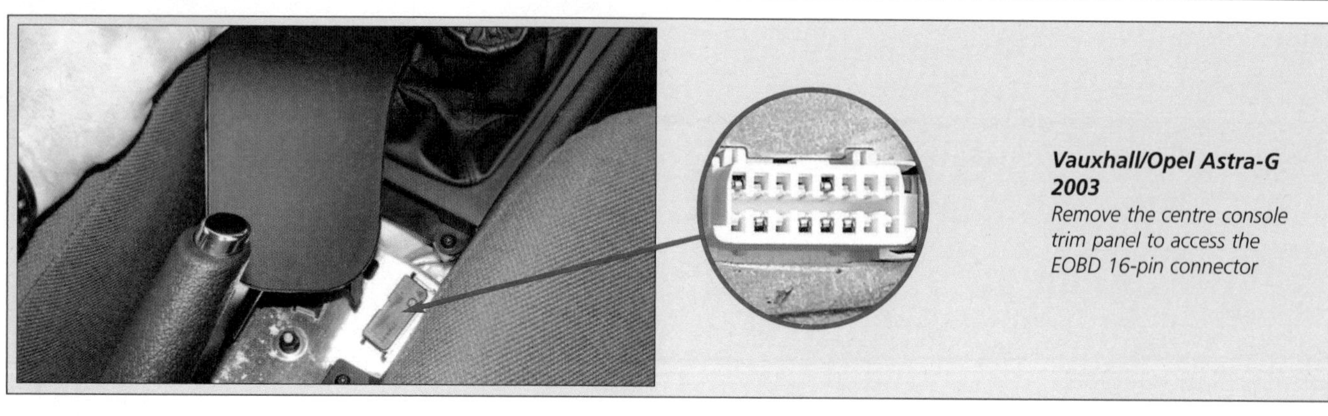

Vauxhall/Opel Astra-G 2003

Remove the centre console trim panel to access the EOBD 16-pin connector

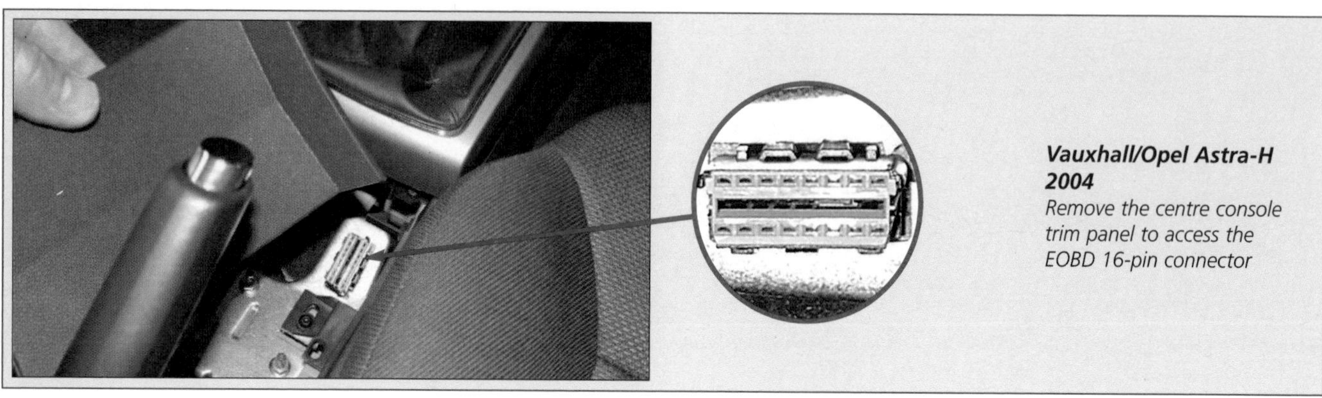

Vauxhall/Opel Astra-H 2004

Remove the centre console trim panel to access the EOBD 16-pin connector

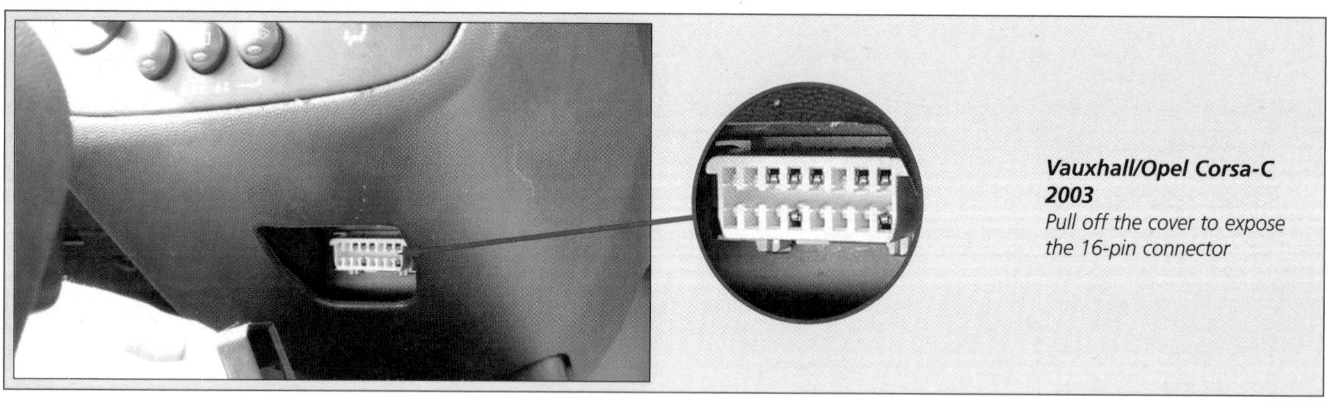

Vauxhall/Opel Corsa-C 2003

Pull off the cover to expose the 16-pin connector

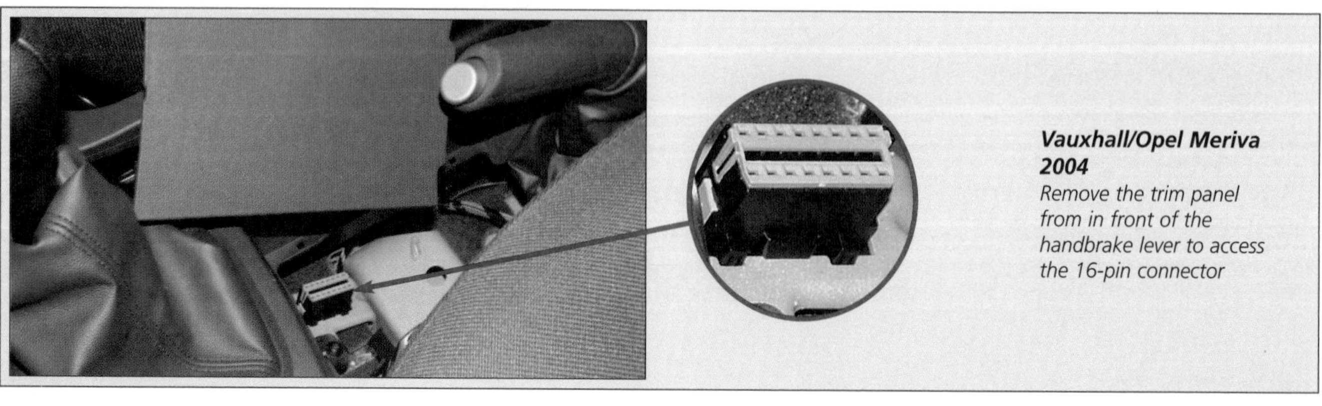

Vauxhall/Opel Meriva 2004

Remove the trim panel from in front of the handbrake lever to access the 16-pin connector

Vauxhall/Opel Omega 2001
Open the fusebox cover beneath the steering column to access the 16-pin connector

Vauxhall/Opel Vectra 2003
Remove the ashtray to expose the 16-pin connector

Vauxhall/Opel Zafira 2001
Remove the centre console trim panel to access the EOBD 16-pin connector

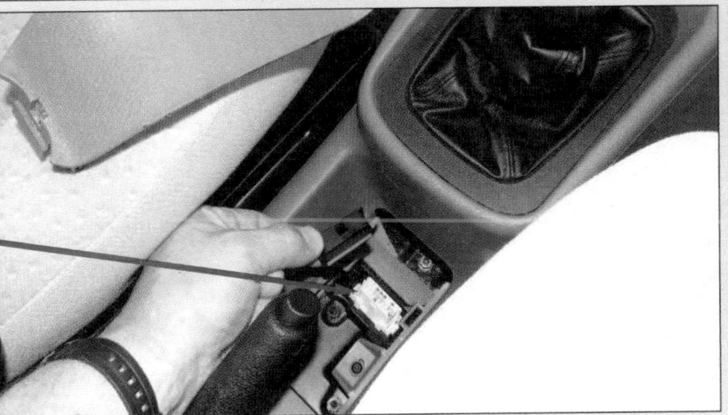

VW Beetle 2003
The 16-pin connector is located under the facia, adjacent to the bonnet release handle

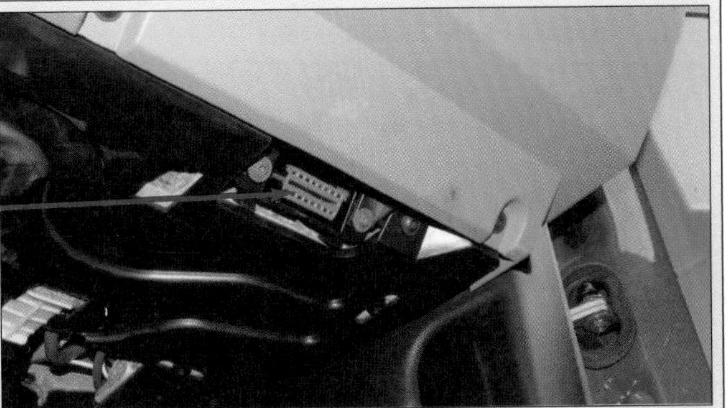

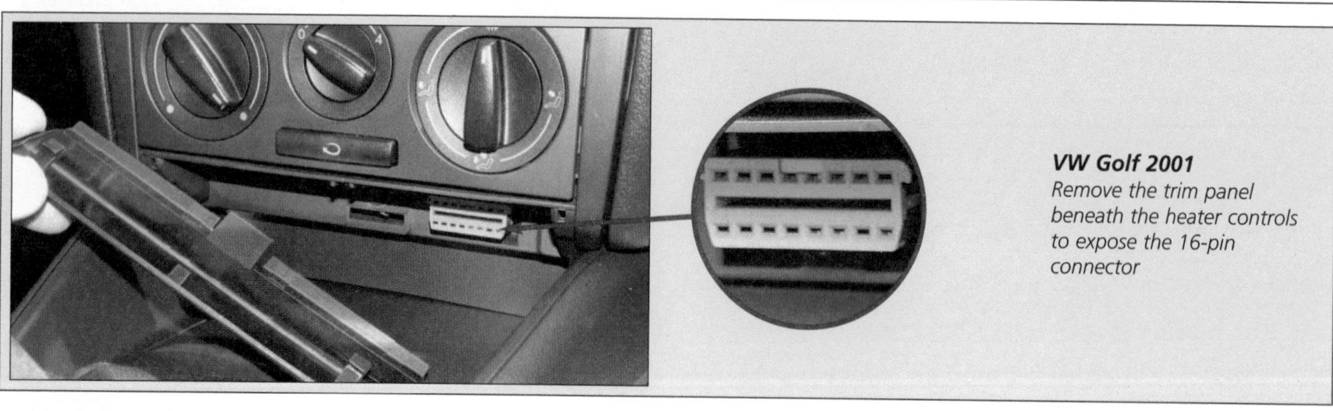

VW Golf 2001
Remove the trim panel beneath the heater controls to expose the 16-pin connector

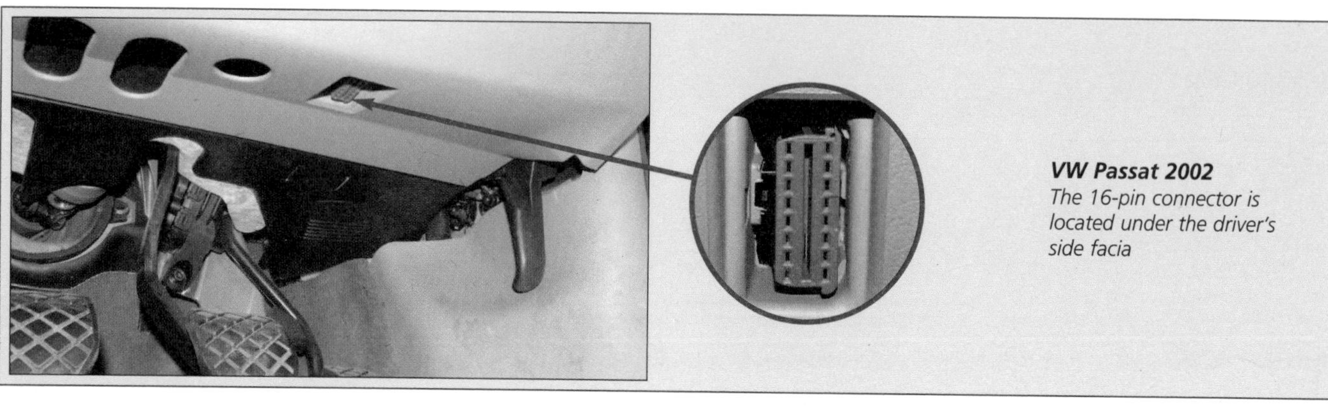

VW Passat 2002
The 16-pin connector is located under the driver's side facia

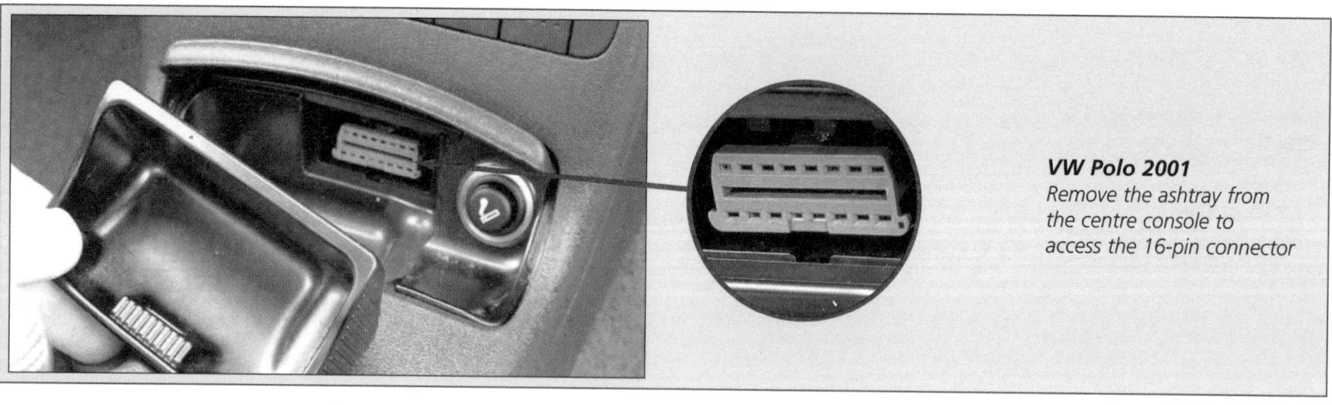

VW Polo 2001
Remove the ashtray from the centre console to access the 16-pin connector

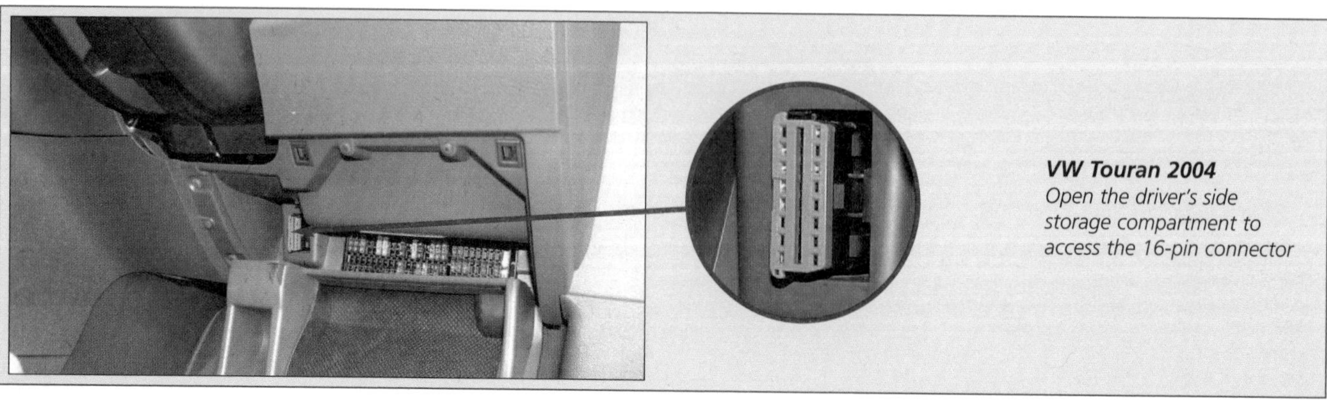

VW Touran 2004
Open the driver's side storage compartment to access the 16-pin connector

BMW 3-Series (E46) '98 – '03

System type

N42 4-cylinder 16 valve DOHC engines	DME (Digital Motor Electronics) ME9 engine management
M43TU 4-cylinder SOHC engines	DME (Digital Motor Electronics) BMS46 engine management
M52TU 6-cyinder engines	DME (Digital Motor Electronics) MS42 engine management
M54 6-cylinder engines	DME (Digital Motor Electronics) MS43 engine management

Fuel system data

Fuel pump type	Electric, immersed in tank
Fuel pressure regulator rating	3.0 ± 0.06 bar
Specified idle speed:	
N42 engine	700 ± 50 rpm (not adjustable – controlled by ECM)
M43TU engine	800 ± 50 rpm (not adjustable – controlled by ECM)
M52TU engine	750 ± 50 rpm (not adjustable – controlled by ECM)
M54 engine	No information available
Specified idle mixture CO content	Not adjustable – controlled by ECM
Tank capacity	68 litres

Torque wrench settings

	Nm	lbf ft
Camshaft position sensor bolt	7	5
Coolant temperature sensor	13	10
Crankshaft position sensor bolt	10	7
Exhaust manifold nuts*:		
M6 nuts	10	7
M7 nuts	20	15
M8 nuts	22	16
Front axle reinforcement plate bolts*		
Stage 1	59	44
Stage 2	Angle-tighten through a further 90°	
Fuel rail-to-inlet manifold bolts	10	7
Fuel tank mounting bolts	23	17
Fuel tank retaining strap bolts	8	6
Inlet manifold nuts:		
M6 nuts	10	7
M7 nuts	15	11
M8 nuts	22	16
Oxygen sensor	50	37

** Do not re-use*

General information

The fuel supply system consists of a fuel tank (which is mounted under the rear of the vehicle, with an electric fuel pump immersed in it), a fuel filter, fuel feed and return lines. The fuel pump supplies fuel to the fuel rail, which acts as a reservoir for the four fuel injectors which inject fuel into the inlet tracts. The fuel filter incorporated in the feed line from the pump to the fuel rail ensures that the fuel supplied to the injectors is clean. On M43TU and M52TU engines, a fuel pressure regulator is fitted to the fuel injection rail, where the fuel then returns to the tank. On N42 and M54 engines, the pressure regulator is incorporated into the fuel filter assembly.

Fuel injection system – general information

1 An integrated engine management system known as DME (Digital Motor Electronics) is fitted to all models, and the system controls all fuel injection and ignition system functions using a central ECM (Electronic Control Module).

2 On all models, the system incorporates a closed-loop catalytic converter and an evaporative emission control system, and complies with the very latest emission control standards.

3 The fuel pump (which is immersed in the fuel tank) supplies fuel from the tank to the fuel rail, via a filter. Fuel supply pressure is controlled by the pressure regulator in the fuel rail on M43TU and M52TU engines, and on integral with the fuel filter on N42 and M54 engines. When the optimum operating pressure of the fuel system is exceeded, the regulator allows excess fuel to return to the tank.

4 The electrical control system consists of the ECM, along with the following sensors:

a) Hot film air mass meter – informs the ECM of the quantity and temperature of air entering the engine.
b) Throttle position sensor (M43TU engine) – informs the ECM of the throttle position, and the rate of throttle opening/closing.
c) Coolant temperature sensor(s) – informs the ECM of engine temperature.
d) Crankshaft position sensor – informs the ECM of the crankshaft position and speed of rotation.
e) Camshaft position sensor(s)s – informs the ECM of the camshaft(s) positions.
f) Oxygen sensor(s) – informs the ECM of the oxygen content of the exhaust gases (explained in greater detail in Part C of this Chapter).
g) Vehicle speed sensor – informs the ECM of the vehicles road speed.
h) Intake air temperature sensor – informs the ECM of the temperature of the air entering the engine (M52TU and M54 engines)
i) Oil temperature – informs the ECM of the engine oil temperature.

5 All the above signals are analysed by the ECM which selects the fuelling response appropriate to those values. The ECM controls the fuel injectors (varying the pulse width – the length of time the injectors are held open – to provide a richer or weaker mixture, as appropriate). The mixture is constantly varied by the ECM, to provide the best setting for cranking, starting (with a hot or cold engine), warm-up, idle, cruising and acceleration.

6 The ECM also has full control over the engine idle speed, via an auxiliary air valve which bypasses the throttle valve. When the throttle valve is closed, the ECM controls the opening of the valve, which in turn regulates the amount of air entering the manifold, and so controls the idle speed.

7 The ECM also controls the exhaust and evaporative emission control systems.

8 On all engines, a Differential Air Inlet System (DISA) is fitted. Variable length inlet tracts incorporated in the inlet manifold are operated by a butterfly valve according to engine speed and load. This improves engines torque at low and medium engine speeds. The butterfly valve

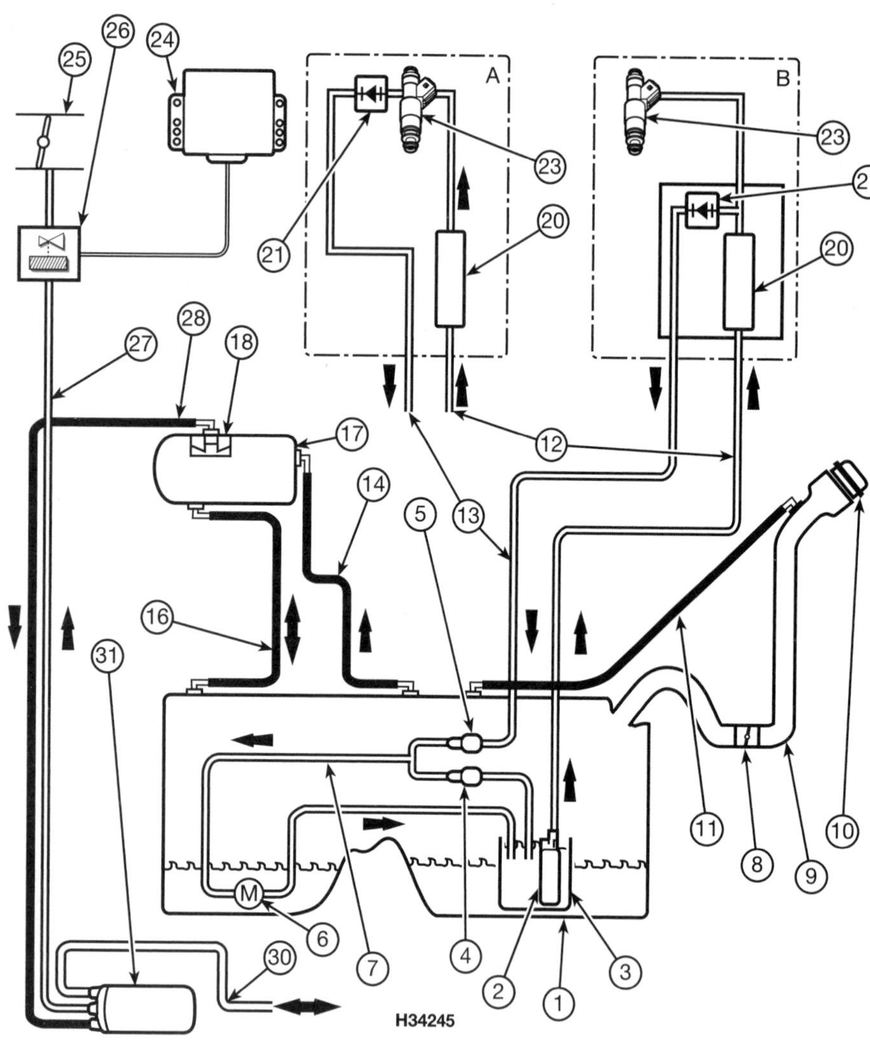

Fuel system

1 Fuel tank
2 Electric pump
3 Surge chamber
4 Pressure limiting valve
5 Outlet protection valve
6 Suction jet pump
7 Tank expansion pipe
8 Non-return valve
9 Filler neck
10 Filler cap
11 Breather hose
12 Fuel supply pipe
13 Fuel return pipe
14 Breather hose
16 Breather hose
17 Expansion tank
18 Roll-over valve
20 Fuel filter
21 Pressure regulator
23 Fuel rail
24 ECM
25 Intake manifold
26 Tank vent valve
27 Purge pipe
28 Vent pipe
30 Evaporation pipe
31 Carbon canister
A = M43TU and M52 TU engines
B = N42 and M54 engines

is operated by a vacuum actuator fitted under the manifold.

9 If there is an abnormality in any of the readings obtained from the sensors, the ECM enters its back-up mode. In this event, it ignores the abnormal sensor signal and assumes a pre-programmed value which will allow the engine to continue running (albeit at reduced efficiency). If the ECM enters this back-up mode, the relevant fault code will be stored in the ECM memory.

10 If a fault is suspected, the vehicle should be taken to a BMW dealer at the earliest opportunity. A complete test of the engine management system can then be carried out, using a special electronic diagnostic test unit which is simply plugged into the system's diagnostic connector. The 16-pin OBD socket is located under the facia on the drivers side, whilst the BMW diagnostic socket is located in the right-hand corner of the engine compartment or under the facia on the drivers side.

Fuel injection system – depressurisation and priming

Depressurisation

1 Remove the fuel pump fuse from the fusebox. The fuse is located in the main fusebox in the passenger glovebox, and the exact location is given on the fusebox cover.

2 Start the engine, and wait for it to stall. Switch off the ignition.

3 Remove the fuel filler cap.

4 The fuel system is now depressurised. **Note:** *Place a wad of rag around fuel lines before disconnecting, to prevent any residual fuel from spilling onto the engine.*

5 Disconnect the battery negative lead before working on any part of the fuel system.

Priming

6 Refit the fuel pump fuse, then switch on the ignition and wait for a few seconds for the fuel pump to run, building up fuel pressure. Switch off the ignition unless the engine is to be started.

Fuel injection system components – removal and refitting

Electronic control module (ECM)

1 Disconnect the battery negative lead. **Note:** *Disconnecting the battery will erase any fault codes stored in the ECM. It is recommended that the fault code memory of the module is interrogated using special test equipment prior to battery disconnection. Entrust this task to a BMW dealer or suitably equipped specialist.*

2 Working in the left-hand corner of the engine compartment, undo the four screws, and remove the cover from the electrical box.

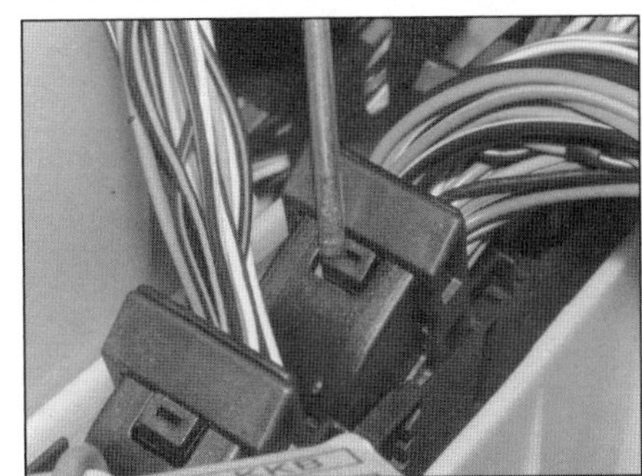

3 Disconnect the module wiring plugs, and remove it from the box.

4 Refitting is a reversal of removal. **Note:** *If a new module has been fitted, it will need to be coded using special test equipment. Entrust this task to a BMW dealer or suitably equipped specialist. After reconnecting the battery, the vehicle must be driven for several miles so that the ECM can learn its basic settings. If the engine still runs erratically, the basic settings may be reinstated by an BMW dealer or specialist using special diagnostic equipment. After reconnecting the wiring, the vehicle must be driven for several miles so that the ECM can learn its basic settings.*

Fuel rail and injectors

M43TU engines

Note: *New fuel injector O-rings should be used on refitting.*

5 Depressurise the fuel system, then disconnect the battery negative lead.

6 Remove the upper section of the inlet manifold.

7 Slacken the retaining clip and disconnect the fuel return pipe, then undo the union and disconnect the fuel supply pipe from the fuel rail. Be prepared for fuel spillage, and take adequate fire precautions. Plug the open ends of the pipes and hoses to prevent dirt entry and further fuel spillage. Discard the fuel supply pipe union O-ring, a new one must be fitted.

8 Disconnect the vacuum hose from the fuel pressure regulator.

9 Disconnect the injector wiring plugs, and release the wiring harness from the retaining clips.

10 Unscrew the two bolts securing the fuel rail to the inlet manifold.

11 Carefully pull the fuel rail upwards to release the fuel injectors from the cylinder head, then withdraw the complete fuel rail/fuel injector assembly.

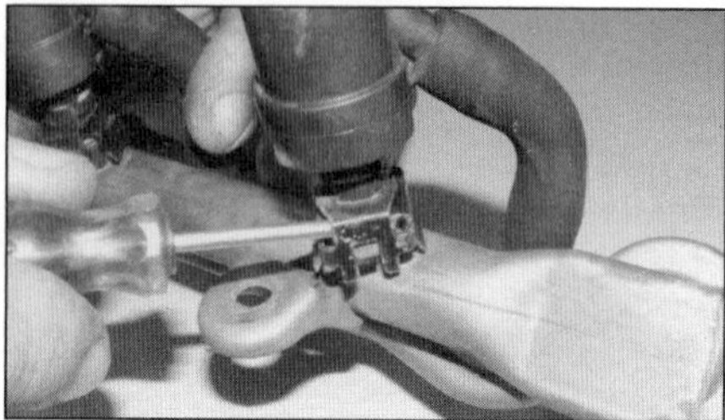

12 To remove a fuel injector from the fuel rail, proceed as follows.

a) Prise off the metal securing clip, using a screwdriver.
b) Pull the fuel injector from the fuel rail.

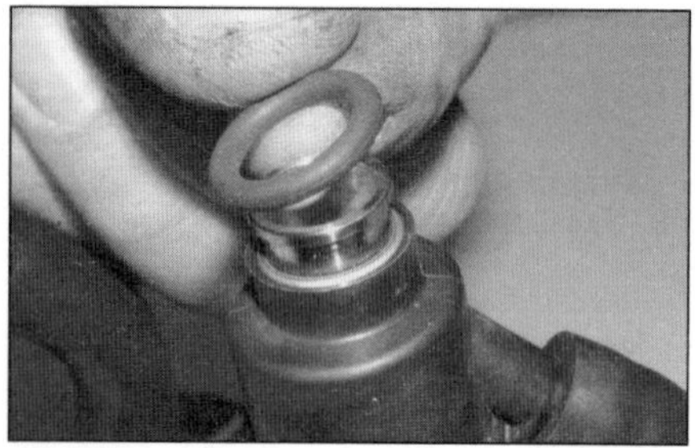

13 Before refitting, it is wise to renew all the fuel injector O-rings as a matter of course.

14 Check that the plastic washer at the bottom of each injector is positioned above the lower O-ring.

15 Lightly lubricate the fuel injector O-rings with a little petroleum jelly or SAE 90 gear oil.

16 Where applicable, refit the fuel injectors to the fuel rail, ensuring that the securing clips are correctly fitted. Note that the injectors should be positioned so that the wiring sockets are uppermost when the assembly is refitted.

17 Slide the fuel rail/fuel injector assembly into position, ensuring that the injectors engage with their bores in the inlet manifold.

18 Further refitting is a reversal of removal, but refit the upper section of the inlet manifold, and pressurise the fuel system (refit the fuel pump fuse and switch on the ignition) and check for leaks before starting the engine.

N42 engines

19 Undo the two nuts, raise the front edge of the plastic cover on top of the engine, and pull it out towards the front.

20 Working at the rear of the engine compartment, turn the fasteners 90° anti-clockwise and remove the pollen filter cover. Slide the filter from the housing.

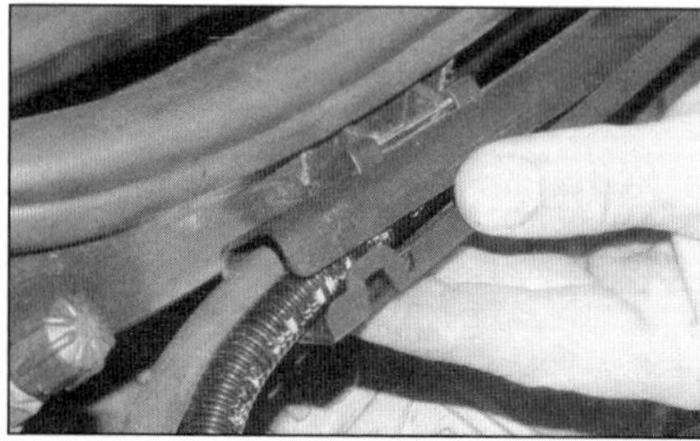

21 Release the retaining clips, and remove the cable from the ducting on the air intake housing.

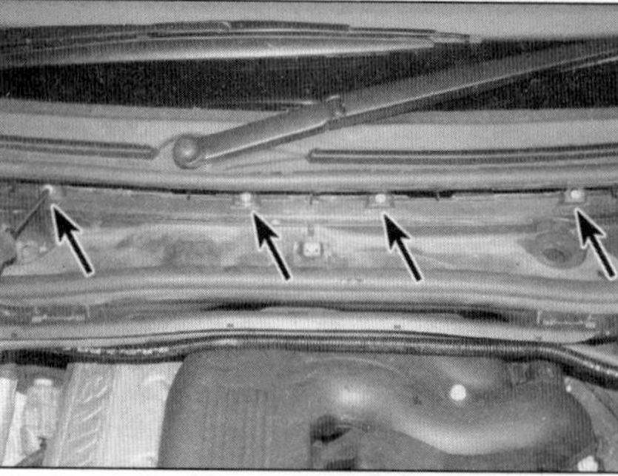

22 Undo the four screws and pull the pollen filter housing to the front.

23 In the left-hand rear corner of the engine compartment, pull up the sealing strip, undo the two fasteners and pull the trim panel forwards a little.

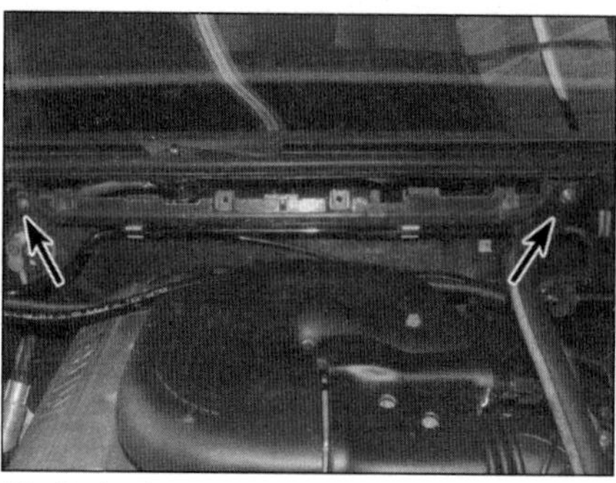

24 Undo the two Torx screws, and lift out the heater intake housing.

25 Press in the locking collar, and detach the fuel supply pipe from the fuel rail. Be prepared for fuel spillage, and take adequate fire precautions. Plug the open ends of the pipe and hose to prevent dirt entry and further fuel spillage.

26 Unclip the cable, undo the two nuts securing the bracket to the cylinder head, then unscrew the two retaining bolts, and remove the fuel rail complete with injectors.

27 Disconnect the injector wiring plugs, and if required, cut through the cable ties to free the rail.

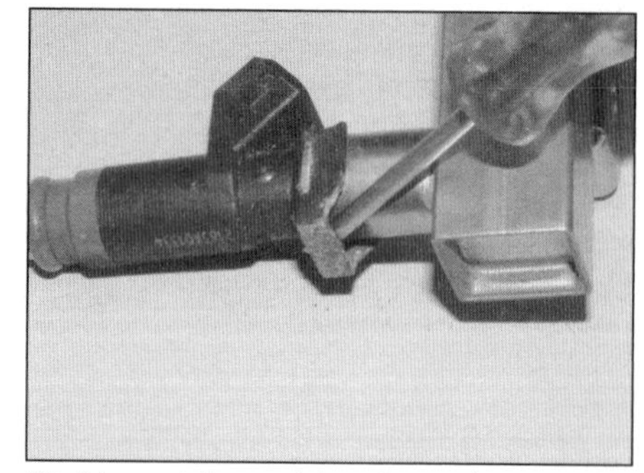

28 Prise out the retaining clips and remove the injectors from the fuel rail. Check the condition of the O-ring seals, and replace if necessary.

29 Lightly lubricate the fuel injector O-rings with a little petroleum jelly, or acid-free grease.

30 Refit the injectors to the fuel rail, and retain them in place with the clips pushed into the grooves.

31 Further refitting is a reversal of removal.

M52TU engines

32 Proceed as described in Paragraphs 20 to 24.

33 Prise out the plastic caps, undo the two screws, and remove the plastic cover from the over the injectors *(see paragraph 44)*.

34 Disconnect the fuel regulator vacuum hose.

35 Disconnect the air intake temperature sensor wiring plug, and pull the wiring rail from the fuel injectors.

36 Mark the two oxygen sensor wiring

connectors to aid refitting, unplug them and release them from the retaining clips.

37 Label the fuel supply and return pipes, then disconnect the pipes at the quick release connectors.

38 Undo the four screws and remove the fuel rail complete with the injectors *(see paragraph 48).*

39 To remove a fuel injector from the fuel rail, proceed as follows.

a) Prise off the metal securing clip, using a screwdriver.
b) Pull the fuel injector from the fuel rail.

40 Lightly lubricate the fuel injector O-rings with a little petroleum jelly, or acid-free grease.

41 Refit the injectors to the fuel rail, and retain them in place with the clips pushed into the grooves.

42 Further refitting is a reversal of removal.

M54 engines

43 Proceed as described in Paragraphs 20 to 24.

44 Prise out the plastic caps, undo the two screws, and remove the plastic cover from over the injectors.

45 Mark the two oxygen sensor wiring connectors to aid refitting, unplug them and release them from the retaining clips.

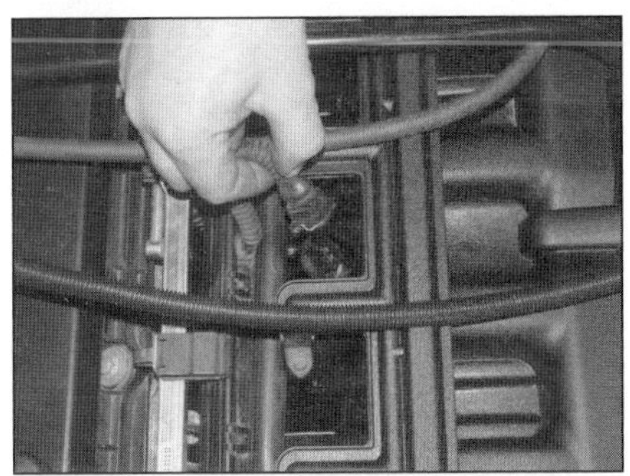

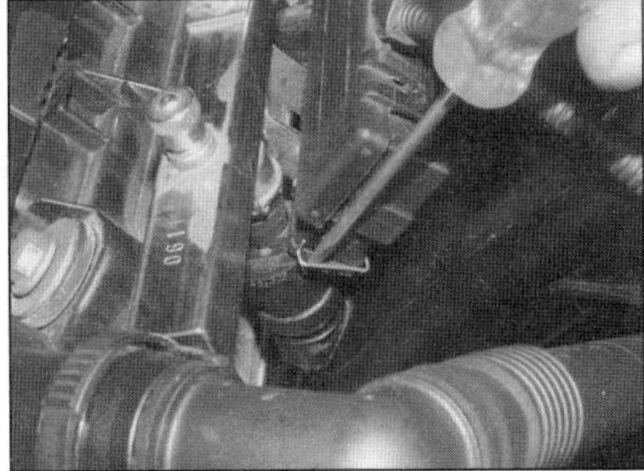

46 Disconnect the air intake temperature sensor wiring plug, the VANOS solenoid wiring plug, and release the retaining clips and pull the wiring rail from the fuel injectors.

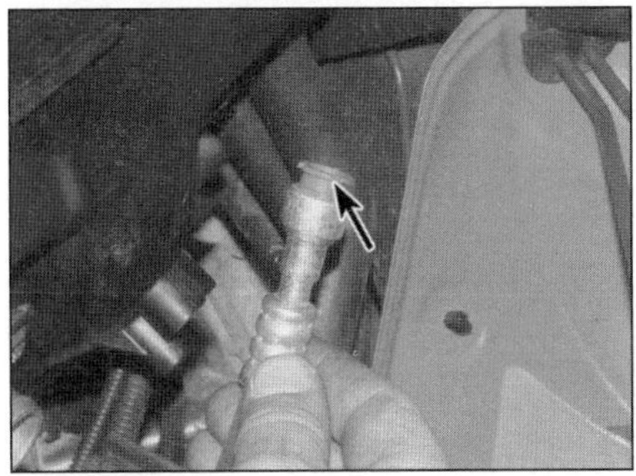

47 Disconnect the fuel supply pipe at the quick release connector.

48 Undo the four screws and remove the fuel rail complete with the injectors.

49 To remove a fuel injector from the fuel rail, proceed as follows.

a) Prise off the metal securing clip, using a screwdriver.
b) Pull the fuel injector from the fuel rail.

50 Lightly lubricate the fuel injector O-rings with a little petroleum jelly, or acid-free grease.

51 Refit the injectors to the fuel rail, and retain them in place with the clips pushed into the grooves.

52 Further refitting is a reversal of removal.

Fuel pressure regulator

M43TU engines

Warning: Refer to the warning notes on page 0•4 before proceeding.

Note: *New O-rings may be required on refitting.*

53 Depressurise the fuel system, then disconnect the battery negative lead.

54 Pull the vacuum hose from the pressure regulator.

55 Where applicable, to improve access, disconnect the breather hose from the front of the cylinder head cover.

56 Note the angle of the vacuum inlet on the pressure regulator, so that it can be refitted in the same position.

57 Where applicable, to give sufficient clearance to remove the pressure regulator, unbolt the engine lifting bracket from the cylinder head.

58 On models where the regulator is secured by a clamping bracket retained by a nut, unscrew the securing nut and withdraw the clamping bracket.

59 On models where the regulator is secured by a locking clip, pull out the clip.

60 Twist and pull the regulator from the fuel rail. Note that it will be difficult to pull the regulator from the fuel rail due to the tight-fitting O-rings.

61 Before refitting, check the O-rings, and renew if necessary.

62 Refitting is a reversal of removal.

N42 engines

63 On these engines, the regulator is integral with the fuel filter assembly. If defective, replace the complete filter housing.

M52TU engines

64 Depressurise the fuel system, then disconnect the battery negative lead.

65 To allow sufficient clearance, remove the heater/ventilation inlet air ducting from the rear of the engine compartment as follows.

a) Rotate the three fasteners 90° anti-clockwise and remove the pollen filter cover from the rear of the engine compartment. Pull the filter forward and remove it.

b) Undo the four retaining clips and thread the cable out of the ducting (see paragraph 21).

c) Unscrew the four screws and pull the filter housing forwards and remove it.

d) Undo the two screws and remove the inlet ducting upwards and out of the engine compartment (see paragraph 24).

66 Prise up the two caps, undo the screws and remove the plastic cover from above the fuel rail.

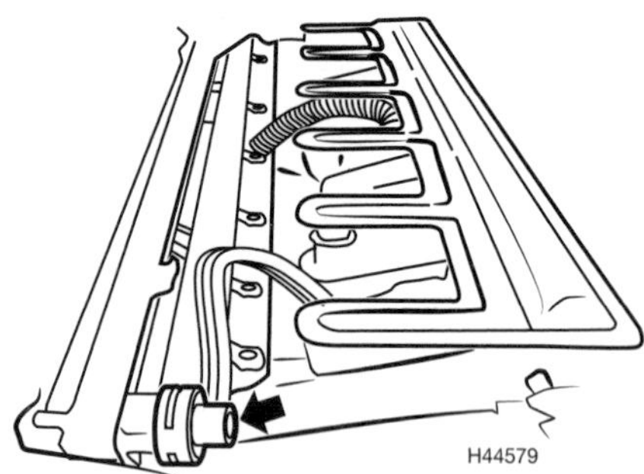

67 Pull the vacuum hose from the fuel regulator.

68 Remove the circlip, twist and pull the regulator from the fuel rail.

69 Before refitting, check the O-rings, and renew if necessary (see paragraph 61).

70 Refitting is a reversal of removal, bearing in mind the following points.

a) Ensure that the regulator is pushed firmly into position in the end of the fuel rail.

b) Make sure that the circlip correctly engages with the recess in the fuel rail..

c) On completion, pressurise the fuel system (refit the fuel pump fuse and switch on the ignition) and check for leaks before starting the engine.

M54 engines

71 On these engines, the regulator is integral with the fuel filter assembly. If defective, replace the complete filter housing.

Mass airflow sensor

M43TU engines

Note: *A new airflow sensor seal may be required on refitting.*

72 Disconnect the battery negative lead.

73 Disconnect the wiring plug from the airflow sensor.

74 Loosen the hose clip, and disconnect the air ducting from the airflow sensor.

75 Release the clips, and remove the sensor from the air cleaner housing. Recover the seal.

76 Refitting is a reversal of removal, but check the condition of the seal and renew if necessary.

N42 engines

77 Remove the air cleaner housing.

78 Slacken the hose clip and remove the intake hose and resonance chamber from the filter housing.

79 Undo the two nuts, and remove the sensor from the filter housing.

80 Refitting is a reversal of removal.

M52TU engines

Note: *A new airflow sensor seal may be required on refitting.*

81 Ensure the ignition is turned off, and disconnect the wiring plug from the sensor **(see illustration 4.84)**.

82 Pull the vacuum hose from the intake ducting, release the retaining clips, and remove the sensor complete with the intake ducting. Recover the seal.

83 Refitting is a reversal of removal, but check the condition of the seal and renew if necessary.

M54 engines

Note: *A new airflow sensor seal may be required on refitting.*

84 Ensure the ignition is turned off, and disconnect the wiring plug from the sensor.

85 Release the retaining clips and detach the intake ducting from the sensor, and the sensor from the sir cleaner housing. Recover the seal.

86 Refitting is a reversal of removal, but check the condition of the seal and renew if necessary.

Throttle position sensor

M43TU engines

Note: *A new O-ring may be required on refitting.*

87 Ensure the ignition is turned off.

88 Disconnect the wiring plug from the sensor.

89 Unscrew the two securing screws, and withdraw the sensor from the throttle body. Where applicable, recover the O-ring

90 Refitting is a reversal of removal, but where applicable, check the condition of the O-ring and renew if necessary, and ensure that the O-ring is correctly positioned. Note that adjustment of the unit is not necessary.

N42 engines

91 On these engines, although the throttle body valve is not used to control engine load, it is used to provide inlet manifold vacuum for purging the carbon canister etc. The throttle positioning motor is integral with the throttle body. The accelerator pedal position sensor is integral with the accelerator pedal.

M52TU and M54 engines

92 The throttle position sensor is integral with the throttle body.

93 Refitting is a reversal of removal. Note that if a new throttle body/position sensor has been fitted, it will be necessary to clear the adaption values stored in the engine management ECM using specialist test equipment. Entrust this task to a BMW dealer or suitably equipped specialist. Once cleared, new adaption values will be 'learnt' as the engine is started and used.

Coolant temperature sensor

M43TU engines

94 The sensor is located in the left-hand side of the cylinder head. Drain the coolant system, and remove the upper section of the intake manifold.

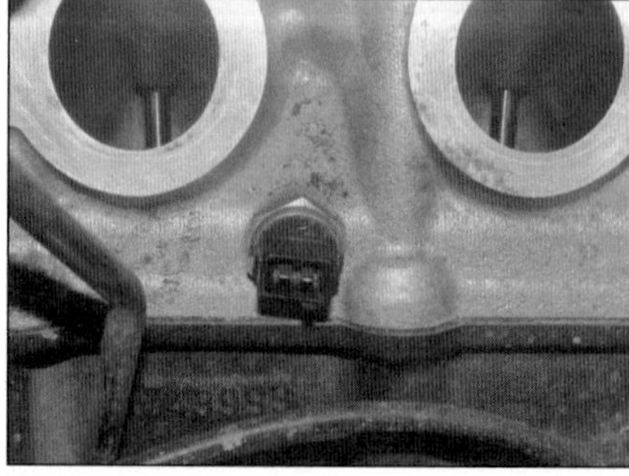

95 Unscrew the sensor from the cylinder head.

96 Refitting is a reversal of removal. Tighten the sensor to the specified torque, and top up the coolant.

N42 engines

97 The sensor is located in the left-hand side of the cylinder block. Drain the coolant system.

98 Remove the alternator.

99 Disconnect the wiring plug and unscrew the sensor from the cylinder block..

100 Refitting is a reversal of removal. Tighten the sensor to the specified torque, and top up the coolant.

M52TU and M54 engines

101 The sensor is located in the left-hand side of the cylinder head under the intake manifold. Drain the coolant system, and remove the intake manifold.

102 Disconnect the sensor wiring plug, and unscrew it from the cylinder head.

103 Refitting is a reversal of removal. Tighten the sensor to the specified torque, and top up the coolant.

Crankshaft position sensor

M43TU engines

104 Jack up the front of the vehicle and support it securely on axle stands. Undo the screws and remove the engine under shield.

105 The sensor is located below the starter motor. Disconnect the sensor wiring plug, undo the retaining screw and remove the sensor. Recover the seal.

106 Refitting is a reversal of removal. Check the condition of the seal and replace if necessary.

N42 engines

107 The sensor is located below the starter motor. Remove the intake manifold.

108 Disconnect the sensor wiring plug, undo the retaining screw and remove the sensor. Recover the seal.

109 Check the condition of the sealing ring and replace if necessary. Refit the sensor and tighten the retaining screw to the specified torque.

110 The remainder of refitting is a reversal of removal.

M52TU engines

111 Proceed as described in Paragraphs 104 to 106.

M54 engines

112 Jack up the front of the vehicle and support it securely on axle stands. Undo the screws and remove the engine under shield.

113 Undo the bolts and remove the reinforcement plate from between the lower control arms. Note that new screws will be needed upon refitting.

114 The sensor is located under the starter motor. Disconnect the sensor wiring plug, undo the retaining screw and remove the sensor. Recover the seal.

115 Refitting is a reversal of removal, but use new bolts when refitting the front axle reinforcement plate.

Camshaft position sensor(s)

M43TU engines

116 The sensor is located in the front of the upper timing chain cover.

117 Ensure the ignition is switched off, and disconnect the sensor wiring plug.

118 To improve access, unscrew the solenoid valve and bracket, then move it to one side.

119 Undo the retaining screw and remove the sensor from the engine. Recover the seal.

120 Check the condition of the sealing ring, and renew if necessary.

121 Refit the sensor to the engine, and tighten the retaining screw securely.

122 The remainder of refitting is a reversal of removal.

N42 engines

123 The sensors are located in the front side of the cylinder head, under their respective camshaft ends. Ensure the ignition is switched off.

124 Prise out the two expanding rivets, and remove the air intake hood from the bonnet slam panel.

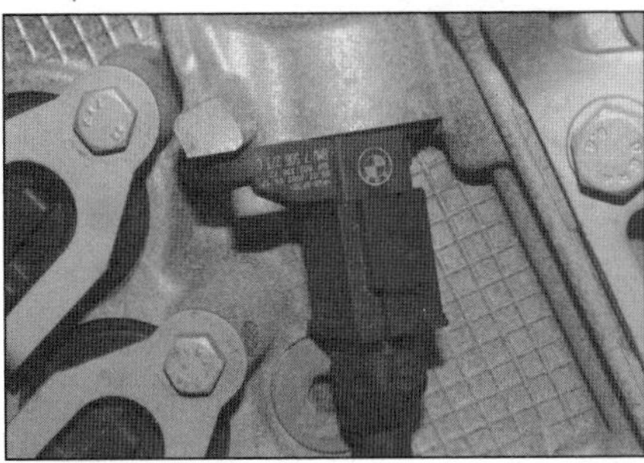

125 Squeeze together the locking lugs, and disconnect the sensor wiring plug.

126 Undo the retaining screw and remove the sensor. Recover the seal.

127 Refitting is a reversal of removal. Check the condition of the seal and renew it if necessary.

Intake camshaft sensor

M52TU and M54 engines

128 Ensure the ignition is turned off, and remove the air cleaner housing.

129 Disconnect the wiring plug, then unscrew the VANOS solenoid valve to access the sensor.

130 Trace the wiring back from the sensor, and disconnect the wiring plug where it clips to the cable ducting behind the alternator.

131 Undo the retaining screw and remove the sensor from the cylinder head. Recover the seal.

132 Refitting is a reversal of removal. Check the condition of the seal and renew it if necessary.

Chapter 6

Exhaust camshaft sensor

M52TU and M54 engines

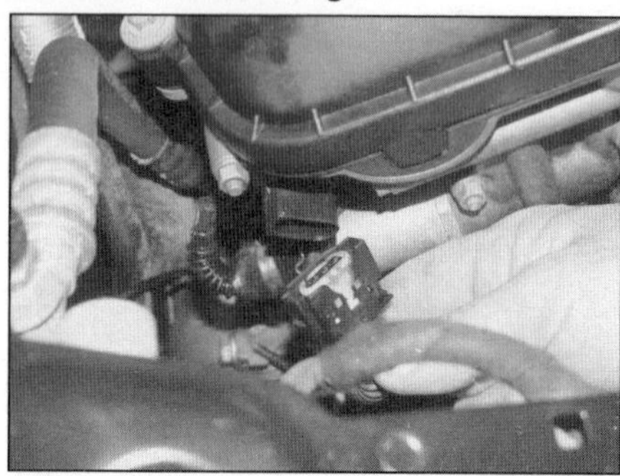

133 Ensure the ignition is switched off. Disconnect the sensor wiring plug, undo the retaining screw and remove the sensor. Recover the seal.

134 Refitting is a reversal of removal. Check the condition of the seal and renew it if necessary.

Oxygen sensor

Note: *Ensure that the exhaust system is cold before attempting to remove the oxygen sensor.*

135 The oxygen sensor(s) is screwed into the following locations.

a) *M43TU engines – in front of and after the catalytic converter.*

b) *N42 engines – in front of and after the catalytic converter*

c) *M52TU and M54 engines – Front exhaust downpipes before and after the catalytic converters*

136 Ensure the ignition is switched off.

137 Apply the handbrake, then jack up the front of the vehicle and support securely on axle stands. Undo the screws and remove the engine undershield.

M43TU and N42 engines

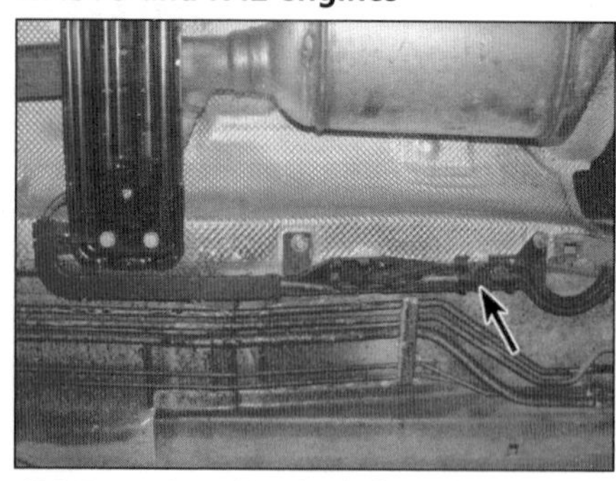

138 Trace the wiring back from the sensor to the wiring connector under the vehicle, and disconnect the connector.

M52TU and M54 engines

Note: *Due to the limited access, if the sensor for cylinders 4 to 6 is to be removed, the complete exhaust system must be removed.*

139 To allow sufficient clearance, remove the heater/ventilation inlet air ducting from the rear of the engine compartment as follows.

a) *Rotate the three fasteners 90° anti-clockwise and remove the pollen filter cover from the rear of the engine compartment. Pull the filter forward and remove it.*

b) *Undo the four retaining clips and thread the cable out of the ducting* (see paragraph 21 earlier in this Section).

c) *Unscrew the four screws and pull the filter housing forwards and remove it.*

d) *Pull up the rubber strip, rotate the two fasteners anti-clockwise, and move the dividing panel in the left-hand corner of the engine compartment forward a little.*

e) *Undo the two screws and remove the inlet ducting upwards and out of the engine compartment* (see paragraph 24 earlier in this Section).

140 Prise out the plastic caps, undo the two screws, and remove the plastic cover from over the injectors.

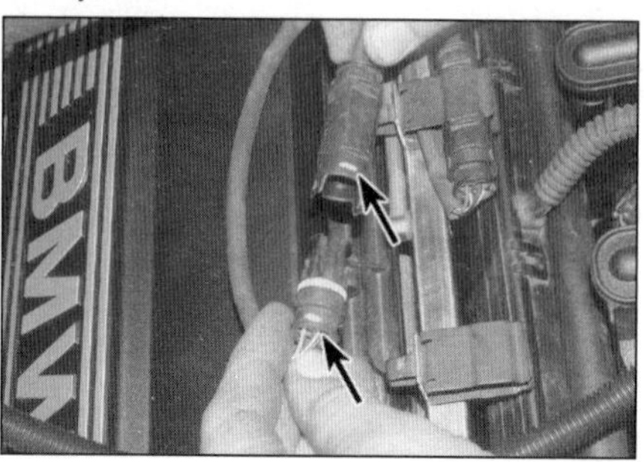

141 Unclip the oxygen sensor cables from the retainer, and disconnect the wiring plugs. Label the connectors to ensure they are refitted to their original locations.

142 Remove the exhaust manifolds.

All models

143 Using a oxygen sensor removal socket, unscrew the sensor and remove it from the exhaust pipe.

144 Refitting is a reverse of the removal procedure, noting the following points:

a) *Tighten the sensor to the specified torque.*

b) *Check that the wiring is correctly routed, and in no danger of contacting the exhaust system.*

c) *Ensure that no lubricant or dirt comes into contact with the sensor probe.*

d) *Apply a smear of copper based, high temperature anti-seize grease to the sensor threads prior to refitting.*

Idle speed control valve

M43TU engines

145 The valve is mounted on the side of the throttle body.

146 Ensure the ignition is switched off.

147 Disconnect the wiring plug from the valve.

148 Disconnect the vacuum hose from the valve.

149 Remove the two securing screws, and withdraw the valve from the throttle body. Recover the seal.

150 Refitting is a reversal of removal, but clean the sealing face of the valve, and check the condition of the seal. Renew the seal if necessary.

N42 engines

151 On these engines, the idle speed is regulated by the lift and duration of the intake valves, controlled by the engine management ECM.

M52TU and M54 engines

152 Disconnect the battery negative lead.

153 The idle speed control valve is located below the intake manifold, and above the throttle body.

154 Remove the air cleaner housing.

155 Pull the throttle cable outer (where fitted) from the support bracket on the throttle body.

156 Undo the two clamps and disconnect the intake ducting from the throttle body and idle speed control valve.

157 Disconnect the wiring plugs from the idle speed control valve and the manifold resonance flap actuator solenoid valve.

158 Undo the nut securing the cable support bracket, and the two screws securing the idle speed control valve bracket. Manoeuvre the valve from the manifold. Discard the seal between the valve and the manifold, a new one must be fitted.

159 Smear the new seal with grease and fit it to the intake manifold. Push the idle speed control valve into place and securely tighten the bracket retaining screws/nuts.

160 The remainder of refitting is a reversal of removal.

Fuel pump relay

161 Disconnect the battery negative lead.

162 Remove the passenger side glovebox.

163 Insert a flat-bladed screwdriver into the slot in the end of the glovebox light and carefully prise it from place. Disconnect the wiring plug as the light is withdrawn.

164 Undo the five screws, release the expanding rivet and remove the glovebox frame from the facia.

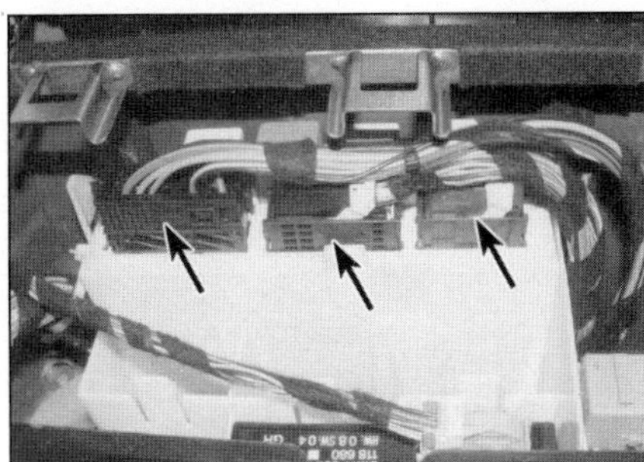
165 Undo the locking levers, and disconnect the wiring plugs from the general control module.

166 Pull the fuel pump relay from the relay plate.

167 Refitting is a reversal of removal.

Main engine management relay

168 Ensure the ignition is switched off.

169 Working in the left-hand corner of the engine compartment, undo the four screws, and remove the cover from the electrical box *(see paragraph 2 earlier in this Section).*

170 Pull the relay from the relay socket.

171 Refitting is a reversal of removal.

Throttle pedal position sensor

N42 engines

172 The throttle pedal position sensor is integral with the pedal assembly.

Intake manifold differential pressure sensor

173 Ensure the ignition is switched off, undo the two nuts, lift the front edge, and pull the plastic engine cover to the front and remove it.

174 Disconnect the wiring plug, undo the screws and remove the sensor from the manifold.

175 Refitting is a reversal of removal.

Notes

Citroën Saxo petrol models '96 – '01

System type		
Single-point injection models:		
954 cc and 1124 cc models	Bosch Monopoint MA3.1	
Multipoint injection models:		
1360 cc models	Magneti Marelli 1 AP	
1587 cc models:		
Up to 97 model year	Bosch Motronic MP5.1	
97 model year onwards	Bosch Motronic MP5.2	
1587 cc 16-valve models	Magneti Marelli 1 AP 41	
Fuel system data		
Fuel pump type	Electric, immersed in tank	
Single-point injection models:		
Fuel pump delivery rate	80 litres per hour	
Fuel pump delivery pressure at full load	1.1 bar	
Regulated fuel pressure	0.7 to 0.9 bar	
Specified idle speed	850 ± 50 rpm (controlled by ECU, non-adjustable)	
Multipoint injection models:		
Fuel pump delivery rate	115 to 120 litres per hour	
Fuel pump delivery pressure at full load	3.0 bar	
Regulated fuel pressure:		
1360 cc models	2.5 to 3.0 bar	
1587 cc (including 16-valve) models	2.0 to 2.5 bar	
Specified idle speed:		
Models without air conditioning	850 ± 50 rpm (not adjustable, controlled by ECU)	
Models with air conditioning	900 ± 50 rpm (not adjustable, controlled by ECU)	
Idle mixture CO content	Less than 0.5 % (controlled by ECU, non-adjustable)	
Recommended fuel		
Minimum octane rating	95 RON unleaded (ie unleaded Premium) only	
Torque wrench setting	**Nm**	**lbf ft**
Knock sensor	20	15

Chapter 7

General information

Single-point injection models

1 The Bosch Monopoint MA3.1 engine management (fuel injection/ignition) system is fitted to all 954 and 1124 cc models. The system is a full engine management system, controlling both the fuel injection and ignition functions. The system employs closed-loop fuelling by means of a catalytic converter and a Lambda sensor, to minimise exhaust gas emissions. An evaporative loss emission control system is also integrated, to minimise the escape of unburned hydrocarbons into the atmosphere from the fuel tank. The overall operation of the system is as follows.

2 The fuel pump, which is immersed in the fuel tank, pumps fuel from the fuel tank to the fuel injector, via a filter mounted underneath the rear of the vehicle. Fuel supply pressure is controlled by the pressure regulator in the throttle body assembly, which lifts to allow excess fuel to return to the tank when the optimum operating pressure of the fuel system is exceeded.

3 The electrical control system consists of the ECU, along with the following sensors and actuators:

a) Throttle potentiometer – informs the ECU of the throttle valve position, and the rate of throttle opening/closing.

b) Coolant temperature sensor – informs the ECU of engine temperature.

c) Intake air temperature sensor – informs the ECU of the temperature of the air passing through the throttle body.

d) Lambda sensor – informs the ECU of the oxygen content of the exhaust gases.

e) Throttle position switch (built into the idle speed stepper motor) – informs the ECU when the throttle valve is closed (ie. if the accelerator pedal is fully released).

f) Crankshaft sensor to inform the ECU of engine speed and crankshaft angular position.

g) Vehicle speed sensor – (1124cc models only) informs the ECU of the vehicle's road speed, allowing manual gear selection to be monitored.

h) Idle control stepper motor – controls the position of the throttle during idling to maintain a constant engine idle speed.

i) Inlet manifold heater – raises the temperature of the inlet manifold to minimise fuel condensation.

j) Instrument panel-mounted warning indicator – illuminated by the ECU when a malfunction is detected.

k) Relay pack – two separate relays mounted in one housing provide separate current supplies to the fuel pump and ECU, sensors and actuators.

4 Signals from each of the sensors are compared by the ECU and, based on this information, the ECU selects the response appropriate to those values, and controls the fuel injector (varying its pulse width – the length of time the injector is held open – to provide a richer or weaker mixture, as appropriate). The mixture and idle speed are constantly varied by the ECU, to provide the best settings for cranking, starting (with either a hot or cold engine) and engine warm-up, idle, cruising and acceleration.

5 The ECU also has full control over the engine idle speed via a stepper motor which is fitted to the throttle body. The motor pushrod rests against a cam on the throttle valve spindle. When the throttle valve is closed (accelerator pedal released), the ECU uses the motor to vary the opening of the throttle valve, and so control the idle speed.

6 If there is any abnormality in any of the readings obtained from either the coolant temperature sensor, the intake air temperature sensor or the Lambda sensor, the ECU enters its back-up mode. If this happens, the sensor signal is overridden, and the ECU assumes a pre-programmed 'back-up' value which will allow the engine to continue running, albeit at reduced efficiency. If the ECU enters this back-up mode, the warning lamp on the instrument panel will be illuminated, and the relevant fault code will be stored in the ECU memory.

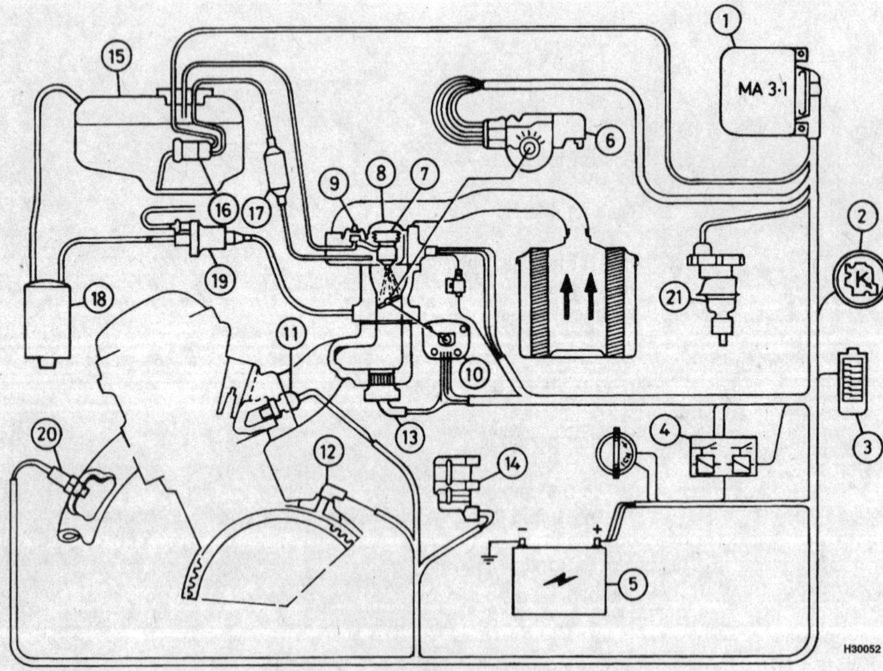

Bosch Monopoint MA3.1 engine management system schematic

1 ECU
2 Warning lamp
3 Diagnostic connector
4 Relay
5 Battery
6 Idle control stepper motor
7 Injector
8 Inlet air temperature sensor
9 Fuel pressure regulator
10 Throttle potentiometer
11 Coolant temperature sensor
12 Crankshaft speed/TDC sensor
13 Manifold heater
14 Ignition coil module
15 Fuel tank
16 Fuel pump
17 Fuel filter
18 Charcoal canister
19 Purge valve
20 Lambda sensor
21 Vehicle speed sensor

Multipoint injection systems

7 All versions of the Magneti Marelli and Bosch Motronic systems covered in this Chapter are full engine management systems, controlling both the fuel injection and ignition functions; refer to the Specifications for system application data. As the layout and overall operation of each of the systems is very similar, a general description is given in the following paragraphs.

8 The systems all employ closed-loop fuelling by means of a catalytic converter and a Lambda sensor, to minimise exhaust gas emissions. An evaporative loss emission control system is also integrated, to minimise the escape of unburned hydrocarbons into the atmosphere from the fuel tank.

9 The fuel pump supplies fuel from the tank to the fuel rail, via a replaceable cartridge filter mounted underneath the rear of the vehicle. The pump itself is mounted inside the fuel tank, the pump motor is permanently immersed in fuel, to keep it cool. The fuel rail is mounted directly above the fuel injectors and acts as a fuel reservoir.

10 The fuel injectors are electromagnetic pintle valves, which spray atomised fuel into the combustion chambers under the control of the engine management system ECU. There are four injectors, one per cylinder, mounted on the fuel rail close to the cylinder head. Each injector is mounted at an angle that allows it to spray fuel directly onto the back of the inlet valve(s). The ECU controls the volume of fuel-injected by varying the length of time for which each injector is held open (the 'duty ratio').

11 The Bosch Motronic systems employ 'banked' fuel injection, where all four injectors are activated simultaneously. Fuel is injected into each cylinder's inlet tract on every engine stroke and is then drawn into the combustion chamber during the induction stroke.

12 The Magneti Marelli systems employ semi-sequential fuel injection, where fuel is injected into each cylinder's inlet tract twice per engine cycle; once during the power stroke and once during the induction stroke.

13 Fuel rail supply pressure is controlled by the pressure regulator, mounted at the end of the fuel rail. The regulator contains a spring-loaded valve, which lifts to allow excess fuel to return to the tank when the optimum operating pressure of the fuel system is exceeded (eg during low speed, light load cruising). The regulator also contains a diaphragm which is supplied with vacuum from the inlet manifold. This allows the regulator to reduce the fuel supply pressure during light load, high manifold depression conditions (e.g. during idling or deceleration) to prevent excess fuel being 'sucked' through the open injectors. Note that on later models, a fuel pressure regulator is not fitted; a fuel pressure monitoring point is provided at the left hand end of the fuel rails.

14 The electrical control system consists of the ECU, along with the following sensors and actuators:

a) Throttle potentiometer – informs the ECU of the throttle valve position, and the rate of throttle opening/closing.

b) Coolant temperature sensor – informs the ECU of engine temperature.

c) Inlet air temperature sensor – informs the ECU of the temperature of the air passing through the throttle body.

d) Lambda sensor – informs the ECU of the oxygen content of the exhaust gases.

e) Inlet manifold pressure sensor – informs the ECU of the engine load.

f) Crankshaft sensor to inform the ECU of engine speed and crankshaft angular position.

g) Vehicle speed sensor – informs the ECU of the vehicle's road speed, allowing manual gear selection to be monitored.

h) Throttle body heater – raises the temperature of the throttle body to minimise fuel condensation and prevent throttle valve icing.

i) Instrument panel-mounted warning indicator – illuminated by the ECU when a malfunction is detected.

j) Relay pack – two separate relays mounted in one housing, providing separate current supplies to the fuel pump and ECU, sensors and actuators.

15 Signals from each of the sensors are compared by the ECU and, based on this information, the ECU selects the response appropriate to those values, and controls the fuel injector duty ratio to provide a richer or weaker air/fuel mixture, as appropriate. The air/fuel mixture is constantly varied by the ECU, to provide the best settings for cranking, starting (with either a hot or cold engine) and engine warm-up, idle, cruising and acceleration.

16 All systems are equipped with a knock sensor, mounted on the front of the cylinder block.

17 The ECU also has full control over the engine idle speed, via a stepper motor fitted to the throttle body (Magneti Marelli systems), or an idle actuator valve (Bosch Motronic systems). On models with Magneti Marelli systems, the stepper motor pushrod controls the amount of air passing through a by-pass drilling at the side of the throttle. When the throttle valve is closed (accelerator pedal released), the ECU uses the motor to alter the position of the pushrod, controlling the amount of air bypassing the throttle valve and so controlling the idle speed. On models with Bosch Motronic systems, the same principle is applied, but the flow of air bypassing the throttle valve is controlled by an idle actuator valve, which is separate from the throttle body. The ECU also carries out 'fine tuning' of the idle speed by varying the ignition timing to increase or reduce the torque of the engine as it is idling. This helps to stabilise the idle speed when electrical or mechanical loads (such as headlights, air conditioning etc) are switched on and off.

18 The throttle body is fitted with an electric heating element. The heater is supplied with current by the ECU, warming the throttle body on cold starts to help prevent icing of the throttle valve.

19 If there is any abnormality in any of the readings obtained from either the coolant temperature sensor, the inlet air temperature sensor or the Lambda sensor, the ECU enters its 'back-up' mode. If this happens, the erroneous sensor signal is overridden, and the ECU assumes a pre-programmed 'back-up' value, which will allow the engine to continue running, albeit at reduced efficiency. If the ECU enters this mode, the warning lamp on the instrument panel will be illuminated, and the relevant fault code will be stored in the ECU memory.

Fuel system – depressurisation

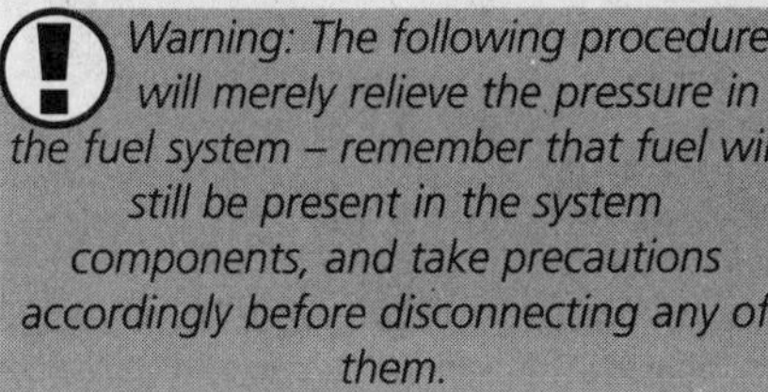

Warning: The following procedure will merely relieve the pressure in the fuel system – remember that fuel will still be present in the system components, and take precautions accordingly before disconnecting any of them.

1 The fuel system referred to in this Section is defined as the tank-mounted fuel pump, the fuel filter, the fuel injector and the pressure regulator in the injector housing, and the metal pipes and flexible hoses of the fuel lines between these components. All these contain fuel, which will be under pressure while the engine is running and/or while the ignition is switched on. The pressure will remain for some time after the ignition has been switched off, and must be relieved before any of these components are disturbed for servicing work.

2 Ensure that the engine has cooled completely before starting work. Disconnect the battery negative cable and position it away from the terminal.

3 Place a suitable container beneath the relevant connection/union to be disconnected, and have a large rag ready to soak up any escaping fuel not being caught by the container.

4 Slowly loosen the connection or union nut (as applicable) to avoid a sudden release of pressure, and position the rag around the connection to catch any fuel spray which may be expelled. Once the pressure is released, disconnect the fuel line. Plug the openings, to minimise fuel loss and prevent the entry of dirt into the fuel system.

Bosch Monopoint system components – removal and refitting

Fuel injector

Note: *If a faulty injector is suspected, before condemning the injector, it is worth trying the effect of one of the proprietary injector-cleaning treatments.*

1 Disconnect the battery negative cable and position it away from the terminal.

2 Remove the air cleaner-to-throttle body duct, then slacken and withdraw the securing screws and remove the plastic adapter ring from the top of the injector body.

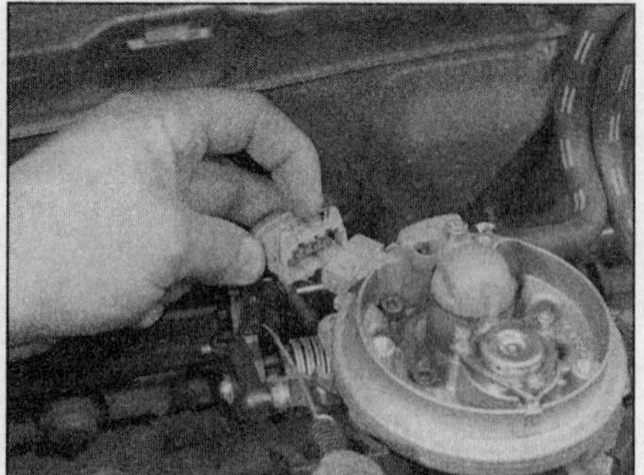

3 Unplug the wiring harness from the fuel injector at the multiway connector.

4 Pad the area around and above the injector cap with absorbent rag, to prevent fuel spray. Slacken and withdraw the securing screw, then remove the injector cap, recover the gasket and lift out the injector.

5 Refitting is a reversal of the removal procedure, ensuring that the injector sealing ring(s) and injector cap O-ring are in good condition. When refitting the injector cap, ensure that the injector pins are correctly aligned with the cap terminals; the cap is shaped so that it can only be refitted the correct way around.

Fuel pressure regulator

Note : *At the time of writing, the fuel pressure regulator assembly is not available separately. If the fuel pressure regulator assembly is faulty, the complete throttle body assembly must be renewed. Refer to a Citroën dealer for further information on parts availability. Although the unit can be dismantled for cleaning, if required, it should not be disturbed unless absolutely necessary.*

6 Disconnect the battery negative cable and position it away from the terminal.

7 Remove the air cleaner-to-throttle body duct.

8 Using a marker pen, make alignment marks between the regulator cover and throttle body. Slacken and remove the cover retaining screws .

9 As the screws are slackened, place an absorbent rag over the cover, to catch any fuel spray which may be released.

10 Lift off the cover, then remove the spring and withdraw the diaphragm, noting its correct fitted orientation.

11 Remove all traces of dirt, and examine the diaphragm for signs of splitting; renew if necessary.

12 Refitting is a reverse of the removal procedure, ensuring that the diaphragm and cover are fitted the correct way around and the retaining screws are securely tightened.

Idle control stepper motor

13 Disconnect the battery negative able and position it away from the terminal.

14 Depress the retaining clip, and disconnect the wiring connector from the idle control stepper motor.

15 Undo the retaining screws, and remove the motor from the throttle body. There are three securing screws; two on the lower edge of the motor casing and one on the upper edge at the right hand side (as viewed from the front of the engine compartment).

16 Refitting is a reverse of the removal procedure, ensuring that the motor retaining screws are securely tightened.

Throttle potentiometer

17 The throttle potentiometer is a sealed unit, and under **no** circumstances should it be disturbed. For this reason, on some models, it is secured to the throttle body assembly by tamperproof screws. If the throttle potentiometer is faulty, the complete throttle body assembly must be renewed. Refer to your Citroën dealer or specialist for further information. The illustration shows the wiring connector being unplugged for the purposes of the throttle body removal.

Intake air temperature sensor

Note : *On some later models, at the time of writing, the intake air temperature sensor is not available separately. If the sensor is faulty, the complete throttle body assembly must be renewed. Refer to your Citroën dealer or specialist for further information on parts availability.*

18 The intake air temperature sensor is an integral part of the throttle body injector cap. To remove the cap, first disconnect the battery negative terminal, then remove the air cleaner-to-throttle body duct.

19 Undo the three retaining screws, and remove the circular plastic ring from the top of the throttle body. Recover its sealing ring.

20 Depress the retaining clip, and unplug the wiring connector from the injector wiring connector.

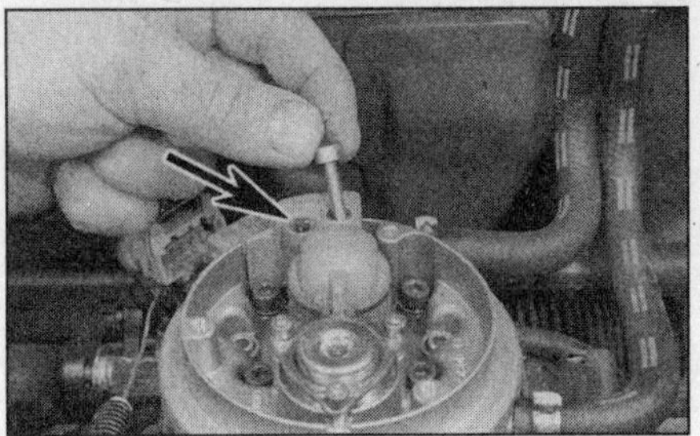

21 Undo the injector cap retaining screw, then lift off the cap and recover the gasket. As the cap screw is slackened, place a rag over the injector, to catch any fuel spray which may be released.

22 Refitting is a reversal of the removal procedure, ensuring that the injector cap gasket and/or O-ring is in good condition. Take care to ensure that the cap terminals are correctly aligned with the injector pin, and securely tighten the cap retaining screw.

Coolant temperature sensor

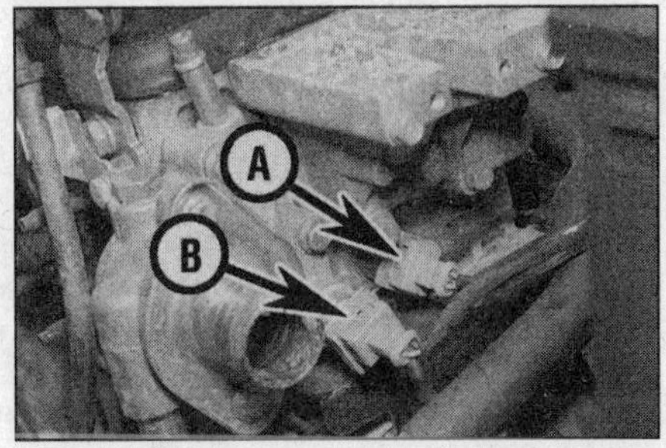

A Temperature warning light switch/temperature gauge sensor
B Fuel injection system coolant temperature sensor

23 The fuel injection system coolant temperature sensor has a green body and is mounted in the left-hand side of the thermostat housing. Partially drain the cooling system to just below the level of the sensor.

24 Unplug the wiring connector from the sensor, then unscrew it from its mounting and recover the sealing ring (where applicable).

25 Refitting is a reversal of removal. If no sealing ring was originally fitted, coat the threads of the new sensor with sealing compound prior to refitting.

Electronic control unit (ECU)

26 The ECU is located on the right-hand side of the engine compartment, underneath a large plastic cover. To remove the ECU, first disconnect the battery negative cable and position it away from the terminal.

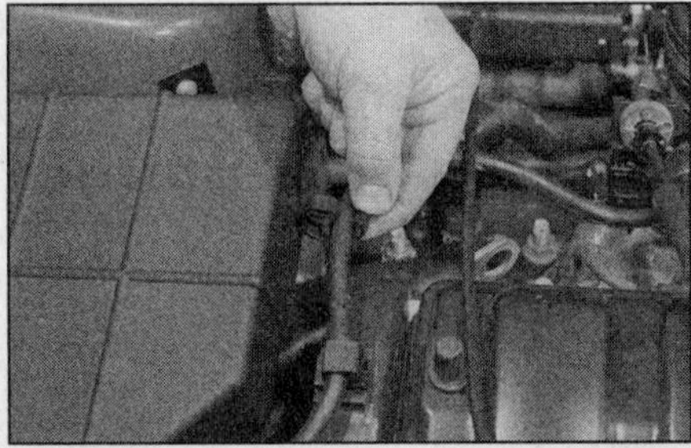

27 Unfasten the securing clips and release the purge valve vacuum hose from the side of the ECU plastic cover. Unclip the cover from the mounting plate, then lift the retaining clip and disconnect the wiring connector from the ECU. Slacken and remove the ECU retaining bolts, and remove it from the vehicle. Alternatively, unbolt the mounting bracket from the bodywork and remove it complete with the ECU.

28 Refitting is a reverse of the removal procedure, ensuring that the wiring connector is securely reconnected.

Fuel injection system relay unit

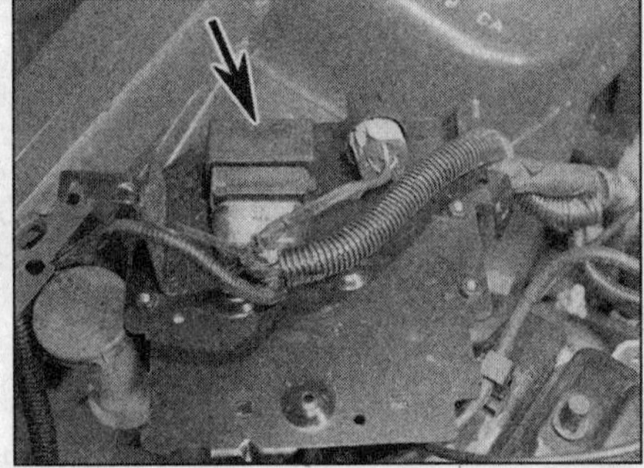

29 The relay unit is clipped onto the underside of the ECU mounting plate, on the right-hand side of the engine compartment. To remove the relay unit, first disconnect the battery negative cable and position it away from the terminal.

30 Remove the ECU mounting bracket as described in the previous sub-Section. Unclip the relay unit from the mounting plate, disconnect the wiring connector and remove the unit from the vehicle.

31 Refitting is the reverse of removal, ensuring that the relay unit is securely held in position by its retaining clip.

Crankshaft/TDC sensor

32 The crankshaft sensor is situated on the front face of the transmission clutch housing.

33 To remove the sensor, first disconnect the battery negative cable and position it away from the terminal.

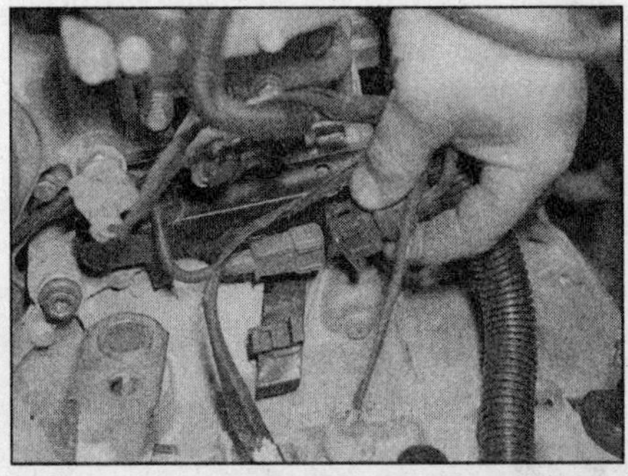

34 Trace the wiring back from the sensor to the wiring connector, and disconnect it from the main harness.

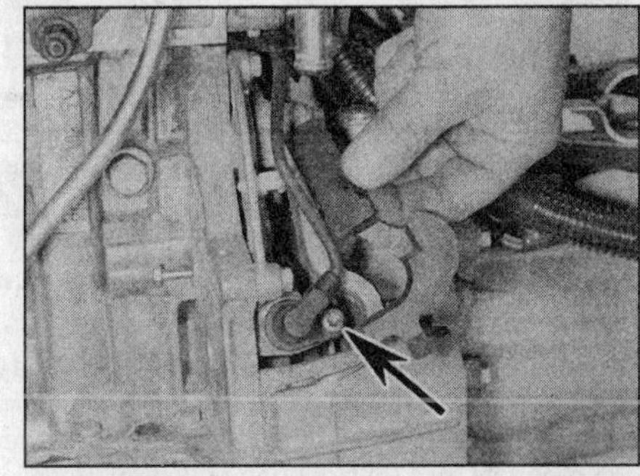

35 Prise out the plastic cover plate, then undo the retaining bolt and withdraw the sensor from the transmission.

36 Refitting is the reverse of the removal procedure. Ensure that the sensor retaining bolt is securely tightened, and that the grommet is correctly seated in the transmission housing.

Vehicle speed sensor

1124 cc models

37 The vehicle speed sensor is an integral part of the speedometer drive housing. Jack up the front of the car and support it securely on axle stands. The speedometer drive is situated on the rear of the transmission housing, next to the inner end of the right-hand driveshaft.

38 Undo the wiring connector from the sensor, then slacken and remove the retaining bolt.

39 Withdraw the sensor from the housing, along with its O-ring.

40 Refitting is a reversal of removal, ensuring that a new O-ring is fitted.

Inlet manifold heater

41 Remove the inlet manifold.

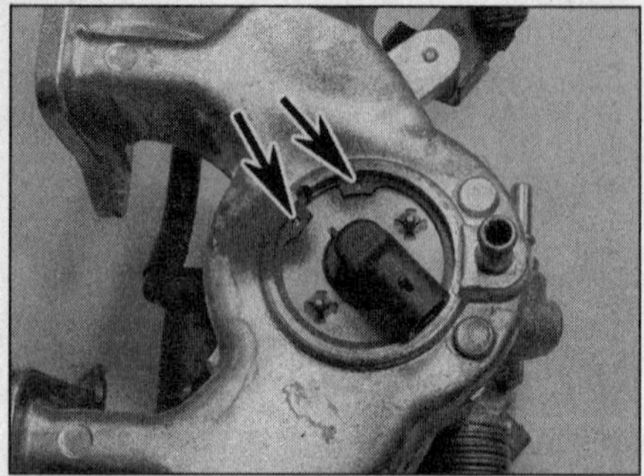

42 Release the circlip and withdraw the heater from the underside of the manifold casting.

43 Refitting is a reversal of removal.

Lambda sensor

44 The Lambda sensor is threaded into the exhaust downpipe, and is easily accessible from the front of the engine bay. Access may be further improved by removing the radiator and cooling fan(s).

45 Unplug the wiring harness from the oxygen sensor at the connector, which is mounted on the bracket at the top of the transmission casing.

46 Note that a flying lead remains connected to the sensor after it has been disconnected; if the appropriate size of an open-ended spanner is not available, a slotted socket will be required to remove the sensor. Slacken and withdraw the sensor, taking care to avoid damaging the sensor probe as it is removed.

47 Apply a little anti-seize grease to the sensor threads – avoid contaminating the probe tip.

48 Refit the sensor to its housing, tightening it securely. Restore the harness connection.

Bosch Motronic system components – removal and refitting

Fuel rail and injectors

Note: *If a faulty injector is suspected, before condemning the injector, it is worth trying the effect of one of the proprietary injector-cleaning treatments.*

1 Disconnect the battery negative cable and position it away from the terminal, then depressurise the fuel system.

2 Disconnect the vacuum pipe from the port at the top of the fuel pressure regulator.

3 Slacken the retaining clips and disconnect the fuel feed and return hoses from the fuel rail. Where the original crimped-type Citroën hose clips are still fitted, cut them off and discard; replace them with standard worm-drive hose clips on refitting.

4 Depress the retaining tangs and disconnect the wiring connectors from the four injectors.

5 Slacken and remove the fuel rail retaining bolts nuts, then carefully ease the fuel rail and injector assembly out from the inlet manifold and remove it from the vehicle. Remove the O-rings from the lower end of each injector, and discard them; they must be renewed whenever they are disturbed.

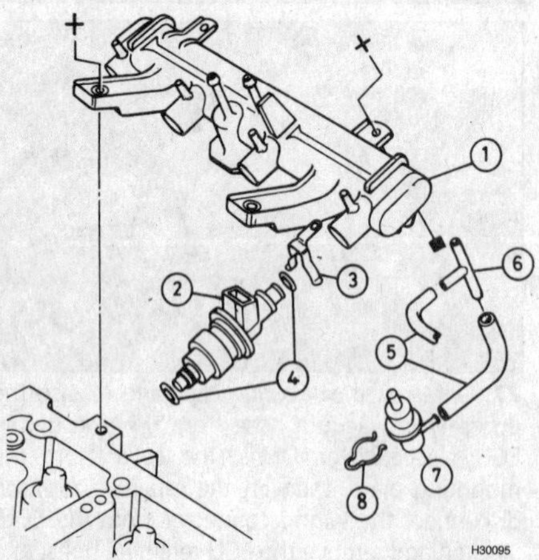
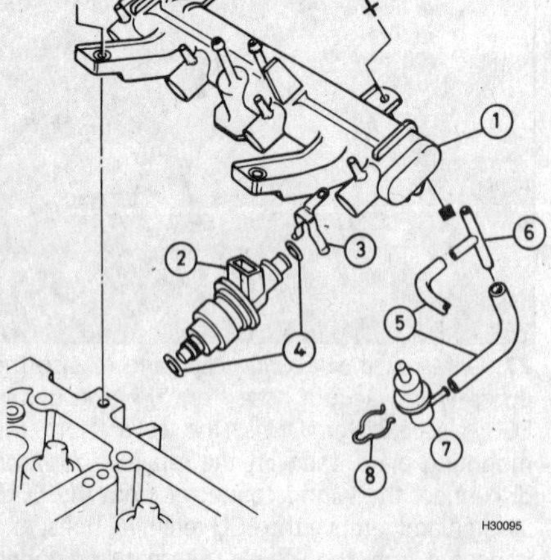

Fuel rail, injector and pressure regulator assembly (early 1.6 litre 8-valve model shown)

1 Fuel rail
2 Injector
3 Retaining clip
4 O-ring seals
5 Fuel hoses
6 T-piece
7 Fuel pressure regulator
8 Retaining clip

6 Slide out the retaining clip(s) and remove the relevant injector(s) from the fuel rail. Remove the upper O-ring from each disturbed injector and discard; any O-rings which are disturbed during removal must be renewed.

7 Refitting is a reversal of the removal procedure, noting the following points:

a) Fit new O-rings to all injector unions disturbed on removal.

b) Apply a smear of engine oil to the O-rings to aid installation, then ease the injectors and fuel rail into position, ensuring that none of the O-rings are displaced.

c) On completion, start the engine and check for fuel leaks.

Fuel pressure regulator

Note: *Later models are not fitted with a fuel pressure regulator.*

8 Disconnect the vacuum pipe from the regulator.

9 Place a wad of rag under and around the regulator, to catch any fuel spray which may be released, then remove the retaining clip and ease the regulator out from the fuel rail.

10 Refitting is a reversal of the removal procedure. Examine the regulator seal for signs of damage or deterioration, and renew if necessary.

Throttle potentiometer

11 Disconnect the battery negative terminal.

12 Depress the retaining clip and disconnect the wiring connector from the throttle potentiometer.

13 Slacken and remove the two retaining screws, then disengage the potentiometer from the throttle valve spindle and remove it from the vehicle.

14 Refitting is a reverse of the removal procedure, ensuring that the potentiometer is correctly engaged with the throttle valve spindle.

Electronic Control Unit (ECU)

15 The ECU is located on the right-hand side of the engine compartment. On some models, a large plastic cover is mounted over the ECU.

16 To remove the ECU, first disconnect the battery negative lead.

17 Unclip the cover from the mounting plate, then lift the retaining clip and disconnect the wiring connector from the ECU. Slacken and remove the ECU retaining bolts, and remove it from the vehicle (refer to the illustrations in the Bosch Monopoint system Section for greater detail).

18 Refitting is a reverse of the removal procedure, ensuring that the wiring connector is securely reconnected by pushing the retaining clip firmly home.

Idle actuator valve

19 On early models, the valve is mounted on a bracket on the underside of the inlet manifold. On later models, the valve is mounted on the right hand end of the inlet manifold casting.

20 To remove it, first disconnect the battery negative cable and position it away from the terminal.

21 Depress the retaining clip, and disconnect the wiring connector from the air valve.

22 Slacken the retaining clips, and disconnect the air hose(s) from the end of the auxiliary air valve.

23 On early models, slide the valve out from its mounting bracket, and remove it from the engine compartment. On later models, remove the securing screw and withdraw the valve from the manifold.

24 Refitting is a reversal of the removal procedure. Examine the mounting bracket rubber lining for signs of deterioration, and renew it if necessary.

Manifold pressure sensor

25 On early models, the manifold pressure sensor is mounted remotely on the right-hand side of the engine compartment bulkhead. On later models, the sensor is mounted directly on the front of the inlet manifold casting. To remove the sensor, first disconnect the battery negative cable and position it away from the terminal.

26 On early models, undo the retaining nut, and free the sensor from its mounting bracket. On later models, remove the securing screw and withdraw the sensor from the manifold.

27 Disconnect the wiring connector and vacuum hose, and remove the sensor from the engine compartment.

28 Refitting is the reverse of the removal procedure.

Coolant temperature sensor

29 The coolant temperature sensor is threaded into the thermostat housing, at the left hand end of the cylinder head. Do not confuse it with the other sensors in the thermostat housing – refer to the information given in the Bosch Monopoint system Section for location and identification details.

30 Allow the engine to cool completely, then partially drain the cooling system.

31 Ensure that the ignition is switched off then unplug the wiring from the sensor connector.

32 Carefully unscrew the sensor from the thermostat housing – be prepared for some coolant spillage. Recover the sensor sealing ring.

33 Refitting is a reversal of removal. Ensure that the sensor wiring is securely reconnected before switching on the ignition and attempting to start the engine.

Inlet air temperature sensor

34 On early 1587 cc models, the inlet air temperature sensor is screwed into the top of the air cleaner housing. On later models, the sensor is threaded into the underside of the throttle body. Before removing the sensor, first ensure that the ignition is switched off.

35 Unplug the wiring from the sensor connector, then unscrew the sensor and remove it from the vehicle. Recover the sealing ring, where applicable.

36 Refitting is the reverse of removal.

Crankshaft/TDC sensor

37 The crankshaft/TDC sensor is situated on the front face of the transmission clutch housing.

38 To remove the sensor, first ensure that the ignition is switched off.

39 Trace the wiring back from the sensor to the wiring connector, and disconnect it from the main harness.

40 Prise out the rubber grommet, then undo the retaining bolt and withdraw the sensor from the transmission. Refer to the illustrations in the Bosch Monopoint system Section for greater detail.

41 Refitting is reverse of the removal procedure, ensuring that the sensor retaining bolt is securely tightened and the grommet is correctly seated in the transmission housing.

Fuel injection system relay unit

42 The relay unit is clipped onto the underside of the ECU mounting plate, on the right-hand side of the engine compartment.

43 To remove the relay unit, first disconnect the battery negative lead.

44 Unclip the relay unit from the mounting plate, disconnect the wiring connector and remove the unit from the vehicle.

45 Refitting is the reverse of removal, ensuring that the relay unit is securely held in position by its retaining clip.

Lambda sensor

46 Refer to the information in the Bosch Monopoint system Section.

Magneti Marelli system components – removal and refitting

Fuel rail and injectors

1 Refer to the information given in the Bosch Motronic system Section.

Fuel pressure regulator

2 Refer to the information given in the Bosch Motronic system Section.

Throttle potentiometer

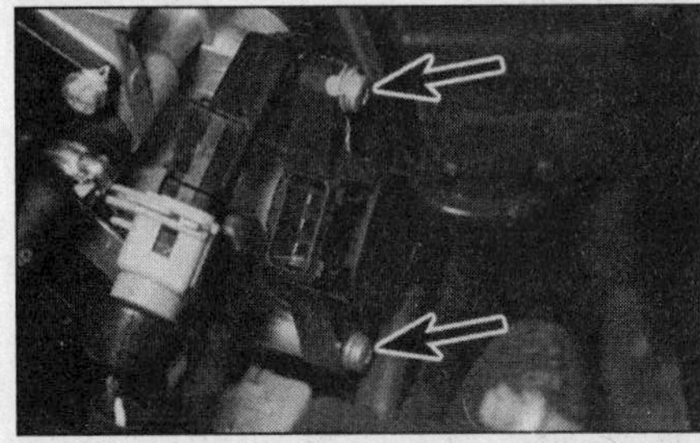

3 Refer to the information given in the Bosch Motronic system Section.

Electronic Control Unit (ECU)

4 Refer to the information given in the Bosch Motronic system Section.

Idle speed control stepper motor

5 The idle speed control stepper motor is located on the side of the throttle housing assembly. To remove the motor, first disconnect the battery negative terminal.

6 Release the retaining clip, and disconnect the wiring connector from the motor.

7 Slacken and remove the two retaining screws, and withdraw the motor from the throttle housing.

8 Refitting is a reversal of the removal procedure.

Manifold pressure sensor

9 On early models, the sensor is mounted on a bracket on the bulkhead at the rear of the engine compartment bulkhead. On later models, the manifold pressure sensor is mounted directly on the inlet manifold casting. To remove it, first ensure that the ignition is switched off.

10 Release the locking tab and unplug the wiring from the sensor connector.

11 On early models, disconnect the vacuum hose, then remove the screws and withdraw the sensor from the mounting bracket. On later models, remove the securing screws and withdraw the sensor from the manifold, recovering the sealing ring.

12 Refitting is a reversal of removal.

Coolant temperature sensor

13 Refer to the information given in the Bosch Monopoint system Section.

Inlet air temperature sensor

14 The sensor is threaded into the underside of the throttle body. Before removing the sensor, first ensure that the ignition is switched off.

15 Unplug the wiring from the sensor connector, then unscrew the sensor and remove it from the vehicle. Recover the sealing ring, where applicable.

16 Refitting is the reverse of removal.

Crankshaft/TDC sensor

17 Refer to the information given in the Bosch Motronic system Section.

Fuel injection system relay unit

18 Refer to the information given in the Bosch Motronic system Section.

Knock sensor

19 The knock sensor is screwed onto the rear face of the cylinder block.

20 To gain access to the sensor, firmly apply the handbrake, then jack up the front of the vehicle and support it on axle stands. Access to the sensor can then be gained from underneath the vehicle.

21 Trace the wiring back from the sensor to its wiring connector, and disconnect it from the main loom.

22 Slacken and remove the bolt securing the sensor to the cylinder block, and remove it from underneath the vehicle.

23 Refitting is a reversal of the removal procedure, ensuring that the sensor wiring is correctly routed and its retaining bolt is tightened to the correct torque.

Note: *The knock sensor will only operate correctly if its securing bolt is tightened accurately to the specified torque setting.*

Throttle housing heating element

24 The throttle housing heating element is fitted to the side of the throttle housing. To remove the element, first disconnect the battery negative cable and position it away from the terminal.

25 To improve access, disconnect the accelerator inner cable from the throttle cam, then withdraw the outer cable from the mounting bracket, along with its flat washer and spring clip.

26 Disconnect the element wiring connector, then undo the retaining screw, and free the wiring connector from the throttle housing.

27 Undo the screws securing the accelerator cable bracket to the side of the throttle housing. Carefully remove the bracket, and recover the spring from the top of the heating element.

28 Ease the heating element out from the throttle housing. Examine the O-ring for signs of damage or deterioration, and renew if necessary.

29 Refitting is a reversal of the removal procedure; where necessary, use a new O-ring.

Vehicle speed sensor

30 The vehicle speed sensor is an integral part of the speedometer drive housing. Refer to The Bosch Monopoint system Section.

Lambda sensor

31 Refer to the information given in the Bosch Monopoint system Section. Note that in some markets, 1360 cc Saxo models conforming to CEE 2000 emissions standards are fitted with two Lambda sensors; one upstream of the catalytic converter, and one downstream.

Citroën Picasso '00 – '02

Petrol models

System type

1.6 litre engines:	
Pre-2001 models (engine code NFZ)	Bosch Motronic MP7.2
2001 models onward (engine code NFV)	Bosch Motronic ME7.4.4
1.8 litre engines	Sagem S2000

Fuel system data

Fuel pump type	Electric, immersed in tank
Idle speed*:	
1.6 litre engines:	
Without air conditioning	850 ± 50 rpm
With air conditioning	900 ± 50 rpm
1.8 litre engines	700 ± 50 rpm
Idle mixture CO content*	Less than 0.5%

** Not adjustable – controlled by ECU*

Recommended fuel

Minimum octane rating	95 RON unleaded

Diesel models

General

System type	HDi (High-pressure Diesel injection) with full electronic control, direct injection and turbocharger
Designation	Bosch EDC 15
Firing order	1-3-4-2 (No 1 at flywheel end)
Fuel system operating pressure	1350 bars

High-pressure fuel pump

Type	Bosch CP 1
Direction of rotation	Clockwise, viewed from sprocket end

Injectors

Type	Electromagnetic

Turbocharger

Type	Garrett GT15 or KKK K03
Boost pressure (approximate)	1 bar at 3000 rpm

Torque wrench settings

	Nm	lbf ft
Petrol models		
Knock sensor retaining bolt	20	15
Diesel models		
Accumulator rail mounting bolts	23	17
Fuel injector clamp nuts	30	22
Fuel pressure sensor to accumulator rail	45	33

Chapter 8

Fuel injection systems – general information

Petrol models

Note: *The fuel injection ECU is of the 'self-learning' type, meaning that as it operates, it also monitors and stores the settings which give optimum engine performance under all operating conditions. When the battery is disconnected, these settings are lost and the ECU reverts to the base settings programmed into its memory at the factory. On restarting, this may lead to the engine running/idling roughly for a short while, until the ECU has relearned the optimum settings. This process is best accomplished by taking the vehicle on a road test (for approximately 15 minutes), covering all engine speeds and loads, concentrating mainly in the 2500 to 3500 rpm region.*

On all engines, the fuel injection and ignition functions are combined into a single engine management system. The systems fitted are manufactured by Bosch and Sagem, and are very similar to each other in most respects, the only significant differences being in the software contained in the system ECU, and specific component location according to engine type. Each system incorporates a closed-loop catalytic converter and an evaporative emission control system, and complies with the latest emission control standards. The system operates as follows.

The fuel pump supplies fuel from the tank to the fuel rail, via a replaceable cartridge filter mounted on the side of the fuel tank. The pump itself is mounted inside the tank, with the pump motor permanently immersed in fuel, to keep it cool. The fuel rail is mounted directly above the fuel injectors and acts as a fuel reservoir.

Fuel rail supply pressure is controlled by the pressure regulator, also located in the fuel tank. The regulator contains a spring-loaded valve, which lifts to allow excess fuel to recirculate within the tank when the optimum operating pressure of the fuel system is exceeded (eg, during low speed, light load cruising).

The fuel injectors are electromagnetic pintle valves, which spray atomised fuel into the combustion chambers under the control of the engine management system ECU. There are four injectors, one per cylinder, mounted in the inlet manifold close to the cylinder head. Each injector is mounted at an angle that allows it to spray fuel directly onto the back of the inlet valve(s). The ECU controls the volume of fuel injected by varying the length of time for which each injector is held open. The fuel injection systems are typically of the sequential type, whereby each injector operates individually in cylinder sequence.

The electrical control system consists of the ECU, along with the following sensors:

a) Throttle potentiometer – informs the ECU of the throttle valve position, and the rate of throttle opening/closing.
b) Coolant temperature sensor – informs the ECU of engine temperature.
c) Inlet air temperature sensor – informs the ECU of the temperature of the air passing through the throttle housing.
d) Lambda sensors – inform the ECU of the oxygen content of the exhaust gases (explained in greater detail in Part C of this Chapter).
e) Manifold pressure sensor – informs the ECU of the load on the engine (expressed in terms of inlet manifold vacuum).
f) Crankshaft sensor – informs the ECU of engine speed and crankshaft angular position.
g) Vehicle speed sensor – informs the ECU of the vehicle speed.
h) Knock sensor – informs the ECU of pre-ignition (detonation) within the cylinders.
i) Camshaft sensor – informs the ECU of which cylinder is on the firing stroke on systems with sequential injection.

Signals from each of the sensors are compared by the ECU and, based on this information, the ECU selects the response appropriate to those values, and controls the fuel injectors (varying the pulse width – the length of time the injectors are held open – to provide a richer or weaker air/fuel mixture, as appropriate). The air/fuel mixture is constantly varied by the ECU, to provide the best settings for cranking, starting (with either a hot or cold engine) and engine warm-up, idle, cruising and acceleration.

The ECU also has full control over the engine idle speed, via a stepper motor fitted to the throttle housing. The stepper motor controls the amount of air passing through a bypass drilling at the side of the throttle. When the throttle valve is closed (accelerator pedal released), the ECU uses the motor to open or close an air passage, controlling the amount of air bypassing the throttle valve and so controlling the idle speed. The ECU also carries out 'fine tuning' of the idle speed by varying the ignition timing to increase or reduce the torque of the engine as it is idling. This helps to stabilise the idle speed when electrical or mechanical loads (such as headlights, air conditioning, etc) are switched on and off.

The throttle housing is also fitted with an electric heating element. The heater is supplied with current by the ECU, warming the throttle housing on cold starts to help prevent icing of the throttle valve.

If there is any abnormality in any of the readings obtained from the coolant temperature sensor, the inlet air temperature sensor or the lambda sensor, the ECU enters its 'back-up' mode. If this happens, the erroneous sensor signal is overridden, and the ECU assumes a pre-programmed 'back-up' value, which will allow the engine to continue running, albeit at reduced efficiency. If the ECU enters this mode, the warning lamp on the instrument panel will be illuminated, and the relevant fault code will be stored in the ECU memory.

Diesel models

The fuel system consists of a centrally-mounted fuel tank and fuel lift pump, a fuel cooler mounted under the car, a fuel filter with integral water separator, and a turbocharged, electronically-controlled High-pressure Diesel injection (HDi) system.

The exhaust system is conventional, but to meet the latest emission levels an unregulated catalytic converter and an exhaust gas recirculation system are fitted to all models.

The HDi system (generally known as a 'common rail' system) derives its name from the fact that a common rail (also referred to as an accumulator rail) or fuel reservoir, is used to supply fuel to all the fuel injectors. Instead of an in-line or distributor type injection pump, which distributes the fuel directly to each injector, a high-pressure pump is used, which generates a very high fuel pressure (approximately 1350 bars) in the accumulator rail. The accumulator rail stores fuel, and maintains a constant fuel pressure, with the aid of a pressure control valve. Each injector is supplied with high-pressure fuel from the accumulator rail, and the injectors are individually controlled via signals from the system electronic control unit (ECU). The injectors are electromagnetically-operated.

In addition to the various sensors used on models with a conventional fuel injection pump, common rail systems also have a fuel pressure sensor. The fuel pressure sensor allows the ECU to maintain the required fuel pressure, via the pressure control valve.

For the purposes of describing the operation of a common rail injection system, the components can be divided into three sub-systems; the low-pressure fuel system, the high-pressure fuel system and the electronic control system.

Low-pressure fuel system

The low-pressure fuel system consists of the following components:

a) Fuel tank.
b) Fuel lift pump.
c) Fuel cooler.
d) Fuel filter/water trap.
e) Low-pressure fuel lines.

The low-pressure system (fuel supply system) is responsible for supplying clean fuel to the high-pressure fuel system.

High-pressure fuel system

The high-pressure fuel system consists of the following components:

a) High-pressure fuel pump with pressure control valve.
b) High-pressure accumulator rail.
c) Fuel injectors.
d) High-pressure fuel lines.

After passing through the fuel filter, the fuel reaches the high-pressure pump, which forces it into the accumulator rail, generating a pressure of 1350 bars. As diesel fuel has a certain elasticity, the pressure in the accumulator rail remains constant, even though fuel leaves the rail each time one of the injectors operates. Additionally, a pressure control valve mounted on the high-pressure pump ensures that the fuel pressure is maintained within pre-set limits.

The pressure control valve is operated by the ECU. When the valve is opened, fuel is returned from the high-pressure pump to the tank, via the fuel return lines, and the pressure in the accumulator rail falls. To enable the ECU to trigger the pressure control valve correctly, the pressure in the accumulator rail is measured by a fuel pressure sensor.

The electromagnetically-controlled fuel injectors are operated individually, via signals from the ECU, and each injector injects fuel directly into the relevant combustion chamber. The fact that high fuel pressure is always available allows very precise and highly flexible injection in comparison to a conventional injection pump: for example combustion during the main injection process can be improved considerably by the pre-injection of a very small quantity of fuel.

Electronic control system

The electronic control system consists of the following components:

a) Electronic control unit (ECU).
b) Crankshaft speed/position sensor.
c) Camshaft position sensor.
d) Accelerator pedal position sensor.
e) Coolant temperature sensor.
f) Fuel temperature sensor.
g) Air mass meter.
h) Fuel pressure sensor.
i) Fuel injectors.
j) Fuel pressure control valve.
k) Preheating control unit.
l) EGR solenoid valve.

The information from the various sensors is passed to the ECU, which evaluates the signals. The ECU contains electronic 'maps' which enable it to calculate the optimum quantity of fuel to inject, the appropriate start of injection, and even pre- and post-injection fuel quantities, for each individual engine cylinder under any given condition of engine operation.

Additionally, the ECU carries out monitoring and self-diagnostic functions. Any faults in the system are stored in the ECU memory, which enables quick and accurate fault diagnosis using appropriate diagnostic equipment (such as a suitable fault code reader).

Fuel injection systems – depressurisation, priming and bleeding

Petrol models

Warning: The following procedure will merely relieve the pressure in the fuel system – remember that fuel will still be present in the system components and take precautions accordingly before disconnecting any of them.

1 The fuel system referred to in this Section is defined as the tank-mounted fuel pump, the fuel filter, the fuel injectors, the fuel rail and the metal pipes and flexible hoses of the fuel lines between these components. All these contain fuel which will be under pressure while the engine is running, and/or while the ignition is switched on. The pressure will remain for some time after the ignition has been switched off, and must be relieved in a controlled fashion when any of these components are disturbed for servicing work.

2 Disconnect the battery negative terminal.

3 On 1.8 litre engines, remove the engine cover.

4 Place a container beneath the connection/union to be disconnected, and have a large rag ready to soak up any escaping fuel not being caught by the container.

5 Slowly loosen the connection or union nut to avoid a sudden release of pressure, and position the rag around the connection to catch any fuel spray which may be expelled. Once the pressure is released, disconnect the fuel line. Plug the pipe ends, to minimise fuel loss and prevent the entry of dirt into the fuel system.

6 On 1.8 litre engines a Schrader valve is fitted to the centre of the fuel rail and can, if desired, be used for depressurisation. The Schrader valve operates like a tyre valve, whereby on depressing the central plunger, the system pressure will be released. Ensure that the valve is protected with rags to soak up escaping fuel as this is done.

Diesel models

7 Although there is no specific requirement to depressurise the fuel system, common sense dictates that the engine should be left to cool prior to commencing any repair procedure. During this cooling period, any residual pressure in the system should have mostly dissipated. Before slackening any fuel pipe union, place a clean rag over the connection to soak up and diesel spray.

8 The fuel system is entirely self-bleeding because the fuel lift pump supplies fuel to the high-pressure pump whenever the ignition is switched on.

9 In the case of running out of fuel, or after disconnecting any part of the fuel supply system, ensure that there is fuel in the tank, then start the engine in the normal way.

Petrol fuel injection system components – removal and refitting

1.6 litre engines

Fuel rail and injectors

1 Disconnect the battery negative terminal.

2 Remove the ignition coil unit.

3 Disconnect the accelerator inner cable from the throttle cam, then withdraw the outer cable from the mounting bracket along with its spring clip.

4 Undo the retaining bolts and remove the accelerator cable mounting bracket.

5 Depress the catch on the fuel feed hose quick-release fitting, and disconnect the hose from the fuel rail. Suitably seal or plug the hose and the fuel rail union after disconnection.

6 Depress the retaining tangs and disconnect the wiring connectors from the four injectors.

7 Unclip the brake servo vacuum hose from the clips on the fuel rail (where applicable).

8 Slacken and remove the three fuel rail retaining bolts, then carefully ease the fuel rail and injector assembly out from the inlet manifold and remove it from the engine. Remove the O-rings from the end of each injector and discard them; they must be renewed whenever they are disturbed.

9 Slide out the retaining clip(s) and remove the relevant injector(s) from the fuel rail. Remove the upper O-ring from each disturbed injector and discard; all disturbed O-rings must be renewed.

10 Refitting is a reversal of the removal procedure, noting the following points.

a) Fit new O-rings to all disturbed injector unions.

b) Apply a smear of engine oil to the O-rings to aid installation, then ease the injectors and fuel rail into position ensuring that none of the O-rings are displaced.

c) Refit the ignition coil unit.

d) Adjust the accelerator cable.

e) On completion, start the engine and check for fuel leaks.

Throttle potentiometer

11 Disconnect the battery negative terminal.

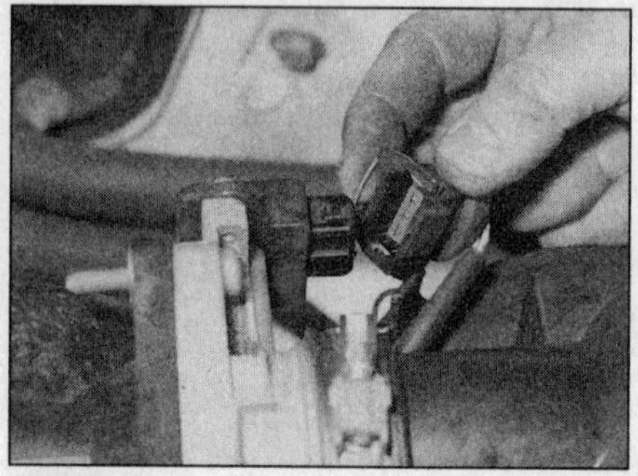

12 Depress the retaining clip and disconnect the wiring connector from the throttle potentiometer.

13 Slacken and remove the two retaining screws, then disengage the potentiometer from the throttle valve spindle and remove it from the engine.

14 Refit in the reverse order of removal. Ensure that the potentiometer is correctly engaged with the throttle valve spindle.

Electronic control unit (ECU)

Note: If a new ECU is to be fitted, this work must be entrusted to a Citroën dealer or suitably equipped specialist. It is necessary to initialise the new ECU after installation which requires the use of dedicated Citroën diagnostic equipment.

15 The ECU is located on the right-hand side of the engine compartment adjacent to the fuse/relay box. The ECU can be withdrawn from its location and moved to one side, without disconnecting its wiring connectors, if this is required for access to other components. To remove the ECU completely, it will be necessary to drill out the pop rivets securing the metal tamperproof cover over the top of the wiring connectors. These rivets are larger than the standard type and it will be necessary to obtain new rivets and a suitable rivet gun for refitting.

16 Disconnect the battery negative terminal.

17 If necessary for improved access, remove the air cleaner assembly.

18 Disconnect the wiring connector from the fuel injection double relay on the side of the ECU case.

19 Release the ECU wiring harness from the retaining clip at the base of the ECU support tray.

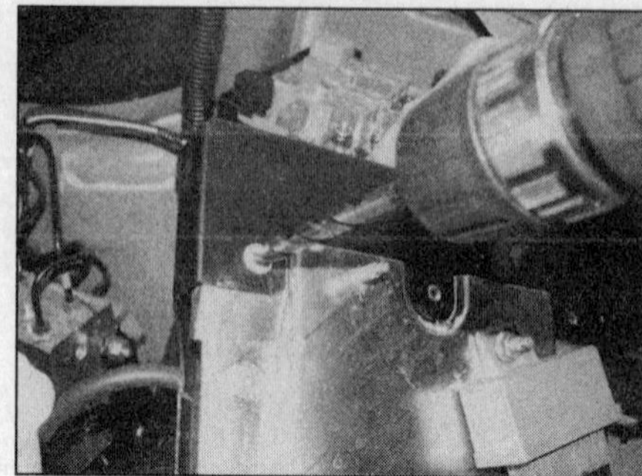

20 Lift the ECU up and out of the support tray, then drill out the pop rivets securing the metal cover over the wiring connectors. Lift off the metal cover and release the protective boot over the connectors.

21 Release the retaining clips on the connector locking catches using a small screwdriver. Rotate the locking catches and disconnect the three wiring connectors from the ECU.

22 If necessary the ECU can be unbolted and removed from its case.

23 To refit the ECU, first attach it to the case (if removed) and secure with the retaining bolts.

24 Locate the ECU in position in its engine compartment support tray.

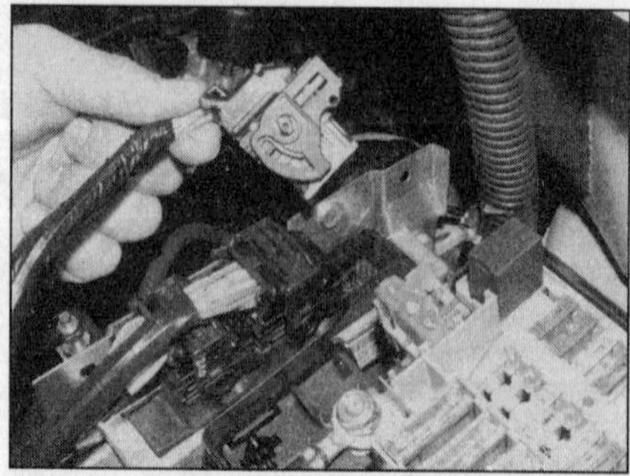

25 Reconnect the three wiring connectors and lock them in place by rotating the retaining catches.

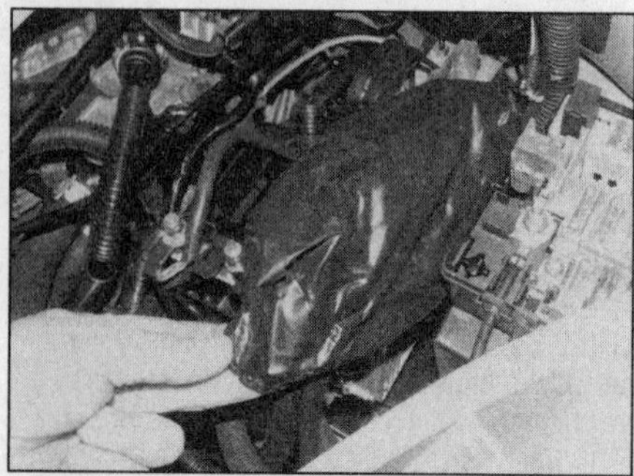

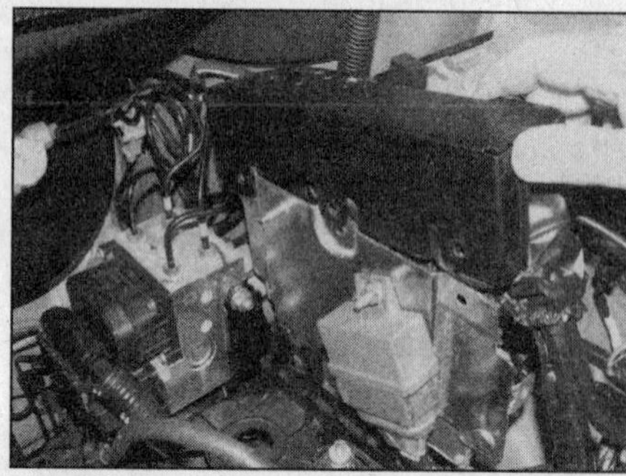

26 Refit the protective boot over the wiring connectors, then locate the metal cover in position. Secure the cover in position using new pop rivets.

27 Reconnect the injection double relay wiring connector and clip the ECU wiring harness to the support tray.

28 Refit the air cleaner assembly (if removed) and reconnect the battery.

Idle speed stepper motor

29 The idle speed stepper motor is located on the right-hand end of the inlet manifold on early engines, and on the side of the throttle housing assembly on later engines.

30 Disconnect the battery negative terminal.

31 Release the retaining clip and disconnect the wiring connector from the motor. On early engines, release the retaining clip and disconnect the air hose.

32 Slacken and remove the two retaining screws, and withdraw the motor from the manifold or throttle housing.

33 Refitting is a reversal of the removal procedure.

Manifold pressure sensor

34 The manifold pressure sensor is located on the front face of the inlet manifold.

35 Disconnect the battery negative terminal.

36 On early engines, remove the securing screw and withdraw the sensor from the manifold. Disconnect the wiring connector, and remove the sensor from the engine.

37 On later engines, disconnect the breather hose quick-release connector and move the hose to one side for improved access to the pressure sensor.

38 Disconnect the wiring connector, then undo the retaining bolt and withdraw the sensor from the manifold.

39 Refitting is the reverse of the removal procedure but fit a new sealing O-ring to the sensor body.

Coolant temperature sensor pre-2001

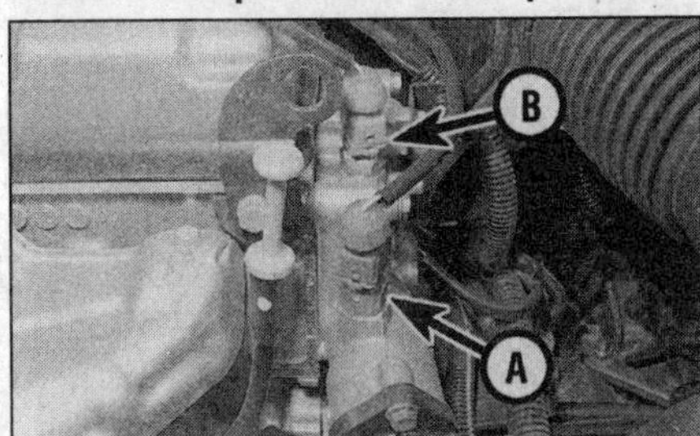

Coolant temperature warning light switch/temperature gauge sensor (A) and engine management coolant temperature sensor (B) – early 1.6 litre engine models

40 The engine management coolant temperature sensor has a green wiring connector and is located above the thermostat in the coolant outlet housing at the left-hand end of the cylinder head.

41 To remove the sensor, partially drain the cooling system to just below the level of the sensor. If necessary, remove the air cleaner and ducts for access to the sensor.

42 Unplug the wiring connector, then unscrew the sensor from its mounting. Recover the sealing ring where fitted.

43 Refitting is a reversal of removal, ensuring that a new sealing ring is fitted where applicable. If no sealing ring was fitted, apply sealing compound to the threads of the sensor prior to refitting.

Coolant temperature sensor from 2001

44 The coolant temperature sensor has a blue or green wiring connector and is located above the thermostat in the coolant outlet housing at the left-hand end of the cylinder head. The temperature signal from this sensor is used by the engine management ECU for fuel injection/ignition regulation, and to control the operation of the cooling fan, air conditioning system and the temperature warning light/gauge.

45 To remove the sensor, follow the procedure described in Paragraphs 41 to 43 above.

Inlet air temperature sensor

46 On early engines, the inlet air temperature sensor is located on the underside of the throttle housing. On later engines the sensor is integral with the manifold pressure sensor. Disconnect the battery negative terminal.

47 Disconnect the wiring connector, then unscrew the sensor and remove it from the manifold.

48 Refitting is the reverse of rémoval.

Crankshaft sensor

49 The crankshaft sensor is situated on the upper face of the transmission bellhousing and is virtually inaccessible. Disconnect the battery negative terminal.

50 Release the cable clips, cable ties and coolant hose support clips for access to the sensor as necessary.

51 Disconnect the sensor to the wiring connector.

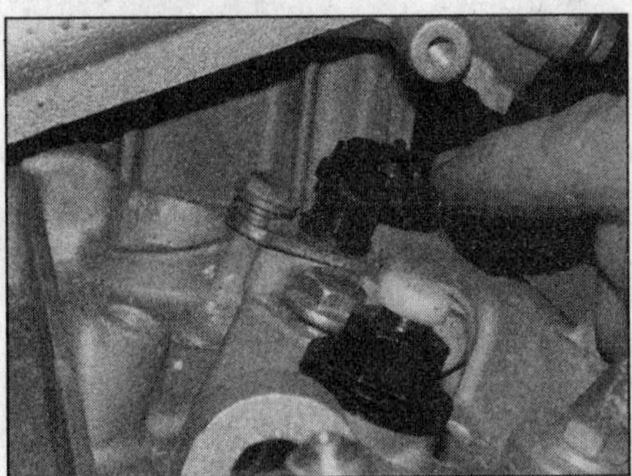

52 Slacken the sensor retaining bolt (there is no need to remove it completely) and rotate the sensor body to disengage it from the retaining bolt. Withdraw the sensor upward and out of its location.

53 Refitting is reverse of the removal procedure, ensuring that all wiring and hoses are correctly clipped back into place.

Knock sensor

54 On 1.6 litre engines, the knock sensor is fitted to the rear face of the cylinder block, whilst on 1.8 litre models, the sensor is fitted to the front face of the block. Disconnect the battery negative terminal.

55 Jack up the front of the vehicle and support it securely on axle stands. Remove the engine undertray.

56 Trace the wiring back from the sensor to its wiring connector, and disconnect it from the main wiring loom.

57 Undo the retaining bolt and remove the sensor from the block.

58 Refitting is a reversal of removal, ensuring the retaining bolt is tightened to the specified torque.

Vehicle speed sensor

59 Jack up the front of the vehicle and support it securely on axle stands. Remove the engine undertray.

60 The sensor if fitted to the rear of the transmission housing, next to the inner end of the right-hand driveshaft. Disconnect the wiring plug from the sensor.

61 Slacken and remove the retaining bolt, along with the heat shield (where fitted), and withdraw the speed sensor and driven pinion assembly from the transmission housing along with its O-ring.

62 Refitting is a reversal of removal, ensuring a new O-ring is fitted.

1.8 litre engines

Fuel rail and injectors

Note: If a faulty injector is suspected, before condemning the injector, it is worth trying the effect of one of the proprietary injector-cleaning treatments which are available from car accessory shops.

63 Disconnect the battery negative terminal.

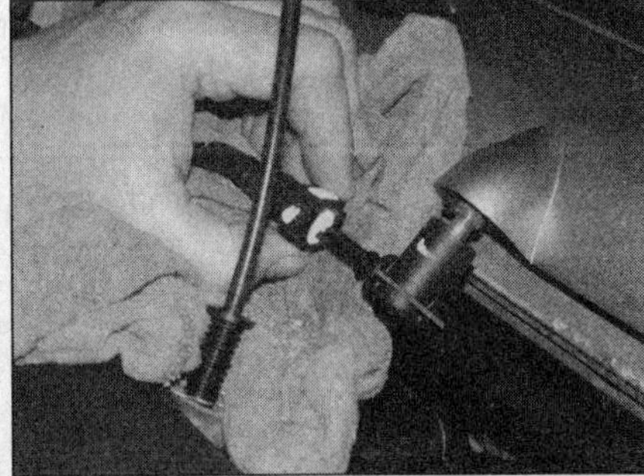

64 Depress the catch on the fuel feed hose quick-release fitting, and disconnect the hose from the fuel rail. Suitably seal or plug the hose and the fuel rail union after disconnection.

65 Using a small screwdriver, release the locking clip and disconnect the wiring connectors from the four fuel injectors.

66 Undo the two fuel rail retaining bolts, then carefully ease the fuel rail and injector assembly out from the inlet manifold and remove it from the engine. Remove the O-rings from the end of each injector and discard them; they must be renewed whenever they are disturbed.

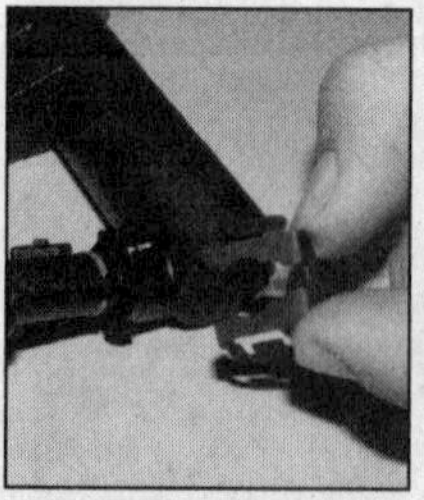

67 Slide out the retaining clip(s) and remove the relevant injector(s) from the fuel rail. Remove the upper O-ring from each disturbed injector and discard; all disturbed O-rings must be renewed.

68 Refitting is a reversal of the removal procedure, noting the following points.

a) Fit new O-rings to all disturbed injector unions.

b) Apply a smear of engine oil to the O-rings to aid installation, then ease the injectors and fuel rail into position ensuring that none of the O-rings are displaced.

c) On completion, start the engine and check for fuel leaks.

Throttle potentiometer

69 Disconnect the battery negative terminal.

70 Remove the air inlet fixed duct from the throttle housing.

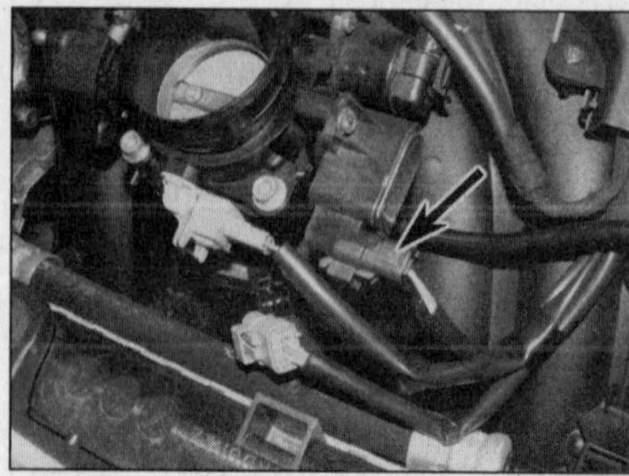

71 Depress the retaining clip and disconnect the wiring connector from the throttle potentiometer, located on the side of the throttle housing.

72 Slacken and remove the two retaining screws, then disengage the potentiometer from the throttle valve spindle and remove it from the engine.

73 Refit in the reverse order of removal. Ensure that the potentiometer is correctly engaged with the throttle valve spindle.

Electronic control unit (ECU)

74 Refer to paragraphs 15 to 28.

Idle speed stepper motor

75 The idle speed stepper motor is located on the side of the throttle housing.

76 Disconnect the battery negative terminal.

77 Remove the air inlet fixed duct from the throttle housing.

78 Release the retaining clip and disconnect the wiring connector from the motor.

79 Slacken and remove the two retaining screws, and withdraw the motor from the throttle housing.

80 Refitting is a reversal of the removal procedure.

Manifold pressure sensor

81 The manifold pressure sensor is located on the front face of the inlet manifold, just below the throttle housing.

82 Disconnect the battery negative terminal.

83 Remove the air inlet fixed duct from the throttle housing.

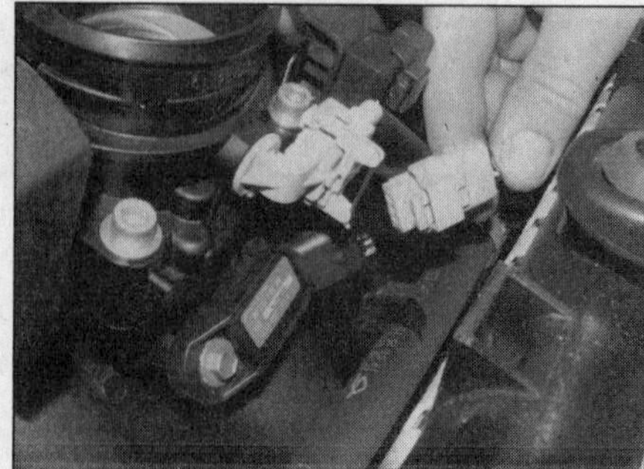

84 Disconnect the wiring connector, then undo the retaining bolt and withdraw the sensor from the manifold.

85 Refitting is the reverse of the removal procedure but fit a new sealing O-ring to the sensor body.

Coolant temperature sensor

86 The coolant temperature sensor has a green wiring connector and is located in the coolant outlet housing at the left-hand end of the cylinder head.

87 The temperature signal from this sensor is used by the engine management ECU for fuel injection/ignition regulation, and to control the operation of the exhaust gas recirculation, secondary air injection, cooling fan, air conditioning system and temperature warning light/gauge.

88 To remove the sensor, follow the procedure described in Paragraphs 41 to 43 above.

Inlet air temperature sensor

89 The inlet air temperature sensor is located on the front face of the throttle housing.

90 Disconnect the battery negative terminal.

91 Remove the air inlet fixed duct from the throttle housing.

92 Disconnect the wiring connector, then withdraw the sensor from the manifold.

93 Refitting is the reverse of removal.

Camshaft position sensor

94 The camshaft position sensor is located at the left-hand end of the exhaust camshaft cylinder head cover.

95 Disconnect the battery negative terminal.

96 Undo the six screws and lift off the engine cover.

97 Disconnect the crankcase breather hose at the quick-fit connector on the rear cylinder head cover.

98 Disconnect the wiring connector at the camshaft position sensor, then undo the bolt and remove the sensor from the rear cylinder head cover.

99 Refitting is the reverse of removal but fit a new sealing O-ring to the sensor body.

Crankshaft sensor

100 Refer to Paragraphs 49 to 53 of this Section.

Knock sensor

101 Refer to Paragraphs 54 to 58 of this Section.

Vehicle speed sensor

102 Refer to Paragraphs 59 to 62 of this Section.

Diesel electronic control system components – removal and refitting

Note: Before carrying out any of the following procedures, disconnect the battery negative terminal. Reconnect the battery on completion of refitting.

Electronic control unit (ECU)

Note: If a new ECU is to be fitted, this work must be entrusted to a Citroën dealer or suitably equipped specialist. It is necessary to initialise the new ECU after installation, which requires the use of dedicated Citroën diagnostic equipment.

1 Refer to the procedures described in the petrol injection Section.

Crankshaft speed/position sensor

2 The crankshaft speed/position sensor is located at the top of the transmission bellhousing, directly above the engine flywheel. To gain access, remove the air cleaner assembly and air inlet ducts.

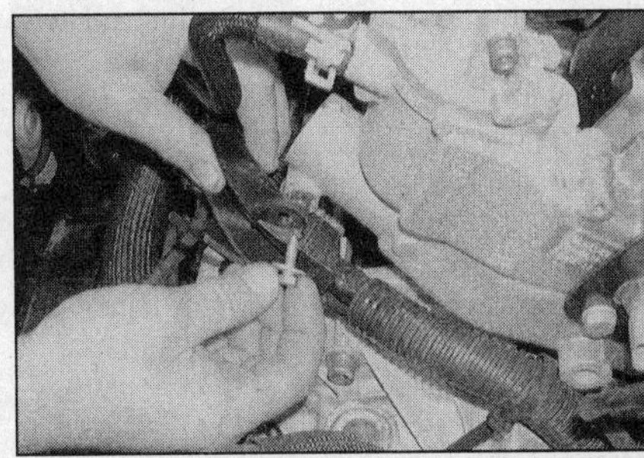

3 Undo the retaining nuts and bolts and release the plastic wiring harness guide from its mountings.

4 Working below the coolant outlet housing, disconnect the wiring connector from the crankshaft speed/position sensor.

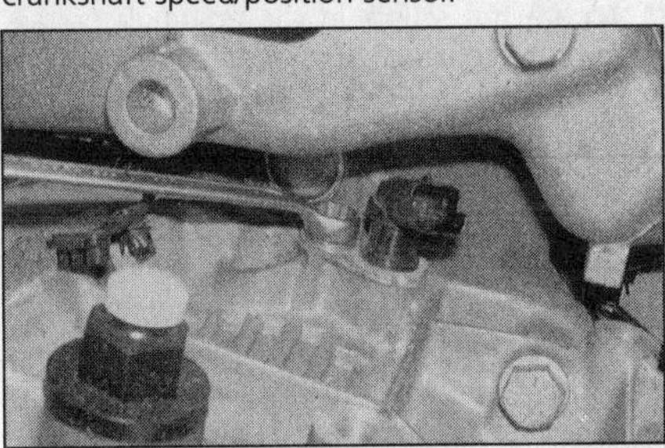

5 Slacken the bolt securing the sensor to the bellhousing. It is not necessary to remove the bolt completely as the sensor mounting flange is slotted.

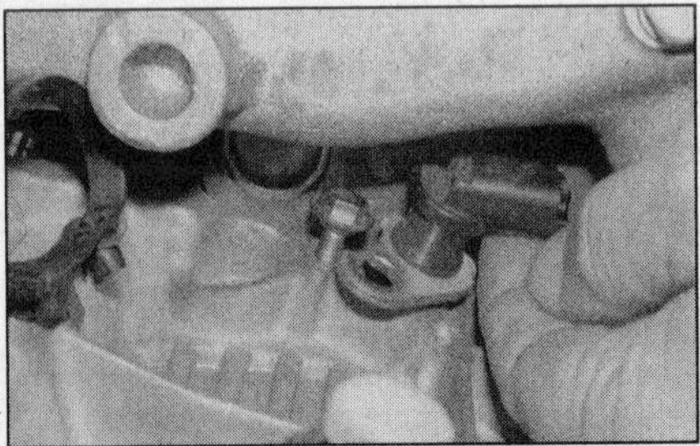

6 Turn the sensor body to clear the mounting bolt, then withdraw the sensor from the bellhousing.

7 Refitting is reverse of the removal procedure ensuring the sensor retaining bolt is securely tightened.

Camshaft position sensor

8 The camshaft position sensor is mounted on the right-hand end of the cylinder head cover, directly behind the camshaft sprocket.

9 Remove the timing belt upper and intermediate covers.

10 Disconnect the sensor wiring connector.

11 Undo the retaining bolt and lift the sensor off the cylinder head cover.

12 To refit and adjust the sensor position, locate the sensor on the cylinder head cover and loosely refit the retaining bolt.

13 The air gap between the tip of the sensor and the target plate at the rear of the camshaft sprocket hub must be set to 1.2 mm, using feeler blades. Clearance for the feeler blades is limited with the timing belt and camshaft sprocket in place, but it is just possible if the feeler blades are bent through 90° so they can be inserted through the holes in the sprocket, to rest against the inner face of the target plate.

14 With the feeler blades placed against the target plate, move the sensor toward the sprocket until it just contacts the feeler blades. Hold the sensor in this position and tighten the retaining bolt.

15 With the gap correctly adjusted, reconnect the sensor wiring connector, then refit the timing belt upper and intermediate covers.

Accelerator pedal position sensor

16 The accelerator pedal position sensor is located on the side of the master cylinder reservoir mounting bracket.

17 Undo the two bolts and remove the sensor assembly from the mounting bracket.

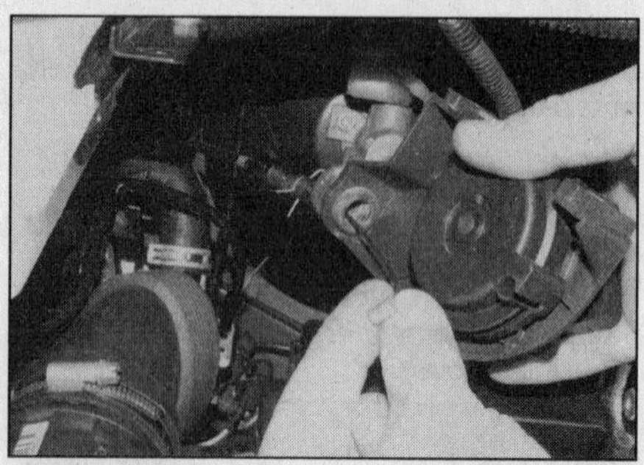

18 Rotate the pedal position sensor quadrant, and release the accelerator inner cable from the quadrant. Withdraw the outer cable from the grommet in the pedal position sensor body.

19 Refitting is reverse of the removal procedure, but adjust the accelerator cable on completion.

Coolant temperature sensor

20 The coolant temperature sensor has a green wiring connector and is mounted in the coolant outlet housing at the left-hand end of the cylinder head. The temperature signal from the sensor is used by the engine management ECU for diesel injection regulation and to control the operation of the exhaust gas recirculation, cooling fans, pre/post heating control unit, air conditioning system, and the temperature warning light-gauge.

21 To remove the sensor, partially drain the cooling system to just below the level of the sensor.

22 Where necessary, remove the air cleaner and air inlet ducts for access to the sensor.

23 Unplug the wiring connector, and unscrew the sensor from the housing. Recover the sealing ring where fitted.

24 Refitting is a reversal of removal, ensuring a new sealing ring is fitted where applicable. Where no sealing ring was fitted, apply sealing compound to the threads of the sensor prior to refitting.

Fuel temperature sensor

Note: *Do not remove the sensor from the accumulator rail unless there is a valid reason to do so. At the time of writing there was no information as to the availability of the sensor seal as a separate item. Consult a Citroën parts stockist for the latest information before proceeding.*

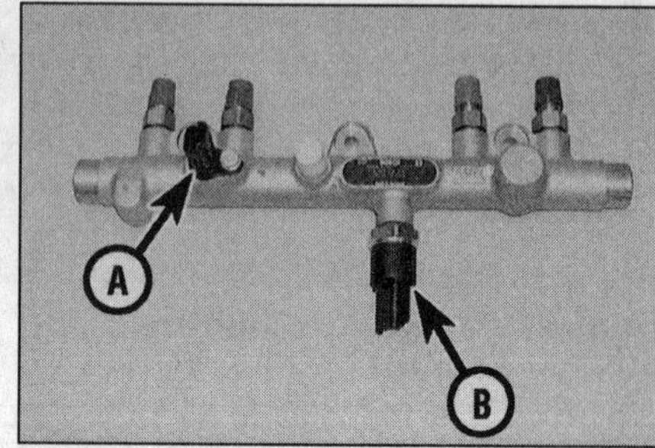

Fuel temperature sensor (A) and fuel pressure sensor (B) locations on the accumulator rail (shown removed for clarity)

25 The fuel temperature sensor is located towards the right-hand end of the accumulator rail. Undo the four plastic nuts and lift off the engine cover.

26 Disconnect the fuel temperature sensor wiring connector.

27 Thoroughly clean the area around the sensor and its location on the accumulator rail.

28 Suitably protect the components below the sensor and have plenty of clean rags handy. Be prepared for considerable fuel spillage.

29 Undo the retaining bolt and withdraw the sensor from the accumulator rail. Plug the opening in the accumulator rail as soon as the sensor is withdrawn.

30 Prior to refitting, if the original sensor is to be refitted, renew the sensor seal, where applicable.

31 Locate the sensor in the accumulator rail and refit the retaining bolt, tightened securely.

32 Refit the sensor wiring connector.

33 Start the engine and allow it to idle. Check for leaks at the fuel temperature sensor with the engine idling. If satisfactory, increase the engine speed to 4000 rpm and check again for leaks. Take the car for a short road test and check for leaks once again on return. If any leaks are detected, obtain and fit a new sensor.

34 Refit the engine cover on completion.

Air mass meter

35 The air mass meter is located in the flexible air inlet duct just behind the air cleaner lid.

36 Undo the four plastic nuts and lift off the engine cover.

37 Disconnect the wiring connector from the underside of the air mass meter.

38 Slacken the hose clips on each side of the air mass meter body.

39 Lift the air cleaner assembly upward to release it from the support bracket then detach the air mass meter from the front inlet duct.

40 Withdraw the air mass meter from the rear inlet duct and remove it from the engine compartment. Suitably plug or cover the rear inlet duct using clean rag to prevent any dirt or foreign material from entering the turbocharger.

41 Refitting is reverse of the removal procedure.

Fuel pressure sensor

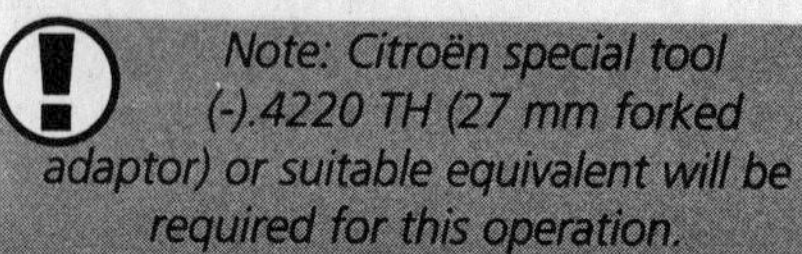

Note: Citroën special tool (-).4220 TH (27 mm forked adaptor) or suitable equivalent will be required for this operation.

42 The fuel pressure sensor is located centrally on the underside of the accumulator rail *(see paragraph 25 earlier in this section)*.

43 Undo the four plastic nuts and lift off the engine cover.

44 Release the retaining clip and disconnect the crankcase ventilation hose from the cylinder head cover.

45 Disconnect the fuel supply and return hose quick-release fittings at the fuel filter, using a small screwdriver to release the locking clip. Suitably plug or cover the open unions to prevent dirt entry. Release the fuel hoses from the relevant retaining clips.

46 Disconnect the fuel pressure sensor wiring connector.

47 Thoroughly clean the area around the sensor and its location on the accumulator rail.

48 Suitably protect the components below the sensor and have plenty of clean rags handy. Be prepared for considerable fuel spillage.

49 Using the Citroën special tool (or suitable alternative 27 mm forked adaptor) and a socket bar, unscrew the fuel pressure sensor from the base of the accumulator rail.

50 Obtain and fit a new sealing ring to the sensor prior to refitting.

51 Locate the sensor in the accumulator rail and tighten it to the specified torque using the special tool (or alternative) and a torque wrench.

52 Refit the sensor wiring connector.

53 Start the engine and allow it to idle. Check for leaks at the fuel pressure sensor with the engine idling. If satisfactory, increase the engine speed to 4000 rpm and check again for leaks. Take the car for a short road test and check for leaks once again on return. If any leaks are detected, obtain and fit another new sensor sealing ring.

54 Refit the engine cover on completion.

Fuel pressure control valve

55 The fuel pressure control valve is integral with the high-pressure fuel pump and cannot be separated.

Fiat Punto
'99 – '03

System type

SOHC (8-valve) engines	Weber-Marelli 59F multi-point fuel injection/ignition system
DOHC (16-valve) engines	Bosch Motronic ME 7.3 H4 multi-point fuel injection/ignition system

Fuel system data

Engine idle speed	900 ± 50 rpm*
CO%	0.35 maximum*
Fuel pump type	Electric, immersed in fuel tank
Fuel pump delivery rate:	
SOHC (8-valve) engines	120 litres/hour minimum
DOHC (16-valve) engines	110 litres/hour minimum
Regulated fuel pressure:	
SOHC (8-valve) engines	3.5 bars
DOHC (16-valve) engines	3.0 bars
Crankshaft TDC sensor resistance at 20°C	1134 to 1386 ohms
Injector electrical resistance:	
SOHC (8-valve) engines	13.8 to 15.2 ohms
DOHC (16-valve) engines	14.0 to 15.0 ohms

** Not adjustable – controlled by ECU*

Recommended fuel

Minimum octane rating	95 RON unleaded

Torque wrench setting

	Nm	lbf ft
Knock sensor bolt	20	15

Chapter 9

General information

The fuel supply system consists of a fuel tank (which is mounted under the centre of the car, with an electric fuel pump immersed in it) and fuel feed line. The fuel pump supplies fuel to the fuel rail, which acts as a reservoir for the four fuel injectors which inject fuel into the inlet tracts.

The fuel injection and ignition functions are combined into a single engine management system. The systems fitted are manufactured by Weber-Marelli and Bosch, and are very similar in terms of construction and operation. The only significant differences being in the software contained in the system Electronic Control Unit (ECU), and certain specific component variations according to engine type. Each system incorporates a closed-loop catalytic converter and an evaporative emission control system, and complies with the latest emission control standards. The fuel side of the system operates as follows.

The fuel pump supplies fuel from the tank to the fuel rail (mounted directly above the fuel injectors) by means of a 'returnless' system. With this arrangement, the fuel filter and fuel pressure regulator are an integral part of the fuel pump assembly located in the fuel tank. The regulator maintains a constant fuel pressure in the supply line to the fuel rail and allows excess fuel to recirculate in the fuel tank, by means of a bypass channel, if the regulated fuel pressure is exceeded. As the fuel filter is an integral part of the pump assembly, fuel filter renewal is no longer necessary as part of the maintenance and servicing schedule.

The fuel injectors are electromagnetic pintle valves which spray atomised fuel into the combustion chambers under the control of the ECU. There are four injectors, one per cylinder, mounted in the inlet manifold close to the cylinder head. Each injector is mounted at an angle that allows it to spray fuel directly onto the back of the inlet valve(s). The ECU controls the volume of fuel injected by varying the length of time for which each injector is held open. The fuel injection systems are of the sequential type, whereby each injector operates individually in cylinder sequence.

The electrical control system consists of the ECU, along with the following sensors:

a) *Throttle potentiometer – informs the ECU of the throttle valve position, and the rate of throttle opening/closing.*
b) *Engine coolant temperature sensor – informs the ECU of engine temperature.*
c) *Intake air temperature/pressure sensor – informs the ECU of intake air temperature and load on the engine (expressed in terms of inlet manifold vacuum).*
d) *Lambda sensors – inform the ECU of the oxygen content of the exhaust gases).*
e) *Crankshaft TDC sensor – informs the ECU of engine speed and crankshaft angular position.*
f) *Vehicle speed sensor – informs the ECU of the vehicle speed.*
g) *Knock sensor – informs the ECU of pre-ignition (detonation) within the cylinders.*
h) *Camshaft position sensor – informs the ECU of which cylinder is on the firing stroke.*

On the Bosch system, the following additional sensors are also used:

a) *Acceleration sensor – enables the ECU to distinguish between engine roughness caused by pre-ignition (detonation) within the cylinders, or engine shake caused by abnormal road conditions.*
b) *Accelerator pedal potentiometer – informs the ECU of accelerator pedal position so that a specific throttle opening can be determined by the throttle valve actuator.*

Signals from each of the sensors are compared by the ECU and, based on this information, the ECU selects the response appropriate to those values, and controls the fuel injectors (varying the pulse width – the length of time the injectors are held open – to provide a richer or weaker air/fuel mixture, as appropriate). The air/fuel mixture is constantly varied by the ECU, to provide the best settings for cranking, starting (with either a hot or cold engine) and engine warm-up, idle, cruising and acceleration.

The ECU also has full control over the engine idle speed, via a stepper motor or throttle valve actuator fitted to the throttle body. On the Weber-Marelli system, the stepper motor controls the amount of air passing through a bypass drilling at the side of the throttle. When the throttle valve is closed (accelerator pedal released), the ECU uses the motor to open or close an air passage, controlling the amount of air bypassing the throttle valve and so controlling the idle speed. The ECU also carries out 'fine tuning' of the idle speed by varying the ignition timing to increase or reduce the torque of the engine as it is idling. This helps to stabilise the idle speed when electrical or mechanical loads (such as headlights, air conditioning, etc) are switched on and off. On the Bosch system, the ECU has total control over the full range of throttle valve movement by means of a throttle valve actuator. The accelerator pedal potentiometer informs the ECU of accelerator pedal position and from this data, the ECU controls the throttle valve actuator so that a corresponding throttle opening can be obtained. This arrangement is often termed 'drive-by-wire' as there is no direct accelerator cable connection between the accelerator pedal and throttle valve.

If there is any abnormality in any of the readings obtained from the main engine sensors, the ECU enters its 'back-up' mode. If this happens, the erroneous sensor signal is overridden, and the ECU assumes a preprogrammed 'back-up' value, which will allow the engine to continue running, albeit at reduced efficiency. If the ECU enters this mode, the warning lamp on the instrument panel will be illuminated, and the relevant fault code will be stored in the ECU memory.

Fuel system – depressurisation

Warning: The following procedure will merely relieve the pressure in the fuel system – remember that fuel will still be present in the system components and take precautions accordingly before disconnecting any of them.

1 The fuel system referred to in this Section is defined as the tank-mounted fuel pump, the fuel rail, the fuel injectors, and the metal pipes and flexible hoses of the fuel lines between these components. All these contain fuel which will be under pressure while the engine is running and/or while the ignition is switched on. The pressure will remain for some time after the ignition has been switched off, and must be relieved before any of these components are disturbed for servicing work.

2 Disconnect the battery negative terminal.

3 Remove the air cleaner assembly.

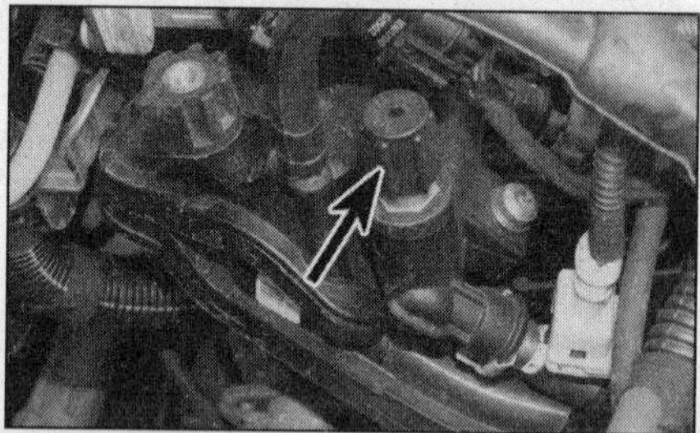

4 Locate the Schrader valve situated at the right-hand end of the fuel rail. The Schrader valve works like a tyre valve whereby on depressing the central plunger, the system fuel pressure will be released.

5 Withdraw the protective plastic cap from the top of the valve.

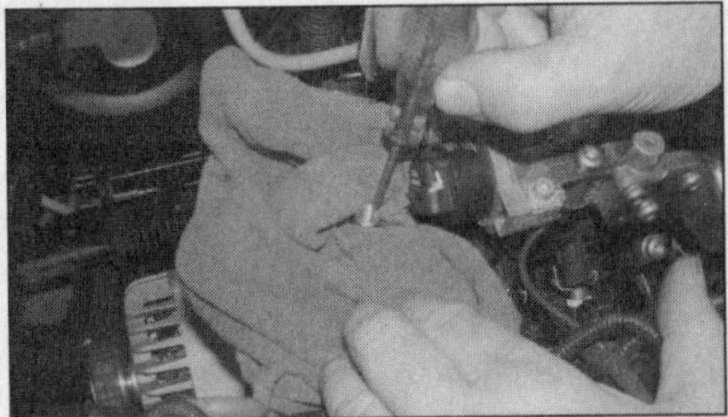

6 Place an absorbent rag around the valve then, using a small screwdriver, slowly depress the central plunger to allow the pressure to be released. Ensure that the rag completely covers the valve to catch the fuel spray which will be expelled.

7 On completion of the operations for which system depressurisation was necessary, refit the Schrader valve cap and reconnect the battery negative terminal. Refit the air cleaner assembly.

Fuel injection components (SOHC 8-valve engines) – removal and refitting

Throttle body assembly

1 Disconnect the battery negative terminal.

2 Remove the air cleaner assembly.

3 Disengage the accelerator inner cable from the throttle cam, and withdraw the outer cable from the rubber grommet on the mounting bracket.

4 Remove the electronic control unit as described later in this Section.

5 Disconnect the wiring connectors from the throttle potentiometer and the idle control stepper motor.

6 Undo the retaining screw and disconnect the ECU earth lead from the front of the throttle body.

7 Slacken and remove the three bolts securing the throttle body assembly to the inlet manifold, then remove the assembly and collect the sealing O-ring.

8 Refitting is a reversal of the removal procedure, bearing in mind the following points:

a) Ensure the throttle body and inlet manifold mating surfaces are clean and dry, then fit the throttle body with a new O-ring, and securely tighten the retaining bolts.

b) Refit the electronic control unit as described later in this Section.

c) Adjust the accelerator cable.

Fuel rail and injectors

Note: *If a faulty injector is suspected, before condemning the injector it is worth trying the effect of one of the proprietary injector-cleaning treatments.*

9 Disconnect the battery negative terminal.

10 Remove the air cleaner assembly.

11 Depressurise the fuel system.

12 Disconnect the wiring connectors at the fuel injectors.

13 Disconnect the fuel supply pipe at the quick-release connector on the left-hand end of the fuel rail.

14 Unscrew the two bolts securing the fuel rail assembly to the inlet manifold, then carefully pull the injectors from their manifold locations. Remove the assembly from the engine and remove the injector lower O-ring seals.

15 The injectors can be removed individually from the fuel rail by extracting the relevant metal clip and easing the injector out of the rail. Remove the injector upper O-ring seals.

16 Check the electrical resistance of the injector using a multimeter and compare it with the Specifications.

17 Refitting is a reversal of the removal procedure, bearing in mind the following points:

a) Renew the injector O-ring seals, and smear them with a little petroleum jelly before assembling. Take care when fitting the injectors to the fuel rail and do not press them in further than required to fit the retaining clip otherwise the O-ring seal may be damaged.

b) Ensure that the injector retaining clips are securely seated.

c) On completion check the fuel rail and injectors for fuel leaks.

Idle control stepper motor

18 The idle control stepper motor is an integral part of the throttle body and cannot be individually renewed.

Throttle potentiometer

19 The idle control stepper motor is an integral part of the throttle body and cannot be individually renewed.

Intake air temperature/pressure sensor

20 Disconnect the battery negative terminal.

21 Remove the air cleaner assembly.

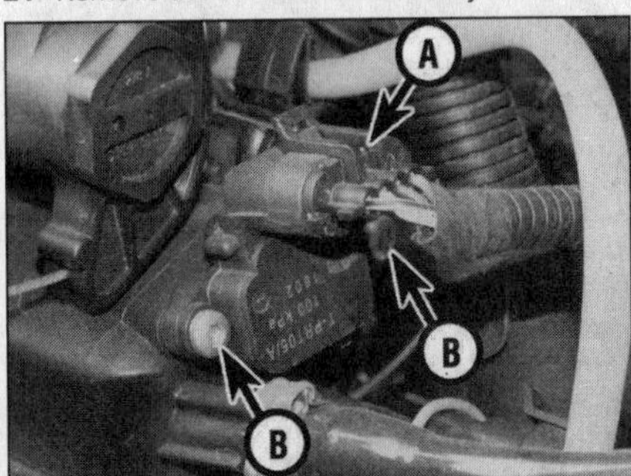

22 Disconnect the wiring connector (A), undo the two retaining screws (B) and remove the sensor from the left-hand end of the inlet manifold.

23 Refitting is a reversal of the removal procedure.

Coolant temperature sensor

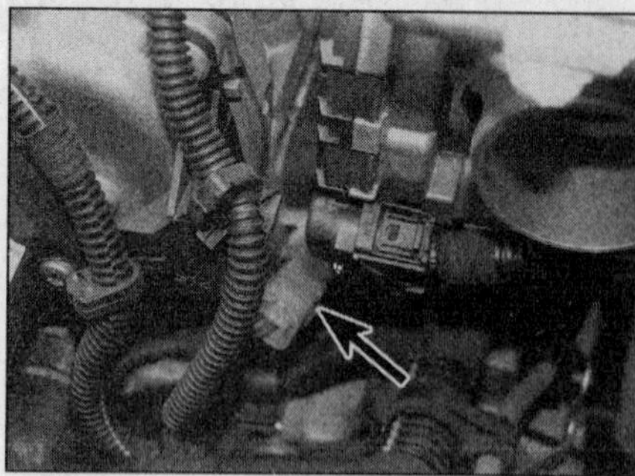

24 The sensor is located at the left-hand end of the inlet manifold, and can be identified by its green wiring connector. The sensor signal is also used to operate the temperature warning light and/or temperature gauge on the instrument panel.

25 To remove the sensor, remove the air cleaner, battery and battery tray.

26 Partially drain the cooling system to just below the level of the sensor.

27 Disconnect the wiring plug, then unscrew the sensor from its mounting and recover the sealing ring.

28 Refitting is a reversal of removal, renewing the sealing ring if necessary.

Lambda sensors

29 Two sensors are fitted into the exhaust system. One 'upstream' of the catalytic converter, and one 'downstream'.

30 To remove the sensors, jack up the front of the vehicle and support it securely on axle stands.

31 Remove the air cleaner assembly.

32 Undo the four bolts securing the heat shield to the exhaust manifold.

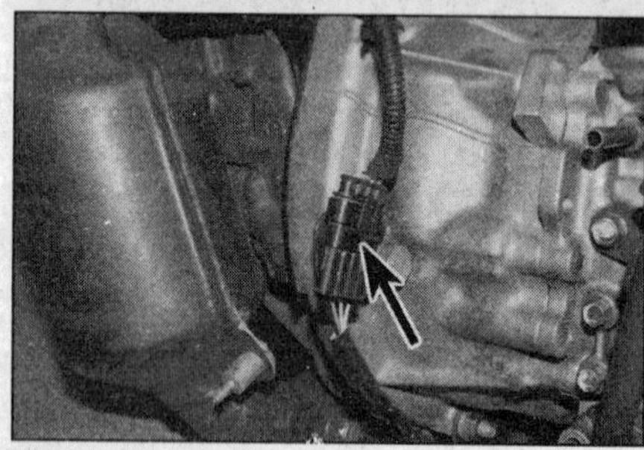

33 Disconnect the relevant sensor wiring connector located on the transmission bellhousing.

34 Unscrew the sensor, taking care to avoid damaging the sensor probe as it is removed. **Note:** *As a flying lead remains connected to the sensor after it has been disconnected, if the correct spanner is not available, a slotted socket will be required to remove the sensor.*

35 Refitting is a reversal of removal, but apply a little anti-seize grease to the sensor threads prior to refitting.

Crankshaft TDC sensor

36 The crankshaft TDC sensor is located adjacent to the front facing side of the crankshaft pulley.

37 Disconnect the battery negative terminal.

38 Remove the timing belt upper and lower covers.

39 Undo the bolt securing the sensor to the oil pump casing and remove the sensor from its location.

40 Locate the sensor in position and secure with the retaining bolt.

41 Engage the sensor wiring harness with the upper and lower timing belt covers, then refit the covers.

Knock sensor

42 The knock sensor is fitted to the rear face of the cylinder block, behind the alternator rear mounting bracket. Jack up the front of the car, and support it securely on axle stands.

43 Trace the wiring back from the sensor to its wiring connector, then disconnect it from the main loom.

44 Undo the sensor securing bolt, and remove it from the cylinder block.

45 Refitting is a reversal of removal, tightening the retaining bolt to the specified torque.

Camshaft position sensor

46 Disconnect the battery negative terminal.

Camshaft position sensor wiring connector (A) and retaining bolt (B)

47 Disconnect the wiring connector from the camshaft position sensor located at the timing belt end of the camshaft cover.

48 Undo the retaining bolt and withdraw the sensor from the camshaft cover.

49 Refitting is a reversal of the removal procedure.

Electronic control unit (ECU)

50 The electronic control unit is attached to the rear of the throttle body.

51 Disconnect the battery negative terminal.

52 Remove the air cleaner assembly.

53 Rotate the wiring connector locking catches and disconnect the two wiring connectors from the top of the ECU.

54 Undo the two mounting bolts and remove the ECU from the throttle body.

55 Refitting is a reversal of the removal procedure.

Inertia safety switch

56 Disconnect the battery negative terminal.

57 Working under the facia on the right-hand side, move away the floor covering and side trim for access to the switch.

58 Disconnect the wiring connector, then undo the two bolts and remove the switch from its location.

59 Refitting is a reversal of the removal procedure.

Vehicle speed sensor

Note: *On cars equipped with ABS, the vehicle speed sensor signal is supplied to the ECU by the ABS wheel speed sensors.*

60 The vehicle speed sensor is located on the top of the transmission differential housing.

61 Disconnect the battery negative terminal.

62 Firmly apply the handbrake, then jack up the front of the car and support it securely on axle stands.

63 Disconnect the sensor wiring connector then unscrew the sensor from the differential housing.

64 Refitting is a reversal of the removal procedure.

Fuel injection components (DOHC 16-valve engines) – removal and refitting

Throttle body assembly

1 Disconnect the battery negative terminal.

2 Remove the air cleaner assembly.

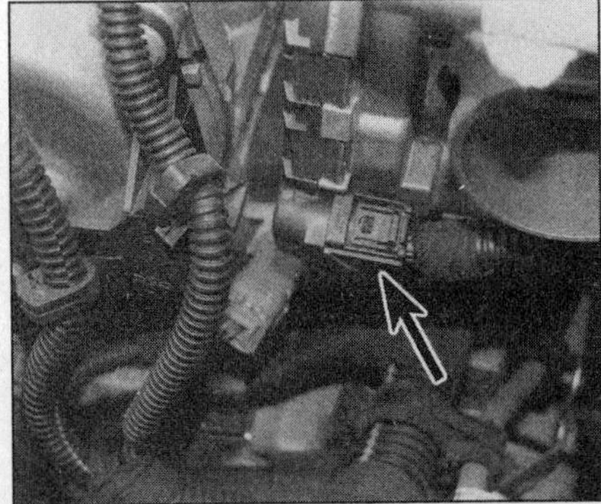

3 Disconnect the wiring connector for the throttle valve actuator.

4 Remove the electronic control unit as described later in this Section.

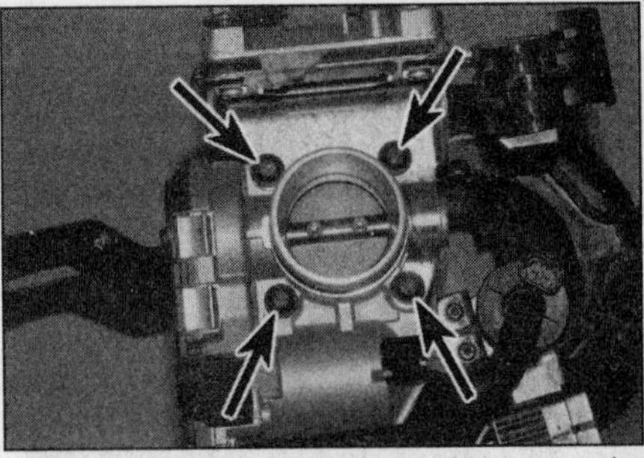

5 Slacken and remove the four bolts securing the throttle body assembly to the inlet manifold. Move the ECU mounting bracket aside, then remove the assembly and collect the sealing O-ring.

6 Refitting is a reversal of the removal procedure, bearing in mind the following points:

a) *Ensure the throttle body and inlet manifold mating surfaces are clean and dry, then fit the throttle body with a new O-ring, and securely tighten the retaining bolts.*

b) *Refit the electronic control unit as described later in this Section.*

Fuel rail and injectors

Note: *If a faulty injector is suspected, before condemning the injector it is worth trying the effect of one of the proprietary injector-cleaning treatments.*

7 Disconnect the battery negative terminal.

8 Remove the air cleaner assembly.

9 Depressurise the fuel system.

10 Disconnect the fuel supply pipe at the quick-release connector in the centre of the fuel rail.

Fuel injector wiring connectors (A) harness clips (B) and fuel rail retaining bolts (C) – shown with inlet manifold removed

11 Disconnect the wiring connectors at the fuel injectors. Release the injector wiring harness from the retaining clips on the fuel rail and move the harness to one side.

12 Unscrew the two bolts securing the fuel rail assembly to the inlet manifold, then carefully pull the injectors from their manifold locations. Remove the assembly from the engine and remove the injector lower O-ring seals.

13 The injectors can be removed individually from the fuel rail by extracting the relevant metal clip and easing the injector out of the rail. Remove the injector upper O-ring seals.

14 Check the electrical resistance of the injector using a multimeter and compare it with the Specifications.

15 Refitting is a reversal of the removal procedure, bearing in mind the following points:

a) *Renew the injector O-ring seals, and smear them with a little petroleum jelly before assembling. Take care when fitting the injectors to the fuel rail and do not press them in further than required to fit the retaining clip otherwise the O-ring seal may be damaged.*

b) *Ensure that the injector retaining clips are securely seated.*

c) *On completion check the fuel rail and injectors for fuel leaks.*

Throttle valve actuator

16 The throttle valve actuator is an integral part of the throttle body and cannot be individually renewed.

Throttle potentiometer

17 The throttle potentiometer is incorporated in the throttle valve actuator and cannot be individually renewed.

Accelerator pedal potentiometer

18 The accelerator pedal potentiometer is an integral part of the accelerator pedal assembly and cannot be individually renewed.

Intake air temperature/pressure sensor

19 Disconnect the battery negative terminal.

20 Remove the air cleaner assembly.

21 Disconnect the wiring connector, then undo the retaining bolt and remove the sensor from the right-hand end of the inlet manifold.

22 Refitting is a reversal of the removal procedure.

Coolant temperature sensor

23 Refer to the SOHC 8-valve engines Section.

Lambda sensors

24 Refer to the SOHC 8-valve engines Section.

Crankshaft TDC sensor

25 The crankshaft TDC sensor is located adjacent to the front facing side of the crankshaft pulley *(see paragraph 36 in the previous section)*.

26 Disconnect the battery negative terminal.

27 Remove the timing belt upper and lower covers.

28 Undo the bolt securing the sensor to the oil pump casing and remove the sensor from its location.

29 Locate the sensor in position and secure with the retaining bolt.

30 Engage the sensor wiring harness with the upper and lower timing belt covers, then refit the covers.

Knock sensor

31 Refer to the SOHC 8-valve engines Section.

Camshaft position sensor

32 Disconnect the battery negative terminal.

33 Remove the air cleaner assembly.

34 Disconnect the wiring connector from the camshaft position sensor located in the centre of the cylinder head extension.

35 Undo the retaining bolt and withdraw the sensor from its location.

36 Refitting is a reversal of the removal procedure.

Acceleration sensor

37 Disconnect the battery negative terminal.

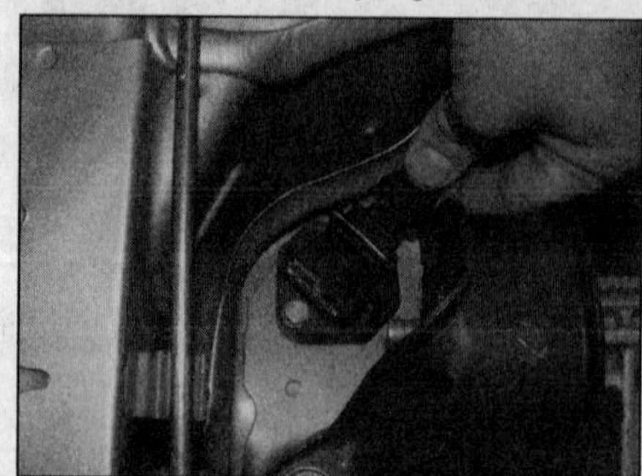

38 Disconnect the wiring connector from the acceleration sensor located adjacent to the right-hand engine mounting.

39 Undo the retaining bolt and remove the sensor from its location.

40 Refitting is a reversal of the removal procedure.

Electronic control unit (ECU)

41 The electronic control unit is attached to the top of the throttle body.

42 Disconnect the battery negative terminal.

43 Remove the air cleaner assembly.

44 Rotate the wiring connector locking catches and disconnect the two wiring connectors from the top of the ECU.

45 Undo the four nuts securing the ECU to the throttle body, noting the location of the earth lead under the front left-hand nut. Lift the unit up and off the throttle body.

46 Refitting is a reversal of the removal procedure.

Inertia safety switch

47 Refer to the SOHC 8-valve engines Section.

Vehicle speed sensor

48 Refer to the SOHC 8-valve engines Section.

Ford Fiesta '95 – '01

10

Petrol models

General

System type	Sequential Electronic Fuel injection (SEFi)
Fuel octane requirement	95 RON unleaded
Regulated fuel pressure:	
Pressure regulator vacuum hose disconnected	2.7 ± 0.2 bar
Engine running and pressure regulator vacuum hose connected	2.1 ± 0.2 bar
Hold pressure – engine stopped after five minutes	1.8 bars minimum

Diesel models

General

System type	Bosch VP-30 combined lift and injection pump, chain-driven from crankshaft. Direct injection via five-hole, pencil-type injectors, electronic pump control unit (PCU) linked to revised EEC-V engine management Powertrain Control Module (PCM)
Application	1.8 litre Endura-DI engine
Firing order	1-3-4-2 (No 1 at timing belt end)
Idle speed	800 ± 50 rpm (regulated by EEC V engine management system – no adjustment possible)
Fuel type	Commercial diesel fuel for road vehicles (DERV)
Injection pump:	
Make and type	Bosch VP-30 electronic, distributor-type
Rotation (viewed from crankshaft pulley end)	Clockwise
Drive	By twin 'gemini' chains from crankshaft
Injection pump timing	By timing pin, at TDC
Injectors:	
Type	Pencil-type, five-hole
Turbocharger:	
Type	Garrett GT15, integral with exhaust manifold.

Chapter 10

Torque wrench settings	Nm	lbf ft
Petrol models		
Camshaft position sensor	10	7
Crankshaft position sensor	7	5
Cylinder head temperature sensor	10	7
Engine coolant temperature sensor	12	9
Fuel injector bolts	6	4
Fuel pressure regulator screws:		
Endura-E engine	10	7
Zetec-SE engine	6	4
Fuel rail to cylinder head (Zetec-SE engine)	23	17
Fuel rail-to-inlet manifold bolts (Endura-E engine)	18	13
Idle air control valve:		
Endura-E engine	8	6
Zetec-SE engine	10	7
Inlet manifold	18	13
Oxygen sensor	42	31
Shrader valve retaining bolts	10	7
Diesel models		
Crankshaft position sensor bolt	10	7
Cylinder head temperature sensor	20	15

General information

Petrol models

The fuel system consists of a fuel tank (mounted under the body, beneath the rear seats), fuel hoses, an electric fuel pump mounted in the fuel tank, and a sequential electronic fuel injection system controlled by a EEC V engine management electronic control unit (Powertrain Control Module).

The electric fuel pump supplies fuel under pressure to the fuel rail, which distributes fuel to the injectors. A pressure regulator controls the system pressure in relation to inlet tract depression. From the fuel rail, fuel is injected into the inlet ports, just above the inlet valves, by four fuel injectors. On the Endura-E engine a plastic moulded fuel rail is fitted, whereas on the Zetec-SE engine the fuel rail is of aluminium alloy. The Endura-E fuel rail is mounted on the inlet manifold, and the Zetec-SE fuel rail is mounted in the cylinder head.

The amount of fuel supplied by the injectors is precisely controlled by the Powertrain Control Module (PCM). The module uses the signals from the crankshaft position sensor and the camshaft position sensor, to trigger each injector separately in cylinder firing order (sequential injection), with benefits in terms of better fuel economy and leaner exhaust emissions.

The Powertrain Control Module is the heart of the entire engine management system, controlling the fuel injection, ignition and emissions control systems. The module receives information from various sensors which is then computed and compared with pre-set values stored in its memory, to determine the required period of injection.

Information on crankshaft position and engine speed is generated by a crankshaft position sensor. The inductive head of the sensor runs just above the engine flywheel and scans a series of 36 protrusions on the flywheel periphery. As the crankshaft rotates, the sensor transmits a pulse to the system's ignition module every time a protrusion passes it. There is one missing protrusion in the flywheel periphery at a point corresponding to 90° BTDC. The ignition module recognises the absence of a pulse from the crankshaft position sensor at this point to establish a reference mark for crankshaft position. Similarly, the time interval between absent pulses is used to determine engine speed. This information is then fed to the Powertrain Control Module for further processing.

On the Zetec-SE engine, the camshaft position sensor is located in the cylinder head so that it registers with a lobe on the camshaft. On the Endura-E engine, it is located on the timing cover and registers with a plate located beneath the camshaft sprocket retaining bolts. The camshaft position sensor functions in the same way as the crankshaft position sensor, producing a series of pulses; this gives the Powertrain Control Module a reference point, to enable it to determine the firing order, and operate the injectors in the appropriate sequence.

The mass air flow sensor is based on a 'hot-wire' system, sending the Powertrain Control Module a constantly-varying (analogue) voltage signal corresponding to the mass of air passing into the engine. Since air mass varies with temperature (cold air being denser than warm), measuring air mass provides the module with a very accurate means of determining the correct amount of fuel required to achieve the ideal air/fuel mixture ratio.

Engine temperature information is supplied by the coolant temperature sensor or, on certain later Zetec-SE engines, by a cylinder head temperature sensor. Both components are NTC (Negative Temperature Coefficient) thermistors – that is, a semi-conductor whose electrical resistance decreases as its temperature increases. The sensors provide the Powertrain Control Module with a constantly-varying (analogue) voltage signal, corresponding to the temperature of the engine coolant or cylinder head, as applicable. This is used to refine the calculations made by the module, when determining the correct amount of fuel required to achieve the ideal air/fuel mixture ratio. While the coolant temperature sensor measures the temperature of the engine coolant, the cylinder head temperature sensor is seated in a blind hole in the cylinder head, and measures the temperature of the metal directly.

Inlet air temperature information is supplied by the inlet air temperature sensor. This component is also an NTC thermistor – see the previous paragraph – providing the module with a signal corresponding to the temperature of air passing into the engine. This is used to refine the calculations made by the module, when determining the correct amount of fuel required to achieve the ideal air/fuel mixture ratio.

On later Endura-E, and 1.25 litre Zetec-SE engines, inlet air temperature and density information for air/fuel mixture ratio calculations is provided by a temperature and manifold absolute pressure (TMAP) sensor. The TMAP sensor is located in the inlet manifold and consists of a pressure transducer and a temperature sensor which directly replaces the mass air flow and inlet air temperature sensors. The TMAP sensor provides information to the Powertrain Control Module relating to inlet manifold vacuum and barometric pressure, and the temperature of the air in the inlet manifold. When the ignition is switched on with the engine stopped, the sensor calculates barometric pressure and, when the engine is running, the sensor calculates inlet manifold vacuum.

A throttle position sensor is mounted on the end of the throttle valve spindle, to provide the Powertrain Control Module with a constantly-varying (analogue) voltage signal corresponding to the throttle opening. This allows the module to register the driver's input when determining the amount of fuel required by the engine.

Road speed is monitored by the vehicle speed sensor. This component is a Hall-effect generator, mounted on the transmission's speedometer drive. It supplies the module with a series of pulses corresponding to the vehicle's road speed, enabling the module to control features such as the fuel shut-off on overrun.

The clutch pedal position is monitored by a switch fitted to the pedal bracket. This sends a signal to the Powertrain Control Module.

Where power steering is fitted, a pressure-operated switch is screwed into the power steering system's high-pressure pipe. The switch sends a signal to the Powertrain Control Module to increase engine speed to maintain idle speed during power steering assistance.

The oxygen sensor in the exhaust system provides the module with constant feedback – 'closed-loop' control – which enables it to adjust the mixture to provide the best possible operating conditions for the catalytic converter.

The air inlet side of the system consists of an air cleaner housing, the mass air flow sensor, an inlet hose and duct, and a throttle housing.

The throttle valve inside the throttle housing is controlled by the driver, through the accelerator pedal. As the valve opens, the amount of air that can pass through the system increases. As the throttle valve opens further, the mass air flow sensor signal alters, and the Powertrain Control Module opens each injector for a longer duration, to increase the amount of fuel delivered to the inlet ports.

Both the idle speed and mixture are under the control of the Powertrain Control Module, and cannot be adjusted.

Diesel models

The fuel system consists of a fuel tank (mounted under the body, beneath the rear seats), fuel filter, electronic fuel injection pump with pump control unit, injectors, fuel lines and hoses, fuel gauge sender unit mounted in the fuel tank, and EEC V engine management electronic control unit (Powertrain Control Module).

Fuel is drawn from the tank by the transfer pump incorporated in the injection pump. It then passes through the fuel filter, located in the engine bay, where foreign matter and water are removed. The injection pump is driven from the crankshaft via a twin-row ('gemini') chain and supplies fuel under very high pressure to each injector in turn as it is needed. The camshaft is driven from the injection pump via a toothed timing belt.

The Endura-DI engine is very much a 'state-of-the-art' unit, in that it features a full electronic engine management system, very similar to that fitted to the Fiesta petrol models. An extensive array of sensors is fitted, which supply information on many different parameters to the Powertrain Control Module (PCM).

Information on crankshaft position and engine speed is generated by a crankshaft position sensor. The inductive head of the sensor runs just above the engine flywheel, and scans a series of 36 protrusions on the flywheel periphery. As the crankshaft rotates, the sensor transmits a pulse every time a protrusion passes it. There is one missing protrusion in the flywheel periphery at a point corresponding to 90° BTDC. The PCM recognises the absence of a pulse from the crankshaft position sensor at this point to establish a reference mark for crankshaft position. Similarly, the time interval between absent pulses is used to determine engine speed.

Information on the quantity and temperature of the inlet air is derived from the MAP sensor and the inlet air temperature sensor. The manifold absolute pressure (or MAP) sensor is connected to the inlet manifold by a vacuum hose, and measures the pressure in the inlet system. The inlet air temperature sensor measures the temperature of the inlet air before it enters the turbocharger. The temperature and quantity of air has a direct bearing on the quantity of fuel to be injected for optimum efficiency.

The traditional coolant temperature sensor has been replaced by a cylinder head temperature sensor. The new sensor is seated in a blind hole in the cylinder head, and measures the temperature of the metal directly. Information on engine temperature is critical for accurate fuelling calculations, and is also used to control the pre-heating system for cold starts.

The clutch pedal sensor informs the PCM whether the clutch is engaged or disengaged. When the clutch pedal is depressed, the quantity of fuel injected is momentarily reduced, to make gearchanging smoother.

The amount of fuel delivered is determined by the pump's internal quantity and timing solenoid valves, which are controlled by the pump control unit (PCU), mounted on top of the pump. The pump is internally equipped with a pulse ring fitted to the main rotor, and an angle sensor determines the pump rotor's position and speed, in much the same way as the crankshaft position sensor and engine flywheel, except that there are four gaps in the pump rotor 'teeth' – one for each cylinder. The pump control unit is supplied with information from the 'main' engine management module (PCM), and from this, is able to calculate the most appropriate values for injection timing and quantity (injection duration). The electronically-controlled pump internals enable these calculated values to be delivered with great accuracy, for improved efficiency and reduced emissions.

No accelerator cable is fitted on the Endura-DI engine – instead, a sensor located next to the accelerator pedal informs the PCM of the accelerator position, and this information is used to determine the most appropriate fuelling requirements from the injection pump. The engine idle speed is also controlled by the PCM, and cannot be adjusted. From the signals it receives from the various sensors, the PCM can control the idle speed very accurately, compensating automatically for additional engine loads or unfavourable ambient/engine temperatures.

Rigid pipes connect the pump to the four injectors. Each injector has five holes, to disperse the fuel evenly, and sprays fuel directly into the combustion chamber as its piston approaches TDC on the compression stroke. This system is known as direct injection. The pistons have an off-centre recess machined into their crowns, the shape of which has been calculated to improve 'swirl' (fuel/air mixing). The injectors open in two stages, to promote smoother combustion – the rate of opening is fixed by two internal springs. Lubrication is provided by allowing a small quantity of fuel to leak back past the injector internal components. The leaked-back fuel is returned to the pump and then to the fuel tank.

Cold-starting performance is automatically controlled by the PCM and pump control unit. Under cold start conditions, the injection pump timing is advanced by the pump control unit, while the PCM operates the glow plug system. The pre-heater or 'glow' plugs fitted to each cylinder are electrically heated before, during and immediately after starting, and are particularly effective during a cold start. A warning light illuminates when the ignition is switched on, showing that the glow plugs are in operation. When the light goes out, pre-heating is complete and the engine can be started. In very cold conditions, the glow plugs remain on after the engine has started (post-heating) – this helps the engine to run more smoothly during warm-up, and reduces exhaust emissions.

Older diesel engines, including the previous Endura-DE, had injection pumps which were equipped with a solenoid valve to cut the fuel supply when the ignition switch is turned off, to stop the engine (the valve was usually known as a 'stop' solenoid). The Endura-DI injection pump does not have a stop solenoid – instead, the PCM is able to 'switch off' the injection pump via the pump control unit, and this forms part of the vehicle immobiliser system.

The fuel system on diesel engines is normally very reliable. Provided that clean fuel is used and the specified maintenance is conscientiously carried out, no problems should be experienced. The injection pump and injectors may require overhaul after a high mileage has been covered, but this cannot be done on a DIY basis.

Chapter 10

Fuel system – depressurisation, priming and bleeding

Warning: *The following procedure will merely relieve the pressure in the fuel system – remember that fuel will still be present in the system components, and take precautions accordingly before disconnecting any of them.*

Petrol models

1 The fuel system referred to in this Chapter is defined as the fuel tank and tank-mounted fuel pump/fuel gauge sender unit, the fuel filter, the fuel injector, fuel pressure regulator, and the metal pipes and flexible hoses of the fuel lines between these components. All these contain fuel, which will be under pressure while the engine is running and/or while the ignition is switched on.

2 The pressure will remain for some time after the ignition has been switched off, and must be relieved before any of these components is disturbed for servicing work.

3 The simplest depressurisation method is to disconnect the fuel pump electrical supply by removing the fuel pump fuse (refer to the wiring diagrams or the label on the relevant fusebox for exact location) and starting the engine; allow the engine to idle until it stops through lack of fuel. Turn the engine over once or twice on the starter to ensure that all pressure is released, then switch off the ignition; do not forget to refit the fuse when work is complete.

4 If an adapter is available to fit the Schrader-type valve on the fuel rail pressure test/release fitting (identifiable by its blue plastic cap, and located on the union of the fuel feed line and the fuel rail), this may be used to release the fuel pressure. The Ford adapter (tool number 23-033) operates similar to a drain tap – turning the tap clockwise releases the pressure. If the adapter is not available, place cloth rags around the valve, then remove the cap and allow the fuel pressure to dissipate. Refit the cap on completion.

5 Note that, once the fuel system has been depressurised and drained (even partially), it will take significantly longer to restart the engine – perhaps several seconds of cranking – before the system is refilled and pressure restored.

Diesel models

6 As this system is intended to be 'self-bleeding', no hand-priming pump or separate bleed screws/nipples are fitted.

7 When any part of the system has been disturbed therefore, air must be purged from the system by cranking the engine on the starter motor until it starts. When it has started, keep the engine running for approximately 5 minutes to ensure that all air has been removed from the system. To minimise the strain on the battery and starter motor when trying to start the engine, crank it in 10-second bursts, pausing for 30 seconds each time, until the engine starts.

8 Depending on the work that has been carried out, it may be possible to partially prime the system so as to spare the battery by reducing the amount of cranking time required to start the engine. To spare the battery, fill the filter with clean fuel via its vent screw opening, but it is essential that no dirt is introduced into the system and that no diesel fuel is poured over vulnerable components when doing this.

9 If a hand-operated vacuum pump is available, this can be connected to the pump's fuel return union and used to suck fuel through the supply lines and filter. This will obviously save the battery a good deal of work. If a long length of clear plastic tubing is used to connect the vacuum pump to the injection pump union, it will be easier to see when fuel emerges free from air bubbles. Turn the ignition switch to position II so that fuel can pass through the pump.

10 If air has entered the injector pipes, slacken each union at the injectors and crank the engine until fuel emerges, then tighten securely all unions and mop up the spilt fuel. Start the engine and keep it running for a few minutes to ensure that all air has been expelled.

Petrol fuel injection system components – removal and refitting

Throttle body housing

1 The housing is located on the left-hand side of the inlet manifold. First disconnect the battery negative (earth) lead.

2 Remove the air inlet duct or air cleaner assembly, as necessary.

3 Disconnect the accelerator cable from the lever on the throttle body housing.

4 Disconnect the throttle position sensor multiplug.

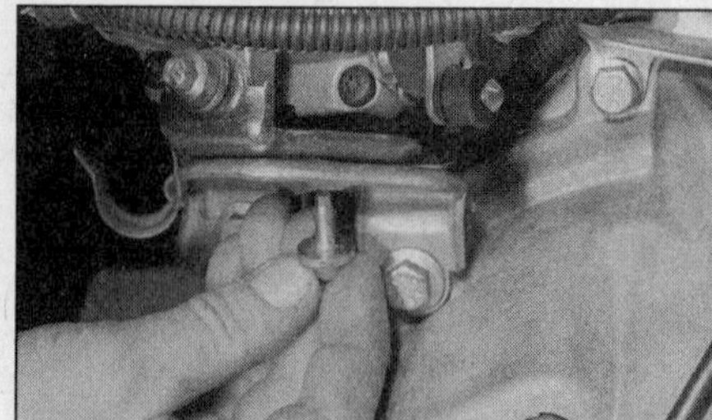

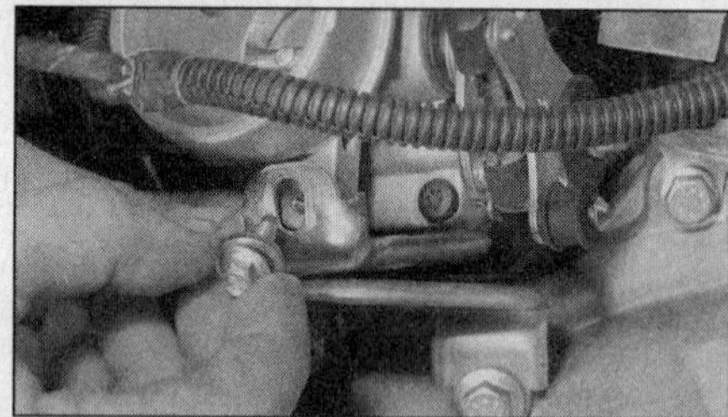

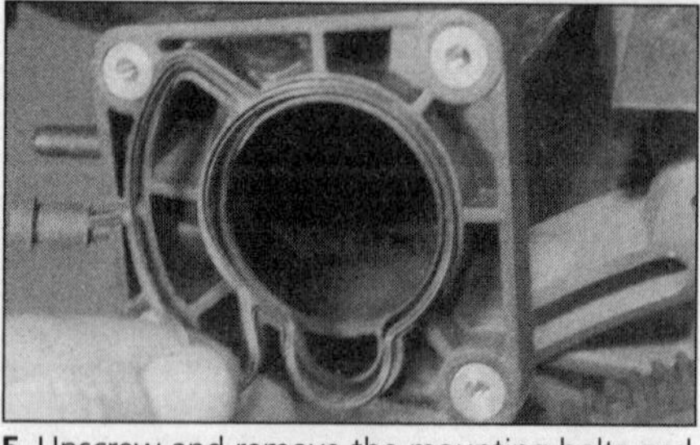

5 Unscrew and remove the mounting bolts and withdraw the throttle housing from the inlet manifold. On the Zetec-SE engine also unbolt and remove the support bracket. Discard the gasket and obtain a new one.

6 Refitting is a reversal of removal, but clean the mating faces and fit a new gasket, and tighten the mounting bolts securely. Check and if necessary adjust the accelerator cable.

Fuel rail and injectors

7 Relieve the residual pressure in the fuel system, and equalise tank pressure by removing the fuel filler cap.

8 Disconnect the battery negative (earth) lead.

Endura-E engine

9 Disconnect the wiring from the idle air control valve. Unscrew the two bolts and remove the idle air control valve.

10 Pull out the accelerator cable adjustment clip, then release the outer cable from the support bracket and disconnect the inner cable from the throttle lever on the throttle housing.

11 Remove the air inlet resonator, then loosen the clips and disconnect the air inlet duct from the air cleaner and throttle housing. Remove the duct from the engine compartment.

12 Disconnect the crankcase ventilation hoses from the inlet manifold.

13 Unscrew the bolt and remove the accelerator cable retaining bracket.

14 Disconnect the HT leads from the spark plugs, then unclip the lead supports and position the leads to one side. When disconnecting the leads, pull on the end fittings and not the leads.

15 Disconnect the wiring from the throttle position sensor. To do this, depress the locking wire and pull off the plug.

16 Disconnect the wiring from the fuel injectors, then undo the injector wiring loom retaining screws, disconnect the loom multiplug, and position the loom to one side.

17 Pull the vacuum hose from the fuel pressure regulator on the end of the fuel rail.

18 Unscrew and remove the fuel rail mounting bolts.

19 Undo the single screw and remove the fuel supply and return pipe retaining bracket.

20 Disconnect the fuel supply and return pipes from the fuel rail by squeezing the lugs on the special quick-release fittings.

21 Carefully pull the injectors from the inlet manifold and withdraw them together with the fuel rail.

22 Remove the clips and carefully pull the injectors from the fuel rail.

23 Using a screwdriver prise the O-rings from the grooves at each end of the injectors. Discard the O-rings and obtain new ones.

24 Refitting is the reverse of the removal procedure, noting the following points:

a) Lubricate the new O-rings with clean engine oil to aid refitting.

b) Ensure that the hoses and wiring are routed correctly, and secured on reconnection by any clips or ties provided.

c) Adjust the accelerator cable.

d) On completion, switch the ignition on to activate the fuel pump and pressurise the system, without cranking the engine. Check for signs of fuel leaks around all disturbed unions and joints before attempting to start the engine.

Zetec-SE engine

25 On later engines with the air cleaner assembly mounted on the engine, remove the air cleaner. On all engines, lift the cover from the centre of the fuel injector wiring loom holder, and disconnect the vent hose.

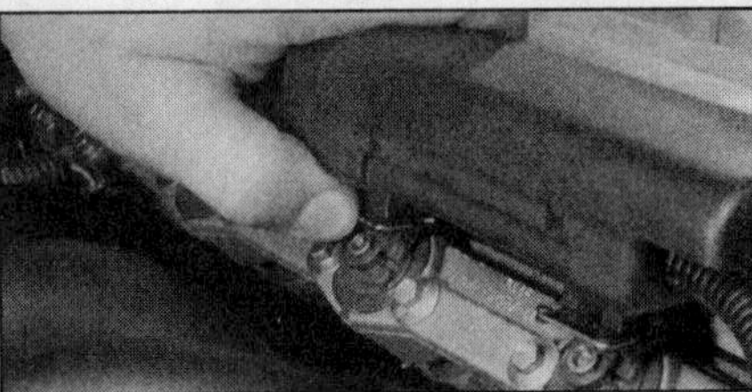

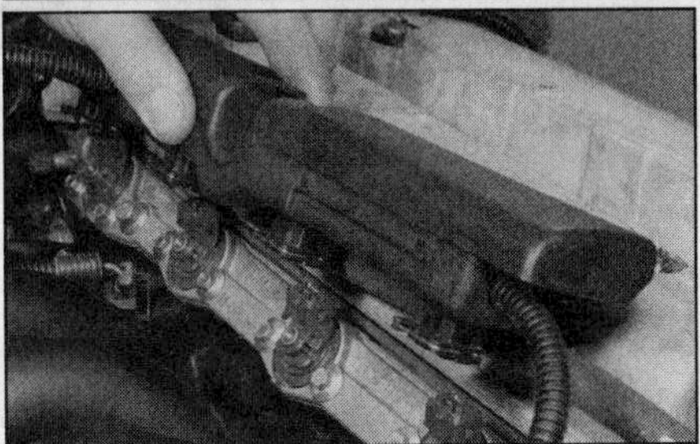

26 The wiring loom adapter must now be disconnected from the injectors. To do this, depress two of the injector clips and lift the adapter up a little until it releases from the injectors. Now depress the remaining two clips and lift the complete adapter from the injectors. Position the adapter to one side.

27 If it is only required to remove the injectors, proceed to paragraph 32.

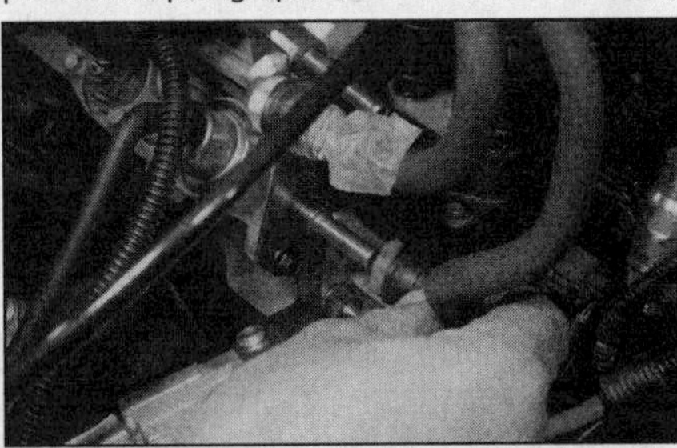

28 To remove the fuel rail, disconnect the feed and return hoses from the left-hand end of the rail by squeezing together the lugs of the quick-release fittings.

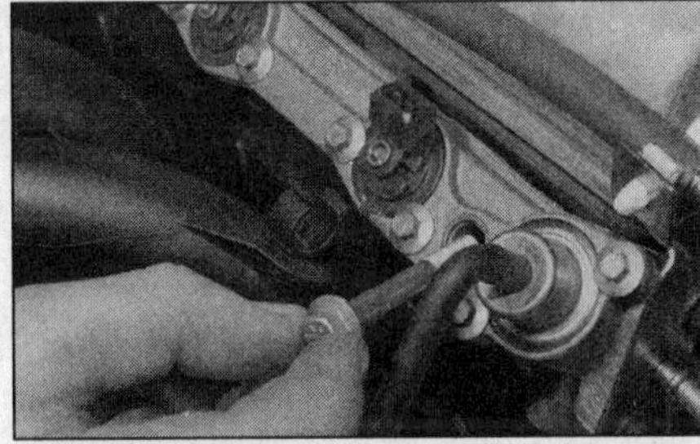

29 Unscrew the mounting bolts and lift the fuel rail from the cylinder head.

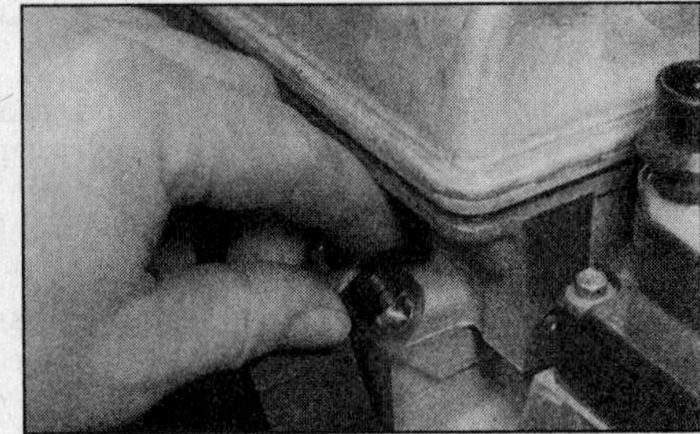

30 Recover the bolt hole adapters from the cylinder head.

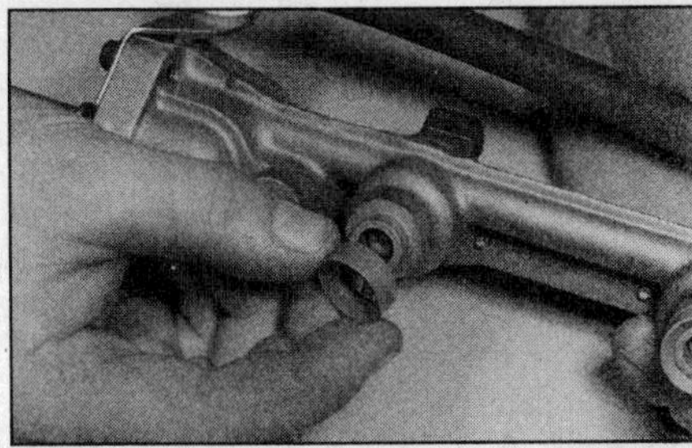

31 Recover the seals from the inner ends of the fuel rail.

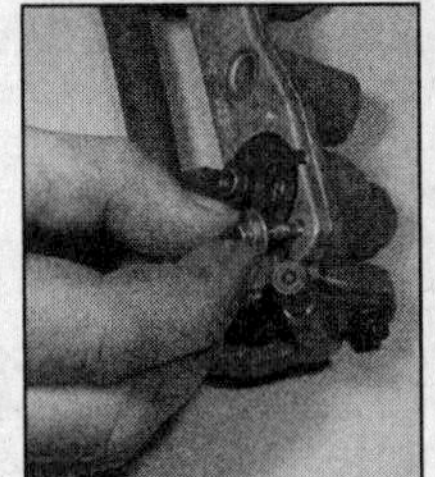

32 To remove the injectors, unscrew the two retaining bolts from each injector and carefully prise the injectors from the fuel rail. If they are tight, use a soft-faced tool inserted from the holes on the inner end of the fuel rail.

33 Remove the O-ring seals from the grooves in the injectors. Discard them and obtain new ones.

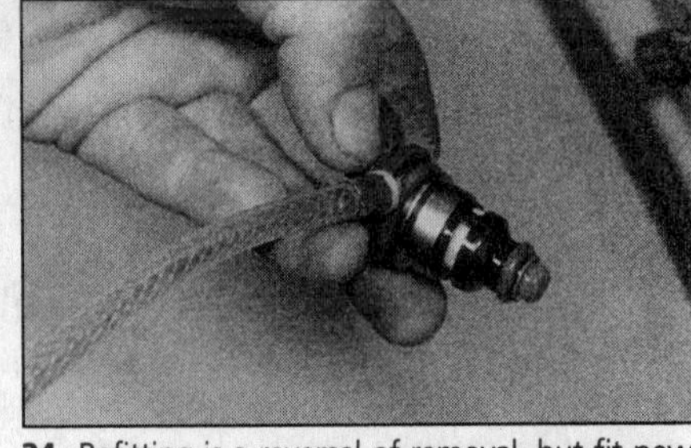

34 Refitting is a reversal of removal, but fit new O-ring seals and lubricate them with a little fresh engine oil before inserting the injectors in the fuel rail. Tighten the bolts to the specified torque.

Fuel pressure regulator

35 Relieve the residual pressure in the fuel system, and equalise tank pressure by removing the fuel filler cap.

36 Disconnect the battery negative (earth) lead.

Endura-E engine

37 Loosen the clips and disconnect the air inlet duct from the air cleaner and throttle housing.

38 Disconnect the vacuum pipe from the fuel pressure regulator.

39 On early models, undo the single screw and remove the fuel supply and return pipe retaining bracket. Disconnect the fuel return pipe from the fuel rail by squeezing the lugs on the special quick-release fittings.

40 Where the regulator is secured by screws, undo the two screws and remove the fuel pressure regulator from the fuel rail.

41 Where the regulator is secured by a spring clip, extract the clip and remove the fuel pressure regulator from the fuel rail.

42 Using a screwdriver, prise the O-ring from the groove in the fuel pressure regulator. Discard the O-ring and obtain a new one.

43 Refitting is a reversal of removal, but lubricate the new O-ring with clean engine oil to aid installation. Where applicable, tighten the mounting screws to the specified torque.

Zetec-SE engine

44 On later engines with the air cleaner assembly mounted on the engine, remove the air cleaner.

45 Disconnect the vacuum pipe from the fuel pressure regulator, then undo the two screws and remove the fuel pressure regulator from the fuel rail.

46 Using a screwdriver, prise the O-rings from the grooves in the fuel pressure regulator. Discard the O-rings and obtain new ones.

47 Refitting is a reversal of removal, but lubricate the new O-rings with clean engine oil to aid installation. Tighten the mounting screws to the specified torque.

Idle air control valve

48 The valve is located on the inlet manifold. On later engines with the air cleaner assembly mounted on the engine, remove the air cleaner for improved access.

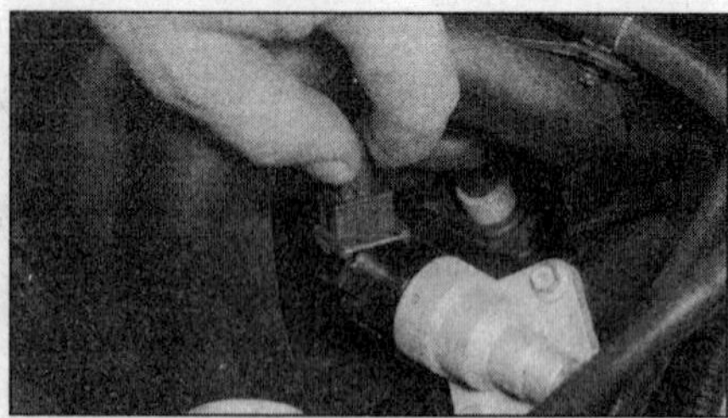

49 Depress the wire clip and disconnect the wiring from the valve.

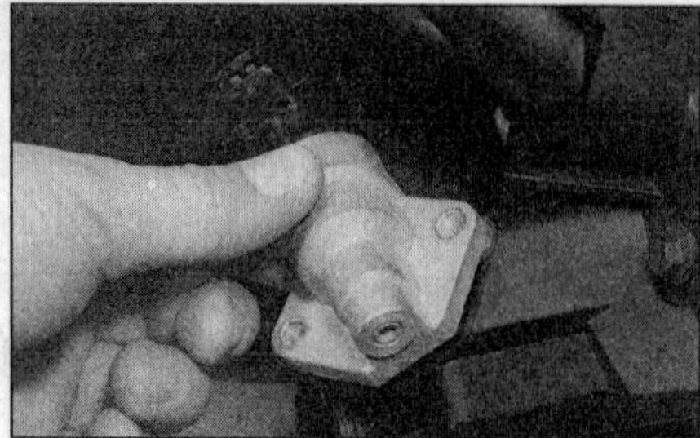

50 Unscrew the mounting bolts and remove the valve from the inlet manifold.

51 Recover the O-ring seals from the inlet manifold and discard them. Obtain new seals.

52 Refitting is a reversal of removal, but note the following points.

a) *Clean the mating surfaces, and fit new O-ring seals.*

b) *Once the wiring and battery are reconnected, start the engine and allow it to idle. When it has reached normal operating temperature, check that the idle speed is stable, and that no induction (air) leaks are evident. Switch on all electrical loads (headlights, heated rear window, etc), and check that the idle speed is still satisfactory.*

Mass air flow sensor

53 Loosen the clip and disconnect the air inlet duct from the mass air flow sensor on the air cleaner cover. If necessary, remove the air cleaner cover for improved access.

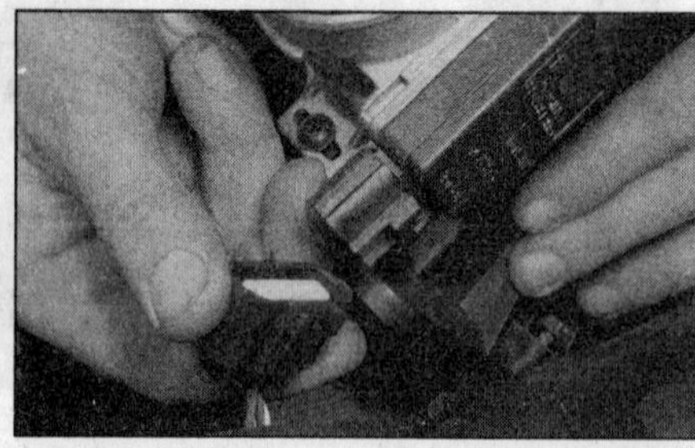

54 Disconnect the wiring from the sensor.

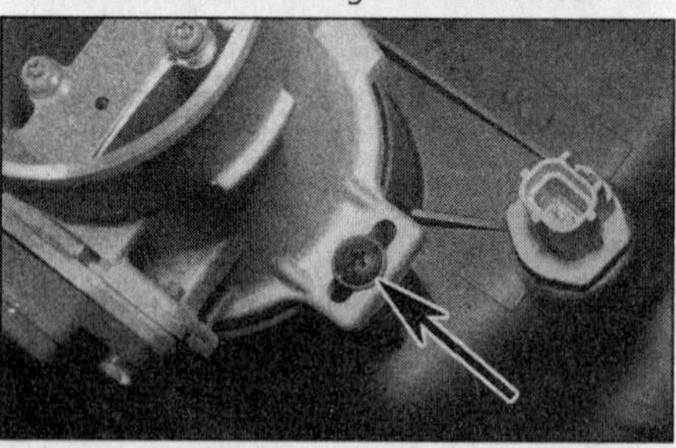

55 Unscrew the crosshead mounting screws and remove the sensor from the air cleaner cover.

56 Refitting is a reversal of removal.

Powertrain Control Module

Note: *The module is fragile. Take care not to drop it, or subject it to any other kind of impact. Do not subject it to extremes of temperature, or allow it to get wet.*

57 The Module (engine management module) is located behind the passenger footwell side trim. First disconnect the battery negative (earth) lead.

58 Remove the fuel cut-off switch.

59 Drill out or cut off the two rivets and remove the security shield from over the module.

60 Release the module from its bracket and withdraw it downwards.

61 Undo the screw and swivel the wiring multiplug away from the module. **Do not** pull on the wiring, only on the multiplug itself.

62 Withdraw the module from inside the vehicle.

63 Refitting is a reversal of removal. Take care when refitting the multiplug, and tighten the retaining bolt by hand first. Use new pop rivets when refitting the security shield.

Crankshaft position sensor

64 The sensor is located on the front left-hand side of the engine. For improved access, apply the handbrake then jack up the front of the vehicle and support it on axle stands.

65 Unbolt (Endura-E) or unclip (Zetec-SE) the cover from the crankshaft position sensor on the front of the engine, then disconnect the wiring.

66 Unscrew the mounting bolt and withdraw the sensor.

67 Refitting is a reversal of removal.

Camshaft position sensor

68 On the Endura-E engine the sensor is located on the timing chain cover. On the Zetec-SE engine it is located on the right-hand rear of the cylinder head.

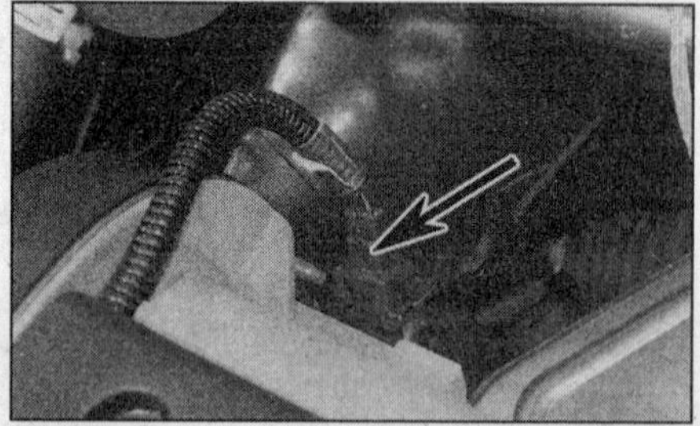

69 Depress the wire clip and disconnect the wiring from the camshaft position sensor. Where applicable on the Endura-E engine, release the fuel feed and return hoses from their clip.

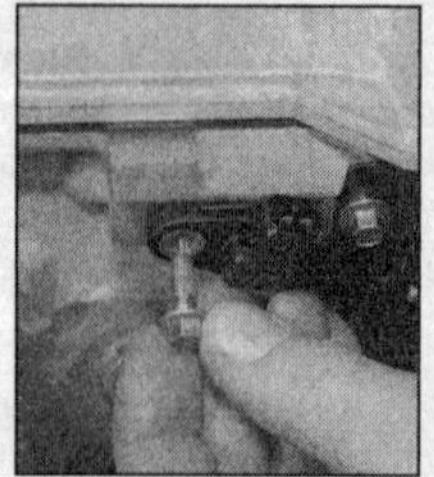

70 Unscrew the mounting bolt and withdraw the sensor from the cylinder head or timing chain cover (as applicable).

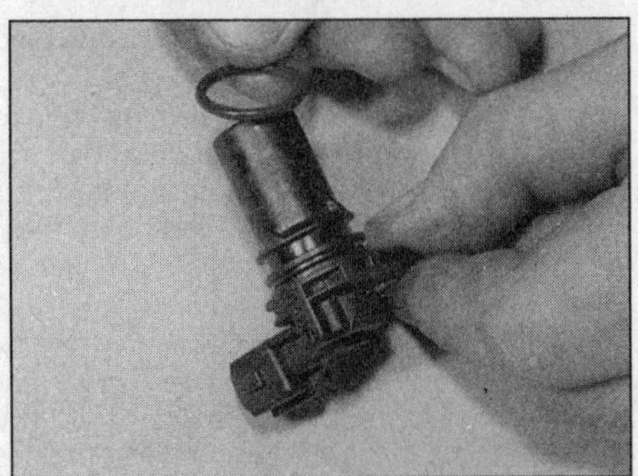

71 Remove the O-ring from the groove in the sensor.

72 Refitting is a reversal of removal but fit a new O-ring. Smear a little engine oil on the seal before fitting the sensor.

Coolant temperature sensor

73 Drain the cooling system.

74 Unbolt the plastic cover from the top of the cylinder head cover.

75 Disconnect the HT leads from the ignition coil, then disconnect the low tension wiring from the coil.

76 Unbolt and remove the ignition coil.

77 Disconnect the wiring from the coolant temperature sensor.

78 Unscrew and remove the sensor.

79 Refitting is a reversal of removal. Clean the threads of the sensor and mounting hole, then refit the sensor and tighten it to the specified torque.

Inlet air temperature sensor

80 The sensor is located in the air cleaner cover. If necessary, remove the air cleaner cover for improved access.

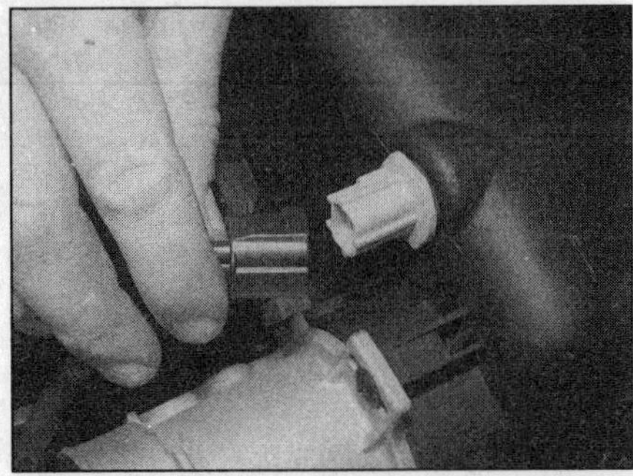

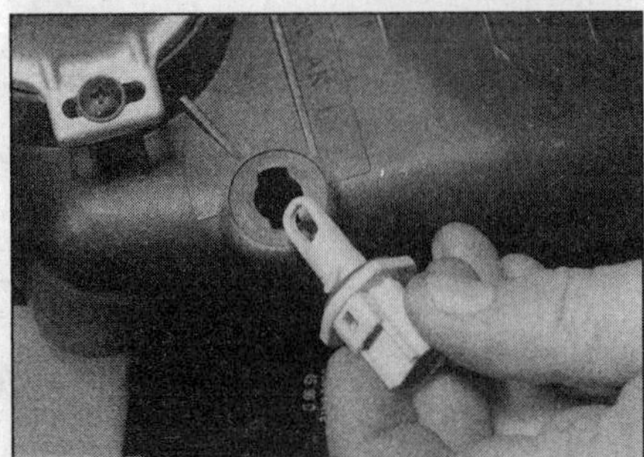

81 Disconnect the wiring from the sensor, then twist the sensor through 90° and remove it.

82 Refitting is a reversal of removal.

Throttle position sensor

83 The sensor is located on the throttle body housing on the left-hand side of the inlet manifold. On later engines with the air cleaner assembly mounted on the engine, remove the air cleaner for improved access.

84 Disconnect the wiring by depressing the retaining clip. Remove the retaining screws, and withdraw the unit from the throttle housing. *Do not* force the sensor's centre to rotate past its normal operating sweep; the unit will be seriously damaged.

85 Refitting is a reversal of removal, but ensure that the sensor is correctly orientated, by locating its centre on the D-shaped throttle shaft (throttle closed), and aligning the sensor body so that the bolts pass easily into the throttle housing.

Vehicle speed sensor

86 The sensor is mounted on the rear of the transmission at the base of the speedometer drive cable.

87 Where applicable, undo the retaining nut, and withdraw the speedometer cable from the vehicle speed sensor. Use two spanners to loosen the nut – one to counterhold the sensor, and the other to unscrew the cable nut.

88 Disconnect the wiring from the vehicle speed sensor, then unscrew the sensor from the top of the drive pinion.

89 Refitting is a reversal of removal.

Temperature and manifold absolute pressure sensor

90 The sensor is mounted in the inlet manifold.

91 Remove the air cleaner for improved access.

92 Disconnect the sensor wiring connector.

93 Undo the sensor mounting screws, lift off the sensor and recover the gasket.

94 Refitting is a reversal of removal, using a new gasket.

Cylinder head temperature sensor

95 The sensor is located in the centre of the head, between Nos 2 and 3 spark plugs.

96 Disconnect the HT leads from the spark plugs as necessary for access – make sure that the leads are numbered for position (with No 1 at the timing belt end). Prise out the rubber bung used to secure the sensor wiring.

97 Disconnect the wiring from the temperature sensor.

98 Unscrew and remove the sensor. **Note:** *The sensor cannot be re-used – a new one must be fitted, since the mating face of the sensor is deformed when fully tightened, to ensure a good contact with the cylinder head. If the old sensor were re-used, it would result in inaccurate readings.*

99 Refitting is a reversal of removal. Clean the sensor mounting hole, then fit the sensor and tighten it to the specified torque.

Clutch pedal position switch

100 Inside the vehicle, reach up under the clutch pedal and disconnect the return spring from the bracket.

101 Disconnect the wiring from the clutch switch, then twist the switch and remove it from the pedal bracket.

102 Refitting is a reversal of removal.

Oxygen sensor

103 The oxygen sensor is located in the exhaust manifold. Trace the wiring back from the oxygen sensor to the connector and disconnect the wiring.

104 Unscrew the sensor and remove it from the exhaust manifold.

105 Refitting is a reversal of removal, by apply a little anti-seize grease to the sensor threads prior to refitting, and tighten the sensor to the specified torque.

Diesel injection system electronic components – removal and refitting

Crankshaft position sensor

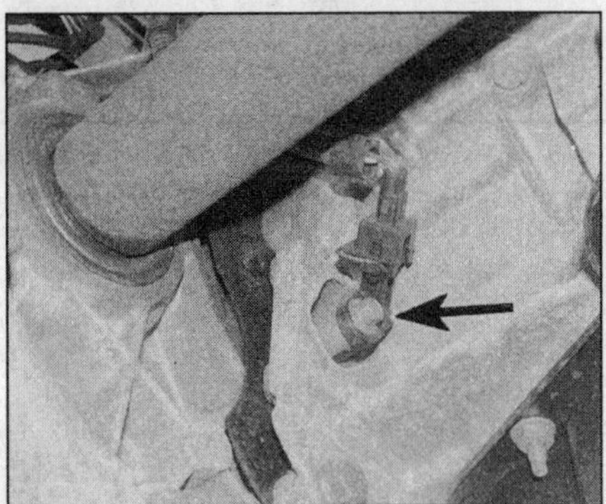

1 The sensor is located at the flywheel end of the engine, low down at the rear. For improved access, apply the handbrake then jack up the front of the vehicle and support it on axle stands.

2 Where fitted, unclip the cover from the crankshaft position sensor, then disconnect the wiring plug.

3 Unscrew the mounting bolt and withdraw the sensor. If a spacer is fitted between the sensor and engine, it is vital that this is refitted when refitting the sensor, or the sensor head will hit the rotor teeth.

4 Refitting is a reversal of removal.

Cylinder head temperature sensor

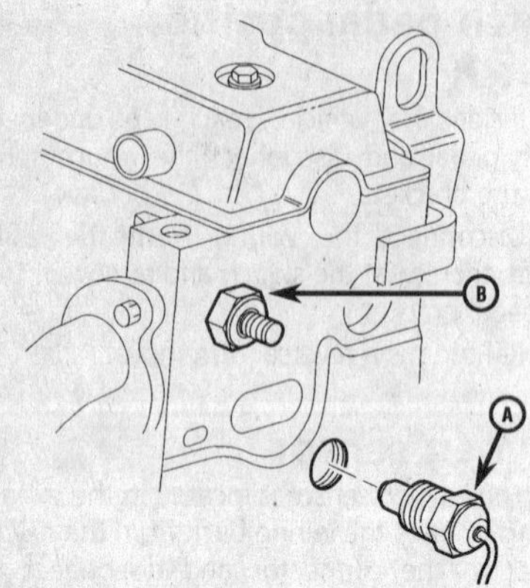

5 The switch (A) is screwed into the left-hand (flywheel) end of the cylinder head, behind the vacuum pump and below the oil pressure warning light switch (B).

6 Disconnect the battery negative (earth) lead.

7 To improve access to the switch, it will be necessary to remove (or partially remove) the air cleaner inlet duct.

8 It will also be helpful to release the hoses and remove the crankcase ventilation system oil separator from the left-hand end of the cylinder head.

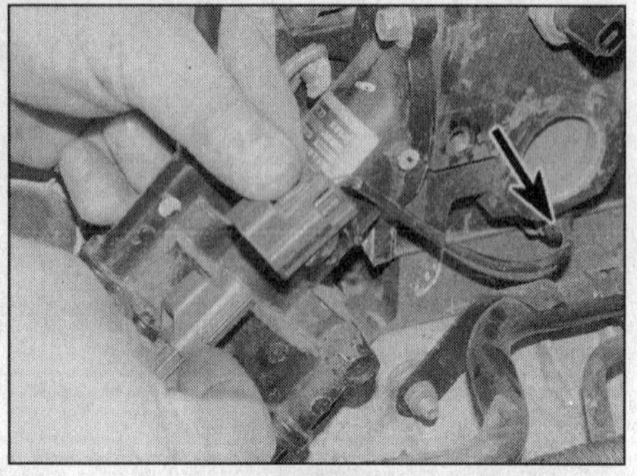

9 Trace the wiring from the sensor, and disconnect it at the plug, which is clipped to the brake vacuum pump at the front of the engine.

10 The sensor can now be unscrewed and removed. However, access to the sensor is such that great difficulty may be experienced in getting a tool to fit onto it. Ultimately, it may be necessary to cut the sensor wiring, unscrew the sensor using a thin-wall socket or box spanner, then re-make the wiring after fitting, using a suitable connector.

11 Refitting is a reversal of removal. Clean the threads of the sensor and mounting hole, then refit the sensor and tighten it to the specified torque.

Inlet air temperature sensor

12 The sensor is located in the air cleaner inlet duct.

13 Disconnect the wiring from the sensor, then twist the sensor anti-clockwise and remove it.

14 Refitting is a reversal of removal.

Vehicle speed sensor

15 Refer to the petrol injection system Section.

Manifold absolute pressure sensor

16 The MAP sensor is located in the centre of the engine compartment bulkhead.

17 Disconnect the vacuum hose from the port at the base of the sensor, then disconnect the wiring plug behind it.

18 Unscrew and remove the two mounting bolts, and withdraw the sensor from the bulkhead.

19 Prior to refitting, remove the vacuum hose from the inlet manifold connection, and check it for signs of perishing or splitting, especially at the pipe ends.

20 Refitting is a reversal of removal, ensuring that the wiring plug and vacuum hose are securely reconnected.

Fuel control valve

21 The control valve is fitted to the top of the fuel filter (at the right-hand rear corner of the engine compartment), and contains a bi-metal strip. Its function is to close the fuel return to the fuel tank at low fuel temperatures, allowing fuel which has been warmed by passing through the injection pump to flow back into the filter, thus warming the fuel being drawn from the tank.

22 Noting their positions for refitting, disconnect the fuel supply and outlet pipes from the connections at the fuel filter.

23 Extract the wire clip securing the control valve and return pipe, and release it from the top of the filter. Once again noting the fitted positions of the fuel pipes, disconnect them from each end of the valve (note that the return pipe may in this case be black, and smaller in diameter than the supply pipe).

24 Refitting is a reversal of removal. Ensure that the pipe connections are correctly and securely remade, then run the engine and check for signs of fuel leakage.

Clutch pedal position switch

25 Remove the trim panel above the driver's footwell to gain access to the clutch pedal.

26 Reach up and disconnect the wiring from the clutch switch at the top of the pedal, then twist the switch anti-clockwise and remove it from the pedal bracket.

27 Refitting is a reversal of removal.

Accelerator pedal sensor

28 The accelerator pedal sensor is integral with the pedal assembly.

Injection pump control unit

29 The pump control unit is integral with the injection pump, and cannot be separated from it. If a new injection pump is fitted, the PCM must be electronically 'matched' to the engine management module (ECU), otherwise the pump will not function correctly (or even at all, if the immobiliser function is not correctly set up) – this is a task for a Ford dealer or suitably equipped specialist, as specialised electronic equipment is required.

Powertrain Control Module

Note: *The module is fragile. Take care not to drop it, or subject it to any other kind of impact. Do not subject it to extremes of temperature, or allow it to get wet. The module wiring plug must never be disconnected while the ignition switch is on.*

30 The Powertrain Control Module is located behind the passenger footwell side trim. First disconnect the battery negative (earth) lead.

31 Drill out or cut off the two rivets and remove the security shield from over the module.

32 Release the module from its bracket and withdraw it downwards.

33 Undo the screw and swivel the wiring multiplug away from the module. **Do not** pull on the wiring, only on the multiplug itself.

34 Withdraw the module from inside the vehicle.

35 Refitting is a reversal of removal. Take care when refitting the multiplug, and tighten the retaining bolt by hand first. Use new pop rivets when refitting the security shield.

Ford Focus '98 – '01

11

Petrol models

General

System type	Sequential Electronic Fuel injection (SEFi)
Fuel octane requirement	95 RON unleaded
Idle speed	700 ± 30 rpm (regulated by EEC-V engine management system – no adjustment possible)
CO content at idle	Less than 0.3%

Fuel system data

Note: *The resistance values quoted below are typical, but may be used for guidance. Generally, a faulty component will be indicated by a zero or infinity reading, rather than a slight deviation from the values given. Always have your findings verified before buying a new component (if possible, perform the same test on a new component, and compare the results).*

Fuel pressure regulator:	
Regulated fuel pressure:	
Pressure regulator hose disconnected	2.7 ± 0.2 bars
Engine running, pressure regulator hose connected	2.1 ± 0.2 bars
Hold pressure – engine stopped after five minutes	1.8 bars minimum
Fuel injectors:	
Resistance	13.7 to 15.2 ohms
Idle speed control valve:	
Resistance	6 to 14 ohms
Idle increase solenoid valve:	
Resistance	50 to 120 ohms
Crankshaft speed/position sensor:	
Resistance	200 to 450 ohms
Camshaft position sensor:	
Resistance	200 to 900 ohms
Intake air temperature sensor:	
Resistance:	
At –40°C	860 to 900 k ohms
At 20°C	35 to 40 k ohms
At 100°C	1.9 to 2.5 k ohms
At 120°C	1.0 to 1.3 k ohms
Throttle potentiometer:	
Resistance	400 to 6000 ohms
Power steering pressure switch:	
Operating pressure – green switch body:	
Contacts open – infinite resistance	31.5 ± 3.5 bars
Contacts close – 0 to 2.5 ohms resistance	Between 13.5 and 24.0 bars

Chapter 11

Diesel models

General

System type	Bosch VP-30 combined lift and injection pump, chain-driven from crankshaft. Direct injection via five-hole, pencil-type injectors, electronic pump control unit (PCU) linked to revised EEC-V engine management module (ECU)
Application	1.8 litre Endura-DI engine
Firing order	1 – 3 – 4 – 2 (No 1 at timing belt end)
Idle speed	800 ± 50 rpm (regulated by EEC-V engine management system – no adjustment possible)
Fuel type	Commercial diesel fuel for road vehicles (DERV)

Injection pump

Make and type	Bosch VP-30 electronic, distributor-type
Rotation (viewed from crankshaft pulley end)	Clockwise
Drive	By twin 'gemini' chains from crankshaft
Injection pump timing	By timing pin, at TDC

Injectors

Type	Pencil-type, five-hole

Turbocharger

Type	Garrett GT15, integral with exhaust manifold. Air-to-air intercooler

Torque wrench settings

	Nm	lbf ft
Petrol models		
Camshaft position sensor:		
1.4 and 1.6 litre models	8	6
1.8 and 2.0 litre	20	15
Crankshaft position sensor	9	7
Cylinder head temperature sensor (renew)	10	7
EGR valve (1.8 and 2.0 litre models)	9	7
Fuel pressure regulator mounting bolts:		
1.4 and 1.6 litre models	10	7
1.8 and 2.0 litre models	6	4
Fuel rail bolts:		
1.4 and 1.6 litre models	15	11
1.8 and 2.0 litre models	10	7
Idle speed control valve	10	7
Knock sensor	20	15
Power steering pressure switch	20	15
Diesel models		
Crankshaft position sensor bolt	10	7
Cylinder head temperature sensor	20	15

General information

Petrol models

The fuel system consists of a fuel tank (mounted under the body, beneath the rear seats), fuel hoses, an electric fuel pump mounted in the fuel tank, and a sequential electronic fuel injection system controlled by a EEC-V engine management control module (also called the ECU, or Electronic Control Unit).

The electric fuel pump supplies fuel under pressure to the fuel rail, which distributes fuel to the injectors. A pressure regulator controls the system pressure in relation to inlet tract depression. From the fuel rail, fuel is injected into the inlet ports, just above the inlet valves, by four fuel injectors. The aluminium alloy fuel rail is mounted to the cylinder head on 1.4 and 1.6 litre models, while the fuel rail on 1.8 and 2.0 litre models is mounted on the inlet manifold.

The amount of fuel supplied by the injectors is precisely controlled by the ECU. The module uses the signals from the crankshaft position sensor, and the camshaft position sensor, to trigger each injector separately in cylinder firing order (sequential injection), with benefits in terms of better fuel economy and leaner exhaust emissions.

The ECU is the heart of the entire engine management system, controlling the fuel injection, ignition and emissions control systems. The module receives information from various sensors which is then computed and compared with pre-set values stored in its memory, to determine the required period of injection.

Information on crankshaft position and engine speed is generated by a crankshaft position sensor. The inductive head of the sensor runs just above the engine flywheel and scans a series of 36 protrusions on the flywheel periphery. As the crankshaft rotates, the sensor transmits a pulse to the system's ignition module every time a protrusion passes it. There is one missing protrusion in the flywheel periphery at a point corresponding to 90° BTDC. The ignition module recognises the absence of a pulse from the crankshaft position sensor at this point to establish a reference mark for crankshaft position. Similarly, the time interval between absent pulses is used to determine engine speed. This information is then fed to the ECU for further processing.

The camshaft position sensor is located in the cylinder head so that it registers with a lobe on the camshaft. The camshaft position sensor functions in the same way as the crankshaft position sensor, producing a series of pulses; this gives the ECU a reference point, to enable it to determine the firing order, and operate the injectors in the appropriate sequence.

The mass airflow sensor is based on a 'hot-wire' system, sending the ECU a constantly-varying (analogue) voltage signal corresponding to the mass of air passing into the engine. Since air mass varies with temperature (cold air being denser than warm), measuring air mass provides the module with a very accurate means of determining the correct amount of fuel required to achieve the ideal air/fuel mixture ratio.

The traditional coolant temperature sensor has been replaced by a cylinder head temperature sensor. The new sensor is seated in a blind hole in the cylinder head, and measures the temperature of the metal directly. This component is an NTC (Negative Temperature Coefficient) thermistor – that is, a semi-conductor whose electrical resistance decreases as its temperature increases. It provides the ECU with a constantly-varying (analogue) voltage signal, corresponding to the temperature of the engine. This is used to refine the calculations made by the module, when determining the correct amount of fuel required to achieve the ideal air/fuel mixture ratio.

Inlet air temperature information is supplied by the inlet air temperature sensor built into the airflow sensor. This component is also an NTC thermistor – see the previous paragraph – providing the module with a signal corresponding to the temperature of air passing into the engine. This is used to refine the calculations made by the module, when determining the correct amount of fuel required to achieve the ideal air/fuel mixture ratio.

A throttle position sensor is mounted on the end of the throttle valve spindle, to provide the ECU with a constantly-varying (analogue) voltage signal corresponding to the throttle opening. This allows the module to register the driver's input when determining the amount of fuel required by the engine.

Road speed is monitored by the vehicle speed sensor. This component is a Hall-effect generator, mounted on the transmission's speedometer drive. It supplies the module with a series of pulses corresponding to the vehicle's road speed, enabling the module to control features such as the fuel shut-off on overrun.

The clutch pedal position is monitored by a switch fitted to the pedal bracket. This sends a signal to the ECU.

Where power steering is fitted, a pressure-operated switch is screwed into the power steering system's high-pressure pipe. The switch sends a signal to the ECU to increase engine speed to maintain idle speed when the power steering pump loads up the engine (typically, during parking manoeuvres).

The oxygen sensor in the exhaust system provides the module with constant feedback – 'closed-loop' control – which enables it to adjust the mixture to provide the best possible operating conditions for the catalytic converter.

The air inlet side of the system consists of an air cleaner housing, the mass airflow sensor, an inlet hose and duct, and a throttle housing.

The throttle valve inside the throttle housing is controlled by the driver, through the accelerator pedal. As the valve opens, the amount of air that can pass through the system increases. As the throttle valve opens further, the mass airflow sensor signal alters, and the ECU opens each injector for a longer duration, to increase the amount of fuel delivered to the inlet ports.

Both the idle speed and mixture are under the control of the ECU, and cannot be adjusted.

Diesel models

The fuel system consists of a fuel tank (mounted under the body, beneath the rear seats), fuel filter, electronic fuel injection pump with pump control unit, injectors, fuel lines and hoses, fuel gauge sender unit mounted in the fuel tank, and EEC-V engine management control module (ECU).

Fuel is drawn from the tank by the transfer pump incorporated in the injection pump. It then passes through the fuel filter, located in the engine bay, where foreign matter and water are removed. The injection pump is driven from the crankshaft via a twin-row ('gemini') chain and supplies fuel under very high pressure to each injector in turn as it is needed. The camshaft is driven from the injection pump via a toothed timing belt.

The Endura-DI engine was very much a 'state-of-the-art' unit, in that it features a full electronic engine management system, very similar to that fitted to the Focus petrol models. An extensive array of sensors is fitted, which supply information on many different parameters to the ECU.

Information on crankshaft position and engine speed is generated by a crankshaft position sensor. The inductive head of the sensor runs just above the engine flywheel, and scans a series of 36 protrusions on the flywheel periphery. As the crankshaft rotates, the sensor transmits a pulse every time a protrusion passes it. There is one missing protrusion in the flywheel periphery at a point corresponding to 90° BTDC. The ECU recognises the absence of a pulse from the crankshaft position sensor at this point to establish a reference mark for crankshaft position. Similarly, the time interval between absent pulses is used to determine engine speed.

Information on the quantity and temperature of the inlet air is derived from the MAP sensor and the inlet air temperature sensors. The manifold absolute pressure (or MAP) sensor is connected to the inlet manifold by a vacuum hose, and measures the pressure in the inlet system. Two air temperature sensors are fitted – one before the turbocharger, and one after the intercooler. The temperature and quantity of air has a direct bearing on the quantity of fuel to be injected for optimum efficiency.

The traditional coolant temperature sensor has been replaced by a cylinder head temperature sensor. The new sensor is seated in a blind hole in the cylinder head, and measures the temperature of the metal directly. Information on engine temperature is critical for accurate fuelling calculations, and is also used to control the pre-heating system for cold starts.

The clutch pedal sensor informs the ECU whether the clutch is engaged or disengaged. When the clutch pedal is depressed, the quantity of fuel injected is momentarily reduced, to make gearchanging smoother.

The stop-light switch and separate brake pedal sensor inform the ECU when the brakes are applied – when this signal is received, the ECU puts the engine into idle mode until a signal is received from the accelerator position sensor.

The amount of fuel delivered is determined by the pump's internal quantity and timing solenoid valves, which are controlled by the pump control unit (PCU), mounted on top of the pump. The pump is internally equipped with a pulse ring fitted to the main rotor, and an angle sensor determines the pump rotor's position and speed, in much the same way as the crankshaft position sensor and engine flywheel, except that there are four gaps in the pump rotor 'teeth' – one for each cylinder. The pump control unit is supplied with information from the 'main' engine management module (ECU), and from this, is able to calculate the most appropriate values for injection timing and quantity (injection duration). The electronically-controlled pump internals enable these calculated values to be delivered with great accuracy, for improved efficiency and reduced emissions.

No accelerator cable is fitted on the Endura-DI engine – instead, a sensor located next to the accelerator pedal informs the ECU of the accelerator position, and this information is used to determine the most appropriate fuelling requirements from the injection pump. The engine idle speed is also controlled by the ECU, and cannot be adjusted. From the signals it receives from the various sensors, the ECU can control the idle speed very accurately, compensating automatically for additional engine loads or unfavourable ambient/engine temperatures.

Rigid pipes connect the pump to the four injectors. Each injector has five holes, to disperse the fuel evenly, and sprays fuel directly into the combustion chamber as its piston approaches TDC on the compression stroke. This system is known as direct injection. The pistons have an off-centre recess machined into their crowns, the shape of which has been calculated to improve 'swirl' (fuel/air mixing). The injectors open in two stages, to promote smoother combustion – the rate of opening is fixed by two internal springs. Lubrication is provided by allowing a small quantity of fuel to leak back past the injector internal components. The leaked-back fuel is returned to the pump and then to the fuel tank.

Cold-starting performance is automatically controlled by the ECU and pump control unit. Under cold start conditions, the injection pump timing is advanced by the pump control unit, while the ECU operates the glow plug system. The pre-heater or 'glow' plugs fitted to each cylinder are electrically heated before, during and immediately after starting, and are particularly effective during a cold start. A warning light illuminates when the ignition is switched on, showing that the glow plugs are in operation. When the light goes out, pre-heating is complete and the engine can be started. In very cold conditions, the glow plugs remain on after the engine has started (post-heating) – this helps the engine to run more smoothly during warm-up, and reduces exhaust emissions.

Older diesel engines, including the previous Endura-DE, had injection pumps which were equipped with a solenoid valve to cut the fuel supply when the ignition switch is turned off, to stop the engine (the valve was usually known as a 'stop' solenoid). The Endura-DI injection pump does not have a stop solenoid – instead, the ECU is able to 'switch off' the injection pump via the pump control unit, and this forms part of the vehicle immobiliser system.

The fuel system has a built-in 'strategy' to prevent it from drawing in air, should the car run low on fuel. The ECU monitors the level of fuel in the tank, via the gauge sender unit. After switching on the low fuel level warning light, it will eventually induce a misfire as a further warning to the driver, and lower the engine's maximum speed until the engine stops.

The fuel system on diesel engines is normally very reliable. Provided that clean fuel is used and the specified maintenance is conscientiously carried out, no problems should be experienced. The injection pump and injectors may require overhaul after a high mileage has been covered, but this cannot be done on a DIY basis.

Fuel system – depressurisation, priming and bleeding

Petrol models

Warning: The following procedure will merely relieve the pressure in the fuel system – remember that fuel will still be present in the system components, and take precautions accordingly before disconnecting any of them.

1 The fuel system referred to in this Chapter is defined as the fuel tank and tank-mounted fuel pump/fuel gauge sender unit, the fuel filter, the fuel injector, fuel pressure regulator, and the metal pipes and flexible hoses of the fuel lines between these components. .All these contain fuel, which will be under pressure while the engine is running and/or while the ignition is switched on.

2 The pressure will remain for some time after the ignition has been switched off, and must be relieved before any of these components is disturbed for servicing work.

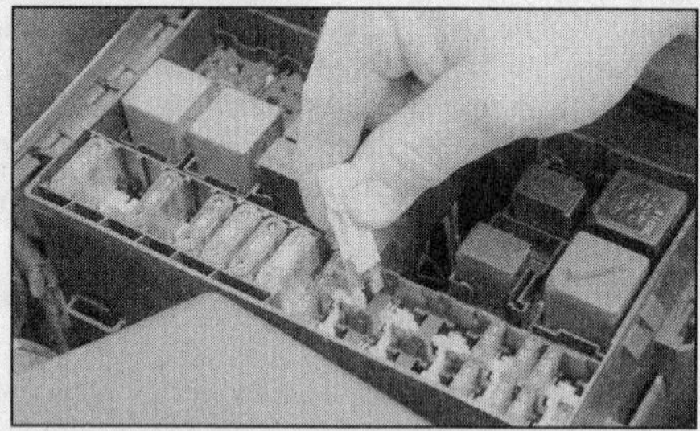

3 The simplest depressurisation method is to disconnect the fuel pump electrical supply by removing the fuel pump fuse (No 12, in the engine compartment fusebox) and starting the engine. If the engine does not start, crank it on the starter for a few seconds, otherwise, allow the engine to idle until it dies through lack of fuel. Turn the engine over once or twice on the starter to ensure that all pressure is released, then switch off the ignition; do not forget to refit the fuse when work is complete.

4 If an adapter is available to fit the Schrader-type valve on the fuel rail pressure test/release fitting (identifiable by its blue plastic cap, and located on the fuel rail), this may be used to release the fuel pressure. The Ford adapter (tool number 23-033) operates similar to a drain tap – turning the tap clockwise releases the pressure. If the adapter is not available, place cloth rags around the valve, then remove the cap and allow the fuel pressure to dissipate. Refit the cap on completion.

5 Note that, once the fuel system has been depressurised and drained (even partially), it will take significantly longer to restart the engine – perhaps several seconds of cranking – before the system is refilled and pressure restored.

Diesel models

6 As this system is intended to be 'self-bleeding', no hand-priming pump or separate bleed screws/nipples are fitted.

7 When any part of the system has been disturbed therefore, air must be purged from the system by cranking the engine on the starter motor until it starts. When it has started, keep the engine running for approximately 5 minutes to ensure that all air has been removed from the system. To minimise the strain on the battery and starter motor when trying to start the engine, crank it in 10-second bursts, pausing for 30 seconds each time, until the engine starts.

8 Depending on the work that has been carried out, it may be possible to partially prime the system so as to spare the battery by reducing as much as possible the amount of cranking time required to start the engine. To spare the battery, fill the filter with clean fuel via its vent screw opening, but it is essential that no dirt is introduced into the system and that no diesel fuel is poured over vulnerable components when doing this.

9 If a hand-operated vacuum pump is available, this can be connected to the pump's fuel return union and used to suck fuel through the supply lines and filter. This will obviously save the battery a good deal of work. If a long length of clear plastic tubing is used to connect the vacuum pump to the injection pump union, it will be easier to see when fuel emerges free from air bubbles. Turn the ignition switch to position II so that fuel can pass through the pump.

10 If air has entered the injector pipes, slacken each union at the injectors and crank the engine until fuel emerges, then tighten securely all unions and mop up the spilt fuel. Start the engine and keep it running for a few minutes to ensure that all air has been expelled.

Petrol fuel injection system components – removal and refitting

Throttle body housing

1 The housing is located on the left-hand side of the inlet manifold. First disconnect the battery negative (earth) lead.

2 Remove the air inlet duct.

3 Disconnect the accelerator cable from the throttle body housing.

4 Disconnect the throttle position sensor multi-plug.

5 Unscrew and remove the four mounting bolts and withdraw the throttle housing from the

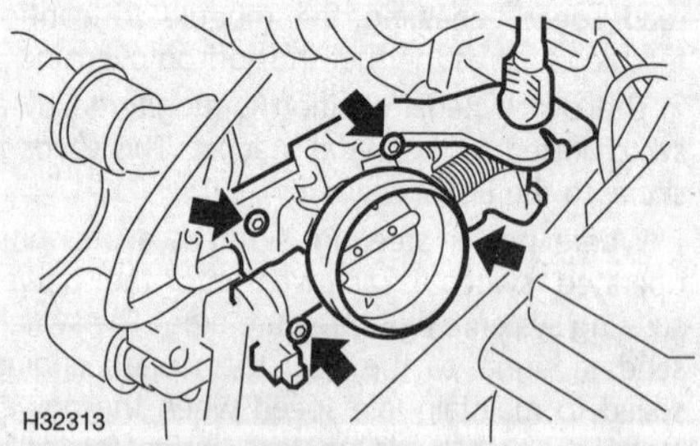

inlet manifold. Discard the rubber gasket and obtain a new one. **Note:** *Take care when cleaning the inside of the housing, as it is treated with a special coating during manufacture, which could be removed by over-enthusiastic cleaning, or by use of powerful solvents.*

6 Refitting is a reversal of removal, but clean the mating faces and fit a new gasket, and tighten the mounting bolts securely.

Fuel rail and injectors

7 Relieve the residual pressure in the fuel system, and equalise tank pressure by removing the fuel filler cap.

> *Warning: This procedure will merely relieve the increased pressure necessary for the engine to run – remember that fuel will still be present in the system components, and take precautions accordingly before disconnecting any of them.*

8 Disconnect the battery negative (earth) lead.

1.4 and 1.6 litre engines

9 Disconnect the breather hose from the cylinder head cover.

10 Disconnect the wiring plugs from the idle speed valve and throttle position sensor, then detach the wiring harness from the clip on the fuel rail.

11 Unclip the fuel lines from the cylinder head cover.

12 Using a small screwdriver, prise up and remove the four wire clips (one on each injector) securing the injector wiring busbar – use a magnetic holding tool to prevent the wire clips falling down the back of the engine as they are removed. Lift off the wiring busbar, and move it to the rear.

13 Disconnect the vacuum hose from the fuel pressure regulator.

14 Disconnect the fuel supply and return lines, noting their fitted positions. The supply line is colour-coded white, and the return line is colour-coded red.

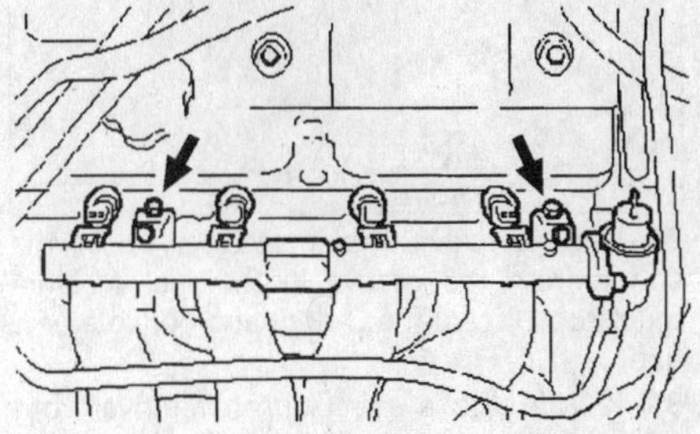

15 Unscrew and remove the two bolts securing the fuel rail. Pulling the fuel rail equally at both ends, remove it squarely from the inlet manifold.

1.8 and 2.0 litre engines

16 Loosen the retaining clips at either end, and remove the air cleaner inlet duct.

17 Disconnect the accelerator cable from the throttle body housing.

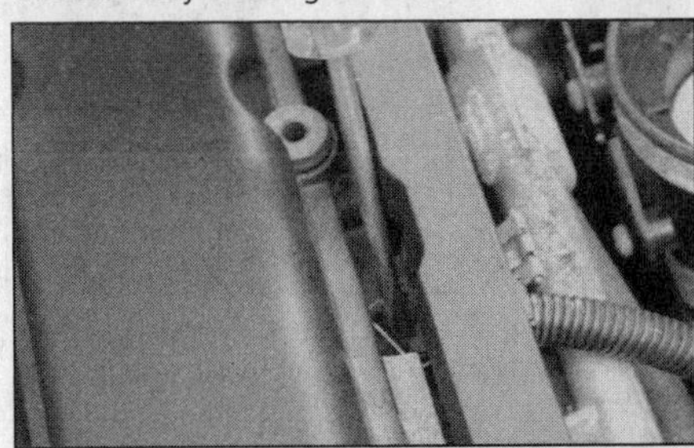

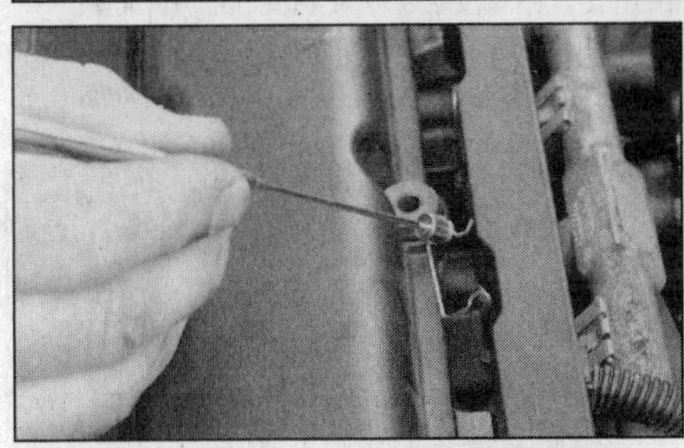

18 Using a small screwdriver, prise up and remove the four wire clips (one on each injector) securing the injector wiring busbar – use a magnetic holding tool to prevent the wire clips falling down the back of the engine as they are removed. Lift off the wiring busbar, and move it to the rear.

19 Disconnect the vacuum hose from the fuel pressure regulator.

20 Disconnect the fuel supply and return lines, noting their fitted positions. The supply line is colour-coded white, and the return line is colour-coded red.

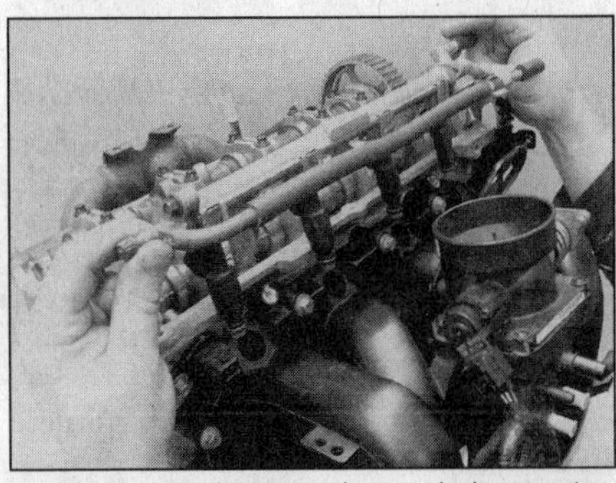

21 Unscrew and remove the two bolts securing the fuel rail. Pulling the fuel rail equally at both ends, remove it squarely from the inlet manifold.

All engines

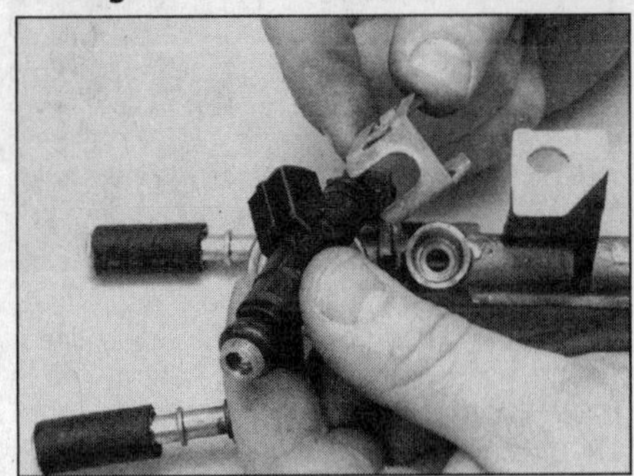

22 If required, remove the clips and carefully pull the injectors from the fuel rail.

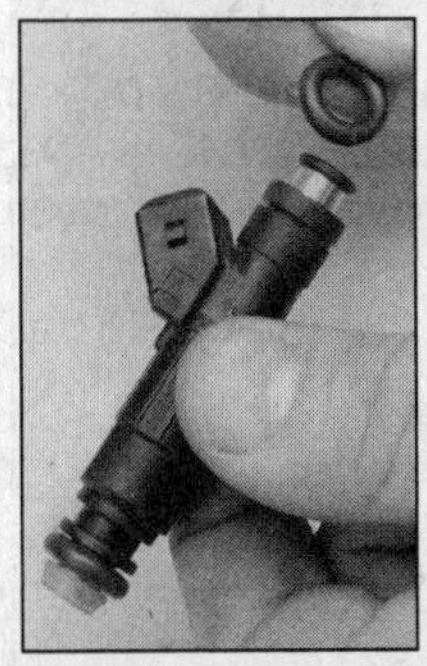

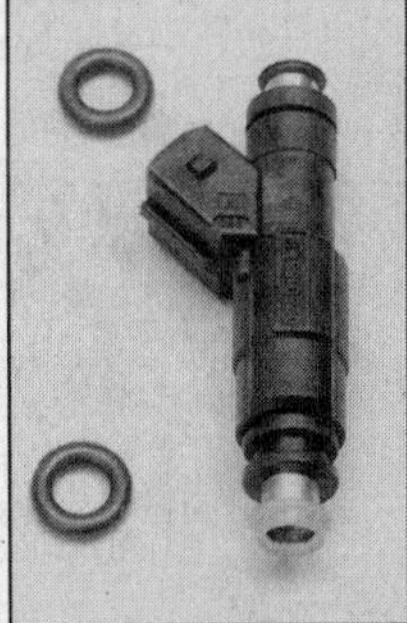

23 Using a screwdriver if necessary, prise the O-rings from the grooves at each end of the injectors. Discard the O-rings and obtain new ones.

24 Refitting is the reverse of the removal procedure, noting the following points:

a) Lubricate the new O-rings with clean engine oil to aid refitting.

b) Ensure that the hoses and wiring are routed correctly, and secured on reconnection by any clips or ties provided.

c) Refit the accelerator cable.

d) On completion, switch the ignition on to activate the fuel pump and pressurise the system, without cranking the engine. Check for signs of fuel leaks around all disturbed unions and joints before attempting to start the engine.

Fuel pressure regulator

25 Relieve the residual pressure in the fuel system, and equalise tank pressure by removing the fuel filler cap.

Warning: This procedure will merely relieve the increased pressure necessary for the engine to run – remember that fuel will still be present in the system components, and take precautions accordingly before disconnecting any of them.

26 Disconnect the battery negative (earth) lead.

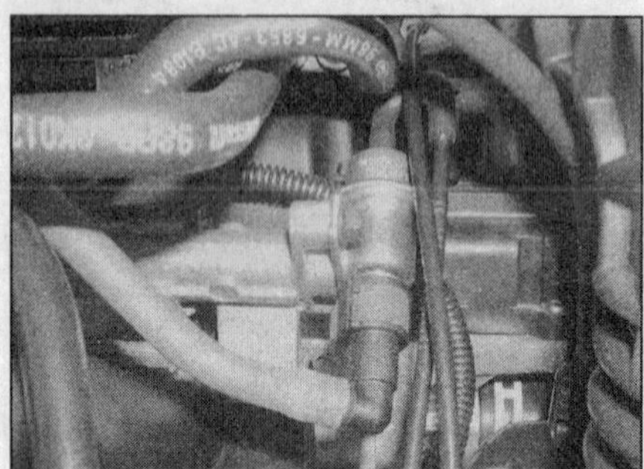

27 Disconnect the fuel supply and return lines, noting their fitted positions. The supply line is colour-coded white, and the return line is colour-coded red.

28 Disconnect the vacuum pipe from the fuel pressure regulator *(refer to paragraph 19)*.

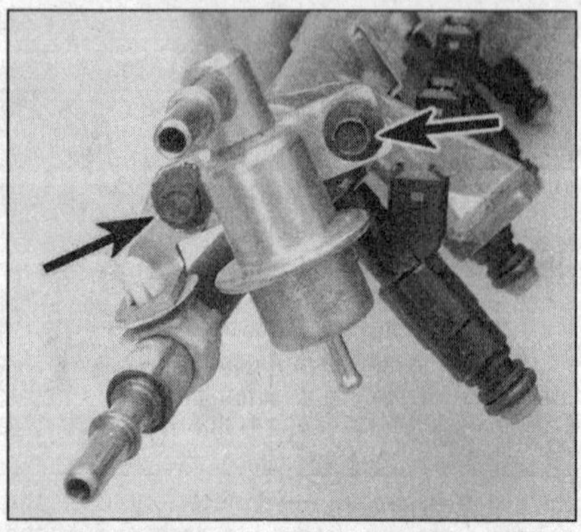

29 Undo the two bolts and remove the fuel pressure regulator from the fuel rail.

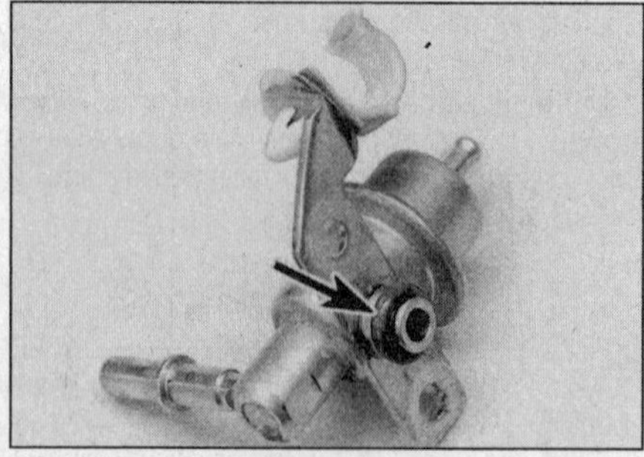

30 Using a screwdriver if necessary, prise the O-ring from the groove in the fuel pressure regulator Discard the O-ring, and obtain a new one.

31 Refitting is a reversal of removal, but lubricate the new O-ring with clean engine oil to aid installation. Tighten the mounting bolts to the specified torque.

Idle speed control valve

32 The valve is located on the inlet manifold, next to the throttle housing. On 1.8 and 2.0 litre models, the valve is on the inside of the inlet manifold, making access especially difficult.

33 On 1.8 and 2.0 litre models, loosen the retaining clips at either end, and remove the air cleaner inlet duct.

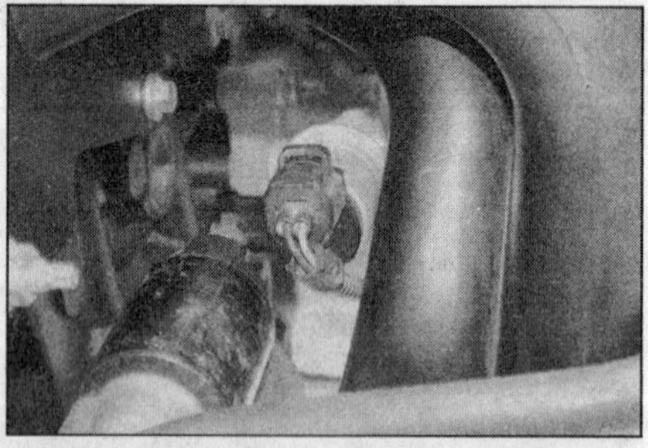

34 Disconnect the wiring plug from the valve.

35 Unscrew the mounting bolts and remove the valve from the inlet manifold. On 1.8 and 2.0 litre models, access to the bolts is less easy; remove the alternator, or ultimately, if suitable tools are not available, it may be simpler to remove the inlet manifold for access.

36 Recover the rubber seal from the valve, and discard it. Obtain a new seal for refitting.

37 Refitting is a reversal of removal, but note the following points.

a) Clean the mating surfaces, and fit a new seal.

b) Tighten the valve bolts to the specified torque.

c) Once the wiring and battery are reconnected, start the engine and allow it to idle. When it has reached normal operating temperature, check that the idle speed is stable, and that no induction (air) leaks are evident. Switch on all electrical loads (headlights, heated rear window, etc), and check that the idle speed is still satisfactory.

Mass airflow sensor

38 Loosen the clip and disconnect the air inlet duct from the mass airflow sensor on the air cleaner cover. If necessary, remove the air cleaner cover for improved access.

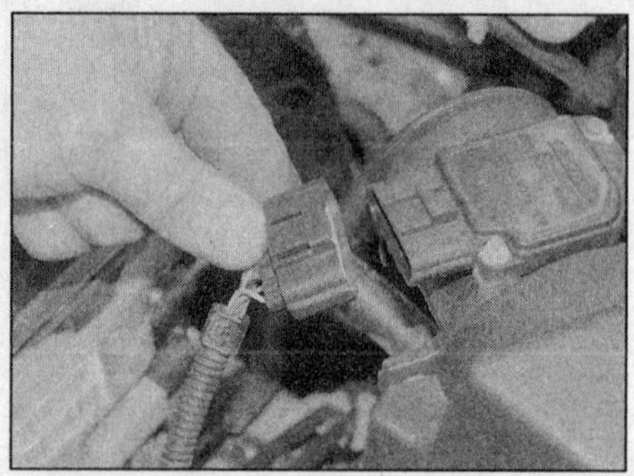

39 Disconnect the wiring from the sensor.

40 Unscrew the crosshead mounting screws and remove the sensor from the air cleaner cover.

41 Refitting is a reversal of removal.

Inlet air temperature sensor

42 The sensor is an integral part of the mass airflow sensor, and cannot be renewed separately.

Engine management module (ECU)

Note: *The module is fragile. Take care not to drop it, or subject it to any other kind of impact. Do not subject it to extremes of temperature, or allow it to get wet. The module wiring plug must never be disconnected while the ignition is switched on.*

43 The engine management module is located behind the facia, on the right-hand side. First disconnect the battery negative (earth) lead.

Right-hand-drive models

44 Pull up the weatherstrip from the driver's door aperture, and release it from the side trim.

45 Prise off the trim panel fitted between the end of the facia panel and the door aperture, approximately half way up the door aperture.

46 Prise out the two screwhead covers, then unscrew and remove the two screws securing the footwell side trim panel. Remove the trim panel for access to the module.

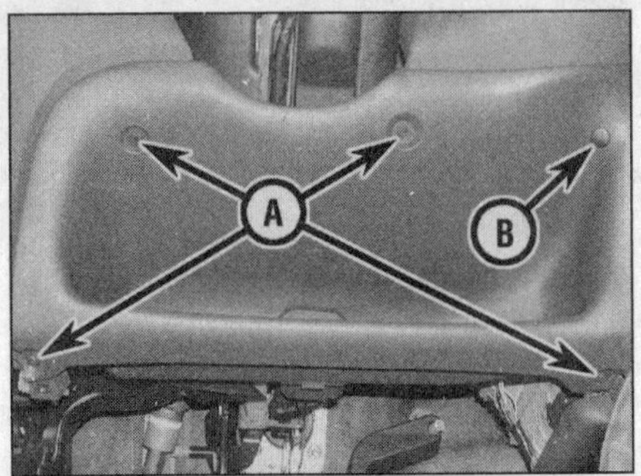

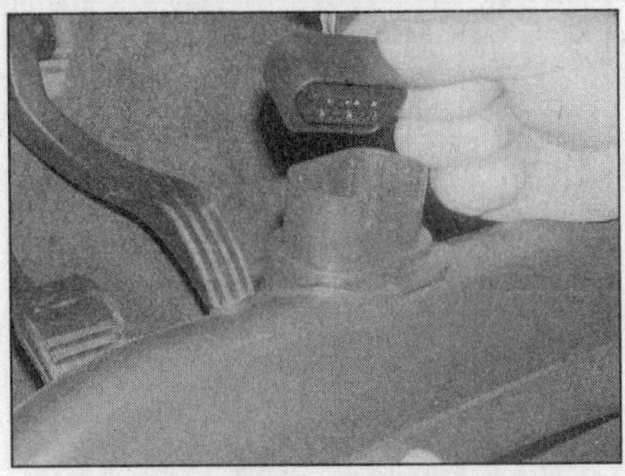

47 Remove the four screws (A) and prise out the plastic clip (B) securing the driver's side lower trim panel. Remove the panel from the facia, and detach the diagnostic connector socket.

Left-hand-drive models

48 Remove the glovebox.

49 Remove the four screws securing the passenger's side lower trim panel. Remove the panel from the facia.

All models

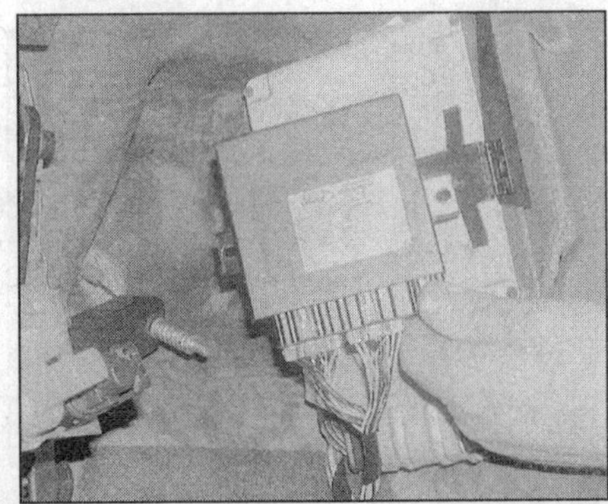

50 Remove the two screws securing the central locking module, and lower the module out of position; there is no need to disconnect the wiring.

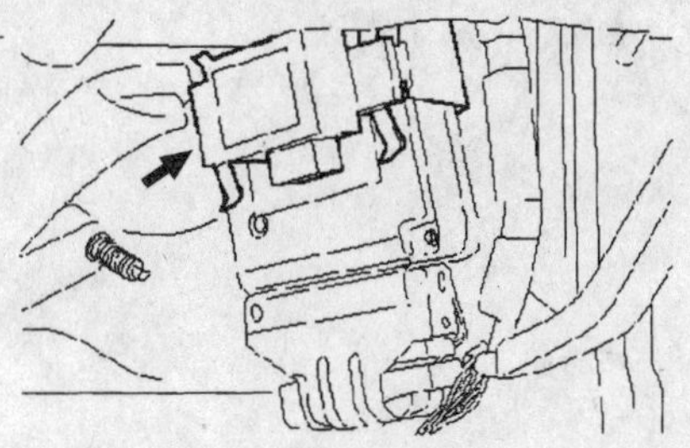

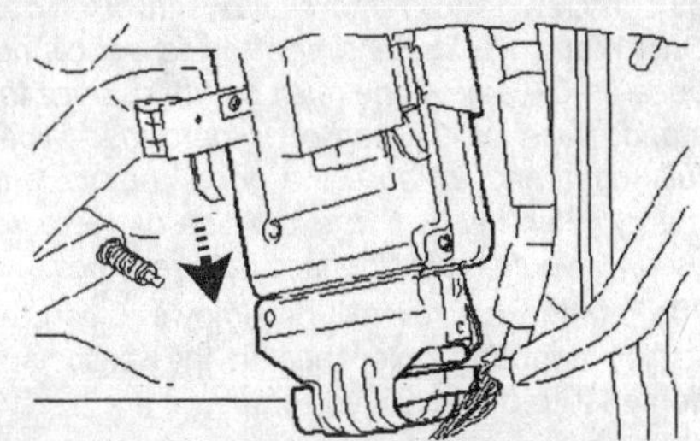

51 Release the spring clips securing the module bracket in position, and remove the module downwards into the footwell.

52 The ECU wiring connector is protected by a tamperproof shield, secured by a shear-bolt and welded nut, which must be drilled to remove the bolt. Great care must be taken not to damage the wiring harness during drilling.

53 First drill a 3 mm pilot hole in the nut, then enlarge the hole using an 8 mm drill until the shear-bolt can be removed.

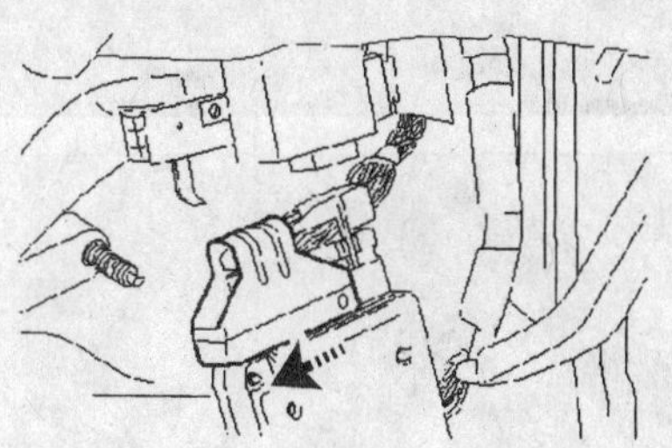

54 With the shear-bolt removed, slide off the tamperproof shield.

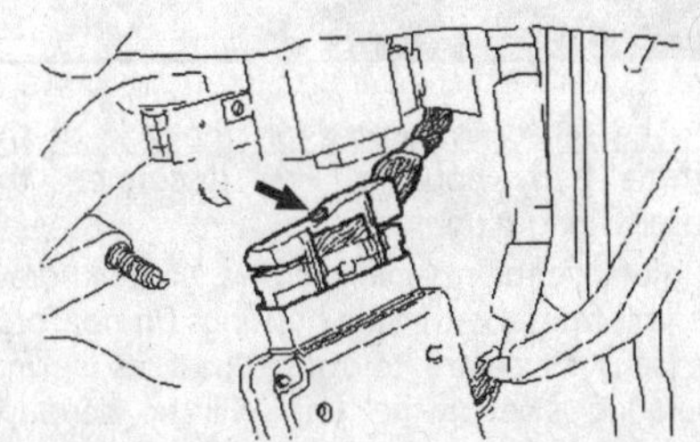

55 Remove the wiring connector securing screw, then disconnect the plug from the ECU, and remove the module from the car.

56 Refitting is a reversal of removal. Use a new shear-bolt, and tighten it until the head shears off.

Crankshaft position sensor

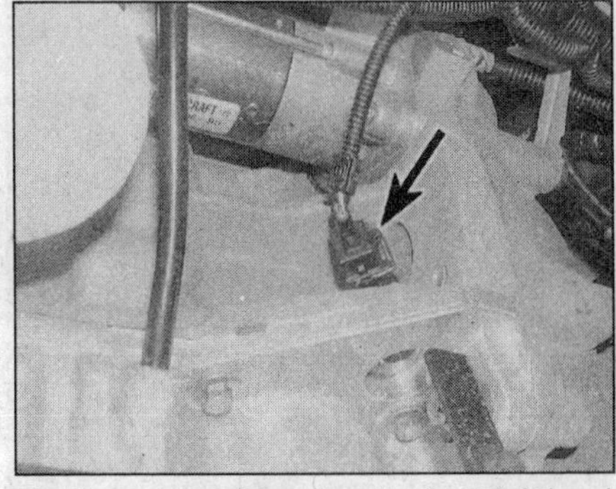

57 The sensor is located on the front left-hand side of the engine. For improved access, apply the handbrake then jack up the front of the vehicle and support it on axle stands.

58 Where fitted, unclip the cover from the crankshaft position sensor on the front of the engine, then disconnect the wiring plug.

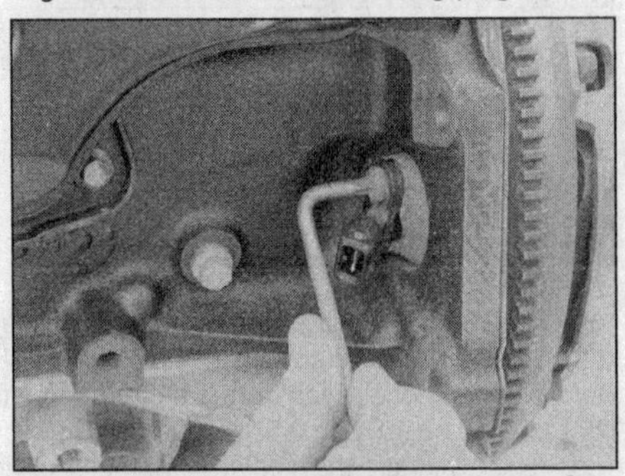

59 Unscrew the mounting bolt and withdraw the sensor. If a spacer is fitted between the sensor and engine, it is vital that this is refitted when refitting the sensor, or the sensor head will hit the rotor teeth.

60 Refitting is a reversal of removal. Tighten the sensor retaining bolt to the specified torque.

Camshaft position sensor

61 The camshaft position sensor is located at the rear of the cylinder head – on the right-hand side on 1.4 and 1.6 litre models, and on the left-hand side on 1.8 and 2.0 litre models (left and right as seen from the driver's seat).

62 On 1.8 and 2.0 litre models, loosen the retaining clips at either end, and remove the air cleaner inlet duct. Disconnect the breather hose from the cylinder head cover.

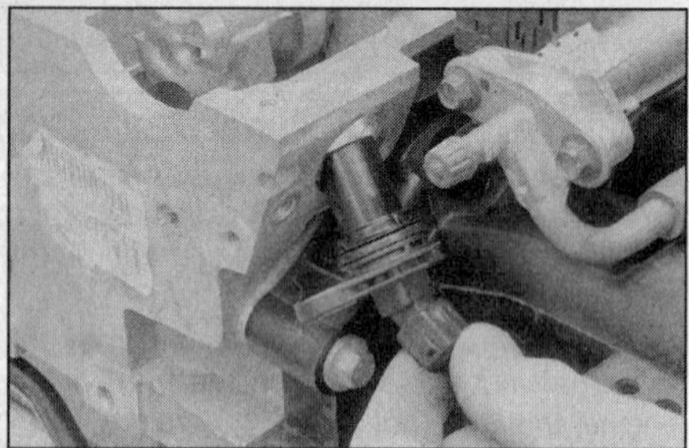

63 Disconnect the wiring connector from the sensor, then unscrew the sensor retaining bolt and withdraw the sensor from the cylinder head.

64 Refitting is a reversal of removal. Tighten the sensor retaining bolt to the specified torque.

Cylinder head temperature sensor

65 The sensor is located in the centre of the head, between Nos 2 and 3 spark plugs, on 1.4 and 1.6 litre models, and on the right-hand end of the head on 1.8 and 2.0 litre models. There is no need to drain the coolant, as the sensor measures the temperature of the metal directly.

66 On 1.4 and 1.6 litre models, disconnect the HT leads from the spark plugs as necessary for access – make sure that the leads are numbered for position (with No 1 at the timing belt end). Prise out the rubber bung used to secure the sensor wiring.

67 On 1.8 and 2.0 litre models, remove the alternator.

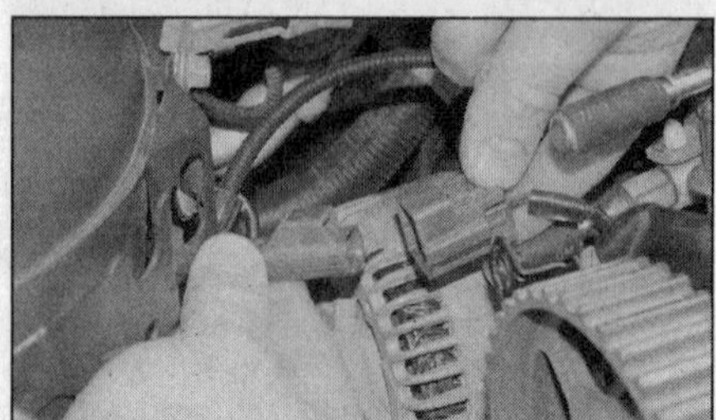

68 Disconnect the wiring from the temperature sensor.

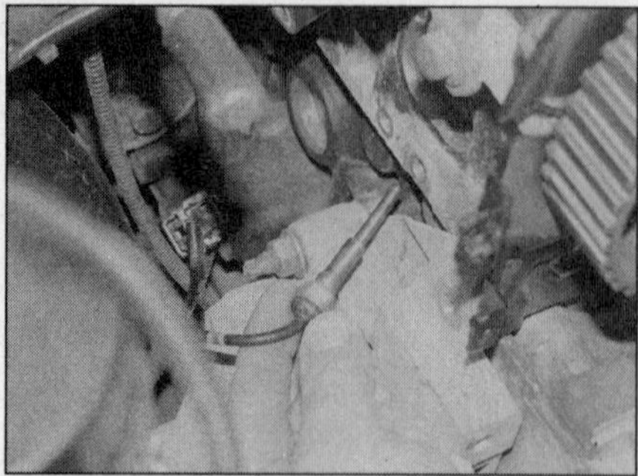

69 Unscrew and remove the sensor. **Note:** *On 1.4 and 1.6 litre engines, the sensor cannot be re-used – a new one must be fitted, since the mating face of the sensor is deformed when fully tightened, to ensure a good contact with the cylinder head. If the old sensor were re-used, it would result in inaccurate readings.*

70 Refitting is a reversal of removal. Clean the sensor mounting hole, then fit the sensor and tighten it to the specified torque.

Throttle position sensor

71 The sensor is located on the side of the throttle body housing. First disconnect the sensor wiring plug.

72 Remove the retaining screws, and withdraw the unit from the throttle housing. *Do not* force the sensor's centre to rotate past its normal operating sweep; the unit will be seriously damaged.

73 Refitting is a reversal of removal, but ensure that the sensor is correctly orientated, by locating its centre on the D-shaped throttle shaft (throttle closed), and aligning the sensor body so that the screws pass easily into the throttle housing.

Vehicle speed sensor

74 The sensor is located next to the right-hand driveshaft at the rear of the transmission. Jack up the front of the vehicle and support it securely on axle stands.

75 Disconnect the wiring plug from the sensor.

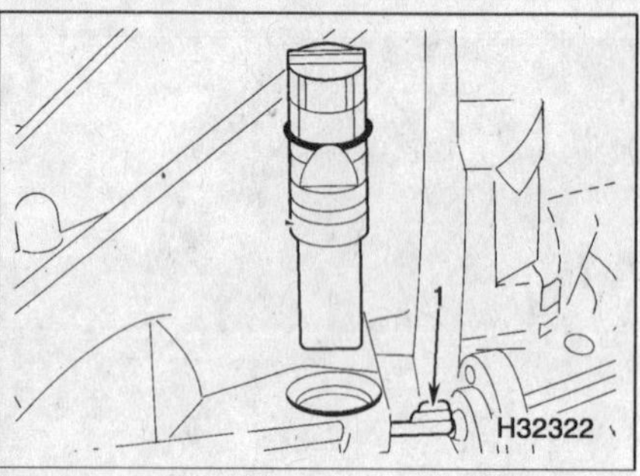

76 Using thin-nosed pliers, pull out the retaining pin at the base of the sensor (1), noting how it is fitted.

77 Pull the sensor from the transmission housing, and recover the O-ring seal.

78 Refitting is a reversal of removal, renewing the O-ring seal where necessary.

Clutch pedal position switch

79 Remove five screws and take out the trim panel above the driver's footwell to gain access to the clutch pedal; unclip the diagnostic connector plug from the panel as it is removed.

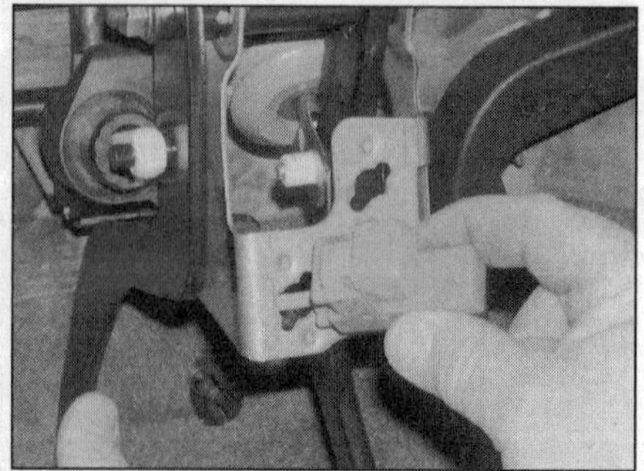

80 Reach up and disconnect the wiring from the clutch switch at the top of the pedal, then twist the switch anti-clockwise and remove it from the pedal bracket.

81 Refitting is a reversal of removal.

Oxygen sensor

Note: *Some later models are fitted with two sensors – one before and one after the catalytic converter.*

82 One 1.4 and 1.6 litre models, access to the sensor in the front downpipe can be gained from above, but access to the second sensor fitted to later models can only be gained with the raised and supported on axle stands.

83 On 1.8 and 2.0 litre models, remove the four bolts securing the heat shield fitted over the exhaust manifold, and lift away the heat shield for access to the sensor.

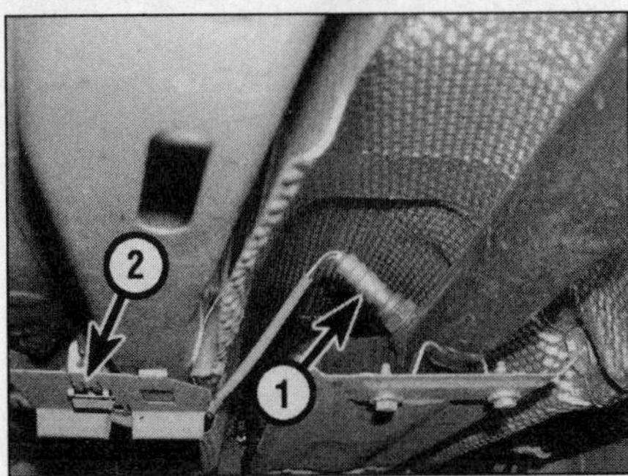

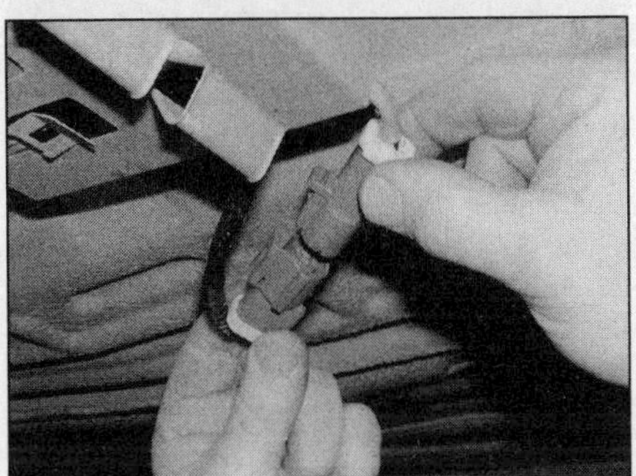

84 Trace the wiring from the sensor body (1) back to its wiring plug (2), and disconnect it. On models with two sensors, the wiring plug the rearmost sensor is clipped to a metal plate attached to the underside of the car. Note how the wiring is routed, as it must not come into contact with hot exhaust components.

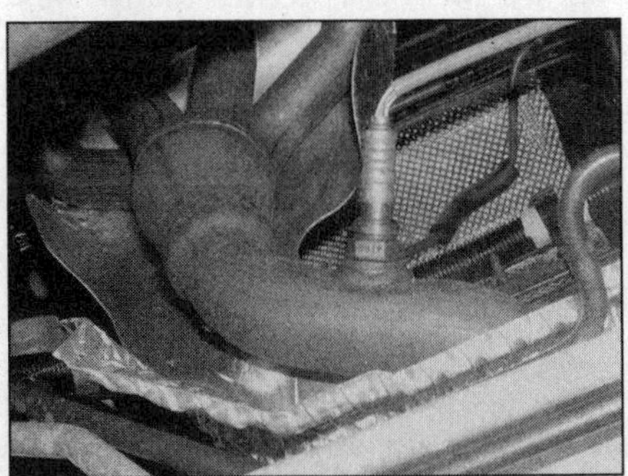

85 Unscrew the sensor from its location, and remove it.

86 Refitting is a reversal of removal, but apply a little anti-seize grease to the sensor threads prior to refitting.

Diesel injection system electronic components – removal and refitting

Crankshaft position sensor

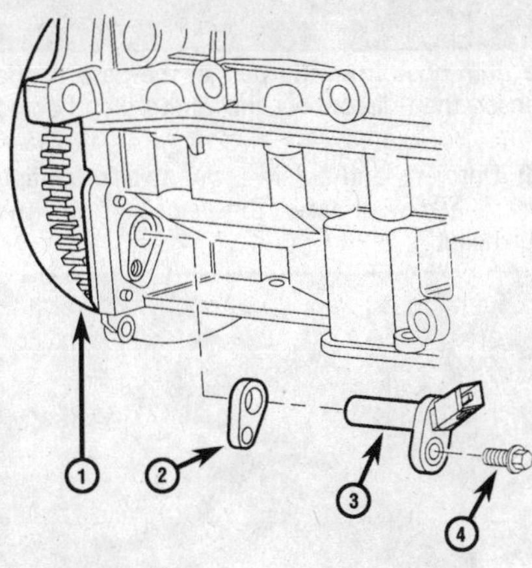

1 *Flywheel*
2 *Spacer*
3 *Crankshaft position sensor*
4 *Mounting bolt*

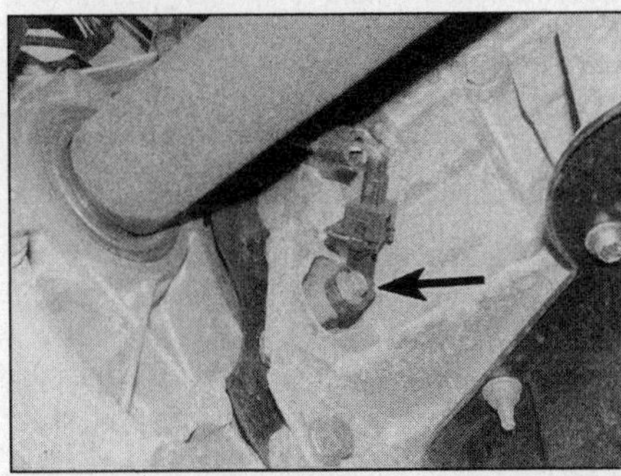

1 The sensor is located at the flywheel end of the engine, low down at the rear. For improved access, apply the handbrake then jack up the front of the vehicle and support it on axle stands.

2 Where fitted, unclip the cover from the crankshaft position sensor, then disconnect the wiring plug.

3 Unscrew the mounting bolt and withdraw the sensor. If a spacer is fitted between the sensor and engine, it is vital that this is refitted when refitting the sensor, or the sensor head will hit the rotor teeth.

4 Refitting is a reversal of removal.

Cylinder head temperature (CHT) sensor

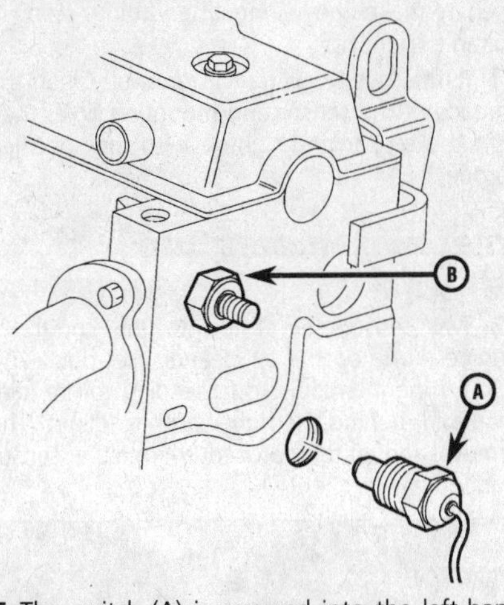

5 The switch (A) is screwed into the left-hand (flywheel) end of the cylinder head, behind the vacuum pump and below the oil pressure warning light switch (B).

6 Open the bonnet and disconnect the battery negative (earth) lead.

7 To improve access to the switch, it will be necessary to remove (or partially remove) the air cleaner inlet duct and the intercooler right-hand air duct.

8 It will also be helpful to release the hoses and remove the crankcase ventilation system oil separator from the left-hand end of the cylinder head.

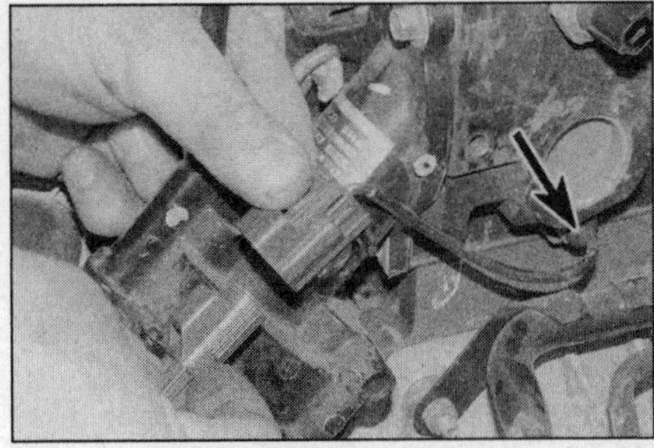

9 Trace the wiring from the sensor, and disconnect it at the plug, which is clipped to the brake vacuum pump at the front of the engine.

10 The sensor can now be unscrewed and removed. However, access to the sensor is such that great difficulty may be experienced in getting a tool to fit onto it. Ultimately, it may be necessary to cut the sensor wiring, unscrew the sensor using a thin-wall socket or box spanner, then re-make the wiring after fitting, using a suitable connector.

11 Refitting is a reversal of removal. Clean the threads of the sensor and mounting hole, then refit the sensor and tighten it to the specified torque.

Inlet air temperature sensors

12 Two sensors are fitted to the Endura-DI engine – one on the air cleaner inlet duct, and one on the intercooler duct leading to the inlet manifold, behind the right-hand headlight. The removal and refitting procedure for either sensor is identical.

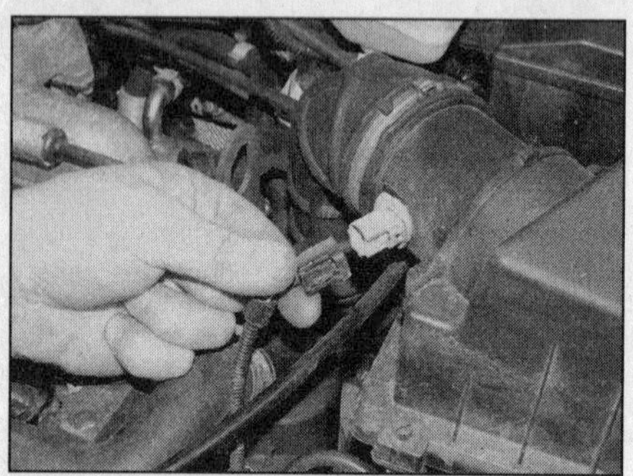

13 Disconnect the wiring from the sensor, then twist the sensor anti-clockwise and remove it.

14 Refitting is a reversal of removal.

Vehicle speed sensor

15 The sensor is located next to the right-hand driveshaft at the rear of the transmission. Jack up the front of the vehicle and support it securely on axle stands.

16 Disconnect the wiring plug from the sensor, then undo the retaining bolt and remove it from the transmission housing. Recover the O-ring seal

17 Refitting is a reversal of removal, renewing the O-ring seal where necessary.

Manifold absolute pressure (MAP) sensor

18 The MAP sensor is mounted on the engine compartment bulkhead, next to the brake/clutch fluid reservoir. Disconnect the

vacuum hose from the port at the base of the sensor, then disconnect the wiring plug behind it.

19 Unscrew and remove the two mounting bolts, and withdraw the sensor from the bulkhead.

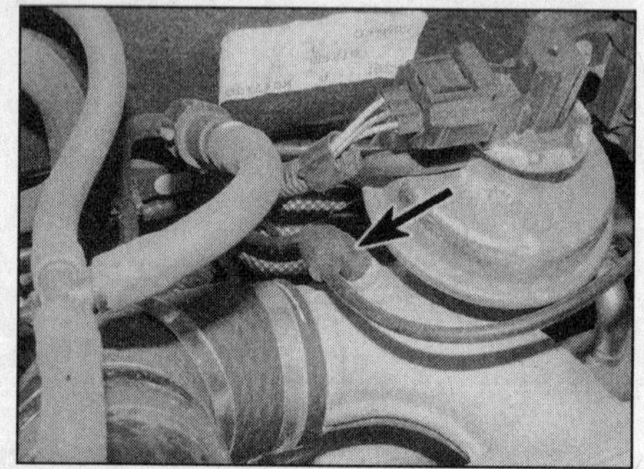

20 Prior to refitting, remove the vacuum hose from the inlet manifold connection, and check it for signs of perishing or splitting, especially at the pipe ends.

21 Refitting is a reversal of removal, ensuring that the wiring plug and vacuum hose are securely reconnected.

Fuel control valve

22 The control valve is fitted to the top of the fuel filter (at the right-hand rear corner of the engine compartment), and contains a bi-metal strip. Its function is to close the fuel return to the fuel tank at low fuel temperatures, allowing fuel which has been warmed by passing through the injection pump to flow back into the filter, thus warming the fuel being drawn from the tank.

23 Noting their positions for refitting, disconnect the fuel supply and outlet pipes from the connections at the fuel filter.

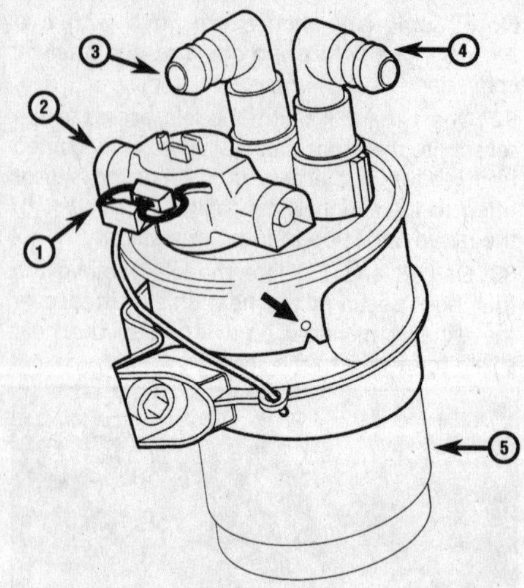

1 Retaining wire clip
2 Fuel control valve
3 Filter outlet
4 Filter inlet
5 Fuel filter

24 Extract the wire clip securing the control valve and return pipe, and release it from the top of the filter. Once again noting the fitted positions of the fuel pipes, disconnect them from each end of the valve (note that the return pipe may in this case be black, and smaller in diameter than the supply pipe).

25 Refitting is a reversal of removal. Ensure that the pipe connections are correctly and securely remade, then run the engine and check for signs of fuel leakage.

Clutch pedal position switch

26 Remove five screws and take out the trim panel above the driver's footwell to gain access to the clutch pedal; unclip the diagnostic connector plug from the panel as it is removed.

27 Reach up and disconnect the wiring from the clutch switch at the top of the pedal, then twist the switch anti-clockwise and remove it from the pedal bracket.

28 Refitting is a reversal of removal.

Stop-light and brake pedal position switches

29 These are two separate switches, both fitted at the top of the brake pedal. Remove five screws and take out the trim panel above the driver's footwell to gain access to the brake pedal; unclip the diagnostic connector plug from the panel as it is removed.

30 Reach up and disconnect the wiring from the switch at the top of the pedal; if both switches are removed at once, take care to note which wiring plug serves which switch. Twist the switch through 90° and remove it from the

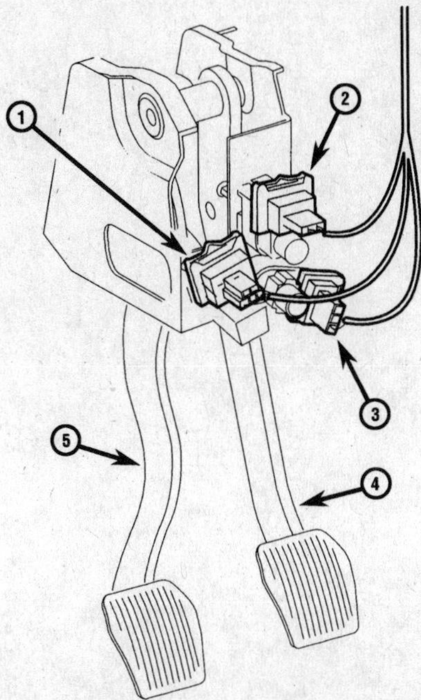

1 Clutch pedal switch
2 Brake stop-light switch
3 Brake pedal position switch
4 Brake pedal
5 Clutch pedal

pedal bracket. The stop-light switch (upper) is removed by twisting anti-clockwise, while the brake pedal position switch (lower) is twisted clockwise to remove.

31 Refitting is a reversal of removal.

Accelerator pedal sensor

32 The accelerator pedal sensor is integral with the pedal assembly.

Injection pump control unit (PCU)

33 The pump control unit is integral with the injection pump, and cannot be separated from it. If a new injection pump is fitted, the PCU must be electronically 'matched' to the engine management module (ECU), otherwise the pump will not function correctly (or even at all, if the immobiliser function is not correctly set up) – this is a task for a Ford dealer or suitably equipped specialist, as specialised electronic equipment is required.

Engine management module (ECU)

Note: *The module is fragile. Take care not to drop it, or subject it to any other kind of impact. Do not subject it to extremes of temperature, or allow it to get wet. The module wiring plug must never be disconnected while the ignition switch is on.*

34 The engine management module is located behind the facia, on the right-hand side. First disconnect the battery negative (earth) lead.

Right-hand-drive models

35 Pull up the weatherstrip from the driver's door aperture, and release it from the side trim.

36 Prise off the trim panel fitted between the end of the facia panel and the door aperture, approximately half way up the door aperture.

37 Prise out the two screwhead covers, then unscrew and remove the two screws securing the footwell side trim panel. Remove the trim panel for access to the module.

38 Remove the four screws securing the driver's side lower trim panel. Remove the panel from the facia, and detach the data link connector socket.

Left-hand-drive models

39 Remove the glovebox.

40 Remove the four screws securing the passenger's side lower trim panel. Remove the panel from the facia.

All models

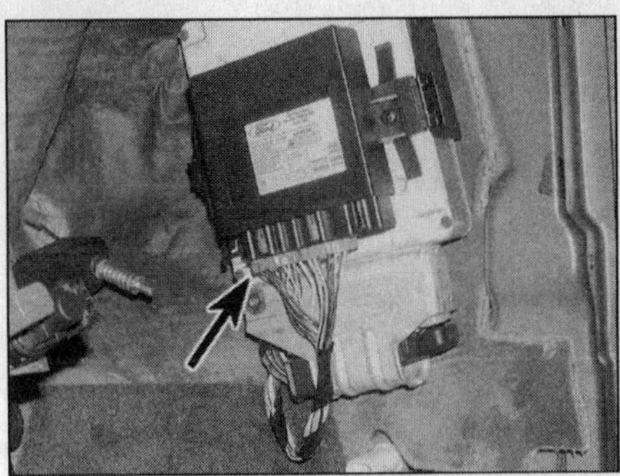

41 Disconnect the small wiring plug from the central security module.

42 Release the spring clips securing the main module in position, and remove the module downwards into the footwell.

43 The ECU wiring connector is protected by a tamperproof shield, secured by a shear-bolt and welded nut, which must be drilled to remove the bolt. Great care must be taken not to damage the wiring harness during drilling.

44 First drill a 3 mm pilot hole in the nut, then enlarge the hole using an 8 mm drill until the shear-bolt can be removed.

45 With the shear-bolt removed, slide off the tamperproof shield.

46 Remove the wiring connector securing screw, then disconnect the plug from the ECU, and remove the module from the car.

47 Refitting is a reversal of removal. Use a new shear-bolt, and tighten it until the head shears off.

Notes

Ford Focus '01 – '04

12

Petrol models

General

System type	Sequential Electronic Fuel injection (SEFi)
Fuel octane requirement	95 RON unleaded
Idle speed	700 ± 30 rpm (regulated by EEC-V engine management system – no adjustment possible)
CO content at idle	Less than 0.3%

Fuel system data

Note: *The resistance values quoted below are typical, but may be used for guidance. Generally, a faulty component will be indicated by a zero or infinity reading, rather than a slight deviation from the values given. Always have your findings verified before buying a new component (if possible, perform the same test on a new component, and compare the results).*

Fuel pressure regulator:	
Regulated fuel pressure:	
Pressure regulator hose disconnected	2.7 ± 0.2 bars
Engine running, pressure regulator hose connected	2.1 ± 0.2 bars
Hold pressure – engine stopped after five minutes	1.8 bars minimum
Fuel injectors:	
Resistance	13.7 to 15.2 ohms
Idle speed control valve:	
Resistance	6 to 14 ohms
Idle increase solenoid valve:	
Resistance	50 to 120 ohms
Crankshaft speed/position sensor:	
Resistance	200 to 450 ohms
Camshaft position sensor:	
Resistance	200 to 900 ohms
Intake air temperature sensor:	
Resistance:	
At –40°C	860 to 900 k ohms
At 20°C	35 to 40 k ohms
At 100°C	1.9 to 2.5 k ohms
At 120°C	1.0 to 1.3 k ohms
Throttle potentiometer:	
Resistance – see text	400 to 6000 ohms
Power steering pressure switch:	
Operating pressure – green switch body:	
Contacts open – infinite resistance	31.5 ± 3.5 bars
Contacts close – 0 to 2.5 ohms resistance	Between 13.5 and 24.0 bars

Chapter 12

Diesel models

General

System type:	
Endura-Di engines	Bosch VP-30 combined lift and injection pump, chain-driven from crankshaft. Direct injection via five-hole, pencil-type injectors, electronic pump control unit (PCU) linked to revised EEC-V engine management module (ECU)
Duratorq-TDCi engines	Dephi DCR1400 common rail system, with a chain-driven high pressure pump, direct injection, with a variable geometry turbocharger, intercooler, six-hole injectors, Injection Driver Module (IDM) linked to the EEC-V engine management module (ECU)
Firing order	1 – 3 – 4 – 2 (No 1 at timing belt end)
Idle speed:	
Endura-Di engines	800 ± 50 rpm (regulated by EEC-V engine management system – no adjustment possible)
DuraTorq-TDCi engines	900 ± 50 rpm (regulated by EEC-V engine management system – no adjustment possible)

Fuel

Fuel type	Commercial diesel fuel for road vehicles (DERV)

Injection pump

Make and type:	
Endura-DI engines	Bosch VP-30 electronic, distributor-type
DuraTorq-TDCi engines	Delphi DCR1400
Rotation (viewed from crankshaft pulley end)	Clockwise
Drive	By twin 'gemini' chains from crankshaft
Injection pump timing:	
Endura-Di engines	By timing pin, at TDC
DuraTorq-TDCi engines	Not applicable (common rail)

Injectors

Type:	
Endura-Di engines	Pencil-type, five-hole
DuraTorq-TDCi engines	Six-hole

Glow plugs

Type	Motorcraft EZD37

Turbocharger

Type:	
Endura-Di engines	Garrett GT15, integral with exhaust manifold. Air-to-air intercooler
DuraTorq-TDCi engines	Integral with the exhaust manifold. Air-to-air intercooler
Boost pressure:	
Endura-Di engines	0.5 bar @ 3000 rpm
DuraTorq-TDCi engines	1.2 bar

Torque wrench settings

	Nm	lbf ft
Petrol models		
Camshaft position sensor:		
1.4 and 1.6 litre models	8	6
1.8 and 2.0 litre	20	15
Crankshaft position sensor	7	6
Cylinder head temperature sensor (renew)	10	7
EGR valve (1.8 and 2.0 litre models)	9	7
Fuel pressure regulator mounting bolts:		
1.4 and 1.6 litre models	10	7
1.8 and 2.0 litre models	6	4
Fuel rail bolts:		
1.4 and 1.6 litre models	15	11
1.8 and 2.0 litre models	10	7
Fuel supply pipe retaining plate bolts	10	7
Idle speed control valve	10	7
Knock sensor	20	15
Power steering pressure switch	20	15

Torque wrench settings	Nm	lbf ft
Diesel models		
Crankshaft position sensor bolt	10	7
Cylinder head temperature sensor	20	15
EGR cooler lower bolts	23	17
EGR tube/cooler clamp nut/bolt*	9	7
EGR valve flange bolts	20	15
Fuel metering valve	6	4
Fuel supply manifold (common rail)	28	21
Fuel temperature sensor bolt	15	11
Glow plugs	15	11
Knock sensor	20	15
** Do not reuse*		

General information

Petrol models

The fuel system consists of a fuel tank (mounted under the body, beneath the rear seats), fuel hoses, an electric fuel pump mounted in the fuel tank, and a sequential electronic fuel injection system controlled by a EEC-V engine management control module (also called the ECU, or Electronic Control Unit).

The electric fuel pump supplies fuel under pressure to the fuel rail, which distributes fuel to the injectors. A pressure regulator controls the system pressure in relation to intake tract depression. From the fuel rail, fuel is injected into the intake ports, just above the intake valves, by four fuel injectors. The aluminium alloy fuel rail is mounted to the cylinder head on 1.4 and 1.6 litre models, while the fuel rail on 1.8 and 2.0 litre models is mounted on the intake manifold.

The amount of fuel supplied by the injectors is precisely controlled by the ECU. The module uses the signals from the crankshaft position sensor, and the camshaft position sensor, to trigger each injector separately in cylinder firing order (sequential injection), with benefits in terms of better fuel economy and leaner exhaust emissions.

The ECU is the heart of the entire engine management system, controlling the fuel injection, ignition and emissions control systems. The module receives information from various sensors which is then computed and compared with pre-set values stored in its memory, to determine the required period of injection.

Information on crankshaft position and engine speed is generated by a crankshaft position sensor. The inductive head of the sensor runs just above the engine flywheel and scans a series of 36 protrusions on the flywheel periphery. As the crankshaft rotates, the sensor transmits a pulse to the system's ignition module every time a protrusion passes it. There is one missing protrusion in the flywheel periphery at a point corresponding to 90° BTDC. The ignition module recognises the absence of a pulse from the crankshaft position sensor at this point to establish a reference mark for crankshaft position. Similarly, the time interval between absent pulses is used to determine engine speed. This information is then fed to the ECU for further processing.

The camshaft position sensor is located in the cylinder head so that it registers with a lobe on the camshaft. The camshaft position sensor functions in the same way as the crankshaft position sensor, producing a series of pulses; this gives the ECU a reference point, to enable it to determine the firing order, and operate the injectors in the appropriate sequence.

The mass airflow sensor is based on a 'hot-wire' system, sending the ECU a constantly-varying (analogue) voltage signal corresponding to the mass of air passing into the engine. Since air mass varies with temperature (cold air being denser than warm), measuring air mass provides the module with a very accurate means of determining the correct amount of fuel required to achieve the ideal air/fuel mixture ratio.

The traditional coolant temperature sensor has been replaced by a cylinder head temperature sensor. The new sensor is seated in a blind hole in the cylinder head, and measures the temperature of the metal directly. This component is an NTC (Negative Temperature Coefficient) thermistor – that is, a semi-conductor whose electrical resistance decreases as its temperature increases. It provides the ECU with a constantly-varying (analogue) voltage signal, corresponding to the temperature of the engine. This is used to refine the calculations made by the module, when determining the correct amount of fuel required to achieve the ideal air/fuel mixture ratio.

Intake air temperature information is supplied by the intake air temperature sensor built into the airflow sensor. This component is also an NTC thermistor – see the previous paragraph – providing the module with a signal corresponding to the temperature of air passing into the engine. This is used to refine the calculations made by the module, when determining the correct amount of fuel required to achieve the ideal air/fuel mixture ratio.

A throttle position sensor is mounted on the end of the throttle valve spindle, to provide the ECU with a constantly-varying (analogue) voltage signal corresponding to the throttle opening. This allows the module to register the driver's input when determining the amount of fuel required by the engine.

Road speed is monitored by the vehicle speed sensor. This component is a Hall-effect generator, mounted on the transmission's speedometer drive. It supplies the module with a series of pulses corresponding to the vehicle's road speed, enabling the module to control features such as the fuel shut-off on overrun.

The clutch pedal position is monitored by a switch fitted to the pedal bracket. This sends a signal to the ECU.

Where power steering is fitted, a pressure-operated switch is screwed into the power steering system's high-pressure pipe. The switch sends a signal to the ECU to increase engine speed to maintain idle speed when the power steering pump loads up the engine (typically, during parking manoeuvres).

The oxygen sensor in the exhaust system provides the module with constant feedback – 'closed-loop' control – which enables it to adjust the mixture to provide the best possible operating conditions for the catalytic converter.

The air intake side of the system consists of an air cleaner housing, the mass airflow sensor, an intake hose and duct, and a throttle housing.

The throttle valve inside the throttle housing is controlled by the driver, through the accelerator pedal. As the valve opens, the amount of air that can pass through the system increases. As the throttle valve opens further, the mass airflow sensor signal alters, and the ECU opens each injector for a longer duration, to increase the amount of fuel delivered to the intake ports.

Both the idle speed and mixture are under the control of the ECU, and cannot be adjusted.

Diesel models

Two different fuel injection systems are fitted to Focus diesel models covered in this manual. The first fuel system, fitted to the Endura-Di engines, consists of a fuel tank (mounted under the body, beneath the rear seats), fuel filter, electronic fuel injection pump with pump control unit, injectors, fuel lines and hoses, fuel gauge sender unit mounted in the fuel tank, and EEC-V engine management control module (ECU). The second, fitted to the DuraTorq-TDCi engines, is a common rail direct injection system. The common-rail system incorporates most of the components used in the first system, with the exception of a high pressure fuel pump, replacing the injection pump, electrically operated injectors replacing the mechanical ones, and a fuel supply manifold.

Endura-Di engines

Fuel is drawn from the tank by the transfer pump incorporated in the injection pump. It then passes through the fuel filter, located in the engine bay, where foreign matter and water are removed. The injection pump is driven from the crankshaft via a twin-row ('gemini') chain and supplies fuel under very high pressure to each injector in turn as it is needed. The camshaft is driven from the injection pump via a toothed timing belt.

The Endura-DI engine features a full electronic engine management system, very similar to that fitted to the Focus petrol models. An extensive array of sensors is fitted, which supply information on many different parameters to the ECU.

DuraTorq-TDCi engines

Fuel is drawn from the tank in the same way, but passes through a filter assembly before being passed to the high-pressure chamber in the pump. The filter assembly incorporates a control valve to close the fuel return circuit at low temperatures, and a water trap. Once inside the high-pressure chamber of the pump, the fuel is pressurised to a maximum of 1200 bar, and delivered to the fuel supply manifold. This manifold stores the pressurised fuel, and supplies all of the injectors – hence the term 'common rail'. The manifold is small enough to hold to retain the pressure required for rapid starting, whilst being large enough the dampen any pressure fluctuation from the pump or injectors. The slimline 6-hole injectors are operated by powerful electrical solenoids integral with their design. On some models (up to August 2002), the IDM (injector driver module) determines the optimum injection timing based on information received from various sensors, and signals the injector solenoids to operate accordingly. On others, the function of the IDM is incorporated into the engine management ECU The advantages of the common rail system are an increase in engine output, accompanied by a decrease in fuel consumption, combustion noise, and harmful exhaust emissions.

All diesel engines

Information on crankshaft position and engine speed is generated by a crankshaft position sensor. The inductive head of the sensor runs just above the engine flywheel, and scans a series of 36 protrusions on the flywheel periphery. As the crankshaft rotates, the sensor transmits a pulse every time a protrusion passes it. There is one missing protrusion in the flywheel periphery at a point corresponding to 90° BTDC. The ECU recognises the absence of a pulse from the crankshaft position sensor at this point to establish a reference mark for crankshaft position. Similarly, the time interval between absent pulses is used to determine engine speed.

Information on the quantity and temperature of the intake air is derived from the MAP sensor and the intake air temperature sensors. The manifold absolute pressure (or MAP) sensor is connected to the intake manifold by a vacuum hose, and measures the pressure in the intake system. Two air temperature sensors are fitted – one before the turbocharger, and one after the intercooler. The temperature and quantity of air has a direct bearing on the quantity of fuel to be injected for optimum efficiency.

The traditional coolant temperature sensor has been replaced by a cylinder head temperature sensor. The new sensor is seated in a blind hole in the cylinder head, and measures the temperature of the metal directly. Information on engine temperature is critical for accurate fuelling calculations, and is also used to control the pre-heating system for cold starts.

The clutch pedal sensor informs the ECU whether the clutch is engaged or disengaged. When the clutch pedal is depressed, the quantity of fuel injected is momentarily reduced, to make gearchanging smoother.

The stop-light switch and separate brake pedal sensor inform the ECU when the brakes are applied – when this signal is received, the ECU puts the engine into idle mode until a signal is received from the accelerator position sensor.

On Endura-Di engines, the amount of fuel delivered is determined by the pump's internal quantity and timing solenoid valves, which are controlled by the pump control unit (PCU), mounted on top of the pump. The pump is internally equipped with a pulse ring fitted to the main rotor, and an angle sensor determines the pump rotor's position and speed, in much the same way as the crankshaft position sensor and engine flywheel, except that there are four gaps in the pump rotor 'teeth' – one for each cylinder. The pump control unit is supplied with information from the 'main' engine management module (ECU), and from this, is able to calculate the most appropriate values for injection timing and quantity (injection duration). The electronically-controlled pump internals enable these calculated values to be delivered with great accuracy, for improved efficiency and reduced emissions.

On DuraTorq-TDCi engines, the system pressure, fuel injection timing and quantity is controlled by the IDM (injector driver module) or engine management ECU, based on information it receives from the knock sensor, camshaft position sensor, crankshaft position sensor, fuel temperature sensor, fuel pressure sensor, and a barometric pressure sensor built into the IDM itself or a manifold absolute pressure sensor on models where no IDM is fitted.

No accelerator cable is fitted on either engines – instead, a sensor located next to the accelerator pedal informs the ECU of the accelerator position, and this information is used to determine the most appropriate fuelling requirements from the injection pump. The engine idle speed is also controlled by the ECU, and cannot be adjusted. From the signals it receives from the various sensors, the ECU can control the idle speed very accurately, compensating automatically for additional engine loads or unfavourable ambient/engine temperatures.

Rigid pipes connect the pump/fuel supply manifold to the four injectors. Each injector has several holes, to disperse the fuel evenly, and sprays fuel directly into the combustion chamber as its piston approaches TDC on the compression stroke. This system is known as direct injection. The pistons have an off-centre recess machined into their crowns, the shape of which has been calculated to improve 'swirl' (fuel/air mixing). The injectors open in several stages, to promote smoother combustion – the rate of opening is fixed by two internal springs and/or the operation of the integral solenoids. Lubrication is provided by allowing a small quantity of fuel to leak back past the injector internal components. The leaked-back fuel is returned to the pump and then to the fuel tank.

Cold-starting performance is automatically controlled by the ECU and pump control unit/IDM. Under cold start conditions, the injection timing is advanced, while the ECU operates the glow plug system. The pre-heater or 'glow' plugs fitted to each cylinder are electrically heated before, during and immediately after starting, and are particularly effective during a cold start. A warning light illuminates when the ignition is switched on, showing that the glow plugs are in operation. When the light goes out, pre-heating is complete and the engine can be started. In very cold conditions, the glow plugs remain on after the engine has started (post-heating) – this helps the engine to run more smoothly during warm-up, and reduces exhaust emissions.

Older diesel engines, including the previous Endura-DE, had injection pumps which were equipped with a solenoid valve to cut the fuel supply when the ignition switch is turned off, to stop the engine (the valve was usually known as a 'stop' solenoid). The Endura-DI injection pump does not have a stop solenoid – instead, the ECU is able to 'switch off' the injection pump via the pump control unit, and this forms part of the vehicle immobiliser system. On DuraTorq-TDCi engines, the ECU switches off the injectors, and operates a vacuum unit fitted to the intake manifold to prevent engine 'shake' as the crankshaft and pistons come to a stop.

The fuel system has a built-in 'strategy' to prevent it from drawing in air, should the car run low on fuel. The ECU monitors the level of fuel in the tank, via the gauge sender unit. After switching on the low fuel level warning light, it will eventually induce a misfire as a further warning to the driver, and lower the engine's maximum speed until the engine stops.

The fuel system on diesel engines is normally very reliable. Provided that clean fuel is used and the specified maintenance is conscientiously carried out, no problems should be experienced. The injection pump and injectors may require overhaul after a high mileage has been covered, but this cannot be done on a DIY basis.

Fuel system – depressurisation, priming and bleeding

Warning: The following procedure will merely relieve the pressure in the fuel system – remember that fuel will still be present in the system components, and take precautions accordingly before disconnecting any of them.

Petrol models

1 The fuel system referred to in this Chapter is defined as the fuel tank and tank-mounted fuel pump/fuel gauge sender unit, the fuel filter, the fuel injector, fuel pressure regulator, and the metal pipes and flexible hoses of the fuel lines between these components. All these contain fuel, which will be under pressure while the engine is running and/or while the ignition is switched on.

2 The pressure will remain for some time after the ignition has been switched off, and must be relieved before any of these components is disturbed for servicing work.

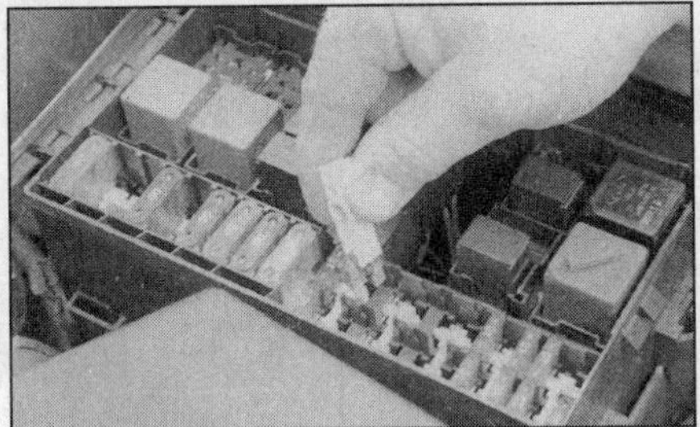

3 The simplest depressurisation method is to disconnect the fuel pump electrical supply by removing the fuel pump fuse (No 12, in the engine compartment fusebox) and starting the engine. If the engine does not start, crank it on the starter for a few seconds, otherwise, allow the engine to idle until it dies through lack of fuel. Turn the engine over once or twice on the starter to ensure that all pressure is released, then switch off the ignition; do not forget to refit the fuse when work is complete.

4 If an adapter is available to fit the Schrader-type valve on the fuel rail pressure test/release fitting (identifiable by its blue plastic cap, and located on the fuel rail), this may be used to release the fuel pressure. The Ford adapter (tool number 23-033) operates similar to a drain tap – turning the tap clockwise releases the pressure. If the adapter is not available, place cloth rags around the valve, then remove the cap and allow the fuel pressure to dissipate. Refit the cap on completion.

5 Note that, once the fuel system has been depressurised and drained (even partially), it will take significantly longer to restart the engine – perhaps several seconds of cranking – before the system is refilled and pressure restored.

Diesel models

Depressurisation (DuraTorq-TDCi engines only)

6 In order to render the system safe to work on, the system pressure must be relieved before commencing any procedures involving the diesel supply, injection or return circuits.

7 Working in the engine compartment, remove fuse No. 12 from the fusebox.

8 Start the engine, and allow it to run until the fuel in the system is exhausted. The system is now depressurised.

Priming and bleeding (all models)

9 As this system is intended to be 'self-bleeding', no hand-priming pump or separate bleed screws/nipples are fitted.

10 When any part of the system has been disturbed therefore, air must be purged from the system by cranking the engine on the starter motor until it starts. When it has started, keep the engine running for approximately 5 minutes to ensure that all air has been removed from the system. To minimise the strain on the battery and starter motor when trying to start the engine, crank it in 10-second bursts, pausing for 30 seconds each time, until the engine starts.

11 Depending on the work that has been carried out, it may be possible to partially prime the system so as to spare the battery by reducing as much as possible the amount of cranking time required to start the engine. To spare the battery, fill the filter with clean fuel via its vent screw opening (Endura-Di engines) or fuel pipe connections ports (DuraTorq-TDCi engines), but it is essential that no dirt is introduced into the system and that no diesel fuel is poured over vulnerable components when doing this.

12 If a hand-operated vacuum pump is available, this can be connected to the pump's fuel return union and used to suck fuel through the supply lines and filter. This will obviously save the battery a good deal of work. If a long length of clear plastic tubing is used to connect the vacuum pump to the injection pump union, it will be easier to see when fuel emerges free from air bubbles. Turn the ignition switch to position II so that fuel can pass through the pump.

13 Start the engine and keep it running for a few minutes to ensure that all air has been expelled.

Petrol injection system components – removal and refitting

Throttle body housing

1 The housing is located on the left-hand side of the intake manifold. First disconnect the battery negative (earth) lead.

2 Remove the air intake duct.

3 Disconnect the accelerator cable from the throttle body housing.

4 Disconnect the throttle position sensor multi-plug.

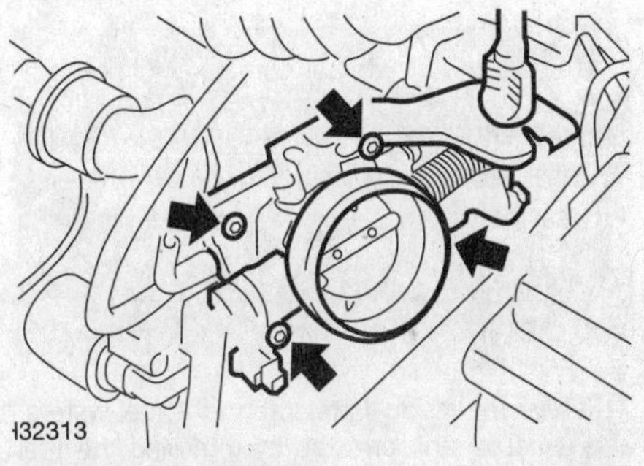

5 Unscrew and remove the four mounting bolts and withdraw the throttle housing from the intake manifold. Discard the rubber gasket and obtain a new one. **Note:** *Take care when cleaning the inside of the housing, as it is treated with a special coating during manufacture, which could be removed by over-enthusiastic cleaning, or by use of powerful solvents.*

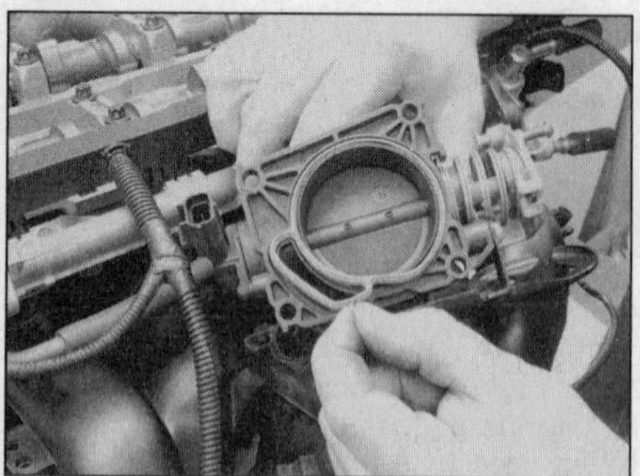

6 Refitting is a reversal of removal, but clean the mating faces and fit a new gasket, and tighten the mounting bolts securely.

Fuel rail and injectors

Warning: This procedure will merely relieve the increased pressure necessary for the engine to run – remember that fuel will still be present in the system components, and take precautions accordingly before disconnecting any of them.

7 Relieve the residual pressure in the fuel system, and equalise tank pressure by removing the fuel filler cap.

8 Disconnect the battery negative (earth) lead.

1.4 and 1.6 litre engines

9 Disconnect the breather hose from the cylinder head cover.

10 Disconnect the wiring plugs from the idle speed valve and throttle position sensor, then detach the wiring harness from the clip on the fuel rail.

11 Unclip the fuel lines from the cylinder head cover.

12 Using a small screwdriver, prise up and remove the four wire clips (one on each injector) securing the injector wiring busbar – use a magnetic holding tool to prevent the wire clips falling down the back of the engine as they are removed. Lift off the wiring busbar, and move it to the rear.

13 Disconnect the vacuum hose from the fuel pressure regulator.

14 Disconnect the fuel supply and return lines, noting their fitted positions. The supply line is colour-coded white, and the return line is colour-coded red.

15 Unscrew and remove the two bolts securing the fuel rail. Pulling the fuel rail equally at both ends, remove it squarely from the intake manifold.

1.8 and 2.0 litre engines

16 Loosen the retaining clips at either end, and remove the air cleaner intake duct.

17 Disconnect the accelerator cable from the throttle body housing.

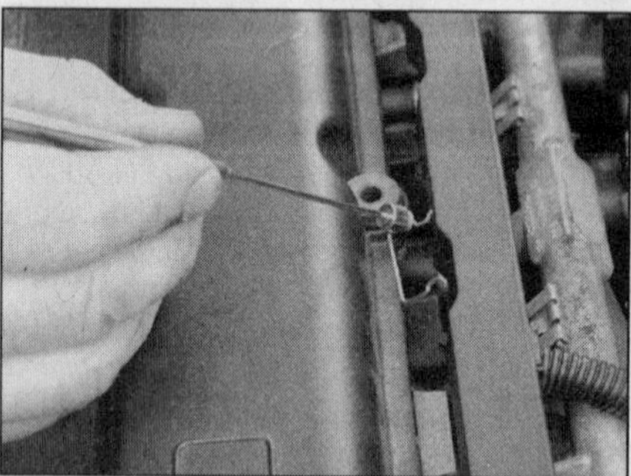

18 Using a small screwdriver, prise up and remove the four wire clips (one on each injector) securing the injector wiring busbar – use a magnetic holding tool to prevent the wire clips falling down the back of the engine as they are removed. Lift off the wiring busbar, and move it to the rear.

19 Disconnect the vacuum hose from the fuel pressure regulator.

20 Disconnect the fuel supply and return lines, noting their fitted positions. The supply line is colour-coded white, and the return line is colour-coded red.

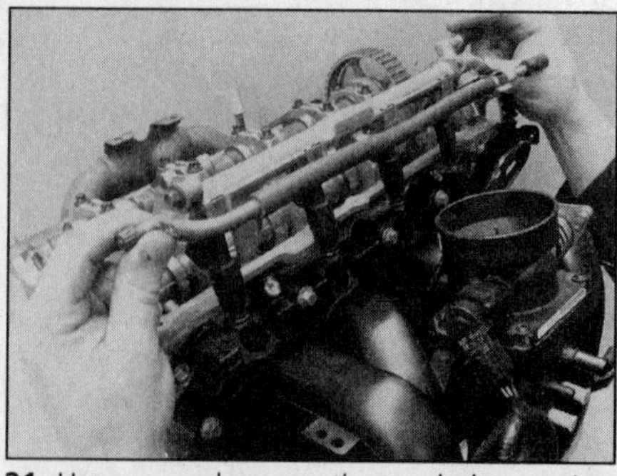

21 Unscrew and remove the two bolts securing the fuel rail. Pulling the fuel rail equally at both ends, remove it squarely from the intake manifold.

All engines

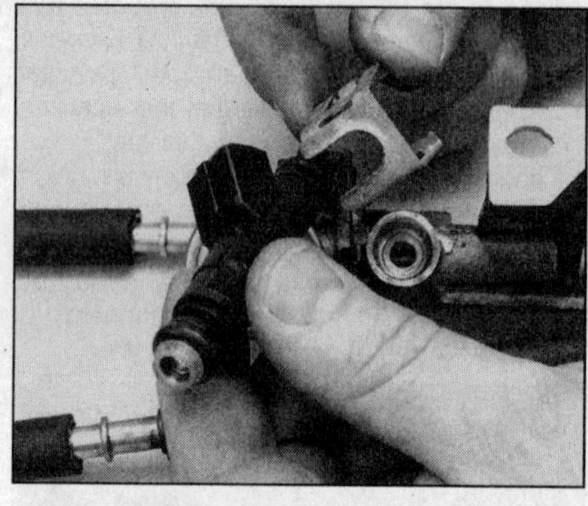

22 If required, remove the clips and carefully pull the injectors from the fuel rail.

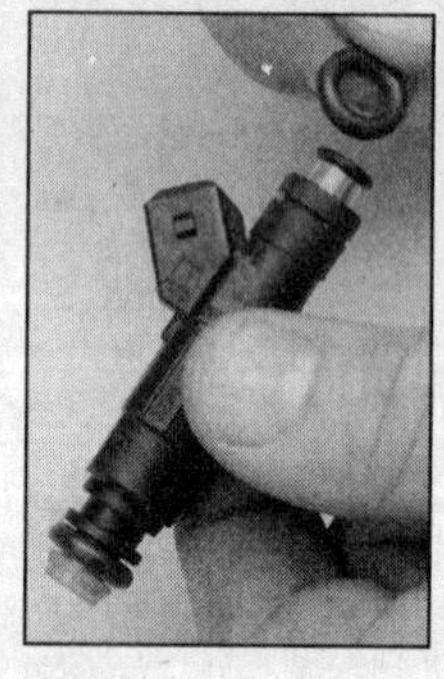

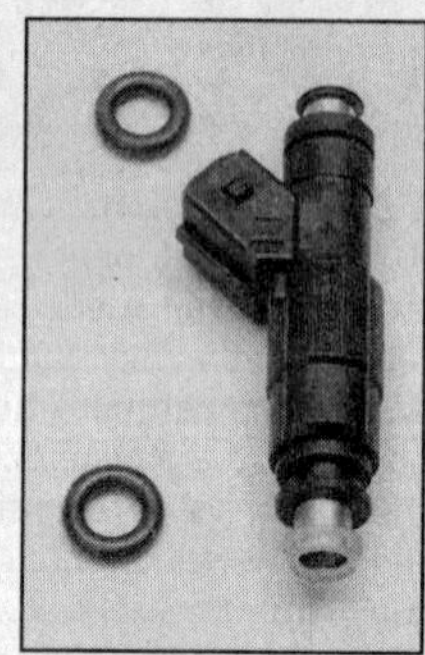

23 Using a screwdriver if necessary, prise the O-rings from the grooves at each end of the injectors. Discard the O-rings and obtain new ones.

24 Refitting is the reverse of the removal procedure, noting the following points:

a) Lubricate the new O-rings with clean engine oil to aid refitting.

b) Ensure that the hoses and wiring are routed correctly, and secured on reconnection by any clips or ties provided.

c) Refit the accelerator cable.

d) On completion, switch the ignition on to activate the fuel pump and pressurise the system, without cranking the engine. Check for signs of fuel leaks around all disturbed unions and joints before attempting to start the engine.

Fuel pressure regulator

25 Relieve the residual pressure in the fuel system, and equalise tank pressure by removing the fuel filler cap.

> *Warning: This procedure will merely relieve the increased pressure necessary for the engine to run – remember that fuel will still be present in the system components, and take precautions accordingly before disconnecting any of them.*

26 Disconnect the battery negative (earth) lead.

27 Disconnect the fuel supply and return lines, noting their fitted positions. The supply line is colour-coded white, and the return line is colour-coded red.

28 Disconnect the vacuum pipe from the fuel pressure regulator *(refer to paragraph 19)*.

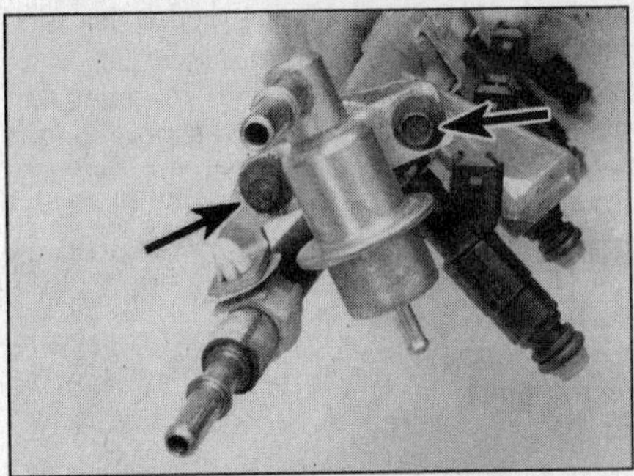

29 Undo the two bolts and remove the fuel pressure regulator from the fuel rail.

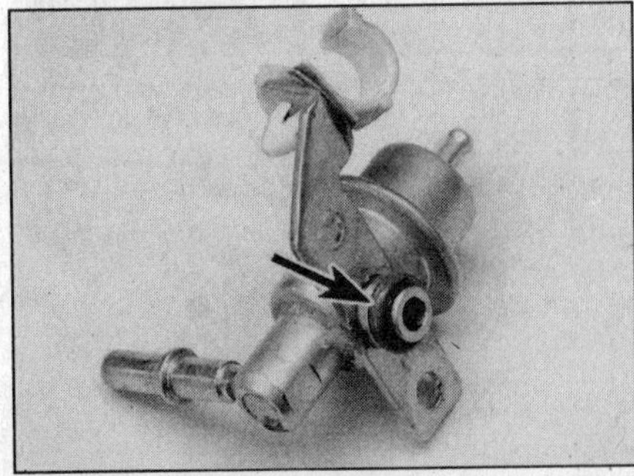

30 Using a screwdriver if necessary, prise the O-ring from the groove in the fuel pressure regulator. Discard the O-ring, and obtain a new one.

31 Refitting is a reversal of removal, but lubricate the new O-ring with clean engine oil to aid installation. Tighten the mounting bolts to the specified torque.

Idle speed control valve

32 The valve is located on the intake manifold, next to the throttle housing. On 1.8 and 2.0 litre models, the valve is on the inside of the intake manifold, making access especially difficult.

33 On 1.8 and 2.0 litre models, loosen the retaining clips at either end, and remove the air cleaner intake duct.

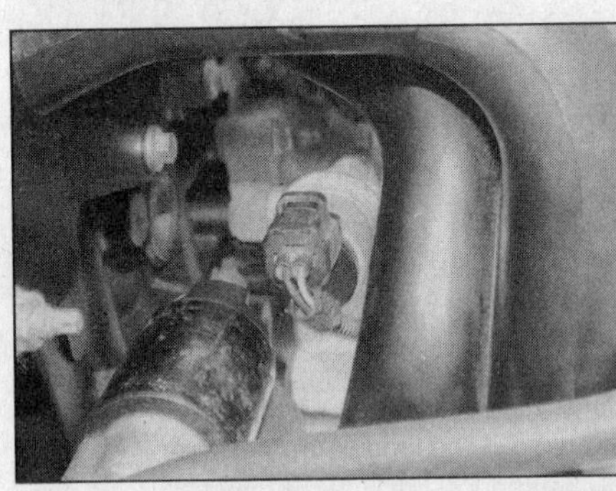

34 Disconnect the wiring plug from the valve.

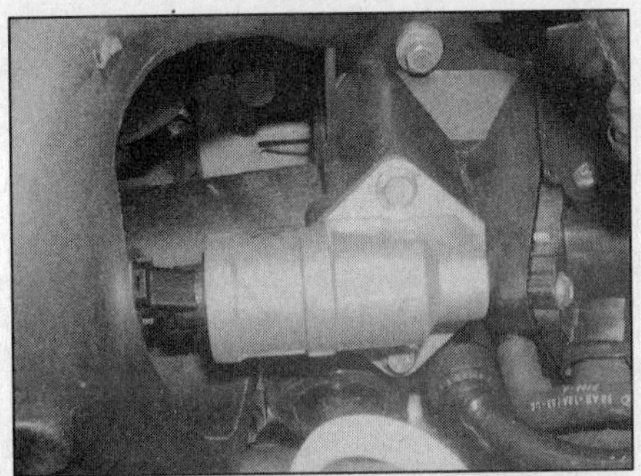

35 Unscrew the mounting bolts and remove the valve from the intake manifold. On 1.8 and 2.0 litre models, access to the bolts is less easy; remove the alternator, or ultimately, if suitable tools are not available, it may be simpler to remove the intake manifold for access.

36 Recover the rubber seal from the valve, and discard it. Obtain a new seal for refitting.

37 Refitting is a reversal of removal, but note the following points.

a) Clean the mating surfaces, and fit a new seal.
b) Tighten the valve bolts to the specified torque.
c) Once the wiring and battery are reconnected, start the engine and allow it to idle. When it has reached normal operating temperature, check that the idle speed is stable, and that no induction (air) leaks are evident. Switch on all electrical loads (headlights, heated rear window, etc), and check that the idle speed is still satisfactory.

Mass airflow sensor

38 Loosen the clip and disconnect the air intake duct from the mass airflow sensor on the air cleaner cover. If necessary, remove the air cleaner cover for improved access.

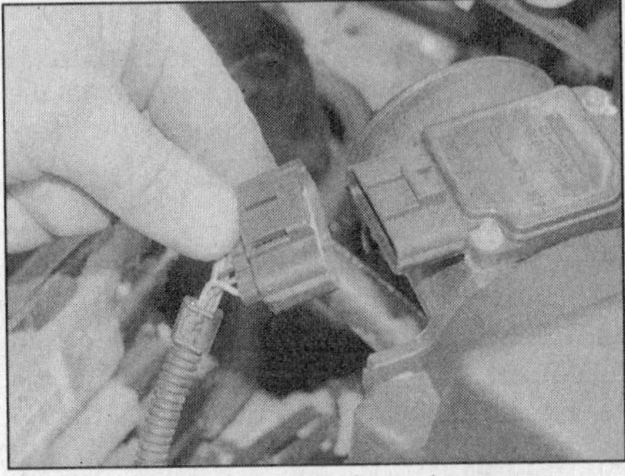

39 Disconnect the wiring from the sensor.

40 Unscrew the crosshead mounting screws and remove the sensor from the air cleaner cover.

41 Refitting is a reversal of removal.

Intake air temperature sensor

42 The sensor is an integral part of the mass airflow sensor, and cannot be renewed separately.

Engine management module (ECU)

Note: *The module is fragile. Take care not to drop it, or subject it to any other kind of impact. Do not subject it to extremes of temperature, or allow it to get wet. The module wiring plug must never be disconnected while the ignition is switched on.*

43 The engine management module is located behind the facia, on the right-hand side. First disconnect the battery negative (earth) lead.

Right-hand-drive models

44 Pull up the weatherstrip from the driver's door aperture, and release it from the side trim.

45 Prise off the trim panel fitted between the end of the facia panel and the door aperture, approximately half way up the door aperture.

46 Prise out the two screw head covers, then unscrew and remove the two screws securing the footwell side trim panel. Remove the trim panel for access to the module.

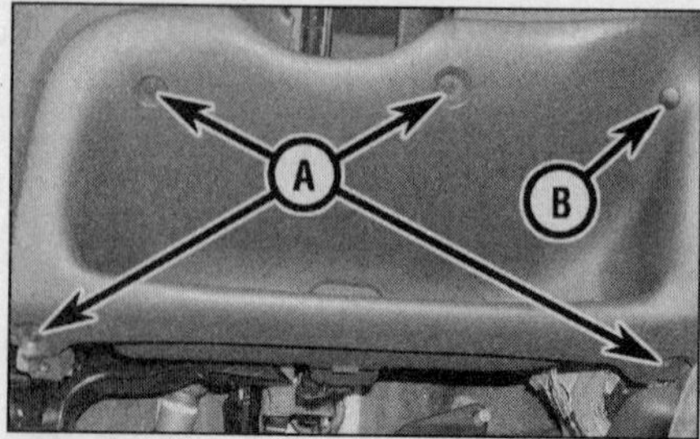

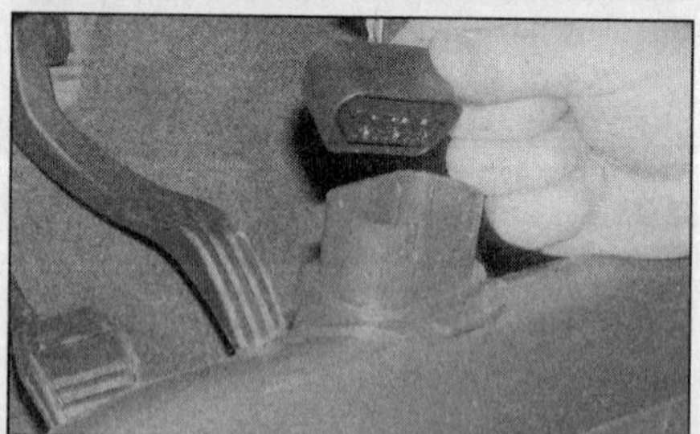

47 Remove the four screws (A) and prise out the plastic clip (B) securing the driver's side lower trim panel. Remove the panel from the facia, and detach the diagnostic connector socket.

Left-hand-drive models

48 Remove the glovebox.

49 Remove the four screws securing the passenger's side lower trim panel. Remove the panel from the facia.

All models

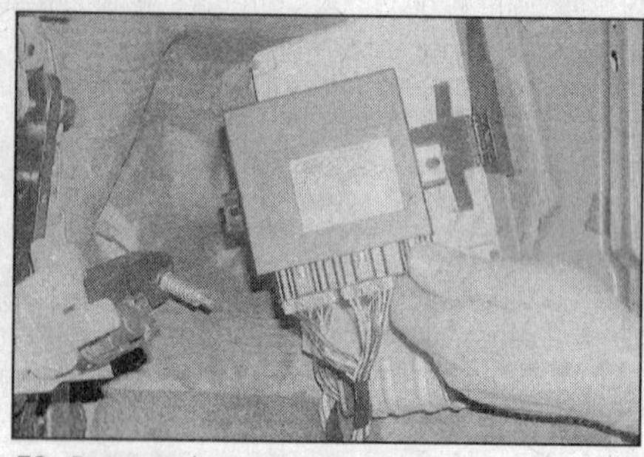

50 Remove the two screws securing the central locking module, and lower the module out of position; there is no need to disconnect the wiring.

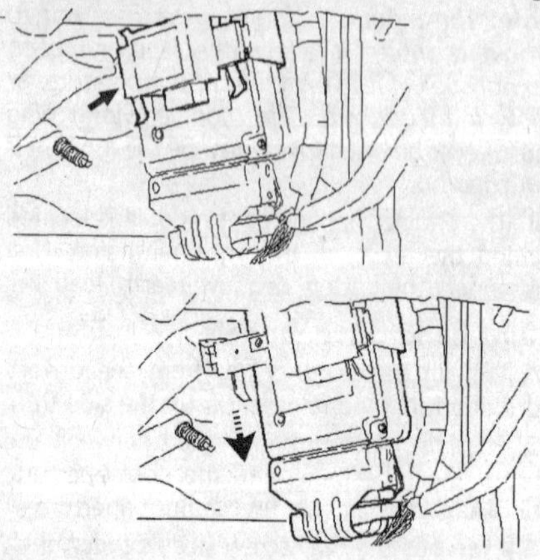

51 Release the spring clips securing the module bracket in position, and remove the module downwards into the footwell.

52 The ECU wiring connector is protected by a tamperproof shield, secured by a shear-bolt and welded nut, which must be drilled to remove the bolt. Great care must be taken not to damage the wiring harness during drilling.

53 First drill a 3 mm pilot hole in the nut, then enlarge the hole using an 8 mm drill until the shear-bolt can be removed.

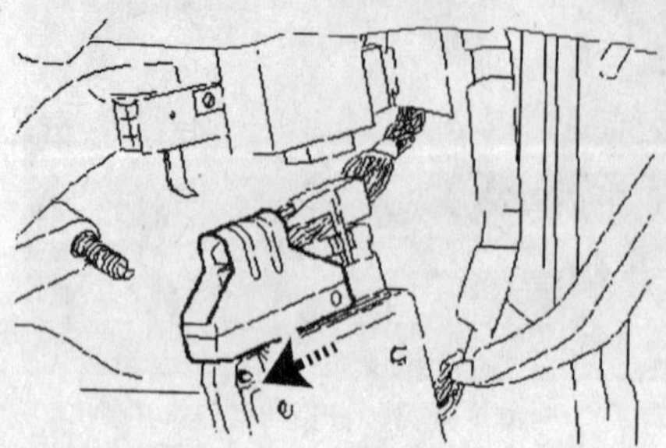

54 With the shear-bolt removed, slide off the tamperproof shield.

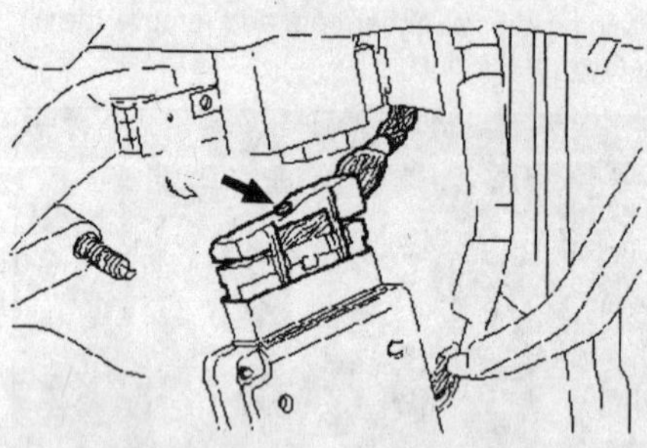

55 Remove the wiring connector securing screw, then disconnect the plug from the ECU, and remove the module from the car.

56 Refitting is a reversal of removal. Use a new shear-bolt, and tighten it until the head shears off.

Crankshaft position sensor

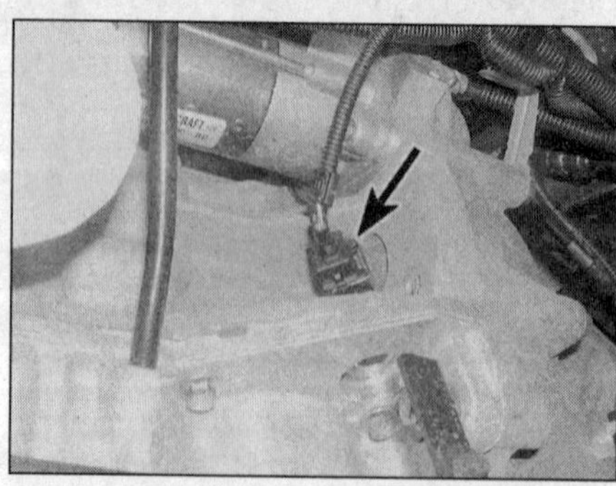

57 The sensor is located on the front left-hand side of the engine. For improved access, apply the handbrake then jack up the front of the vehicle and support it on axle stands.

58 Where fitted, unclip the cover from the crankshaft position sensor on the front of the engine, then disconnect the wiring plug.

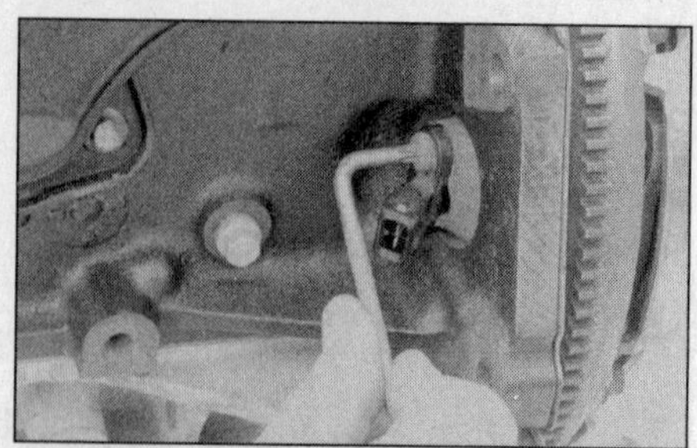

59 Unscrew the mounting bolt and withdraw the sensor. If a spacer is fitted between the sensor and engine, it is vital that this is refitted when refitting the sensor, or the sensor head will hit the rotor teeth.

60 Refitting is a reversal of removal. Tighten the sensor retaining bolt to the specified torque.

Camshaft position sensor

61 The camshaft position sensor is located at the rear of the cylinder head – on the right-hand side on 1.4 and 1.6 litre models, and on the left-hand side on 1.8 and 2.0 litre models (left and right as seen from the driver's seat).

62 On 1.8 and 2.0 litre models, loosen the retaining clips at either end, and remove the air cleaner intake duct. Disconnect the breather hose from the cylinder head cover.

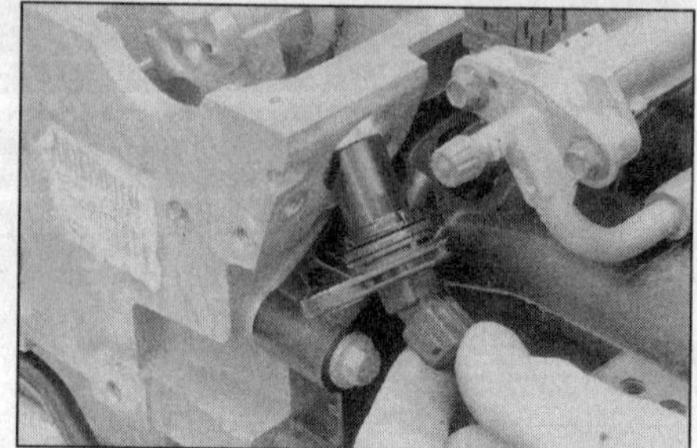

63 Disconnect the wiring connector from the sensor, then unscrew the sensor retaining bolt and withdraw the sensor from the cylinder head.

64 Refitting is a reversal of removal. Tighten the sensor retaining bolt to the specified torque.

Cylinder head temperature sensor

65 The sensor is located in the centre of the head, between Nos 2 and 3 spark plugs, on 1.4 and 1.6 litre models, and on the right-hand end of the head on 1.8 and 2.0 litre models. There is no need to drain the coolant, as the sensor measures the temperature of the metal directly.

66 On 1.4 and 1.6 litre models, disconnect the HT leads from the spark plugs as necessary for access – make sure that the leads are numbered for position (with No 1 at the timing belt end). Prise out the rubber bung used to secure the sensor wiring.

67 On 1.8 and 2.0 litre models, remove the alternator.

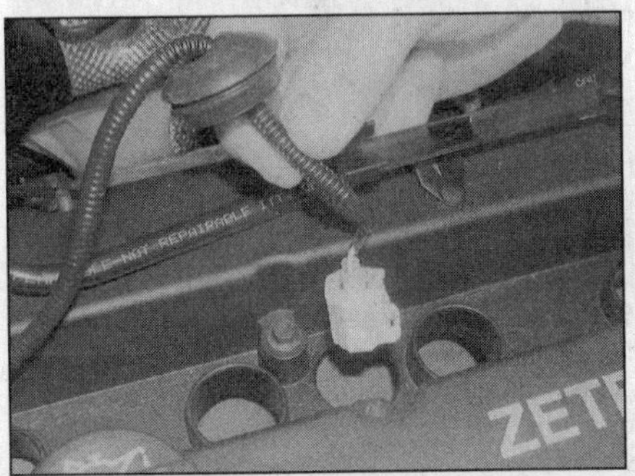

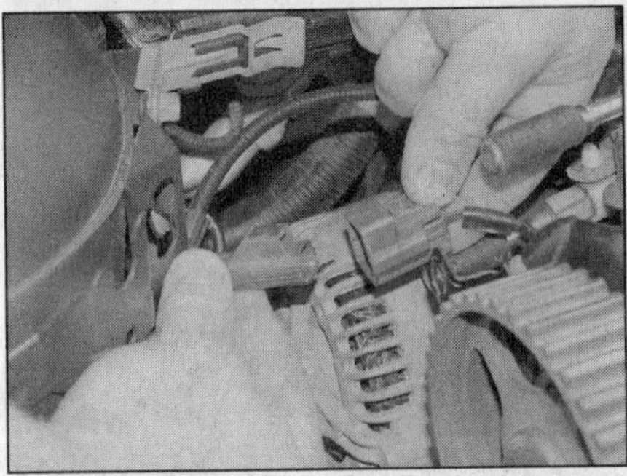

68 Disconnect the wiring from the temperature sensor.

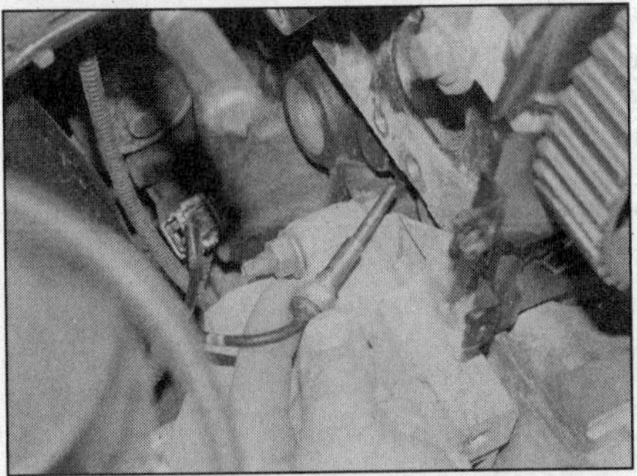

69 Unscrew and remove the sensor. **Note:** *On 1.4 and 1.6 litre engines, the sensor cannot be re-used – a new one must be fitted, since the mating face of the sensor is deformed when fully tightened, to ensure a good contact with the cylinder head. If the old sensor were re-used, it would result in inaccurate readings.*

70 Refitting is a reversal of removal. Clean the sensor mounting hole, then fit the sensor and tighten it to the specified torque.

Throttle position sensor

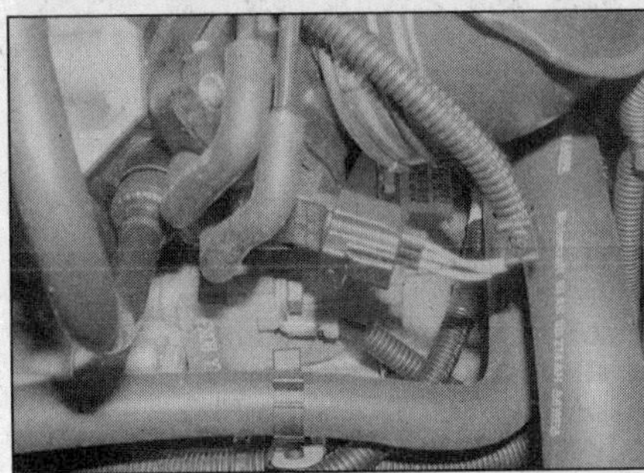

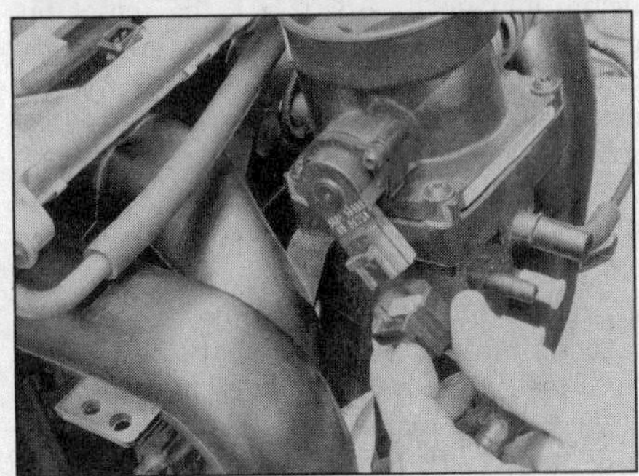

71 The sensor is located on the side of the throttle body housing. First disconnect the sensor wiring plug.

72 Remove the retaining screws, and withdraw the unit from the throttle housing. *Do not* force the sensor's centre to rotate past its normal operating sweep; the unit will be seriously damaged.

73 Refitting is a reversal of removal, but ensure that the sensor is correctly orientated, by locating its centre on the D-shaped throttle shaft (throttle closed), and aligning the sensor body so that the screws pass easily into the throttle housing.

Vehicle speed sensor

74 Access to the speed sensor is easiest from below. Apply the handbrake, then loosen the left-hand front wheel nuts. Jack up the front of the car and support it on axle stands.

75 Remove the left-hand front wheel and the wheelarch liner – the sensor is located next to the right-hand driveshaft, at the rear of the transmission.

76 Disconnect the wiring plug from the top of the sensor.

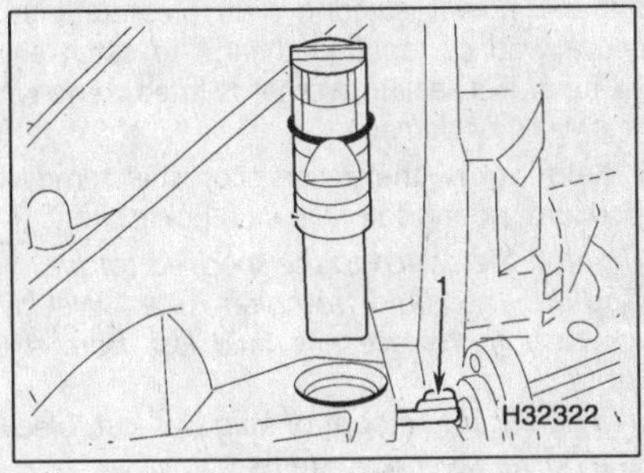

77 Using thin-nose pliers, pull out the retaining pin at the base of the sensor (1), noting how it is fitted.

78 Pull the sensor out of its location in the transmission – be prepared for a little oil spillage. Recover the O-ring seal – fit a new seal if the old one is in poor condition.

79 Refitting is a reversal of the removal procedure, but lightly oil the O-ring before inserting the assembly in the transmission casing.

Clutch pedal position switch

80 Remove five screws and take out the trim panel above the driver's footwell to gain access to the clutch pedal; unclip the diagnostic connector plug from the panel as it is removed.

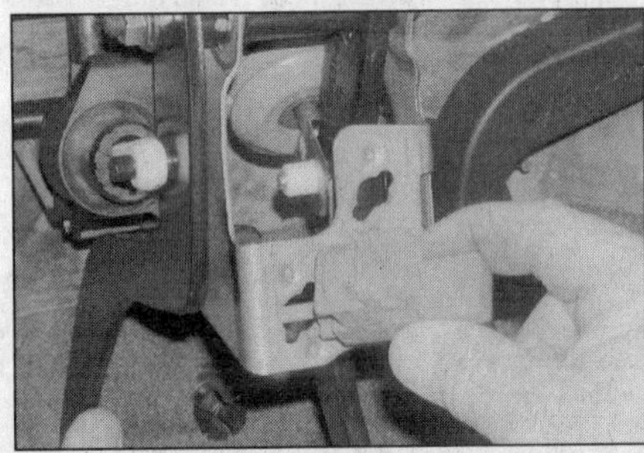

81 Reach up and disconnect the wiring from the clutch switch at the top of the pedal, then twist the switch anti-clockwise and remove it from the pedal bracket.

82 Refitting is a reversal of removal.

Power steering pressure switch

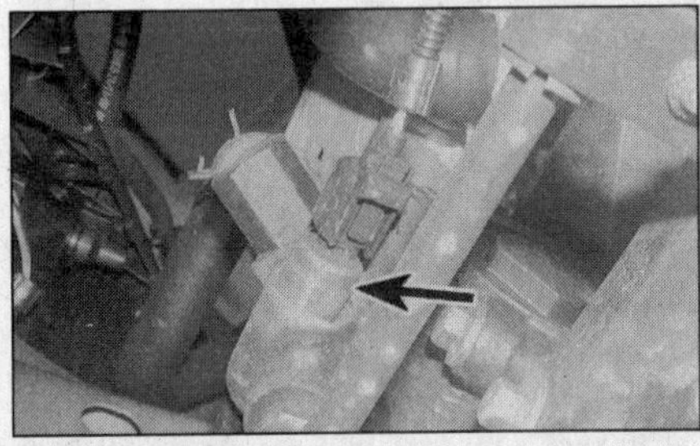

83 The switch is screwed into the power steering system's high-pressure pipe at the front right-hand side of the engine compartmen. On some models, access is improved by removing the radiator grille.

84 Releasing its clip, unplug the switch's electrical connector, then unscrew the switch from the power steering high pressure pipe. Place a wad of rag underneath, to catch any spilt fluid. If a sealing washer is fitted, renew it if it is worn or damaged.

85 Refitting is the reverse of the removal procedure, noting the following points:

a) Tighten the switch to the specified torque.
b) Top-up the fluid reservoir (see 'Weekly checks') to replace any fluid lost from the system.
c) If a significant amount of fluid was lost, bleed the power steering system.

Oxygen sensor

Note: *Some later models are fitted with two sensors – one before and one after the catalytic converter. This enables more efficient monitoring of the exhaust gas, allowing a faster response time. The overall efficiency of the converter itself can also be checked.*

86 On 1.4 and 1.6 litre models, access to the sensor in the front downpipe can be gained from above, but access to the second sensor fitted to later models can only be gained with the car raised and supported.

87 On 1.8 and 2.0 litre models, remove the four bolts securing the heat shield fitted over the exhaust manifold, and lift away the heat shield for access to the sensor.

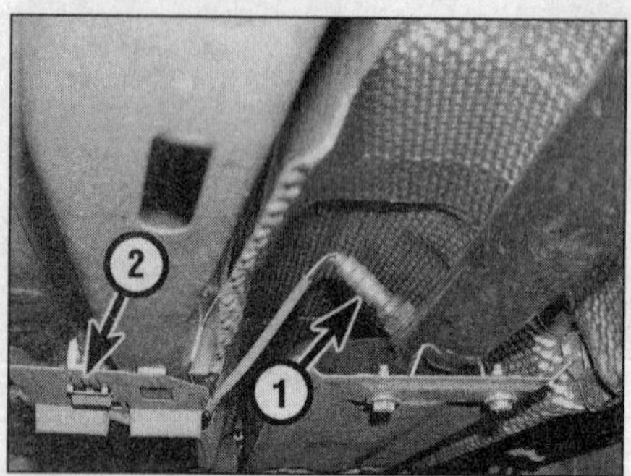

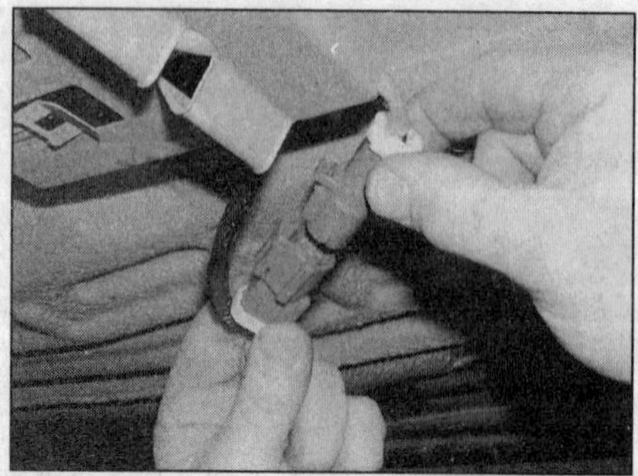

88 Trace the wiring from the sensor body (1) back to its wiring plug (2), and disconnect it. On models with two sensors, the wiring plug for the rearmost sensor is clipped to a metal plate attached to the underside of the car. Note how the wiring is routed, as it must not come into contact with hot exhaust components.

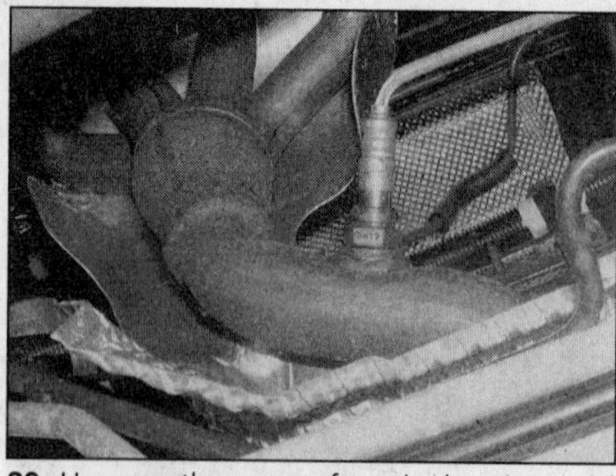

89 Unscrew the sensor from its location, and remove it. Once removed, take great care that the sensor is not dropped or damaged. Keep the sensor tip clean while it is removed.

90 It may be beneficial to clean the sensor before refitting it, especially if the sensor tip appears to be contaminated. However, great care must be exercised, as the tip will be damaged by any abrasives, and by certain solvents. Seek the advice of a Ford dealer before cleaning the sensor.

91 Refitting is a reversal of removal, noting the following points:

a) Apply a little anti-seize compound to the sensor threads, taking care not to allow any on the sensor tip, and tighten the sensor to the specified torque.
b) Reconnect the wiring, ensuring that it is routed clear of any hot exhaust components.
c) Remember that a faulty sensor will have generated a fault code – if this code is still logged in the ECU electronic memory, the engine management system may still be in LOS.

Diesel injection system electronic components – removal and refitting

Crankshaft position sensor

Endura-Di engines

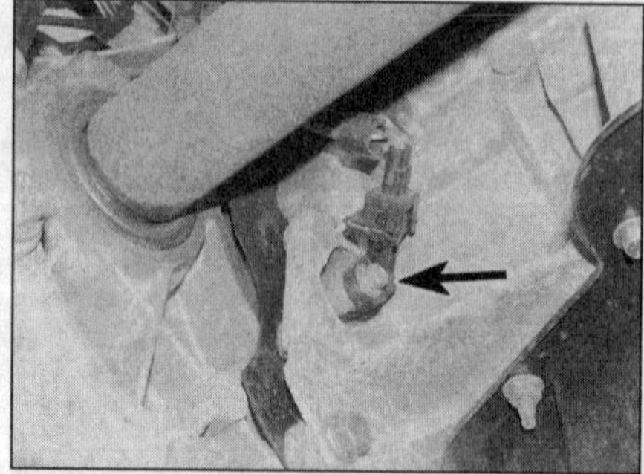

1 The sensor is located at the flywheel end of the engine, low down at the rear. For improved access, apply the handbrake then jack up the front of the vehicle and support it on axle stands.

2 Where fitted, unclip the cover from the crankshaft position sensor, then disconnect the wiring plug.

3 Unscrew the mounting bolt and withdraw the sensor. If a spacer is fitted between the sensor and engine, it is vital that this is refitted when refitting the sensor, or the sensor head will hit the rotor teeth.

4 Refitting is a reversal of removal, tightening the retaining bolt to the specified torque.

DuraTorq-TDCi engines

5 Before removing the sensor, set the crankshaft at TDC on No. 1 cylinder.

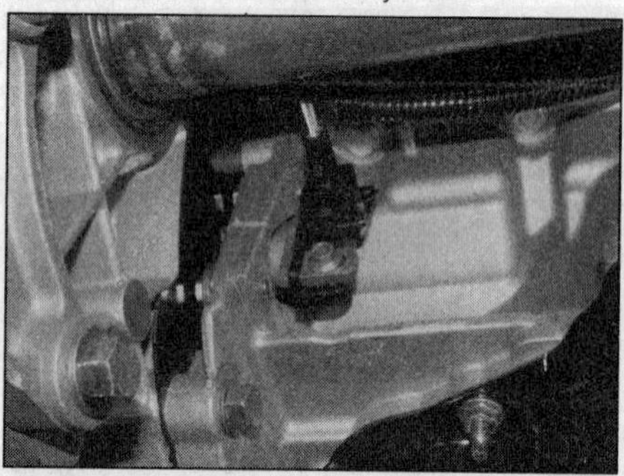

6 The sensor is located at the flywheel end of the engine, low down at the rear. For improved access, apply the handbrake then jack up the front of the vehicle and support it on axle stands.

7 Disconnect the wiring plug from the sensor.

8 Unscrew the mounting bolt, then slide the sensor to towards the timing belt end of the engine and withdrawn it.

9 Refitting is a reversal of removal, but ensure that the sensor and the block mounting face is clean and free from debris, and tighten the bolt to the specified torque.

Cylinder head temperature (CHT) sensor

10 The sensor is screwed into the left-hand (flywheel) end of the cylinder head, behind the vacuum pump and below the oil pressure warning light switch.

11 Open the bonnet and disconnect the battery negative (earth) lead.

12 To improve access to the sensor, it will be necessary to remove (or partially remove) the air cleaner intake duct and the intercooler right-hand air duct.

13 It will also be helpful to release the hoses and remove the crankcase ventilation system oil separator from the left-hand end of the cylinder head.

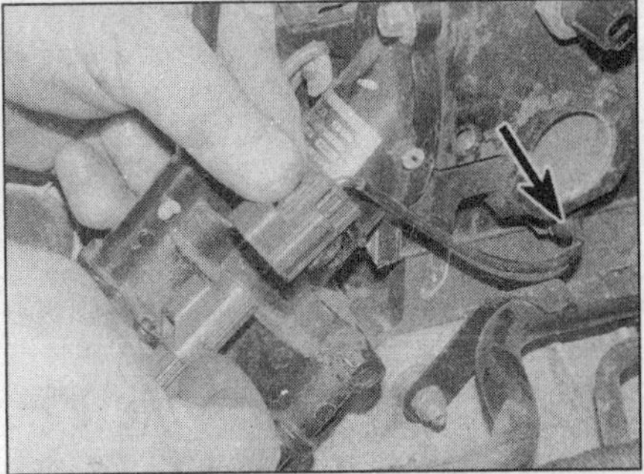

14 Trace the wiring from the sensor, and disconnect it at the plug, which is clipped to the brake vacuum pump at the front of the engine.

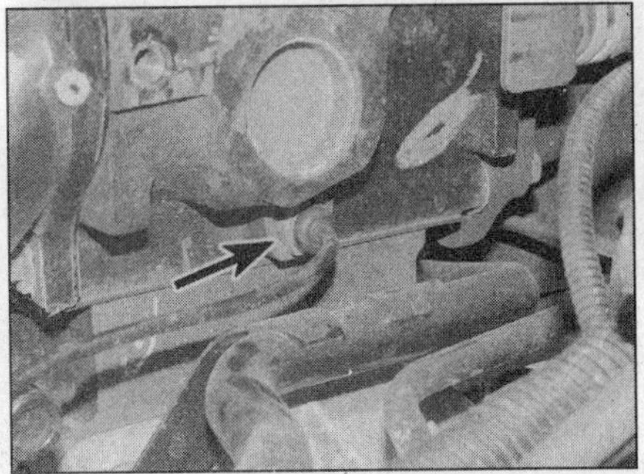

15 The sensor can now be unscrewed and removed. However, access to the sensor is such that great difficulty may be experienced in getting a tool to fit onto it. Ultimately, it may be necessary to cut the sensor wiring, unscrew the sensor using a thin-wall socket or box spanner, then re-make the wiring after fitting, using a suitable connector.

16 Refitting is a reversal of removal. Clean the threads of the sensor and mounting hole, then refit the sensor and tighten it to the specified torque.

Intake air temperature sensors

17 Two sensors are fitted – one on the air cleaner intake duct, and one on the intercooler duct leading to the intake manifold, behind the right-hand headlight. On some TDCi models, the sensor is incorporated into the mass air flow sensor at the air filter housing outlet. On TDdi and some TDCi models, removal and refitting procedure for either sensor is identical (see paragraph 18), whilst on some TDCi models, the clamp must be released and the air filter-to-turbo pipe disconnected, then the sensor unscrewed from the housing.

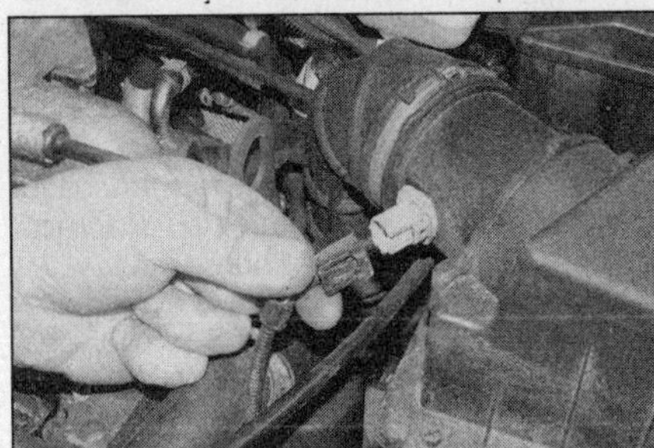

18 Disconnect the wiring from the sensor, then twist the sensor anti-clockwise and remove it.

19 Refitting is a reversal of removal.

Vehicle speed sensor

20 Refer to the petrol injection system Section.

Manifold absolute pressure (MAP) sensor

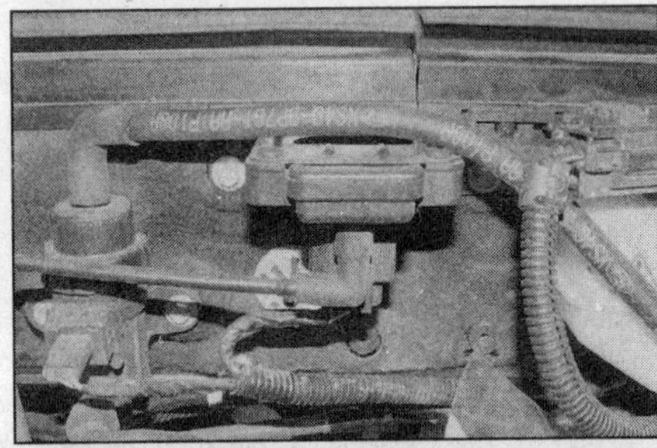

21 The MAP sensor is mounted on the engine compartment bulkhead, next to the brake/clutch fluid reservoir on TDdi models, and adjacent to the engine compartment fusebox on TDCi models.

22 Disconnect the vacuum hose from the port at the base of the sensor, then disconnect the wiring plug behind it.

23 Unscrew and remove the two mounting bolts, and withdraw the sensor from the bulkhead.

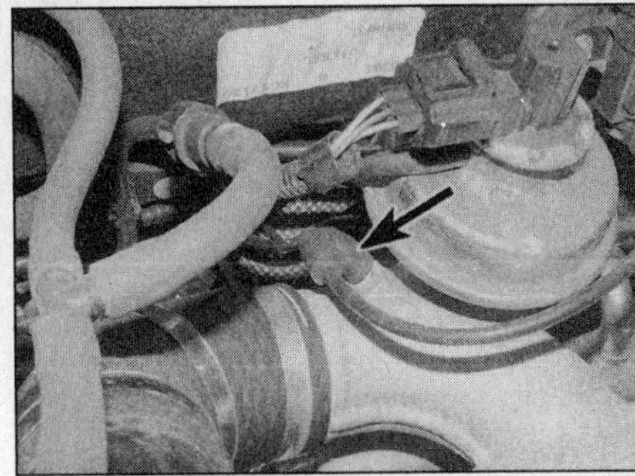

24 Prior to refitting, remove the vacuum hose from the intake manifold connection, and check it for signs of perishing or splitting, especially at the pipe ends.

25 Refitting is a reversal of removal, ensuring that the wiring plug and vacuum hose are securely reconnected.

Fuel control valve

26 The control valve is fitted to the top of the fuel filter (at the right-hand rear corner of the engine compartment), and contains a bi-metal strip. Its function is to close the fuel return to the fuel tank at low fuel temperatures, allowing fuel which has been warmed by passing through the injection pump to flow back into the filter, thus warming the fuel being drawn from the tank. **Note:** *On DuraTorq-TDCi engines, the valve is integral with the filter assembly and cannot be renewed separately.*

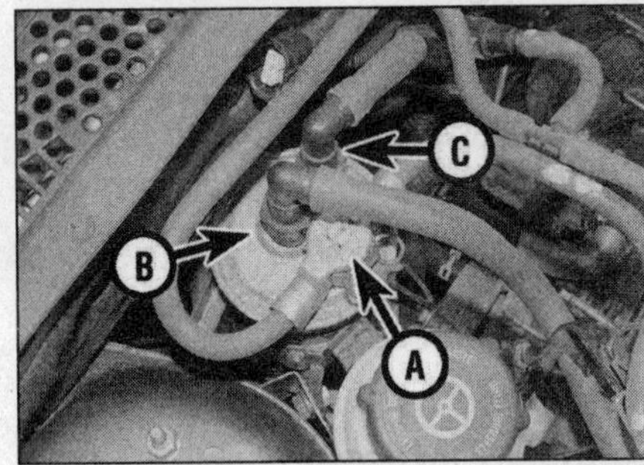

A Control valve B Outlet C Inlet

27 Noting their positions for refitting, disconnect the fuel supply and outlet pipes from the connections at the fuel filter.

28 Extract the wire clip securing the control valve and return pipe, and release it from the top of the filter. Once again noting the fitted positions of the fuel pipes, disconnect them from each end of the valve (note that the return pipe may in this case be black, and smaller in diameter than the supply pipe).

29 Refitting is a reversal of removal. Ensure that the pipe connections are correctly and securely remade, then run the engine and check for signs of fuel leakage.

Clutch pedal position switch

30 Remove five screws and take out the trim panel above the driver's footwell to gain access to the clutch pedal; unclip the diagnostic connector plug from the panel as it is removed.

31 Reach up and disconnect the wiring from the clutch switch at the top of the pedal.

32 Twist the switch anti-clockwise and remove it from the pedal bracket.

33 Refitting is a reversal of removal.

Stop-light and brake pedal position switches

34 These are two separate switches, both fitted at the top of the brake pedal. Remove five screws and take out the trim panel above the driver's footwell to gain access to the brake pedal; unclip the diagnostic connector plug from the panel as it is removed.

35 Reach up and disconnect the wiring from the switch at the top of the pedal; if both switches are removed at once, take care to note which wiring plug serves which switch. Twist the switch through 90° and remove it from the pedal bracket. The stop-light switch (upper) is removed by twisting anti-clockwise, while the brake pedal position switch (lower) is twisted clockwise to remove.

36 Refitting is a reversal of removal.

Accelerator pedal sensor

37 The accelerator pedal sensor is integral with the pedal assembly.

Injection pump control unit (PCU)

Endura-Di engines only

38 The pump control unit is integral with the injection pump, and cannot be separated from it. If a new injection pump is fitted, the PCU must be electronically 'matched' to the engine management module (ECU), otherwise the pump will not function correctly (or even at all, if the immobiliser function is not correctly set up) – this is a task for a Ford dealer, as specialised electronic equipment is required.

Injector driver module (IDM)

DuraTorq-TDCi engines only

Note: *In order to successfully carry out the IDM renewal procedure, the configuration data within the existing unit must be uploaded and stored into Ford's WDS diagnostic unit, ready to be downloaded later into the new unit. Do not attempt to renew the IDM without access to this facility.*

39 Undo the two screws and move the air cleaner intake pipe on the bonnet slam panel forward.

40 Remove the battery, and battery tray.

41 Disconnect the breather pipe from the IDM cover .

42 Undo the four screws and remove the IDM cover.

43 Release the locking catches and disconnect the wiring connectors from the IDM.

44 Undo the fasteners and remove the IDM.

45 Refitting is a reversal of removal, ensuring that the original configuration data is downloaded using Ford's WDS equipment.

Engine management module (ECU)

Note: *The module is fragile. Take care not to drop it, or subject it to any other kind of impact. Do not subject it to extremes of temperature, or allow it to get wet. The module wiring plug must never be disconnected while the ignition switch is on.*

46 The engine management module is located behind the facia, on the right-hand side. First disconnect the battery negative (earth) lead.

Right-hand-drive models

47 Pull up the weatherstrip from the driver's door aperture, and release it from the side trim.

48 Prise off the trim panel fitted between the end of the facia panel and the door aperture, approximately half way up the door aperture.

49 Prise out the two screw head covers, then unscrew and remove the two screws securing the footwell side trim panel. Remove the trim panel for access to the module.

50 Remove the four screws securing the driver's side lower trim panel. Remove the panel from the facia, and detach the data link connector socket.

Left-hand-drive models

51 Remove the glovebox.

52 Remove the four screws securing the passenger's side lower trim panel. Remove the panel from the facia.

All models

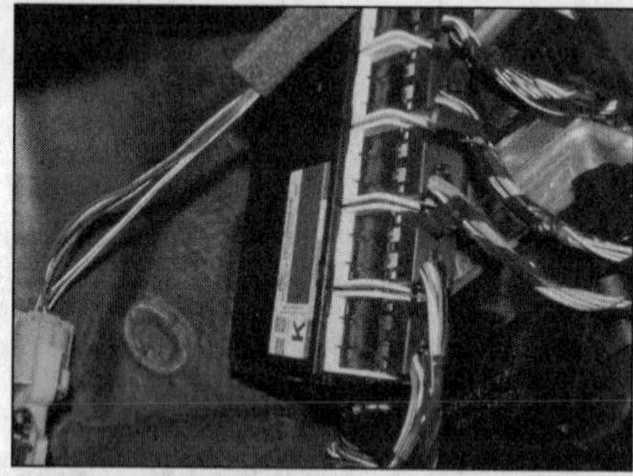

53 Release the clips and separate the generic electronic module (GEM) from the ECU. Position the GEM to one side without disconnecting its wiring plugs.

54 Release the spring clips securing the ECU in position, and remove it downwards into the footwell.

55 The ECU wiring connector is protected by a tamperproof shield, secured by a shear-bolt and welded nut, which must be drilled to remove the bolt. Great care must be taken not to damage the wiring harness during drilling.

56 First drill a 3 mm pilot hole in the nut, then enlarge the hole using an 8 mm drill until the shear-bolt can be removed.

57 With the shear-bolt removed, slide off the tamperproof shield.

58 Remove the wiring connector securing screw, then disconnect the plug from the ECU, and remove it from the car.

59 Refitting is a reversal of removal. Use a new shear-bolt, and tighten it until the head shears off.

Fuel temperature sensor

DuraTorq-TDCi engines only

60 Undo the two fasteners and remove the plastic cover from the top of the engine.

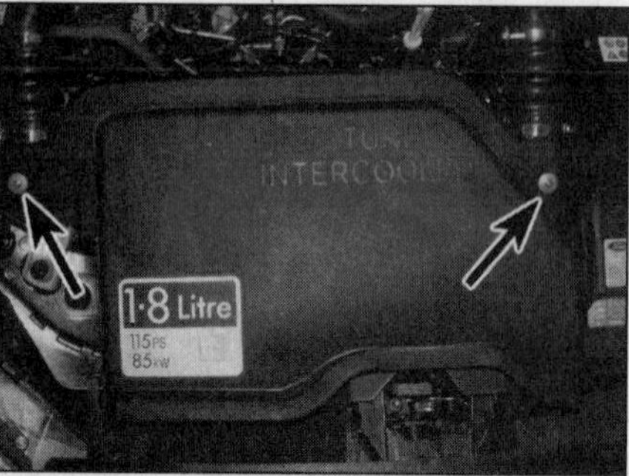

61 Undo the two fasteners and remove the air deflector from the front of the engine.

62 Slacken the clamps, remove the bolts, disconnect the charge air temperature sensor wiring plug, and remove the intercooler pipes.

63 Place absorbent rags over the top of the alternator to protect it from any split fuel, then working at the rear of the high-pressure fuel pump, disconnect the wiring plug from the temperature sensor.

64 Ensure that the area around the sensor is clean, and have a blanking plug or cap ready to cover the exposed port in the pump.

65 Unscrew the sensor from the pump – be prepared for fuel spillage. Plug or cap the exposed port in the pump to prevent dirt ingress.

66 Fit a new sealing O-ring to the sensor, and spray the O-ring with a lubricant spray (WD40 etc.). Remove the plug or cap from the pump, and refit the sensor, tightening it to the specified torque.

67 The remainder of refitting is a reversal of removal.

Fuel metering valve

Duratorq-TDCi engines only

68 Undo the two fasteners and remove the plastic cover from the top of the engine.

69 Undo the two fasteners and remove the air deflector from the front of the engine *(see paragraph 61)*.

70 Slacken the clamps, remove the bolts, disconnect the charge air temperature sensor wiring plug, and remove the intercooler pipes.

71 Place absorbent rags over the top of the alternator to protect it from any split fuel, then working at the rear of the high-pressure fuel pump, disconnect the wiring plug from the metering valve.

72 Ensure that the area around the sensor is clean, and have a blanking plug or cap ready to cover the exposed port in the pump.

73 Undo the two retaining screws and remove the metering valve from the pump – be prepared for fuel spillage. Plug or cap the exposed port in the pump to prevent dirt ingress.

74 Fit a new sealing O-ring to the valve, and spray the O-ring with a lubricant spray (WD40 etc.). Remove the plug or cap from the pump, and refit the valve, tightening the retaining screws finger tight at first, then to the specified torque.

75 The remainder of refitting is a reversal of removal.

Fuel pressure sensor

DuraTorq-TDCi engines only

76 The pressure sensor is fitted to the rear of the high-pressure pump, but is not available separately, and no attempt should be made to remove it.

Camshaft position sensor

DuraTorq-TDCi engines only

77 Undo the two nuts and remove the plastic cover from the top of the engine.

78 Disconnect the camshaft position sensor wiring plug.

79 Undo the bolt and remove the sensor from the camshaft cover.

80 Refitting is a reversal of removal, remembering to renew the O-ring and tighten the retaining bolt securely.

Knock sensor

DuraTorq-TDCi engines only

81 The knock sensor is located at the rear of the cylinder block.

82 Jack up the front of the vehicle and support it securely on axle stands. Undo the fasteners and remove the engine undertray.

83 Reach up behind the engine block and disconnect the wiring plug from the sensor.

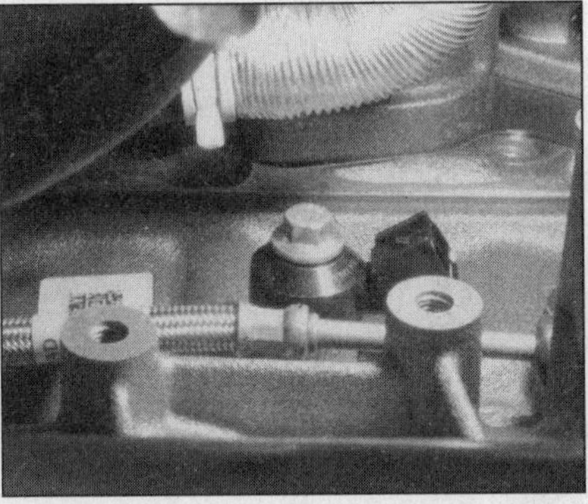

84 Undo the bolt and remove the sensor from the cylinder block.

85 Refitting is a reversal of removal, tightening the sensor to the specified torque.

Notes

Ford Galaxy '95 – '00

13

Petrol models

General

Type	Ford EEC V sequential multipoint injection

Fuel system data

Fuel pump type	Electric, immersed in fuel tank
Regulated fuel pressure at idling speed:	
Vacuum hose fitted	3.5 bar (approx)
Vacuum hose disconnected	4.0 bar (approx)
Minimum holding pressure (after 10 minutes)	2.0 bar
Engine idle speed (non-adjustable, ECM controlled):	
2.0 litre 8V engine	865 rpm
2.3 litre 16V engine	875 rpm
Engine maximum speed:	
All models	6175 rpm
Idle CO content	Not adjustable, ECM controlled
Fuel tank capacity	70 litres

Diesel models

General

Type	Direct injection, turbocharged and intercooled

Fuel system data

Maximum engine speed (all models)	5100 rpm (not adjustable ECM controlled)
Engine idle speed (all models)	900 rpm (not adjustable ECM controlled)
Injection pressure	170 bar minimum
Turbocharger type	Garrett
Maximum boost pressure:	
Engine code 1Z, AHU and AVG	1.7 to 1.9 bar
Engine code AFN	1.7 to 2.2 bar

Chapter 13

Torque wrench settings

	Nm	lbf ft
Petrol models		
Camshaft position sensor	5	4
Crankshaft position sensor	4	3
Engine coolant temperature sensor	23	17
Fuel injector bolts	6	5
Fuel pressure regulator	6	5
Fuel rail	10	7
Idle air control valve-to-intake manifold	10	7
Intake manifold:		
Studs	14	10
Nuts and bolts	23	17
Throttle body-to-intake manifold	10	7
Diesel models		
Accelerator pedal position sender	10	7
Crankshaft position sensor	10	7
Fuel cut-off solenoid	40	30
Fuel pipe unions to injection pump and injectors	25	18

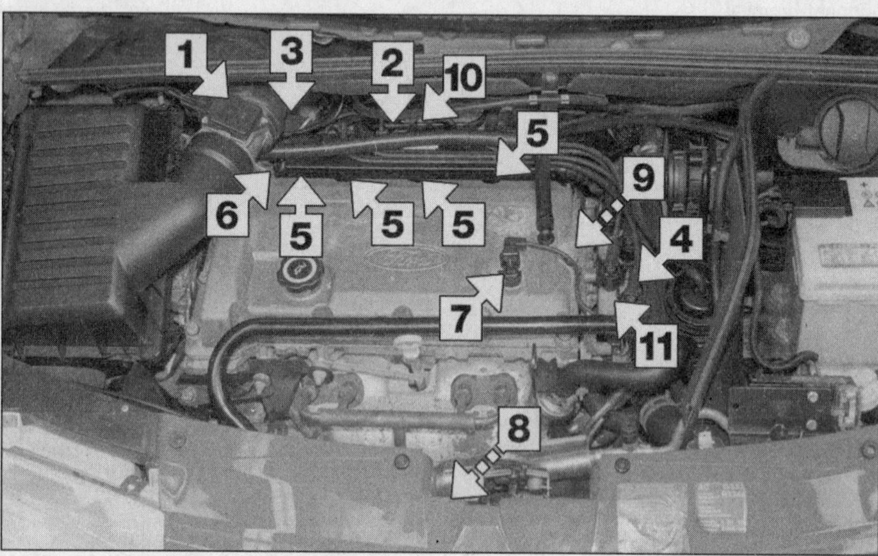

2.0 litre engine management components

1 Mass airflow sensor
2 Throttle position sensor
3 Intake air temperature sensor
4 Coolant temperature sensor
5 Fuel injectors and fuel rail
6 Fuel pressure regulator
7 Camshaft position sensor
8 Oxygen sensor
9 Crankshaft position sensor
10 Idle air control valve
11 Ignition coils

2.3 litre engine management components

1 Mass airflow sensor
2 Throttle position sensor
3 Coolant temperature sensor
4 Fuel injectors wiring rail
5 Camshaft position sensor
6 Radio suppressor
7 Ignition coils/spark plugs cover
8 Crankshaft position sensor (rear of engine)
9 Oxygen sensor
10 Idle air control valve

General information

Petrol models

The Ford EEC V sequential multipoint petrol injection system described in this Chapter is a self-contained engine management system, which controls both the fuel injection and, via a separate module, ignition.

The fuel injection system comprises a fuel tank, an electric fuel pump, a fuel filter, fuel supply and return lines, a throttle body, an air mass sensor, a fuel rail and four electronic injectors, a fuel pressure regulator, and an electronic control module (ECM), together with its associated sensors, actuators and wiring.

The air mass sensor is located on the air cleaner outlet to the throttle body. Fuel is supplied under pressure to a fuel rail, and then passes to four electronic injectors. The duration of the injection period is determined by the ECM which switches the injectors on and off as required.

The tank immersed fuel pump delivers a constant supply of fuel through a cartridge filter. The fuel is supplied to a fuel rail, and the fuel pressure regulator maintains a constant fuel pressure to the fuel injectors and returns excess fuel to the tank via the return line. The constant fuel flow system helps to reduce fuel temperature and prevents vaporisation.

The ECM controls starting and warm-up enrichment together with idle speed regulation and Lambda control. Idle speed control is achieved by an idle air control valve on the side of the throttle body, and partly by the ignition system. Manual adjustment of the idle speed is not possible.

Intake air is drawn into the engine through the air cleaner, which contains a renewable paper filter element.

The exhaust gas oxygen content is constantly monitored by the ECM via the Oxygen sensor, which is mounted in the front section of the exhaust pipe, before the catalytic converter. On models for some markets, a second Oxygen sensor is mounted after the catalytic converter. The ECM then uses this information to adjust the air/fuel ratio. Manual adjustment of the idle speed exhaust CO content is not possible. A catalytic converter is fitted to the exhaust system on all models. A secondary air injection system is fitted, where air is introduced into the exhaust manifold during the warm-up phase, to shorten the amount of time taken for the catalytic converter to reach operating temperature, thus reducing harmful exhaust emissions. A fuel evaporative control system is fitted, and the ECM controls the operation of the activated charcoal canister.

It should be noted that fault diagnosis of all the engine management systems described in this Chapter is only possible with dedicated electronic test equipment.

Diesel models

The fuel system comprise of a fuel tank, an engine-bay mounted fuel filter with an integral water separator, fuel supply and return lines, fuel injection pump and four fuel injectors. The commencement of injection is controlled by the engine management electronic control module (ECM) and a solenoid valve on the injection pump. The pump is driven at half crankshaft speed by the camshaft timing belt. Fuel is drawn from the fuel tank, through the filter by the injection pump, which then distributes the fuel under very high pressure to the injectors via separate delivery pipes. All engines are fitted with a turbocharger and an intercooler. However, on engine code AFN the turbocharger is equipped with variable vanes which change the shape of the inlet port. This increases the speed of the incoming exhaust gases at low engine speed, and the result is that full boost pressure is available over almost all of the engine speed range, decreasing fuel consumption and improving torque/power output. The operation of the guide vanes is controlled by the engine management ECM via a charge pressure sensor, and vacuum operated actuator.

Basic injection timing is set mechanically by the position of the pump on its mounting bracket. Dynamic timing and injection duration are controlled by the ECM and are dependant on engine speed, throttle position and rate of opening, intake air flow, intake air temperature, coolant temperature, fuel temperature, ambient pressure (altitude) and manifold depression information, received from sensors mounted on and around the engine. Closed loop control of the injection timing is achieved by means of an injector needle lift sensor. Note that injector No 3 is fitted with the needle lift sensor. Two-stage injectors are used, which improve the engine's combustion characteristics, leading to quieter running and better exhaust emissions. The direct-injection fuelling system is controlled electronically by a diesel engine management system, comprising an ECM and its associated sensors, actuators and wiring

On all engines, the ECM also manages the operation of the Exhaust Gas Recirculation (EGR) emission control system, the turbocharger boost pressure control system and the glow plug control system.

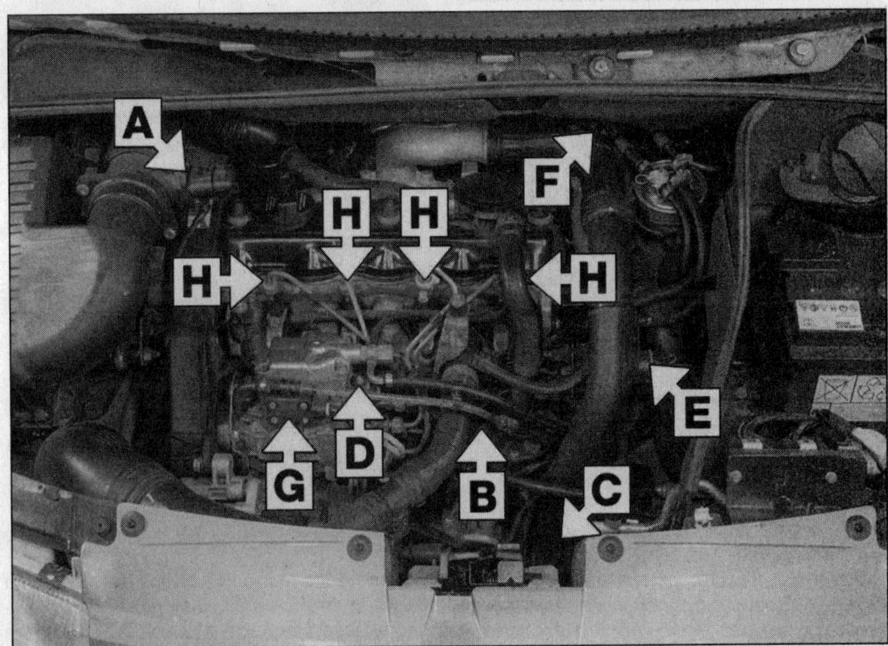

Diesel injection components

A Mass air flow sensor
B Coolant temperature sensor
C Engine speed sensor
D Fuel cut-off solenoid
E Charge pressure/temperature sensor
F Charge pressure control valve
G Fuel injection pump
H Fuel injectors

Chapter 13

Fuel system depressurisation

Petrol models

> **Warning:** *The following procedure will merely relieve the pressure in the fuel system – remember that fuel will still be present in the system components and take precautions accordingly before disconnecting any of them.*

1 The fuel system referred to in this Section comprises the tank-mounted fuel pump and sender, the fuel filter, the fuel rail and injectors, the fuel pressure regulator and the metal pipes and flexible hoses of the fuel lines between these components. All these contain fuel which will be under pressure while the engine is running and/or while the ignition is switched on. The pressure will remain for some time after the ignition has been switched off and must be relieved before any of these components are disturbed for servicing work.

2 Disconnect the battery negative lead.

3 Open the fuel filler flap and briefly remove the filler cap to relieve any pressure in the fuel tank. Refit the cap.

4 Place some cloth rags around the Schrader valve on the right-hand end of the fuel rail.

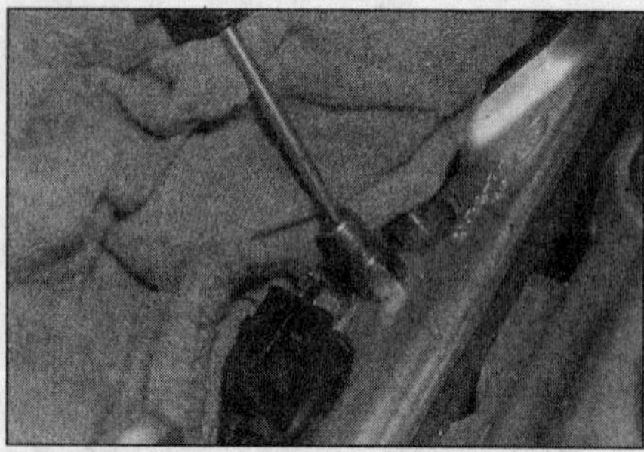

5 Unscrew the valve cap, and depress the valve pin using a small screwdriver (or similar). Try to absorb the fuel spray with the rags.

6 On completion, refit the valve cap.

Diesel models

7 Although there is no specific requirement to depressurise the fuel system, common sense dictates that it would be wise to wait until the engine has cooled sufficiently prior to attempting to dismantle any part of the engine. If this advice is followed, most of the residual pressure in the system will have dissipated before any fuel connections are disturbed. However, to be on the safe side, place a clean rag over the fuel pipe connection as the pipe union is slackened to soak up any fuel spray.

Petrol models engine management systems – component removal and refitting

Note: *The ignition must be switched off at all times.*

Mass airflow sensor

1 The mass airflow sensor is located on the air ducting from the air cleaner housing.

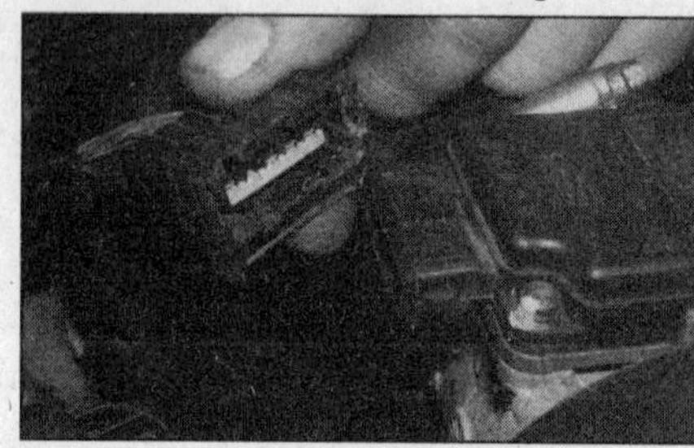

2 Disconnect the wiring plug from the sensor.

3 Slacken the retaining clip, and disconnect the intake hose from the sensor.

4 Release the retaining clips and remove the sensor from the air ducting. Check that the sensor sealing ring is in good condition.

5 Refitting is a reversal of removal.

Throttle position sensor

> *Note: On automatic transmission models, the ECM 'basic set-up' procedure must be carried out following disconnection of the sensor. This necessitates the use of FDS2000 dedicated diagnostic equipment, and should be entrusted to your local Ford dealer or suitably equipped specialist.*

6 Remove the mass airflow sensor, as described earlier in this Section. Undo the three screws and remove the bulkhead panel trim. Slide the panel to the right-hand side and manoeuvre it from behind the coolant expansion tank.

7 Disconnect the wiring plug from the intake air temperature sensor *(see paragraph 3.15)*.

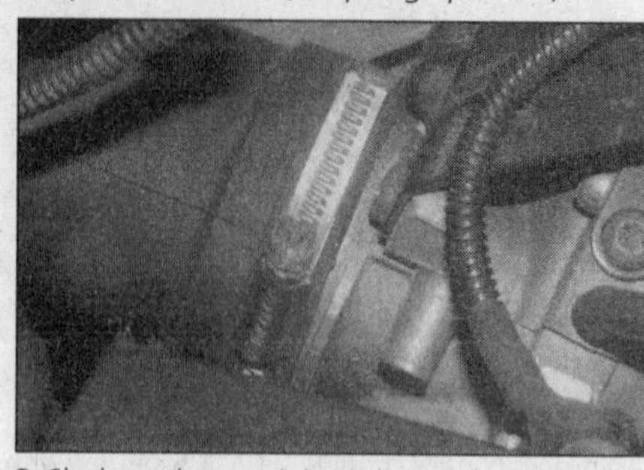

8 Slacken the retaining clip and remove the intake ducting from the throttle body.

9 On 2.0 litre DOHC 8V models, disconnect the wiring multiplugs from the injectors and position to one side.

10 Unplug the throttle position sensor wiring, and disconnect the inner accelerator cable from the throttle valve. On 2.3 litre DOHC 16V models, undo the retaining bolts and position the accelerator cable bracket to one side.

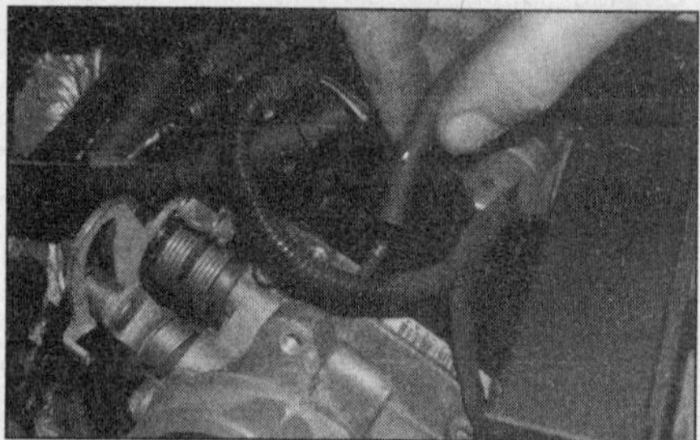

11 Disconnect the vacuum hose from the throttle body, undo the retaining Torx screws, and remove the throttle body.

12 With the throttle body removed, undo the two Torx screws, and remove the throttle position sensor.

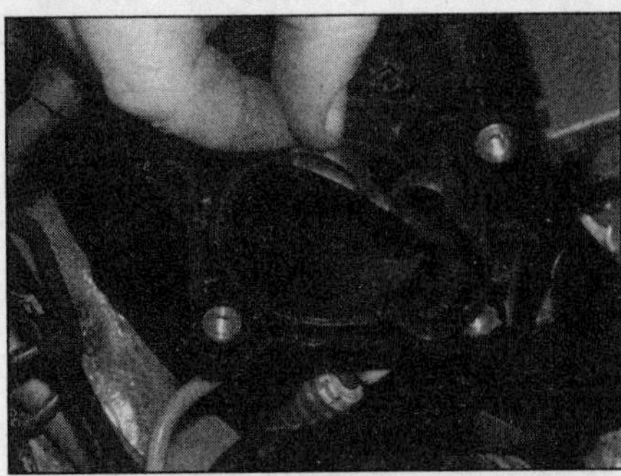

13 Refitting is a reversal of removal, ensuring that the sensor is correctly located on the throttle valve shaft. Before refitting the throttle body, check the sealing gasket is in good condition. On automatic models, it will now be necessary to carry out the 'basic set-up' procedure using FDS2000 diagnostic equipment.

Intake air temperature sensor

14 The intake air temperature sensor is located on the intake ducting between the mass airflow sensor and the throttle body. On 2.0 litre DOHC 8V models, remove the mass airflow sensor as described earlier in this Section.

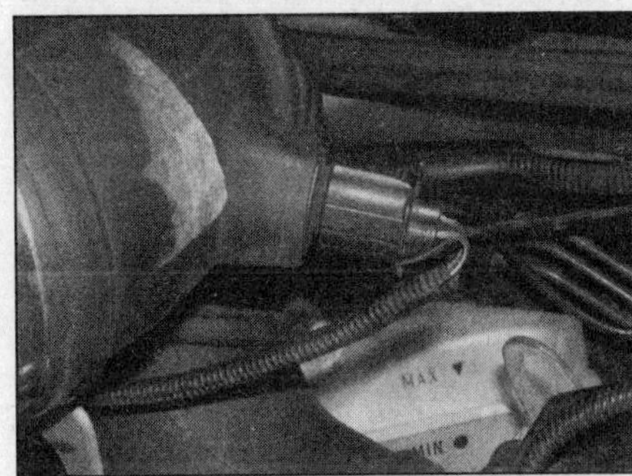

15 Disconnect the wiring plug from the intake air temperature sensor.

16 Carefully prise the sensor from the ducting. If required, the sensor can be tested by checking the output voltage and resistance, and comparing the values obtained with those given below.

Air temperature °C	Resistance kOhms	Resistance V
-10	183	4.20
0	95	3.87
10	66	3.55
20	38	3.10
30	27	2.70
40	16	2.10

17 Refitting is a reversal of removal, tightening the sensor securely.

Vehicle speed sensor

18 Where fitted, the road speed sensor is fitted into the back of the gearbox.

19 Chock the rear wheels, apply the handbrake, then jack up the front of the vehicle and support it on axle stands. Undo the two retaining nuts and four bolts (two each side), and remove the undertray from under the engine compartment.

20 The sensor is fitted on the rear of the transmission casing, above the driveshafts. Lift up the heat shield (where fitted), and disconnect the sensor wiring plug.

21 Unscrew and remove the sensor.

22 If required, withdraw the roll pin and pull the sensor drive from the casing, complete with the drive pinion. Discard the O-ring seal, a new one must be fitted.

23 If the sensor drive was previously removed, fit the new O-ring to the drive body, and insert it, complete with pinion, into the transmission casing.

24 Ensure that the drive is fully seated, then insert the roll pin.

25 Screw the sensor onto the drive, tightening it securely.

26 Reconnect the sensor wiring plug, and slide the heat shield (where fitted) over the sensor.

27 Refit the engine/transmission undertray, and lower the vehicle to the ground.

Any fault with the sensor must be checked by a Ford dealer or suitably equipped specialist, and if necessary renewed.

Coolant temperature sensor

28 The engine coolant temperature sensor is located on the rear of the thermostat/coolant outlet housing on the left-hand end of the cylinder head. The upper sensor is the engine coolant temperature sensor, and the lower sensor is the coolant temperature gauge sensor.

29 Disconnect the wiring plug from the sensor.

30 Either drain the cooling system to below the level of the sensor, or be prepared for coolant spillage.

31 Unscrew the sensor from the housing. If required, the sensor can be tested by checking the output voltage and resistance. The values obtained should be identical to those given for the intake air temperature sensor in Paragraph 16.

32 Refitting is a reversal of removal, tightening the sensor to the specified torque. Top-up the cooling system.

Crankshaft position sensor

33 The engine speed sensor is mounted on the rear of the cylinder block, adjacent to the mating surface of the block and transmission bellhousing, just behind the oil filter. Apply the handbrake, then jack up the front of the vehicle and support it on axle stands (see *Jacking and vehicle support*). Where applicable, remove the splash guard from under the engine compartment.

34 Disconnect the wiring plug from the sensor.

35 Unscrew the retaining bolt and withdraw the sensor from the cylinder block. If required, the sensor can be tested as follows. Ensure the ignition is turned off, and disconnect the sensor wiring plug. Connect a multi-meter between the two sensor terminals, and set the meter to measure resistance. A sensor in good condition should register a resistance of 200 to 450 ohms. If the resistance is outside the range specified, it is likely the sensor is defective.

36 Refitting is a reversal of removal. Tighten the securing bolt to the specified torque.

Throttle body

37 The throttle body removal procedure is incorporated into the throttle position sensor removal procedure described in Paragraph 6 of this Section.

Fuel injectors and fuel rail

2.0 litre DOHC 8V models

38 Disconnect the battery negative lead.

39 Carefully release the wiring multiplug rail from the top of the injectors, and place it to one side.

40 Depressurise the fuel system.

41 Unscrew the fuel pressure regulator retaining bolts at the right-hand end of the fuel rail, and disconnect the regulator from the fuel rail. Be prepared for fuel spillage.

42 Undo the fuel supply hose union bolts at the left hand end of the fuel rail, disconnect the hose and recover any sealing washers/ O-rings.

43 Unscrew the mounting bolts, then carefully lift the fuel rail together with the injectors from the intake manifold.

44 With the assembly on the bench, undo the retaining bolts and release each of the injectors from the fuel rail. Recover the O-ring seals. Check the availability of new O-ring seals before discarding the old ones.

45 Refit the injectors, regulator (where applicable) and fuel rail by following the removal procedure in reverse, noting the following points:

a) The No. 4 injector electrical socket must point to the rear of the engine, whilst the remainder face the front.

b) Where available, renew the injector O-ring seals, and smear them with a little clean engine oil before fitting them.

c) Check that the fuel supply and return pipes is reconnected correctly.

d) Check that all vacuum and electrical connections are remade correctly and securely.

e) Reconnect the battery.

f) On completion, start the engine and check for fuel leaks.

2.3 litre DOHC 16V models

46 Disconnect the battery negative lead. Remove the air cleaner housing.

47 Disconnect the intake air temperature sensor wiring plug.

48 Slacken the retaining clip and remove the air intake ducting from the throttle body.

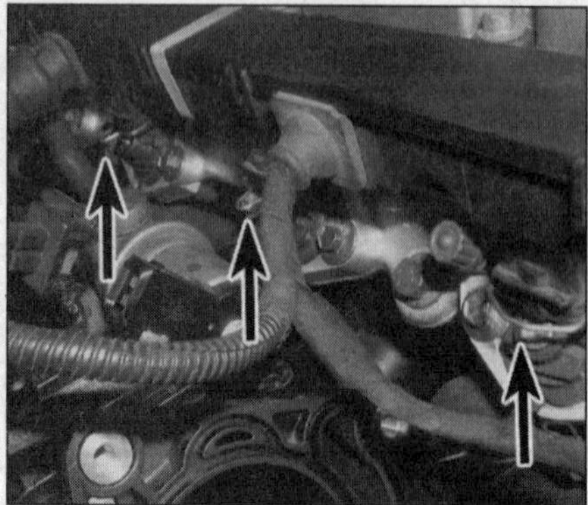

49 Carefully release the wiring multiplug rail from the top of the injectors, and place it to one side.

50 Depressurise the fuel system.

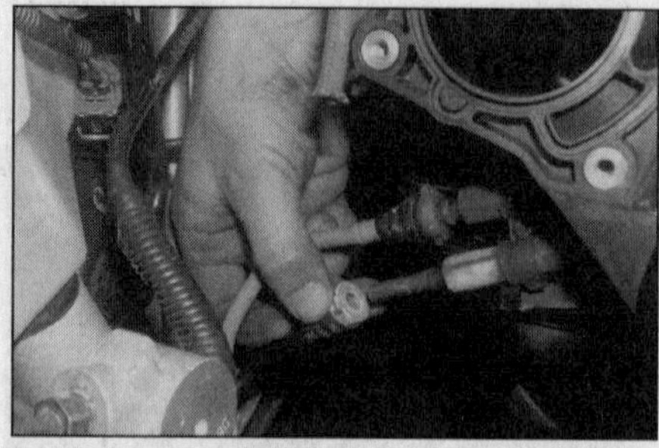

51 Squeeze together the locking tabs and disconnect the fuel supply and return pipes.

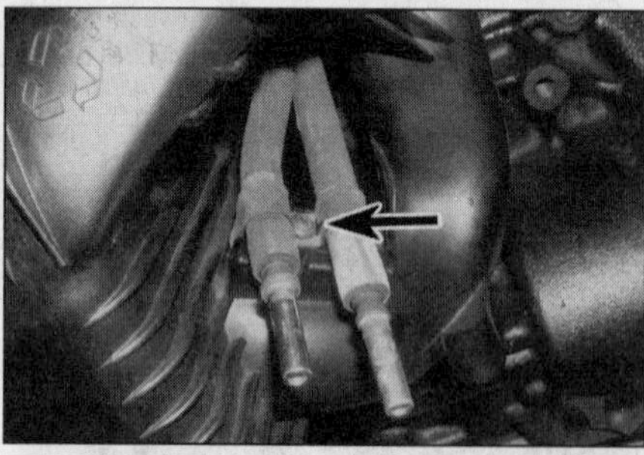

52 Undo the Torx screw and remove the fuel pipes retaining clip.

53 Disconnect the vacuum pipe from the fuel pressure regulator.

54 Undo the retaining bolts and remove the fuel rail complete with pressure regulator and injectors.

55 If required, the pressure regulator can be removed by unscrewing the two retaining bolts. Check the availability of a new O-ring seal before discarding the old one.

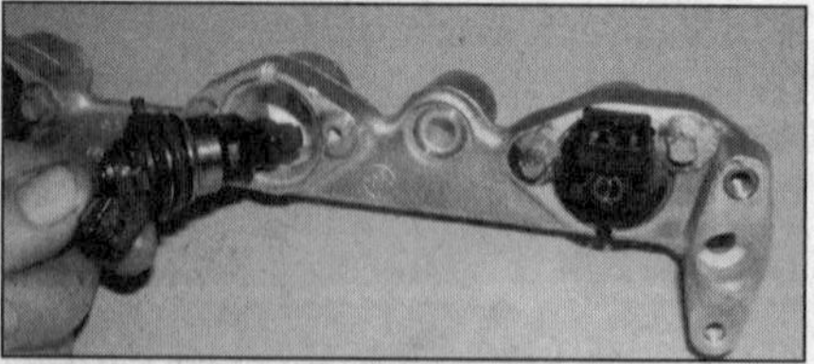

56 Undo the retaining bolts and carefully prise the injectors from the fuel rail. Check the availability of new O-ring seals before discarding the old ones.

57 Refit the injectors, regulator (where applicable) and fuel rail by following the removal procedure in reverse, noting the following points:

a) The No. 4 injector electrical socket must point to the rear of the engine, whilst the remainder face the front.

b) Where available, renew the injector O-ring seals, and smear them with a little clean engine oil before fitting them.

c) Check that the fuel supply and return pipes is reconnected correctly.

d) Check that all vacuum and electrical connections are remade correctly and securely.

e) Reconnect the battery.

f) On completion, start the engine and check for fuel leaks.

Fuel pressure regulator

2.0 litre DOHC 8V models

58 Depressurise the fuel system.

59 Disconnect the vacuum hose from the pressure regulator at the right-hand end of the fuel rail.

60 Undo the two regulator retaining bolts, and remove the pressure regulator from the fuel rail. Be prepared for fuel spillage.

61 Disconnect the fuel hose from the regulator. Check the availability of a new O-ring seal before discarding the old one

62 Refit the fuel pressure regulator by following the removal procedure in reverse, tightening the retaining bolts to the specified torque.

2.3 litre DOHC 16V models

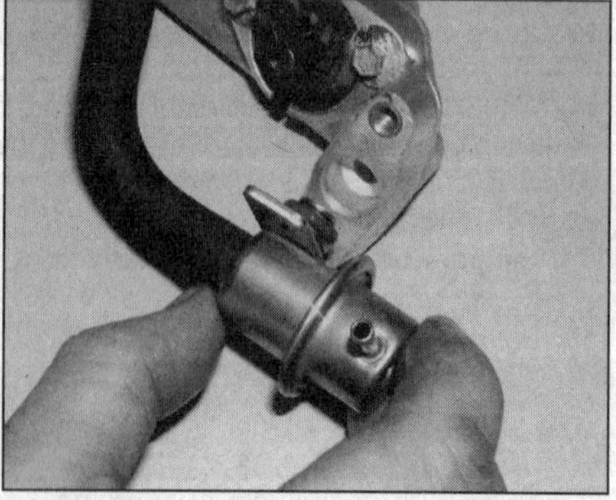

63 Remove the fuel rail assembly as described earlier in this Section. Undo the two retaining bolts, and remove the regulator from the fuel rail. Check the availability of a new O-ring seal before discarding the old one.

64 Refit the fuel pressure regulator by following the removal procedure in reverse, tightening the retaining bolts to the specified torque.

Camshaft position sensor

65 Disconnect the battery negative lead.

66 Disconnect the wiring plug from the sensor.

67 Unscrew the mounting bolt and withdraw the sensor from the camshaft cover. Recover the seal. If required, the sensor can be tested as follows. Ensure the ignition is turned off, and disconnect the sensor wiring plug. Connect a multi-meter between the two sensor terminals, and set the meter to measure resistance. A sensor in good condition should register a resistance of 200 to 900 ohms. If the resistance is outside the range specified, it is likely the sensor is defective.

68 Refitting is a reversal of removal, tightening the mounting bolt to the specified torque.

Oxygen sensor(s)

69 Disconnect the battery negative lead.

70 Apply the handbrake, then jack up the front of the vehicle and support it on axle stands (see *Jacking and vehicle support*). Undo the two retaining nuts and four bolts (two each side) and remove the engine undershield.

71 Disconnect the wiring plug from the sensor.

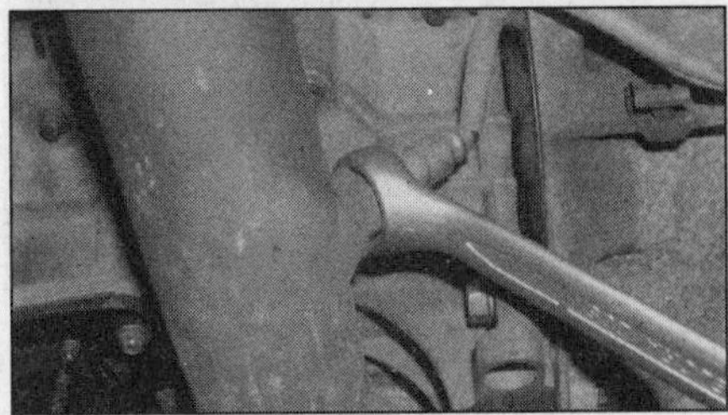

72 Unscrew and remove the sensor, taking care to avoid damaging the sensor probe as it is removed.

73 Apply a little anti-seize grease to the sensor threads, but avoid contaminating the probe tip. **Note:** *New Oxygen sensors may be supplied with fitting paste on the threads.*

74 Refit the sensor and tighten it to the correct torque.

75 Reconnect the wiring plug, refit the engine undershield, and lower the vehicle to the ground.

Electronic control module (ECM)

Caution: Always wait at least 30 seconds after switching off the ignition before disconnecting the wiring from the ECM. When the wiring is disconnected, all the learned values may be erased, however any contents of the fault memory are retained. After reconnecting the wiring, the vehicle must be driven for several miles so that the ECM can learn its basic settings. If the engine still runs erratically, the basic settings may be reinstated by an Ford dealer or specialist using special diagnostic equipment. Note also that if the ECM is renewed, the identification of the new ECM must be transferred to the immobiliser control unit by a Ford dealer or specialist. If a new ECM is fitted, or on automatic transmission models even if the original ECM is refitted, the 'basic set-up' procedure must be carried out using FDS2000 diagnostic equipment, which should be entrusted to a Ford dealer or suitably equipped specialist.

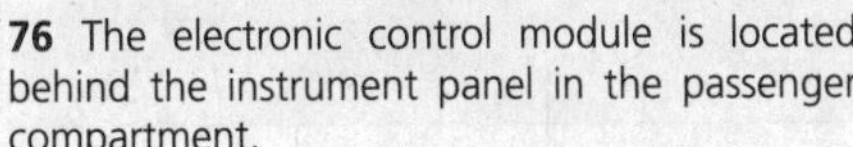

76 The electronic control module is located behind the instrument panel in the passenger compartment.

77 Disconnect the battery negative (earth) lead.

78 Release the steering column adjustment handle, and position the steering wheel as far away from the instrument panel as possible.

79 Undo the two retaining screws and pull the trim surround from the facia. Note the retaining clip at the lower edge of the trim.

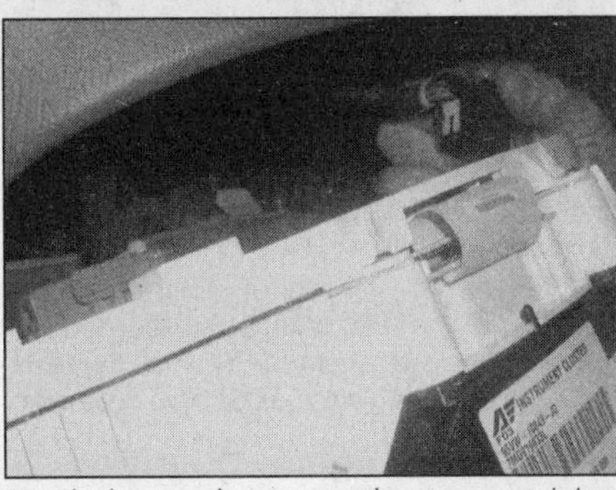

80 Slacken and remove the two retaining screws, tilt the top of the instrument panel towards the steering wheel, and manoeuvre it through the facia aperture. Disconnect the wiring plug(s) as the unit is withdrawn.

Vehicles up to 1998 model year

81 Remove the driver's side lower facia trim panel.

82 Where fitted, unscrew the bulb failure module retaining nut, disconnect the wiring plugs, and remove the module.

83 Slacken the ECM retaining band bolt. Unscrew the retaining bolt and disconnect the ECM wiring plug. Slide the ECM from the retaining band, and manoeuvre it through the instrument panel aperture *(see paragraph 86)*.

84 Refitting is a reversal of removal. If a new ECM has been fitted, or on automatic models, it will now be necessary to carry out the 'basic set-up' procedure using FDS2000 diagnostic equipment.

Vehicles from 1998 model year

85 Working through the instrument panel aperture, undo the two ECM plug bracket screws.

86 Unscrew the retaining bolt and disconnect the ECM wiring plug.

87 Slacken the ECM retaining bracket bolt, and remove the ECM.

88 Refitting is a reversal of removal. If a new ECM has been fitted, or on automatic models, it will now be necessary to carry out the 'basic set-up' procedure using FDS2000 diagnostic equipment.

Idle air control valve

89 Disconnect the battery negative lead.

90 Release the retaining clips, carefully pull the wiring multiplug rail from the top of the injectors, and position it to one side.

91 Disconnect the wiring plug from the idle air control valve.

92 Unscrew the two retaining bolts and remove the valve.

93 Refitting is a reversal of removal. Tighten the valve retaining bolts to the specified torque.

Diesel models engine management systems – component removal and refitting

Coolant temperature sensor

1 The coolant temperature sensor is located on the front of the cylinder head. Prise out the cover caps, undo the retaining nuts and remove the engine cover.

2 Drain approximately one quarter of the coolant from the engine.

3 Disconnect the wiring, pull out the retaining clip, and remove the sensor. The sensor can be tested using a multimeter. Connect the multimeter leads to the terminals of the sensor and set the meter to measure resistance (Ohms). As the temperature increases, the resistance of the sensor decreases. So at 30°C the resistance should be 1500 to 2000 Ohms, and at 80°C the resistance should be 275 to 375 Ohms. If the resistance value of the sensor does not match these values, or fails to change, it must be renewed.

4 Refitting is a reversal of removal. Top-up the cooling system.

Crankshaft position sensor

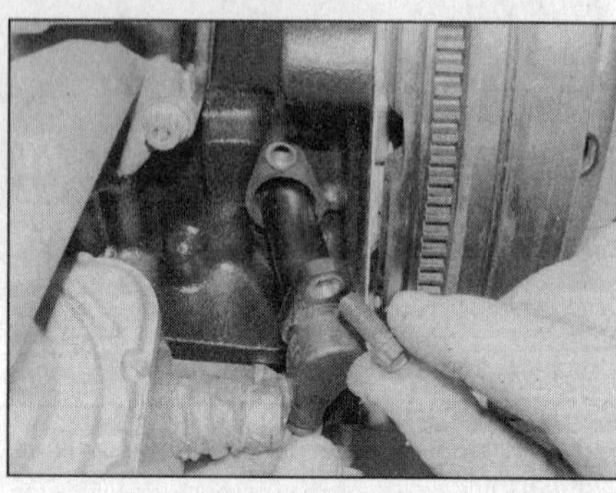

5 The crankshaft position sensor is mounted on the front left-hand side of the cylinder block, adjacent to the mating surface of the block and transmission bellhousing. Prise out the cover caps, undo the retaining nuts and remove the engine cover.

6 Trace the wiring back from the sensor to the connector and disconnect it.

7 Undo the retaining screw and withdraw the sensor from the cylinder block. The sensor can be tested using a multimeter. Connect the multimeter leads to terminals 1 (adjacent to the square side of the plug) and 2 (centre terminal) of the sensor plug, and set the meter to measure resistance (Ohms). If the sensor is usable, the resistance should be 1000 to 1500 ohms. If the sensor resistance is outside this range, renew the sensor.

8 Refitting is a reversal of removal. Tighten the sensor retaining bolt to the specified torque.

Fuel cut-off solenoid

9 The fuel cut-off solenoid is located on top of the distributor head on the fuel injection pump. Prise out the cover caps, undo the retaining nuts and remove the engine cover. Clean the area around the solenoid to prevent dust and dirt entering the fuel system.

10 To remove the solenoid, unscrew the nut and disconnect the wiring.

11 Unscrew the solenoid and remove the O-ring, spring and plunger.

12 Refitting is a reversal of removal, but clean all components before fitting and tighten the solenoid to the specified torque.

Start of injection valve

13 The start of injection valve is located just below the distributor head on the fuel injection pump. Prise out the cover caps, undo the retaining nuts and remove the engine cover. Clean the area around the valve to prevent dust and dirt entering the fuel system.

14 Undo the screw and withdraw the valve from the injection pump. Be prepared for some loss of fuel.

15 Recover the outer O-ring, strainer and inner O-ring.

16 Disconnect the wiring at the connector.

17 Refitting is a reversal of removal, but clean all components before fitting and tighten the screw securely.

Charge pressure control valve

18 The charge pressure control valve is located on the bulkhead in the left-hand rear corner of the engine compartment.

19 Disconnect the wiring from the valve.

20 Remove the vacuum hoses, noting their order of connection carefully to aid correct refitting.

21 Unscrew the mounting nuts and withdraw the valve.

22 Refitting is a reversal of removal.

Air intake charge pressure/temperature sensor

Note: *The charge pressure sensor pipe take-off port is incorporated into the temperature sensor assembly. However, the charge pressure sensor is incorporated into the engine management ECM, and is not available separately.*

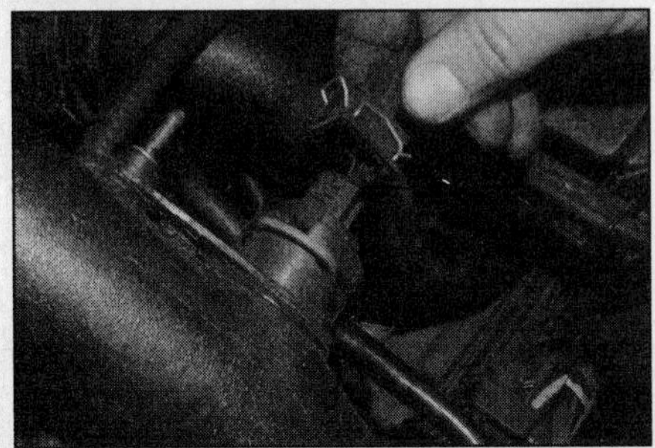

23 The air intake charge pressure port/temperature sensor is located on the air duct leading from the intercooler to the intake manifold, on the left-hand side of the engine compartment. First disconnect the wiring and the pressure hose.

24 Undo the screws and remove the sensor from the air duct.

25 Refitting is a reversal of removal.

Mass airflow sensor

26 The mass airflow sensor is located in the ducting from the air cleaner.

27 Disconnect the wiring plug from the sensor.

28 Loosen the clip and disconnect the air ducting from the airflow sensor.

29 Undo the two bolts and remove the sensor. Recover the O-ring seal.

30 Refitting is a reversal of removal. The O-ring seal can be reused, providing it is in good condition. Tighten the retaining bolts securely.

Electronic control module (ECM)

31 The ECM is located behind the instrument panel.

Caution: Always wait at least 30 seconds after switching off the ignition before disconnecting the wiring from the ECM. When the wiring is disconnected, all the learned values may be erased, however any contents of the fault memory are retained. After reconnecting the wiring, the values will be re-learned after approximately 10 to 25 miles. During this period, engine performance may be erratic. as Note also that if the ECM is renewed, the identification of the new ECM must be transferred to the immobiliser control unit by an Ford dealer or specialist.

32 Removal of the ECM is described in the petrol injection system Section.

33 Refitting is a reversal of removal. Reconnect the battery.

Fuel temperature sensor

34 The fuel temperature sender is located in the top of the injection pùmp, below the cover.

35 Undo the screws and lift the top cover from the injection pump. Recover the gasket.

36 Undo the screws and remove the fuel temperature sensor.

37 Refitting is a reversal of removal.

Altitude sensor

38 The altitude sensor is built into the engine management ECM. It is not available as a separate part. If the sensor is defective, the ECM must be replaced.

Notes

Ford Ka '96 – '02

14

General information

Fuel system – depressurisation

Petrol fuel injection system components – removal and refitting

General

System type	Sequential Electronic Fuel injection (SEFi)
Application	1.3 litre JJB, J4D, JJF, J4M, JJD, and J4K Endura-E engines

Fuel grade

Fuel octane requirement	95 RON unleaded

Fuel system data

Regulated fuel pressure:	
Pressure regulator vacuum hose disconnected	2.7 ± 0.2 bar
With engine running and pressure regulator vacuum hose connected	2.1 ± 0.2 bar
Hold pressure – engine stopped after five minutes	1.8 bars minimum

Torque wrench settings

Torque wrench settings	Nm	lbf ft
Camshaft position sensor	10	7
Crankshaft position sensor	7	5
Engine coolant temperature sensor	12	9
Fuel pressure regulator screws	10	7
Fuel rail-to-inlet manifold bolts	18	13
Idle air control valve	10	7
Oxygen sensor	42	31
Shrader valve retaining bolts	10	7
Throttle body housing	10	7
TMAP sensor	4	3

Chapter 14

General information

The fuel system consists of a steel fuel tank (mounted under the body, beneath the rear seats), fuel hoses, an electric fuel pump mounted in the fuel tank, and a sequential electronic fuel injection system controlled by an EEC V engine management control module (Powertrain Control Module).

The electric fuel pump supplies fuel under pressure to the plastic moulded fuel rail, which distributes fuel to the injectors. A pressure regulator controls the system pressure in relation to inlet tract depression. From the fuel rail, fuel is injected into the inlet ports, just above the inlet valves, by four fuel injectors.

The amount of fuel supplied by the injectors is precisely controlled by the Powertrain Control Module (PCM). The module uses the signals from the crankshaft position sensor and the camshaft position sensor, to trigger each injector separately in cylinder firing order (sequential injection), with benefits in terms of better fuel economy and leaner exhaust emissions.

The Powertrain Control Module is the heart of the entire engine management system, controlling the fuel injection, ignition and emissions control systems. The module receives information from various sensors which is then computed and compared with pre-set values stored in its memory, to determine the required period of injection.

Information on crankshaft position and engine speed is generated by a crankshaft position sensor. The inductive head of the sensor runs just above the engine flywheel and scans a series of 36 protrusions on the flywheel periphery. As the crankshaft rotates, the sensor transmits a pulse to the system's ignition module every time a protrusion passes it. There is one missing protrusion in the flywheel periphery at a point corresponding to 90° BTDC. The ignition module recognises the absence of a pulse from the crankshaft position sensor at this point to establish a reference mark for crankshaft position. Similarly, the time interval between absent pulses is used to determine engine speed. This information is then fed to the Powertrain Control Module for further processing.

The camshaft position sensor is located on the timing cover and registers with a plate located beneath the camshaft sprocket retaining bolts. The camshaft position sensor functions in the same way as the crankshaft position sensor, producing a series of pulses; this gives the Powertrain Control Module a reference point, to enable it to determine the firing order, and operate the injectors in the appropriate sequence.

On models manufactured up to mid-1999 model year, the mass air flow sensor is based on a 'hot-wire' system, sending the Powertrain Control Module a constantly-varying (analogue) voltage signal corresponding to the mass of air passing into the engine. Since air mass varies with temperature (cold air being denser than warm), measuring air mass provides the module with a very accurate means of determining the correct amount of fuel required to achieve the ideal air/fuel mixture ratio. Together with the mass air flow sensor, pre-1999 models are fitted with an inlet air temperature sensor which provides the Powertrain Control Module with a signal corresponding to the temperature of air passing into the engine. This is used to refine the calculations made by the module, when determining the correct amount of fuel required to achieve the ideal air/fuel mixture ratio.

Models manufactured from mid-1999 model year onwards are fitted with a temperature manifold absolute pressure (TMAP) sensor which replaces the mass air flow and inlet air temperature sensors fitted to earlier models. The TMAP sensor consists of a pressure transducer and a temperature sensor fitted directly into the inlet manifold, and the system provides the Power Control Module with information on the inlet manifold vacuum and inlet air temperature. When the ignition is switched on with the engine stationary, the system provides the PCM with barometric pressure information.

Engine temperature information is supplied by the coolant temperature sensor located in the inlet manifold. This component is an NTC (Negative Temperature Coefficient) thermistor – that is, a semi-conductor whose electrical resistance decreases as its temperature increases. It provides the Powertrain Control Module with a constantly-varying (analogue) voltage signal, corresponding to the temperature of the engine coolant. This is used to refine the calculations made by the module, when determining the correct amount of fuel required to achieve the ideal air/fuel mixture ratio.

A throttle position sensor is mounted on the end of the throttle valve spindle, to provide the Powertrain Control Module with a constantly-varying (analogue) voltage signal corresponding to the throttle opening. This allows the module to register the driver's input when determining the amount of fuel required by the engine.

Road speed is monitored by the vehicle speed sensor. This component is a Hall-effect generator, mounted on the transmission's speedometer drive. It supplies the module with a series of pulses corresponding to the vehicle's road speed, enabling the module to control features such as the fuel shut-off on overrun.

The clutch pedal position is monitored by a switch fitted to the pedal bracket. This sends a signal to the Powertrain Control Module.

Where power steering is fitted, a pressure-operated switch is screwed into the power steering system's high-pressure pipe. The switch sends a signal to the Powertrain Control Module to increase engine speed to maintain idle speed during power steering assistance.

The oxygen sensor (located on the exhaust manifold) provides the module with constant feedback – 'closed-loop' control – which enables it to adjust the mixture to provide the best possible operating conditions for the catalytic converter.

The air inlet side of the system consists of an air cleaner housing, the mass air flow sensor (pre 1999 models only), an inlet hose and duct, and a throttle housing.

The throttle valve inside the throttle housing is controlled by the driver, through the accelerator pedal. As the valve opens, the amount of air that can pass through the system increases. As the throttle valve opens further, the mass air flow sensor signal alters, and the Powertrain Control Module opens each injector for a longer duration, to increase the amount of fuel delivered to the inlet ports.

Both the idle speed and mixture are under the control of the Powertrain Control Module, and cannot be adjusted.

Fuel system – depressurisation

1 The fuel system described in this Chapter includes the fuel tank and tank-mounted fuel pump/fuel gauge sender unit, the fuel filter, the fuel injectors and the pressure regulator in the injector rail, and the metal pipes and flexible hoses of the fuel lines between these components. All these contain fuel, which will be under pressure while the engine is running and/or while the ignition is switched on.

2 The pressure will remain for some time after the ignition has been switched off, and must be relieved before any of these components is disturbed for servicing work.

3 On pre-1999 models, the Ford method of depressurisation is to use service tool 23-033 fitted to the fuel rail pressure test/release fitting. The fitting consists of a Schrader-type valve with a plastic cap located on the fuel rail, and the tool acts as a tap by depressing the valve core. Access to the valve is gained by removing the air inlet duct from between the air cleaner and throttle housing.

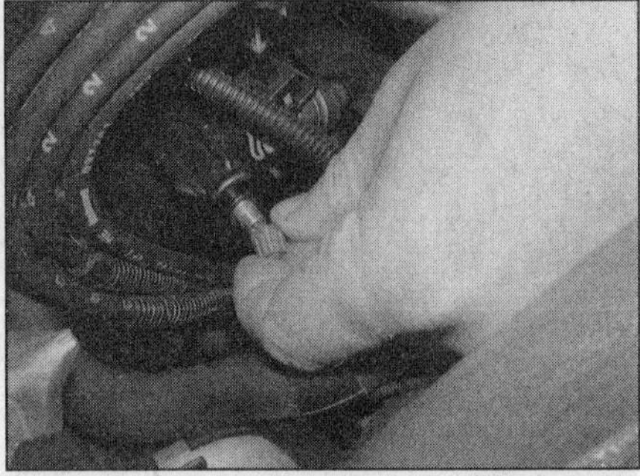

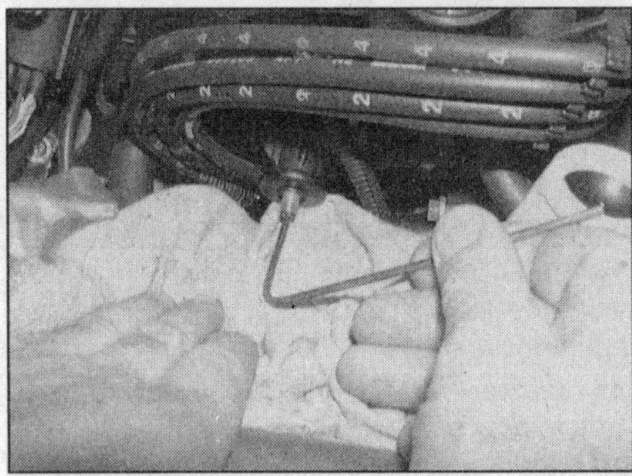

4 To release the pressure without using the service tool, use a suitable container and rag wrapped around the fitting to catch the fuel then use a piece of rod to depress the valve. **Do not** simply depress the valve core without wrapping rag around it, as fuel will spray out, with the consequent risk of fire, and personal injury.

5 The alternative method, and the only method to use on 1999-on models, is simply to disconnect the fuel pump's electrical supply while the engine is running, by removing the fuel pump fuse (number 19), and to allow the engine to idle until it dies through lack of fuel. Turn the engine over once or twice on the starter to ensure that all pressure is released, then switch off the ignition. Do not forget to refit the fuse when work is complete.

6 Note that, once the fuel system has been depressurised, it may take a little longer to restart the engine – perhaps several seconds of cranking – before the system is refilled and pressure restored.

Petrol fuel injection system components – removal and refitting

Throttle body housing

1 The housing is located on the left-hand side of the inlet manifold.

2 Loosen the clips and disconnect the air inlet duct from between the air cleaner (or airflow meter) and throttle housing.

3 Disconnect the accelerator inner cable from the quadrant on the throttle housing by pivoting the quadrant then extracting the retaining clip.

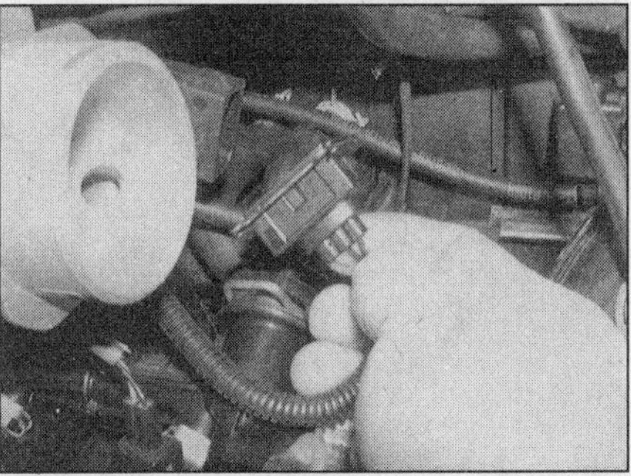

4 Disconnect the throttle position sensor multi-plug.

5 Unscrew and remove the mounting bolts and withdraw the throttle housing from the inlet manifold. Discard the gasket and obtain a new one.

6 Refitting is a reversal of removal, but clean the mating faces and fit a new gasket, and tighten the mounting bolts to the specified torque. Check and if necessary adjust the accelerator cable.

Fuel rail and injectors

7 Relieve the residual pressure in the fuel system, and equalise the tank pressure by removing the fuel filler cap.

8 Disconnect the wiring from the idle air control valve. Unscrew the two bolts and remove the idle air control valve from the inlet manifold.

9 Pull out the accelerator cable adjustment clip, then release the outer cable from the support bracket and disconnect the inner cable from the quadrant on the throttle housing. Position the accelerator cable to one side.

10 Withdraw the air inlet resonator (where fitted) from the inlet manifold, then loosen the clips and disconnect the air inlet duct from the air cleaner and throttle housing. Remove the duct from the engine compartment.

11 Disconnect the crankcase ventilation hoses from the inlet manifold.

12 Unscrew the bolt and remove the accelerator cable support bracket from the inlet manifold.

13 Disconnect the HT leads from the spark plugs, then unclip the lead supports and position the leads to one side. When disconnecting the leads, Twist the plug caps slightly to break the seal, then pull on the end fittings and not the leads.

14 Disconnect the wiring from the throttle position sensor. To do this, depress the locking wire and pull off the plug.

15 Disconnect the wiring from the fuel injectors, then undo the injector wiring loom retaining screws, disconnect the loom multiplug on the bulkhead, and position the loom to one side.

16 Pull the vacuum hose from the fuel pressure regulator on the left-hand end of the fuel rail.

17 Unscrew and remove the fuel rail mounting bolts.

18 On pre-mid 1998 models, undo the single screw and remove the fuel supply and return pipe retaining bracket.

19 Disconnect the fuel supply and return pipes from the fuel rail by squeezing the lugs on the special quick release fittings. Note that the supply pipe is coloured white and the return pipe is coloured red.

20 Carefully pull the injectors from the inlet manifold and withdraw them together with the fuel rail.

21 Remove the clips and carefully pull the injectors from the fuel rail.

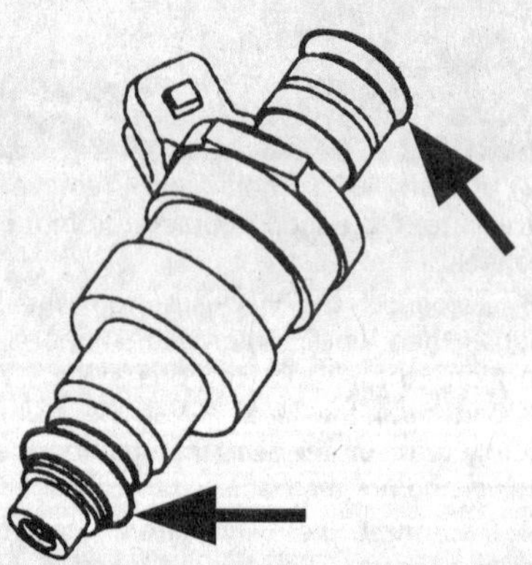

22 Using a screwdriver prise the O-rings from the grooves at each end of the injectors. Discard the O-rings and obtain new ones.

23 Refitting is a reversal of removal, but note the following points:

a) Lubricate the new injector O-rings with clean engine oil to aid refitting.

b) Tighten the mounting bolts to the specified torque.

c) Ensure that the hoses and wiring are routed correctly, and secured on reconnection by any clips or ties provided.

d) Adjust the accelerator cable.

e) On completion, switch the ignition on to activate the fuel pump and pressurise the system, without cranking the engine. Check for signs of fuel leaks around all disturbed unions and joints before attempting to start the engine.

Fuel pressure regulator

24 Relieve the residual pressure in the fuel system, and equalise tank pressure by removing the fuel filler cap.

25 Loosen the clips and disconnect the air inlet duct from the air cleaner and throttle housing.

26 Disconnect the vacuum hose from the fuel pressure regulator.

27 On pre-mid 1998 models, undo the single screw and remove the fuel supply and return pipe retaining bracket.

28 Disconnect the fuel return pipe from the fuel rail by squeezing the lugs on the special quick release fittings, then undo the two screws and remove the fuel pressure regulator from the fuel rail.

29 On mid 1998-on models, prise out the retaining clip from the fuel pressure regulator, then pull the regulator from the fuel rail.

30 Using a screwdriver, prise out the O-ring(s) from the groove(s) in the fuel pressure regulator. Discard the O-ring(s).

31 Refitting is a reversal of removal, but lubricate the new O-ring(s) with clean engine oil to aid installation. On pre-mid 1998 models, tighten the mounting screws to the specified torque.

Idle air control (IAC) valve

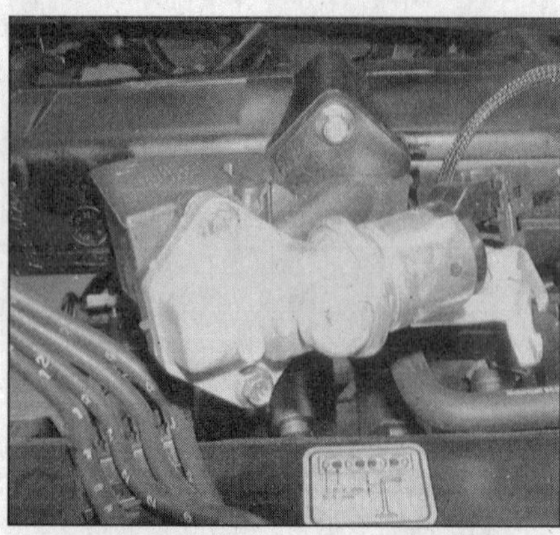

32 The idle air control valve is located on the inlet manifold.

33 Depress the wire clip and disconnect the wiring from the valve.

34 Unscrew the mounting bolts and remove the valve from the inlet manifold. If necessary, also remove the air inlet resonator.

35 Recover the O-ring seals and discard them. Obtain new seals.

36 Refitting is a reversal of removal, but note the following points.

a) Clean the mating surfaces, and fit new O-ring seals.

b) On completion, start the engine and allow it to idle. When it has reached normal operating temperature, check that the idle speed is stable, and that no induction (air) leaks are evident. Switch on all electrical loads (headlights, heated rear window, etc), and check that the idle speed is still satisfactory.

Mass air flow sensor

Pre 1999 models

37 Loosen the clip and disconnect the air inlet duct from the mass air flow sensor on the air cleaner cover. If necessary, remove the air cleaner cover for improved access.

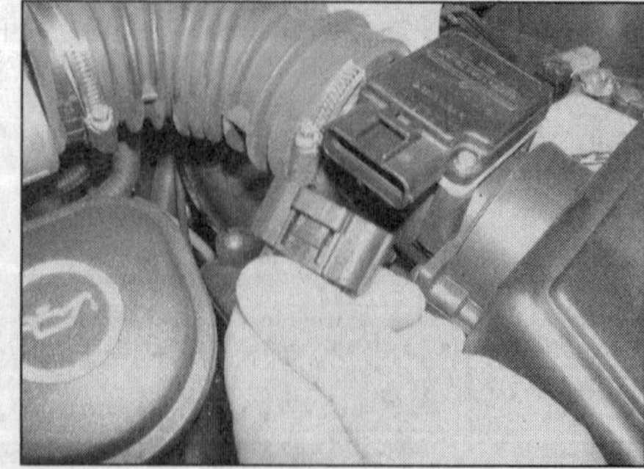

38 Disconnect the wiring from the sensor.

39 Unscrew the crosshead mounting screws and remove the sensor from the air cleaner cover.

40 Refitting is a reversal of removal.

Powertrain Control Module

Note: *The module is fragile. Take care not to drop it, or subject it to any other kind of impact. Do not subject it to extremes of temperature, or allow it to get wet.*

41 The Module (engine management module) is located behind the front left-hand footwell side trim. First disconnect the battery negative (earth) lead.

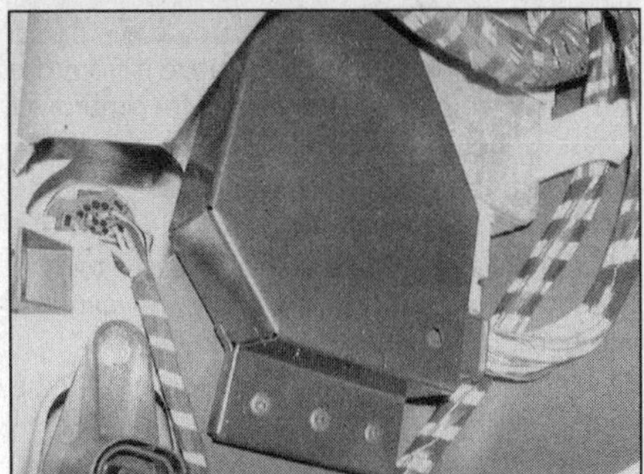

42 Carefully pull back the footwell floor covering for access to the powertrain control module.

43 Drill out the three rivets from the security shield, then unhook the shield and withdraw it downwards from over the module.

44 Undo the bolt and disconnect the wiring multiplug from the module. **Do not** pull on the wiring, only on the multiplug itself.

Caution: Do not touch the module terminals, as there is the possibility of static electricity damaging the internal components.

45 Carefully ease the module forwards then downwards from the mounting bracket. Withdraw the module from inside the vehicle.

46 Refitting is a reversal of removal. Take care when refitting the multiplug, and tighten the retaining bolt by hand first. Use new pop rivets when refitting the security shield.

Crankshaft position sensor

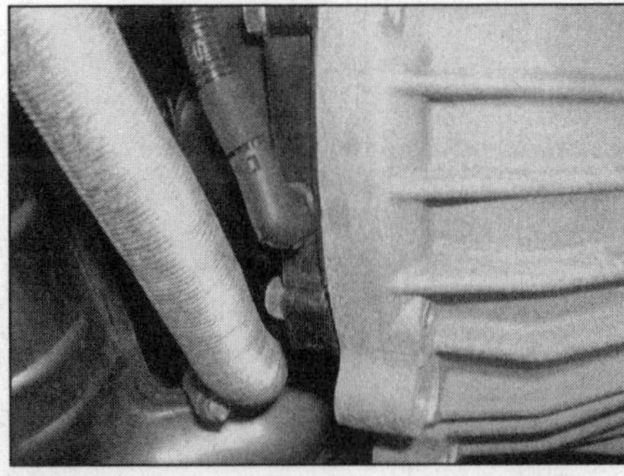

47 The crankshaft position sensor is located on the front left-hand side of the engine cylinder block. For improved access, apply the handbrake then jack up the front of the vehicle and support it on axle stands.

48 Trace the wiring back from the sensor, and disconnect the connector plug.

49 Unscrew the mounting bolts and withdraw the sensor.

50 Refitting is a reversal of removal.

Camshaft position sensor

51 The sensor is located on the rear face of the timing chain cover.

52 Depress the wire clip and disconnect the wiring from the camshaft position sensor.

53 Unscrew the mounting bolt and withdraw the sensor from the timing chain cover.

54 Remove the O-ring seal. Tape over the hole in the timing chain cover while the sensor is removed.

55 Refitting is a reversal of removal but fit a new O-ring. Smear a little engine oil on the seal before fitting the sensor.

Engine coolant temperature (ECT) sensor

56 The engine coolant temperature sensor is located on the lower right-hand end of the inlet manifold.

57 With the engine cold, unscrew the filler cap from the coolant expansion reservoir to release any residual pressure, then refit and tighten the cap.

58 Disconnect the wiring from the sensor.

59 Have ready a suitable plug (or a new sensor). Unscrew the sensor from the bottom of the inlet manifold, and quickly fit the plug.

60 Refitting is a reversal of removal. If necessary, clean the threads of the sensor and mounting hole, then refit the sensor and tighten it to the specified torque.

Inlet air temperature sensor

Pre 1999 models

61 The sensor is located on the left-hand side of the air cleaner cover.

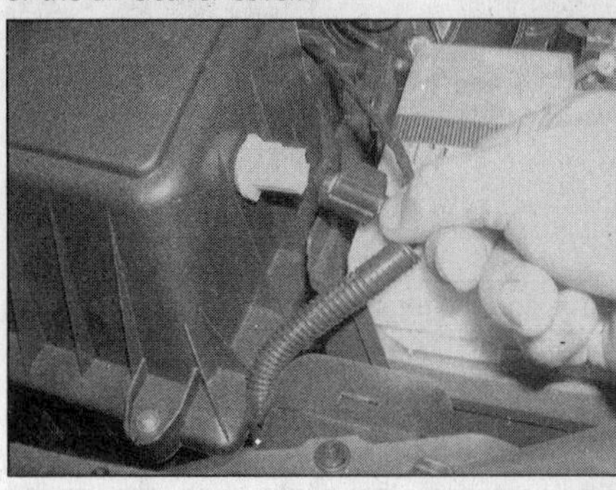

62 Disconnect the wiring from the sensor, then twist the sensor through 90° and remove it.

63 Refitting is a reversal of removal.

Throttle position sensor

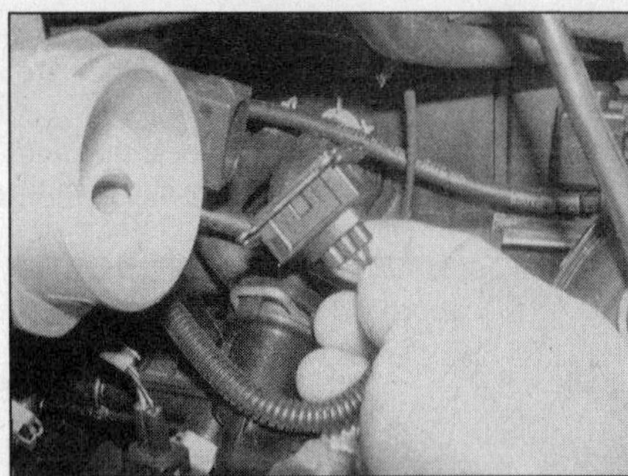

64 The throttle position sensor is located on the rear of the throttle body housing on the left-hand side of the inlet manifold. First disconnect the wiring by depressing the retaining clip.

65 Remove the retaining screws, and withdraw the unit from the throttle housing. ***Do not*** force the sensor's centre to rotate past its normal operating sweep; the unit will be seriously damaged.

66 Refitting is a reversal of removal, but ensure that the sensor is correctly orientated, by locating its centre on the D-shaped throttle shaft (throttle closed), and aligning the sensor body so that the bolts pass easily into the throttle housing.

Vehicle speed sensor

67 The sensor is mounted on the rear of the transmission at the base of the speedometer drive cable.

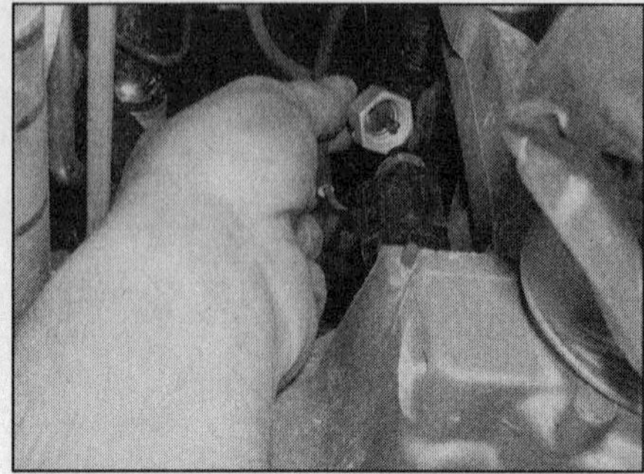

68 Undo the retaining collar and withdraw the speedometer cable from the vehicle speed sensor. Use two spanners to loosen the nut – one to counterhold the sensor, and the other to unscrew the cable nut.

69 Disconnect the wiring from the vehicle speed sensor, then unscrew the sensor from the top of the drive pinion.

70 Refitting is a reversal of removal.

Clutch pedal position switch

71 The switch is located at the top of the clutch pedal.

72 Inside the vehicle, reach up under the clutch pedal and disconnect the return spring from the bracket.

73 Disconnect the wiring from the clutch switch, then twist the switch and remove it from the pedal bracket.

74 Refitting is a reversal of removal.

Power steering pressure switch

75 The switch is located in the high pressure fluid pipe on the steering gear. Apply the handbrake, then jack up the front of the vehicle and support it on axle stands.

76 Place a container beneath the switch location to catch any escaping fluid, then disconnect the wiring. Unscrew and remove the switch from the fluid pipe. Be prepared for fluid spillage, and plug or cover the orifice in the pipe to prevent dirt entry and further fluid loss.

77 Refitting is a reversal of removal, but tighten the switch securely, and on completion bleed the power steering hydraulic circuit.

Oxygen sensor

78 Trace the wiring back from the oxygen sensor to the connector on the radiator cowling and disconnect the wiring.

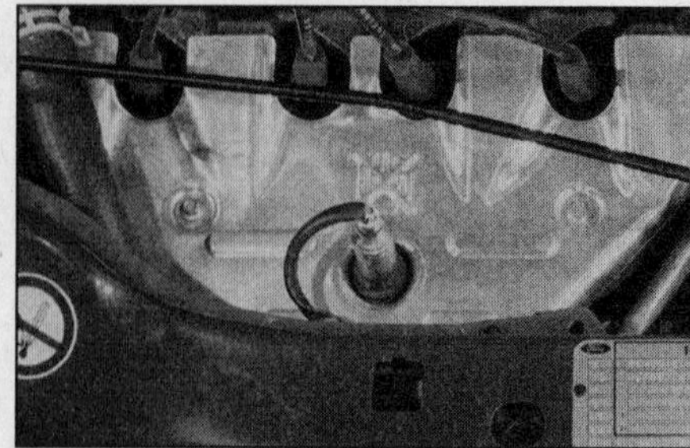

79 Unscrew the sensor and remove it from the exhaust manifold. If necessary for improved access, unbolt the heatshield first.

80 Clean the threads of the sensor and the threads in the exhaust manifold.

81 Apply a little anti-seize grease to the threads, then insert the sensor into the manifold and tighten it to the specified torque.

82 Refit the heatshield, and reconnect the wiring ensuring that it is in no danger of contacting the exhaust manifold. **Note:** *On later models (from 1999), there is a second oxygen sensor fitted after the catalytic converter at the bottom of the exhaust downpipe. The renewal procedure is same for both sensors.*

T-MAP sensor

Models from 1999

83 The sensor is located on the back of the inlet manifold.

84 Disconnect the wiring from the sensor, then undo the retaining screw and remove it from the manifold.

85 Refitting is a reversal of removal.

Ford Mondeo '93 – '00

General

System type	Sequential Electronic Fuel injection (SEFi)
Recommended fuel (Minimum octane rating)	95 RON unleaded
Idle speed:	
4-cylinder (Zetec) engine:	
Regulated – nominal (± 50 rpm):	
Manual transmission models	880 rpm*
Automatic transmission models	710 rpm*
Unregulated – base (to 7/1996 only)	1500 rpm*
V6 (Duratec) engine:	
Regulated – nominal (± 50 rpm)	725 rpm*
Idle mixture (CO level):	
4-cylinder (Zetec) engine	0.5% maximum*
V6 (Duratec) engine	0.5% maximum*

** Given for reference only – not adjustable.*

Rev limiter operation

Fuel injectors shut off at:	
Automatic transmission, position N selected	4100 rpm
Automatic transmission, any other position selected	6800 rpm (approximately)
Manual transmission	6800 to 7100 rpm

Fuel pressure

Regulated fuel pressure – engine running at idle speed:	
Pressure regulator vacuum hose connected	2.1 ± 0.2 bars
Pressure regulator vacuum hose disconnected	2.7 ± 0.2 bars

Note: *When the ignition is switched off, the system should hold 1.8 bars for 5 minutes. If the engine is hot, the pressure may rise to maximum of 2.7 bars during this check. Pressure regulator (when reconnected) should prevent any higher pressure being reached.*

Fuel injectors

Resistance	13.7 to 15.2 ohms

Idle speed control valve

Resistance	6 to 14 ohms

Idle increase solenoid valve

Resistance	50 to 120 ohms

Crankshaft speed/position sensor

Resistance	200 to 450 ohms

Camshaft position sensor

Resistance	200 to 900 ohms

Intake air temperature sensor

Resistance:	
At -40°C	860 to 900 k ohms
At 20°C	35 to 40 k ohms
At 100°C	1.9 to 2.5 k ohms
At 120°C	1.0 to 1.3 k ohms

Throttle potentiometer

Resistance – see text	400 to 6000 ohms

Power steering pressure switch

Operating pressure – green switch body:	
Contacts open – infinite resistance	31.5 ± 3.5 bars
Contacts close – 0 to 2.5 ohms resistance	Between 13.5 and 24.0 bars

Torque wrench settings

	Nm	lbf ft
Plenum chamber-to-inlet manifold fasteners	4	3
Throttle housing-to-inlet manifold screws	10	7
Idle speed control valve bolts	6	4
Fuel pressure regulator bolts:		
4-cylinder engine:		
To mid 1997	6	4
From mid 1997	10	7
V6 engine	6	4
Fuel pressure relief valve (V6)	8	6
Fuel injector bolts	6	4
Fuel rail-to-inlet manifold bolts	10	7
Fuel feed and return line threaded couplings at fuel rail	27	20
Inlet air temperature sensor	23	17
Camshaft position sensor screw	21	16
Crankshaft speed/position sensor:		
Sensor-to-bracket screw	8	6
Bracket-to-cylinder block crankcase screw	21	15

General information

1 All petrol models are equipped with a Sequential Electronically-controlled Fuel Injection (SEFi) system.

Fuel supply and air induction systems

2 An electric fuel pump located inside the fuel tank supplies fuel under pressure to the fuel rail, which distributes fuel to the injectors. A filter between the fuel pump and the fuel rail protects the components of the system. A pressure regulator controls the system pressure in relation to inlet depression. From the fuel rail, fuel is injected into the inlet ports, just above the inlet valves, by four fuel injectors.

3 The amount of fuel supplied by the injectors is precisely controlled by an Electronic Control Unit (ECU). The ECU uses the signals from the engine speed/crankshaft position sensor and the camshaft position sensor, to trigger each injector separately in cylinder firing order (sequential injection).

4 The air induction system consists of an air filter housing, an air mass meter, intake ducting, and a throttle housing. The air mass meter is an information-gathering device for the ECU; it uses a 'hot-wire' system to send the ECU a constantly-varying (analogue) voltage signal corresponding to the volume of air passing into the engine. Another sensor in the air mass meter measures intake air temperature. The ECU uses these signals to calculate the mass of air entering the engine.

5 The throttle valve inside the throttle housing is controlled by the driver, through the accelerator pedal. As the valve opens, the quantity of air entering the engine increases. The throttle potentiometer is attached to the throttle valve and informs the ECU of the throttle position. The ECU calculates the relevant period of injection, and controls the injector opening times.

Electronic control system

6 The EEC-IV and EEC-V (Electronic Engine Control) engine management systems control fuel injection by means of a microcomputer known as the ECU (Electronic Control Unit). The ECU receives signals from a number of sensors, which monitor the intake air mass and temperature, coolant temperature, engine speed and crankshaft position, acceleration/deceleration, and exhaust gas oxygen content. The signals are processed by the ECU to determine the injection duration necessary for the optimum air/fuel ratio. The sensors and associated ECU-controlled relays are located throughout the engine compartment.

7 In the event of a sensor malfunction, a back-up circuit will take over, to provide driveability until the problem is identified and fixed. The following paragraphs describe the components of the electronic control system.

ECU (Electronic Control Unit)

8 This component is the heart of the entire engine management system, controlling the fuel injection, ignition and emissions control systems. It also controls sub-systems such as the radiator cooling fan, air conditioning and automatic transmission, where appropriate.

Air mass meter

9 This uses a 'hot-wire' system, sending the ECU a constantly-varying (analogue) voltage signal corresponding to the mass of air passing into the engine. Since air mass varies with temperature (cold air being denser than warm), measuring air mass provides the ECU with a very accurate means of determining the correct amount of fuel required to achieve the ideal air/fuel mixture ratio.

Crankshaft speed/position sensor

10 This is an inductive pulse generator bolted to the cylinder block/crankcase (4-cylinder) or timing cover (V6). On 4-cylinder engines it scans ridges between 36 holes machined in the inboard (right-hand) face of the flywheel/driveplate; as each ridge passes the sensor tip, a signal is generated, which is used by the ECU to determine engine speed. On V6 engines, the sensor scans similar cut-outs machined on a timing disc located on the timing end of the crankshaft.

11 The ridge between the 35th and 36th holes (corresponding to 90° BTDC) is missing – this step in the incoming signals is used by the ECU to determine crankshaft (ie, piston) position.

Camshaft position sensor

12 On 4-cylinder models the camshaft position sensor is bolted to the rear left-hand end of the cylinder head, and is triggered by a high-point on the inlet camshaft. On V6 models, the camshaft position sensor is mounted on the right-hand side of the front cylinder head and is triggered by a high-point on the front inlet camshaft. The sensor functions in the same way as the crankshaft speed/position sensor, producing a series of pulses (corresponding to No 1 cylinder at 46° ATDC). This gives the ECU a reference point, to enable it to determine the firing order, and operate the injectors in the appropriate sequence.

Coolant temperature sensor

13 This component, which is screwed into the top of the thermostat housing on 4-cylinder models or into the coolant crossover on V6 models, is an NTC (Negative Temperature Coefficient) thermistor – that is, a semi-conductor whose electrical resistance decreases as its temperature increases. It provides the ECU with a constantly-varying (analogue) voltage signal, corresponding to the temperature of the engine coolant. This is used to refine the calculations made by the ECU, when determining the correct amount of fuel required to achieve the ideal air/fuel mixture ratio.

Intake air temperature sensor

14 On 4-cylinder models manufactured up to 1997, this component is screwed into the underside of the air intake resonator. On 1998-on models it is located in the air inlet duct. On V6 models it is located in the air cleaner cover. The sensor is an NTC thermistor – see the previous paragraph – providing the ECU with a signal corresponding to the temperature of air passing into the engine. This is used to refine the calculations made by the ECU, when determining the correct amount of fuel required to achieve the ideal air/fuel mixture ratio.

Throttle potentiometer

15 This is mounted on the end of the throttle valve spindle, to provide the ECU with a constantly-varying (analogue) voltage signal corresponding to the throttle opening. This allows the ECU to register the driver's input when determining the amount of fuel required by the engine.

Vehicle speed sensor

16 This component is a Hall-effect generator, mounted on the transmission's speedometer drive. It supplies the ECU with a series of pulses corresponding to the vehicle's road speed, enabling the ECU to control features such as the fuel shut-off on the overrun, and to provide information for the trip computer, adaptive damping and cruise control systems (where fitted).

Power steering pressure switch

17 This is a pressure-operated switch, screwed into the power steering system's high-pressure pipe. Its contacts are normally closed, opening when the system reaches the specified pressure – on receiving this signal, the ECU increases the idle speed, to compensate for the additional load on the engine.

Air conditioning system

18 Two pressure-operated switches and the compressor clutch solenoid are connected to the ECU, to enable it to determine how the system is operating. The ECU can increase idle speed or switch off the system, as necessary, so that normal vehicle operation and driveability are not impaired.

Idle speed control valve

19 The idle speed control valve maintains a stable idle speed by varying the quantity of air entering the engine through an auxiliary air passage. The valve is activated by a signal from the ECU.

Automatic transmission sensors

20 In addition to the driver's controls, the transmission has a speed sensor, a fluid temperature sensor (built into the solenoid valve unit), and a selector lever position sensor. All of these are connected to the ECU, to enable it to control the transmission through the solenoid valve unit.

Exhaust gas oxygen sensor

21 The oxygen sensor in the exhaust system provides the ECU with constant feedback – 'closed-loop' control – which enables it to adjust the mixture to provide the best possible conditions for the catalytic converter to operate.

Heated front windscreen solenoid

22 When the heated front windscreen is switched on, the solenoid admits additional air to the inlet manifold in order to counteract the extra load on the alternator.

Fuel system – depressurisation

1 The fuel system referred to in this Chapter is defined as the fuel tank and tank-mounted fuel pump/fuel gauge sender unit, the fuel filter, the fuel injectors and the pressure regulator in the injector rail, and the metal pipes and flexible hoses of the fuel lines between these components. All these contain fuel, which will be under pressure while the engine is running and/or while the ignition is switched on.

2 The pressure will remain for some time after the ignition has been switched off, and must be relieved before any of these components is disturbed for servicing work.

3 The Ford method of depressurisation is to use service tool 23-033 fitted to the fuel rail pressure test/release fitting. The fitting consists of a Schrader-type valve with a plastic cap located on the fuel rail, and the tool acts as a tap by depressing the valve core. On 1997-on models it will be necessary to remove the air inlet duct from between the air cleaner and airflow meter.

4 To release the pressure without using the service tool, use a suitable container and rag wrapped around the fitting to catch the fuel. Do not simply depress the valve core without wrapping rag around it, as fuel will spray out, with the consequent risk of fire, and personal injury.

5 An alternative method is simply to disconnect the fuel pump's electrical supply while the engine is running, by removing the fuel pump fuse (number 14), and to allow the engine to idle until it dies through lack of fuel. Turn the engine over once or twice on the starter to ensure that all pressure is released, then switch off the ignition. Do not forget to refit the fuse when work is complete.

6 Note that, once the fuel system has been depressurised, it may take longer to restart the engine – perhaps several seconds of cranking – before the system is refilled and pressure restored.

Petrol fuel injection system components – removal and refitting

ECU (Electronic Control Unit)

Caution: The ECU is fragile. Take care not to drop it or subject it to any other kind of impact, and do not subject it to extremes of temperature, or allow it to get wet. Do not touch the ECU terminals as there is a chance that static electricity may damage the internal electronic components.

Note: *If renewing an EEC-V fuel injection system ECU, note that it must be re-programmed for the specific model by a Ford dealer using the FDS 2000 diagnostic equipment. Failure to do so will result in the ECU assuming its limited operating strategy (LOS) settings giving poor performance and economy.*

1 Disconnect the battery negative (earth) lead.

2 Carefully lift the power steering fluid reservoir upwards out of its clip on the suspension mounting. The ECU connector is secured with a bolt, however on later models a cover is riveted over the ECU and it will be necessary to drill out the rivets to remove the cover.

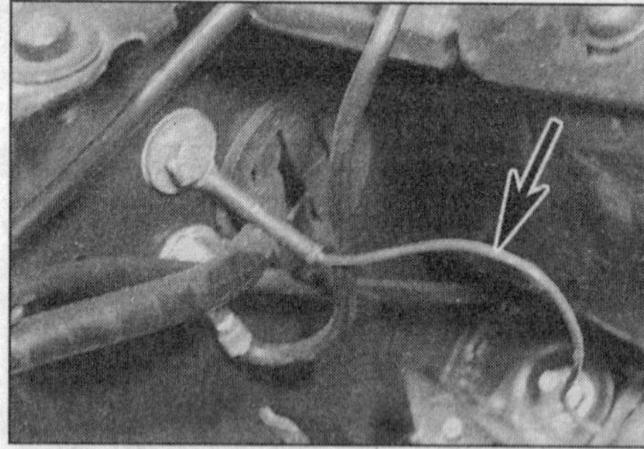

3 on later models, unbolt the earth cable from the inner wing panel. Unscrew the retaining bolt, and unplug the connector – to prevent dust and

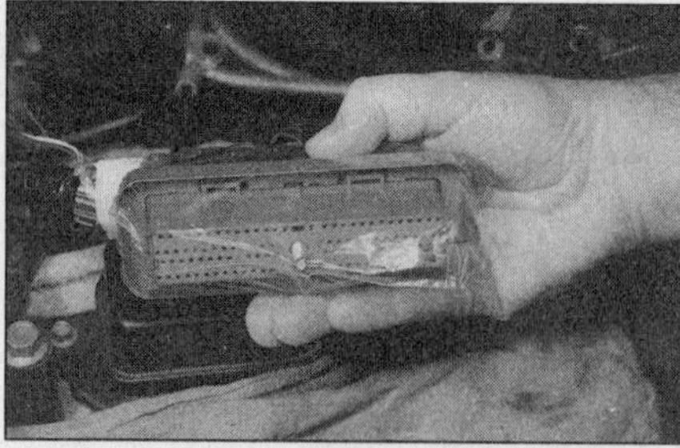

dirt entering the multiplug, cover it with a polythene bag.

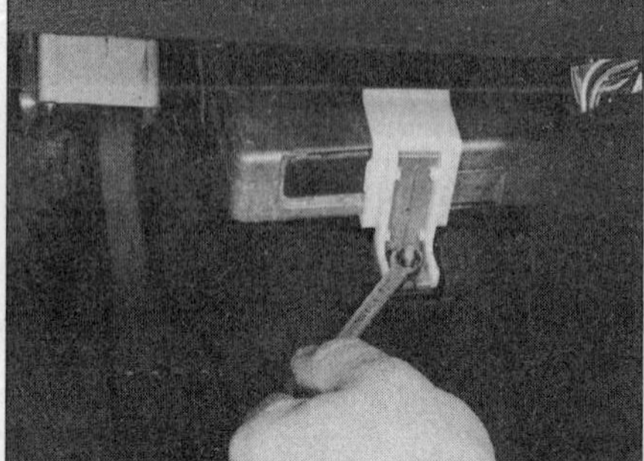

4 Working inside the car, unscrew the retaining bolt and withdraw the mounting bracket. Lift the ECU to release it from the bulkhead carrier bracket, then withdraw the unit.

5 Refitting is the reverse of the removal procedure. Whenever the ECU (or battery) is disconnected, the information relating to idle speed control and other operating values will be lost from its memory until the unit has re-programmed itself; until then, there may be surging, hesitation, erratic idle and a generally-inferior level of performance. To allow the ECU to re-learn these values, start the engine and run it as close to idle speed as possible until it reaches its normal operating temperature, then run it for approximately two minutes at 1200 rpm. Next, drive the vehicle as far as necessary – approximately 5 miles of varied driving conditions is usually sufficient – to complete the re-learning process. Refer to the note at the beginning of this Section for models fitted with the EEC-V fuel injection system.

Air mass meter

6 Disconnect the battery negative (earth) lead.

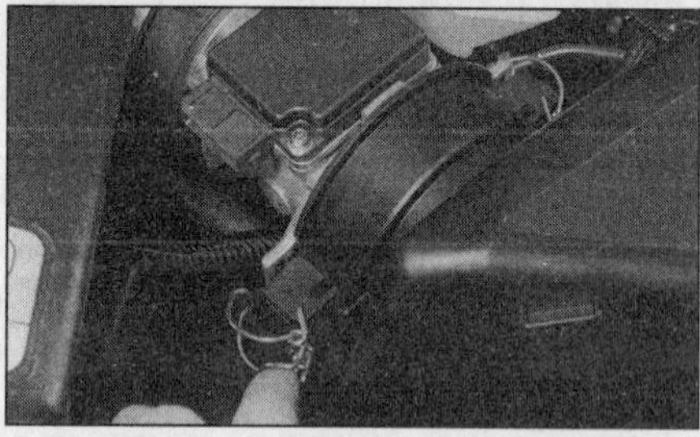

7 On 4-cylinder (Zetec) engines, release the wire clip, and unplug the meter's electrical connector. Release the clips and lift the air cleaner cover, then release the two smaller clips and detach the meter from the cover. Slacken the clamp securing the meter to the resonator hose, and withdraw the meter.

8 On V6 (Duretec) engines, remove the air cleaner assembly. Unscrew and remove the meter mounting bolts and withdraw the meter from the air cleaner cover.

9 Refitting is the reverse of the removal procedure. Ensure that the meter and air cleaner cover are seated correctly and securely fastened, so that there are no air leaks.

Crankshaft speed/position sensor

10 Disconnect the battery negative (earth) lead.

11 Apply the handbrake, then jack up the front of the vehicle and support it on axle stands.

12 On 4-cylinder models remove the left-hand wheel arch liner. On V6 models, remove the right-hand wheel arch liner.

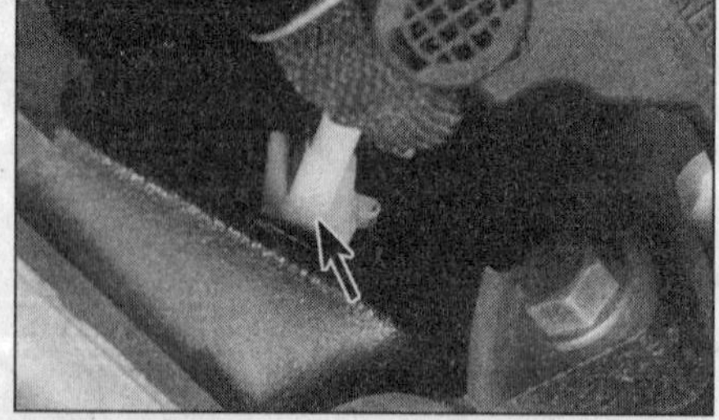

13 Disconnect the wiring from the sensor.

14 Undo the retaining screw and withdraw the sensor.

15 Refitting is the reverse of the removal procedure.

Camshaft position sensor

16 On 4-cylinder models, the camshaft sensor is located at the left-hand rear of the cylinder head. On V6 models, the sensor is located on the right-hand end of the timing cover.

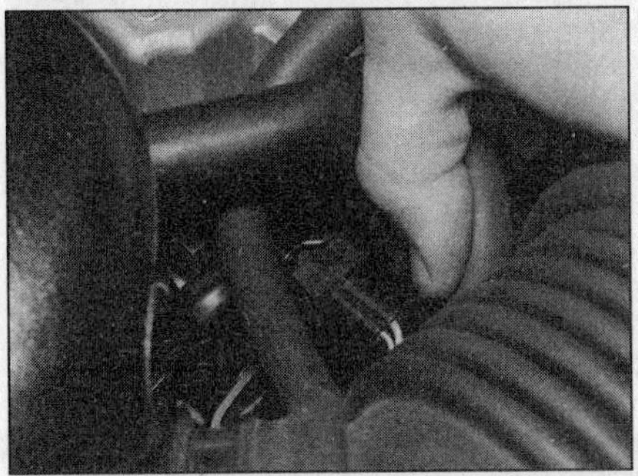

17 Remove the air mass meter and on 4-cylinder models the resonator to gain access to the sensor. Release the fuel feed and return hoses from their clips.

18 Release the wire clip and disconnect the sensor's electrical connector. Remove the retaining screw, and withdraw the sensor from the cylinder head; be prepared for slight oil loss.

19 Refitting is the reverse of the removal procedure, noting the following points:

a) Apply petroleum jelly or clean engine oil to the sensor's sealing O-ring.

b) Locate the sensor fully in the cylinder head, and wipe off any surplus lubricant before securing it.

c) Tighten the screw to the specified torque wrench setting.

Coolant temperature sensor

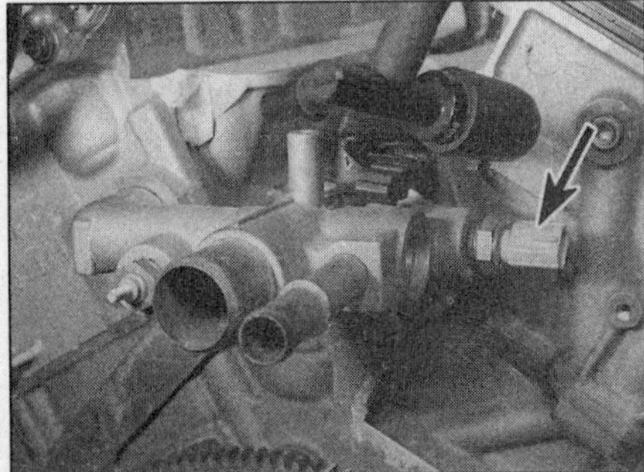

20 On 4-cylinder models, the sensor is screwed into the top of the thermostat. On V6 engine models, then sensor is fitted into the rear of the bypass housing at the left-hand end of the engine.

21 Drain the cooling system.

22 On 4-cylinder models, unbolt the resonator support bracket from the engine compartment front crossmember. Slacken the two clamp screws securing the resonator to the air mass meter and plenum chamber hoses, then swing the resonator up clear of the thermostat housing.

23 On V6 engines, access to the sensor is poor. Access can be improved by removing the air inlet hose from the air cleaner – if preferred, remove the air cleaner housing completely. Move the wiring harness around the sensor out of the way as far as possible.

24 On all engines, unplug the wiring connector from the sensor, and unscrew it from its location.

25 Refitting is a reversal of removal, but apply a little sealant to the sensor threads prior to refitting.

Intake air temperature sensor

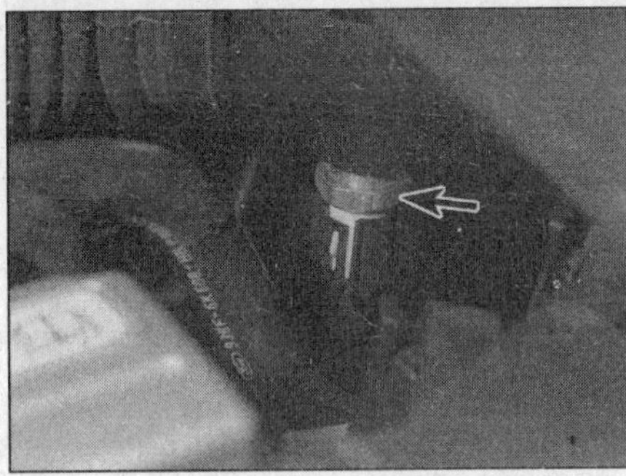

26 Remove the air mass meter (and resonator on 4-cylinder models) to gain access to the sensor.

27 Releasing its clip, unplug the sensor's electrical connector, then unscrew the sensor from the resonator, air inlet duct or air cleaner housing.

28 Refitting is the reverse of the removal procedure. Tighten the sensor to the specified torque wrench setting. If it is over-tightened, its tapered thread may crack the resonator, duct or air cleaner housing.

Throttle position sensor (potentiometer)

29 On 4-cylinder models, remove the plenum chamber. On V6 models, remove the water pump pulley shield. If necessary, remove the PCV hose from the air inlet duct.

30 Disconnect the wiring plug. Remove the retaining screws, and withdraw the unit from the throttle housing. Do not force the sensor's centre to rotate past its normal operating sweep, otherwise the unit will be seriously damaged.

31 Refitting is the reverse of the removal procedure, noting the following points:

a) Ensure that the sensor is correctly orientated, by locating its centre on the D-shaped throttle shaft (throttle closed), and aligning the sensor body so that the bolts pass easily into the throttle housing.

b) Tighten the screws to the Specified torque (but do not overtighten them, or the sensor body will be cracked).

Vehicle speed sensor

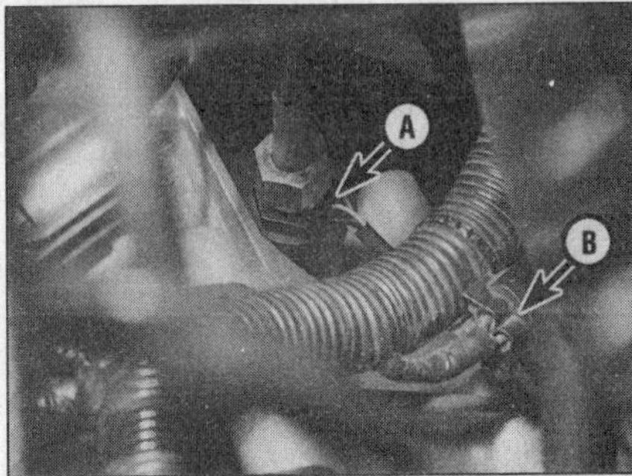

Vehicle speed snesor (A) with its electrical connector (B)

32 On models manufactured up to 1997, the sensor is mounted at the base of the speedometer drive cable, and is removed with the speedometer drive pinion. On later models it is located in the same position, however the cable is not fitted as the speedometer is operated electronically. Jack up the front of the vehicle and support is securely on axle stands.

34 On early models, unscrew the nut and disconnect the speedometer cable from the vehicle speed sensor on the transmission.

35 Disconnect the wiring from the speed sensor, and unscrew the sensor from the top of the drive pinion.

36 Refitting is a reversal of removal, renewing the O-ring seal if necessary.

Power steering pressure switch

37 Disconnect the wiring from the switch, then position a cloth rag beneath the switch then unscrew it from the pipe. Where fitted, recover the sealing washer.

38 Refitting is the reverse of the removal procedure; tighten the switch securely, then top-up the fluid reservoir to replace any fluid lost from the system, and bleed out any trapped air.

Throttle housing (4-cylinder models)

39 Disconnect the battery negative (earth) lead.

40 Remove the plenum chamber.

41 Disconnect the accelerator cable from the throttle linkage. Where fitted, also disconnect the cruise control actuator cable.

42 Disconnect the wiring from the large connector next to the fuel pressure regulator. Similarly release and unplug the throttle position sensor's electrical connector.

43 Clearly label, then detach, all vacuum hoses from the throttle housing.

44 Remove the throttle housing mounting screws, then detach the throttle housing and gasket from the inlet manifold. Discard the gasket – this must be renewed whenever it is disturbed.

45 Using a soft brush and a suitable liquid cleaner, thoroughly clean the exterior of the throttle housing, then blow out all passages with compressed air.

> *Caution: Do not clean the throttle housing's bore, the throttle valve, or the throttle position sensor, either by scraping or with a solvent. Just wipe them over carefully with a clean soft cloth.*

46 Refitting is the reverse of the removal procedure. Fit a new gasket, and tighten the housing screws to the specified torque.

Throttle housing (V6 models)

47 Disconnect the battery negative (earth) lead.

48 Remove the air cleaner and air inlet duct . Also remove the water pump pulley shield.

49 Disconnect the wiring from the throttle position sensor and detach the wiring from the stud. Position the wiring to one side.

50 Disconnect the accelerator cable and speed control cable (if fitted) from the throttle housing.

51 Unscrew the mounting bolts and nut, and withdraw the throttle housing from the upper inlet manifold. Recover the gasket.

52 Using a soft brush and a suitable liquid cleaner, thoroughly clean the exterior of the throttle housing, then blow out all passages with compressed air *(see Caution above)*.

53 Refitting is the reverse of the removal procedure. Fit a new gasket, and tighten the housing bolts and nut to the specified torque.

Fuel rail and injectors (4-cylinder models)

Note: *The following procedure describes the removal of the fuel rail assembly, complete with the injectors and pressure regulator, to enable the injectors to be serviced individually on a clean work surface. It is possible to remove and refit an individual injector once the fuel system has been depressurised and the battery has been disconnected.*

54 Depressurise the fuel system. Also equalise tank pressure by removing the fuel filler cap.

55 Disconnect the battery negative (earth) lead.

56 Remove the plenum chamber.

57 If the additional clearance is required, disconnect the accelerator cable from the throttle linkage. Where fitted, also disconnect the cruise control actuator cable.

58 Disconnect the wiring from the four fuel injector electrical connectors. On mid 1997-on models the wiring is contained in a holder.

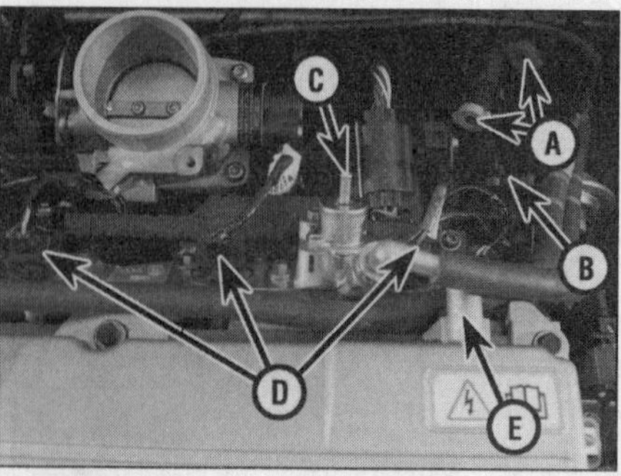

Injector removal – disconnect fuel lines at quick-release couplings (A), unclip hoses (B), disconnect vacuum hose from regulator (C), unplug electrical connectors (D) – three of four shown – and disconnect breather hose from union (E)

59 Disconnect the fuel feed and return lines at the quick-release couplings next to the braking system vacuum servo unit, then unclip the fuel hoses from the inlet manifold; use rag to soak up any spilt fuel. **Note:** *Do not disturb the threaded couplings at the fuel rail unions unless absolutely necessary; these are sealed at the factory. The quick-release couplings will suffice for all normal service operations.*

60 Disconnect the crankcase breather hose from the cylinder head cover union (pre mid 1997 models), and the vacuum hose from the fuel pressure regulator.

61 Unscrew the bolts securing the fuel rail, and withdraw the rail, carefully prising it out of the inlet manifold, and draining any remaining fuel into a suitable clean container. Note that mid 1997-on models are fitted with a different fuel rail, and the injectors locate directly in the inlet manifold. Recover the seals from the manifold and obtain new ones.

62 On pre mid 1997 models, unscrew the two bolts securing each injector, and withdraw the injectors. On later models, pull the clips apart and withdraw the injectors from the fuel rail, then remove the injectors from the clips. Place the injectors in a clean storage container.

63 Discard the nose seals/O-rings and obtain new ones.

64 Further testing of the injectors is beyond the scope of the home mechanic. If you are in doubt as to the status of any injector, it can be tested at a dealer service department.

65 Refitting is the reverse of the removal procedure, noting the following points:

a) Lubricate each (new) seal/O-ring with clean engine oil before refitting.

b) On pre mid 1997 models, locate each injector carefully in the fuel rail recess, ensuring that the locating tab on the injector head fits into the slot provided in the rail. Tighten the bolts to the specified torque.

c) Tighten the fuel rail bolts to the torque wrench setting specified.

d) Fasten the fuel feed and return quick-release coupling.

e) Ensure that the breather hose, vacuum hose and wiring are routed correctly, and secured on reconnection by any clips or ties provided.

f) On completion, switch the ignition on to activate the fuel pump and pressurise the system, without cranking the engine. Check for signs of fuel leaks around all disturbed unions and joints before attempting to start the engine.

Fuel rail and injectors (V6 models)

66 Depressurise the fuel system. Also equalise tank pressure by removing the fuel filler cap.

67 Disconnect the battery negative (earth) lead.

68 Remove the upper inlet manifold.

69 Disconnect the wiring harnesses from the injectors, and position to one side.

Pre-1997 models

70 The injectors can be removed at this stage by unscrewing the clamp bolts, removing the clamps and removing the injectors from the fuel rail. Place them in a clean container. Alternatively, the injectors can be removed together with the fuel rail.

71 To remove the fuel rail, loosen the clips and disconnect the fuel hoses from the fuel rail, and also disconnect the vacuum hose from the regulator, then unscrew the fuel rail mounting bolts and remove the rail from the lower inlet manifold.

1997-on models

72 Disconnect the vacuum hose from the pressure regulator.

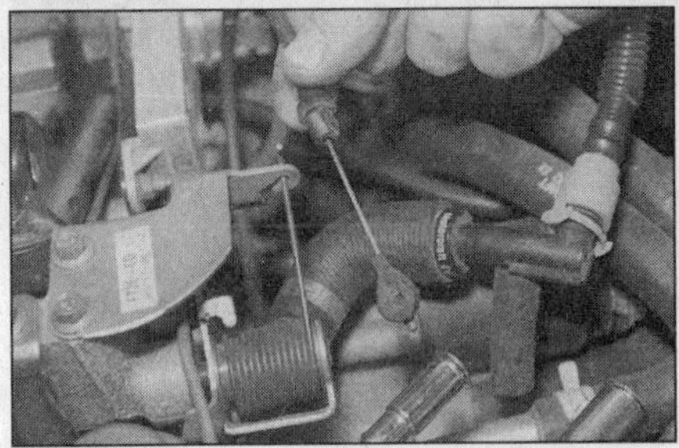

73 Disconnect the inlet manifold runner control (IMRC) actuator rod, by disconnecting the cable and cable end.

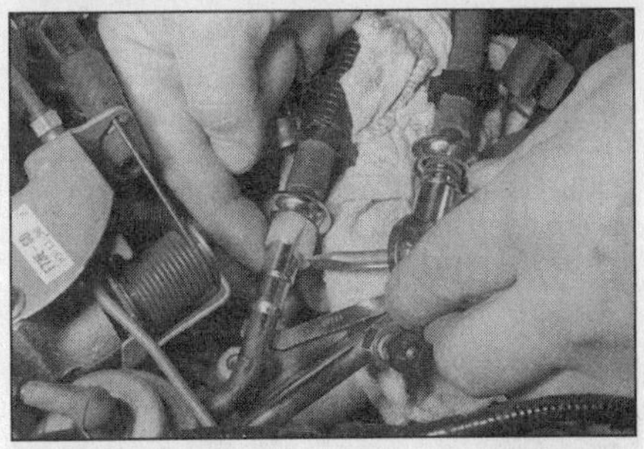

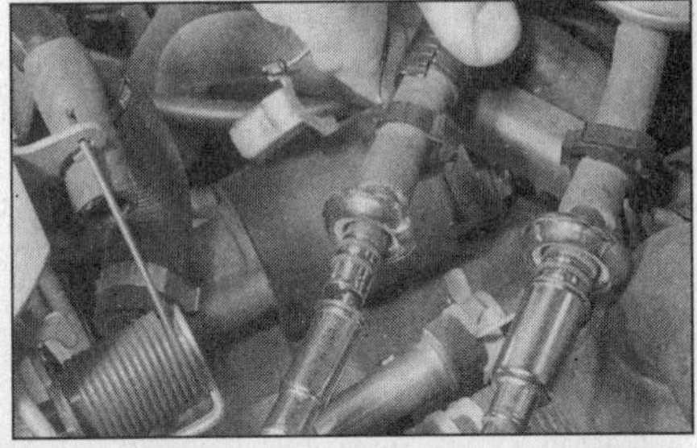

74 Disconnect the fuel feed and return lines by first releasing the clips. The lines have fittings that require a special tool to release them, however it is possible to use a home-made tool made from a coil of thin plastic about 15 mm wide. Insert the coil into the open end of the fitting so that the internal coiled spring is lifted over the retaining lip, then separate the lines.

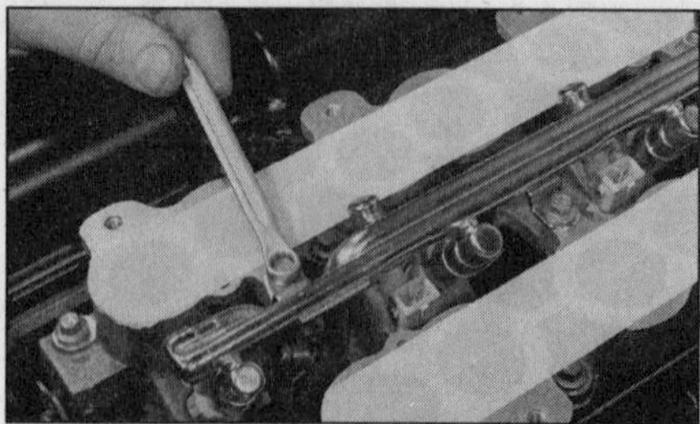

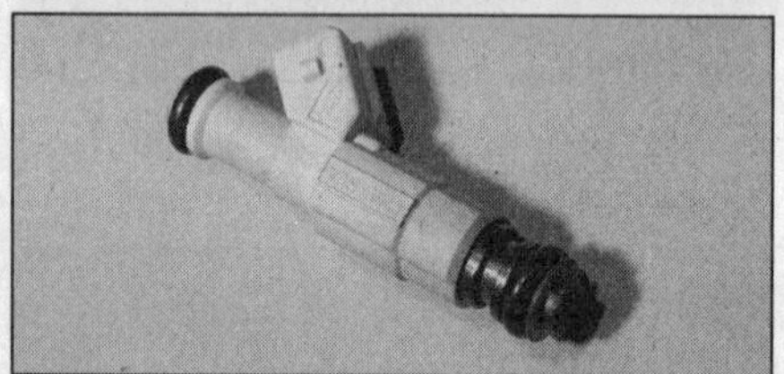

75 Unscrew the mounting bolts and remove the fuel rail from the lower inlet manifold. The injectors may come away with the rail or they may remain in the inlet manifold.

76 Carefully ease the injectors from the lower inlet manifold or fuel rail and place them in a clean container.

All models

77 Remove and discard all O-ring seals. Further testing of the injectors is beyond the scope of the home mechanic. If you are in doubt as to the status of any injector, it can be tested at a dealer service department.

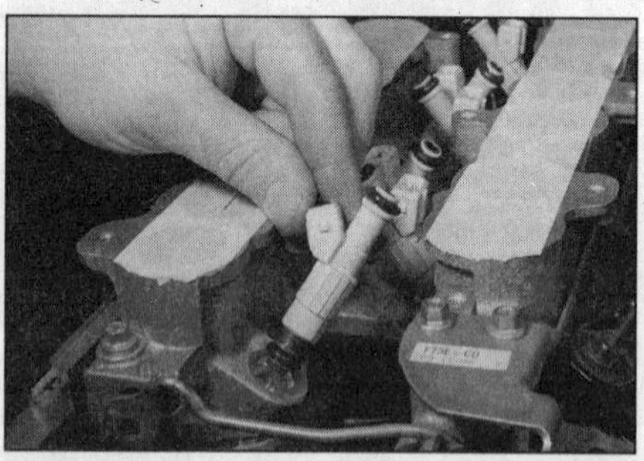

78 Refitting is the reverse of the removal procedure, noting the following points:

a) Lubricate each (new) O-ring seal with clean engine oil before refitting, and refit the injectors in the lower inlet manifold before fitting the fuel rail to them.

b) Tighten the fuel rail bolts to the specified torque.

c) On completion, switch the ignition on to activate the fuel pump and pressurise the system, without cranking the engine. Check for signs of fuel leaks around the unions and joints before attempting to start the engine.

Fuel pressure regulator

79 Depressurise the fuel system. Also equalise tank pressure by removing the fuel filler cap.

80 Disconnect the battery negative (earth) lead.

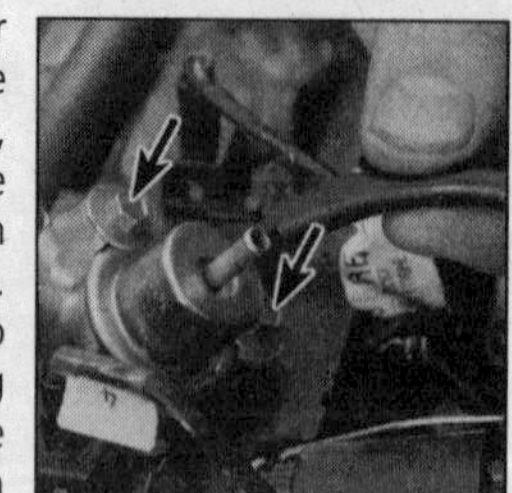

81 On 4-cylinder models, remove the plenum chamber, then disconnect the vacuum hose from the regulator. Unscrew the two regulator mounting screws and remove the regulator. On mid 1997-on models it will also be necessary to disconnect the fuel return hose and release it from the clip. Recover the O-ring seal. Soak up spilled fuel using a clean rag.

82 On pre mid 1997 V6 models, remove the fuel rail as described earlier. On later models, remove the upper inlet manifold. Disconnect the vacuum hose from the regulator. Remove the two regulator screws and remove the regulator. Recover the O-ring seal. Soak up spilled fuel using a clean rag.

83 Refitting is the reverse of the removal procedure, noting the following points:

a) Renew the regulator sealing O-ring whenever the regulator is disturbed. Lubricate the new O-ring with clean engine oil on installation.

b) Locate the regulator carefully in the fuel rail recess, and tighten the bolts to the specified torque wrench setting.

c) On completion, switch the ignition on and off five times, to activate the fuel pump and pressurise the system, without cranking the engine. Check for signs of fuel leaks around all disturbed unions and joints before attempting to start the engine.

Idle speed control valve

84 Disconnect the battery negative (earth) lead.

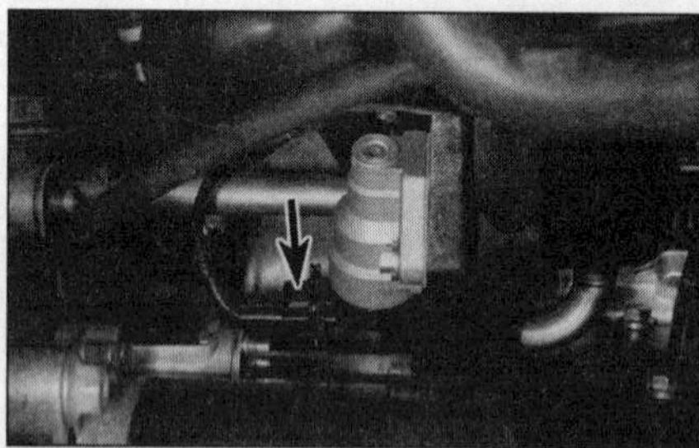

85 On 4-cylinder models, apply the handbrake, then jack up the front of the vehicle and support it on axle stands. Unplug the connector.

86 Disconnect the wiring from the idle speed control valve.

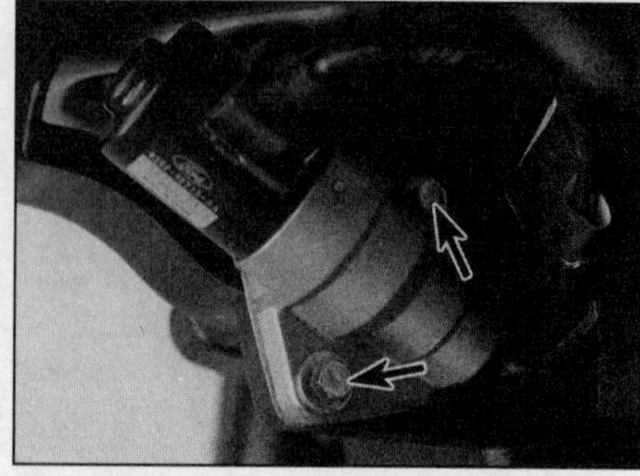

87 Unscrew the two retaining bolts (4-cylinder) or nuts (V6), and withdraw the valve from the inlet manifold. Recover the gasket.

88 Since the valve's individual components are not available separately, and the complete assembly must be renewed if it is thought to be faulty, there is nothing to be lost by attempting to flush out the passages, using carburettor cleaner or similar solvent. This won't take much time or effort, and may well cure the fault.

89 Refitting is the reverse of the removal procedure, noting the following points:

a) Clean the mating surfaces carefully, and always fit a new gasket whenever the valve is disturbed.

b) Tighten the bolts evenly and to the specified torque wrench setting.

c) Once the wiring and battery are reconnected, start the engine and allow it to idle. When it has reached normal operating temperature, check that the idle speed is stable, and that no induction (air) leaks are evident. Switch on all electrical loads (headlights, heated rear window, etc), and check that the idle speed is still correct.

Idle-increase solenoid valve

90 If better access is required, remove the plenum chamber.

91 Disconnect the battery negative (earth) lead.

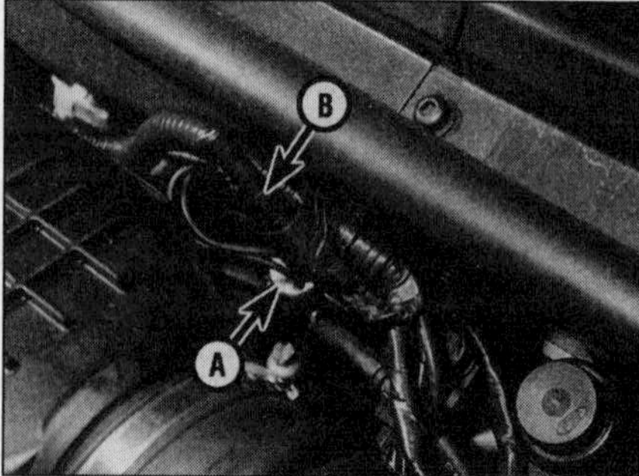

Location of idle-increase solenoid valve (A) and diode (B)

92 Unplug the valve's electrical connector. Unclip the valve from the bulkhead, then disconnect the vacuum hoses and withdraw it.

93 Refitting is the reverse of the removal procedure.

Ford Mondeo '00 – '03

16

Petrol models

General

System type	Sequential multi-point Fuel injection (SFi)
Recommended fuel (minimum octane rating)	95 RON unleaded
Idle speed:	
4-cylinder engines	700± 50 rpm
V6 engine	725± 50 rpm
Idle mixture (CO level):	
4-cylinder engines	0.5% maximum
V6 engine	0.5% maximum

Intake air temperature sensor

Resistance:	
At –40°C	approximately 880 k ohms
At –20°C	approximately 272 k ohms
At 0°C	approximately 95 k ohms
At 20°C	approximately 38 k ohms
At 40°C	approximately 16 k ohms
At 60°C	approximately 7.5 k ohms
At 80°C	approximately 3.8 k ohms
At 100°C	approximately 2.2 k ohms
At 120°C	approximately 1.2 k ohms

Diesel models

General

System type	Electronic fuel injection, controlled by an EEC V PCM
Application:	
2.0 litre Duratorq – DI engine	D5BA (66kW)
2.0 litre Duratorq – DI engine	D6BA (85kW)
2.0 litre Duratorq – DCI (common rail) engine	FMBA (96kW)
Firing order	1 – 3 – 4 – 2 (No 1 at timing chain end)
Idle speed	900 rpm (regulated by 'Black Oak' engine management system – no adjustment possible)

Diesel models (continued)

Glow plugs		
Maximum preheat time	8 seconds	
Maximum after-glow phase (less than 2500rpm)	30 seconds	
Engine above 80°C	No preheating	
Fuel		
Fuel type	Commercial diesel fuel for road vehicles (DERV)	
Injection pump		
Make and type:		
D5BA	Bosch VP-30	
D6BA	Bosch VP-44	
FMBA	Bosch VP-44	
Rotation (viewed from crankshaft pulley end)	Clockwise	
Drive	Twin chain via crankshaft and camshaft sprockets	
Turbocharger		
Type:		
D5BA	Fixed vane turbocharger – integral with exhaust manifold	
D6BA and FMBA	Variable vane turbocharger – integral with exhaust manifold	
Torque wrench settings	**Nm**	**lbf ft**
Petrol models		
Camshaft position sensor bolts (4-cylinder engines)	6	4
Crankshaft speed/position sensor:		
Sensor-to-bracket screw	8	6
Bracket-to-cylinder block crankcase screw	21	15
Fuel pressure regulator bolts	6	4
Fuel pressure relief valve (V6 engine)	8	6
Fuel rail:		
4-cylinder engines	25	18
V6 engine	10	7
Idle air control valve bolts	10	7
Intake air temperature sensor (V6 engine)	23	17
Intake manifold tuning valve	10	7
Throttle housing-to-inlet manifold screws	10	7
Diesel models		
Catalytic monitor sensor	46	34
Crankshaft position sensor bolt	7	5
Cylinder head temperature sensor	11	8
Oxygen sensor	48	35

General information

Petrol models

1 All petrol models are equipped with a Sequential Electronically-controlled Fuel Injection system. The engine management system on the 1.8 and 2.0 litre engines, control the sequential multi-point fuel injection (SFI), the electrically-heated thermostat and the electronically-controlled water-cooled exhaust gas recirculation (EGR) valve.

Fuel supply and air induction

2 An electric fuel pump located inside the fuel tank supplies fuel under pressure to the fuel rail, which distributes fuel to the injectors. A filter between the fuel pump and the fuel rail protects the components of the system. A pressure regulator controls the system pressure in relation to inlet depression. From the fuel rail, fuel is injected into the inlet ports, just above the inlet valves, by four fuel injectors.

3 The amount of fuel supplied by the injectors is precisely controlled by the Powertrain Control Module (PCM). The PCM uses the signals from the engine speed/crankshaft position sensor and the camshaft position sensor, to trigger each injector separately in cylinder firing order (sequential injection).

4 The air induction system consists of an air filter housing, a Mass Air Flow (MAF) sensor on V6 engines, a Temperature and Manifold Absolute Pressure (T-MAP) sensor on 4-cylinder engines, intake ducting, and a throttle housing. The MAF and T-MAP sensors are information-gathering devices for the PCM. The MAF sensor measures the volume and temperature of the air passing through the air intake pipe into the engine. The T-MAP sensor measures the air pressure and temperature in the intake manifold. The PCM uses these signals to calculate the mass/pressure of the air entering the engine.

5 The throttle valve inside the throttle housing is controlled by the driver, through the accelerator pedal. As the valve opens, the quantity of air entering the engine increases. The throttle potentiometer is attached to the throttle valve and informs the PCM of the throttle position. The PCM calculates the relevant period of injection, and controls the injector opening times.

Electronic control system

6 The 'Black Oak' engine management system controls the fuel injection by means of a microcomputer known as the PCM (Power-train Control Module) (see illustrations). The PCM receives signals from a number of sensors, which monitor the intake air mass/ pressure and temperature, coolant tempera-ture, camshaft and crankshaft position, throttle position, and exhaust gas oxygen content. The signals are processed by the PCM to determine the injection duration necessary for the optimum air/fuel ratio. The sensors and associated PCM-controlled relays are located throughout the engine compartment.

7 In the event of a sensor malfunction, a back-up circuit will take over, to provide driveability until the problem is identified and fixed. The following paragraphs describe the components of the electronic control system.

Powertrain Control Module

8 This component is the heart of the entire engine management system, controlling the fuel injection, ignition and emissions control systems. It also controls sub-systems such as the radiator cooling fan, air conditioning and automatic transmission, where appropriate.

Mass Air Flow sensor

V6 engine

9 This uses a 'hot-wire' system, sending the PCM a constantly-varying (analogue) voltage signal corresponding to the mass of air passing into the engine. Since air mass varies with temperature (cold air being denser than warm), measuring air mass provides the PCM with a very accurate means of determining the correct amount of fuel required to achieve the ideal air/fuel mixture ratio.

Temperature and Manifold Absolute Pressure sensor

4-cylinder engines

10 The T-MAP sensor measures the temperature and air pressure in the intake manifold. The PCM then uses these signals to calculate the pressure of the air in the manifold. Manifold absolute pressure is calculated from the formula:

Atmospheric Pressure minus Manifold Pressure = Manifold Absolute Pressure

When the manifold vacuum is high (eg, engine at idle), manifold absolute pressure is low and the PCM provides less fuel. When the manifold vacuum is low (eg, throttle wide open), manifold absolute pressure is high and the PCM provides more fuel.

Crankshaft speed/position sensor

11 This is an inductive pulse generator bolted to the timing chain cover. The sensor scans cut-outs machined on a timing disc located on the timing chain end of the crankshaft. The ridge between the 35th and 36th holes (corresponding to 50° BTDC) is missing – this step in the incoming signals is used by the PCM to determine crankshaft (ie, piston) position.

Camshaft position sensor

12 On 4-cylinder models the camshaft position sensor is bolted to the front left-hand end of the camshaft cover, and is triggered by a high-point on the inlet camshaft. On V6 models, the camshaft position sensor is mounted on the right-hand side of the front cylinder head and is triggered by a high-point on the front inlet camshaft. The sensor functions in the same way as the crankshaft speed/position sensor, producing a series of pulses. This gives the PCM a reference point, to enable it to determine the firing order, and operate the injectors in the appropriate sequence.

Coolant temperature sensor

13 This component, which is screwed into the coolant housing on the left-hand end of the cylinder head on 4-cylinder models or into the coolant crossover on V6 models, is an NTC (Negative Temperature Coefficient) thermistor – that is, a semi-conductor whose electrical resistance decreases as its temperature increases. It provides the PCM with a constantly-varying (analogue) voltage signal, corresponding to the temperature of the engine coolant. This is used to refine the calculations made by the PCM, when determining the correct amount of fuel required to achieve the ideal air/fuel mixture ratio.

Intake air temperature sensor

14 On 4-cylinder models, this component is part of the temperature and manifold absolute pressure (T-MAP) sensor which is fitted into the intake manifold. On V6 models it is located in the air cleaner housing cover with the mass air flow (MAF) sensor. The sensor is an NTC thermistor – see the previous paragraph – providing the PCM with a signal corresponding to the temperature of air passing into the engine. This is used to refine the calculations made by the PCM, when determining the correct amount of fuel required to achieve the ideal air/fuel mixture ratio.

Throttle potentiometer

15 This is mounted on the end of the throttle valve spindle, to provide the PCM with a constantly-varying (analogue) voltage signal corresponding to the throttle opening. This allows the PCM to register the driver's input when determining the amount of fuel required by the engine.

Vehicle speed sensor

16 This component is a Hall-effect generator, mounted on the transmission's speedometer drive. It supplies the PCM with a series of pulses corresponding to the vehicle's road speed, enabling the PCM to control features such as the fuel shut-off on the overrun, and to provide information for the trip computer, adaptive damping and cruise control systems (where fitted).

Power steering pressure switch

17 This is a pressure-operated switch, screwed into the power steering system's high-pressure pipe. Its contacts are normally closed, opening when the system reaches the specified pressure – on receiving this signal, the PCM increases the idle speed, to compensate for the additional load on the engine.

Air conditioning system

18 Two pressure-operated switches and the compressor clutch solenoid are connected to the PCM, to enable it to determine how the system is operating. The PCM can increase idle speed or switch off the system, as necessary, so that normal vehicle operation and driveability are not impaired. Fault diagnosis and repair should be left to a dealer service department or air conditioning specialist.

Idle air control valve

19 The idle air control (IAC) valve maintains a stable idle speed by varying the quantity of air entering the engine through an auxiliary air passage. The valve is activated by a signal from the PCM.

Intake manifold runner control actuator

4-cylinder engines

20 The intake manifold runner control (IMRC) actuator, operates four (one for each cylinder) plates in the passageways of the intake manifold. These plates are called swirl plates and reduce the cross-section of the intake passages at low speeds, this increases the intake air swirl to make the mixture ignite more efficiently around the spark plug area. This in turn will reduce exhaust emissions and fuel consumption.

Intake manifold tuning valve

4-cylinder engines

21 The intake manifold tuning (IMT) valve controls the intake manifold runner control actuator via a vacuum pipe, this will then give the correct position of the swirl plates depending on the vacuum in the intake manifold.

Automatic transmission sensors

22 In addition to the driver's controls, the 4-speed transmission has a speed sensor, a fluid temperature sensor (built into the solenoid valve unit), and a selector lever position sensor. All of these are connected to the PCM, to enable it to control the transmission through the solenoid valve unit. On 5-speed transmissions, there is a selector lever position sensor and a Transmission Control Module (TCM).

Exhaust gas oxygen sensor

23 The oxygen sensor in the exhaust system provides the PCM with constant feedback – 'closed-loop' control – which enables it to adjust the mixture to provide the best possible conditions for the catalytic converter to operate.

Diesel models

The fuel system consists of a fuel tank (mounted under the body, beneath the rear seats), fuel filter, electronic fuel injection pump with pump control unit, injectors, fuel lines and hoses, fuel gauge sender unit mounted in the fuel tank, and EEC-V PCM (powertrain control module).

Fuel is drawn from the tank (via an electrically-operated lift pump, on the common rail engines), it then passes through the fuel filter located in the engine bay, where foreign matter and water are removed. The injection pump is driven from the crankshaft via a twin-row chain and supplies fuel under very high pressure to each injector in turn as it is needed. The camshafts are also driven from the same chain.

Chapter 16

The Duratorq-DI engine is very much a 'state-of-the-art' unit, in that it features a full electronic engine management system. An extensive array of sensors are fitted, which supply information on many different parameters to the PCM.

Information on crankshaft position and engine speed is generated by a crankshaft position sensor. The inductive head of the sensor runs just above the engine flywheel, and scans a series of 36 protrusions on the flywheel periphery. As the crankshaft rotates, the sensor transmits a pulse every time a protrusion passes it. There is one missing protrusion in the flywheel periphery at a point corresponding to 50° BTDC. The PCM recognises the absence of a pulse from the crankshaft position sensor at this point to establish a reference mark for crankshaft position. Similarly, the time interval between absent pulses is used to determine engine speed.

Information on the quantity and temperature of the intake air is derived from the T-MAP sensor. The temperature and manifold absolute pressure (or T-MAP) sensor is located in the top of the air intake duct that runs along the front of the engine subframe below the radiator. It measures the temperature and the pressure of the air in the intake system. The temperature and quantity of air has a direct bearing on the quantity of fuel to be injected for optimum efficiency.

The traditional coolant temperature sensor has been superseded by a cylinder head temperature sensor. The new sensor is seated in a blind hole in the cylinder head, and measures the temperature of the metal directly. Information on engine temperature is critical for accurate fuelling calculations, and is also used to control the preheating system for cold starts.

The clutch pedal sensor informs the PCM whether the clutch is engaged or disengaged. When the clutch pedal is depressed, the quantity of fuel injected is momentarily reduced, to make gearchanging smoother.

The stop-light switch and separate brake pedal sensor inform the PCM when the brakes are applied – when this signal is received, the PCM puts the engine into idle mode until a signal is received from the accelerator position sensor.

The amount of fuel delivered is determined by the pump's internal quantity and timing solenoid valves, which are controlled by the pump control unit (PCU), mounted on top of the pump. The pump is internally equipped with a pulse ring fitted to the main rotor, and an angle sensor determines the pump rotor's position and speed, in much the same way as the crankshaft position sensor and engine flywheel, except that there are four gaps in the pump rotor 'teeth' – one for each cylinder. The pump control unit is supplied with information from the 'main' powertrain control module (PCM), and from this, is able to calculate the most appropriate values for injection timing and quantity (injection duration). The electronically-controlled pump internals enable these calculated values to be delivered with great accuracy, for improved efficiency and reduced emissions.

No accelerator cable is fitted on the Duratorq engines – instead, a sensor located next to the accelerator pedal informs the PCM of the accelerator position, and this information is used to determine the most appropriate fuelling requirements from the injection pump. The engine idle speed is also controlled by the PCM, and cannot be adjusted. From the signals it receives from the various sensors, the PCM can control the idle speed very accurately, compensating automatically for additional engine loads or unfavourable ambient/engine temperatures.

Rigid pipes connect the pump to the four injectors, on the D5BA and D6BA engines. On the FMBA engine there are four pipes from the injectors to the fuel supply manifold and one from the supply manifold down to the fuel pump. Each injector disperses the fuel evenly, and sprays fuel directly into the combustion chamber as its piston approaches TDC on the compression stroke. This system is known as direct injection. The pistons have a recess machined into their crowns, the shape of which has been calculated to improve 'swirl' (fuel/air mixing). Lubrication is provided by allowing a small quantity of fuel to leak back past the injector internal components. The leaked-back fuel is returned to the pump and then to the fuel tank.

Cold-starting performance is automatically controlled by the PCM. Under cold start conditions, the cylinder head temperature (CHT) sensor informs the PCM on the engine temperature, this determines the preheat time. The glow plugs are located in the side of the cylinder head, one to each cylinder, and are electrically-heated. A warning light illuminates when the ignition is switched on, showing that the glow plugs are in operation. When the light goes out, preheating is complete and the engine can be started. The glow plugs have an after-glow phase which only operates under 2500 rpm, and below temperatures of 50°C . This helps the engine to run more smoothly during idling, and reduces exhaust emissions through more efficient combustion just after starting.

The fuel system has a built-in 'strategy' to prevent it from drawing in air, should the car run low on fuel. The PCM monitors the level of fuel in the tank, via the gauge sender unit. After switching on the low fuel level warning light, it will eventually induce a misfire as a further warning to the driver, and lower the engine's maximum speed until the engine stops.

The fuel system on diesel engines is normally very reliable. Provided that clean fuel is used and the specified maintenance is conscientiously carried out, no problems should be experienced. The injection pump and injectors may require overhaul after a high mileage has been covered, but this cannot be done on a DIY basis.

Fuel system – depressurisation, priming and bleeding

Petrol models

1 The fuel system referred to in this Chapter is defined as the fuel tank and tank-mounted fuel pump/fuel gauge sender unit, the fuel filter, the fuel injectors and the pressure regulator in the injector rail, and the metal pipes and flexible hoses of the fuel lines between these components. All these contain fuel, which will be under pressure while the engine is running and/or while the ignition is switched on.

2 The pressure will remain for some time after the ignition has been switched off, and must be relieved before any of these components is disturbed for servicing work.

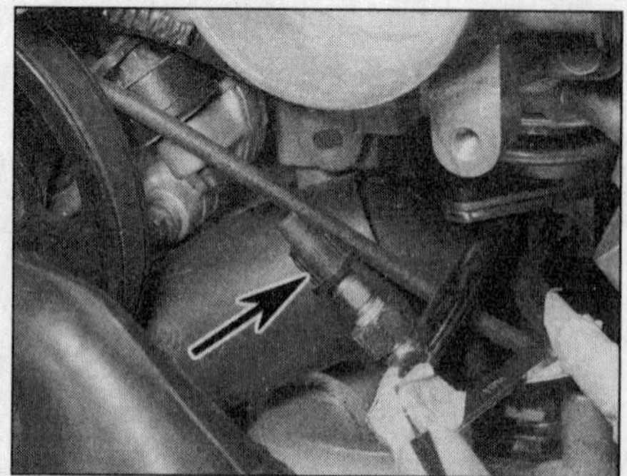

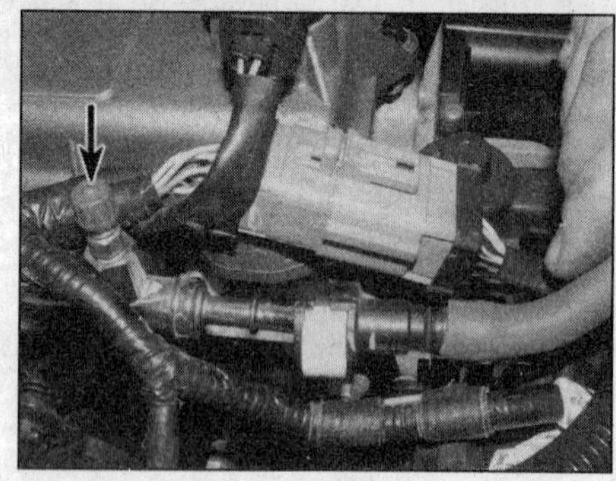

3 The Ford method of depressurisation is to use service tool 23-033 fitted to the fuel rail pressure test/release fitting. The fitting consists of a Schrader-type valve with a plastic cap located on the fuel rail, and the tool acts as a tap by depressing the valve core.

4 To release the pressure without using the service tool, use a suitable container and rag wrapped around the fitting to catch the fuel. Do not simply depress the valve core without wrapping rag around it, as fuel will spray out, with the consequent risk of fire, and personal injury.

5 An alternative method is simply to disconnect

the fuel pump's electrical supply while the engine is running, by removing the fuel pump fuse, and to allow the engine to idle until it dies through lack of fuel. Turn the engine over once or twice on the starter to ensure that all pressure is released, then switch off the ignition. Do not forget to refit the fuse when work is complete.

6 Note that, once the fuel system has been depressurised, it may take longer to restart the engine – perhaps several seconds of cranking – before the system is refilled and pressure restored.

Diesel models

7 As this system is intended to be 'self-bleeding', no hand-priming pump or separate bleed screws/nipples are fitted.

8 When any part of the system has been disturbed therefore, air must be purged from the system by cranking the engine on the starter motor until it starts. When it has started, keep the engine running for approximately 5 minutes to ensure that all air has been removed from the system. To minimise the strain on the battery and starter motor when trying to start the engine, crank it in 10 second bursts, pausing for 30 seconds each time, until the engine starts.

9 Depending on the work that has been carried out, it may be possible to partially prime the system so as to spare the battery by reducing as much as possible the amount of cranking time required to start the engine. To spare the battery, fill the filter with clean fuel via its vent screw opening (where applicable), but it is essential that no dirt is introduced into the system and that no diesel fuel is poured over vulnerable components when doing this.

10 On models with a Bosch VP-30 injection pump, if a hand-operated vacuum pump is available, this can be connected to the pump's fuel return union and used to suck fuel through the supply lines and filter. This will obviously save the battery a good deal of work. If a long length of clear plastic tubing is used to connect the vacuum pump to the injection pump union, it will be easier to see when fuel emerges free from air bubbles. Turn the ignition switch to position II so that fuel can pass through the pump.

11 On models with a Bosch VP-44 injection pump, a clear plastic T-piece will need to be connected into the return line (across the front of the timing chain cover) with one piece going into a container. Remove the left-hand side front trim panel to access the inertia fuel cut-off switch (IFS). Disconnect the 3-pin wiring connector and using a fused (15 amp) jumper wire, supply a battery feed to pin No 1 in the wiring connector. This will start the fuel lift pump and supply fuel through the circuit, check for air in the clear plastic T-piece. When no more air is visible, disconnect the jumper wire, remove the T-piece and reconnect the return pipe.

Petrol fuel injection system components – removal and refitting

Powertrain Control Module

> *Caution: The PCM is fragile. Take care not to drop it or subject it to any other kind of impact, and do not subject it to extremes of temperature, or allow it to get wet. Do not touch the PCM terminals as there is a chance that static electricity may damage the internal electronic components.*

Note: *If renewing the powertrain control module, note that it must be reprogrammed for the specific model by a Ford dealer using the FDS2000 diagnostic equipment. Failure to do so will result in the PCM assuming its limited operating strategy (LOS) settings giving poor performance and economy.*

1 Disconnect the battery negative (earth) lead.

2 Unclip the plastic cover and remove the three retaining nuts from the right-hand strut top mounting. Note: DO NOT jack up the front of the vehicle, while these nuts are removed.

3 Undo the four retaining bolts from the right-hand strut top mounting brace and remove it from the vehicle.

4 A security shield is riveted over the PCM and it will be necessary to drill out the rivet(s) to remove the shield and earth cable.

> *Caution: The PCM earth cable is connected behind the security shield rivet. Take care not to damage the earth cable when drilling through the rivet.*

5 Unclip the wiring harness from along the right-hand inner wing panel.

6 Undo the retaining bolt and release the wiring connector from the PCM – to prevent dust and dirt entering the multi-plug, cover it with a polythene bag.

7 From inside the vehicle, unclip the generic electronic module (GEM) from the PCM bracket.

8 Undo the retaining nut and withdraw the PCM and mounting bracket from the driver's side footwell.

9 Refitting is the reverse of the removal procedure. Whenever the PCM (or battery) is disconnected, the information relating to idle speed control and other operating values will be lost from its memory until the unit has reprogrammed itself; until then, there may be surging, hesitation, erratic idle and a generally-inferior level of performance. To allow the PCM to relearn these values, start the engine and run it as close to idle speed as possible until it reaches its normal operating temperature, then run it for approximately two minutes at 1200 rpm. Next, drive the vehicle as far as necessary – approximately 5 miles of varied driving conditions is usually sufficient – to complete the relearning process.

Mass air flow sensor

V6 engine

10 Disconnect the battery negative (earth) lead.

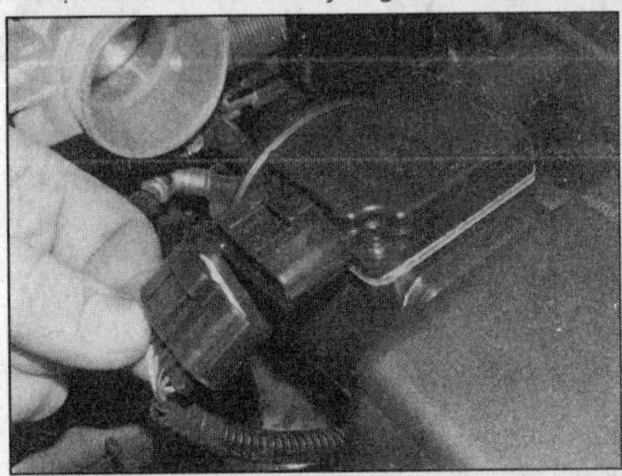

11 Disconnect the wiring connector from the mass air flow sensor.

12 Remove the air cleaner assembly. Unscrew and remove the sensor mounting bolts and withdraw the sensor from the air cleaner cover.

13 Refitting is the reverse of the removal procedure. Ensure that the sensor and air cleaner cover are seated correctly and securely fastened, so that there are no air leaks.

Temperature and Manifold Absolute Pressure sensor

4-cylinder engines

14 Disconnect the battery negative (earth) lead. Unclip the breather hose from the engine upper plastic cover, then unclip the cover from the top of the engine.

15 Disconnect the wiring connector from the temperature and manifold absolute pressure sensor.

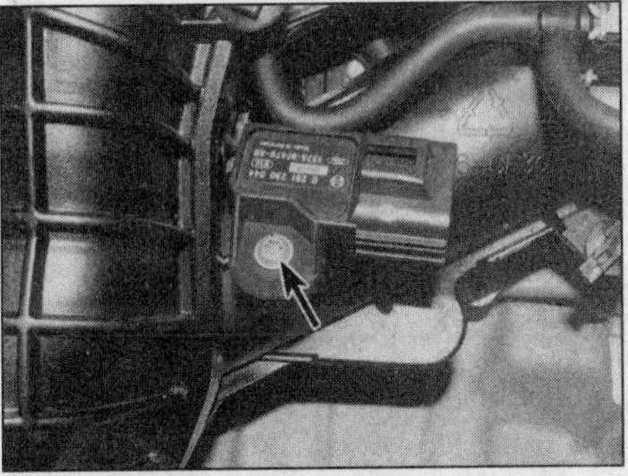

16 Undo the retaining screw and withdraw the sensor from the intake manifold.

Crankshaft speed/position sensor

17 Disconnect the battery negative (earth) lead.

18 Apply the handbrake, then jack up the front of the vehicle and support it on axle stands.

19 Remove the right-hand front roadwheel, undo the retaining bolts and withdraw the wheel arch liner.

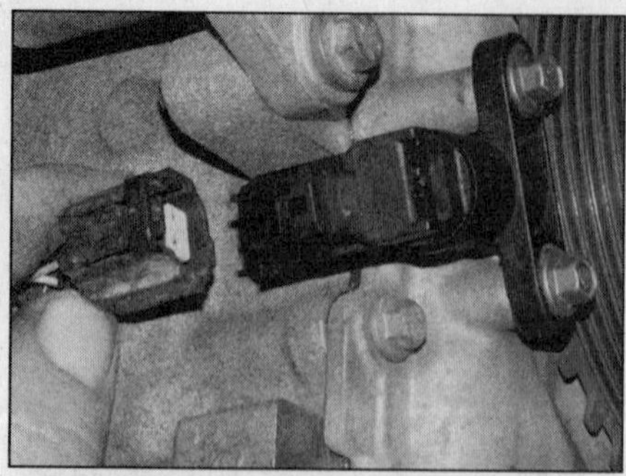

20 Disconnect the wiring from the sensor.

21 Undo the retaining bolt and withdraw the sensor.

22 Refitting is the reverse of the removal procedure.

Camshaft position sensor

23 The camshaft sensor is located on the front left-hand end of the camshaft cover on 4-cylinder models. On V6 models, the sensor is located in the timing chain cover, adjacent to the front intake camshaft sprocket.

24 On 4-cylinder models, unclip the breather hose from the engine upper plastic cover, then unclip the cover from the top of the engine.

25 Disconnect the sensor's electrical connector. Remove the retaining bolt, and withdraw the sensor from the camshaft cover (4-cylinder engines) or timing chain cover (V6 engines); be prepared for slight oil loss.

26 Refitting is the reverse of the removal procedure, noting the following points:

a) Apply petroleum jelly or clean engine oil to the sensor's sealing O-ring.

b) Locate the sensor fully in the camshaft cover/timing chain cover, and wipe off any surplus lubricant before securing it.

c) Tighten the screw to the specified torque wrench setting.

Coolant temperature sensor

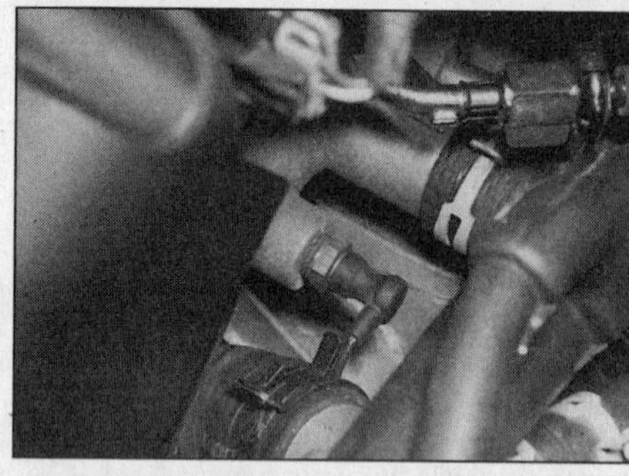

27 On 1.8 and 2.0 litre petrol and diesel models, then sensor is screwed into the outlet housing on the left-hand end of the cylinder head, whilst on 2.5 litre V6 engines, the sensor is at the front of the bypass housing on the left-hand end of the engine. Drain the coolant system.

28 On 1.8 and 2.0 litre petrol engine models, remove the ignition coil pack.

29 On V6 and diesel models, remove the water pump drivebelt cover, and disconnect the air inlet hose from the air cleaner housing. Access will be improved by removing the battery.

30 Disconnect the wiring plug and unscrew the sensor.

31 Refitting is a reversal of removal. Apply a little sealing to the sensor threads, and refill the coolant system.

Intake air temperature sensor

V6 engine

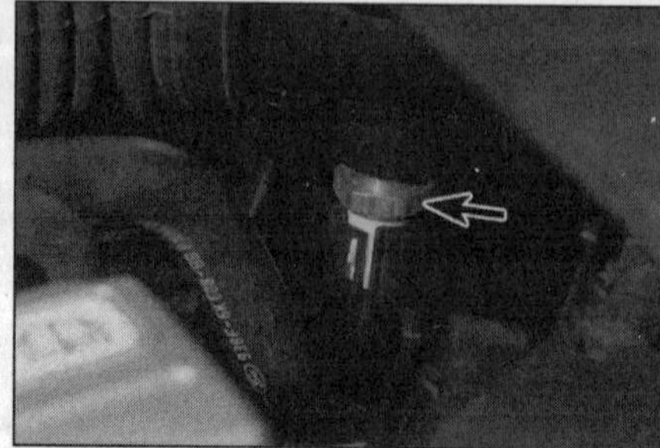

32 Remove the mass air flow sensor to gain access to the temperature sensor.

33 Releasing its clip, unplug the sensor's electrical connector, then unscrew the sensor from the resonator, air inlet duct or air cleaner housing.

34 Refitting is the reverse of the removal procedure. Tighten the sensor to the specified torque wrench setting. If it is over-tightened, its tapered thread may crack the resonator, duct or air cleaner housing.

Throttle position sensor (potentiometer)

35 On 4-cylinder models, unclip the breather hose from the engine upper plastic cover, then unclip the cover from the top of the engine. On V6 models, remove the water pump pulley shield. If necessary, remove the PCV hose from the air inlet duct.

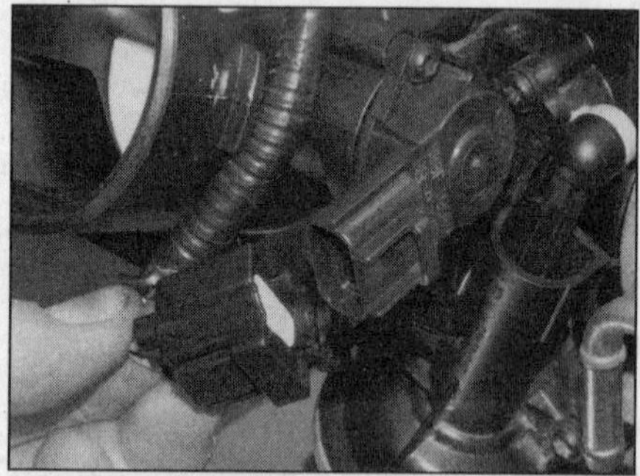

36 Disconnect the wiring plug. Remove the retaining screws, and withdraw the unit from the throttle housing. Do not force the sensor's centre to rotate past its normal operating sweep, otherwise the unit will be seriously damaged.

37 Refitting is the reverse of the removal procedure, noting the following points:

a) Ensure that the sensor is correctly orientated, by locating its centre on the D-shaped throttle shaft (throttle closed), and aligning the sensor body so that the bolts pass easily into the throttle housing.

b) Tighten the screws securely (but do not overtighten them, or the sensor body will be cracked).

Vehicle speed sensor

38 The speed sensor is located in the transmission casing above the right-hand driveshaft, it is operated electronically. Jack up the front of the vehicle and support it securely on axle stands.

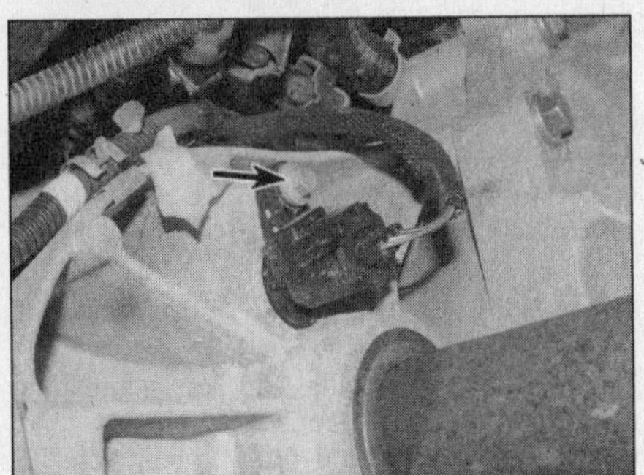

39 Disconnect the wiring plug, then undo the retaining bolt and remove the sensor from the casing. **Note:** *On some models then sensor may be retained by a roll-pin, not a bolt. Use a pair of pliers to pull the pin from the casing.*

40 Refitting is a reversal of removal, ensuring a new O-ring seal is fitted where necessary.

Throttle housing

4-cylinder engines

41 Disconnect the battery negative (earth) lead, then unclip the breather hose from the engine upper plastic cover, and unclip the cover from the top of the engine. Slacken the retaining clips and remove the air intake pipe from the throttle housing.

42 Disconnect the accelerator cable from the throttle linkage. Where fitted, also disconnect the cruise control actuator cable.

43 Disconnect the wiring plug from the throttle position sensor's electrical connector.

44 Remove the throttle housing mounting screws, then detach the throttle housing and gasket from the inlet manifold. Discard the gasket – this must be renewed whenever it is disturbed.

45 Using a soft brush and a suitable liquid cleaner, thoroughly clean the exterior of the throttle housing, then blow out all passages with compressed air.

> *Caution: Do not clean the throttle housing's bore, the throttle valve, or the throttle position sensor, either by scraping or with a solvent. Just wipe them over carefully with a clean soft cloth.*

46 Refitting is the reverse of the removal procedure. Fit a new gasket, and tighten the housing screws to the specified torque.

V6 engine

47 Disconnect the battery negative (earth) lead.

48 Remove the air cleaner and air inlet duct. Also remove the water pump pulley shield.

49 Disconnect the wiring from the throttle position sensor and detach the wiring from the stud. Position the wiring to one side.

50 Disconnect the accelerator cable and speed control cable (if fitted) from the throttle housing.

51 Unscrew the mounting bolts and nut, and withdraw the throttle housing from the upper inlet manifold. Recover the gasket.

52 Using a soft brush and a suitable liquid cleaner, thoroughly clean the exterior of the throttle housing, then blow out all passages with compressed air.

53 Refitting is the reverse of the removal procedure. Fit a new gasket, and tighten the housing bolts and nut to the specified torque.

Fuel rail and injectors

4-cylinder engines

> *Note: The following procedure describes the removal of the fuel rail assembly, complete with the injectors and pulse damper, to enable the injectors to be serviced individually on a clean work surface.*

54 Depressurise the fuel system. Also equalise tank pressure by removing the fuel filler cap.

55 Disconnect the battery negative (earth) lead, then unclip the breather hose from the engine upper plastic cover, and unclip the cover from the top of the engine.

56 Disconnect the wiring from the four fuel injector electrical connectors.

57 Disconnect the fuel supply line at the quick-release coupling from the end of the fuel rail, use rag to soak up any spilt fuel.

58 Disconnect the vacuum pipe from the fuel pulse damper.

59 Unscrew the bolts securing the fuel rail, and withdraw the rail, complete with the injectors out of the cylinder head. Drain any remaining fuel into a suitable clean container. Recover the fuel rail spacers from the cylinder head.

60 Clamp the rail carefully in a vice fitted with soft jaws. On models up to 03/2002, slide the retaining clips sideways from the top of the injectors, to withdraw the injectors from the fuel rail. On models built after 03/2002 release ,the locking tangs at the top of the injector, to withdraw the injectors from the fuel rail. Place the injectors in a clean storage container.

61 Discard the seals/O-rings from the fuel injectors and obtain new ones.

62 Refitting is the reverse of the removal procedure, noting the following points:

a) Lubricate each (new) seal/O-ring with clean engine oil before refitting.

b) Locate each injector carefully in the fuel rail recess, ensuring that the retaining clips are located correctly on the top of the injectors.

c) Tighten the fuel rail bolts to the torque wrench setting specified.

d) Reconnect the fuel feed quick-release coupling.

e) Ensure that the breather hose, vacuum hose and wiring are routed correctly, and secured on reconnection by any clips or ties provided.

f) On completion, switch the ignition on to activate the fuel pump and pressurise the system, without cranking the engine. Check for signs of fuel leaks around all disturbed unions and joints before attempting to start the engine.

V6 engine

63 Depressurise the fuel system. Also equalise tank pressure by removing the fuel filler cap.

64 Disconnect the battery negative (earth) lead.

65 Remove the upper intake manifold.

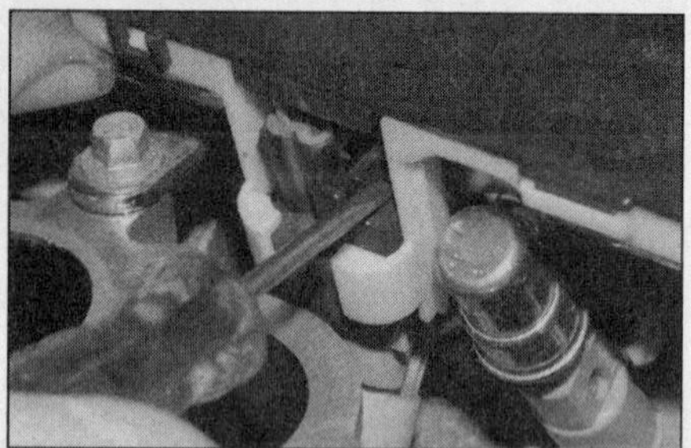

66 Disconnect the wiring harnesses from the injectors, and position to one side.

67 Disconnect the fuel supply line at the quick-release coupling from the end of the fuel rail.

68 Unscrew the mounting bolts and remove the fuel rail from the lower intake manifold. The

injectors may come away with the rail or they may remain in the intake manifold.

69 Where applicable, carefully ease the injectors from the lower intake manifold or fuel rail and place them in a clean container.

70 Remove and discard all O-ring seals. Further testing of the injectors is beyond the scope of the home mechanic. If you are in doubt as to the status of any injector, it can be tested at a dealer service department.

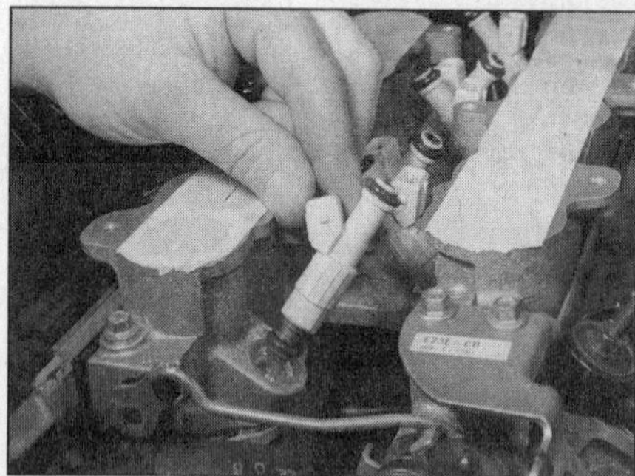

71 Refitting is the reverse of the removal procedure, noting the following points:

a) Lubricate each (new) O-ring seal with clean engine oil before refitting, and refit the injectors in the lower inlet manifold before fitting the fuel rail to them.

b) Tighten the fuel rail bolts to the specified torque.

c) On completion, switch the ignition on to activate the fuel pump and pressurise the system, without cranking the engine. Check for signs of fuel leaks around the unions and joints before attempting to start the engine.

Fuel pressure regulator

72 Depressurise the fuel system. Also equalise tank pressure by removing the fuel filler cap.

73 Disconnect the battery negative (earth) lead.

74 On V6 models, remove the upper inlet manifold.

75 Disconnect the vacuum hose from the regulator. Remove the two regulator screws and remove the regulator. Recover the O-ring seal. Soak up spilled fuel using a clean rag.

76 Refitting is the reverse of the removal procedure, noting the following points:

a) Renew the regulator sealing O-ring whenever the regulator is disturbed. Lubricate the new O-ring with clean engine oil on installation.

b) Locate the regulator carefully, and tighten the bolts to the specified torque wrench setting.

c) On completion, switch the ignition on and off five times, to activate the fuel pump and pressurise the system, without cranking the engine. Check for signs of fuel leaks around all disturbed unions and joints before attempting to start the engine.

Idle air control valve

77 Disconnect the battery negative (earth) lead.

78 On 4-cylinder engines, unclip the breather hose from the engine upper plastic cover, then unclip the cover from the top of the engine.

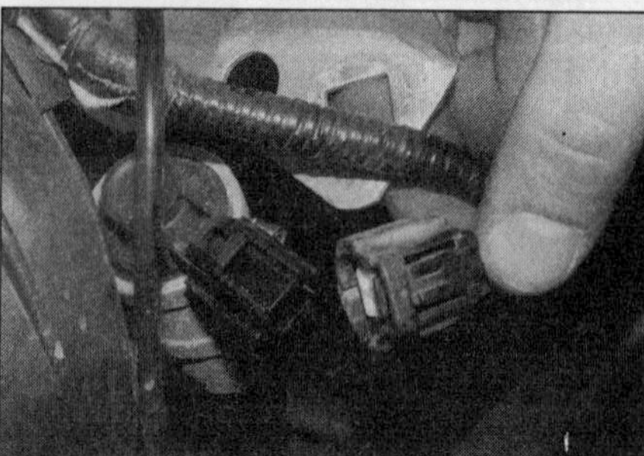

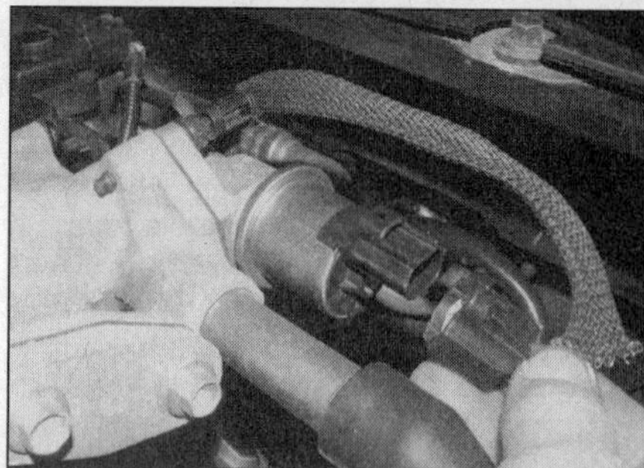

79 Disconnect the wiring from the idle air control valve.

80 Unscrew the two retaining bolts (4-cylinder) or nuts (V6), and withdraw the valve from the intake manifold. Recover the gasket.

81 Since the valve's individual components are not available separately, and the complete assembly must be renewed if it is thought to be faulty, there is nothing to be lost by attempting to flush out the passages, using carburettor cleaner or similar solvent. This won't take much time or effort, and may well cure the fault.

82 Refitting is the reverse of the removal procedure, noting the following points:

a) Clean the mating surfaces carefully, and always fit a new gasket whenever the valve is disturbed.

b) Tighten the bolts evenly and to the specified torque wrench setting.

c) Once the wiring and battery are reconnected, start the engine and allow it to idle. When it has reached normal operating temperature, check that the idle speed is stable, and that no induction (air) leaks are evident. Switch on all electrical loads (headlights, heated rear window, etc), and check that the idle speed is still correct.

Intake manifold runner control actuator

4-cylinder engines

83 Remove the throttle housing as described in paragraphs 41 to 46 in this Section.

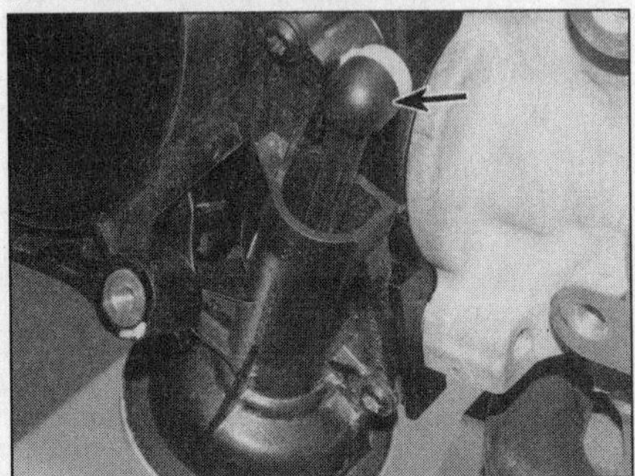

84 Disconnect the actuator arm from the intake manifold swirl plate linkage.

85 Disconnect the vacuum pipe from the intake manifold runner control actuator.

86 Undo the three retaining screws and withdraw the intake manifold runner control actuator from the intake manifold.

87 Refitting is the reverse of the removal procedure, noting the following points:

a) Clean the mating surfaces carefully, and always fit a new gasket.

b) Tighten the bolts evenly.

c) Once the wiring and battery are reconnected, start the engine and allow it to idle. When it has reached normal operating temperature, check that the idle speed is stable, and that no induction (air) leaks are evident. Switch on all electrical loads (headlights, heated rear window, etc), and check that the idle speed is still correct.

V6 engine

88 Undo the retaining bolts and remove the plastic cover from the front camshaft cover.

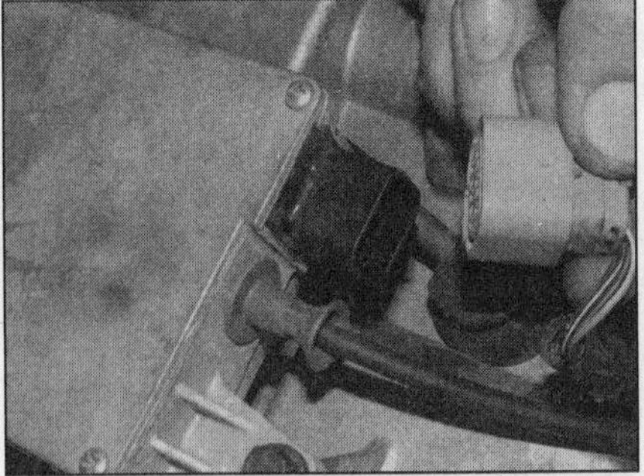

89 Disconnect the wiring connector from the intake manifold runner control unit.

90 Undo the three retaining bolts and withdraw the intake manifold runner control unit from the camshaft cover.

91 Disconnect the operating cable from the throttle housing linkage.

92 Refitting is the reverse of the removal procedure

Intake manifold tuning valve

4-cylinder engines

93 Disconnect the battery negative (earth) lead.

94 Unclip the breather hose from the engine upper plastic cover, then unclip the cover from the top of the engine.

95 Disconnect the wiring connector from the intake manifold tuning valve.

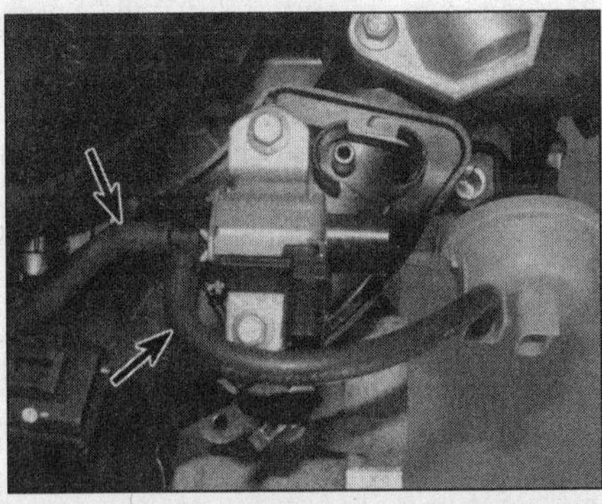

96 Disconnect the vacuum pipes from the intake manifold tuning valve.

97 Undo the two retaining bolts and withdraw the valve from the intake manifold.

98 Refitting is the reverse of the removal procedure, noting the following points:

a) Clean the mating surfaces carefully, and always fit a new gasket.

b) Tighten the bolts evenly and to the specified torque wrench setting.

c) Once the wiring and battery are reconnected, start the engine and allow it to idle. When it has reached normal operating temperature, check that the idle speed is stable, and that no induction (air) leaks are evident. Switch on all electrical loads (headlights, heated rear window, etc), and check that the idle speed is still correct.

Diesel injection system electronic components – removal and refitting

1 Disconnect the battery negative (earth) lead.

2 Withdraw the oil level dipstick and unclip the engine plastic cover.

Crankshaft position sensor

3 The sensor is located in the top of the bellhousing to the rear.

4 For better access remove the air cleaner assembly.

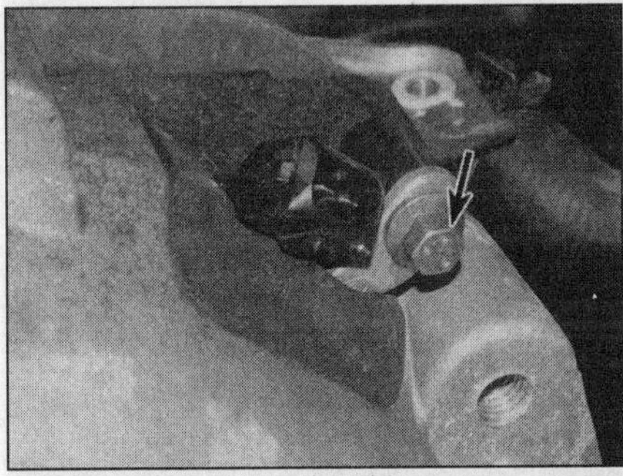

5 Disconnect the wiring plug, then unscrew the mounting bolt and withdraw the sensor. **Note:** *Before removing the sensor, mark the position of the retaining bolt, so that it can be refitted in the same place.*

6 Refitting is a reversal of removal.

Cylinder head temperature sensor

7 The switch is screwed into the left-hand (flywheel) end of the cylinder head, behind the power steering pump pulley.

8 To improve access to the switch, it will be necessary to remove the power steering pump pulley.

9 It may also be helpful to release the air intake hoses and move them to one side.

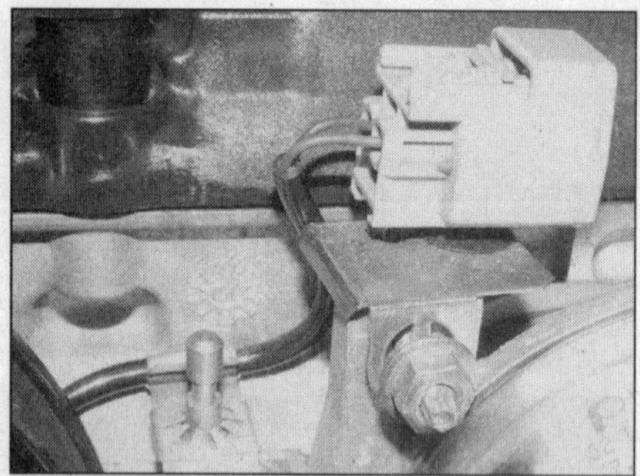

10 Trace the wiring from the sensor, and disconnect it at the plug, which is clipped to the brake vacuum pump.

11 The sensor can now be unscrewed and removed. However, access to the sensor because of the length of wiring, may make it difficult getting a tool to fit onto it. Ultimately, it may be necessary to cut the sensor wiring, unscrew the sensor using a thin-wall socket or box spanner, then remake the wiring after fitting, using a suitable connector.

12 Refitting is a reversal of removal. Clean the threads of the sensor and mounting hole, then refit the sensor and tighten it to the specified torque.

Vehicle speed sensor

13 Refer to the petrol injection system Section.

Temperature and manifold absolute pressure sensor

14 The T-MAP sensor is mounted on the air intake duct which runs along the front of the engine subframe, below the radiator.

15 Disconnect the wiring plug from the sensor.

16 Unscrew and remove the two mounting bolts, and withdraw the sensor from the intake duct.

17 Refitting is a reversal of removal, ensuring that the wiring plug is securely reconnected and the sensor is fitted securely to the intake duct.

Fuel metering valve

18 The fuel metering valve is integral with the injection pump, and cannot be separated from it. If a new injection pump is fitted, the pump control unit (PCU) must be electronically 'matched' to the powertrain control module (PCM), otherwise the pump will not function correctly (or even at all, if the immobiliser function is not correctly set up) – this is a task for a Ford dealer, as specialised electronic equipment is required.

Clutch pedal position switch

19 Remove the trim panel above the driver's footwell to gain access to the clutch pedal; unclip the diagnostic connector plug from the panel as it is removed.

20 Reach up and disconnect the wiring from the clutch switch at the top of the pedal bracket.

21 Slide the retaining clip (where fitted) down to release the switch, then twist the switch anti-clockwise and remove it from the pedal bracket.

22 Refitting is a reversal of removal. **Note:** *A new retaining clip will be required.*

Brake pedal position switch

23 Remove the trim panel above the driver's footwell to gain access to the brake pedal; unclip the diagnostic connector plug from the panel as it is removed.

24 Reach up and disconnect the wiring from the switch at the top of the pedal; there are two switches and if both switches are removed at once, take care to note which wiring plug serves which switch.

25 Slide the retaining clip down to release the switch, then twist the switch through 90° and remove it from the pedal bracket.

26 Refitting is a reversal of removal. **Note:** *A new retaining clip will be required.*

Accelerator pedal sensor

27 The accelerator pedal sensor is integral with the pedal assembly.

Injection pump control unit

28 The pump control unit is integral with the injection pump, and cannot be separated from it. If a new injection pump is fitted, the PCU must be electronically 'matched' to the powertrain control module (PCM), otherwise the pump will not function correctly (or even at all, if the immobiliser function is not correctly set up) – this is a task for a Ford dealer or authorise repairer, as specialised electronic equipment is required.

Powertrain Control Module

Caution: The PCM is fragile. Take care not to drop it or subject it to any other kind of impact, and do not subject it to extremes of temperature, or allow it to get wet. Do not touch the PCM terminals as there is a chance that static electricity may damage the internal electronic components.

Note: *If renewing the powertrain control module, note that it must be reprogrammed for the specific model by a Ford dealer or authorised repairer using the FDS2000 diagnostic equipment. Failure to do so will result in the PCM assuming its 'limited operating strategy' (LOS) settings giving poor performance and economy.*

29 Disconnect the battery negative (earth) lead.

30 Unclip the plastic cover and remove the three retaining nuts from the right-hand strut top mounting. **Note:** ***DO NOT*** *jack up the front of the vehicle, while these nuts are removed.*

31 Unclip the cover from the right-hand strut top mounting brace and undo the retaining nuts and bolts, then remove it from the vehicle.

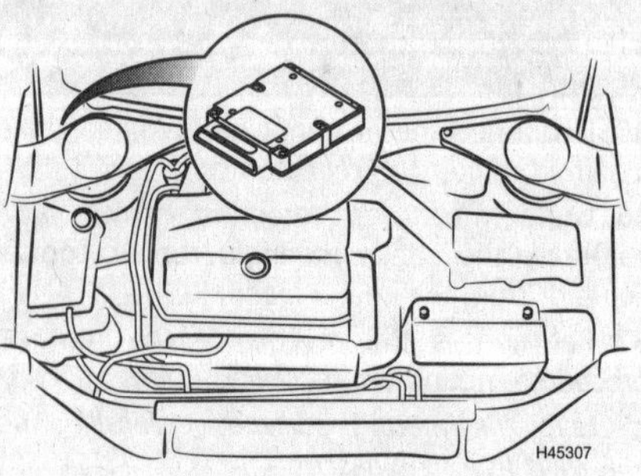

32 A security shield is riveted over the PCM and it will be necessary to drill out the rivet(s) to remove the shield and earth cable.

Caution: The PCM earth cable is connected behind the security shield rivet. Take care not to damage the earth cable when drilling through the rivet.

33 Unclip the wiring harness from along the right-hand inner wing panel.

34 Undo the retaining bolt and release the wiring connector from the PCM – to prevent dust and dirt entering the multi-plug, cover it with a polythene bag.

35 From inside the vehicle, unclip the generic electronic module (GEM) from the PCM bracket.

36 Undo the retaining nut and withdraw the PCM and mounting bracket from the driver's side footwell.

37 Refitting is the reverse of the removal procedure. Whenever the PCM (or battery) is disconnected, the information relating to idle speed control and other operating values will be lost from its memory until the unit has reprogrammed itself; until then, there may be surging, hesitation, erratic idle and a generally-inferior level of performance. To allow the PCM to relearn these values, start the engine and run it as close to idle speed as possible until it reaches its normal operating temperature, then run it for approximately two minutes at 1200 rpm. Next, drive the vehicle as far as necessary – approximately 5 miles of varied driving conditions is usually sufficient – to complete the relearning process.

Injector driver module

38 The module is located under the front, of the left-hand front wing (as viewed from the driver's seat).

39 Remove the front bumper.

40 Undo the retaining bolts and remove the intake pipe from under the left-hand wing. Where applicable, disconnect the foglight's wiring block connector from the clip on the intake pipe.

41 Remove the shear bolts and detach the IDM unit from the housing cover.

42 Support the IDM unit and disconnect the wiring connectors from the unit.

43 The IDM must be electronically 'configured' to the powertrain control module (PCM), this must be done by your Ford Dealer or suitably equipped specialist, as specialised electronic equipment is required.

Mass air flow (MAF) sensor

44 Disconnect the battery negative (earth) lead, and the mass air flow sensor wiring plug.

45 Remove the air cleaner assembly. Unscrew and remove the sensor mounting bolts and withdraw the sensor from the air cleaner cover.

46 Refitting is the reverse of the removal procedure. Ensure that the sensor and air cleaner cover are seated correctly and securely fastened, so that there are no air leaks.

Notes

Land Rover Freelander '97 – '02

17

Petrol models

General

System type:	
Models up to 2001	Rover/Motorola multi-point injection Modular Engine Management system MEMS 1.9
Models 2001-on	Rover/Motorola multi-point injection Modular Engine Management system MEMS 3

Fuel system data

Fuel pump type	Electric, immersed in tank
Fuel pump pressure (approximate):	
Regulated	3.5 ± 0.2 bar
Specified idle speed	775 ± 50 rpm not adjustable
Idle mixture CO content	Regulated by ECM not adjustable

Recommended fuel

Minimum octane rating	95 RON unleaded (UK unleaded premium). Leaded fuel must **not** be used

Diesel models

General

System type:	
L Series	Direct injection with Bosch Electronic Diesel Control (EDC) system
TD4	Direct injection common rail with Bosch high pressure delivery pump and Electronic Diesel Control with DDE 4.0 ECM

Fuel system data	**L Series**	**TD4**
Fuel tank pump pressure	Not applicable	2.5 bar
Fuel injection pump	Bosch VP37	Bosch Cp1 3 x radial piston pump
Injection pressure	Not available	250 to 1350 bar
Turbocharger type	Garrett GT 1549	Mitsubishi MR1 TD025L3-08T-3.3 or Garrett GT1749
Turbocharger boost pressure	1.2 bar (max)	Mitsubishi 1.765 ± 0.03 bar (max) Garrett– Not available
Idle speed*	800 ± 50 rpm	780 ± 30 rpm

** Not adjustable – controlled by engine control module (ECM)*

Torque wrench settings

	Nm	lbf ft
Petrol models		
Camshaft sensor bolt (MEMS 3 only)	6	4
Coolant temperature sensor	15	11
Crankshaft position sensor bolt	6	4
Electronic Control Module (ECM) nuts	9	6
Fuel rail bolts	10	7
Idle air control valve screws	1.5	1
Intake air temperature sensor	7	5
Oxygen sensor	50	36
Spark plugs	27	19
Spark plug cover	10	7
Throttle housing retaining bolts	7	5
Throttle position sensor screws*	1.5	1
L Series diesel engine		
Coolant temperature sensor	5	3
Crankshaft sensor bolt	6	4
Engine electronic control module (ECM) nuts/bolts	9	6
Fuel cut-off solenoid	20	15
Fuel hose union bolt	25	18
Injector pipe union nuts	20	15
Intake air temperature sensor	12	9
Manifold absolute pressure (MAP) sensor bolt	5	4
Throttle position sensor-to-bracket	4	3
Throttle position sensor lever	10	7
TD4 diesel engine		
Camshaft position sensor	8	6
Coolant temperature sensor	15	11
Crankshaft position sensor	8	6
Fuel pressure regulator Torx screws	9	6
Fuel high-pressure sensor	38	28
Fuel rail-to-cylinder head bolts	25	18
Fuel rail-to-coolant rail bolts	10	7
Mass air flow/inlet air temperature sensor	6	4
Throttle position sensor-to-bracket	10	7
Throttle position sensor lever	10	7
Turbocharger boost pressure sensor	8	6

* *Do not reuse*

Engine management components – MEMS 1.8

1 Air cleaner
2 Distributor
3 Ignition coil
4 Spark plugs
5 Intake air temperature sensor
6 Crankshaft position sensor
7 Oxygen sensor
8 Engine electronic control module (with integral MAP sensor)
9 Fuel injectors
10 Idle air control valve
11 Throttle position sensor

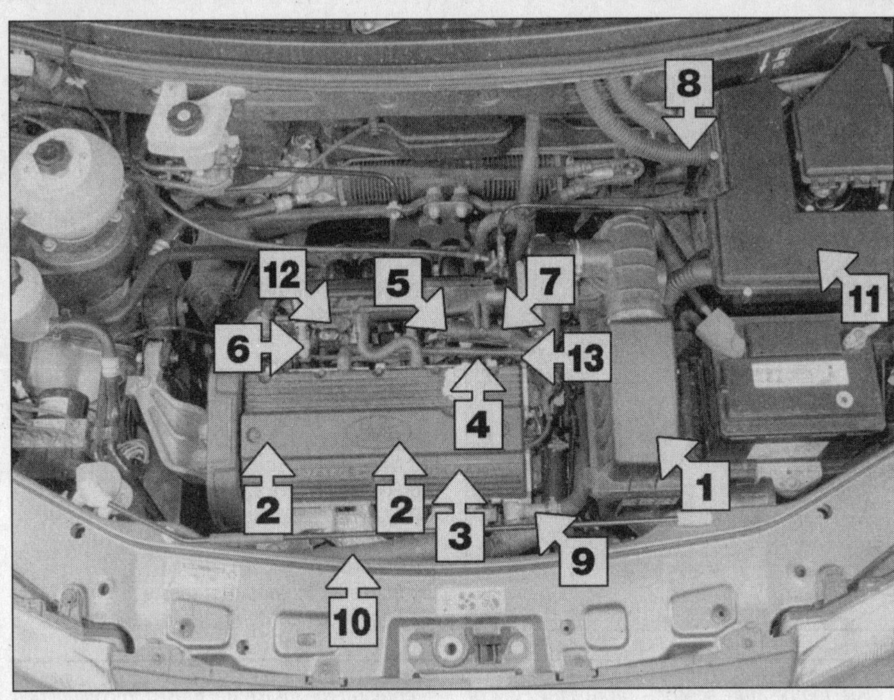

Engine management components – MEMS 3.0

1 Air cleaner
2 Ignition coils
3 Camshaft position sensor
4 Intake air temperature sensor
5 Idle air control valve
6 Fuel injectors
7 Throttle position sensor
8 Fuel inertia shut-off valve
9 Coolant temperature sensor
10 Oxygen sensor
11 Engine electronic control module
12 Manifold absolute pressure (MAP) sensor
13 Crankshaft position sensor

General information and precautions

Petrol models

The 1.8 litre K Series petrol engines are equipped with a multi-point Rover/Motorola modular engine management (fuel injection/ignition) system (MEMS). The earlier models (upto 2001 model year) are fitted with MEMS 1.8, and the later models (2001 model year-on) are fitted with MEMS 3. The main difference between the two systems is that MEMS 1.8 incorporates a distributor and rotor arm to deliver the HT voltage to the spark plugs, and the injectors are fired semi-sequentially (injectors 1 and 4, 2 and 3 are triggered together). Whilst MEMS 3 has fully sequential injection, with spark plugs 1 and 4, 2 and 3 fired together by two ignition coils mounted directly above Nos.1 and 3 spark plugs (no distributor). MEMS 3 relies on the camshaft position sensor to inform the ECM of the engine firing sequence position. Both systems incorporate a closed-loop catalytic converter and an evaporative emission control system, and complies with the latest emission control standards. The fuel injection side of the system operates as follows.

The fuel pump, immersed in the fuel tank, pumps fuel from the fuel tank to the fuel rail, via a filter which is located in the engine compartment. Fuel supply pressure is controlled by the pressure regulator, which is incorporated as part of the pump assembly in the tank. A pressure damper is fitted to the fuel rail, to absorb pressure fluctuations.

The electrical control system consists of the electronic control module (ECM), along with the following sensors.

a) Throttle position sensor – informs the ECM of the throttle position, and the rate of throttle opening or closing.
b) Coolant temperature sensor – informs the ECM of engine temperature.
c) Intake air temperature sensor – informs the ECM of the temperature of the air passing through the inlet manifold.
d) Oxygen sensor – informs the ECM of the oxygen content of the exhaust gases (explained in greater detail in Part C of this Chapter).
e) Crankshaft sensor – informs the ECM of engine speed and crankshaft position.
f) Manifold absolute pressure (MAP) sensor (contained within the ECM on MEMS 1.8) – informs the ECM of the engine load by monitoring the pressure in the inlet manifold.
g) Camshaft position sensor (MEMS 3 only) – informs the ECM of the camshaft position.
h) Idle air control valve

All the above information is analysed by the ECM and, based on this, the ECM determines the appropriate ignition and fuelling requirements for the engine. The ECM controls the fuel injector by varying its pulse width – the length of time the injector is held open – to provide a richer or weaker mixture, as appropriate. The mixture is constantly varied by the ECM, to provide the best setting for cranking, starting (with either a hot or cold engine), warm-up, idle, cruising, and acceleration.

The ECM also has full control over the engine idle speed. The idle speed is controlled via the idle air control valve; the valve controls the opening of an air passage which bypasses the throttle valve. When the throttle valve is closed, the ECM controls the opening of the valve, which in turn regulates the amount of air entering the manifold, and so controls the idle speed.

The ECM also controls the exhaust and evaporative emission control systems.

If there is an abnormality in any of the readings obtained from any sensor, the ECM enters its back-up mode. In this event, the ECM ignores the abnormal sensor signal, and assumes a pre-programmed value which will allow the engine to continue running (albeit at reduced efficiency). If the ECM enters this back-up mode, the warning light on the instrument panel will come on, and the relevant fault code will be stored in the ECM memory.

An inertia switch is incorporated into the fuel system to cut off the fuel supply in the event of an accident. The switch is located behind the front of the centre console, on the driver's side, and can be reset by depressing the button on the top of the switch.

Diesel models

L Series engines

The system is under the overall control of the Electronic Diesel Control (EDC) system, which also controls the pre-heating system.

Fuel is supplied from the rear-mounted fuel tank, via a fuel filter, to the fuel injection pump. The fuel injection pump supplies the exact amount of fuel required by the engine, according to the prevailing engine operating conditions.

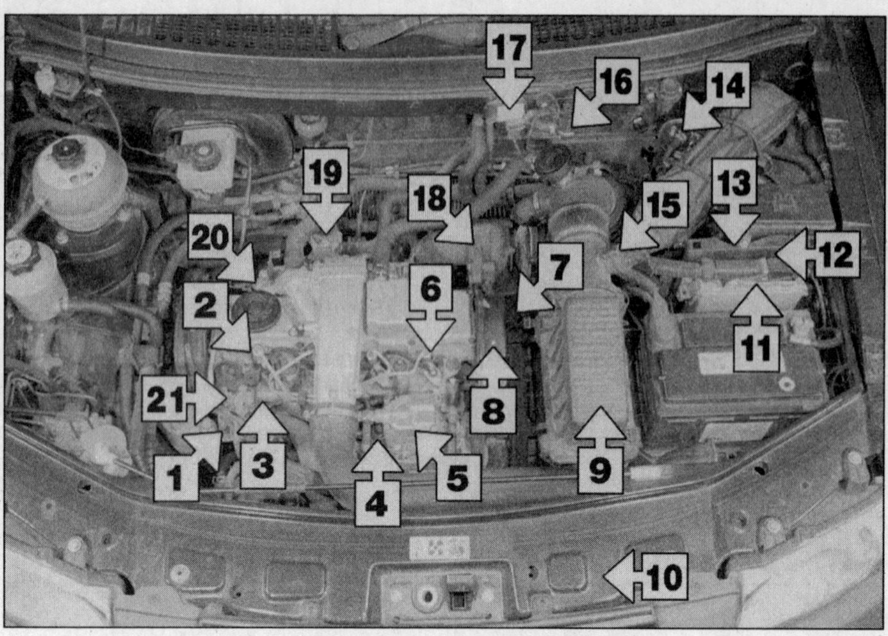

L-Series electronic diesel control components

1 Coolant temperature gauge sender
2 Needle lift sensor
3 Glow plugs
4 Fuel shut-off solenoid
5 Fuel injection pump
6 Injectors
7 Vehicle speed sensor
8 Crankshaft position sensor
9 Air cleaner
10 Intercooler
11 Engine control module
12 Glow plug relay
13 Fuel injection pump relay
14 Fuel filter
15 Mass airflow sensor
16 EGR modulator valve
17 Manifold absolute pressure sensor (MAP)
18 Turbocharger
19 EGR valve
20 Intake air temperature sensor
21 Engine coolant temperature sensor

The engine is fitted with various sensors, which monitor the engine operating conditions, and transmit data to the engine control module (ECM). The control module processes the data from the various sensors, and determines the optimum amount of fuel required, and the injection timing for the prevailing running conditions. Additionally, the control module activates the fuel injection pump stop solenoid, the pre-heating system, and the exhaust gas recirculation (EGR) system.

The system uses the following sensors.

a) Crankshaft sensor – informs the ECM of the crankshaft speed and position.

b) Coolant temperature sensor – informs the ECM of engine temperature.

c) Fuel temperature sensor – informs the ECM of fuel temperature (in the injection pump).

d) Airflow meter – informs the ECM of the mass of air entering the intake tract.

e) Fuel injector needle lift sensor – informs the ECM of the start of the injection sequence.

f) Vehicle speed sensor – informs the ECM of the vehicle speed (non-ABS models only).

g) Fuel quantity servo position sensor – informs the ECM of the quantity of fuel supplied to the injectors by the fuel injection pump.

h) Throttle position sensor – informs the ECM of throttle position, and the rate of throttle opening/closing.

i) Manifold absolute pressure (MAP) sensor – informs the ECM of the pressure of air entering the intake tract (used in conjunction with the intake air temperature sensor to calculate the volume of oxygen in the air entering the engine).

j) Intake air temperature sensor – informs the ECM of the temperature of air entering the engine.

k) Stop-light switch – informs the ECM when the brakes are being applied.

TD4 engines

The system is under the overall control of the Electronic Diesel Control (EDC) system, which also controls the pre-heating system.

Fuel is supplied from the rear-mounted fuel tank, via an electrically powered lift pump, and

TD4 electronic diesel control components

1 Air cleaner
2 Crankshaft position sensor
3 Coolant temperature sensor
4 Fuel high-pressure sensor
5 Mass airflow/intake air temperature sensor
6 Engine electronic control module
7 Fuel low-pressure sensor
8 Fuel lift pump
9 EGR control valve
10 Turbocharger boost pressure sensor
11 High pressure fuel pump
12 Fuel pressure regulator
13 Fuel injectors

fuel filter, to the fuel injection pump. The fuel injection pump supplies fuel under high pressure to the common fuel rail. The fuel rail provides a reservoir of fuel under pressure ready for the injectors to deliver direct to the combustion chamber. The individual fuel injectors incorporate solenoids, which when operated, allow the high pressure fuel to be injected. The solenoids are controlled by the ECM. The fuel injection pump purely provides high pressure fuel. The timing and duration of the injection is controlled by the ECM based, on the information received from the various sensors. In order to increase combustion efficiency and reduce combustion noise (diesel 'knock'), a small amount of fuel is injected before the main injection takes place – this is known as Pre- or Pilot-injection.

Additionally, the control module activates the pre-heating system, and the exhaust gas recirculation (EGR) system.

The system uses the following sensors.

a) Crankshaft sensor – informs the ECM of the crankshaft speed and position.
b) Coolant temperature sensor – informs the ECM of engine temperature.
c) Mass airflow/intake temperature sensor – informs the ECM of the mass and temperature of air entering the intake tract.
d) Wheel speed sensor – informs the ECM of the vehicle speed.
e) Throttle position sensor – informs the ECM of throttle position, and the rate of throttle opening/closing.
f) Fuel high-pressure sensor – informs the ECM of the pressure of the fuel in the common rail.
g) Fuel low pressure sensor – informs the ECM of the pressure of the fuel in the filter.
h) Fuel pressure regulator – controls the pressure produced by the high pressure fuel pump.
i) Camshaft position sensor – informs the ECM of the camshaft position so that the engine firing sequence can be established.
j) Stop-light switch – informs the ECM when the brakes are being applied
k) Turbocharger boost pressure sensor – informs the ECM of the boost pressure generated by the turbocharger.

All diesel enignes

On all models, a 'drive-by-wire' throttle control system is used. The accelerator pedal is not physically connected to the fuel injection pump, but instead is connected by a cable to a throttle position sensor, mounted in the engine compartment, which provides the engine control module (ECM) with a signal relating to accelerator pedal movement.

The signals from the various sensors are processed by the ECM, and the optimum fuel quantity and injection timing settings are selected for the prevailing engine operating conditions.

A catalytic converter and an exhaust gas recirculation (EGR) system is fitted, to reduce harmful exhaust gas emissions.

If there is an abnormality in any of the readings obtained from any sensor, the ECM enters its back-up mode. In this event, the ECM ignores the abnormal sensor signal, and assumes a pre-programmed value which will allow the engine to continue running (albeit at reduced efficiency). If the ECM enters this back-up mode, the warning light on the instrument panel will come on, and the relevant fault code will be stored in the ECM memory.

The diagnostic connector is located behind the driver's side of the facia; to gain access to the connector unclip the storage pocket and reach in through the facia aperture.

Fuel injection system – depressurisation, priming and bleeding

Petrol models

Warning: Refer to the warning notes in Chapter 1 before proceeding. The following procedure will merely relieve the pressure in the fuel system – remember that fuel will still be present in the system components, and take precautions accordingly before disconnecting any of them.

1 The fuel system referred to in this Section is defined as the tank-mounted fuel pump, the fuel filter, the fuel injector(s) and the pressure regulator, and the metal pipes and flexible hoses of the fuel lines between these components. All these contain fuel which will be under pressure while the engine is running, and/or while the ignition is switched on. The pressure will remain for some time after the ignition has been switched off, and it must be relieved in a controlled fashion when any of these components are disturbed for servicing work.

2 The fuel system is depressurised by disconnecting the fuel rail quick release coupling.

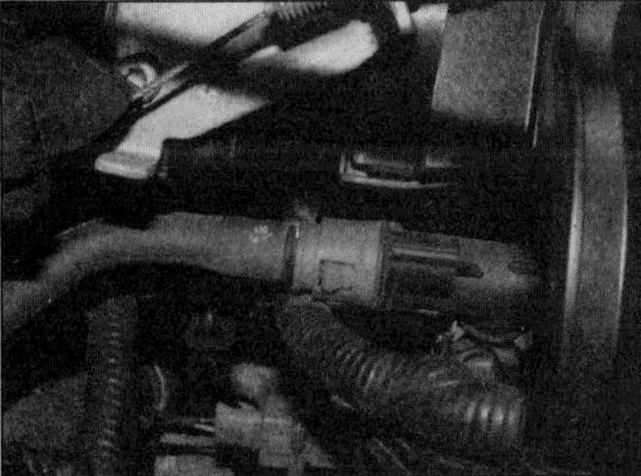

3 Ensure the ignition is switched off and position wads of rag around the fuel feed pipe connection to the fuel rail to catch the spilled fuel. Squeeze together the two tabs of the quick release coupling, and slowly release the pipe from the rail.

4 Ensure all fuel pressure has been released, then push the feed pipe back into the fuel rail.

Diesel models

Depressurisation

5 Although there is no specific requirement to depressurise the fuel system, common sense dictates that it would be wise to wait until the engine has cooled sufficiently prior to attempting to dismantle any part of the engine. If this advice is followed, most of the residual pressure in the system will have dissipated before any fuel connections are disturbed. However, to be on the safe side, place a clean rag over the fuel pipe connection as the pipe union is slackened to soak up any fuel spray.

Priming and bleeding

6 On L Series engines, after any operation which requires the disconnection of any fuel hose, it is necessary to prime and bleed the fuel system; the priming pump is located in the left-hand rear corner of the engine compartment, where it is fitted to the fuel filter inlet pipe. On models fitted with the TD4 engine, the system is self priming.

7 Position wads of absorbent rag around the fuel filter then slacken the bleed screw which is fitted to the top of the fuel filter. Gently squeeze and release the pump until fuel which is free of air bubbles is flowing out of the filter. Once all traces of air have been removed, squeeze and hold the pump then securely tighten the bleed screw before releasing the pump. Remove the rag from around the filter and mop up any spilt fuel.

8 Turn on the ignition switch and gently squeeze and release the pump until resistance is felt. Once the lines are full of fuel (indicated by the resistance felt when the pump is squeezed), stop pumping and turn off the ignition.

9 On all models, depress the accelerator pedal to the floor then start the engine as normal (this may take longer than usual, especially if the fuel system has been allowed to run dry – operate the starter in ten second bursts with 5 seconds rest in between each operation). Run the engine at a fast idle speed for a minute or so to purge any remaining trapped air from the fuel lines. After this time the engine should idle smoothly at a constant speed.

10 If the engine idles roughly, then there is still some air trapped in the fuel system. Increase the engine speed again for another minute or so then recheck the idle speed. Repeat this procedure as necessary until the engine is idling smoothly.

Petrol injection system components – removal and refitting

Fuel rail and injectors

Note: *If a faulty injector is suspected, before condemning the injector, it is worth trying the effect of one of the proprietary injector-cleaning treatments.*

1 Disconnect the battery negative terminal, then depressurise the fuel system.

2 On models equipped with MEMS 1.8, remove the air cleaner assembly

3 Disconnect the wiring plug from the IAC (idle air control) valve.

4 Release the retaining clip, and disconnect the vacuum pipe from the fuel pressure damper at the right-hand end of the fuel rail.

5 Slide the engine oil dipstick out of the guide tube.

6 On models equipped with MEMS 1.8, release the injector wiring harness wiring plug from its retaining bracket, and disconnect the plug.

7 Undo the two retaining bolts, and release the fuel rail and injectors from the inlet manifold.

8 On models equipped with MEMS 3, release and remove the injector spacer, and disconnect the wiring plugs from each injector, and remove the fuel rail with the injectors from the engine.

9 On models equipped with MEMS 1.8, remove the fuel rail, complete with injectors and wiring harness from the engine. Disconnect the wiring plugs from each injector.

10 On all models, remove the lower sealing rings from the injectors and discard them; they must be renewed whenever they are disturbed.

11 Slide off the retaining clips and withdraw the injectors from the fuel rail. Remove the upper sealing ring from the each injector and discard it; all disturbed sealing rings must be renewed.

12 Refitting is a reversal of the removal procedure, noting the following points:

a) Renew all disturbed sealing rings and apply a smear of engine oil to them to aid installation.

b) Ease the injector(s) into the fuel rail, ensuring that the sealing ring(s) remain correctly seated. Secure in position with the retaining clips and reconnect the wiring connectors.

c) On refitting the fuel rail, take care not to damage the injectors and ensure that all sealing rings remain in position. Once the fuel rail is correctly seated, tighten its retaining bolts to the specified torque.

d) On completion start the engine and check for fuel leaks.

Fuel pressure damper

13 The damper is an integral part of the fuel rail and cannot be renewed separately. If the damper is faulty, renew the fuel rail as described earlier in this Section.

Throttle position sensor

Note: *New retaining screws must be used on refitting.*

14 Disconnect and remove the hose from the IAC (idle air control) valve to the throttle body.

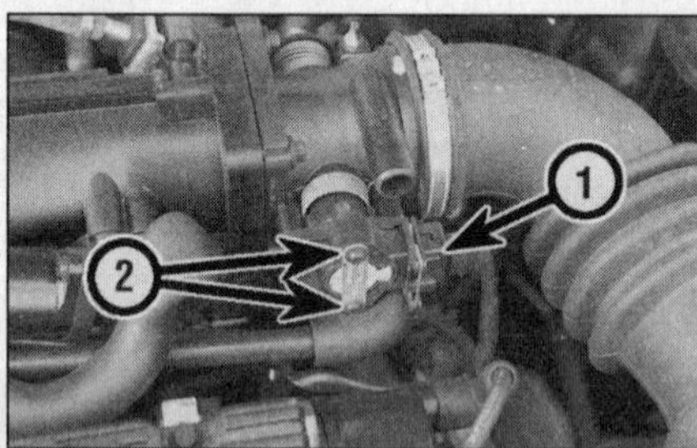

15 Ensure the ignition is switched off then disconnect the wiring connector (1) from the sensor which is fitted to the side of the throttle housing.

16 Slacken and remove the two Torx screws (2) securing the sensor to the throttle body. Discard the screws and wave washers – new ones must be fitted. Remove the sensor and specification plate. **Note:** *Do not twist or lever the sensor.*

17 On refitting, carefully align the throttle valve spindle with the sensor slot and seat the sensor on the housing. Gently push the sensor onto the spindle, applying pressure only to the centre portion of the sensor.

18 Rotate the sensor in an anti-clockwise direction to align the fixing holes. Do **not** turn the sensor in a clockwise direction, and ensure that it is not turned beyond its internal stops – the sensor is easily damaged.

19 Fit the retaining plate and new retaining screws and washers, tightening them to the specified torque, and reconnect the wiring connector.

Caution: Do not overtighten the retaining screws as the sensor is easily damaged.

20 Reconnect the wiring connector to the sensor, and refit the IAC hose. Note that after replacing the sensor, the ECM may take a little while to 're-learn' the throttle valve closed position. During this time, the engine may suffer from a raised, or erratic, idle.

Intake air temperature sensor

21 The sensor is screwed into the left-hand end of the inlet manifold, in the side of the No.4 cylinder inlet tract.

22 Ensure the ignition is switched off then disconnect the wiring connector from the sensor.

23 Unscrew the sensor and remove it from the manifold along with its sealing washer (where fitted).

24 On refitting, ensure the manifold and sensor threads are clean and dry. If the sensor was originally fitted with a sealing washer, use a new sealing washer. Where no sealing washer was fitted, clean the threads and apply a smear of sealant to them.

25 Refit the sensor to the manifold, tightening it to the specified torque, and reconnect the wiring connector.

Idle air control valve

26 Ensure the ignition is switched off then disconnect the wiring connector (1) from the valve which is fitted to the top of the inlet manifold.

27 On models equipped with MEMS 1.8, undo the two retaining screws (2) then carefully remove the motor from its mounting bracket and lift it away from the engine. Recover the sealing ring and discard it, a new one should be used on refitting.

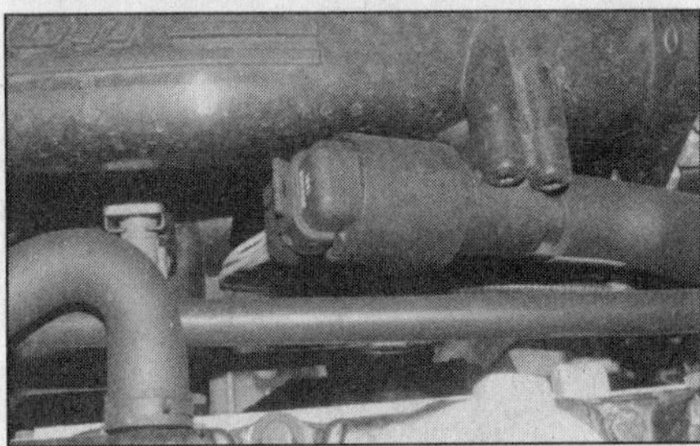

28 On models equipped with MEMS 3, disconnect the hose from the valve to the throttle body, undo the four Torx screws, and remove the valve. Discard the O-ring seal, a new must be fitted.

29 On refitting, ensure the mating surfaces are clean and dry. Fit a new sealing ring then ease the motor into position and tighten its retaining screws to the specified torque. Reconnect the wiring connector, and where applicable, throttle body hose.

Coolant temperature sensor

30 The sensor is located in the coolant outlet elbow at the front left-hand end of the cylinder head. Disconnect the wiring plug, and unscrew the sensor from the elbow. Be prepared for coolant spillage.

31 Refitting is a reversal of removal, but apply suitable sealant (Land Rover recommend Loctite 577) to the sensor threads and tighten it to the specified torque. Top up the cooling system.

Manifold absolute pressure (MAP) sensor

32 On models equipped with MEMS 1.8, the sensor is an integral part of the ECM and cannot be renewed separately. details. On models equipped with MEMS 3, the sensor is located on the right-hand side of the inlet manifold. Slide off the cover, and disconnect the wiring plug from the sensor.

33 Unscrew the two Torx screws and remove the sensor.

34 On refitting, ensure the sensor and manifold mating surfaces are clean.

35 Refit the sensor to the manifold, and tighten the Torx screws securely. Reconnect the wiring plug, and fit the plug cover.

Crankshaft position sensor

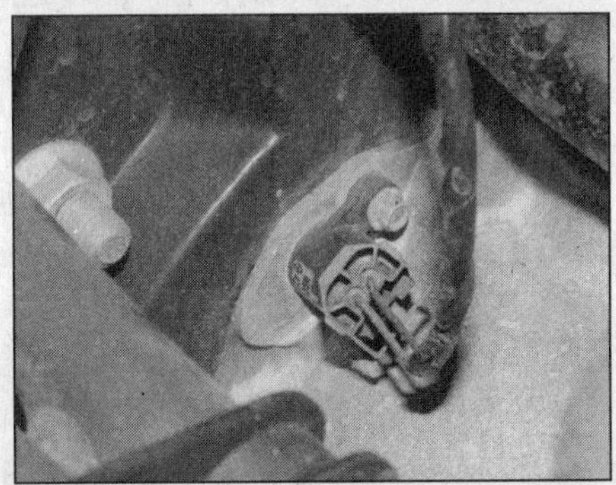

36 The sensor is fitted into the transmission mounting plate at the rear of the engine. Ensure the ignition is switched off, then disconnect the wiring plug from the sensor.

37 Unscrew the retaining bolt and remove the sensor from the engine.

38 Refitting is the reverse of removal, tightening the retaining bolt to the specified torque.

Vehicle speed sensor

Non ABS models

39 The vehicle speed sensor is mounted on the top of the speedometer drive. As all later models were equipped with ABS, the speedometer drive is only fitted to pre-June 2000 PG1 gearboxes. The speedometer drive is situated on the top of the gearbox housing, next to the inner end of the right-hand driveshaft. Access to the drive is poor from above, to gain access from below, firmly apply the handbrake then jack up the front of the vehicle and support it on axle stands . Release the retaining screws and remove the engine/transmission undertray.

40 Disconnect the wiring connector from the vehicle speed sensor then unscrew the sensor and remove it from the top of the speedometer drive.

41 Refitting is a reversal of removal, ensuring that the drive pin is engages correctly with the pinion.

Electronic control module (ECM)

Models equipped with MEMS 1.8

42 The ECM is mounted onto the engine compartment bulkhead. Prior to removal, first disconnect the battery negative terminal.

Note: If a new ECM is to be fitted, it will be necessary to entrust the task to a Land Rover dealer. After fitting, it will be necessary to programme the anti-theft system code into the ECM to enable it function correctly. This can only be done using the special Land Rover equipment which is plugged into the diagnostic connector.

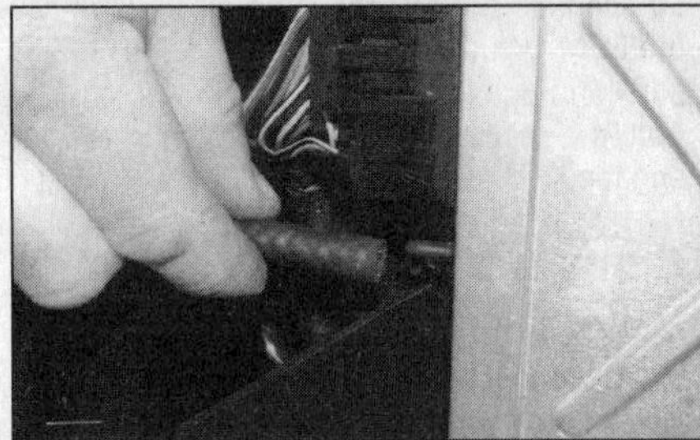

43 Disconnect the wiring connector and vacuum hose from the ECM.

44 Undo the retaining nut, then free the ECM from its lower mounting and remove it from the engine compartment.

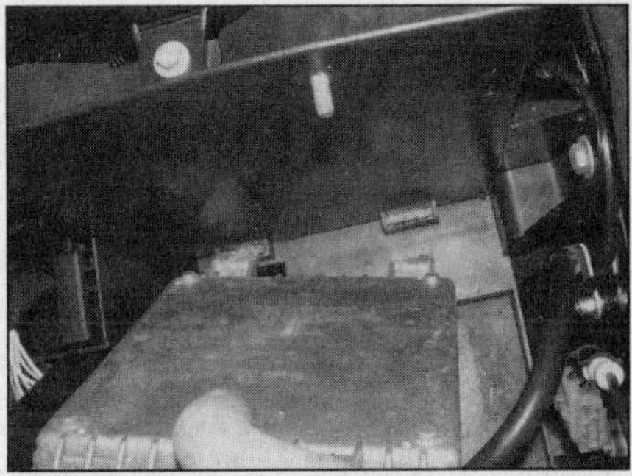

45 Refitting is the reverse of removal ensuring the wiring connector and vacuum hose are securely reconnected. Tighten the module nut to the specified torque setting. Note that the ECM lugs must engage with the slots in the bracket.

Models equipped with MEMS 3

46 Disconnect the battery negative terminal.

47 Unscrew the five Allen bolts and remove the electrical box cover, located behind the battery.

48 Undo the two screws securing the ECM harness clamp.

49 Pull out the locking latches and disconnect the wiring plugs from the ECM.

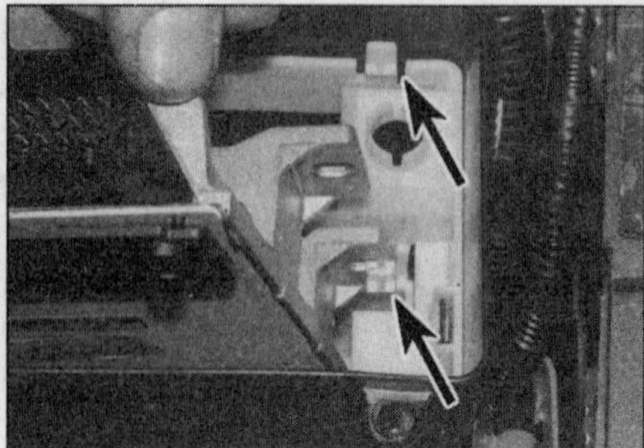

50 Depress the two retaining clips and remove the ECM.

51 Refitting is the reverse of removal, ensuring that the wiring plug locking catches are fully engaged.

Camshaft position sensor

MEMS 3 only

52 Disconnect the battery negative lead.

53 Slacken and remove the three retaining bolts, and remove the spark plug cover from the engine.

54 Release the sensor harness from the retaining bracket, and disconnect the sensor wiring plug.

55 Unscrew the sensor retaining bolt, and remove the sensor, freeing the harness from the retaining clips as the sensor is withdrawn.

56 On refitting, ensure the sensor and mating face are clean and dry. Refit the sensor, tighten the retaining bolt to the specified torque, and reconnect the wiring connector.

57 Ensure that the harness and connector are refitted to their retaining bracket/clips, and refit the spark plug cover. Tighten the bolts to the specified torque, and reconnect the battery negative lead.

Fuel cut-off inertia switch

58 The fuel cut-off inertia switch is located on the engine compartment bulkhead, in the left-hand corner of the engine compartment. Ensure the ignition is switched off, then disconnect the wiring plug from the switch.

59 Undo the retaining screws and remove the switch.

60 Refitting is the reverse of removal. On completion, reset the switch by depressing the button on the top of the switch.

Electronic Diesel Control (EDC) system components – removal and refitting

Crankshaft sensor

1 The sensor is located at the rear of the engine where it is mounted onto the engine/transmission mounting plate, at the rear of the flywheel on L Series engines. To gain access to the sensor, remove the inlet and exhaust manifolds. On TD4 engines, the sensor is mounted on the front face of the engine block. To gain access, remove the starter motor.

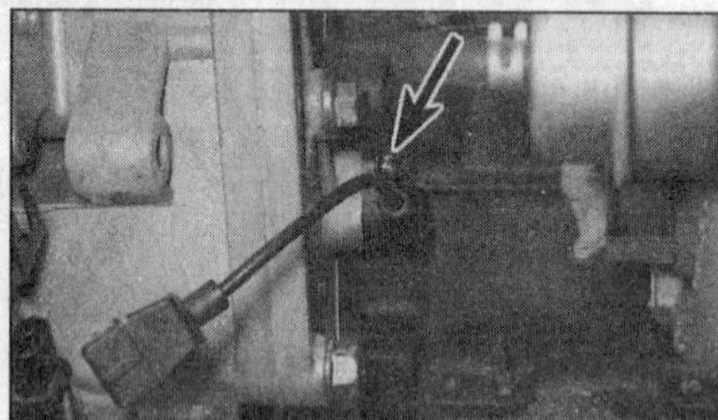

2 Ensure the ignition is switched off, then disconnect the wiring connector from the sensor.

3 Slacken and remove the retaining bolt and carefully remove the sensor from the engine.

4 Refitting is the reverse of removal, tightening the retaining bolt to the specified torque.

Airflow meter

L Series engines only

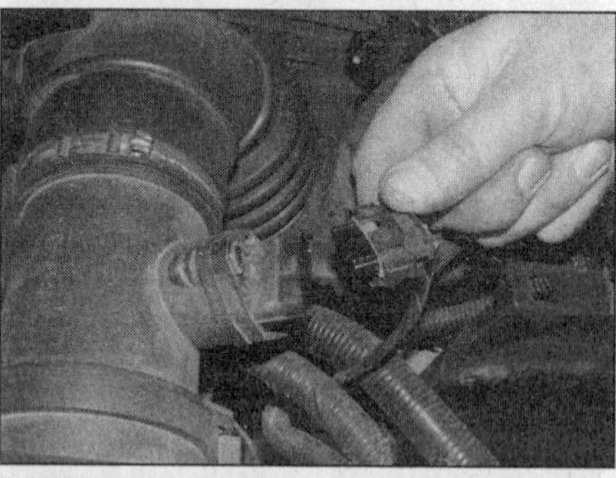

5 Ensure the ignition is switched off then release the retaining clip and disconnect the wiring connector from the airflow meter.

6 Slacken the retaining clip and detach the intake duct from the airflow meter.

7 Release the retaining clips then remove the airflow meter from the air cleaner housing, along with its sealing ring.

8 Refitting is the reverse of removal, using a new sealing ring. Make sure the meter is clipped securely in position and the intake duct is securely retained by its clip.

Fuel injector needle lift sensor

L Series engines only

9 The fuel injector needle lift sensor is an integral part of No 1 cylinder fuel injector and cannot be renewed separately.

Coolant temperature sensor

2.0 litre L Series engine

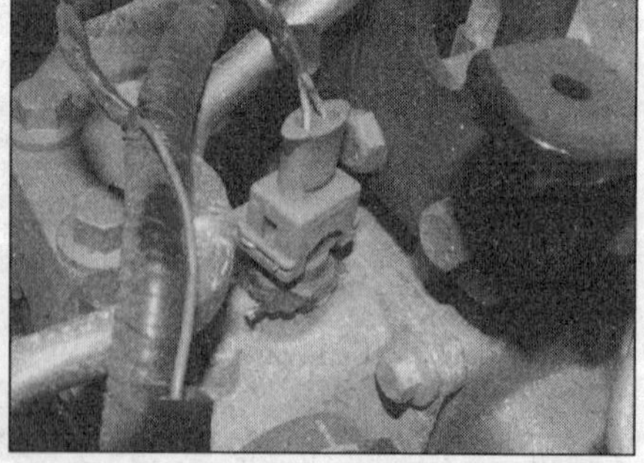

10 The sensor is located in the coolant outlet elbow at the front right-hand end of the cylinder head. Disconnect the wiring plug, and unscrew the sensor from the outlet. Be prepared for coolant spillage. Note that on L Series engines, the temperature gauge sender unit is also fitted to the coolant elbow.

11 Refitting is a reversal of removal, but apply suitable sealant (Land Rover recommend Loctite 577) to the sensor threads and tighten it to the specified torque. Top up the cooling system.

2.0 litre TD4 engine

12 Drain the cooling system. Remove the intake manifold.

13 Disconnect the wiring plug from the sensor, and using a deep socket, unscrew it from the cylinder head. Discard the seal – a new one must be fitted.

14 Refitting is a reversal of removal, but apply suitable sealant (Land Rover recommend Loctite 577) to the sensor threads and tighten it to the specified torque. Top up the cooling system.

Fuel temperature sensor

L Series engines only

15 The sensor is an integral part of the fuel injection pump, and cannot be renewed separately.

Vehicle speed sensor

Non-ABS models only

16 The sensor is driven by the speedometer drive on the transmission unit. Refer to the petrol injection system Section for removal and refitting details.

Fuel quantity servo position sensor

L Series engines only

17 The sensor is integral with the fuel injection pump, and cannot be renewed separately.

Throttle pedal position sensor

18 The sensor is secured to the accelerator pedal bracket. Disconnect the battery negative lead, and on LHD models, it is necessary to remove the facia.

19 Working underneath the drivers side facia, release the throttle sensor wiring plug from its bracket, then disconnect the wiring plug.

20 Unscrew the nut securing the lever to the sensor, then unscrew the retaining bolts/nuts and remove the sensor.

21 Refitting is the reverse of removal. **Note:** *If the pedal operation is unsatisfactory after fitting a new sensor, have the sensor operation confirmed by means of dedicated test equipment – consult your Land Rover dealer or specialist.*

Manifold absolute pressure (MAP) sensor

L Series engines only

22 The sensor is mounted onto the engine compartment bulkhead, where it is located left of centre adjacent to the EGR modulator valve.Ensure the ignition is switched off then disconnect the wiring connector and vacuum hose from the sensor.

23 Slacken and remove the retaining bolt and remove the MAP sensor from the vehicle.

24 Refitting is the reverse of removal, tightening the sensor retaining bolt to the specified torque.

Intake air temperature sensor

L Series engines

25 Ensure the ignition is switched off then disconnect the wiring connector from the intake air temperature sensor which is screwed into the right-hand end of the inlet manifold.

26 Unscrew the sensor and remove it from the manifold.

27 On refitting, remove all traces of sealant from the sensor and manifold threads and apply a smear of fresh sealant (Land Rover recommend the use of Loctite 577) to the sensor threads. Refit the sensor, tightening it to the specified torque, and reconnect the wiring connector. Refit the cover to the engine.

Intake air temperature/Mass air flow sensor

TD4 engines

28 Unscrew the retaining bolts and remove the plastic cover from the top of the engine, taking care not to lose the spacers which are fitted to the cover mounting rubbers.

29 Slacken the retaining clips and release the intake ducting assembly. Unscrew the two retaining screws and remove the assembly from the camshaft cover.

30 Remove the engine oil filler cap, unscrew the five Allen screws and remove the air cleaner cover. Refit the oil filler cap.

31 Disconnect the wiring plug from the sensor.

32 Free the vacuum hose and wiring harness from the turbocharger duct.

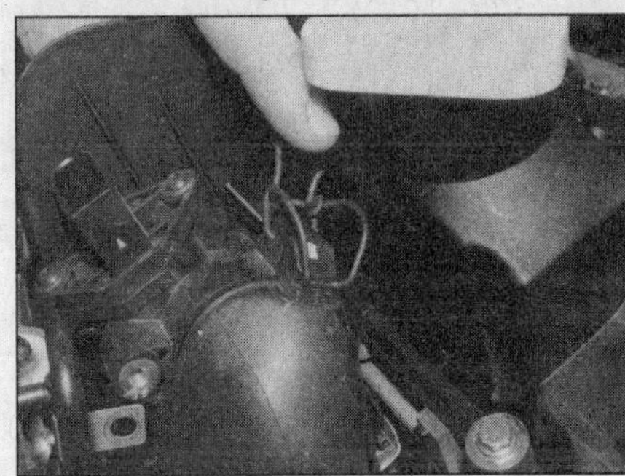

33 Release the two retaining clips and disconnect the turbocharger duct from the sensor. Recover the duct seal and engine breather tube connector.

34 Unscrew the breather pipe retaining bracket bolt and move the pipe to one side.

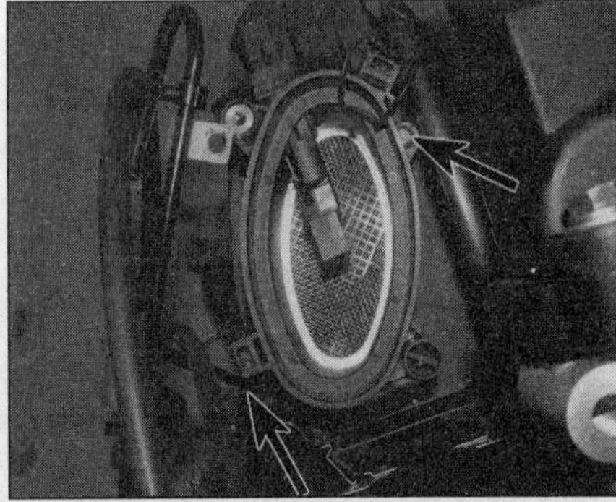

35 Slacken and remove the two Torx screws, and withdraw the sensor.

36 Refitting is a reversal of removal.

Stop-light switch

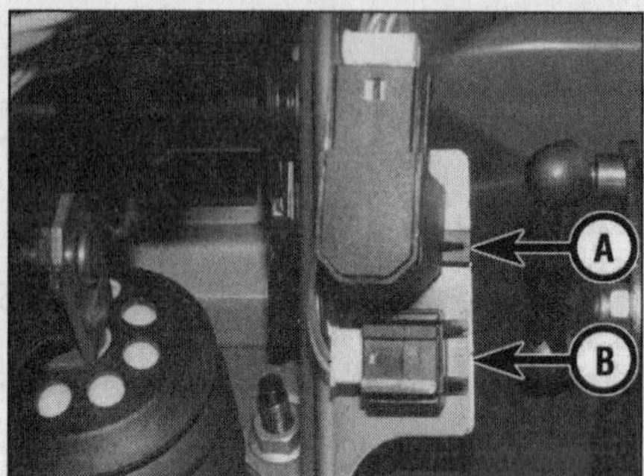

Stop-light switch (A) and brake pedal position switch (B)

37 The stop-light switch is located on the pedal bracket behind the facia. Disconnect the wiring connector plugs from the stop-light switch. On models with cruise control, the stop-light switch is the upper-mounted of the two switches.

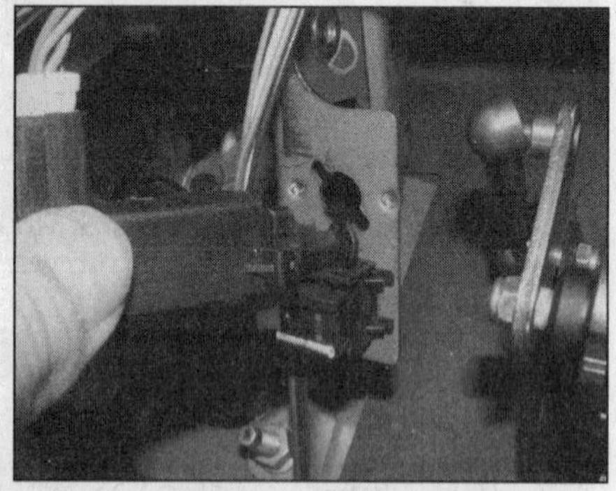

38 Twist the switch body clockwise slightly, to disengage its bayonet fitting, and remove it from the pedal bracket.

39 Refitting is a reversal of removal, but adjust the switch as follows. With the switch refitted in the pedal bracket, depress and hold the brake pedal, then pull out the switch plunger to its full extent. Release the pedal, and the switch plunger will automatically be set to the correct position. Test the operation of the switch on completion.

Fuel cut-off solenoid

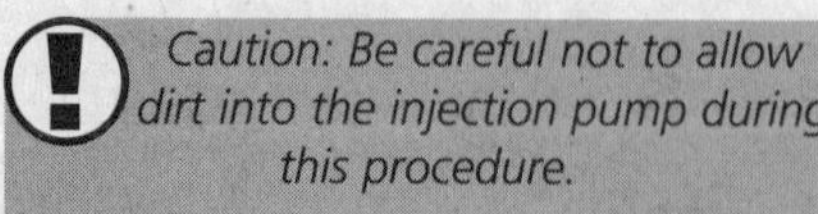

Caution: Be careful not to allow dirt into the injection pump during this procedure.

L Series engines only

40 The fuel cut-off solenoid is fitted to the top of the fuel injection pump. Unscrew the retaining bolts and remove the plastic cover from the top of the engine, taking care not to lose the spacers which are fitted to the cover mounting rubbers. Slacken the retaining clip and disconnect the intercooler duct from the manifold to improve access to the pump.

41 Ensure the ignition is switched off then unscrew the retaining nut and disconnect the wire from the solenoid terminal.

42 Wipe clean the area around the solenoid and position a wad of absorbent cloth around the pump to absorb escaping fuel, then slacken and remove the solenoid from the injection pump and recover the solenoid plunger and spring, noting each components correct fitted location. Remove the sealing ring and discard it; a new one must be used on refitting.

43 Refitting is the reverse of removal using a new sealing ring. Ensure the plunger and spring are fitted the correct way around and tighten the solenoid to the specified torque.

Engine electronic control module (ECM)

L Series engines

44 The electronic control module (ECM) is mounted on the left-hand side of the engine compartment, behind the battery. Remove the battery.

45 Using a flat-bladed screwdriver, release the locking clamp and disconnect the wiring plug from the ECM.

Note: If a new ECM is to be fitted, it will be necessary to entrust the task to a Land Rover dealer or specialist. After fitting, it will be necessary to programme the anti-theft system code into the ECM to enable it function correctly. This can only be done using the special Land Rover equipment which is plugged into the diagnostic connector.

46 Undo the two bolts and move the engine compartment fusebox to one side.

47 Slacken and remove the four retaining nuts/bolts and withdraw the ECM from its mounting bracket.

48 Refitting is the reverse of removal ensuring the wiring connector is securely reconnected. Tighten the ECM retaining. nuts/bolts to the specified torque setting.

TD4 engines

49 Disconnect the battery negative lead.

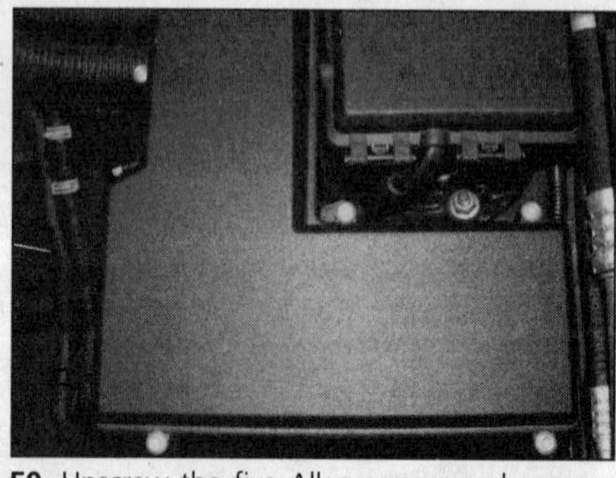

50 Unscrew the five Allen screws and remove the electrical box cover, located on the left-hand side of the engine compartment, behind the battery.

51 Note their fitted locations, and disconnect the five wiring plugs from the ECM. Some of the

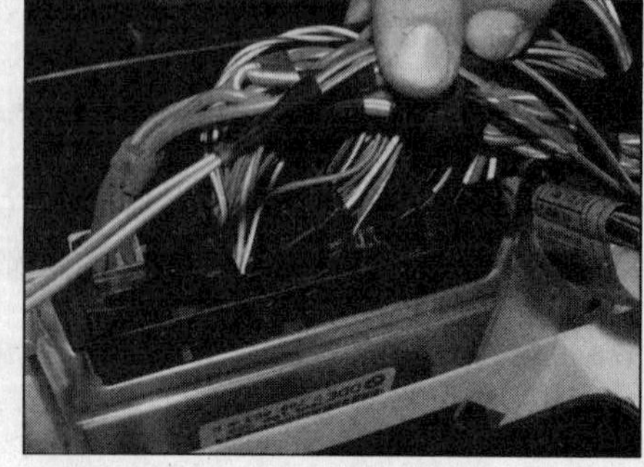

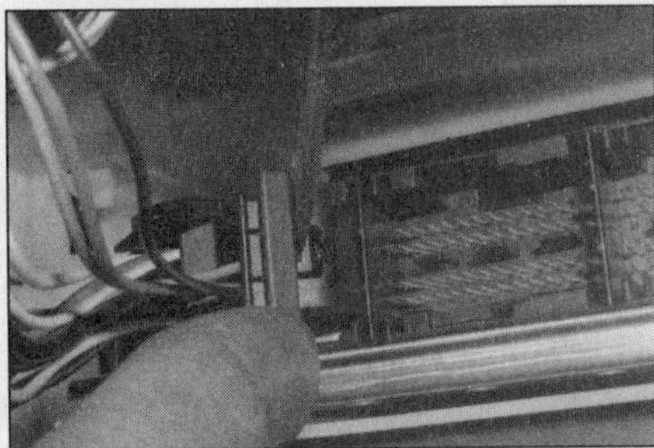

plugs are locked in place. To release these plugs, depress the lug and pivot the locking lever over.

52 Release the two retaining clips and remove the ECM.

53 Refitting is the reverse of removal.

Relays

L-Series engines

54 On these models, the main (fuel injection pump) relay and the glow plug relay are mounted adjacent to the main engine compartment fusebox, located in the left-hand side corner; the glow plug relay is the one nearest the wing and the main relay is situated next to the glow plug relay.

55 In order to improve access, unscrew the two screws and position the fusebox to one side.

56 Release the relay from the mounting bracket, and disconnect the two wiring plugs.

57 Refitting is a reversal of removal.

TD4 engines

58 The fuel pump relay is located in the fuse box in the drivers' side glove box. Open the glove box, and unclip the fuse box cover.

59 Carefully pull the relay from the holder. The fuel pump relay is lower of the two on the far right-hand side of the fuse box.

60 Refitting is a reversal of removal.

Fuel temperature switch

TD4 engines only

61 Fitted to the fuel return circuit, this purely mechanical switch diverts hot fuel through the fuel cooler matrix, prior to it being returned to the plastic tank. The switch is located in the left-hand front area of the engine compartment.

62 Note their fitted locations, and disconnect the fuel hoses from the switch. Some of the connections are of the 'quick release' type. On these, squeeze the two plastic lugs together and pull the connection apart.

63 Refitting is a reversal of removal.

Inertia fuel cut-off switch

TD4 engines only

64 The inertia fuel cut-off switch is designed to cut-off the fuel supply from the tank, should the vehicle be involved in an accident. The switch is located on the engine compartment bulkhead, adjacent to the brake fluid reservoir.

65 Ensure that the ignition is switched off, and disconnect the wiring plug from the switch.

66 Unscrew the securing bolt and remove the switch from the bulkhead. If required the switch can be separated from the mounting bracket by unscrewing the two nuts.

67 Refitting is a reversal of removal. **Note:** *If the fuel cut-off switch has been activated by an accident or similar, reset the switch by pressing the top down.*

Turbocharger boost pressure sensor

TD4 engines only

Note: *On L-Series engines the turbo boost pressure is 'sensed' by the wastegate actuator. The mechanism contains no serviceable parts.*

68 Undo the retaining bolts and remove the plastic cover from the top of the engine. Ensure that the ignition is switched off.

69 Release the two retaining clips and remove the intercooler outlet hose.

70 Release the wiring harness and the fuel return hose from the retaining clips on the inlet manifold, adjacent to the right-hand engine mounting bracket.

71 Disconnect the wiring plug, undo the retaining bolt, and remove the sensor.

72 Ensure that the mating surfaces of the sensor and manifold are clean and dry. With a new seal, refit the sensor to the manifold, and tighten the securing bolt to the specified torque.

73 The remainder of refitting is a reversal of removal.

Turbocharger boost pressure control solenoid valve

TD4 engines only

74 Unscrew the retaining bolts and remove the plastic cover from the top of the engine, taking care not to lose the spacers which are fitted to the cover mounting rubbers.

75 Slacken the retaining clips and release the intake ducting assembly. Unscrew the two retaining screws and remove the assembly from the camshaft cover.

76 Remove the engine oil filler cap, unscrew the five Allen screws and remove the air cleaner cover. Refit the oil filler cap.

77 Free the vacuum hose and wiring harness from the turbocharger duct.

78 Release the two retaining clips, and disconnect the turbocharger duct from the mass air flow/intake air temperature sensor. Recover the duct seal and breather pipe connector.

79 Chock the rear wheels then jack up the front of the vehicle and support it securely on axle stands. Undo the screws and remove the engine undertray.

80 Ensure that the ignition is switched off, and disconnect the wiring plug from the solenoid valve.

81 Note their fitted locations, and disconnect the vacuum hoses from the solenoid valve.

82 Unscrew the two retaining nuts and remove the solenoid from the mounting bracket.

83 Refitting is a reversal of removal.

Fuel high-pressure sensor

TD4 engines only

84 Unscrew the retaining bolts and remove the plastic cover from the top of the engine, taking care not to lose the spacers which are fitted to the cover mounting rubbers.

85 Slacken the retaining clips and release the intake ducting assembly. Unscrew the two retaining screws and remove the assembly from the camshaft cover.

86 The fuel high-pressure sensor is located in the left-hand end of the common fuel rail. Ensure that the ignition is switched off, and disconnect the wiring plug from the sensor.

87 Using a suitable deep socket, unscrew the sensor, and discard the seal. A new one must be fitted. Plug the port of the common rail to prevent contamination.

88 Ensure that the mating surfaces of the sensor and common rail are clean and dry, and refit the sensor with a new seal, tightening the sensor to the specified torque.

89 The remainder of refitting is a reversal of removal.

Fuel low-pressure sensor

TD4 engines only

90 The low pressure sensor is integral with the fuel filter head, and cannot be replaced separately. If faulty the complete filter should be replaced.

Fuel pressure regulator

TD4 engines only

91 Remove the inlet manifold.

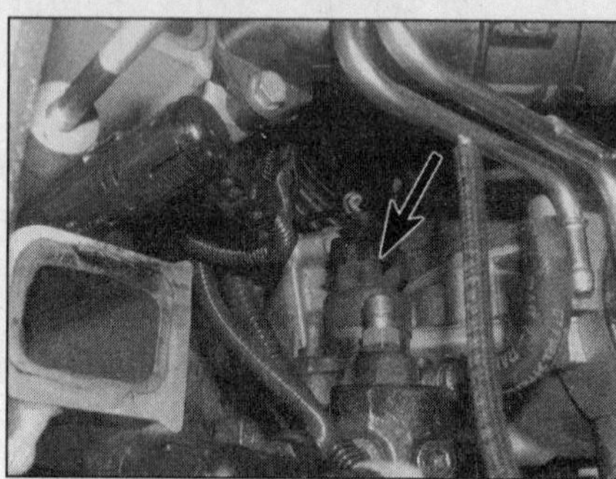

92 Disconnect the wiring plug from the regulator, located in the left-hand end of the fuel pump.

93 Undo the two Torx screws and remove the regulator. Discard the two seals, new ones must be fitted.

94 Ensure the pump and regulator mating faces are clean, and with new seals fitted, refit the regulator. Tighten the Torx screws to the specified torque.

95 Refit the manifold.

Camshaft position sensor

TD4 engines only

96 Undo the three retaining bolts and remove the plastic cover from the top of the engine.

97 Ensure that the ignition is switched off, and disconnect the wiring plug from the sensor, located on the right-hand side of the camshaft cover.

98 Undo the Torx screw and remove the sensor. Discard the seal, a new one must be fitted.

99 To refit the sensor, ensure the mating face of the camshaft cover and sensor are clean, and fit the new seal.

100 Fit the sensor into the cover, and tighten the Torx screw to the specified torque.

Peugeot 206 '98 – '01

Petrol models

General

1.1 litre models	Magneti Marelli 1AP
1.4 litre models:	
Later models with pulse air injection	Bosch Motronic MP7.3
All other models	Magneti Marelli 1AP
1.6 litre models	Bosch Motronic MP7.2

Fuel system data

Fuel pump type	Electric, immersed in tank
Fuel pump regulated constant pressure	3.5 ± 0.2 bars
Specified idle speed	850 ± 100 rpm (not adjustable – controlled by ECU)
Idle mixture CO content	Less than 1.0 % (not adjustable- controlled by ECU)

Recommended fuel

Minimum octane rating	95 RON unleaded (UK unleaded premium). Leaded/lead replacement fuel (LRP) must **not** be used

Diesel models

General

System type:	
1.9 litre engine:	
WJZ engine	Indirect injection system incorporating a distributor fuel injection pump
WJY engine	Indirect injection system incorporating a semi-electronically controlled distributor fuel injection pump
2.0 litre engine	High-pressure direct injection (HDI) 'common-rail' system incorporating an electronically controlled fuel injection pump and injectors
Injection pump type:	
1.9 litre engine:	
WJZ engine	Lucas DWLP11
WJY engine	Lucas DWLP12
2.0 litre engine	Bosch EDC 15C2

	1.9 litre engine	**2.0 litre engine**
Adjustment data		
Idle speed	875 ± 25 rpm	800 ± 20 rpm – controlled by ECU
Maximum speed	Not available	5000 rpm – controlled by ECU
Injectors		
Opening pressure	135 ± 5 bars	Controlled by ECU
Turbocharger		
Type	KKK K03	
Boost pressure (approximate)	1 bar at 3000 rpm	

Chapter 18

Torque wrench settings	Nm	lbf ft
Petrol models		
Knock sensor	20	15
Diesel models		
Fuel pressure sensor	35	26

General information

Petrol models

Note: *The engine management ECU is of the 'auto-adaptive' type, meaning that as it operates, it also monitors and stores the settings which give optimum engine performance under all operating conditions. When the battery is disconnected, these settings are lost and the ECU reverts to the base settings programmed into its memory at the factory. On restarting, this may lead to the engine running/idling roughly for a short while, until the ECU has re-learned the optimum settings during the first few miles of driving. To aid this re-learning process, initialise the ECU as follows once the battery has been reconnected.*

a) Turn on the ignition switch then turn it off for 10 seconds. Turn the ignition on and wait for 10 seconds before starting the engine without touching the accelerator pedal.

b) On engines with a Bosch system, warm the engine up to to normal operating temperature. When the cooling fan cuts in, switch off the engine. Restart the engine, allow it to idle for a few minutes, then take it on a short road test driving at varying engine speeds/loads. At least once during the road test, decelerate from an engine speed of above 4000 rpm for at least 5 seconds.

c) On engines with a Magneti Marelli system, warm the engine up to normal operating temperature. When the cooling fan cuts in, take the vehicle on a short road test driving at varying engine speeds/loads. At least once during the road test, when it is safe to do so, fully depress the accelerator pedal and accelerate for at least 2 seconds.

Bosch Motronic MP7.2 system

1 The Bosch Motronic MP7.2 engine management (fuel injection/ignition) system is fitted to all 1.6 litre models. The system incorporates a closed-loop catalytic converter and an evaporative emission control system, and complies with the very latest emission control standards.

2 The fuel pump (which is immersed in the fuel tank) supplies fuel from the tank to the fuel rail, via a filter mounted on the front of the fuel tank. Fuel supply pressure is controlled by the pressure regulator which is incorporated in the fuel pump assembly. The outlet of the fuel filter is connected to a T-piece, one end of which is connected to the pressure regulator and the other to the fuel rail. When the optimum operating pressure of the fuel system is exceeded, the regulator allows excess fuel to return directly from the filter to the fuel tank. This system means that there is only a single fuel pipe running from the fuel tank to the fuel rail.

3 The electrical control system consists of the ECU, along with the following sensors:

a) Throttle potentiometer – informs the ECU of the throttle position, and the rate of throttle opening/closing.

b) Coolant temperature sensor – informs the ECU of engine temperature.

c) Intake air temperature sensor – informs the ECU of the temperature of the air passing through the throttle housing.

d) Oxygen sensor – informs the ECU of the oxygen content of the exhaust gases.

e) Crankshaft sensor – informs the ECU of the crankshaft position and speed of rotation.

f) Manifold absolute pressure (MAP) sensor – informs the ECU of the load on the engine (expressed in terms of inlet manifold vacuum).

g) Knock sensor – informs the ECU when pre-ignition ('pinking') is occurring.

h) Vehicle speed sensor – informs the ECU of the vehicle speed.

i) Power steering pressure switch – informs the ECU when the power steering pump is under load.

j) Air conditioning system pressure switch – informs ECU when the air conditioning compressor is under load.

4 All the above signals are analysed by the ECU which selects the fuelling response appropriate to those values. The ECU controls the fuel injectors (varying the pulse width – the length of time the injectors are held open – to provide a richer or weaker mixture, as appropriate). The mixture is constantly varied by the ECU, to provide the best setting for cranking, starting (with either a hot or cold engine), warm-up, idle, cruising and acceleration. The injectors are operated 'semi-sequentially', injectors No 1 and 4 being operated as one pair and injectors No 2 and 3 as the other.

5 The ECU also has full control over the engine idle speed, via a stepper motor fitted to the throttle housing. When the throttle valve is closed, the ECU controls the opening of the valve, which in turn regulates the amount of air entering the manifold, and so controls the idle speed.

6 The ECU also controls the exhaust and evaporative emission control systems.

7 On early models, an electric heating element is fitted to the aluminium throttle housing; the heater is supplied with current by the ECU, and warms the throttle housing on cold starts to prevent possible icing of the throttle valve. Later models have a plastic throttle valve housing and no longer require a heating element.

8 If there is an abnormality in any of the readings obtained from the various sensors, the ECU enters its back-up mode. In this event, it ignores the abnormal sensor signal and assumes a pre-programmed value which will allow the engine to continue running (albeit at reduced efficiency). If the ECU enters this back-up mode, the warning light on the instrument panel will come on, and the relevant fault code will be stored in the ECU memory.

9 If the warning light comes on, the vehicle should be taken to a Peugeot dealer at the earliest opportunity. A complete test of the engine management system can then be carried out, using a special electronic diagnostic test unit which is simply plugged into the system's diagnostic connector (located next to the fusebox).

Magneti Marelli 1AP system

10 The Magneti Marelli 1AP engine management (fuel injection/ignition) system is fitted to all 1.1 litre models and all 1.4 litre models not equipped with pulse air injection.

11 The system is very similar in operation to the Bosch MP7.2 system described above, apart from the idle speed control system.

12 On the Magneti Marelli system, the idle speed is controlled by the ECU via a stepper motor fitted to the throttle housing. The motor has a pushrod controlling the opening of an air passage which bypasses the throttle valve. When the throttle valve is closed, the ECU controls the movement of the motor pushrod, which regulates the amount of air which flows through the throttle housing passage, so controlling the idle speed. The bypass passage is also used as an additional air supply during cold starting.

Bosch Motronic MP7.3 system

13 Some later 1.4 litre engines are fitted with a Bosch Motronic MP7.3 engine management system which incorporates a pulse air injection system. The system is a development of the MP7.2 system designed to meet the next level of emission standards. The changes include the following.

a) The system is fully 'sequential' with all injectors being controlled individually.

b) A body accelerometer sensor is fitted to measure the vertical movement of the vehicle body. The sensor informs the ECU of the quality of the road surface on which the vehicle is being driven. This ensures the ECU does not diagnose variations in engine speed due to an uneven road surface as an ignition misfire.

c) A camshaft sensor is fitted to inform the ECU of the position of No 1 piston.

d) An additional oxygen sensor is fitted down stream of the catalytic converter.

e) Pulse air system is fitted to improve exhaust emissions during warm-up.

f) The idle speed is controlled by a stepper motor fitted to the throttle housing (see paragraph 12).

Diesel models

1.9 litre (WJZ) engine

14 The fuel system consists of a fuel tank (which is mounted under the rear of the car), a fuel filter with integral water separator, a fuel injection pump, injectors and associated components.

15 The injection pump draws fuel from the tank through the fuel filter, which is mounted on the thermostat housing on the left-hand end of the cylinder head. The fuel filter removes all foreign matter and water, and ensures that the fuel supplied to the injection pump is clean. Excess fuel is returned from the bleed outlet on the filter housing lid to the tank. The filter housing incorporates a thermostat. When the temperature of the fuel in the filter housing is below 15°C, the filter housing thermostat opens and allows the fuel to circulate between the filter housing and thermostat housing which effectively warms the fuel. When the fuel in the filter housing reaches 35°C, the thermostat closes.

16 The fuel injection pump is driven at half-crankshaft speed by the timing belt. The high pressure required to inject the fuel into the compressed air in the swirl chambers is achieved by two opposed pistons forced together by rollers running in a cam ring. The fuel passes through a central rotor with a single outlet drilling which aligns with ports leading to the injector pipes.

17 Fuel metering is controlled by a centrifugal governor, which reacts to accelerator pedal position and engine speed. The governor is linked to a metering valve, which increases or decreases the amount of fuel delivered at each pumping stroke.

18 Basic injection timing is determined when the timing belt is fitted. When the engine is running, it is varied automatically to suit the prevailing engine speed by a mechanism which turns the cam plate or ring.

19 The four fuel injectors produce a homogeneous spray of fuel into the swirl chambers located in the cylinder head. The injectors are calibrated to open and close at critical pressures to provide efficient and even combustion. Each injector needle is lubricated by fuel, which accumulates in the spring chamber and is channelled to the injection pump return hose by leak-off pipes.

20 Cold starting is assisted by preheater or 'glow' plugs fitted to each swirl chamber. A thermostatic sensor in the cooling system operates a fast idle lever on the injection pump to increase the idling speed when the engine is cold.

21 A stop solenoid cuts the fuel supply to the injection pump rotor when the ignition is switched off, and there is also a hand-operated stop lever for use in an emergency.

22 Provided that the specified maintenance is carried out, the fuel injection equipment will give long and trouble-free service. The injection pump itself may well outlast the engine. The main potential cause of damage to the injection pump and injectors is dirt or water in the fuel.

23 A catalytic converter and exhaust gas recirculation (EGR) system are fitted to the engine to reduce exhaust emissions.

24 Servicing of the injection pump and injectors is very limited for the home mechanic, and any dismantling or adjustment other than that described in this Chapter must be entrusted to a Peugeot dealer or fuel injection specialist.

1.9 litre (WJY) engine

25 Later 1.9 litre (WJY) engines are fitted with an electronically-controlled fuel injection pump to improve emissions in order to meet the next level of emission standards being introduced. The fuel system is very similar to that described in paragraphs 14 to 24, with the following changes to the injection pump.

26 The fuel injection pump electrical control system consists of the ECU, along with the following sensors:

a) Throttle potentiometer – informs the injection pump accelerator lever position, and the rate of throttle opening/closing.

b) Coolant temperature sensor – informs the ECU of engine temperature.

c) Crankshaft sensor – informs the ECU of the crankshaft position and speed of rotation.

d) Vehicle speed sensor – informs the ECU of the vehicle speed.

e) Injector needle lift sensor – informs the ECU when the start of injection occurs at No 1 injector.

f) Atmospheric pressure sensor (incorporated in the ECU) – measures the atmospheric pressure to prevent problems when driving at high-altitude.

27 All the above signals are analysed by the ECU which controls the injection timing via the advance solenoid valve which is fitted to the injection pump. By opening and closing the solenoid valve, the ECU can advance and retard the injection timing as necessary. When the advance solenoid is open, the hydraulic pressure on the pump piston is reduced which results in the injection timing being retarded. To advance the injection timing, the ECU closes the solenoid valve which increases the pressure on the piston.

28 The ECU also controls the exhaust gas recirculation (EGR) system, and the engine cooling fan.

29 If there is an abnormality in any of the readings obtained from the various sensors, the ECU enters its back-up mode. In this event, it ignores the abnormal sensor signal and assumes a pre-programmed value which will allow the engine to continue running (albeit at reduced efficiency). If the ECU enters this back-up mode, the warning light on the instrument panel will come on, and the relevant fault code will be stored in the ECU memory.

2.0 litre engine

30 All 2.0 litre engines are fitted with a high-pressure direct injection (HDI) system which incorporates the very latest in diesel injection technology. On the HDI system, the injection pump is used purely to provide the pressure required for the injection system and has no control over the injection timing (unlike conventional diesel injection systems). The injection timing is controlled by the electronic control unit (ECU) via the electrically-operated injectors. The system operates as follows.

31 The fuel system consists of a fuel tank (which is mounted under the rear of the car, with an electric fuel pump immersed in it), a fuel filter with integral water separator, a fuel injection pump, injectors and associated components.

32 The fuel pump supplies fuel to the fuel filter housing which is located at the front of the engine. The fuel filter removes all foreign matter and water and ensures that the fuel supplied to the injection pump is clean. Excess fuel is returned from the outlet on the filter housing lid to the tank via the fuel cooler. The fuel cooler is fitted to the underside of the vehicle and is cooled by the passing airflow to ensure the fuel is cool before it enters the fuel tank.

33 The fuel is heated to ensure no problems occur when the ambient temperature is very low. On early models the filter housing is connected to the coolant outlet housing on the left-hand end of the cylinder head and is fitted with a thermostat. When the temperature of the fuel in the filter housing is below 15°C, the filter housing thermostat opens and allows the fuel to circulate around the coolant outlet housing which effectively warms the fuel. When the fuel in the filter housing reaches 25°C, the thermostat closes. On later models an electrically-operated fuel heater is fitted to the fuel feed pipe to the filter housing, the heater is controlled by the ECU.

34 The fuel injection pump is driven at half-crankshaft speed by the timing belt. The high pressure required in the system (up to 1350 bar) is produced by the three pistons in the pump. The injection pump supplies high pressure fuel to the fuel rail, which acts as a reservoir for the four injectors. Since the injection pump has no control over the injection timing (unlike conventional diesel injection systems), this means that there is no need to time the injection pump when installing the timing belt.

35 The electrical control system consists of the ECU, along with the following sensors:

a) Accelerator pedal position sensor – informs the ECU of the accelerator pedal position, and the rate of throttle opening/closing.

b) Coolant temperature sensor – informs the ECU of engine temperature.

c) Airflow meter (incorporating the intake air temperature sensor – informs the ECU of the *amount and temperature of air passing through the intake duct.*

d) Crankshaft sensor – informs the ECU of the crankshaft position and speed of rotation.

e) Camshaft position sensor – informs the ECU of the positions of the pistons.

f) Fuel temperature sensor (where fitted) – informs the ECU of the temperature of the fuel in the fuel rail.

g) Fuel pressure sensor – informs the ECU of the fuel pressure present in the fuel rail.

h) Atmospheric pressure sensor (incorporated in the ECU) – measures the atmospheric pressure to prevent problems when driving at high-altitude.

i) Vehicle speed sensor – informs the ECU of the vehicle speed.

j) Power steering pressure switch – informs the ECU when the power steering pump is under load.

k) Air conditioning system relay – informs ECU when the air conditioning compressor is under load.

36 All the above signals are analysed by the ECU which selects the fuelling response appropriate to those values. The ECU controls the fuel injectors to provide the best setting for cranking, starting (with either a hot or cold engine), warm-up, idle, cruising and acceleration.

37 The ECU also has full control over the fuel pressure present in the fuel rail via the high-pressure fuel regulator and third piston deactivator solenoid valve which are fitted to the injection pump. To reduce the pressure, the ECU opens the high-pressure fuel regulator which allows the excess fuel to return direct to the tank from the pump. The third piston deactivator is used mainly to reduce the load on the engine, but can also be used to lower the fuel pressure. The deactivator solenoid valve relieves the fuel pressure from the third piston of the pump which results in only two of the pistons pressurising the fuel system.

38 The ECU also controls the exhaust gas recirculation (EGR) system, and the engine cooling fan.

39 A turbocharger is fitted to increases engine efficiency. It does this by raising the pressure in the inlet manifold above atmospheric pressure. Instead of the air simply being sucked into the cylinders, it is forced in.

40 Energy for the operation of the turbocharger comes from the exhaust gas. The gas flows through a specially-shaped housing (the turbine housing) and in so doing, spins the turbine wheel. The turbine wheel is attached to a shaft, at the end of which is another vaned wheel known as the compressor wheel. The compressor wheel spins in its own housing, and compresses the inlet air on the way to the inlet manifold. The turbo shaft is pressure-lubricated by an oil feed pipe from the main oil gallery. The shaft 'floats' on a cushion of oil. A drain pipe returns the oil to the sump. Boost pressure (the pressure in the inlet manifold) is limited by a wastegate, which diverts the exhaust gas away from the turbine wheel in response to a pressure-sensitive actuator.

41 If there is an abnormality in any of the readings obtained from the various sensors, the ECU enters its back-up mode. In this event, it ignores the abnormal sensor signal and assumes a pre-programmed value which will allow the engine to continue running (albeit at reduced efficiency). If the ECU enters this back-up mode, the warning light on the instrument panel will come on, and the relevant fault code will be stored in the ECU memory.

Fuel system – depressurisation, pressurising and priming

Petrol models

Depressurisation

Warning: The following procedure will merely relieve the pressure in the fuel system – remember that fuel will still be present in the system components and take precautions accordingly before disconnecting any of them.

1 The fuel system referred to in this Section is defined as the tank-mounted fuel pump, the fuel filter, the fuel injectors, the fuel rail and the pipes of the fuel lines between these components. All these contain fuel which will be under pressure while the engine is running, and/or while the ignition is switched on. The pressure will remain for some time after the ignition has been switched off, and must be relieved in a controlled fashion when any of these components are disturbed for servicing work.

2 Disconnect the battery negative terminal.

3 Some models are equipped with a pressure relief valve on the fuel rail. On these models, unscrew the cap from the valve and position a container beneath the valve. Hold a wad of rag over the valve and relieve the pressure in the system by depressing the valve core with a suitable screwdriver. Be prepared for the squirt of fuel as the valve core is depressed and catch it with the rag. Hold the valve core down until no more fuel is expelled from the valve. Once the pressure is relieved, securely refit the valve cap.

4 Where no valve is fitted to the fuel rail, it will be necessary to release the pressure as the fuel pipe is disconnected. Place a container beneath the union and position a large rag around the union to catch any fuel spray which may be expelled. Slowly release and disconnect the fuel pipe and catch any spilt fuel in the container. Plug the pipe/union to minimise fuel loss and prevent the entry of dirt into the fuel system.

Pressurising

5 After any work is carried out on the fuel system, the system should be pressurised as follows.

6 Depress the accelerator pedal fully then switch on the ignition. Hold the pedal depressed for approximately 1 second then release it. The ECU should then operate the fuel pump for between 20 and 30 seconds to refill the fuel system. Once the fuel pump stops the ignition can be switched off.

Diesel models

7 After disconnecting part of the fuel supply system or running out of fuel, it is necessary to prime the fuel system as follows.

1.9 litre engine

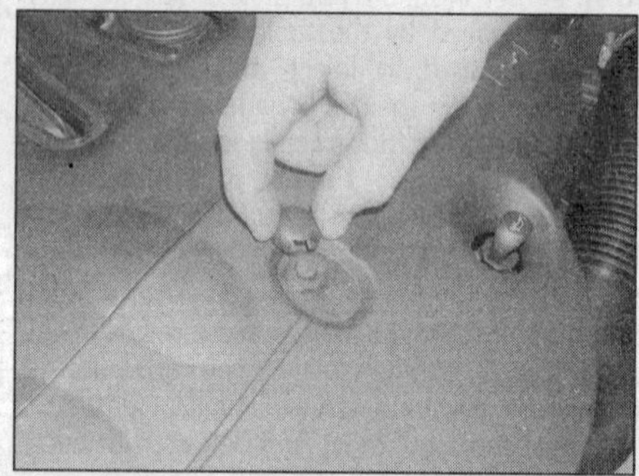

8 All models are fitted with a hand-operated priming pump which is built into the fuel filter housing. To gain access to the pump, release the fasteners from the right-hand side and top of the engine cover then lift off the cover, taking care not to lose its mounting rubbers.

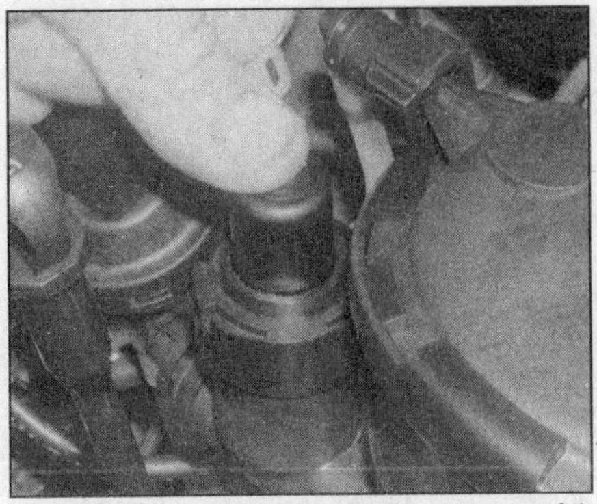

9 Pump the priming pump until resistance is felt then pump a few more times. This will prime the fuel system components and remove all air from the system.

10 Start the engine as normal. If difficulty is encountered, pump the priming pump a few times with the ignition switched on.

11 Once the engine has started, ensure the mounting rubbers are all correctly fitted then install the engine cover, securing it in position with the fasteners.

2.0 litre engine

12 On completion of work on the fuel system slacken, or leave loose (as appropriate), the **new** fuel filter drain screw. Operate the low pressure pump 4 or 5 times by switching on the ignition each time for a period of 5 seconds. Switch off the ignition and wait for a period of 5 to 10 seconds to allow the pressure to fall in the fuel supply circuit. Tighten the drain screw and wipe away all spilt fuel. Start the engine and check that there is no sign whatever of fuel seepage from the filter drain screw once the engine is running.

Petrol engine management system components – removal and refitting

Fuel rail and injectors

Note: *If a faulty injector is suspected, before condemning the injector, it is worth trying the effect of one of the proprietary injector-cleaning treatments which are available from car accessory shops.*

1 Unclip and disconnect the breather hose(s) from the manifold end of the duct connecting the air cleaner housing to the inlet manifold. Slacken the retaining clips then free the duct from the manifold and air cleaner housing and remove it from the engine compartment.

2 Remove the ignition HT coil.

3 Free the accelerator inner cable from the throttle housing cam, then pull the outer cable out from its mounting bracket rubber grommet, complete with its spring clip.

4 Unscrew the bolts and remove the accelerator cable bracket from the manifold/cylinder head.

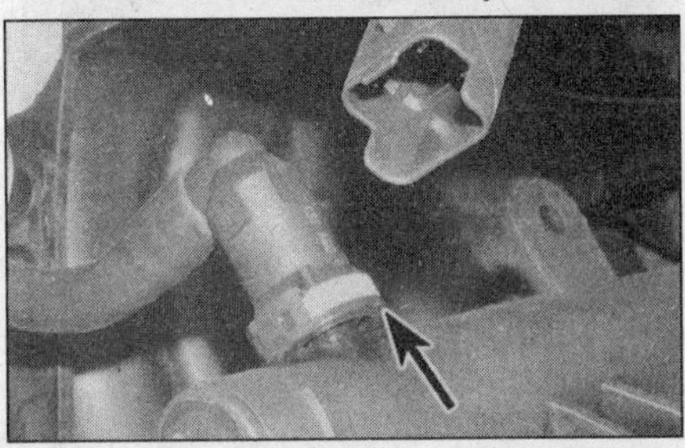

5 Depress the retaining clip and disconnect the fuel pipe from the right-hand end of the fuel rail, bearing in mind the information given in the Section on depressurising the fuel system.

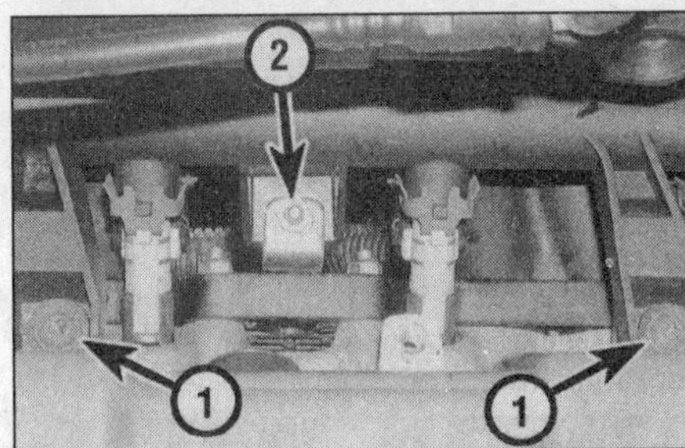

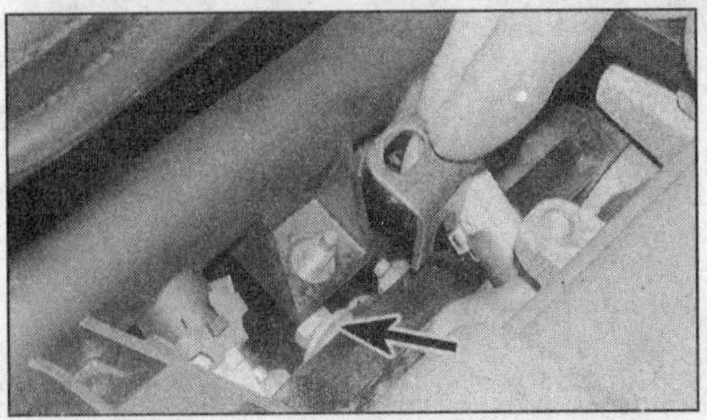

6 Slacken and remove the two bolts (1) securing the fuel rail to the cylinder head, and the nut (2) securing the rail to the manifold. Loosen the bolt securing the fuel rail centre bracket to the inlet manifold, then lift off the bracket (the bracket is slotted to ease removal).

7 Disconnect the injector wiring harness connector then unclip the connector from the rear of the inlet manifold. Also disconnect the wiring connectors from the throttle housing and power steering switch and position the wiring harness clear of the manifold so that it does not hinder fuel rail removal.

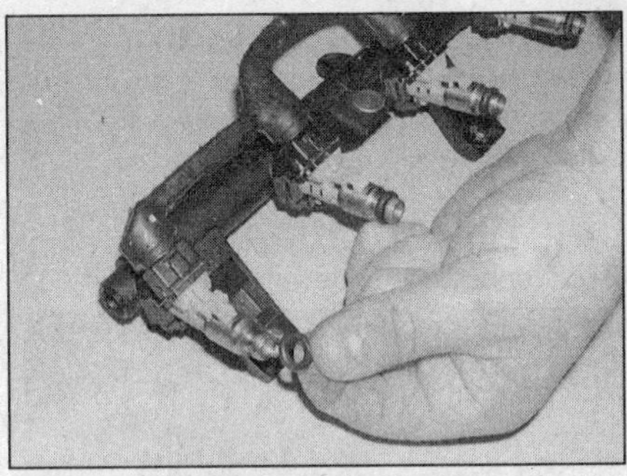

8 Carefully ease the fuel rail and injector assembly out from the cylinder head and manoeuvre it out of position. Remove the seals from the end of each injector and discard them; they must be renewed whenever they are disturbed.

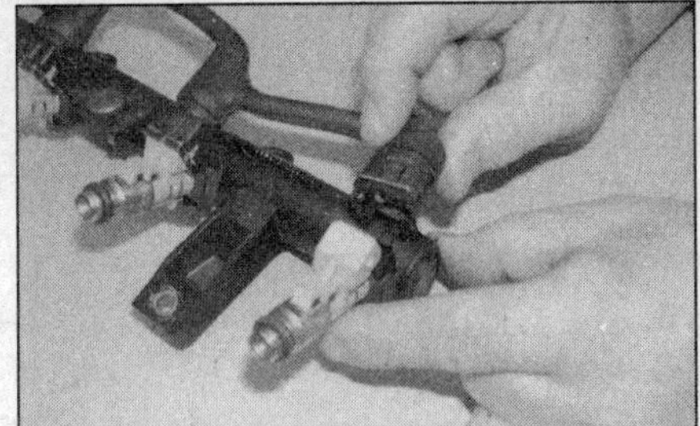

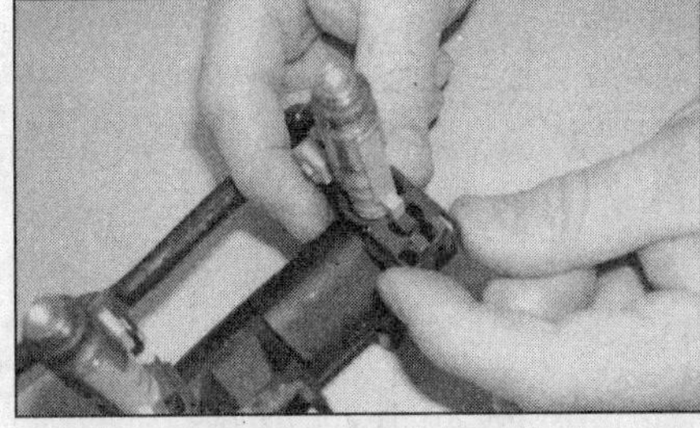

9 Disconnect the wiring connector(s) then slide out the retaining clip(s) and remove the relevant injector(s) from the fuel rail. Remove the upper seal from each disturbed injector and discard; all disturbed seals must be renewed.

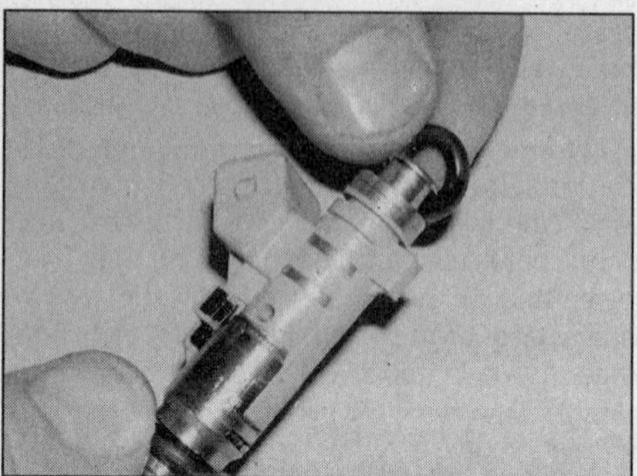

10 Refitting is a reversal of the removal procedure, noting the following points.

a) Fit new seals to all disturbed injector unions.

b) Apply a smear of engine oil to the seals to aid installation, then ease the injectors and fuel rail into position ensuring that none of the seals are displaced.

c) On completion, reconnect the battery and pressurise the fuel system. Start the engine and check for fuel leaks.

Fuel pressure regulator

11 The fuel pressure regulator is an integral part of the fuel pump assembly and is not available separately.

Throttle potentiometer

12 Disconnect the battery negative terminal.

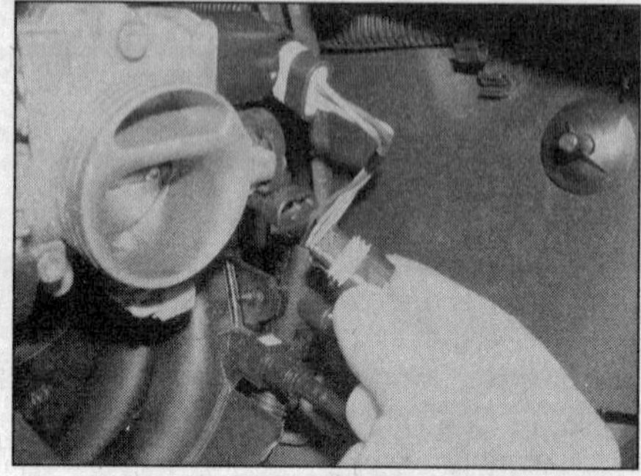

13 Depress the retaining clip and disconnect the wiring connector from the throttle potentiometer.

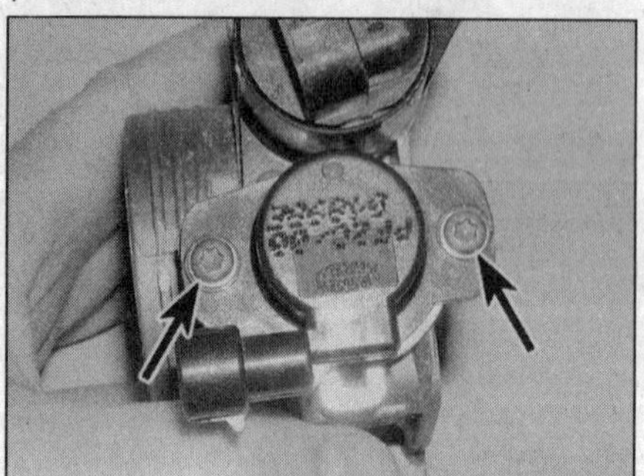

14 Slacken and remove the two retaining screws, then disengage the potentiometer from the throttle valve spindle and remove it from the vehicle.

15 Refit in the reverse order of removal.

16 Ensure that the potentiometer is correctly engaged with the throttle valve spindle.

Electronic Control Unit (ECU)

17 The ECU is located on the right-hand side of the engine compartment.

18 To remove the ECU, first disconnect the battery.

19 Unclip the wiring harness then slacken and remove the ECU mounting plate nut.

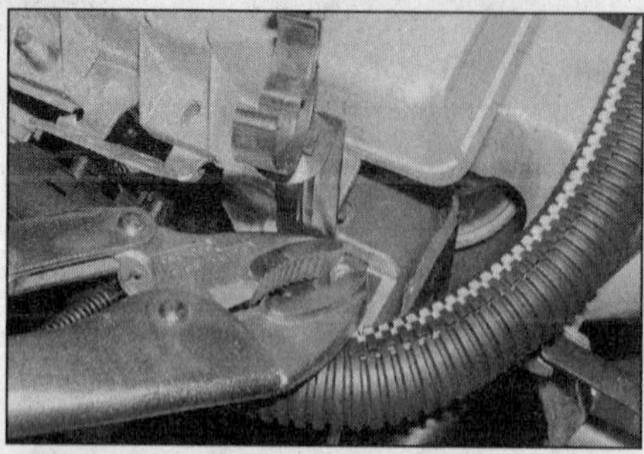

20 The plate is secured in position with an anti-tamper nut, unscrew the nut using a pair of grips then lift off the wiring bracket.

21 Free the mounting plate from the body, then lift the retaining clip and disconnect the ECU wiring connector.

22 Unscrew the ECU mounting nuts (these are also anti-tamper nuts) then remove the ECU from the mounting plate.

23 Refitting is a reverse of the removal procedure ensuring the wiring connector is securely reconnected.

24 Secure the ECU and mounting plate in position with new anti-tamper nuts or standard nuts.

Idle speed stepper motor

25 The idle speed stepper motor is fitted to the rear of the throttle housing.

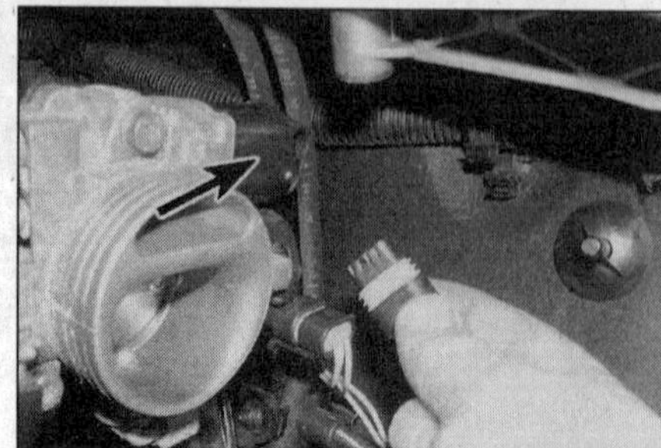

26 Disconnect the battery negative terminal then disconnect the wiring connector from the motor.

27 Slacken and remove the retaining screws then remove the motor from the throttle housing. If necessary, remove the throttle potentiometer to improve access to the motor lower screw.

28 Refitting is a reversal of the removal procedure ensuring the seal is in good condition.

Manifold absolute pressure (MAP) sensor

29 The MAP sensor is mounted on the inlet manifold. To remove it, first disconnect the battery negative terminal.

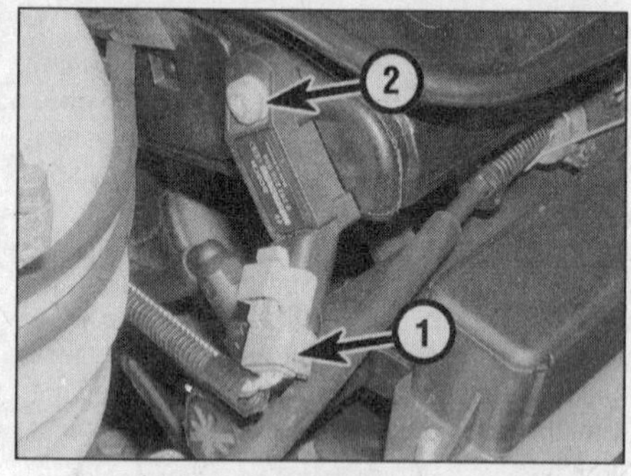

30 Disconnect the wiring connector (1) then undo the screw (2) and remove the sensor from the manifold.

31 Refitting is a reversal of the removal procedure ensuring the sensor seal is in good condition.

Coolant temperature sensor

Engine management system coolant sensor (1) and air conditioning control system (Bitron) temperature sensor (2) locations – petrol engines

32 On petrol and diesel models, the engine management coolant temperature sensor is screwed into the coolant outlet housing on the left-hand end of the cylinder head, and can be identified by its green-coloured wiring connector. Drain the coolant system.

33 Disconnect the wiring plug and unscrew the sensor from the coolant outlet housing. **Note:** *On 2.0 litre diesel engines fitted with a plastic outlet housing, prise out the sensor retaining clip, then remove the sensor and sealing ring from the housing.*

34 Refitting is a reversal of removal. Apply a little sealing compound to the threads of the sensor prior to refitting. On 2.0 litre diesel engines with a plastic outlet housing, renew the sealing ring.

Intake air temperature sensor

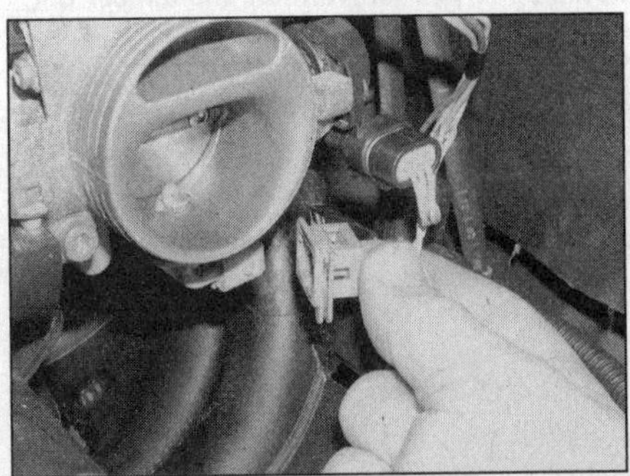

35 The intake air temperature sensor is fitted to the base of the throttle housing. To remove the sensor first disconnect the battery negative terminal, and the wiring connector from the sensor.

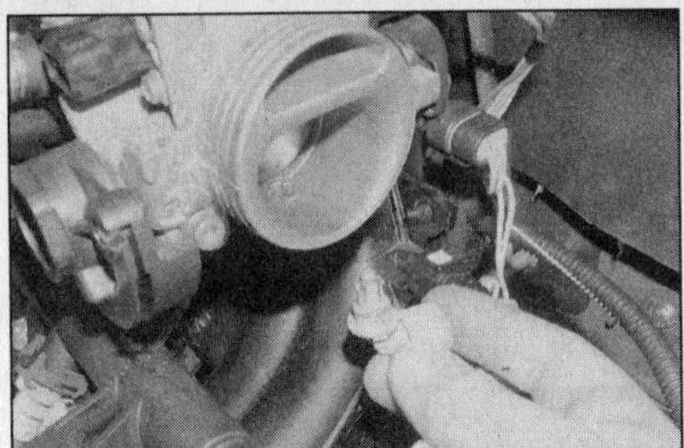

36 Release the retaining clips by passing a close-fitting ring spanner over the end of the sensor, and ease the sensor out from the throttle housing.

37 Refitting is the reverse of removal ensuring the sensor seal is in good condition.

Crankshaft sensor

38 The crankshaft sensor is situated on the front face of the transmission clutch housing. To remove the sensor, first disconnect the battery negative terminal.

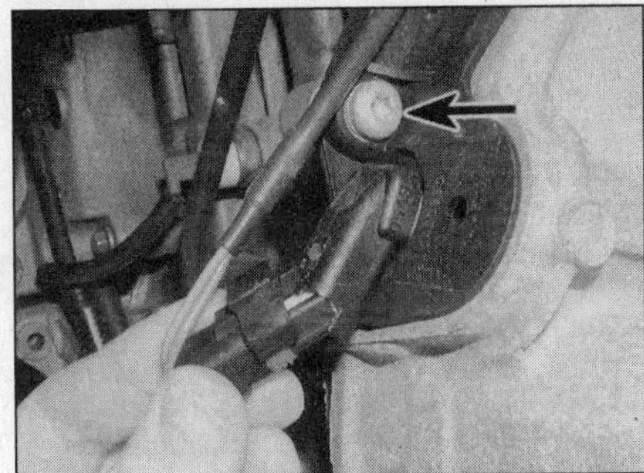

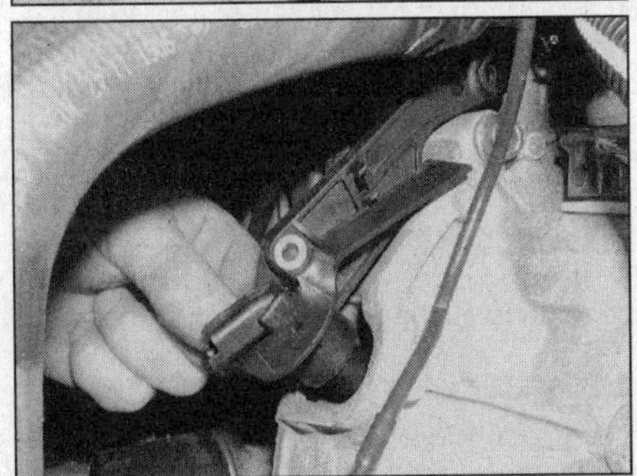

39 Disconnect the sensor wiring connector and unclip the wiring. Undo the retaining bolt and remove the sensor and bracket assembly from the transmission unit.

40 Refitting is reverse of the removal procedure.

Engine management system relay unit

41 The relay unit is mounted onto the underside of the ECU mounting plate. To remove the sensor, first disconnect the battery negative terminal.

42 Unclip the wiring harness then slacken and remove the ECU mounting plate nut. The plate is secured in position with an anti-tamper nut, unscrew the nut using a pair of grips and lift off the wiring bracket *(see paragraph 20)*.

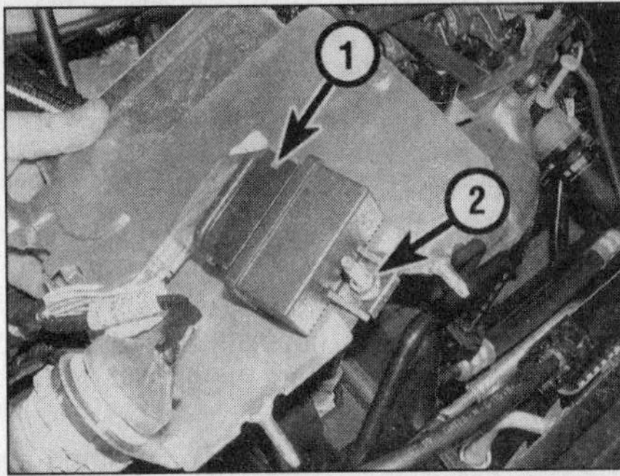

43 Free the mounting plate assembly from the body. Disconnect the wiring connector (1) from the relay then unscrew the nut (2) and remove the relay from the plate.

44 Refitting is a reverse of the removal procedure ensuring the wiring connector is securely reconnected. Secure the ECU mounting plate in position with a new anti-tamper nut or a standard nut.

Throttle housing heating element

Note: *The heating element is only fitted to aluminium throttle housings. Plastic housings do not need a heater.*

45 The heating element is fitted to the top of the throttle housing. To remove it, first disconnect the battery negative terminal.

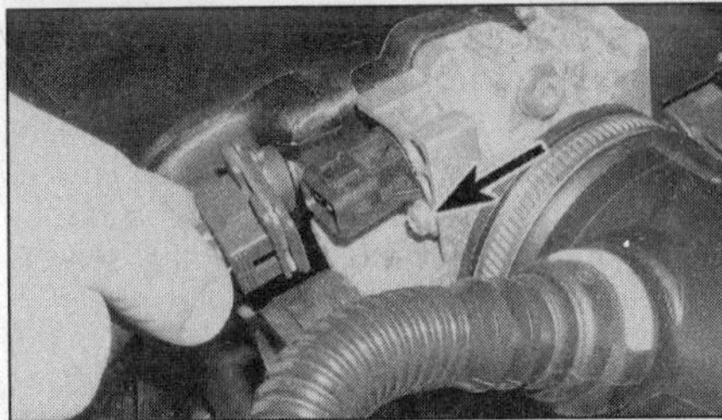

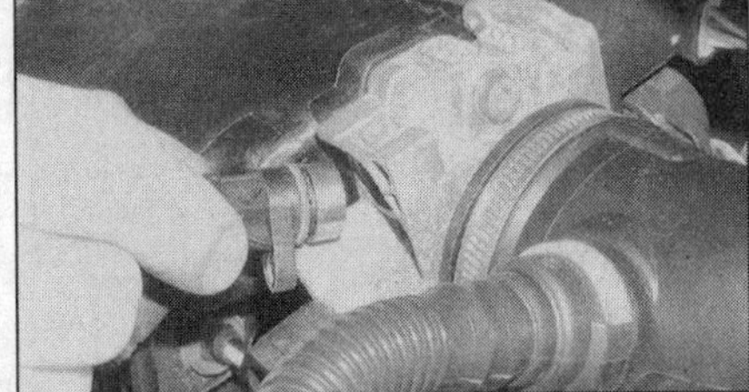

46 Disconnect the wiring connector then unscrew the retaining screw and remove the heating element from the throttle housing.

47 Refitting is the reverse of removal.

Vehicle speed sensor

48 The vehicle speed sensor is an integral part of the speedometer drive, fitted to the transmission housing above the right-hand driveshaft. Jack up the front of the vehicle and support it securely on axle stands.

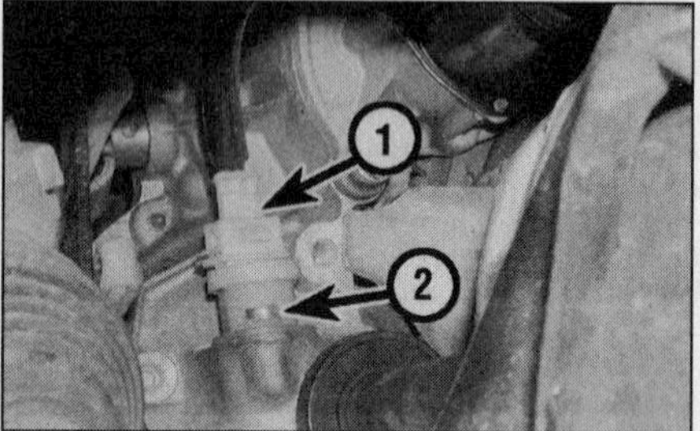

49 Disconnect the wiring connector (1) from the speed sensor

50 Unscrew the retaining bolt (2) and remove the heatshield (where fitted), then withdrawn the speedometer drive from the housing along with the sealing ring.

51 Refitting is a reversal of removal, fitting a new sealing ring.

Knock sensor

52 The knock sensor is screwed onto the rear of the cylinder block. To gain access to the sensor, firmly apply the handbrake then jack up the front of the vehicle and support it on axle stands.

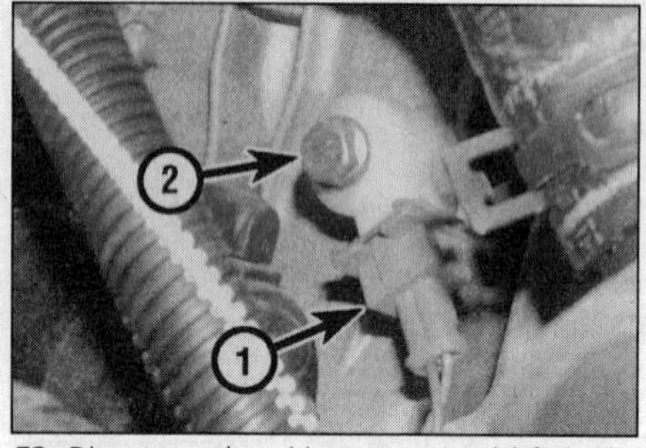

52 Disconnect the wiring connector (1) from the sensor then unscrew the retaining bolt (2) and remove the sensor from underneath the vehicle.

53 Refitting is the reverse of removal. Ensure the sensor and cylinder block mating surfaces are clean and dry and tighten the sensor mounting bolt to the specified torque.

Fuel cut-off inertia switch

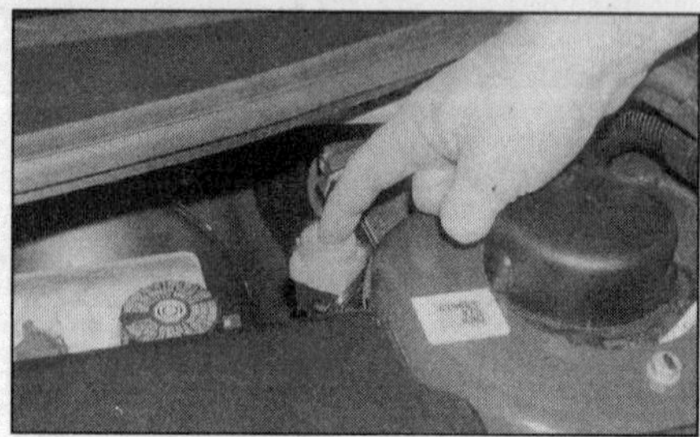

54 The fuel cut-off inertia switch is located in the left-hand rear corner of the engine compartment. To remove it, first disconnect the battery negative terminal.

55 Cut the retaining clip (where fitted) then unscrew the retaining nuts and remove the switch from the vehicle, disconnecting it from the wiring connector.

56 Refitting is the reverse of removal. On completion, reset the switch by firmly depressing its button.

Power steering pressure switch

Caution: Ensure no dirt is allowed to enter into the hydraulic system during the following procedure.

57 The power steering pressure switch is fitted to the power steering pipe located on the left-hand end of the cylinder head. To remove it, first disconnect the battery negative terminal.

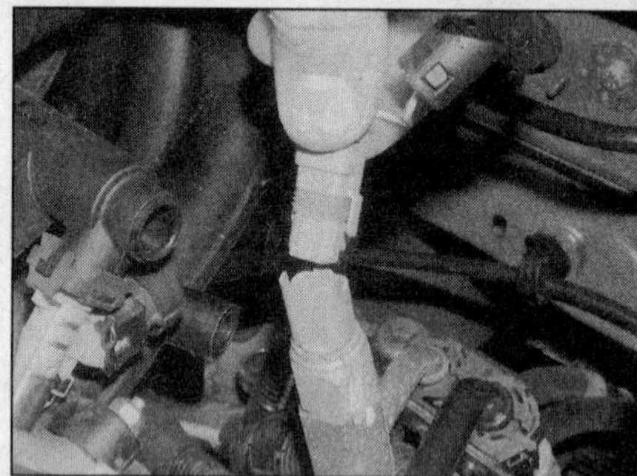

58 Disconnect the wiring connector from the switch then wipe the area around the switch clean. Unscrew the switch and remove it from the pipe, along with its sealing washer. Plug the pipe whilst the switch is removed to minimise fluid loss and prevent the entry of dirt into the hydraulic system.

59 Refitting is the reverse of removal, using a new sealing washer. On completion, check the power steering fluid level and, if necessary, bleed the hydraulic system.

Camshaft position sensor

Note: *1.4 litre engine with pulse air injection*

60 The camshaft position sensor is fitted to the rear of the coolant outlet housing on the left-hand end of the cylinder head. To remove it, first disconnect the battery negative terminal.

61 Disconnect the wiring connector then unscrew the retaining bolt and remove the sensor from the housing.

62 Refitting is the reverse of removal ensuring the sensor seal is in good condition.

Body accelerometer

Note: *1.4 litre engines with pulse air injection*

63 The body accelerometer is located in the engine compartment. To remove it, first disconnect the battery negative terminal.

64 Disconnect the wiring connector then unscrew the mounting bolt and remove the sensor from the vehicle.

65 Refitting is the reverse of removal.

Diesel injection system electronic components – removal and refitting

1.9 litre (WJZ) engine

Stop solenoid

1 The stop solenoid is part of the immobiliser unit which is located on the top of the fuel injection pump, its purpose being to cut the fuel supply when the ignition is switched off. Renewal of the immobiliser/solenoid unit is a complex operation which should be entrusted to a Peugeot dealer or Lucas injection specialist. The immobiliser unit is secured in position with shear bolts which have to drilled out (a high-risk operation which could lead to damage if carried out carelessly) and the new unit will have to be initialised on refitting.

Fuel cut-off inertia switch

Note: *A fuel cut-off inertia switch is not fitted to all models.*

2 The fuel cut-off inertia switch is located in the left-hand rear corner of the engine compartment. To remove it, first disconnect the battery negative terminal.

3 Unscrew the retaining bolts then disconnect the wiring connector and remove the switch from the vehicle.

4 Refitting is the reverse of removal. On completion, reset the switch by firmly depressing its button.

1.9 litre (WJY) engine

Stop solenoid

5 See paragraph 1.

Electronic control unit

6 The ECU is located on the right-hand side of the engine compartment.

7 To remove the ECU, first disconnect the battery.

8 Unclip the wiring harness then slacken and remove the ECU mounting plate nut. The plate is secured in position with an anti-tamper nut, unscrew the nut using a pair of grips.

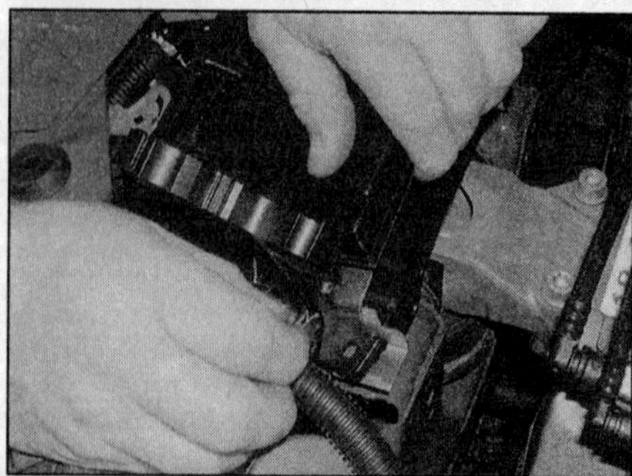

9 Remove the cable/hose guide bracket then free the mounting plate from the body.

10 Lift the retaining clips and disconnect the ECU wiring connectors then unscrew the mounting nuts (these are also anti-tamper nuts) and remove the ECU from the mounting plate.

11 Refitting is a reverse of the removal procedure, ensuring the wiring connectors are securely reconnected. Secure the ECU and mounting plate in position with new anti-tamper nuts or standard nuts.

Crankshaft sensor

12 The crankshaft sensor is situated on the top of the transmission unit.

13 To remove the sensor, first disconnect the battery negative terminal.

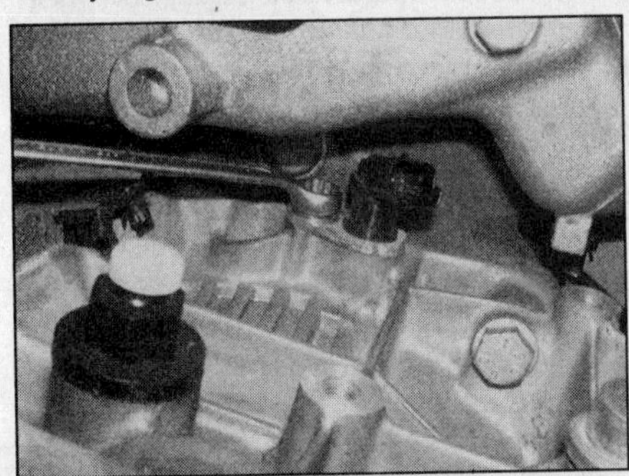

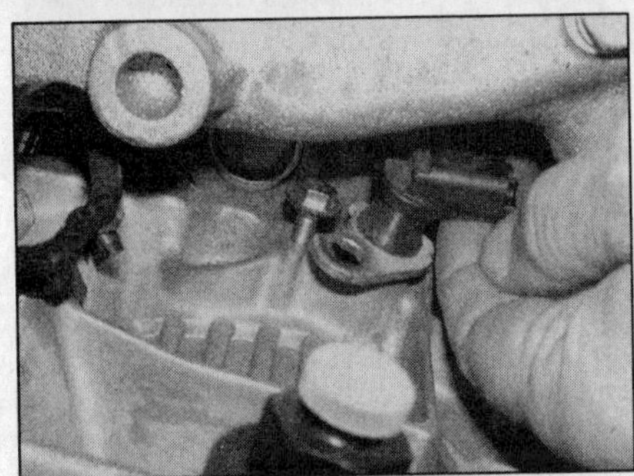

14 Disconnect the wiring connector from the sensor then slacken the mounting bolt and remove the sensor from the transmission unit (the sensor is slotted to ease removal).

15 Refitting is reverse of the removal procedure.

Vehicle speed sensor

16 The vehicle speed sensor is an integral part of the speedometer drive. Refer to the petrol injection system Section for removal and refitting details.

Injector needle lift sensor

17 The needle lift sensor is an integral part of No 1 cylinder injector.

Injection system relay unit

18 The relay unit is mounted onto the underside of the ECU mounting plate. To remove the sensor, first disconnect the battery negative terminal.

19 Unclip the wiring harness then slacken and remove the ECU mounting plate nut. The plate is secured in position with an anti-tamper nut, unscrew the nut using a pair of grips.

20 Remove the cable/hose guide bracket then free the mounting plate from the body.

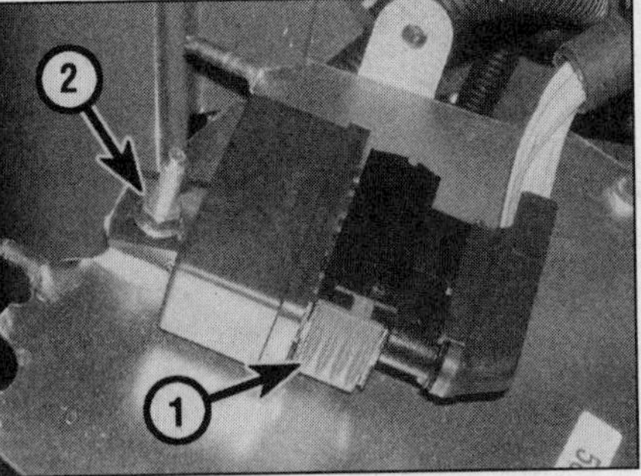

21 Lift the retaining clip (1) and disconnect the wiring connector from the relay unit. Unscrew the mounting nut (2) and remove the relay from the mounting plate.

22 Refitting is a reverse of the removal procedure, ensuring the wiring connector is securely reconnected. Secure the mounting plate in position with new anti-tamper nuts or standard nuts.

Coolant temperature sensor

23 The coolant temperature sensor is screwed into the coolant outlet housing. Refer to the petrol injection system Section for removal and refitting information.

Throttle potentiometer and advance solenoid

24 These components are both fitted to the injection pump. If either of the above are faulty, replacement should be entrusted to a Peugeot dealer who will have the necessary special equipment to adjust and calibrate the replacements on refitting.

Fuel cut-off inertia switch

Note: *A fuel cut-off inertia switch is not fitted to all models.*

25 Refer to paragraphs 2 to 4.

2.0 litre engine

Electronic control unit

26 The ECU is located on the right-hand side of the engine compartment.

27 To remove the ECU, first disconnect the battery.

28 Unclip the wiring harness then slacken and remove the ECU mounting plate nut. The plate is secured in position with an anti-tamper nut, unscrew the nut using a pair of grips.

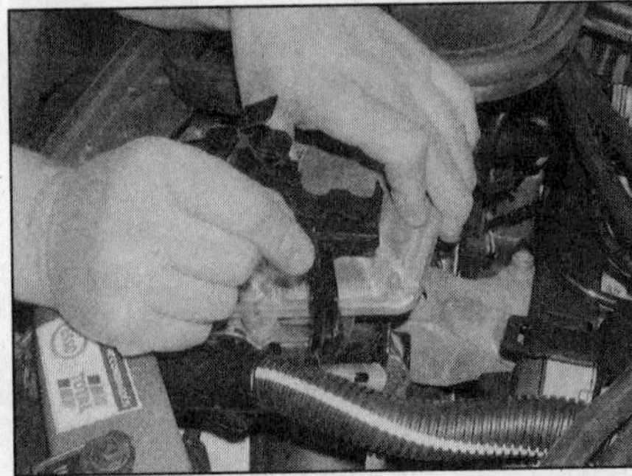

29 Remove the bracket then free the mounting plate from the body.

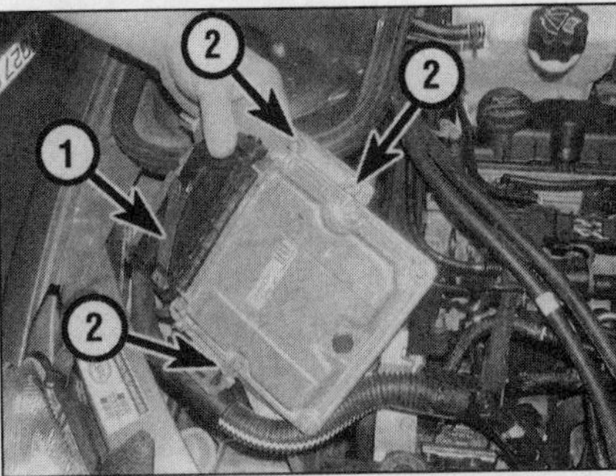

30 Lift the retaining clip and disconnect the ECU wiring connector (1) then unscrew the mounting nuts (2) - these are also anti-tamper nuts - and remove the ECU from the mounting plate.

31 Refitting is a reverse of the removal procedure, ensuring the wiring connector is securely reconnected. Secure the ECU and mounting plate in position with new anti-tamper nuts or standard nuts.

Crankshaft sensor

32 Refer to paragraphs 12 to 15. Access to the sensor is poor; remove the battery and battery tray to improve access.

Vehicle speed sensor

33 The vehicle speed sensor is an integral part of the speedometer drive. Refer to the petrol injection system Section for removal and refitting details.

Injection system relay unit

34 The relay unit is mounted onto the underside of the ECU mounting plate. To remove the sensor, first disconnect the battery negative terminal.

35 Unclip the wiring harness then slacken and remove the ECU mounting plate nut. The plate is secured in position with an anti-tamper nut, unscrew the nut using a pair of grips.

36 Free the mounting plate from the body.

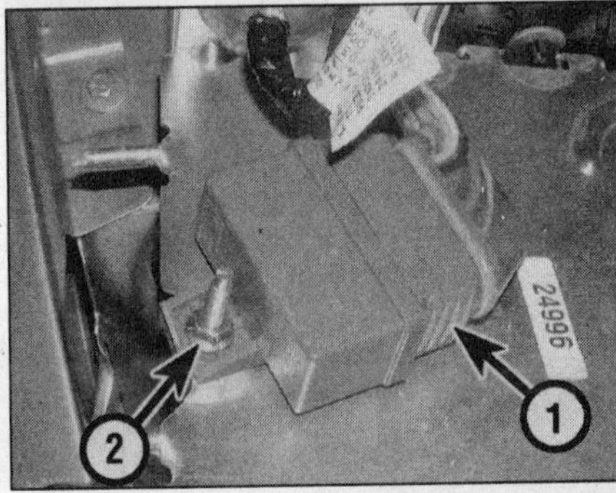

37 Release the retaining clip and disconnect the wiring connector (1) from the relay unit. Unscrew the mounting nut (2) and remove the relay from the mounting plate.

38 Refitting is a reverse of the removal procedure, ensuring the wiring connector is securely reconnected. Secure the mounting plate in position with a new anti-tamper nut or standard nut.

Coolant temperature sensor

39 The coolant temperature sensor is fitted to the side of the coolant outlet housing on the left-hand end of the cylinder. Refer to the petrol injection system Section for removal and refitting information.

Airflow meter

40 The airflow meter is fitted to intake duct connecting the air cleaner housing to the turbocharger intake pipe. Disconnect the battery negative terminal prior to removal.

41 Slacken the retaining clips and remove the intake duct connecting the airflow meter to the intake pipe.

42 Slacken the retaining clip and free the airflow meter from the air cleaner housing. Disconnect the wiring connector and remove the airflow meter from the vehicle, noting which way around it is fitted.

Caution: Never touch the air temperature sensor or metallic plate of the airflow meter as they are easily damaged.

43 Refitting is the reverse of removal, ensuring the airflow meter is fitted the correct way around with the arrow moulded on its body pointing in the direction of airflow (towards the turbocharger pipe).

Camshaft position sensor

44 The camshaft position sensor is fitted to the top of the cylinder head cover right-hand end.

45 Remove the fasteners (rotate them 90° to release them) and remove the engine cover.

46 Remove the camshaft sprocket timing belt cover.

47 Disconnect the battery negative terminal then disconnect the camshaft sensor wiring connector.

48 Unscrew the retaining bolt and remove the sensor from the engine.

49 On refitting, align the camshaft hub timing hole with the cylinder head hole . Insert the flywheel locking tool to ensure the camshaft is correctly positioned.

50 Fit the sensor to its mounting bracket and lightly tighten its bolt.

51 If a new sensor is being fitted, position the sensor so its fitting lug is in contact with the rear of the camshaft hub then securely tighten the sensor bolt. The lug automatically sets the sensor air gap to the correct distance and will be knocked off the first time the engine is started.

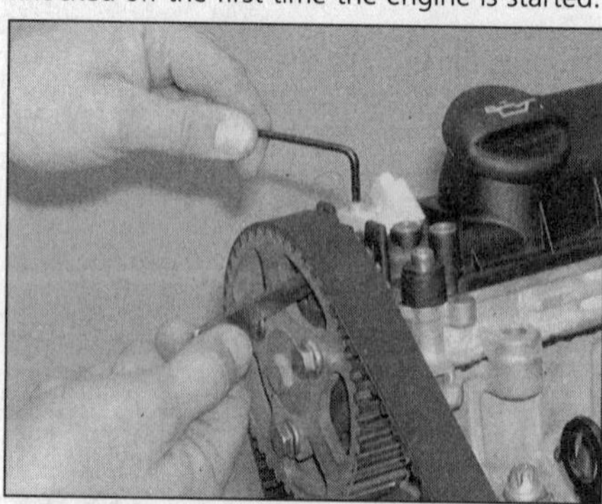

52 If the original sensor is being refitted, using feeler gauges, set the gap between the sensor tip and camshaft hub to 1.2 mm. Ensure the sensor is correctly positioned then securely tighten its retaining bolt. Check the air gap and, if necessary, readjust.

53 Once the sensor is correctly positioned, refit the timing belt cover and remove the flywheel locking tool (where fitted).

54 Reconnect the sensor wiring connector then securely refit the engine cover.

Fuel temperature sensor

55 The fuel temperature sensor (where fitted) is fitted to the fuel rail.

56 Disconnect the battery negative terminal then remove the fasteners (rotate them 90° to release them) and remove the engine cover *(see paragraph 45)*.

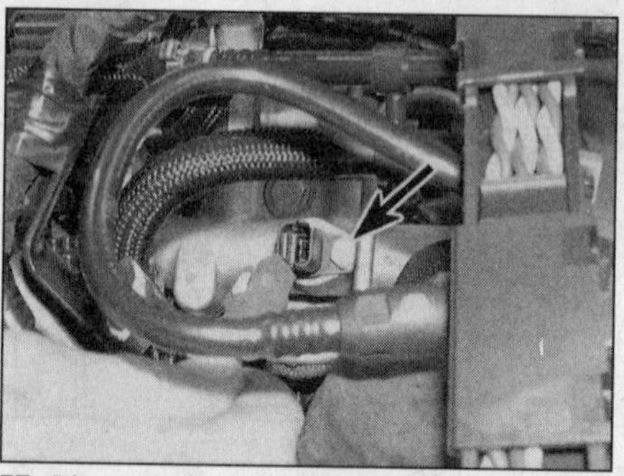

57 Disconnect the wiring connector then unscrew the retaining bolt and remove the sensor from the fuel rail.

58 Refitting is the reverse of removal.

Fuel pressure sensor

Warning: Always wait at least 30 seconds after switching off the engine before disconnecting any fuel hoses/pipes. This will allow the pressure in the system to return to atmospheric pressure.

Caution: Be careful not to allow dirt into the fuel rail/sensor during this procedure.

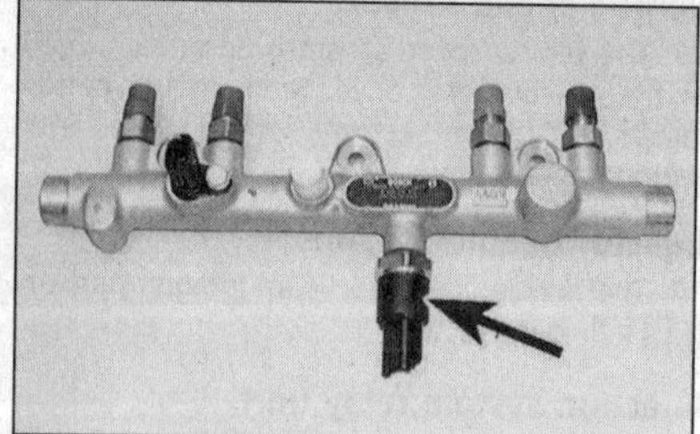

59 The fuel pressure sensor is fitted to the fuel rail.

60 Disconnect the battery negative terminal then release the fasteners (rotate them 90° to release them) and remove the engine cover *(see paragraph 45 opposite)*.

61 Disconnect the wiring connector from the pressure sensor.

62 Remove all traces of dirt from around the

sensor; the sensor and its surrounding area must be clean and dry before proceeding. Position a wad of rag beneath the sensor to catch any spilt fuel.

63 Unscrew the sensor from the fuel rail and remove it along with its sealing washer. Discard the washer; a new one must be used on refitting. Plug the fuel rail to prevent the entry of dirt into the fuel system.

64 Refitting is the reverse of removal, using a new sealing washer and tightening the sensor to the specified torque. On completion, prime the fuel system.

Third piston deactivator solenoid valve

65 The solenoid valve is mounted on the top of the injection pump; it is an integral part of the pump and cannot be renewed. If the valve is faulty the injection pump will have to be renewed or repaired. Never attempt to remove the solenoid from the pump.

High-pressure fuel regulator

66 The high-pressure fuel regulator is mounted on the rear of the injection pump; it is an integral part of the pump and cannot be renewed. If the regulator is faulty the injection pump will have to be renewed or repaired. Never attempt to remove the regulator from the pump.

Fuel cut-off inertia switch

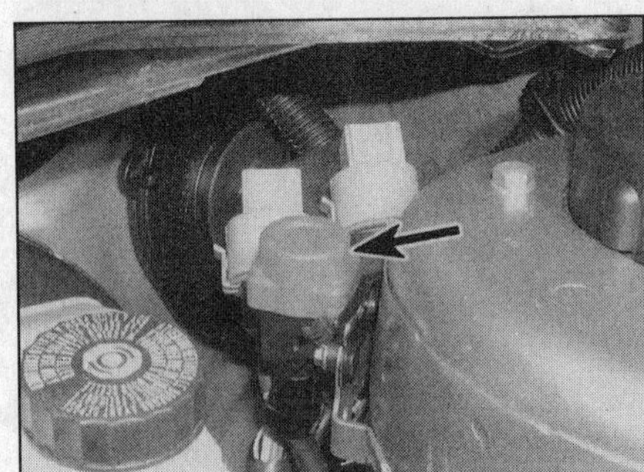

67 Refer to paragraphs 2 to 4.

Electrically-operated fuel heater

Warning: Always wait at least 30 seconds after switching off the engine before disconnecting any fuel hoses/pipes. This will allow the pressure in the system to return to atmospheric pressure.

Caution: Be careful not to allow dirt into the fuel pipes/heater during this procedure.

68 The fuel heater is fitted to the fuel feed pipe to the filter housing.

69 Disconnect the battery negative terminal then release the fasteners (rotate them 90° to release them) and remove the engine cover *(see paragraph 45)*.

70 Disconnect the wiring connector from the heater.

71 Remove all traces of dirt from around the heater; the heater and its surrounding area must be clean and dry before proceeding. Position a wad of rag beneath the heater to catch any spilt fuel.

72 Release the clips and disconnect the fuel pipes from the heater. Plug the fuel pipes to prevent the entry of dirt into the fuel system.

73 Unscrew the retaining bolt and remove the fuel heater from the engine.

74 Refitting is the reverse of removal. On completion, prime the fuel system.

Accelerator pedal position sensor

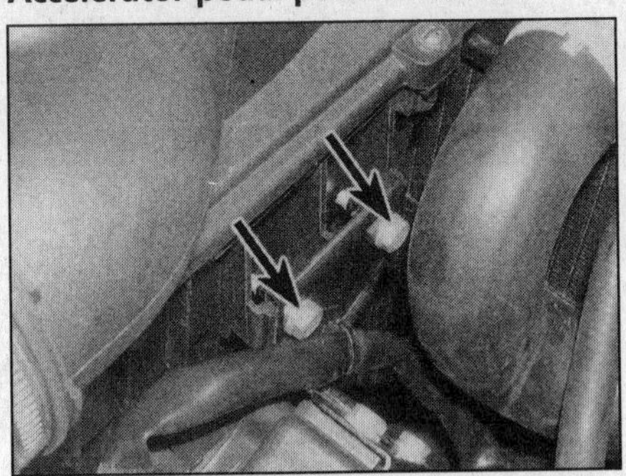

75 Working in the engine compartment, unscrew the two bolts and free the accelerator

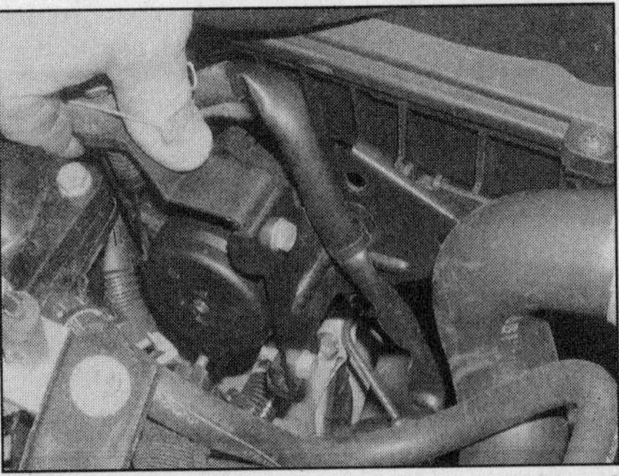

pedal position sensor bracket from the side of the air cleaner housing.

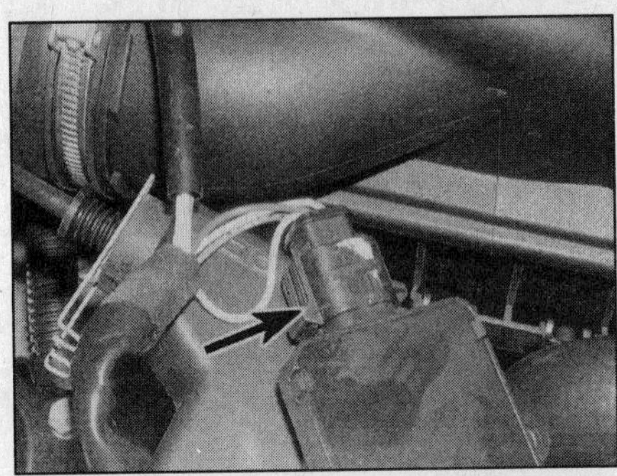

76 Disconnect the sensor wiring connector.

77 Free the accelerator inner cable from the sensor then free the outer cable from sensor bracket, complete with the spring clip, and remove the sensor from the vehicle.

78 Refitting is the reverse of removal. Where necessary, adjust the throttle cable.

Notes

Peugeot 307 '01 – '04

Petrol models

General

System type:

1.4 litre models	Sagem S2000
1.6 litre models	Bosch Motronic ME7.4.4
2.0 litre models	Magneti Marelli 4.8P

Fuel system data

Fuel pump type	Electric, immersed in tank
Fuel pump regulated constant pressure	3.5 ± 0.2 bars
Specified idle speed	850 ± 100 rpm (not adjustable – controlled by ECU)
Idle mixture CO content	Less than 1.0 % (not adjustable- controlled by ECU)

Recommended fuel

Minimum octane rating	95 RON unleaded (UK unleaded premium). Leaded/lead replacement fuel (LRP) must **not** be used

Diesel models

General

System type	HDi (High-pressure Diesel injection) with full electronic control, direct injection and turbocharger
Designation:	
1.4 litre	Bosch EDC 16
2.0 litre	Bosch EDC 15C2
Firing order	1-3-4-2 (No 1 at flywheel end)
Fuel system operating pressure	200 to 1350 bars (according to engine speed)
Idle speed:	
1.4 litre	800 ± 20 rpm (controlled by ECU)
2.0 litre engine	800 ± 20 rpm (controlled by ECU)
Engine cut-off speed:	
1.4 litre	5000 rpm (controlled by ECU)
2.0 litre	5000 rpm (controlled by ECU)

High-pressure fuel pump

Type:	
1.4 litre	CP3.2
2.0 litre	Bosch CP 1
Direction of rotation	Clockwise, viewed from sprocket end

Injectors

Type	Electromagnetic

Turbocharger

Type:	
1.4 litre	KKK
2.0 litre	Garrett GT15 or KKK K03
Boost pressure (approximate)	1 bar at 3000 rpm

Chapter 19

Torque wrench settings	Nm	lbf ft
Petrol models		
Exhaust manifold-to-cylinder head nuts:	20	15
Exhaust manifold-to-catalytic converter	15	11
Inlet manifold nuts:		
M6	10	7
M8	20	15
Diesel models		
Camshaft position sensor bolt (1.4 litre engine)	5	4
Crankshaft speed/position sensor (1.4 litre engine)	5	4
Fuel pressure sensor to accumulator rail	45	33

General information

Petrol models

Note: *The fuel injection ECU is of the 'self-learning' type, meaning that as it operates, it also monitors and stores the settings which give optimum engine performance under all operating conditions. When the battery is disconnected, these settings are lost and the ECU reverts to the base settings programmed into its memory at the factory. On restarting, this may lead to the engine running/idling roughly for a short while, until the ECU has relearned the optimum settings. This process is best accomplished by taking the vehicle on a road test (for approximately 15 minutes), covering all engine speeds and loads, concentrating mainly in the 2500 to 3500 rpm region.*

On all engines, the fuel injection and ignition functions are combined into a single engine management system. The systems fitted are manufactured by Bosch, Magneti Marelli and Sagem, and are very similar to each other in most respects, the only significant differences being in the software contained in the system ECU, and specific component location according to engine type. Each system incorporates a closed-loop catalytic converter and an evaporative emission control system, and complies with the latest emission control standards.

The fuel pump supplies fuel from the tank to the fuel rail, via a replaceable cartridge filter mounted on the side of the fuel tank. The pump itself is mounted inside the tank, with the pump motor permanently immersed in fuel, to keep it cool. The fuel rail is mounted directly above the fuel injectors and acts as a fuel reservoir.

Fuel rail supply pressure is controlled by the pressure regulator, also located in the fuel tank. The regulator contains a spring-loaded valve, which lifts to allow excess fuel to recirculate within the tank when the optimum operating pressure of the fuel system is exceeded (eg, during low speed, light load cruising).

The fuel injectors are electromagnetic pintle valves, which spray atomised fuel into the combustion chambers under the control of the engine management system ECU. There are four injectors, one per cylinder, mounted in the inlet manifold close to the cylinder head. Each injector is mounted at an angle that allows it to spray fuel directly onto the back of the inlet valve(s). The ECU controls the volume of fuel injected by varying the length of time for which each injector is held open. The fuel injection systems are typically of the sequential type, whereby each injector operates individually in cylinder sequence.

The electrical control system consists of the ECU, along with the following sensors:

**a) Throttle potentiometer – informs the ECU of the throttle valve position, and the rate of throttle opening/closing (Sagem S2000).*
b) Coolant temperature sensor – informs the ECU of engine temperature.
c) Inlet air temperature sensor – informs the ECU of the temperature of the air passing through the throttle housing.
d) Oxygen sensors – inform the ECU of the oxygen content of the exhaust gases (explained in greater detail in Part C of this Chapter).
e) Manifold pressure sensor – informs the ECU of the load on the engine (expressed in terms of inlet manifold vacuum).
f) Crankshaft position sensor – informs the ECU of engine speed and crankshaft angular position.
g) Vehicle speed sensor – informs the ECU of the vehicle speed.
h) Knock sensor – informs the ECU of pre-ignition (detonation) within the cylinders.
i) Camshaft sensor – informs the ECU of which cylinder is on the firing stroke on systems with sequential injection (MM 4.8.P).
**J) Accelerator pedal position sensor – informs the ECU of the pedal position and rate of change (ME7.4.4, MM 4.8.P).*
**H) Throttle valve positioner motor – allows the ECU to control the throttle valve position (ME7.4.4, MM 4.8.P).*
** 2.0 litre petrol engines may be fitted with a manual throttle body with an accelerator cable, or a motorised throttle body with no cable.*

Signals from each of the sensors are compared by the ECU and, based on this information, the ECU selects the response appropriate to those values, and controls the fuel injectors (varying the pulse width – the length of time the injectors are held open – to provide a richer or weaker air/fuel mixture, as appropriate). The air/fuel mixture is constantly varied by the ECU, to provide the best settings for cranking, starting (with either a hot or cold engine) and engine warm-up, idle, cruising and acceleration.

The ECU also has full control over the engine idle speed, via a stepper motor fitted to the throttle housing. The stepper motor either controls the amount of air passing through a bypass drilling at the side of the throttle or controls the position of the throttle valve itself, depending on model. On some models, a sensor informs the ECU of the position, and rate of change, of the accelerator pedal. The ECU them controls the throttle valve by means of a throttle positioning motor integral with the throttle body – no accelerator cable is fitted. On 1.6 litre models, the accelerator cable is connected to a sensor in the left-hand front corner of the engine compartment, which informs the ECU of the accelerator pedal position. The ECU also carries out 'fine tuning' of the idle speed by varying the ignition timing to increase or reduce the torque of the engine as it is idling. This helps to stabilise the idle speed when electrical or mechanical loads (such as headlights, air conditioning, etc) are switched on and off.

The throttle housing is also fitted with an electric heating element. The heater is supplied with current by the ECU, warming the throttle housing on cold starts to help prevent icing of the throttle valve.

If there is any abnormality in any of the readings obtained from the coolant temperature sensor, the inlet air temperature sensor or the lambda sensor, the ECU enters its 'back-up' mode. If this happens, the erroneous sensor signal is overridden, and the ECU assumes a pre-programmed 'back-up' value, which will allow the engine to continue running, albeit at reduced efficiency. If the ECU enters this mode, the warning lamp on the instrument panel will be illuminated, and the relevant fault code will be stored in the ECU memory.

The system's diagnostic connector is located behind the trim panel to the right of the steering column.

Diesel models

The fuel system consists of a rear-mounted fuel tank and fuel lift pump, a fuel filter with integral water separator, on some models a fuel cooler mounted under the car, and an electronically-controlled High-pressure Diesel injection (HDi) system, together with a turbocharger.

The exhaust system is conventional, but to meet the latest emission levels an unregulated catalytic converter and an exhaust gas recirculation system are fitted to all models. On some 2.0 litre models, an exhaust emission particulate filter is fitted.

The HDi system (generally known as a 'common rail' system) derives its name from the fact that a common rail (referred to as an accumulator rail), or fuel reservoir, is used to supply fuel to all the fuel injectors. Instead of an in-line or distributor type injection pump, which distributes the fuel directly to each injector, a high-pressure pump is used, which generates a very high fuel pressure (1350 bars at high engine speed) in the accumulator rail. The accumulator rail stores fuel, and maintains a constant fuel pressure, with the aid of a pressure control valve. Each injector is supplied with high-pressure fuel from the accumulator rail, and the injectors are individually controlled via signals from the system electronic control unit (ECU). The injectors are electro-magnetically-operated.

In addition to the various sensors used on models with a conventional fuel injection pump, common rail systems also have a fuel pressure sensor. The fuel pressure sensor allows the ECU to maintain the required fuel pressure, via the pressure control valve.

System operation

For the purposes of describing the operation of a common rail injection system, the components can be divided into three sub-systems; the low-pressure fuel system, the high-pressure fuel system and the electronic control system.

Low-pressure fuel system

The low-pressure fuel system consists of the following components:

a) Fuel tank.
b) Fuel lift pump.
c) Fuel cooler (not all models).
d) Fuel heater (not all models).
e) Fuel filter/water trap.
f) Low-pressure fuel lines.

The low-pressure system (fuel supply system) is responsible for supplying clean fuel to the high-pressure fuel system.

High-pressure fuel system

The high-pressure fuel system consists of the following components:

a) High-pressure fuel pump with pressure control valve.
b) High-pressure fuel accumulator rail.
c) Fuel injectors.
d) High-pressure fuel lines.

After passing through the fuel filter, the fuel reaches the high-pressure pump, which forces it into the accumulator rail. As diesel fuel has a certain elasticity, the pressure in the accumulator rail remains constant, even though fuel leaves the rail each time one of the injectors operates. Additionally, a pressure control valve mounted on the high-pressure pump ensures that the fuel pressure is maintained within preset limits.

The pressure control valve is operated by the ECU. When the valve is opened, fuel is returned from the high-pressure pump to the tank, via the fuel return lines, and the pressure in the accumulator rail falls. To enable the ECU to trigger the pressure control valve correctly, the pressure in the accumulator rail is measured by a fuel pressure sensor.

The electromagnetically-controlled fuel injectors are operated individually, via signals from the ECU, and each injector injects fuel directly into the relevant combustion chamber. The fact that high fuel pressure is always available allows very precise and highly flexible injection in comparison to a conventional injection pump: for example combustion during the main injection process can be improved considerably by the pre-injection of a very small quantity of fuel.

Electronic control system

The electronic control system consists of the following components:

a) Electronic control unit (ECU).
b) Crankshaft speed/position sensor.
c) Camshaft position sensor.
d) Accelerator pedal position sensor.
e) Coolant temperature sensor.
f) Fuel temperature sensor.
g) Air mass meter.
h) Fuel pressure sensor.
i) Fuel injectors.
j) Fuel pressure control valve.
k) Preheating control unit.
l) EGR solenoid valve.

The information from the various sensors is passed to the ECU, which evaluates the signals. The ECU contains electronic 'maps' which enable it to calculate the optimum quantity of fuel to inject, the appropriate start of injection, and even pre- and post-injection fuel quantities, for each individual engine cylinder under any given condition of engine operation.

Additionally, the ECU carries out monitoring and self-diagnostic functions. Any faults in the system are stored in the ECU memory, which enables quick and accurate fault diagnosis using appropriate diagnostic equipment (such as a suitable fault code reader).

Fuel system – depressurisation, pressurising, priming and bleeding

Petrol models

Depressurisation

Warning: The following procedure will merely relieve the pressure in the fuel system – remember that fuel will still be present in the system components and take precautions accordingly before disconnecting any of them.

1 The fuel system referred to in this Section is defined as the tank-mounted fuel pump, the fuel filter, the fuel injectors, the fuel rail and the pipes of the fuel lines between these components. All these contain fuel which will be under pressure while the engine is running, and/or while the ignition is switched on. The pressure will remain for some time after the ignition has been switched off, and must be relieved in a controlled fashion when any of these components are disturbed for servicing work.

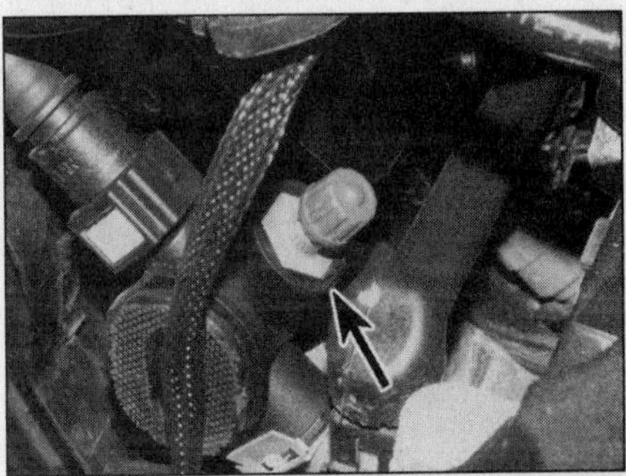

2 Some models are equipped with a pressure relief valve on the fuel rail. On these models, unscrew the cap from the valve and position a container beneath the valve. Hold a wad of rag over the valve and relieve the pressure in the system by depressing the valve core with a suitable screwdriver. Be prepared for the squirt of fuel as the valve core is depressed and catch it with the rag. Hold the valve core down until no more fuel is expelled from the valve. Once the pressure is relieved, securely refit the valve cap.

3 Where no valve is fitted to the fuel rail, it will be necessary to release the pressure as the fuel pipe is disconnected. Place a container beneath the union and position a large rag around the union to catch any fuel spray which may be expelled. Slowly release and disconnect the fuel pipe and catch any spilt fuel in the container. Plug the pipe/union to minimise fuel loss and prevent the entry of dirt into the fuel system.

Pressurising

4 After any work is carried out on the fuel

system, the system should be pressurised as follows.

5 Depress the accelerator pedal fully then switch on the ignition. Hold the pedal depressed for approximately 1 second then release it. The ECU should then operate the fuel pump for between 20 and 30 seconds to refill the fuel system. Once the fuel pump stops the ignition can be switched off.

Diesel models

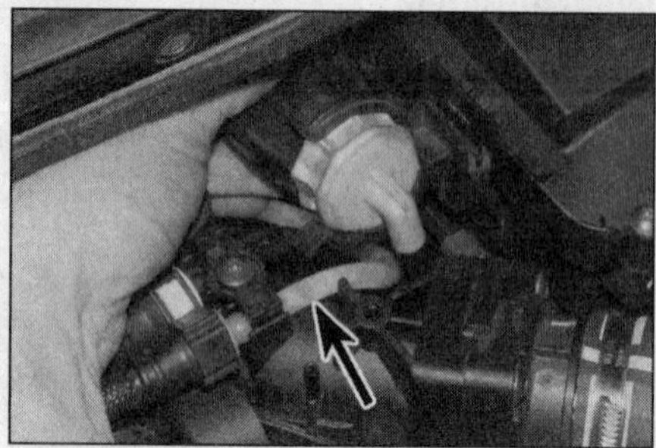

6 Should the fuel supply system be disconnected between the fuel tank and high pressure pump, it is necessary to prime the fuel system. This is achieved by operating the hand priming pump until resistance is felt (2.0 litre models) or fuel appears in the transparent fuel supply pipe in the engine compartment (1.4 litre models). Remove the plastic cover from the top of the engine to access the priming pump. **Note:** *A hand priming pump is not fitted to all 2.0 litre models.*

7 Once this is achieved, operate the starter until the engine starts.

Petrol engine management system components – removal and refitting

Fuel rail and injectors

Note: *If a faulty injector is suspected, before condemning the injector, it is worth trying the effect of one of the proprietary injector-cleaning treatments which are available from car accessory shops.*

1 Slacken the retaining clips then free the duct from the throttle housing and air cleaner housing and remove it from the engine compartment. **Note:** In order to improve access,

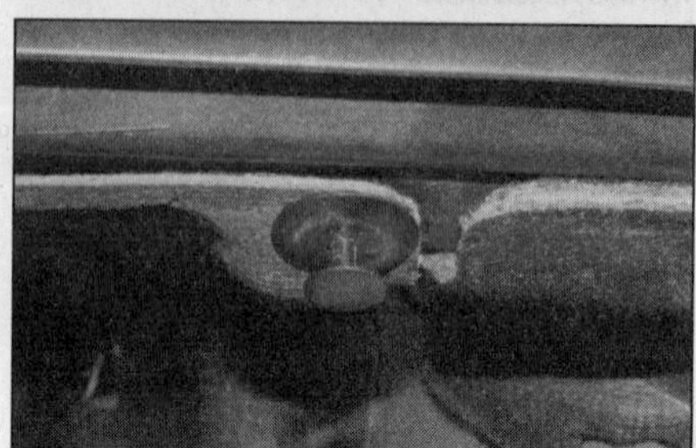

remove the wiper blades and plastic scuttle trim. The trim is secured by plastic expanding rivets at each end – push in the centre pins a little, then prise out the complete rivet. Undo the two screws and move the master cylinder upper reservoir to one side. Release the sound insulation trim from the scuttle crossmember, undo the bolt at each end, and remove the crossmember from the vehicle

1.4 litre models

2 Remove the ignition HT coil.

3 Free the accelerator inner cable (where fitted) from the throttle housing cam, then pull the outer cable out from its mounting bracket rubber grommet, complete with its spring clip.

4 Unscrew the bolts and remove the accelerator cable bracket (where fitted) from the manifold/cylinder head.

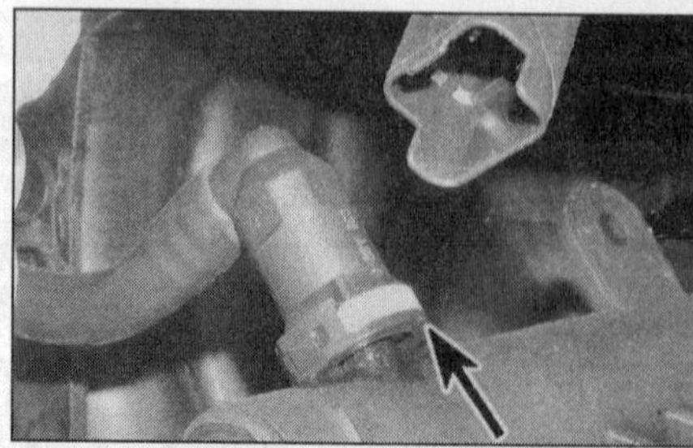

5 Depress the retaining clip and disconnect the fuel pipe from the right-hand end of the fuel rail, bearing in mind the information on depressurising the fuel system.

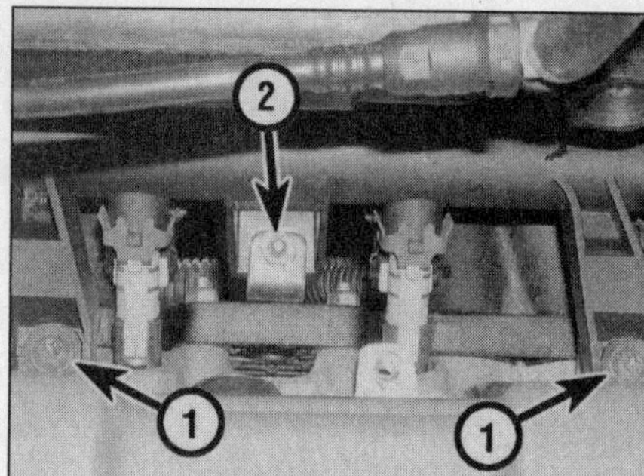

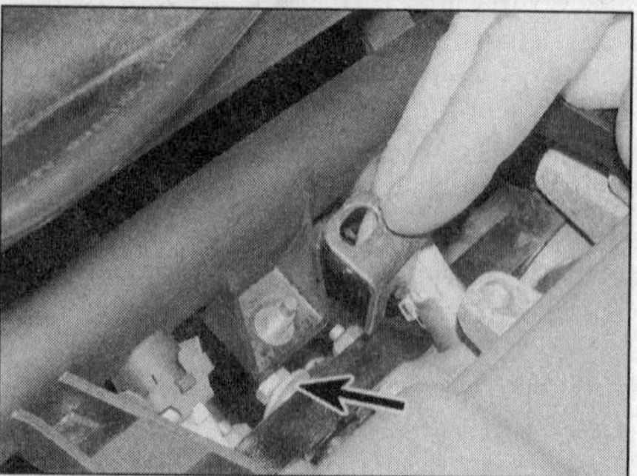

6 Slacken and remove the two bolts (1) securing the fuel rail to the cylinder head, and the nut (2) securing the rail to the manifold. Loosen the bolt securing the fuel rail centre bracket to the inlet manifold, then lift off the bracket (the bracket is slotted to ease removal).

7 Disconnect the injector wiring harness connector, then unclip the connector from the rear of the inlet manifold. Also disconnect the wiring connectors from the throttle housing and position the wiring harness clear of the manifold so that it does not hinder fuel rail removal.

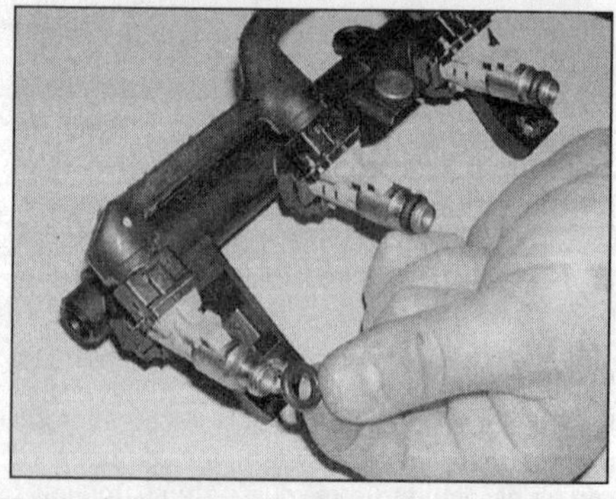

8 Carefully ease the fuel rail and injector assembly out from the cylinder head and manoeuvre it out of position. Remove the seals from the end of each injector and discard them; they must be renewed whenever they are disturbed.

9 Refitting is a reversal of the removal procedure, noting the following points.

a) Fit new seals to all disturbed injector unions.

b) Apply a smear of engine oil to the seals to aid installation, then ease the injectors and fuel rail into position ensuring that none of the seals are displaced.

c) On completion, pressurise the fuel system. Start the engine and check for fuel leaks.

1.6 litre models

10 Remove the inlet manifold.

11 Undo the two bolts and remove the fuel rail with injectors from the manifold.

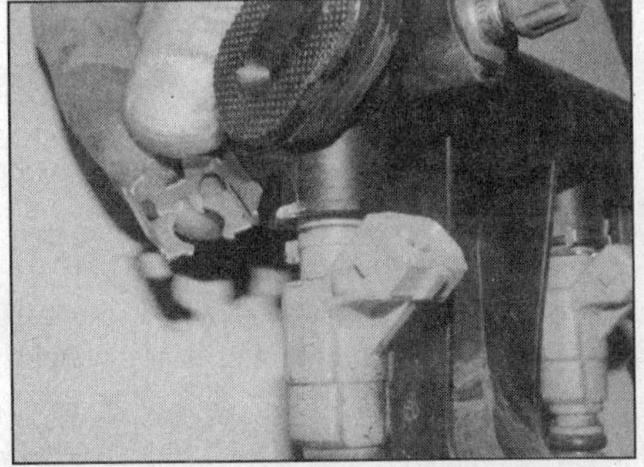

12 Disconnect the wiring connector(s) then slide out the retaining clip(s) and remove the relevant injector(s) from the fuel rail. Remove the seals from each disturbed injector and discard; all disturbed seals must be renewed.

13 Refitting is a reversal of the removal procedure, noting the following points.

a) Fit new seals to all disturbed injector unions.

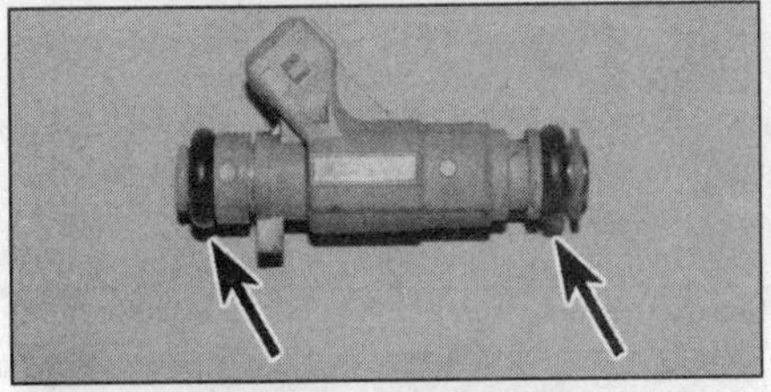

b) Apply a smear of engine oil to the seals to aid installation, then ease the injectors and fuel rail into position ensuring that none of the seals are displaced.

c) On completion, pressurise the fuel system. Start the engine and check for fuel leaks.

2.0 litre models

14 Unbolt and remove the engine top cover, then disconnect the air duct from between the air cleaner and throttle body.

15 Disconnect the wiring from the inlet air temperature sensor.

16 Unbolt the wiring tray from the top of the inlet manifold, and position to one side.

17 Depress the release button and disconnect the fuel supply hose from the fuel rail.

18 Depress the retaining clip(s), and disconnect the wiring connector(s) from the injector(s).

19 Unscrew the mounting bolts and carefully ease the fuel rail complete with injectors from the inlet manifold. Remove the O-rings from the end of each injector, and discard them; these must be renewed whenever they are disturbed.

20 Slide out the retaining clip(s) and remove the relevant injector(s) from the fuel rail. Remove the upper O-ring from each disturbed injector and discard; all disturbed O-rings must be renewed.

21 Refitting is a reversal of the removal procedure, noting the following points.

a) Fit new O-rings to all disturbed injector unions.

b) Apply a smear of engine oil to the O-rings to aid installation then ease the injectors and fuel rail into position ensuring that none of the O-rings are displaced.

c) On completion start the engine and check for fuel leaks.

Fuel pressure regulator

22 The fuel pressure regulator is an integral part of the fuel pump assembly and is not available separately.

Throttle potentiometer

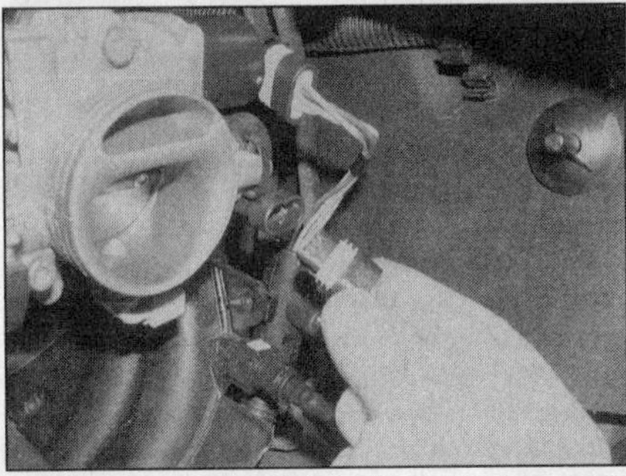

23 Depress the retaining clip and disconnect the wiring connector from the throttle potentiometer.

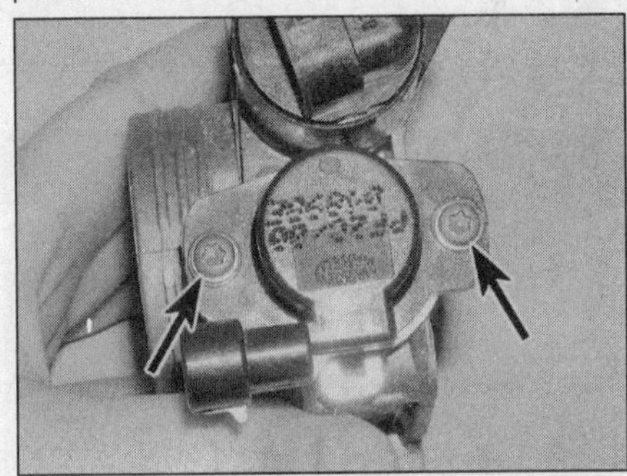

24 Slacken and remove the two retaining screws, then disengage the potentiometer from the throttle valve spindle and remove it from the vehicle.

25 Refit in the reverse order of removal.

26 Ensure that the potentiometer is correctly engaged with the throttle valve spindle.

Electronic Control Unit (ECU)

27 The ECU is located on the left-hand side of the engine compartment.

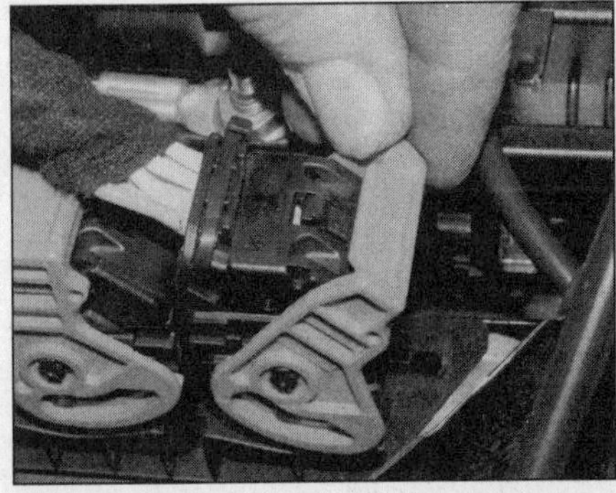

28 Remove the electrical box lid, and disconnect the wiring plugs.

29 The ECU lifts up from the electrical box.

30 Refitting is a reverse of the removal procedure ensuring the wiring connectors are securely reconnected.

Note: *If a replacement ECU has been fitted, the vehicle will not start until the immobiliser ECU has been matched to the engine management ECU. This can only be performed using dedicated test equipment. Consequently, entrust the procedure to a Peugeot dealer or suitably equipped specialist.*

Idle speed stepper motor

31 The idle speed stepper motor is fitted to the rear of the throttle housing.

32 Disconnect the wiring connector from the motor.

33 Slacken and remove the retaining screws then remove the motor from the throttle housing. If necessary, remove the throttle potentiometer to improve access to the motor lower screw.

34 Refitting is a reversal of the removal procedure ensuring the seal is in good condition.

Manifold pressure sensor

35 The MAP sensor is mounted on the inlet manifold.

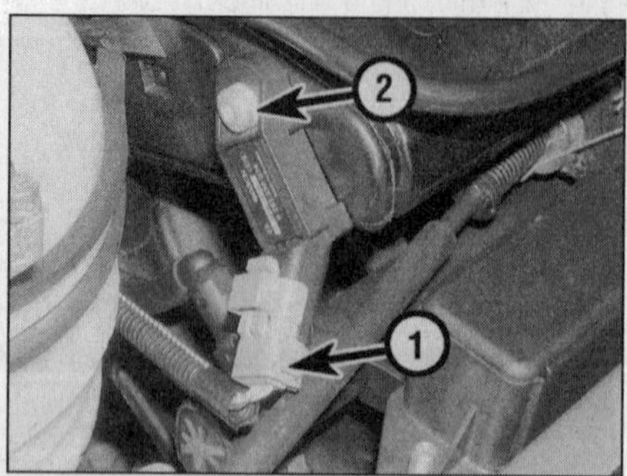

Wiring connector (1) and retaining screw (2) – 1.4 litre engine

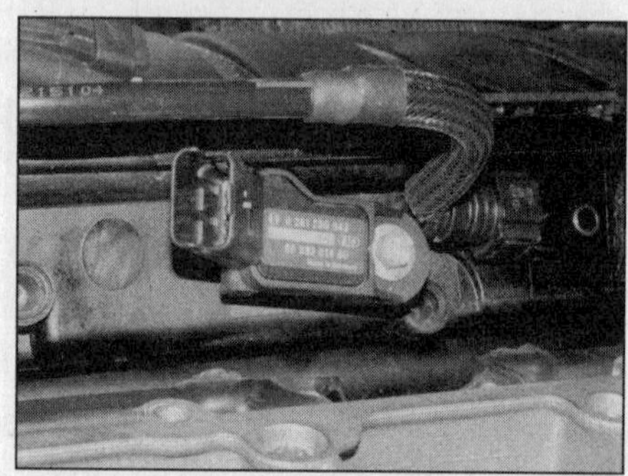

1.6 litre engine

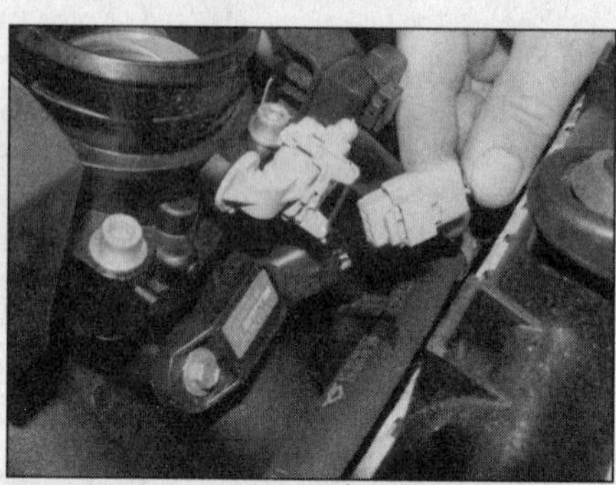

2.0 litre engine

36 Disconnect the wiring connector then undo the screw and remove the sensor from the manifold. .

37 Refitting is a reversal of the removal procedure ensuring the sensor seal is in good condition.

Coolant temperature sensor

38 The coolant temperature sensor is screwed into the coolant outlet housing on the left-hand end of the cylinder head. Partially drain the cooling system to just below the level of the sensor. On diesel models, to improve access, remove the battery.

39 Disconnect the wiring connector from the sensor.

40 On some engines, the sensor is clipped in place. Prise out the sensor retaining circlip then remove the sensor and sealing ring from the housing. On other engines, unscrew the sensor and recover the sealing washer (where applicable).

41 Refitting is a reversal of removal, ensuring a new sealing ring is used.

Intake air temperature sensor

42 The intake air temperature sensor is integral with the throttle housing and is not available separately.

Crankshaft position sensor

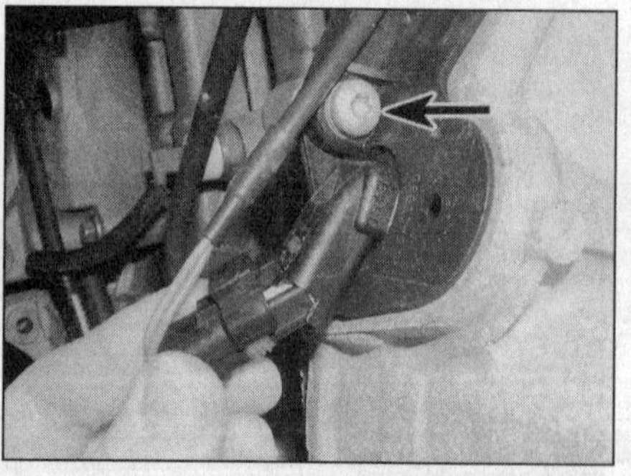

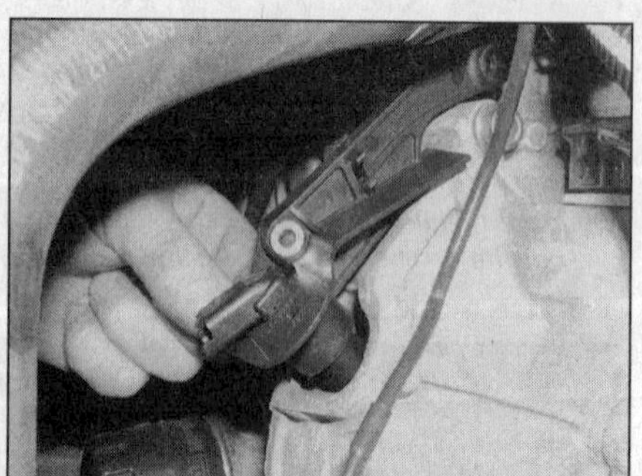

43 The crankshaft sensor is situated on the front face of the transmission clutch housing. Disconnect the sensor wiring connector and unclip the wiring. Undo the retaining bolt and remove the sensor and bracket assembly from the transmission unit.

44 Refitting is reverse of the removal procedure.

Throttle housing heating element

Note: *The heating element is only fitted to aluminium throttle housings. Plastic housings do not need a heater.*

45 The heating element is fitted to the top of the throttle housing. Disconnect the wiring connector then unscrew the retaining screw and remove the heating element from the throttle housing.

46 Refitting is the reverse of removal.

Vehicle speed sensor

47 The vehicle speed sensor is an integral part of the speedometer drive on 1.4 litre models. On other models, the ECU receives vehicle speed data from the wheel speed sensors, via the ABS ECU.

Air conditioning system pressure switch

48 The air conditioning pressure switch is fitted to the refrigerant pipe located on the right-hand side of the engine compartment. Switch renewal requires the air conditioning system to be discharged and drained.

Camshaft position sensor

49 The camshaft position sensor is located on the left-hand end of the exhaust camshaft cylinder head cover on 2.0 litre models only.

50 Rotate the 6 fasteners 90° anti-clockwise and remove the plastic cover from the top of the engine.

51 Disconnect the wiring plug, then undo the bolt and remove the sensor from the cylinder head cover.

52 Refitting is the reverse of removal ensuring the sensor seal is in good condition.

Throttle valve positioner motor

53 The throttle valve positioner motor (where fitted) is integral with the throttle body, and is not available separately.

Accelerator pedal position sensor

54 The sensor is integral with the accelerator pedal assembly.

Diesel electronic control system components – removal and refitting

Before carrying out any of the following procedures, disconnect the battery. Reconnect the battery on completion of refitting.

Electronic control unit (ECU)

Note: *If a new ECU is to be fitted, this work must be entrusted to a Peugeot dealer or suitable equipped specialist. It is necessary to initialise the new ECU after installation, which requires the use of dedicated Peugeot diagnostic equipment.*

1 The ECU is located in a plastic box which is mounted on the left-hand front wheelarch.

2 Lift off the ECU module box lid.

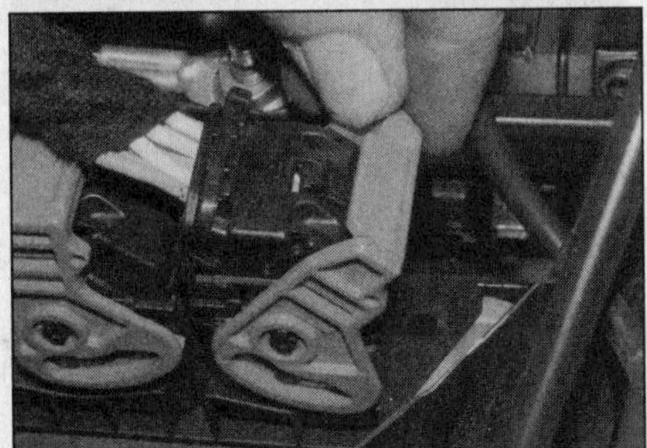

3 Release the wiring connector(s) by depressing the tab and moving the locking lever on top of the connector from the vertical to the fully horizontal position. Carefully withdraw the connector from the ECU pins.

4 Lift the ECU upwards and remove it from its location.

5 To remove the ECU module box, undo the internal and external retaining bolts and remove the module box.

6 Refitting is a reversal of removal.

Crankshaft speed/position sensor

1.4 litre models

7 The crankshaft position sensor is located adjacent to the crankshaft pulley on the right-hand end of the engine. Slacken the right-hand front road wheel bolts, then jack the front of the vehicle up and support it on axle stands. Remove the right-hand front road wheel.

8 Push in the centre pins a little, then prise out the rivets and remove the wheelarch liner.

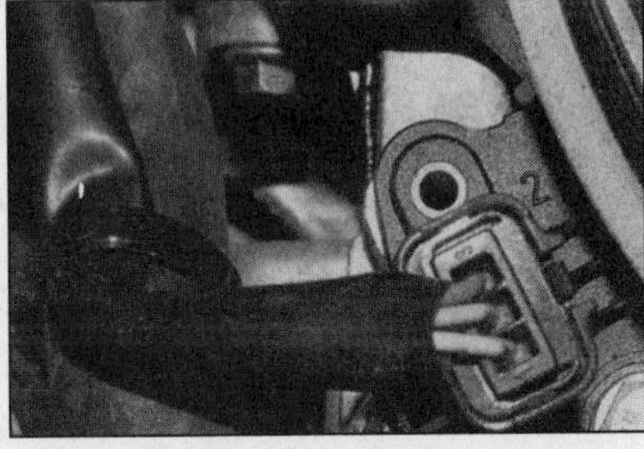

9 Disconnect the sensor wiring plug.

10 Undo the bolt and remove the sensor.

11 Refitting is a reversal of removal, tightening the sensor retaining bolt securely.

2.0 litre models

12 The crankshaft speed/position sensor is located at the top of the transmission bell-housing, directly above the engine flywheel. To gain access, remove the air cleaner assembly, then remove the battery and battery tray.

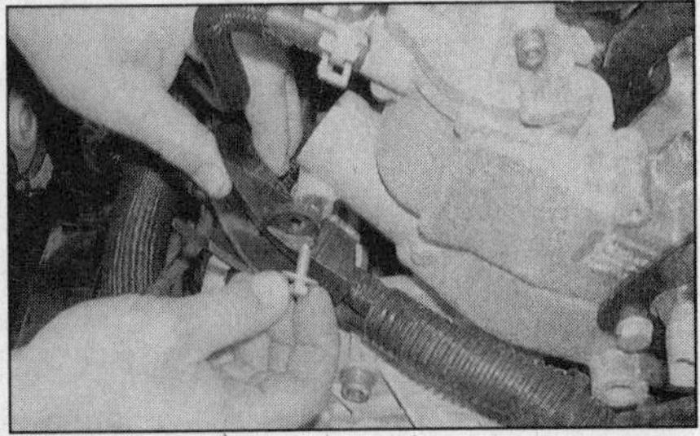

13 Undo the retaining nuts and bolts and release the plastic wiring harness guide from its mountings.

14 Working below the thermostat housing, disconnect the wiring connector from the crankshaft speed/position sensor.

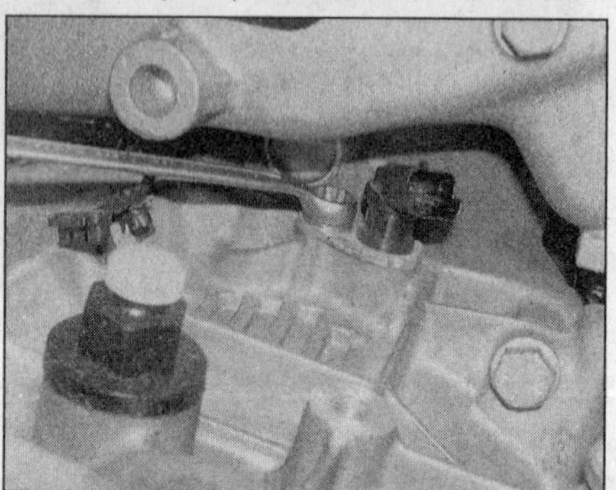

15 Slacken the bolt securing the sensor to the bellhousing. It is not necessary to remove the bolt completely as the sensor mounting flange is slotted.

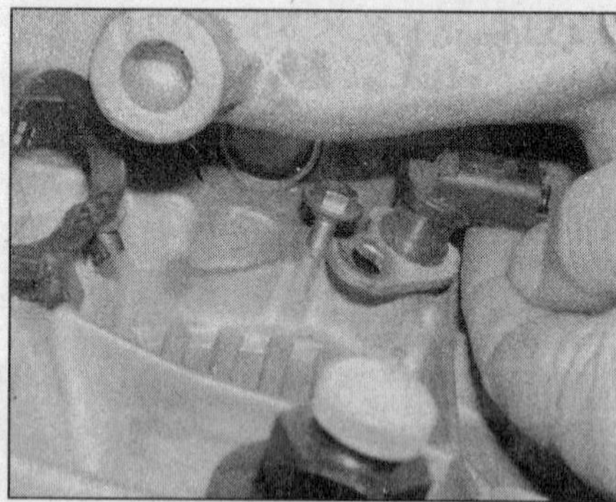

16 Turn the sensor body to clear the mounting bolt, then withdraw the sensor from the bellhousing.

17 Refitting is reverse of the removal procedure ensuring the sensor retaining bolt is securely tightened.

Camshaft position sensor

18 The camshaft position sensor is mounted on the right-hand end of the cylinder head cover, directly behind the camshaft sprocket.

1.4 litre models

19 Remove the upper timing belt cover.

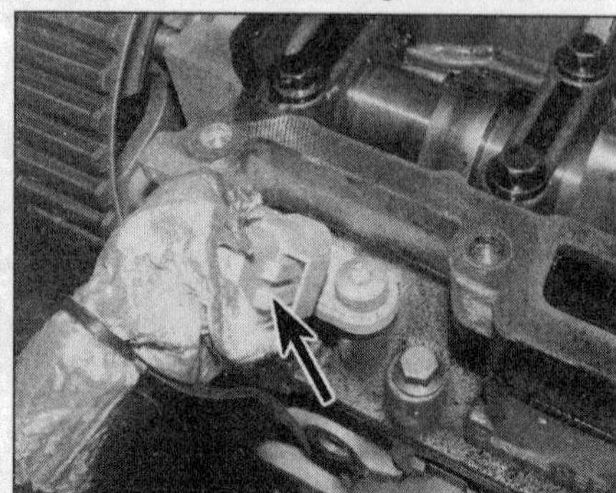

20 Unplug the sensor wiring connector.

21 Undo the bolt and pull the sensor from position.

22 Upon refitting, position the sensor so that the air gap between the sensor end and the webs of the signal wheel is 1.2 mm, measured with feeler gauges, for a used sensor. If fitting a new sensor, the small tip of the sensor must be just touching one of the three webs of the signal wheel. Tighten the sensor retaining bolt to the specified torque.

23 The remainder of refitting is a reversal of removal.

2.0 litre models

24 Remove the timing belt upper and intermediate covers.

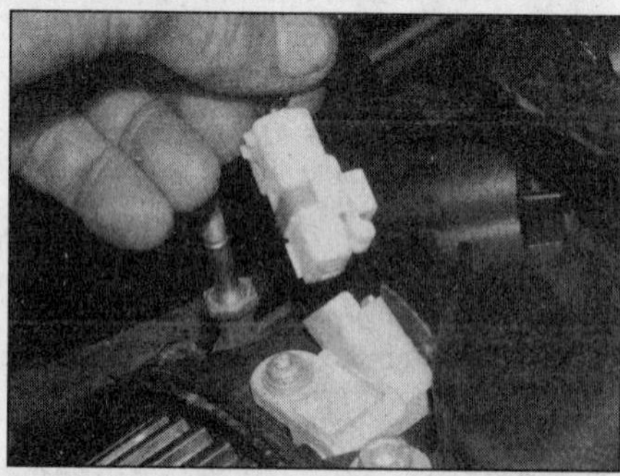

25 Disconnect the sensor wiring connector.

26 Undo the retaining bolt and lift the sensor off the cylinder head cover.

27 To refit and adjust the sensor position, locate the sensor on the cylinder head cover and loosely refit the retaining bolt.

28 The air gap between the tip of the sensor and the signal wheel at the rear of the camshaft sprocket hub must be set to 1.2 mm, using feeler blades. Clearance for the feeler blades is limited with the timing belt and camshaft sprocket in place, but it is just possible if the feeler blades are bent through 90° so they can be inserted through the holes in the sprocket, to rest against the inner face of the signal wheel.

29 With the feeler blades placed against the target plate, move the sensor toward the sprocket until it just contacts the feeler blades. Hold the sensor in this position and tighten the retaining bolt.

30 With the gap correctly adjusted, reconnect the sensor wiring connector, then refit the timing belt upper and intermediate covers.

Accelerator pedal position sensor

Models without accelerator cable

31 On these models, the pedal sensor is integral with the accelerator pedal assembly.

Models with accelerator cable

32 The accelerator pedal position sensor is located in the left-hand front corner of the engine compartment.

33 Remove the air cleaner assembly.

34 Undo the two nuts and bolts and remove the sensor assembly from the mounting bracket on the side of the air cleaner housing.

35 Refitting is reverse of the removal procedure.

Coolant temperature sensor

36 Refer to the petrol engine system Section.

Fuel temperature sensor

1.4 litre models

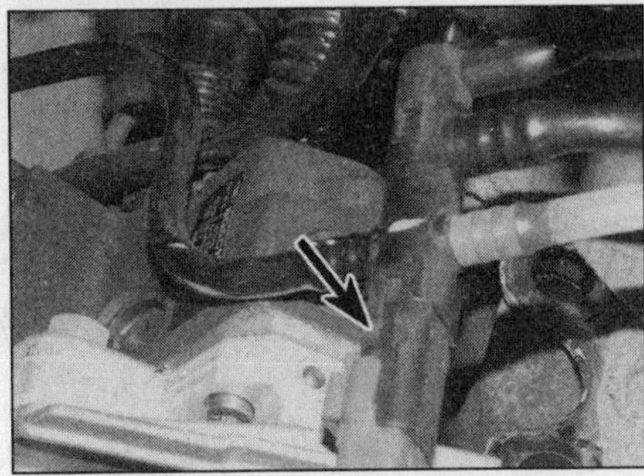

37 The sensor is clipped in to the plastic fuel manifold at the right-hand rear end of the cylinder head. To remove the sensor, disconnect the wiring plug, then unclip the sensor from the manifold. Be prepared for fuel spillage.

38 Refitting is a reversal of removal.

2.0 litre models

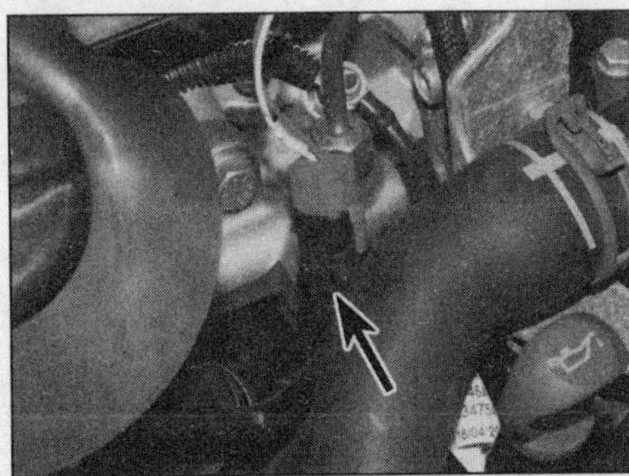

39 The fuel temperature sensor is located to the left-hand side of the oil filler cap.

40 Undo the four plastic fasteners and lift off the engine cover.

41 Disconnect the fuel temperature sensor wiring connector.

42 Thoroughly clean the area around the sensor and its location.

43 Suitably protect the components below the sensor and have plenty of clean rags handy. Be prepared for considerable fuel spillage.

44 Release the retaining clips and detach the sensor from the fuel pipes.

45 Refit the sensor to the fuel pipes, ensuring the clips fully engage.

46 Reconnect the sensor wiring plug.

47 Refit the engine plastic cover.

All models

48 Start the engine and allow it to idle. Check for leaks at the fuel temperature sensor with the engine idling. If satisfactory, increase the engine speed to 4000 rpm and check again for leaks. Take the car for a short road test and check for leaks once again on return. If any leaks are detected, obtain and fit a new sensor.

Air mass meter

49 Air mass meter is located in the intake ducting from the air cleaner housing. On 1.4 litre models, pull the plastic cover from its mountings on top of the engine.

50 Disconnect the meter wiring plug.

Air mass meter wiring plug (arrowed) – 1.4 litre models

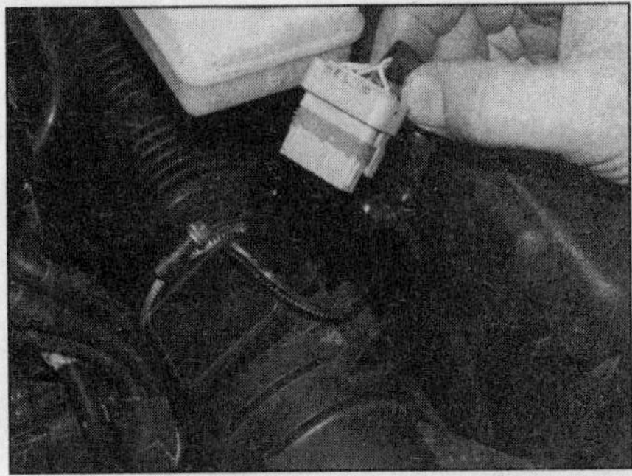

Air mass meter wiring plug – 2.0 litre models

51 Slacken the retaining clips and disconnect the air inlet ducting from either side of the air mass meter. Suitably plug or cover the turbocharger rigid inlet duct, using clean rag to prevent any dirt or foreign material from entering. On 1.4 litre models, the air cleaner outlet is bolted to the air mass meter.

52 Refitting is reverse of the removal procedure.

Fuel pressure sensor

53 The fuel pressure sensor is integral with the accumulator rail, and is not available separately. Peugeot insist that the sensor is not removed from the rail.

Fuel pressure control valve

54 The fuel pressure control valve is integral with the high-pressure fuel pump and cannot be separated.

Vehicle speed sensor

55 The engine management ECU receives the vehicle speed signal from the wheels speed sensors via the ABS ECU.

Notes

Peugeot 406 '99 – '02

20

Petrol models

General

System type:	
1.8 litre XU7JP4 models	Bosch Motronic MP5.1.1, MP7.3 or Sagem SL96
1.8 litre EW7J4 models	Sagem S2000 MPI
2.0 litre EW10J4 models	Magneti Marelli 4.8P

Fuel system data

Fuel pump type	Electric, immersed in tank
Specified idle speed:	
1.8 litre engine	700 ± 50 rpm (not adjustable – controlled by ECU)
2.0 litre engine	850 ± 50 rpm (not adjustable – controlled by ECU)
Idle mixture CO content	Less than 0.4 % (not adjustable- controlled by ECU)

Recommended fuel

Minimum octane rating	95 RON unleaded (UK unleaded premium)

Diesel models

General

System type	HDi (High-pressure Diesel injection) with full electronic control, direct injection and turbocharger
Designation	Bosch EDC 15C2
Firing order	1-3-4-2 (No 1 at flywheel end)
Fuel system operating pressure	200 to 1350 bars (according to engine speed)

High-pressure fuel pump

Type	Bosch CP 1
Direction of rotation	Clockwise, viewed from sprocket end

Injectors

Type	Electromagnetic

Turbocharger

Type:	
DW10 engine	Garrett GT15 or KKK K03
DW12 engine	Allied Signal GT 1549P
Boost pressure (approximate)	1 bar at 3000 rpm

Torque wrench settings

	Nm	lbf ft
Petrol models		
Knock sensor	20	15
Diesel models		
Fuel pressure sensor to accumulator rail	45	33

Chapter 20

General information

Petrol models

Note: *The fuel injection ECU is of the 'self-learning' type, meaning that as it operates, it also monitors and stores the settings which give optimum engine performance under all operating conditions. When the battery is disconnected, these settings are lost and the ECU reverts to the base settings programmed into its memory at the factory. On restarting, this may lead to the engine running/idling roughly for a short while, until the ECU has relearned the optimum settings. This process is best accomplished by taking the vehicle on a road test (for approximately 15 minutes), covering all engine speeds and loads, concentrating mainly in the 2500 to 3500 rpm region.*

On all engines, the fuel injection and ignition functions are combined into a single engine management system. The systems fitted are manufactured by Bosch, Sagem Lucas and Magneti Marelli, and are very similar to each other in most respects, the only significant differences being within the system ECUs. Each system incorporates a closed-loop catalytic converter and an evaporative emission control system, and complies with the latest emission control standards. Refer to Chapter 5B for information on the ignition side of each system; the fuel side of the system operates as follows.

The fuel pump supplies fuel from the tank to the fuel rail, via a renewable cartridge filter mounted underneath the rear of the vehicle. The pump itself is mounted inside the fuel tank, the pump motor is permanently immersed in fuel, to keep it cool. The fuel rail is mounted directly above the fuel injectors and acts as a fuel reservoir.

Fuel rail supply pressure is controlled by the pressure regulator, mounted at the end of the fuel rail or, on later Bosch systems, in front of the fuel tank. The regulator contains a spring-loaded valve, which lifts to allow excess fuel to return to the tank when the optimum operating pressure of the fuel system is exceeded (eg, during low speed, light load cruising). The regulator also contains a diaphragm which is supplied with vacuum from the inlet manifold. This allows the regulator to reduce the fuel supply pressure during light load, high manifold depression conditions (eg, during idling or deceleration) to prevent excess fuel being 'sucked' through the open injectors.

The fuel injectors are electromagnetic valves, which spray atomised fuel into the inlet manifold tracts under the control of the engine management system ECU. There are four injectors, one per cylinder, mounted in the inlet manifold close to the cylinder head. Each injector is mounted at an angle that allows it to spray fuel directly onto the back of the inlet valve(s). The ECU controls the volume of fuel injected by varying the length of time for which each injector is held open.

The fuel injection systems are typically of the simultaneous injection type, whereby all four injectors open at the same time and fuel is injected into each cylinder's inlet tract twice per engine cycle; once during the power stroke and once during the induction stroke. On later Bosch

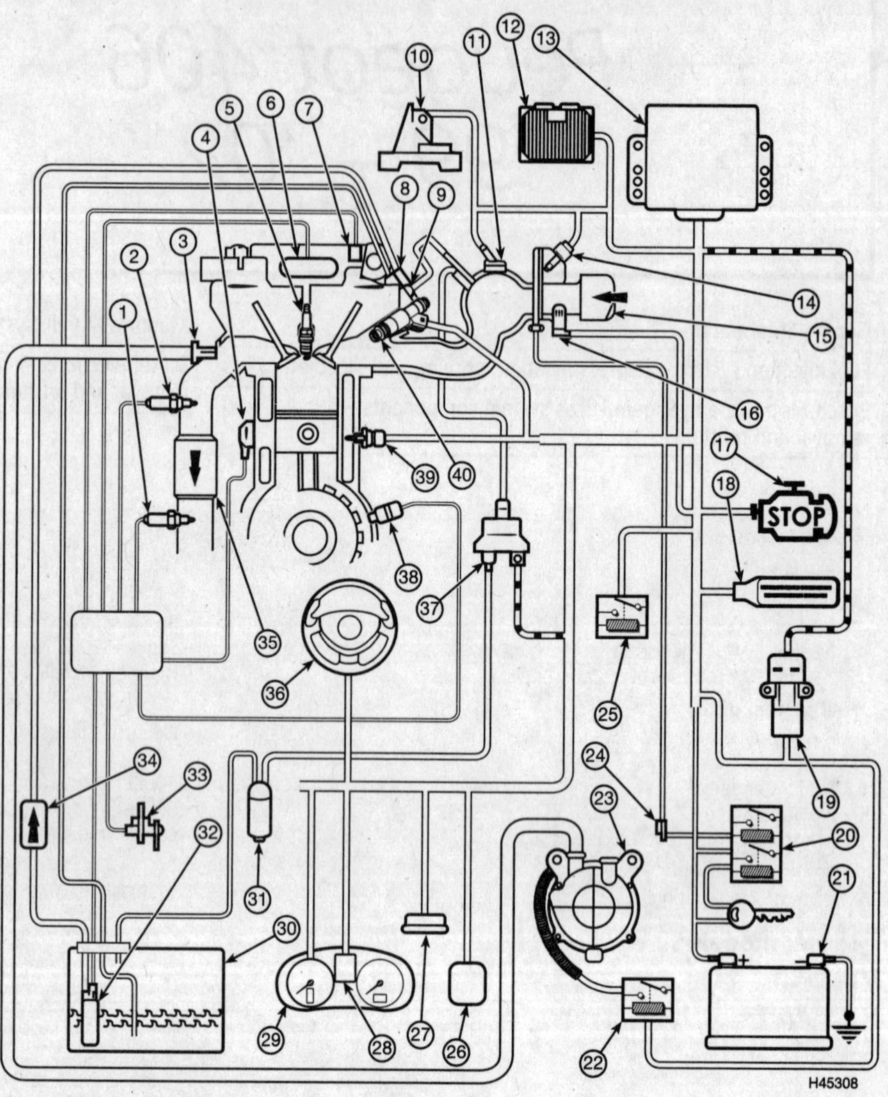

Bosch MP7.3 fuel injection/ignition system

1 Downstream oxygen sensor
2 Upstream oxygen sensor
3 Exhaust air injection valve
4 Knock sensor
5 Spark plugs
6 Ignition coil
7 Cylinder reference sensor
8 Pressure regulator
9 Fuel rail
10 Body accelerometer
11 Induction air pressure sensor
12 Automatic transmission ECU
13 Ignition/injection ECU
14 Idle control stepper motor
15 Throttle housing
16 Inlet air temperature sensor
17 Engine diagnosis warning light
18 Diagnostic connector
19 Inertia switch
20 Engine management multi-function double relay
21 Battery
22 Pulsair relay
24 Throttle housing heating resistor
25 Heating/ventilation
26 Pulsair pump
27 Throttle position sensor
28 Fuel consumption data
29 Tachometer
30 Fuel tank
31 Fuel evaporative system charcoal canister
32 Fuel pump
32 Power steering fluid pressure sensor
33 Vehicle speed sensor
34 Fuel filter
35 Catalytic converter
36 Electronic immobiliser
37 Purge canister solenoid valve
38 Engine speed sensor
39 Coolant temperature sensor
40 Fuel injectors

systems, however, sequential fuel injection is used whereby each injector operates individually in cylinder sequence.

The electrical control system consists of the ECU, along with the following sensors:

a) *Throttle potentiometer – informs the ECU of the throttle valve position, and the rate of throttle opening/closing.*
b) *Coolant temperature sensor – informs the ECU of engine temperature.*
c) *Inlet air temperature sensor – informs the ECU of the temperature of the air passing through the throttle housing.*
d) *Lambda sensor – informs the ECU of the oxygen content of the exhaust gases.*
e) *Manifold pressure sensor – informs the ECU of the load on the engine (expressed in terms of inlet manifold vacuum).*
f) *Crankshaft sensor – informs the ECU of engine speed and crankshaft angular position.*
g) *Vehicle speed sensor – informs the ECU of the vehicle speed.*
h) *Knock sensor – informs the ECU of pre-ignition (detonation) within the cylinders. Not all systems utilise this sensor.*
i) *Camshaft sensor – informs the ECU of which cylinder is on the firing stroke on later systems with sequential injection.*

Signals from each of the sensors are compared by the ECU and, based on this information, the ECU selects the response appropriate to those values, and controls the fuel injectors (varying the pulse width – the length of time the injectors are held open – to provide a richer or weaker air/fuel mixture, as appropriate). The air/fuel mixture is constantly varied by the ECU, to provide the best settings for cranking, starting (with either a hot or cold engine) and engine warm-up, idle, cruising and acceleration.

The ECU also has full control over the engine idle speed, typically via a stepper motor fitted to the throttle housing. The stepper motor pushrod controls the amount of air passing through a bypass drilling at the side of the throttle. When the throttle valve is closed (accelerator pedal released), the ECU uses the motor to alter the position of the pushrod, controlling the amount of air bypassing the throttle valve and so controlling the idle speed. The ECU also carries out 'fine tuning' of the idle speed by varying the ignition timing to increase or reduce the torque of the engine as it is idling. This helps to stabilise the idle speed when electrical or mechanical loads (such as headlights, air conditioning, etc) are switched on and off.

On certain Bosch systems, ECU control of the engine idle speed is by means of an auxiliary air valve which bypasses the throttle valve. When the throttle valve is closed, the ECU controls the opening of the air valve, which in turn regulates the amount of air entering the manifold, and so controls the idle speed.

The throttle housing on most engines is fitted with an electric heating element. The heater is supplied with current by the ECU, warming the throttle housing on cold-starts to help prevent icing of the throttle valve.

If there is any abnormality in any of the readings obtained from the coolant temperature sensor, the inlet air temperature sensor or the lambda sensor, the ECU enters its 'back-up' mode. If this happens, the erroneous sensor signal is overridden, and the ECU assumes a pre-programmed 'back-up' value, which will allow the engine to continue running, albeit at reduced efficiency. If the ECU enters this mode, the warning lamp on the instrument panel will be illuminated, and the relevant fault code will be stored in the ECU memory.

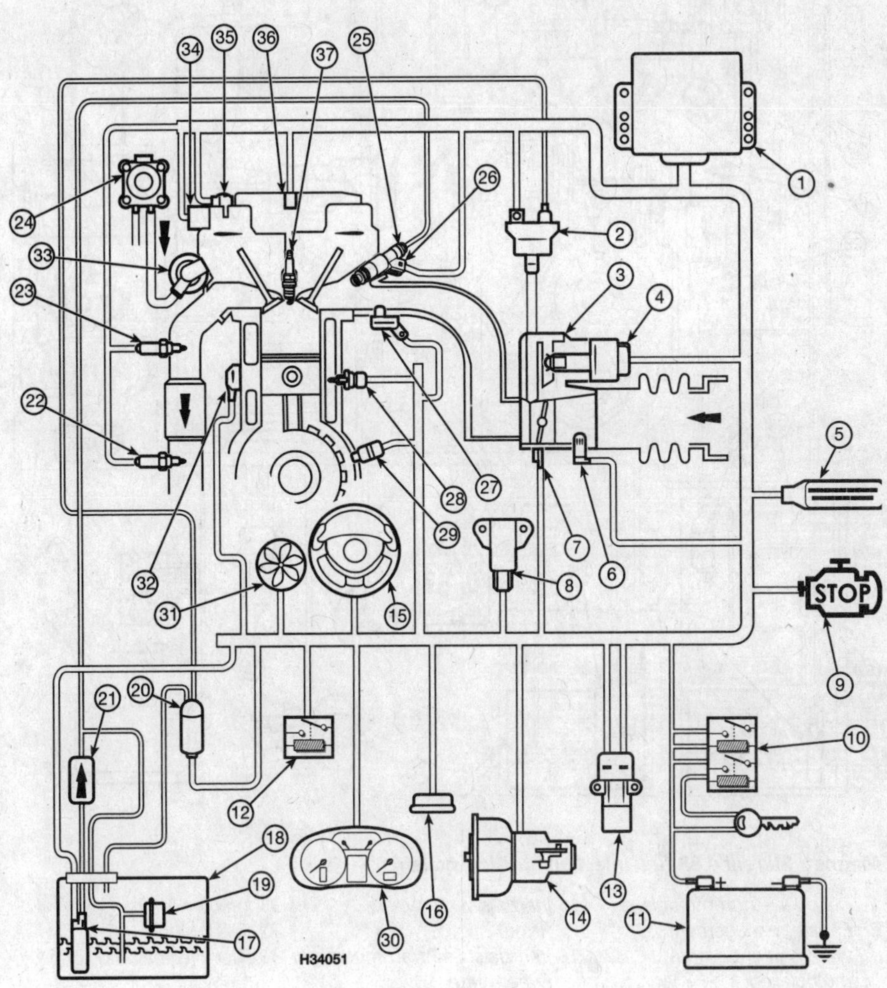

Sagem 2000 fuel injection/ignition system

1 Ignition/injection ECU
2 Purge canister solenoid valve
3 Throttle housing
4 Idle control stepper motor
5 Diagnostic connector
6 Inlet air temperature sensor
7 Throttle housing heating resistor
8 Throttle position sensor
9 Engine diagnosis warning light
10 Engine management multi-function double relay
11 Battery
12 Air conditioning relay
13 Inertia switch
14 Vehicle speed sensor
15 Power steering fluid pressure sensor
16 Electronic immobiliser
17 Fuel pump
18 Fuel tank
19 Fuel pressure regulator
20 Fuel evaporative system charcoal canister
21 Fuel filter
22 Upstream oxygen sensor
23 Downstream oxygen sensor
24 Pulsair pump
25 Fuel rail
26 Fuel injectors
27 Induction air pressure sensor
28 Coolant temperature sensor
29 Crankshaft position sensor
30 Instrument panel
31 Electric cooling fans
32 Knock sensor
33 Secondary air valve
34 Camshaft position sensor
35 EGR valve
36 Ignition coil module
37 Spark plugs

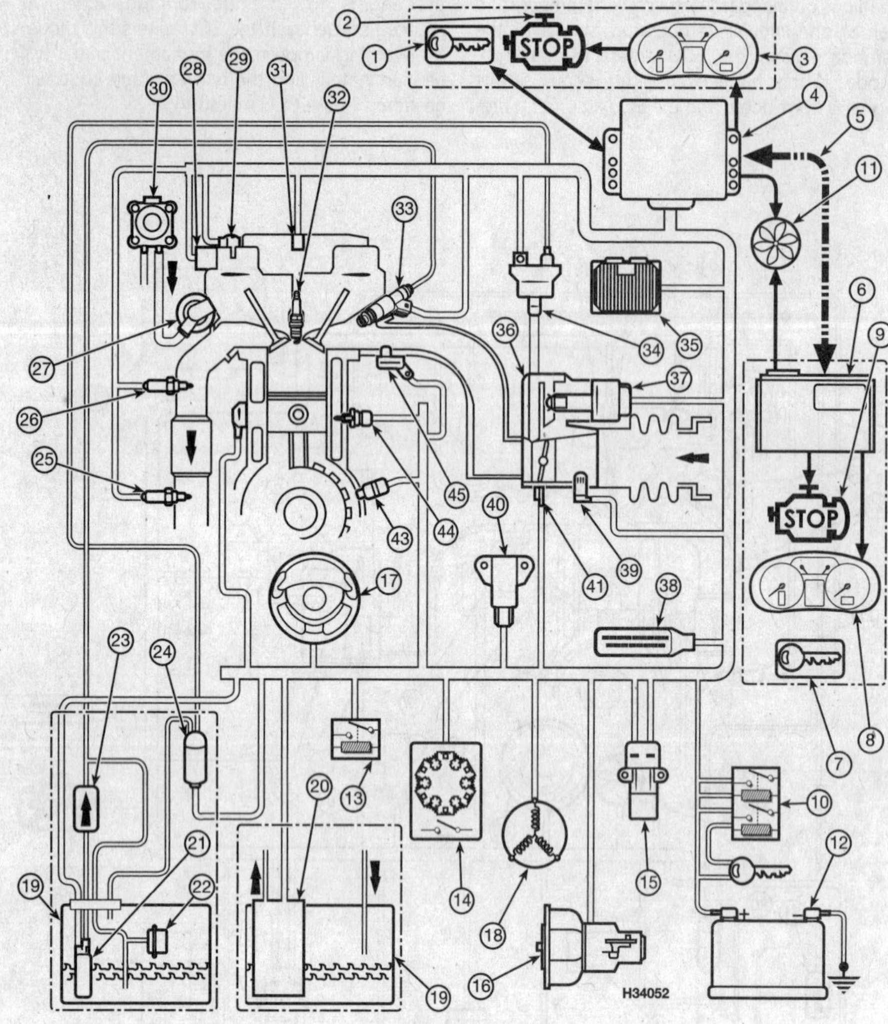

Magneti Marelli 4.8P fuel injection/ignition system

1 Decoder – non-multiplexed
2 Engine management warning light – non-multiplexed
3 Rev counter and instrument panel – non-multiplexed
4 ECU
5 Multiplex network
6 Control box – multiplexed
7 Immobiliser – multiplexed
8 Instrument panel – multiplexed
9 Engine management warning light – non-multiplexed
10 Double relay or built-in systems interface
11 Fans
12 Battery
13 Air conditioning relay
14 Air conditioning pressure switch
15 Inertia switch (where fitted)
16 Vehicle speed transmitter (where fitted)
17 Power steering fluid pressure switch
18 Alternator
19 Fuel tank
20 Fuel pump
21 Fuel pump
22 Fuel pressure regulator
23 Fuel filter
24 Charcoal canister
25 Lambda (oxygen) sensor (downstream)
26 Lambda (oxygen) sensor (upstream)
27 Air inlet valve
28 Exhaust gas recirculation (EGR) valve
29 Camshaft position sensor
30 Secondary air pump
31 Ignition coil module
32 Spark plugs
33 Fuel injectors and supply rail
34 Purge valve
35 Automatic transmission ECU
36 Throttle body
37 Idle control stepper motor
38 Diagnostic socket
39 Inlet air temperature sensor
40 Throttle position potentiometer
41 Throttle body heating element
42 Knock sensor
43 Engine speed sensor
44 Coolant temperature sensor
45 Inlet air pressure sensor

Diesel models

The fuel system consists of a rear-mounted fuel tank and fuel lift pump, a fuel filter with integral water separator, on some models a fuel cooler mounted under the car, and an electronically-controlled High-pressure Diesel injection (HDi) system, together with a turbocharger.

The exhaust system is conventional, but to meet the latest emission levels an unregulated catalytic converter and an exhaust gas recirculation system are fitted to all models.

The HDi system (generally known as a 'common rail' system) derives its name from the fact that a common rail (referred to as an accumulator rail), or fuel reservoir, is used to supply fuel to all the fuel injectors. Instead of an in-line or distributor type injection pump, which distributes the fuel directly to each injector, a high-pressure pump is used, which generates a very high fuel pressure (1350 bars at high engine speed) in the accumulator rail. The accumulator rail stores fuel, and maintains a constant fuel pressure, with the aid of a pressure control valve. Each injector is supplied with high-pressure fuel from the accumulator rail, and the injectors are individually controlled via signals from the system electronic control unit (ECU). The injectors are electromagnetically-operated.

In addition to the various sensors used on models with a conventional fuel injection pump, common rail systems also have a fuel pressure sensor. The fuel pressure sensor allows the ECU to maintain the required fuel pressure, via the pressure control valve.

System operation

For the purposes of describing the operation of a common rail injection system, the components can be divided into three sub-systems; the low-pressure fuel system, the high-pressure fuel system and the electronic control system.

Low-pressure fuel system

The low-pressure fuel system consists of the following components:

a) Fuel tank.
b) Fuel lift pump.
c) Fuel cooler (not all models).
d) Fuel filter/water trap.
e) Low-pressure fuel lines.

The low-pressure system (fuel supply system) is responsible for supplying clean fuel to the high-pressure fuel system.

High-pressure fuel system

The high-pressure fuel system consists of the following components:

a) High-pressure fuel pump with pressure control valve.
b) High-pressure fuel accumulator rail.
c) Fuel injectors.
d) High-pressure fuel lines.

After passing through the fuel filter, the fuel reaches the high-pressure pump, which forces it into the accumulator rail. As diesel fuel has a certain elasticity, the pressure in the accumulator rail remains constant, even though fuel leaves

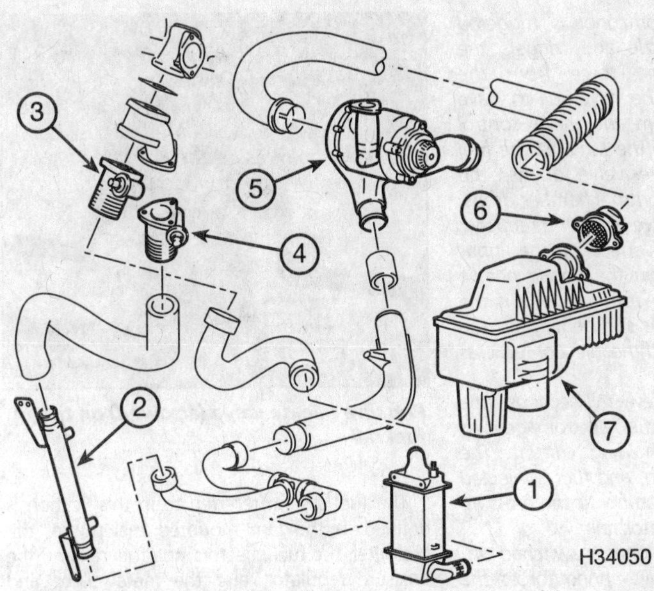

Fuel inlet system on the 2.2 litre DW12 engine

1 Intercooler (air-to-air heat exchanger)
2 Water-to-air heat exchanger
3 Cold air inlet
4 Hot air inlet
5 Turbocharger
6 Airflow sensor
7 Air filter

Fuel cooler mounted under the vehicle

the rail each time one of the injectors operates. Additionally, a pressure control valve mounted on the high-pressure pump ensures that the fuel pressure is maintained within preset limits.

The pressure control valve is operated by the ECU. When the valve is opened, fuel is returned from the high-pressure pump to the tank, via the fuel return lines, and the pressure in the accumulator rail falls. To enable the ECU to trigger the pressure control valve correctly, the pressure in the accumulator rail is measured by a fuel pressure sensor.

The electromagnetically-controlled fuel injectors are operated individually, via signals from the ECU, and each injector injects fuel directly into the relevant combustion chamber. The fact that high fuel pressure is always available allows very precise and highly flexible injection in comparison to a conventional injection pump: for example combustion during the main injection process can be improved considerably by the pre-injection of a very small quantity of fuel.

Electronic control system

The electronic control system consists of the following components:

a) Electronic control unit (ECU).
b) Crankshaft speed/position sensor.
c) Camshaft position sensor.
d) Accelerator pedal position sensor.
e) Coolant temperature sensor.
f) Fuel temperature sensor.
g) Air mass meter.
h) Fuel pressure sensor.
i) Fuel injectors.
j) Fuel pressure control valve.
k) Preheating control unit.
l) EGR solenoid valve.

The information from the various sensors is passed to the ECU, which evaluates the signals. The ECU contains electronic 'maps' which enable it to calculate the optimum quantity of fuel to inject, the appropriate start of injection, and even pre- and post-injection fuel quantities, for each individual engine cylinder under any given condition of engine operation.

Additionally, the ECU carries out monitoring and self-diagnostic functions. Any faults in the system are stored in the ECU memory, which enables quick and accurate fault diagnosis using appropriate diagnostic equipment (such as a suitable fault code reader).

Fuel lift pump

The fuel lift pump and integral fuel gauge sender unit is electrically-operated, and is mounted in the fuel tank.

High-pressure pump

The high-pressure pump is mounted on the engine in the position normally occupied by the conventional distributor fuel injection pump. The pump is driven at half engine speed by the timing belt, and is lubricated by the fuel which it pumps.

The fuel lift pump forces the fuel into the high-pressure pump chamber, via a safety valve.

The high-pressure pump consists of three radially-mounted pistons and cylinders. The pistons are operated by an eccentric cam mounted on the pump drive spindle. As a piston moves down, fuel enters the cylinder through an inlet valve. When the piston reaches bottom dead centre (BDC), the inlet valve closes, and as the piston moves back up the cylinder, the fuel is compressed. When the pressure in the cylinder reaches the pressure in the accumulator rail, an outlet valve opens, and fuel is forced into the accumulator rail. When the piston reaches top dead centre (TDC), the outlet valve closes, due to the pressure drop, and the pumping cycle is repeated. The use of multiple cylinders provides a steady flow of fuel, minimising pulses and pressure fluctuations.

As the pump needs to be able to supply sufficient fuel under full-load conditions, it will supply excess fuel during idle and part-load conditions. This excess fuel is returned from the high-pressure circuit to the low-pressure circuit (to the tank) via the pressure control valve.

The pump incorporates a facility to effectively switch off one of the cylinders to improve efficiency and reduce fuel consumption when maximum pumping capacity is not required. When this facility is operated, a solenoid-operated needle holds the inlet valve in the relevant cylinder open during the delivery stroke, preventing the fuel from being compressed.

Accumulator rail

As its name suggests, the accumulator rail acts as an accumulator, storing fuel and preventing pressure fluctuations. Fuel enters the rail from the high-pressure pump, and each injector has its own connection to the rail. The fuel pressure sensor is mounted in the rail, and the rail also has a connection to the fuel pressure control valve on the pump.

Pressure control valve

The pressure control valve is operated by the ECU, and controls the system pressure. The valve is integral with the high-pressure pump and cannot be separated.

If the fuel pressure is excessive, the valve opens, and fuel flows back to the tank. If the pressure is too low, the valve closes, enabling the high-pressure pump to increase the pressure.

The valve is an electromagnetically-operated ball valve. The ball is forced against its seat, against the fuel pressure, by a powerful spring, and also by the force provided by the electromagnet. The force generated by the electromagnet is directly proportional to the current applied to it by the ECU. The desired pressure can therefore be set by varying the current applied to the electromagnet. Any pressure fluctuations are damped by the spring.

Fuel pressure sensor

The fuel pressure sensor is mounted in the accumulator rail, and provides very precise information on the fuel pressure to the ECU.

Fuel injector

The injectors are mounted on the engine in a similar manner to conventional diesel fuel injectors. The injectors are electro-magnetically-

The control vacuum servos on the bulkhead

operated via signals from the ECU, and fuel is injected at the pressure existing in the accumulator rail. The injectors are high-precision instruments and are manufactured to very high tolerances.

Fuel flows into the injector from the accumulator rail, via an inlet valve and an inlet throttle, and an electromagnet causes the injector nozzle to lift from its seat, allowing injection. Excess fuel is returned from the injectors to the tank via a return line. The injector operates on a hydraulic servo principle: the forces resulting inside the injector due to the fuel pressure effectively amplify the effects of the electromagnet, which does not provide sufficient force to open the injector nozzle directly. The injector functions as follows. Five separate forces are essential to the operation of the injector.

a) *A nozzle spring forces the nozzle needle against the nozzle seat at the bottom of the injector, preventing fuel from entering the combustion chamber.*
b) *In the valve at the top of the injector, the valve spring forces the valve ball against the opening to the valve control chamber. The fuel in the chamber is unable to escape through the fuel return.*
c) *When triggered, the electromagnet exerts a force which overcomes the valve spring force, and moves the valve ball away from its seat. This is the triggering force for the start of injection. When the valve ball moves off its seat, fuel enters the valve control chamber.*
d) *The pressure of the fuel in the valve control chamber exerts a force on the valve control plunger, which is added to the nozzle spring force.*
e) *A slight chamfer towards the lower end of the nozzle needle causes the fuel in the control chamber to exert a force on the nozzle needle.*

When these forces are in equilibrium, the injector is in its rest (idle) state, but when a voltage is applied to the electromagnet, the forces work to lift the nozzle needle, injecting fuel into the combustion chamber. There are four phases of injector operation as follows:

a) *Rest (idle) state – all forces are in equilibrium. The nozzle needle closes off the nozzle opening, and the valve spring forces the valve ball against its seat.*
b) *Opening – the electromagnet is triggered which opens the nozzle and triggers the injection process. The force from the electromagnet allows the valve ball to leave its seat. The fuel from the valve control chamber flows back to the tank via the fuel return line. When the valve opens, the pressure in the valve control chamber drops, and the force on the valve plunger is reduced. However, due to the effect of the input throttle, the pressure on the nozzle needle remains unchanged. The resulting force in the valve control chamber is sufficient to lift the nozzle from its seat, and the injection process begins.*
c) *Injection – within a few milliseconds, the triggering current in the electromagnet is reduced to a lower holding current. The nozzle is now fully open, and fuel is injected into the combustion chamber at the pressure present in the accumulator rail.*
d) *Closing – the electromagnet is switched off, at which point the valve spring forces the valve ball firmly against its seat, and in the valve control chamber, the pressure is the same as that at the nozzle needle. The force at the valve plunger increases, and the nozzle needle closes the nozzle opening. The forces are now in equilibrium once more, and the injector is once more in the idle state, awaiting the next injection sequence.*

ECU and sensors

The ECU and sensors are described earlier in this Section – see Electronic control system.

Air inlet sensor and turbocharger

An airflow sensor is fitted downstream of the air filter to monitor the quantity of air supplied to the turbocharger. On models with the 2.2 litre diesel engine (DW12), air from the high-pressure side of the turbocharger is channelled either through the intercooler and cold air throttle body, or through the coolant heat exchanger and hot air throttle body. The throttle bodies are controlled by the engine management ECU via vacuum servos located on the bulkhead.

Fuel injection system – depressurisation, priming and bleeding

Petrol models

Warning: The following procedure will merely relieve the pressure in the fuel system – remember that fuel will still be present in the system components and take precautions accordingly before disconnecting any of them.

Pressure release valve (arrowed) on the fuel rail

1 The fuel system referred to in this Section is defined as the tank-mounted fuel pump, the fuel filter, the fuel injectors, the fuel rail and the pressure regulator, and the metal pipes and flexible hoses of the fuel lines between these components. All these contain fuel which will be under pressure while the engine is running, and/or while the ignition is switched on. The pressure will remain for some time after the ignition has been switched off, and must be relieved in a controlled fashion when any of these components are disturbed for servicing work.

2 Peugeot technicians connect a special tube to the Schrader valve on the fuel rail in order to depressurize the fuel system. The tube incorporates a union nut which is screwed onto the valve, and an inner cable which is used to depress the valve core. If this tube is not available, cover the valve and surrounding area with cloth rag, and depress the valve with a screwdriver through the rag. Make sure that enough rag is used to soak up the fuel. Access to the valve is gained by first removing the engine top cover.

3 With the pressure released, refit the cap to the valve.

4 Note that pressure may increase again in the fuel system due to an increase in ambient temperature, so any work required on the system should be started immediately after releasing the pressure.

Diesel models

5 The fuel system is entirely self-bleeding because the fuel lift pump supplies fuel to the high-pressure pump whenever the ignition is switched on.

6 In the case of running out of fuel, or after disconnecting any part of the fuel supply system, ensure that there is fuel in the tank, then start the engine in the normal way. **Note:** *To avoid possible damage, do not operate the starter motor for more than 10 seconds at a time, and allow a 'cooling down' period of 30 seconds between starting attempts.*

Bosch Motronic and Sagem Lucas system components – removal and refitting

Fuel rail and injectors

Note: *If a faulty injector is suspected, before condemning the injector, it is worth trying the effect of one of the proprietary injector-cleaning treatments which are available from car accessory shops.*

1 Disconnect the battery negative lead.

2 Disconnect the vacuum pipe from the fuel pressure regulator.

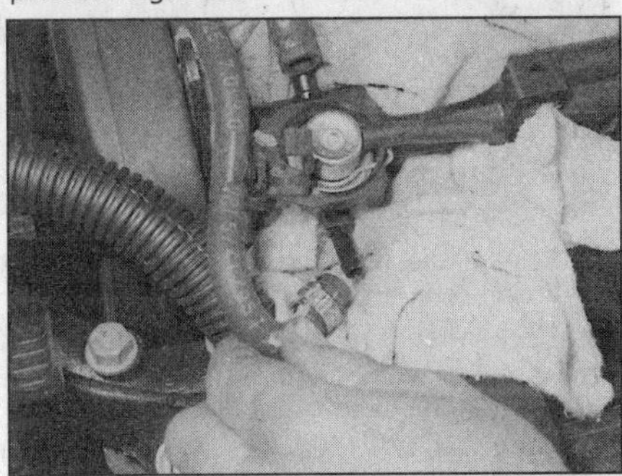

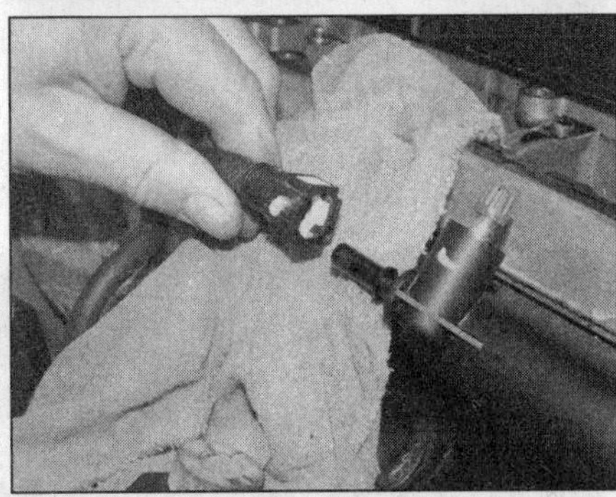

3 Depress the catch on the fuel feed hose quick-release fitting, and disconnect the fuel feed (and return hoses, where applicable) from the fuel rail. Suitably seal or plug the hose and the fuel rail union(s) after disconnection.

4 Open the retaining clips and release the wiring and hoses running along the front of the fuel rail.

5 Depress the retaining tangs and disconnect the wiring connectors from the four injectors.

6 Slacken and remove the fuel rail retaining bolts then carefully ease the fuel rail and injector assembly out from the inlet manifold and remove it from the engine. Remove the O-rings from the end of each injector and discard them; they must be renewed whenever they are disturbed.

7 Slide out the retaining clip(s) and remove the relevant injector(s) from the fuel rail. Remove the upper O-ring from each disturbed injector and discard; all disturbed O-rings must be renewed.

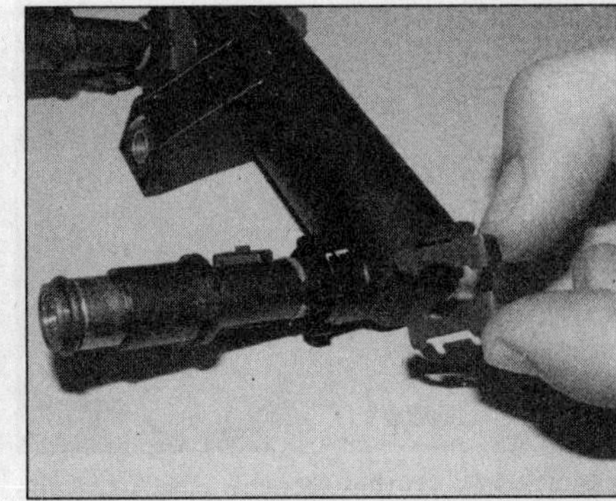

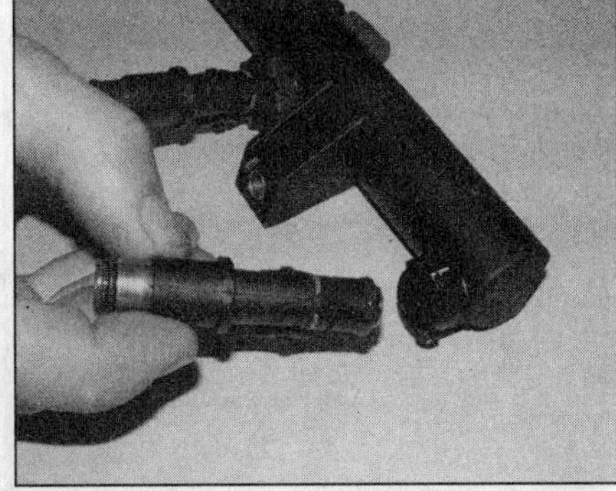

8 Refitting is a reversal of the removal procedure, noting the following points.

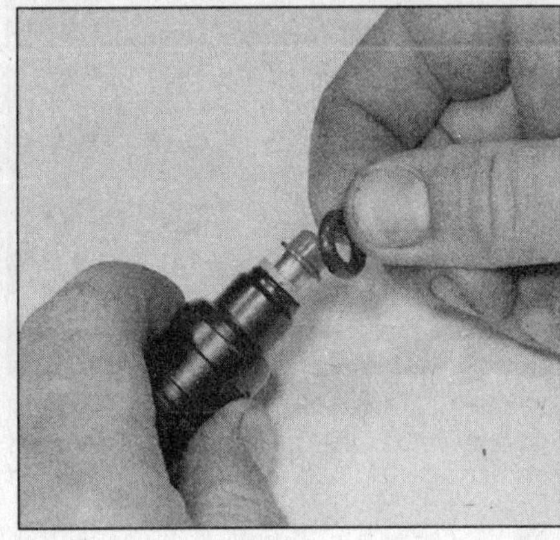

a) Fit new O-rings to all disturbed injector unions.

b) Apply a smear of engine oil to the O-rings to aid installation then ease the injectors and fuel rail into position ensuring that none of the O-rings are displaced.

c) On completion start the engine and check for fuel leaks.

Fuel pressure regulator

Note 2: *On the S2000 MPI system, the pressure regulator is located in the fuel tank.*

9 Disconnect the battery negative lead.

10 Slacken the retaining clips and disconnect the fuel feed and return hoses from the fuel rail and pressure regulator.

11 Disconnect the vacuum pipe from the regulator.

12 Place some rags under the regulator, to catch any spilt fuel. Remove the retaining clip and ease the regulator out from the fuel rail.

13 Refitting is a reversal of the removal procedure. Examine the regulator seal for signs of damage or deterioration and renew if necessary.

Throttle potentiometer

14 Depress the retaining clip and disconnect the wiring connector from the throttle

potentiometer located beneath the throttle housing.

15 Slacken and remove the two retaining screws then disengage the potentiometer from the throttle valve spindle and remove it from the vehicle.

16 Refitting is a reverse of the removal procedure ensuring that the potentiometer is correctly engaged with the throttle valve spindle.

Electronic Control Unit

17 The ECU is located in a plastic box which is mounted on the right-hand front wheelarch.

18 Ensure that the ignition is switched off then lift off the ECU module box lid. On automatic transmission models there will be two ECUs in the box; the fuel injection/ignition ECU is the unit nearest to the engine.

19 Release the wiring connectors by lifting the locking lever on top of the connector upwards. Lift the connector at the rear, disengage the tag at the front and carefully withdraw the connector from the ECU pins.

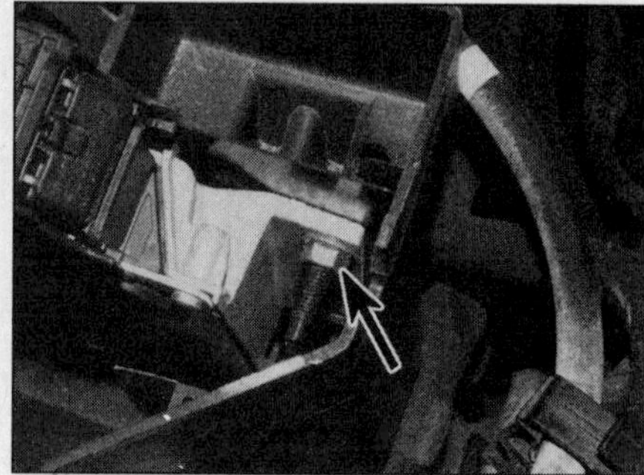

20 Unscrew the mounting bolts, then lift the ECU upwards and remove it from its location.

21 Refitting is a reversal of removal. Note that if a new ECU has been fitted, the vehicle should be taken on an extensive road test. Initially, engine performance may be less than acceptable, but should improve as the ECU control circuitry adapts to the engine parameters.

Idle speed stepper motor

22 The idle speed control stepper motor is located on the side of the throttle housing assembly.

23 Release the retaining clip, and disconnect the wiring connector from the motor.

24 Slacken and remove the two retaining screws, and withdraw the motor from the throttle housing.

25 Refitting is a reversal of the removal procedure.

Manifold absolute pressure sensor

26 The MAP sensor is situated on the underside of the inlet manifold.

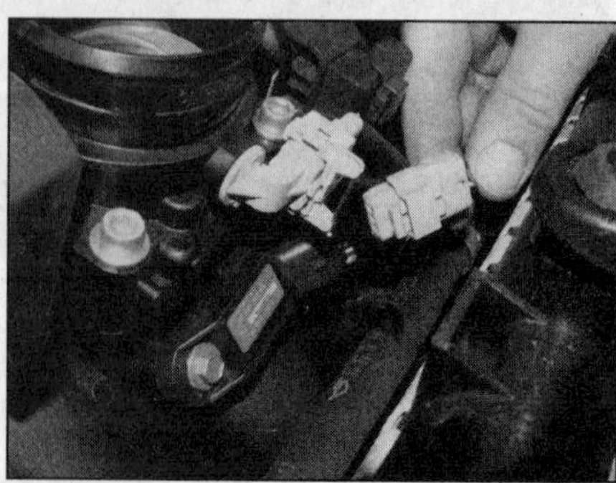

27 Disconnect the wiring connector from the sensor.

28 Undo the securing screw, then pull the sensor out of the manifold.

29 Refitting is the reverse of the removal procedure.

Coolant temperature sensor

30 The coolant temperature sensor is located in the coolant/thermostat housing bolted to the left-hand end of the cylinder head. On models with an alloy housing the sensor is screwed into place, whilst on those with a plastic housing the sensor is retained by a clip. The sensor can be identified by its green wiring connector. Partially drain the cooling system.

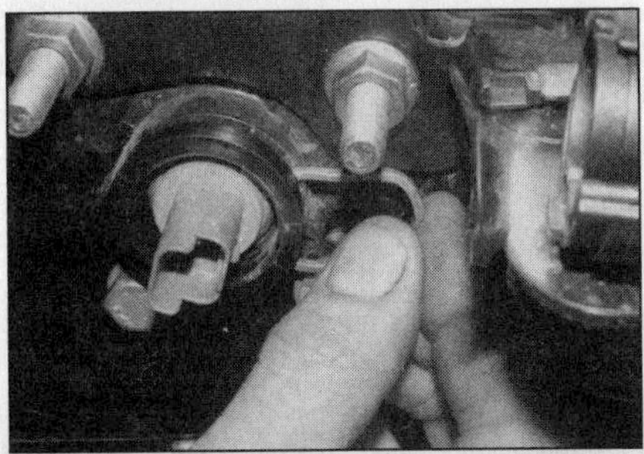

31 Disconnect the wiring plug and unscrew or unclip the sensor. Where fitted, recover the sealing ring.

32 Refitting is a reversal of removal. Renew the sealing ring where necessary. Where no seal was fitted, apply a little sealant to the threads of the sensor prior to refitting.

Inlet air temperature sensor

33 The inlet air temperature sensor is located on the underside or front of the throttle housing. Loosen the retaining clip, and release the air inlet duct from the throttle housing. The inlet air temperature sensor is visible in the top of the housing.

34 Trace the wiring back from the sensor to its wiring connector on the throttle housing, and unplug the connector.

35 The sensor itself can be pressed out of the throttle housing. Note that it is sealed in place, to prevent air leaks; a suitable sealant will be required for refitting.

36 Refitting is the reverse of removal.

Camshaft position sensor

37 The camshaft position sensor is located at the left-hand end of the exhaust camshaft cylinder head cover. Undo the six screws and lift off the engine cover.

38 Disconnect the crankcase breather hose at the quick-fit connector on the rear cylinder head cover.

39 Disconnect the wiring connector at the camshaft position sensor, then undo the bolt and remove the sensor from the cylinder head cover.

40 Refitting is the reverse of removal but fit a new sealing O-ring to the sensor body.

Crankshaft (RPM) sensor

41 The crankshaft sensor is situated on the front face of the transmission clutch housing. Trace the wiring back from the sensor to the wiring connector and disconnect it from the main harness.

42 Undo the retaining bolt and withdraw the sensor from the transmission.

43 Refitting is the reverse of the removal procedure.

Vehicle speed sensor

44 The vehicle speed sensor is an integral part of the transmission speedometer drive assembly, fitted to the transmission housing next to the inner end of the right-hand driveshaft. Jack up the front of the vehicle and support it securely on axle stands.

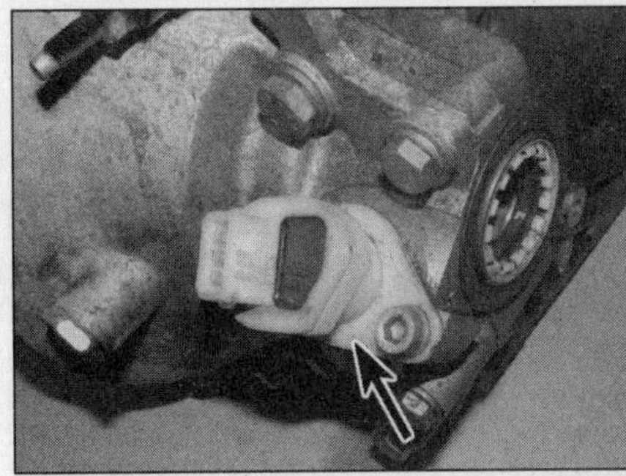

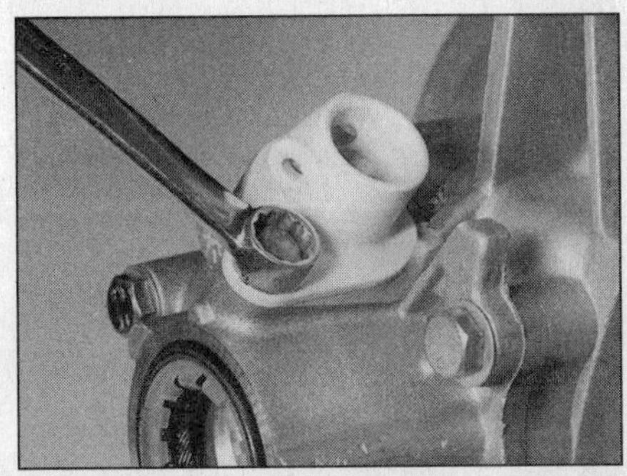

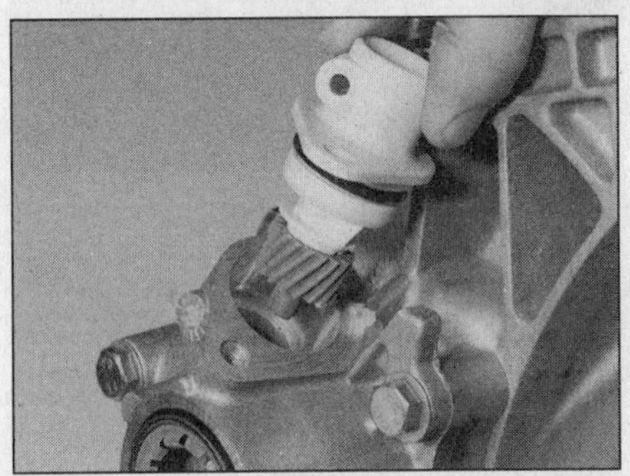

45 Disconnect the wiring plug, then unscrew the retaining bolt, and remove the drive/sensor along with the heatshield (where fitted). Recover the drive O-ring.

46 Refitting is a reversal of removal. Apply a little transmission oil to the new O-ring seal prior to refitting.

Knock sensor

47 The knock sensor is fitted to the front face of the cylinder block. To gain access, jack up the front of the vehicle, and support it securely on axle stands. Remove the engine under tray.

48 Disconnect the wiring plug, unscrew the retaining bolt and remove the sensor.

49 Refitting is a reversal of removal. Ensure the mating faces are clean and the retaining bolt is tighten to the specified torque.

Fuel injection relay unit

50 The relay unit is located in the ECU module box which is mounted on the right-hand front wheelarch.

51 Ensure that the ignition is switched off then lift off the ECU module box lid.

52 Disconnect the wiring connector and remove the relay unit from the mounting plate.

53 Refitting is the reverse of removal.

Magneti Marelli system components – removal and refitting

Fuel rail and injectors

Note: *If a faulty injector is suspected, before condemning the injector, it is worth trying the effect of one of the proprietary injector-cleaning treatments.*

1 Disconnect the battery negative lead.

2 Unbolt and remove the engine top cover.

3 Disconnect the air duct from between the air cleaner and throttle body elbow.

4 Disconnect the wiring from the inlet air temperature sensor.

5 Loosen the clamp and remove the inlet elbow from the throttle body.

6 Disconnect the accelerator cable from the throttle body.

7 Unbolt the wiring tray from the top of the inlet manifold, and position to one side.

8 Depress the retaining clip(s), and disconnect the wiring connector(s) from the injector(s).

9 Unscrew the mounting bolts and carefully ease the fuel rail complete with injectors from the inlet manifold. Remove the O-rings from the end of each injector, and discard them; these must be renewed whenever they are disturbed.

10 Slide out the retaining clip(s) and remove the relevant injector(s) from the fuel rail. Remove the upper O-ring from each disturbed injector and discard; all disturbed O-rings must be renewed.

11 Refitting is a reversal of the removal procedure, noting the following points.

a) Fit new O-rings to all disturbed injector unions.

b) Apply a smear of engine oil to the O-rings to aid installation then ease the injectors and fuel rail into position ensuring that none of the O-rings are displaced.

c) On completion start the engine and check for fuel leaks.

Fuel pressure regulator

12 The fuel pressure regulator is located in the fuel tank, and is removed together with the fuel pump and gauge sender.

Throttle potentiometer

13 The throttle potentiometer is fitted to the right-hand side of the throttle housing.

14 Release the retaining clip, and disconnect the potentiometer wiring connector.

15 Slacken and remove the two retaining screws, and remove the potentiometer from the throttle housing.

16 Refitting is the reverse of removal, ensuring that the potentiometer is correctly engaged with the throttle valve spindle.

Electronic Control Unit

17 Refer to the Bosch Motronic system Section.

Idle speed stepper motor

18 The idle speed control stepper motor is located on the front of the throttle housing assembly.

19 Release the retaining clip, and disconnect the wiring connector from the motor.

20 Slacken and remove the two retaining screws, and withdraw the motor from the throttle housing.

21 Refitting is a reversal of the removal procedure.

Inlet air pressure sensor

22 The inlet air pressure sensor is located on the inlet manifold, below the throttle housing.

23 Disconnect the wiring from the sensor.

24 Undo the screw and withdraw the sensor from the inlet manifold.

25 Refitting is the reverse of the removal procedure.

Coolant temperature sensor

26 Refer to the Bosch Motronic system Section.

Inlet air temperature sensor

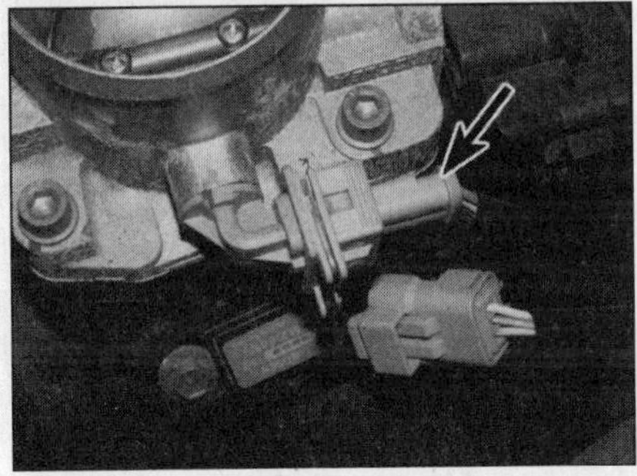

27 The inlet air temperature sensor is located on the front of the throttle housing and cannot be removed separately.

Throttle housing heater

28 The throttle housing heating element is integral with the throttle housing and cannot be removed separately.

Crankshaft (RPM) sensor

29 Refer to the Bosch Motronic system Section.

Camshaft position sensor

30 Refer to the Bosch Motronic system Section.

Vehicle speed sensor

31 Refer to the Bosch Motronic system Section.

Knock sensor

32 Refer to the Bosch Motronic system Section.

Fuel injection relay unit

33 Refer to the Bosch Motronic system Section.

Diesel electronic control system components – removal and refitting

1 Before carrying out any of the following procedures, disconnect the battery negative terminal. Reconnect the battery on completion of refitting.

Electronic control unit (ECU)

Note: *If a new ECU is to be fitted, this work must be entrusted to a Peugeot dealer or specialist. It is necessary to initialise the new ECU after installation, which requires the use of dedicated Peugeot diagnostic equipment.*

2 The ECU is located in a plastic box which is mounted on the right-hand front wheelarch.

3 Lift off the ECU module box lid.

4 Release the wiring connector by lifting the locking lever on top of the connector upwards. Lift the connector at the rear, disengage the tag at the front and carefully withdraw the connector from the ECU pins.

5 Lift the ECU upwards and remove it from its location.

6 To remove the ECU module box, undo the internal and external retaining bolts and remove the module box.

7 Refitting is a reversal of removal.

Crankshaft speed/position sensor

8 The crankshaft speed/position sensor is located at the top of the transmission bell-housing, directly above the engine flywheel. To gain access, remove the air cleaner assembly.

9 Remove the battery and battery tray.

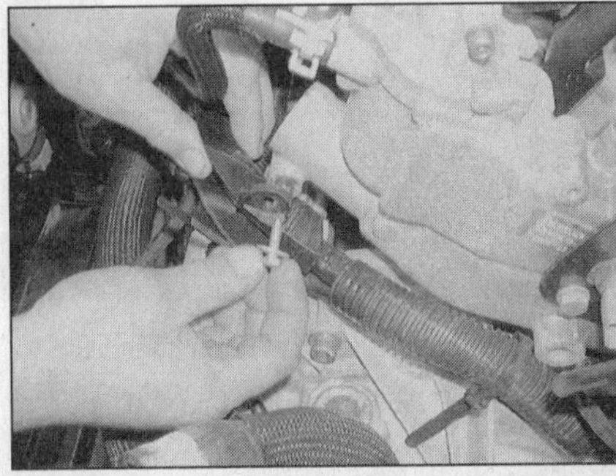

10 Undo the retaining nuts and bolts and release the plastic wiring harness guide from its mountings.

11 Working below the thermostat housing, disconnect the wiring connector from the crankshaft speed/position sensor.

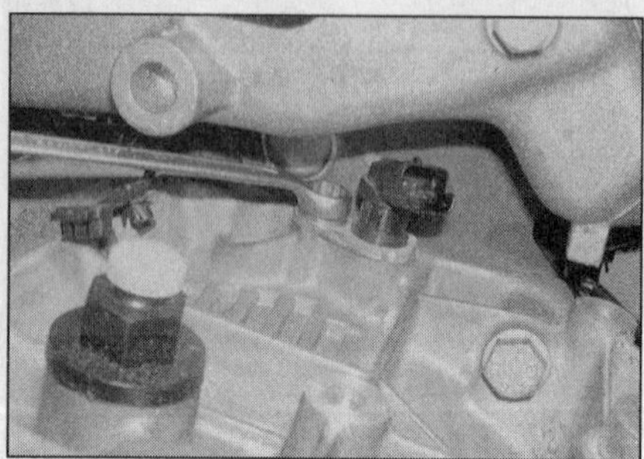

12 Slacken the bolt securing the sensor to the bellhousing. It is not necessary to remove the bolt completely as the sensor mounting flange is slotted.

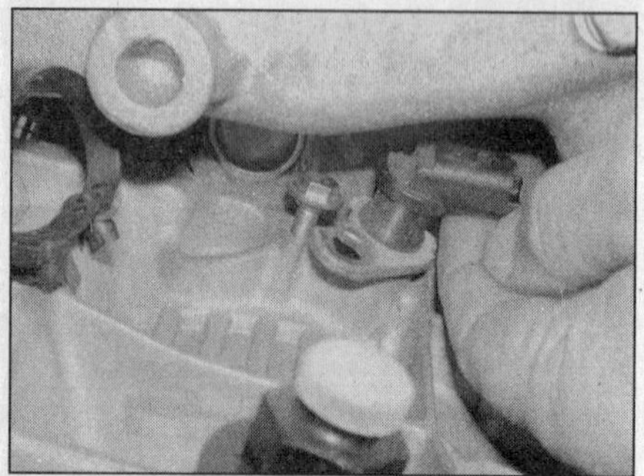

13 Turn the sensor body to clear the mounting bolt, then withdraw the sensor from the bellhousing.

14 Refitting is reverse of the removal procedure ensuring the sensor retaining bolt is securely tightened.

Camshaft position sensor

15 The camshaft position sensor is mounted on the right-hand end of the cylinder head cover, directly behind the camshaft sprocket.

16 Remove the timing belt upper and intermediate covers.

17 Disconnect the sensor wiring connector.

18 Undo the retaining bolt and lift the sensor off the cylinder head cover.

19 To refit and adjust the sensor position, locate the sensor on the cylinder head cover and loosely refit the retaining bolt.

20 The air gap between the tip of the sensor and the target plate at the rear of the camshaft sprocket hub must be set to 1.2 mm, using feeler blades.

21 Clearance for the feeler blades is limited with the timing belt and camshaft sprocket in place, but it is just possible if the feeler blades are bent through 90° so they can be inserted through the holes in the sprocket, to rest against the inner face of the target plate.

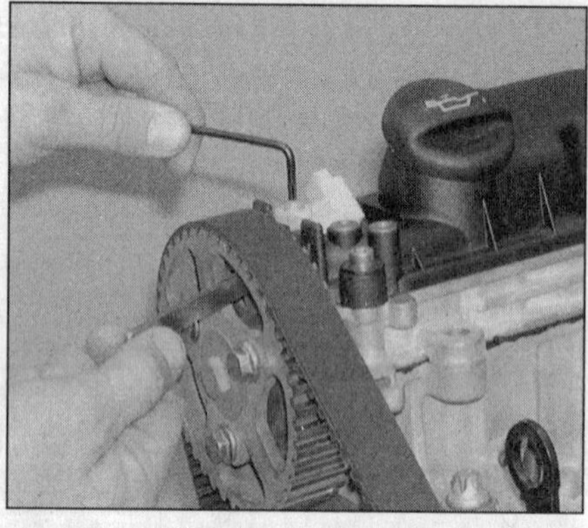

22 With the feeler blades placed against the target plate, move the sensor toward the sprocket until it just contacts the feeler blades. Hold the sensor in this position and tighten the retaining bolt.

23 With the gap correctly adjusted, reconnect the sensor wiring connector, then refit the timing belt upper and intermediate covers.

Accelerator pedal position sensor

24 The accelerator pedal position sensor is located on the left-hand side of the engine compartment, adjacent to the air cleaner housing.

25 Remove the air cleaner assembly.

26 Undo the two nuts and bolts and remove the sensor assembly from the mounting bracket on the side of the air cleaner housing.

27 Refitting is reverse of the removal procedure.

Coolant temperature sensor

28 Refer to the Bosch Motronic system Section.

Fuel temperature sensor

Note: *Do not remove the sensor from the accumulator rail unless there is a valid reason to do so. At the time of writing there was no information as to the availability of the sensor seal as a separate item. Consult a Peugeot parts stockist for the latest information before proceeding.*

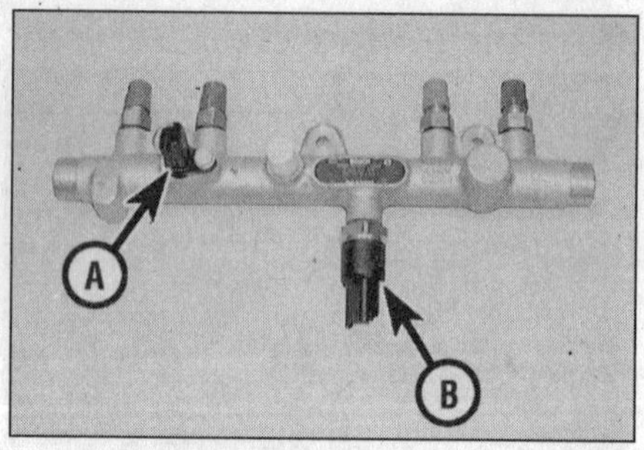

Fuel temperature sensor (A) and fuel pressure sensor (B) locations on the accumulator rail (shown removed for clarity)

29 The fuel temperature sensor is located towards the right-hand end of the accumulator rail.

30 Undo the four plastic nuts and lift off the engine cover.

31 Disconnect the fuel temperature sensor wiring connector.

32 Thoroughly clean the area around the sensor and its location on the accumulator rail.

33 Suitably protect the components below the sensor and have plenty of clean rags handy. Be prepared for considerable fuel spillage.

34 Undo the retaining bolt and withdraw the sensor from the accumulator rail. Plug the opening in the accumulator rail as soon as the sensor is withdrawn.

35 Prior to refitting, if the original sensor is to be refitted, renew the sensor seal, where applicable.

36 Locate the sensor in the accumulator rail and refit the retaining bolt, tightened securely.

37 Refit the sensor wiring connector.

38 Start the engine and allow it to idle. Check for leaks at the fuel temperature sensor with the engine idling. If satisfactory, increase the engine speed to 4000 rpm and check again for leaks. Take the car for a short road test and check for leaks once again on return. If any leaks are detected, obtain and fit a new sensor.

39 Refit the engine cover on completion.

Air mass meter

40 The air mass meter is attached to the lid of the air cleaner housing.

41 Undo the four plastic nuts and lift off the engine cover.

42 Slacken the retaining clips and disconnect the flexible air inlet duct from the air mass meter and turbocharger rigid inlet duct. Suitably plug or cover the turbocharger rigid inlet duct, using clean rag to prevent any dirt or foreign material from entering.

43 Disconnect the wiring connector from the air mass meter.

44 Undo the screws securing the lid to the air cleaner housing and lift off the lid, complete with air mass meter.

45 Undo the two screws and withdraw the air mass meter from the air cleaner lid.

46 Refitting is reverse of the removal procedure.

Fuel pressure sensor

Note: *Peugeot special tool (-).4220 TH (27 mm forked adaptor) or suitable equivalent will be required for this operation.*

47 The fuel pressure sensor is located centrally on the underside of the accumulator rail.

48 Undo the four plastic nuts and lift off the engine cover.

49 Release the retaining clip and disconnect the crankcase ventilation hose from the cylinder head cover.

50 Disconnect the fuel supply and return hose quick-release fittings at the fuel filter, using a small screwdriver to release the locking clip. Suitably plug or cover the open unions to prevent dirt entry. Release the fuel hoses from the relevant retaining clips.

51 Disconnect the fuel pressure sensor wiring connector.

52 Thoroughly clean the area around the sensor and its location on the accumulator rail.

53 Suitably protect the components below the sensor and have plenty of clean rags handy. Be prepared for considerable fuel spillage.

54 Using the Peugeot special tool (or suitable alternative 27 mm forked adaptor) and a socket bar, unscrew the fuel pressure sensor from the base of the accumulator rail.

55 Obtain and fit a new sealing ring to the sensor prior to refitting.

56 Locate the sensor in the accumulator rail and tighten it to the specified torque using the special tool (or alternative) and a torque wrench.

57 Refit the sensor wiring connector.

58 Start the engine and allow it to idle. Check for leaks at the fuel pressure sensor with the engine idling. If satisfactory, increase the engine speed to 4000 rpm and check again for leaks. Take the car for a short road test and check for leaks once again on return. If any leaks are detected, obtain and fit another new sensor sealing ring.

59 Refit the engine cover on completion.

Fuel pressure control valve

60 The fuel pressure control valve is integral with the high-pressure fuel pump and cannot be separated.

Vehicle speed sensor

61 Refer to the Bosch Motronic system Section.

Throttle bodies (DW12 engine)

62 Remove the air cleaner and inlet air ducts.

63 Disconnect the vacuum hoses from the servos, and disconnect the air inlet ducts from the throttle bodies.

64 Unbolt the throttle body assembly from the inlet manifold, and withdraw from the engine compartment. Recover the gaskets.

65 If necessary, unbolt the throttle bodies from the elbow.

66 Refitting is a reversal of removal, but fit new gaskets.

Notes

Renault Clio '98 – '01

21

Petrol models

General

System type:	
1.2 litre models (D7F engine):	
Engine code D7F 720	Sagem or Magneti Marelli semi-sequential multi-point injection with Magneti Marelli throttle body
Engine code D7F 702 and D7F 726	Siemens-Sirius sequential multi-point injection
1.4 litre models (E7J engine):	
Engine code E7J 780	Siemens-Fenix 5 semi-sequential multi-point injection with Pierburg throttle body
Engine code E7J 634	Siemens-Sirius sequential multi-point injection
1.4 litre models (K4J engine)	Siemens-Sirius sequential multi-point injection
1.6 litre models (K7M engine)	Siemens-Fenix 5 semi-sequential multi-point injection with Pierburg throttle body
1.6 litre models (K4M engine)	Siemens-Sirius sequential multi-point injection

Fuel system data

Fuel pressure regulator control pressure (all models):	
Zero vacuum	3.0 ± 0.2 bars
500 mbars vacuum	2.5 ± 0.2 bars
Air temperature sensor resistance:	
D7F 720 engine:	
At 0°C	5000 to 7000 ohms
At 20°C	1700 to 3300 ohms
At 40°C	800 to 1550 ohms
E7J 780 and K7M engines:	
At 0°C	7470 to 11 970 ohms
At 20°C	3060 to 4045 ohms
At 40°C	1315 to 1600 ohms
E7J 634 engine:	
At –10°C	10 450 to 8525 ohms
At 25°C	2120 to 1880 ohms
At 50°C	860 to 760 ohms
D7F 702/726, K4J and K4M engines:	
At 0°C	5290 to 6490 ohms
At 20°C	2400 to 2600 ohms
At 40°C	1070 to 1270 ohms

Fuel system data (continued)

Coolant temperature sensor resistance:	
D7F 720, E7J 780 and K7M engines:	
At 20°C	2600 to 3000 ohms
At 40°C	1100 to 1300 ohms
At 80°C	270 to 300 ohms
At 90°C	200 to 215 ohms
E7J 634 engine:	
At 25°C	2360 to 2140 ohms
At 50°C	770 to 850 ohms
At 80°C	275 to 290 ohms
At 110°C	112 to 117 ohms
D7F 702/726, K4J and K4M engines:	
At 20°C	3060 to 4045 ohms
At 40°C	1315 to 1600 ohms
At 80°C	300 to 370 ohms
At 90°C	210 to 270 ohms
Throttle potentiometer:	
D7F 720 engine:	
Voltage	5.0 volts
Resistance:	
Track A-B	1300 ohms (no load), 1300 ohms (full load)
Track A-C	1360 ohms (no load), 2350 ohms (full load)
Track B-C	2300 ohms (no load), 1260 ohms (full load)
D7F 702/726 engine:	
Voltage	5.0 volts
Resistance:	
Track A-B	1200 ohms (no load), 1200 ohms (full load)
Track A-C	1260 ohms (no load), 2200 ohms (full load)
Track B-C	2200 ohms (no load), 1260 ohms (full load)
E7J 780 engine:	
Voltage	5.0 volts
Resistance:	
Track 1-2	5400 ohms (no load), 2200 ohms (full load)
Track 1-3	4500 ohms (no load), 4460 ohms (full load)
Track 2-3	2160 ohms (no load), 5340 ohms (full load)
K7M engine:	
Voltage	5.0 volts
Resistance:	
Track 1-2	5440 ohms (no load), 2200 ohms (full load)
Track 1-3	4500 ohms (no load), 4460 ohms (full load)
Track 2-3	2160 ohms (no load), 5340 ohms (full load)
K4J and K4M engines:	
Voltage	5.0 volts
Resistance:	
Track A-B	1250 ohms (no load), 1250 ohms (full load)
Track A-C	1245 ohms (no load), 2230 ohms (full load)
Track B-C	2230 ohms (no load), 1245 ohms (full load)
E7J 634 engine:	
Type	Incorporated in throttle housing
Track resistance	4100 ± 800 ohms
Cursor resistance	1500 ± 150 ohms
Stepper motor:	
D7F engine:	
Voltage	12 volts
Resistance:	
Track A-D	100 ± 10 ohms
Track B-C	100 ± 10 ohms
E7J 780, K7M engines:	
Voltage	12 volts
Resistance:	
Track A-D	52.0 ± 5 ohms
Track B-C	52.0 ± 5 ohms
E7J 634 engine:	
Resistance	50.0 ohms

Fuel system data (continued)

Stepper motor: (continued)	
K4J and K4M engines:	
Resistance	53.0 ± 5 ohms
Injector resistance	14.5 ± 1.0 ohms
TDC sensor resistance	220 ohms
Specified idle speed (non-adjustable):	
D7F engine	740 ± 50 rpm
E7J, K7M, K4J, K4M engines	750 ± 50 rpm
Idle mixture CO content (non-adjustable)	0.5% maximum (0.3% at 2500 rpm)

Diesel models

General

System type	Rear-mounted fuel tank, fuel injection pump with integral transfer pump, indirect injection
Firing order	1-3-4-2 (number 1 at flywheel end)
Idle speed:	
F8Q 630 engine	850 ± 25 rpm
F8Q 632 engine	825 ± 50 rpm
F8Q 662 engine	850 ± 50 rpm
Maximum no-load speed:	
F8Q 630 engine	5100 ± 100 rpm
F8Q 632 engine	5175 ± 50 rpm
F8Q 662 engine	4600 ± 100 rpm

Injection pump

Type:	
F8Q 630 engine	Lucas 8448B 171 A/231A
F8Q 632 engine	Lucas EPIC
F8Q 662 engine	Lucas DPCN
Direction of rotation	Clockwise viewed from sprocket end

Sensor resistances

Air temperature sensor (F8Q 630/662 engine):	
0° C	7470 to 11 970 ohms
20° C	3060 to 4045 ohms
40° C	1315 to 1600 ohms
Air temperature sensor (F8Q 632 engine):	
–10° C	10 454 to 8623 ohms
25° C	2175 to 1928 ohms
50° C	857 to 763 ohms
80° C	325 to 292 ohms
TDC sensor (F8Q 630/662 engine)	220 ohms
Engine speed sensor (F8Q 632 engine)	760 ohms at 20° C
Positive flow solenoid valve (F8Q 632 engine)	31 ± 2 ohms
Negative flow solenoid valve (F8Q 632 engine)	31 ± 2 ohms
Advance solenoid valve (F8Q 632 engine)	31 ± 2 ohms
Electrical solenoid valve (F8Q 632 engine)	1.39 ± 0.1 ohms
Advance cam position sensor (F8Q 632 engine)	52 ± 4 ohms
Flow valve position sensor (F8Q 632 engine)	41 ± 4 ohms
Pump temperature thermistor (F8Q 632 engine)	2716 ± 60 ohms at 20° C

Torque wrench settings

	Nm	lbf ft
Petrol models		
Oxygen sensor	45	33
Fuel rail	10	7
Knock sensor	20	15
Diesel models		
Advance solenoid valve	30	22
Altimetric solenoid valve	30	22

Chapter 21

General information

Petrol models

Sagem and Magneti Marelli systems

1.2 litre models (D7F engines) are equipped with a Sagem or Magneti Marelli semi-sequential multi-point injection system with a Magneti Marelli throttle body. The system incorporates a closed-loop catalytic converter and an evaporative emission control system, and complies with the latest emission control standards. The system is of semi-sequential design with the injectors operating in pairs (1 and 4, then 2 and 3). There is one injection per revolution of the engine for each cylinder, ie, during the complete four-stroke cycle these take place on the induction and combustion strokes. The fuel injection side of the system operates as follows;

The fuel pump, immersed in the fuel tank, pumps fuel from the fuel tank to the fuel rail, via a filter mounted underneath the rear of the vehicle. Fuel supply pressure is controlled by the pressure regulator in the throttle body assembly. The regulator operates by allowing excess fuel to return to the tank. There are four injectors (one per cylinder) located in the inlet manifold downstream of the throttle valve. All the injectors are fed from the fuel rail.

The electrical control system consists of the ECU, along with the following sensors:

a) Throttle potentiometer – informs the ECU of the throttle position, and the rate of throttle opening or closing.
b) Coolant temperature sensor – informs the ECU of engine temperature.
c) Inlet air temperature sensor – informs the ECU of the temperature of the air passing through the throttle body.
d) Lambda sensor – informs the ECU of the oxygen content of the exhaust gases.
e) Idle speed regulation stepper motor – controls the idle speed.
f) Crankshaft TDC sensor – informs the ECU of engine speed and crankshaft position.
g) Power steering pressure switch – informs the ECU when the power steering pump is working so the engine idle speed can be increased to prevent stalling.
h) Knock sensor – informs the ECU when pre-ignition ('pinking') is occurring.
i) Manifold absolute pressure (MAP) sensor – informs the ECU of the engine load by monitoring the pressure in the inlet manifold.
j) Fuel vapour recirculation valve – operates the fuel evaporative control system.

All the above information is analysed by the ECU and, based on this, the ECU determines the appropriate ignition and fuelling requirements for the engine. The ECU controls the fuel injector by varying its pulse width – the length of time the injector is held open – to provide a richer or weaker mixture, as appropriate. The mixture is constantly varied by the ECU, to provide the best setting for cranking, starting (with either a hot or cold engine), warm-up, idle, cruising, and acceleration. On automatic transmission models, information from sensors on the transmission is sent to the ECU for processing to determine the most efficient settings for the engine.

The ECU also has full control over the engine idle speed, via a stepper motor which is fitted to the throttle body. The motor pushrod rests against a cam on the throttle spindle. When the throttle is closed (accelerator pedal released), the ECU uses the motor to vary the opening of the throttle valve and so control the idle speed.

The ECU also controls the exhaust and evaporative emission control systems, which are described in detail in Part C of this Chapter.

If there is an abnormality in any of the readings obtained from either the coolant temperature sensor, the inlet air temperature sensor or the Lambda sensor, the ECU enters its back-up mode. In this event, the ECU ignores the abnormal sensor signal, and assumes a pre-programmed value which will allow the engine to continue running (albeit at reduced efficiency). If the ECU enters this back-up mode, the warning light on the instrument panel will come on, and the relevant fault code will be stored in the ECU memory.

A fuel cut-off inertia switch is incorporated into the fuel injection system. In the event of an impact the switch cuts off the electrical supply to the fuel pump and so prevents fuel being expelled should the fuel pipes/hoses be damaged in an accident. The switch is located in the left-hand rear corner of the engine compartment.

Siemens-Fenix 5 system

1.4 and 1.6 litre models with SOHC engines (E7J and K7M) are equipped with a Siemens-Fenix 5 semi-sequential multi-point fuel injection/ignition system with a Pierburg throttle housing. This system operates as for the system described for the 1.2 litre engine.

Siemens-Sirius system

1.4 and 1.6 litre models with DOHC engines (K4J and K4M) are equipped with a Siemens-Sirius sequential multi-point fuel injection/ignition system. The system is of closed-loop type incorporating two oxygen sensors, one located upstream and the other downstream of the catalytic converter. An evaporative emission control system is fitted.

The multi-point injection system uses one injector and one ignition coil for each cylinder, and the injectors are operated individually and sequentially at the beginning of the inlet stroke. The electronic control unit (ECU) is able to determine which cylinder is on its inlet stroke without the use of a camshaft position sensor, however if the unit is renewed, the car must be taken for a road test lasting at least 25 minutes to enable the ECU to reprogram itself; the stepper motor must also be reset.

The system operates as for the system described for the 1.2 litre engine.

Diesel models

The fuel system consists of a rear-mounted fuel tank, a fuel filter with integral water separator, a fuel injection pump, injectors and associated components *(see illustrations)*. Before passing through the filter, the fuel is heated by an electric heating element which is fitted to the filter housing.

Fuel is drawn from the fuel tank to the fuel injection pump by a vane-type transfer pump incorporated in the fuel injection pump. Before reaching the pump, the fuel passes through a fuel filter, where foreign matter and water are removed. Excess fuel lubricates the moving components of the pump, and is then returned to the tank.

The fuel injection pump is driven at half crankshaft speed by the timing belt. The high pressure required to inject the fuel into the compressed air in the swirl chambers is achieved by two opposed pistons forced together by rollers running in a cam ring. The fuel passes through a central rotor with a single outlet drilling which aligns with ports leading to the injector pipes.

The four fuel injectors inject a homogeneous spray of fuel into the swirl chambers located in the cylinder head. The injectors are calibrated to open and close at critical pressures to provide efficient and even combustion. Each injector needle is lubricated by fuel, which accumulates in the spring chamber.

To enable the engine to meet stringent exhaust emission regulations, fuel metering and injection timing is controlled electronically by an injection electronic control unit (ECU) located in the left-hand rear corner of the engine compartment. On early models, the injection pump has semi-electronic controls, however on later models fitted with the EPIC (Electronically Programmed Injection Control) injection pump, the controls are fully electronic. This highly sophisticated system is similar in operation to a full engine management system as used on petrol engine vehicles and uses similar sensors to provide data to the ECU on engine operating conditions. The sensors typically monitor coolant temperature, air temperature, fuel flow, engine speed, vehicle speed, atmospheric

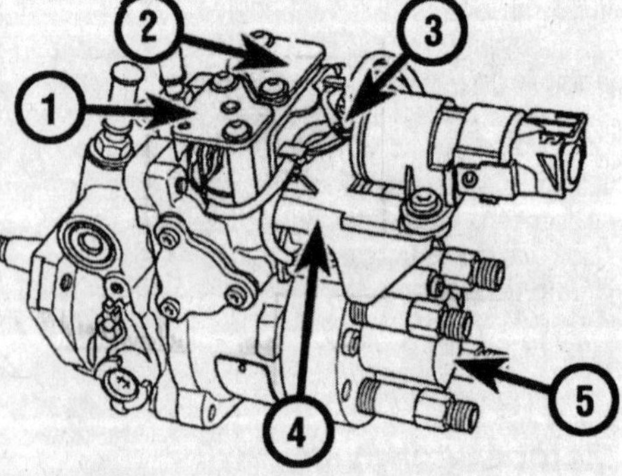

Lucas EPIC fuel injection pump components
1 Advance solenoid valve
2 Positive flow solenoid valve
3 Negative flow solenoid valve
4 Electrical solenoid valve
5 Flow valve position sensor (cannot be removed)

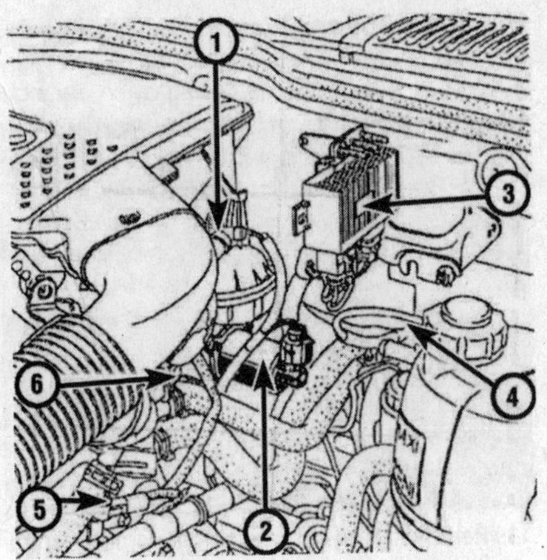

Lucas EPIC system components

1 *Inlet air temperature sensor*
2 *EGR valve*
3 *Pre-post heating unit*
4 *Accelerator pedal position potentiometer*
5 *TDC sensor*
6 *Coolant temperature sensor*

pressure, fuel temperature and accelerator pedal position. From the data received, the ECU controls injection pump fuel metering and injection advance, pre/post-heating system, idle speed, exhaust gas recirculation, the anti-theft system engine immobiliser, and the electric stop control. This allows precise control of all engine fuelling requirements providing optimum engine operation and minimal exhaust emissions under all engine operating conditions. Idle speed is dependent on coolant temperature, gear selected, battery voltage and electrical consumption. If one of the tracks in the accelerator pedal potentiometer is faulty, the idle speed is set at 1000 rpm, if both tracks are faulty, it is set to 1300 rpm. In 1st gear the idle speed is set to 850 rpm, in 2nd gear 875 rpm and in all other gears 900 rpm.

If a fault appears in the injection system, first ensure that all the system wiring connectors are securely connected and free of corrosion. Ensure that the fault is not due to poor maintenance; ie, check that the air cleaner filter element is clean, the cylinder compression pressures are correct, and that the engine breather hoses are clear and undamaged.

Fuel injection system – depressurisation, priming and bleeding

Petrol models

Note: *The fuel system referred to in this Section includes the tank-mounted fuel pump, the fuel filter, the fuel injector(s) and the pressure regulator in the throttle body/fuel rail, and the metal pipes and flexible hoses of the fuel lines between these components. All these contain fuel which will be under pressure while the engine is running, and/or while the ignition is switched on. The pressure will remain for some time after the ignition has been switched off, and it must be relieved when any of these components are disturbed for servicing work.*

Method 1

1 Disconnect the battery negative lead.

2 Place a suitable container beneath the connection or union to be disconnected, and have a large rag ready to soak up any escaping fuel not being caught by the container.

3 Slowly loosen the connection or union nut to avoid a sudden release of pressure, and position the rag around the connection, to catch any fuel spray which may be expelled. Once the pressure is released, disconnect the fuel line. Plug the pipe ends, to minimise fuel loss and prevent the entry of dirt into the fuel system.

Method 2

4 Remove the fuel pump relay located in the engine compartment fuse/relay box.

5 Start the engine and allow it to idle until it stops due to lack of fuel. Operate the starter motor a couple more times, to ensure that all fuel pressure has been relieved.

6 Switch off the ignition and refit the fuel pump relay.

Diesel models

7 After disconnecting part of the fuel supply system or running out of fuel, it is necessary to prime the system and bleed off any air which may have entered the system components.

8 All models are fitted with a hand-operated priming bulb located next to the fuel filter, in the left-hand front corner of the engine compartment.

9 To prime the system, loosen the bleed screw located on the filter head or on the injection pump inlet union.

10 Fit a piece of pipe to the bleed screw and place the open end in a container to catch the fuel. Pump the priming bulb until fuel free from air bubbles emerges from the bleed screw. Retighten the bleed screw, and remove the pipe and container. Continue pumping until firm resistance is felt.

11 Attempt to start the engine at this stage, by fully depressing the accelerator pedal and operating the starter motor. Do not operate the heater plugs. If the engine refuses to start after 15 seconds, operate the heater plugs as normal then attempt to start the engine again.

12 If air has entered the injector pipes, it may be necessary to bleed them as follows. Place wads of rag around the injector pipe unions at the injectors (to absorb spilt fuel), then slacken the unions. Crank the engine on the starter motor until fuel emerges from the unions, then stop cranking the engine and retighten the unions. Mop up spilt fuel.

Warning: Be prepared to stop the engine if it should fire, to avoid fuel spray and spillage.

13 Attempt to start the engine by fully depressing the accelerator pedal and operating the starter motor. Initially do not operate the heater plugs, however, if it refuses to start after 15 seconds, operate the heater plugs as normal then start the engine.

Petrol fuel injection system components – removal and refitting

Fuel rail and injectors

Note: *If a faulty injector is suspected, before condemning the injector, it is worth trying the effect of one of the proprietary injector-cleaning treatments.*

D7F engine

Note: *New O-rings will be required on refitting.*

1 Depressurise the fuel system, then disconnect the battery negative lead.

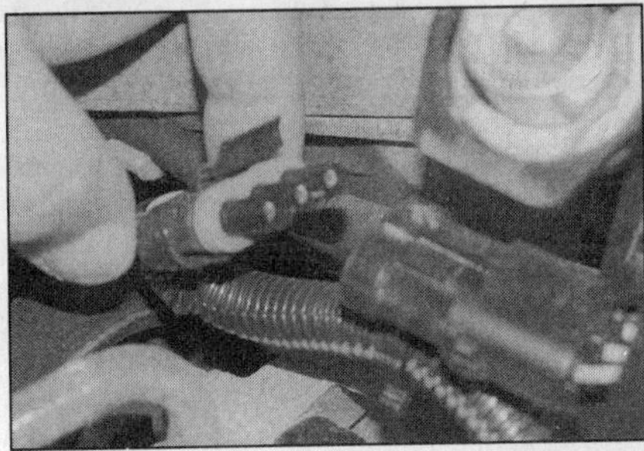

2 Separate the two halves of the fuel injector harness wiring connector. The connector is located in front of the fuel pressure regulator.

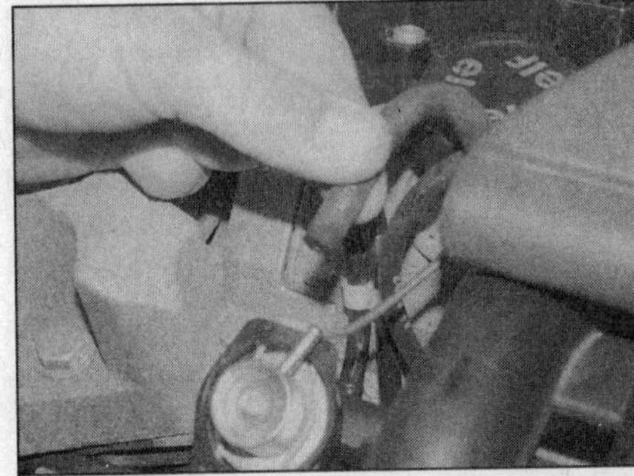

3 Disconnect the vacuum pipe from the fuel pressure regulator.

4 Disconnect the fuel return pipe from the end of the fuel rail. Be prepared for fuel spillage. Plug or clamp the hose to prevent dirt entry and further fuel loss.

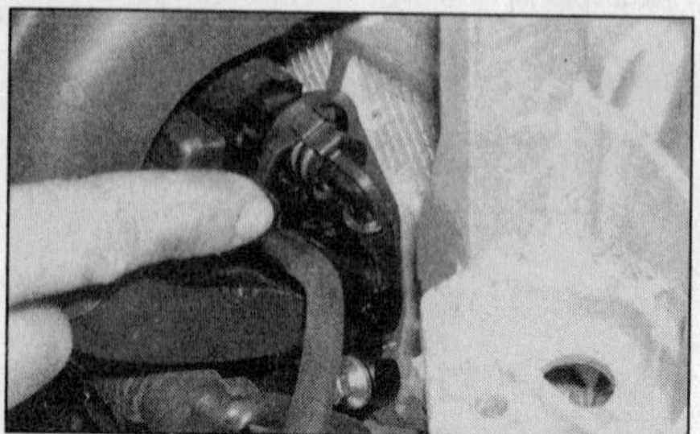

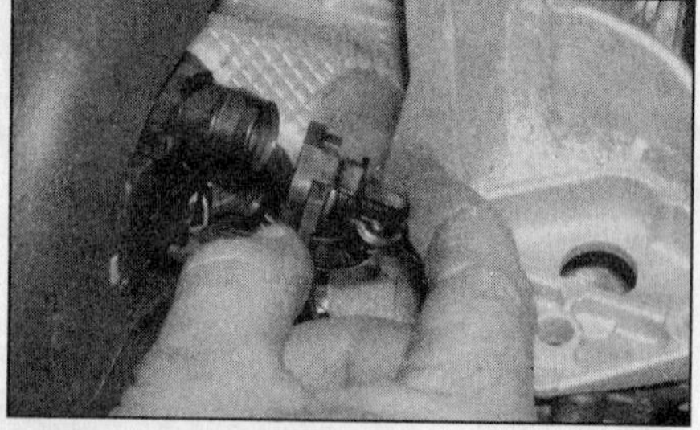

5 It is now necessary to disconnect the fuel feed hose from the fuel rail. To do this, a special tool will be required to release the connector. The appropriate Renault special tool (Mot. 1311-06) slides through the connector collar to release the securing lugs, but the same effect can be achieved using a small flat-bladed screwdriver to release the lugs. Note that some models have the Renault special tool built into the hose connection. Once the connector securing lugs have been released, the hose can be pulled from the end of the fuel rail. Plug or clamp the hose to prevent dirt entry and further fuel loss.

6 Working underneath the manifold, unscrew the two bolts securing the fuel rail to the manifold.

7 Carefully slide the fuel rail and injector assembly towards the right-hand side of the vehicle, between the manifold and cylinder head.

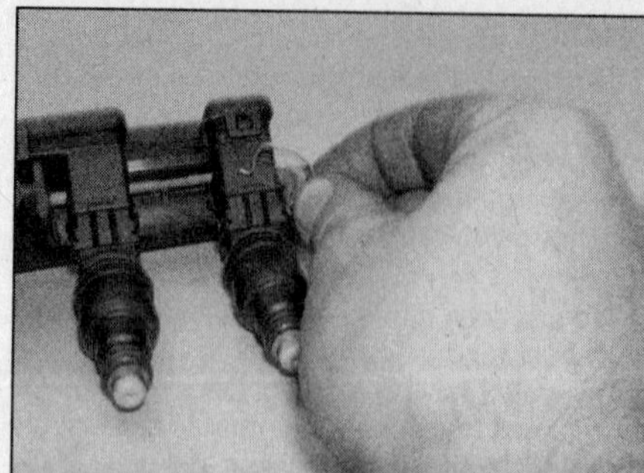

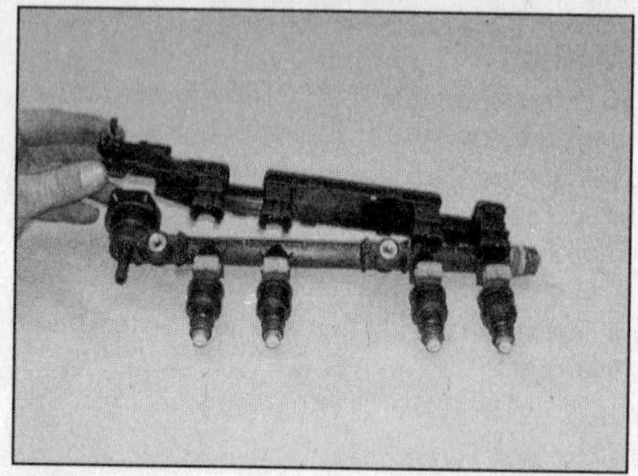

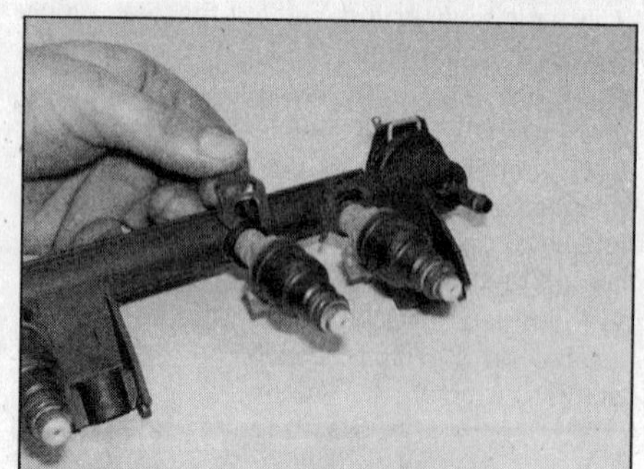

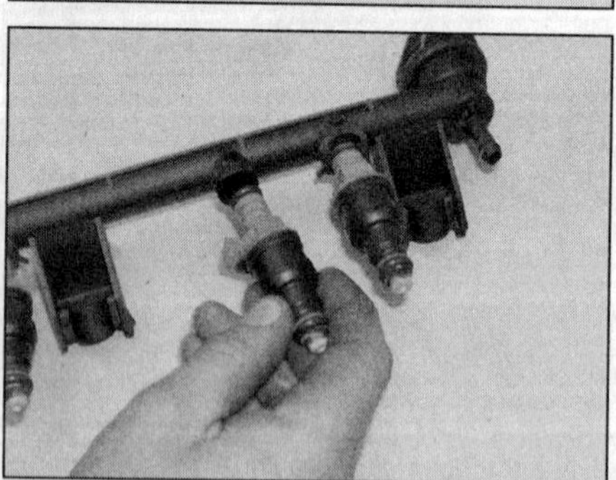

8 To remove a fuel injector from the fuel rail, release the wiring clips from the injectors, and pull the wiring tube assembly from the top of the injectors. Release the relevant injector securing clip, and remove the fuel injector.

9 Refitting is a reversal of removal, but renew the O-rings at the top and bottom of each injector (check on availability before removing the old O-rings), and ensure that the fuel feed hose is securely reconnected (the connector should click securely into position).

E7J and K7M engines

10 Depressurise the fuel system, then disconnect the battery negative lead.

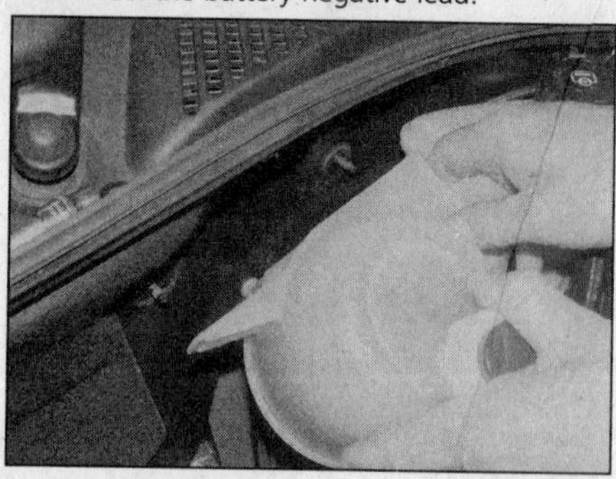

11 Remove the air cleaner assembly, then unbolt the coolant expansion tank from the bulkhead and position it to one side.

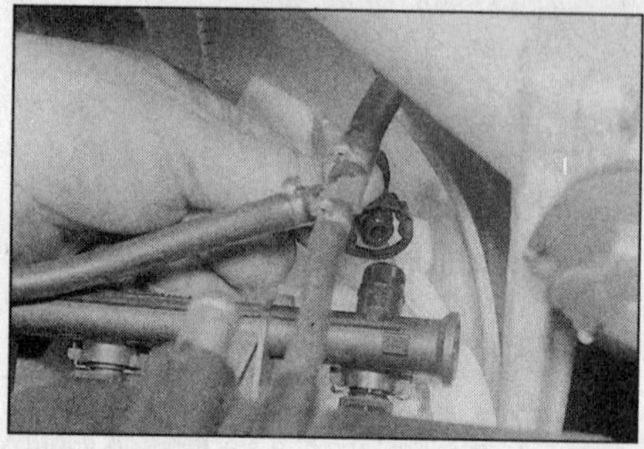

12 Disconnect the fuel feed and return hoses from the fuel rail. The hoses have the same quick-release connectors as fitted to the fuel pump.

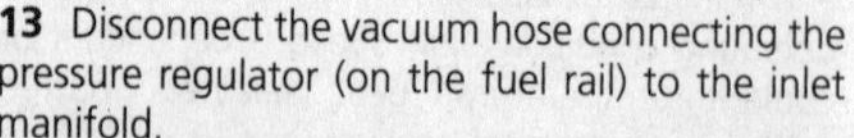

13 Disconnect the vacuum hose connecting the pressure regulator (on the fuel rail) to the inlet manifold.

14 Disconnect the wiring from the injectors and move the loom to one side.

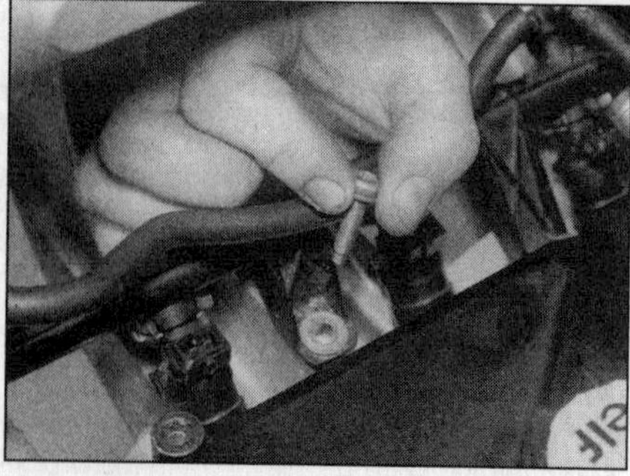

15 Unscrew and remove the mounting bolts and carefully ease the fuel rail together with the injectors from the inlet manifold.

16 Note the fitted positions of the injectors, then remove the clips and ease the injectors from the fuel rail.

17 Remove the sealing rings from the grooves at each end of the injectors and obtain new ones.

18 Refitting is a reversal of the removal procedure, noting the following points:

a) Renew all sealing rings, using a smear of engine oil to aid installation.

b) Refit the fuel rail assembly to the manifold, making sure the sealing rings remain correctly positioned, and tighten the retaining bolts to the specified torque.

c) On completion start the engine and check for fuel leaks.

K4J and K4M engines

19 Depressurise the fuel system, then disconnect the battery negative lead.

20 Remove the cover for access to the fuel rail.

21 Disconnect the fuel feed and return hoses from the fuel rail.

22 Disconnect the vacuum hose connecting the pressure regulator (on the fuel rail) to the inlet manifold.

23 Disconnect the wiring from the injectors and move the loom to one side.

24 Unscrew and remove the mounting bolts and carefully ease the fuel rail together with the injectors from the inlet manifold.

25 Note the fitted positions of the injectors, then remove the clips and ease the injectors from the fuel rail.

26 Remove the sealing rings from the grooves at each end of the injectors and obtain new ones.

27 Refitting is a reversal of the removal procedure, noting the following points:

a) Renew all sealing rings, using a smear of engine oil to aid installation.

b) Refit the fuel rail assembly to the manifold, making sure the sealing rings remain correctly positioned, and tighten the retaining bolts to the specified torque.

c) On completion start the engine and check for fuel leaks.

Fuel pressure regulator

28 Disconnect the vacuum pipe from the regulator.

29 Place a wad of rag over the regulator to catch any spilled fuel, then extract the retaining spring and ease the regulator from the fuel rail.

30 Remove the sealing rings from the grooves in the pressure regulator and obtain new ones.

31 On refitting, fit new sealing rings to the regulator grooves and apply a smear of engine oil to them to ease installation. Ease the regulator back into the end of the fuel rail and refit the retaining spring and vacuum pipe.

Throttle potentiometer

32 Remove the throttle housing.

33 Undo the retaining screws and remove the potentiometer from the throttle housing.

34 Refitting is a reverse of the removal procedure ensuring that the potentiometer is correctly engaged with the throttle spindle.

Inlet air temperature sensor

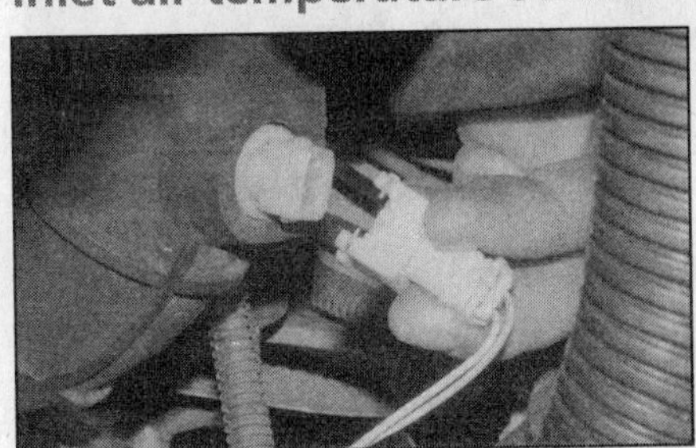

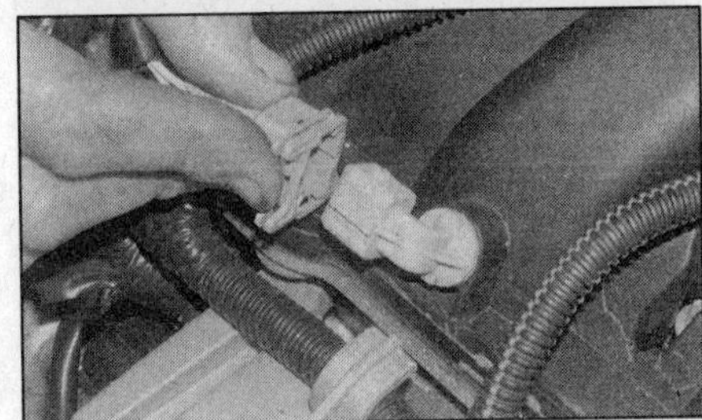

35 The air temperature sensor is located on the air inlet duct to the throttle housing on SOHC engines, and on the upper section of the inlet manifold on DOHC engines. To remove it, first disconnect the wiring from the sensor, then loosen the clips and remove the air inlet duct.

36 Unscrew and remove the inlet air temperature sensor from the air inlet duct.

37 Refitting is a reversal of removal.

Coolant temperature sensor

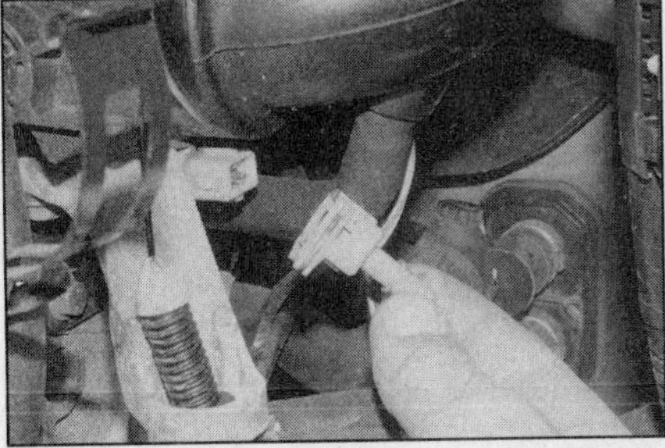

38 The sensor is located on the thermostat housing or at the left-hand end of the cylinder head above the gearbox bellhousing. Drain the coolant system.

39 Disconnect the wiring plug and unscrew the sensor.

40 Refitting is a reversal of removal, but apply a little sealant to the sensor threads prior to refitting.

Power steering pressure switch

41 The switch is screwed into the feed pipe from the power steering pump to the steering gear. To remove the switch, disconnect the wiring connector.

42 Wipe clean the area around the switch then unscrew the switch and remove it from the pipe. Plug the pipe aperture to prevent excess fluid leakage and prevent dirt entry into the hydraulic system.

43 Refitting is the reverse of removal. On completion check the power steering fluid level.

Knock sensor

44 On the D7F, K4J and K4M engines the knock sensor is located on the front of the cylinder block, however on E7J and K7M engines it is located on the rear of the cylinder block.

45 Disconnect the wiring plug and unscrew the sensor from the cylinder block.

46 Refitting is a reversal of removal. Tighten the sensor to the specified torque.

Idle speed control stepper motor

47 On the D7F engine, the idle speed control stepper motor is mounted on the top of the throttle housing on the left-hand side of the engine, however on E7J and K7M engines it is located on the rear of the throttle housing. On K4J and K4M engines the stepper motor is located on the top of the inlet plenum chamber on the right-hand rear side of the engine.

48 To remove the stepper motor, first remove the throttle housing.

49 Undo the retaining screws and remove the stepper motor from the throttle housing. Recover the gasket and discard it; a new one should be used on refitting.

50 Refitting is a reversal of the removal procedure using a new gasket.

Manifold absolute pressure (MAP) sensor

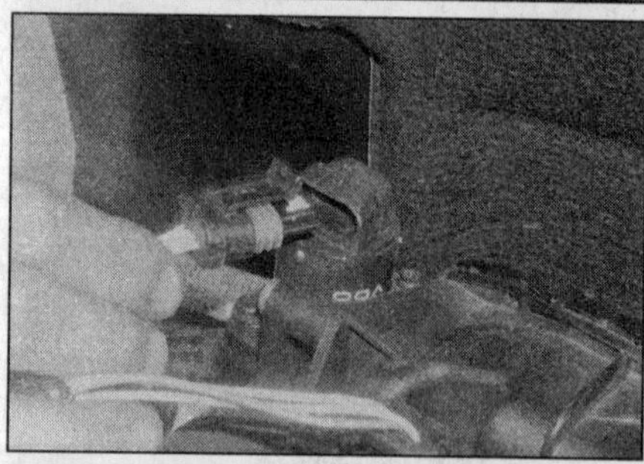

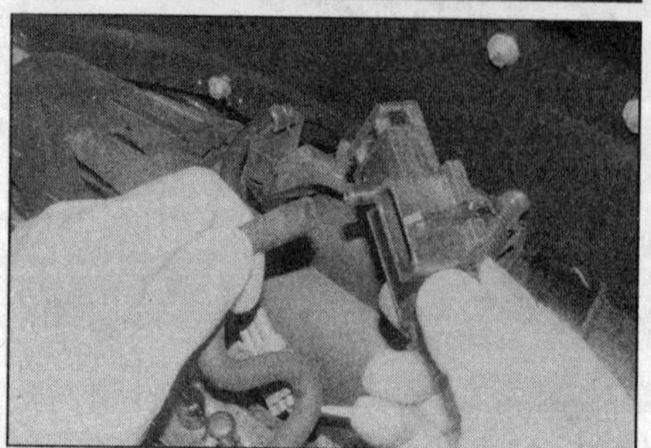

51 The manifold absolute sensor is mounted on the rear of the air cleaner. Disconnect the wiring and vacuum hose from the sensor.

52 Unscrew the mounting nuts and remove the sensor.

53 Refitting is a reversal of removal.

Fuel injection system relay and fuel pump relay

54 These relays are located in the left-hand side of the engine compartment. Remove the cover from the box.

55 Remove the relevant relay from the fuse/relay box.

56 Refitting is the reverse of removal.

Crankshaft TDC sensor

57 The sensor is mounted on the top of the transmission bellhousing at the left-hand end of the cylinder block.

58 To remove the sensor, remove the air cleaner housing.

59 Trace the wiring back from the sensor to the wiring connector, and disconnect it from the main harness.

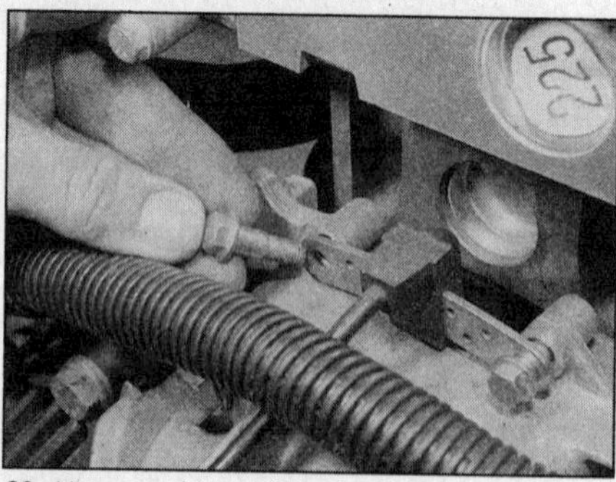

60 Unscrew the retaining bolts and remove the sensor.

61 Refitting is a reversal of removal. Ensure that the sensor retaining bolts are securely tightened – note that only the special shouldered bolts originally fitted must be used to secure the sensor; these bolts locate the sensor precisely to give the correct air gap between the sensor tip and the flywheel/ driveplate.

Fuel cut-off inertia switch

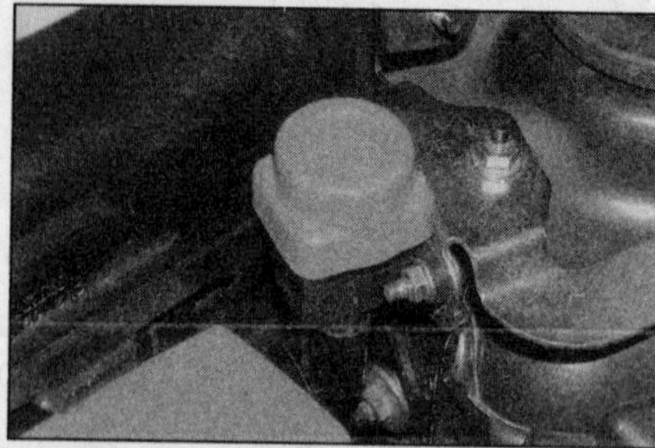

62 The switch is located in the left-hand side of the engine compartment.

63 Unscrew and remove the switch retaining screws then disconnect its wiring connector and remove the switch from the engine compartment.

64 Refitting is the reverse of removal. On completion, reset the switch by depressing its button.

Electronic control unit (ECU)

Note: *The ECU is electronically-coded to match the engine immobiliser. If the ECU is being removed to enable a new unit to be fitted, the new unit must be programmed with the vehicle code as described.*

65 The ECU is located in the left-hand side of the engine compartment, behind the battery. First disconnect the battery negative lead.

66 Unclip the power steering hydraulic fluid reservoir from the bulkhead and position it to one side.

67 Where applicable, unbolt the bracket from the top of the ECU and release the strap. Alternatively the bracket can remain on the ECU until the assembly is removed.

68 Undo the mounting screws and remove the ECU and mounting bracket.

69 Disconnect the wiring connector and remove the ECU from the engine compartment.

70 Refitting is a reverse of the removal procedure ensuring that the wiring connector is securely reconnected. If a new ECU has been fitted, reprogramme it as follows. Turn the ignition on for a few seconds, then turn it off. Now remove the key to operate the immobiliser. After 10 seconds, the red immobiliser warning light should start to flash.

Diesel fuel injection pump components and ECU – removal and refitting

Note: *After refitting the following components it is recommended that the ECU memory is checked and if necessary erased.*

Advance solenoid valve

F8Q 630

1 The advance solenoid valve is located on the front-facing side of the injection pump.

2 The two wires for the valve must be removed from the multi-plug connector for the pump. To do this, first separate the connectors, then press the two tabs and remove the yellow guide. The wires can now be removed using a small screwdriver to prise the retaining tabs to one side. Release the wiring from the main loom.

3 Remove the cover from the valve, then unscrew and remove the valve from the pump.

4 Recover the return connection, seals and filter. Note that the filter must be renewed whenever the valve is removed.

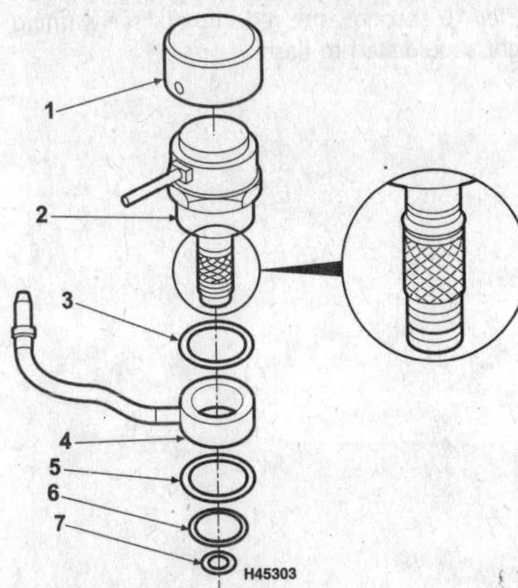

Advance solenoid valve – engine code F8Q 630

1 Protective cover
2 Solenoid valve
3 Seal
4 Return connection
5 Seal
6 Seal
7 Small filter

5 Refitting is a reversal of removal, but tighten the valve to the specified torque and bleed the fuel system.

Altimetric solenoid valve

F8Q 630 and F8Q 662

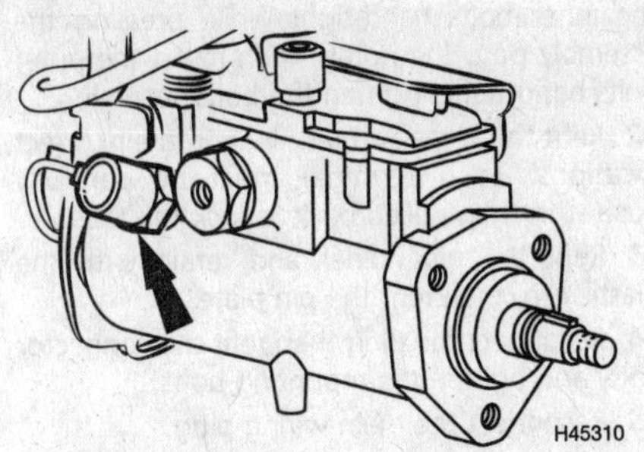

6 The altimetric corrector valve is located on the rear face of the injection pump, and its purpose is to reduce the fuel flow at altitudes above 1000 metres. First unscrew the unions and remove the high pressure pipes.

7 Remove the injector with the sensor.

8 The two wires for the valve must be removed from the multi-plug connector for the pump. To do this, first separate the connectors, then press the two tabs and remove the yellow guide. The wires can now be removed using a small screwdriver to prise the retaining tabs to one side. Release the wiring from the main loom.

9 Remove the cover from the valve, then unscrew and remove the valve from the pump.

10 Recover the small filter. Note that the filter must be renewed whenever the valve is removed.

11 Refitting is a reversal of removal, but tighten the valve to the specified torque and bleed the fuel system.

Load potentiometer

F8Q 630 and F8Q 662

Note: *The following procedure is a delicate operation, and extra care must be taken to prevent damage to components. If a new unit is fitted, the 'full load' position must be reprogrammed into the ECU by a Renault dealer or suitably equipped specialist using specialist equipment.*

12 Disconnect the wiring from the injection pump.

13 Remove the wiring from the multi-plug connector as described in paragraph 8, then remove the wires from the plastic sheath.

14 Mark the potentiometer and bracket in relation to each other as a guide to refitting the unit.

15 Undo the mounting screws, then use a small screwdriver to release the slide contact from the lever. Turn the insert 90° to remove it.

16 With the load lever in the 'full load' position, turn the potentiometer 90° anti-clockwise and remove it.

17 Refitting is a reversal of removal, but if a new unit has been fitted, have the 'full load' position reprogrammed into the ECU by a Renault dealer.

Advance and positive flow solenoid valves

F8Q 632

18 The advance and positive flow solenoid valves are located on the top of the injection pump and are in one integral unit. To remove them, first disconnect the battery negative (earth) lead and position it away from the terminal.

19 Disconnect the main wiring plug from the connector.

20 Unbolt the connector body from its mounting, then pull out the plug by depressing the tabs.

21 Remove the pin plate, then carefully pull out the plastic clip and remove the pin holder.

22 Note the location of the 4 wires for the solenoid valve, then remove them together with the terminals. Renault technicians use a special tool for this, however a paper clip or similar tool may be used instead.

23 Unbolt the solenoid valve assembly from the injection pump and recover the 4 O-ring seals. Discard the seals as new ones must be fitted.

24 Wipe clean the mating surfaces of the valve assembly and injection pump using lint-free cloth. Do not use any solvent.

25 Locate the new O-ring seals on the injection pump. Do not locate them on the solenoid valve assembly.

26 Locate the solenoid valve assembly on the pump, insert the bolts and hand-tighten them.

27 Tighten the bolts to the specified torque, tightening the inner bolts first then the outer bolts.

28 Refit the wires and terminals in their correct locations. Each terminal must be carefully pushed in then pulled back to lock the tabs.

29 Refit the pin holder and retain with the plastic clip, then refit the pin plate.

30 Reconnect the plug then refit the connector body and tighten the mounting bolts.

31 Reconnect the main wiring plug.

32 Reconnect the battery negative lead.

Negative flow and electrical solenoid valves

F8Q 632

33 The negative flow and electrical solenoid valves are located on the lower part of the injection pump and are in one integral unit. To remove them, first disconnect the battery negative (earth) lead and position it away from the terminal.

34 Disconnect the main wiring plug from the connector.

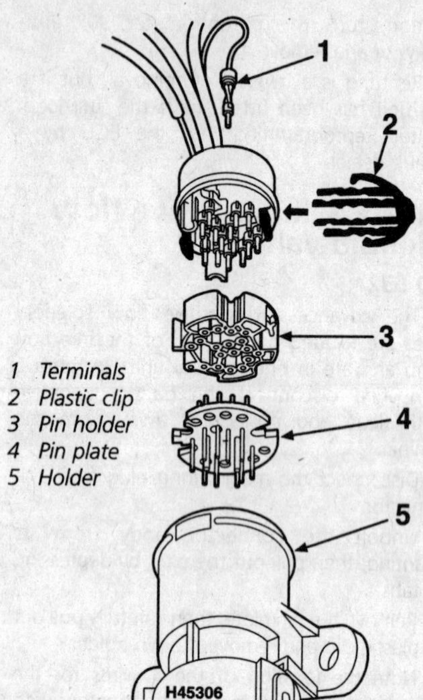

1 Terminals
2 Plastic clip
3 Pin holder
4 Pin plate
5 Holder

35 Unbolt the connector body from its mounting, then pull out the plug by depressing the tabs.

36 Remove the pin plate, then carefully pull out the plastic clip and remove the pin holder.

37 Note the location of the wires for the solenoid valve, then remove them together with the terminals. Renault technicians use a special tool for this, however a paper clip or similar tool may be used instead.

38 Unbolt the solenoid valve assembly and plastic mounting from the injection pump loosening the outer bolts first. As it is being removed note the position of the core and spring. Recover the 3 O-ring seals. Discard the seals as new ones must be fitted.

39 Wipe clean the mating surfaces of the valve assembly and injection pump using lint-free cloth. Do not use any solvent.

40 Locate the new O-ring seals on the injection pump, using diesel fuel to hold them in position. Do not locate them on the solenoid valve assembly.

41 Refit the solenoid valve assembly together with core, spring and plastic mounting, making sure that the O-rings are not disturbed. Insert the inner bolts hand-tight while pressing the assembly onto the pump, then insert the outer bolts hand-tight. Tighten the bolts securely.

42 Refit the wires and terminals in their correct locations. Each terminal must be carefully pushed in then pulled back to lock the tabs.

43 Refit the pin holder and retain with the plastic clip, then refit the pin plate.

44 Reconnect the plug then refit the connector body and tighten the mounting bolts.

45 Reconnect the main wiring plug.

46 Reconnect the battery negative lead.

Accelerator potentiometer

F8Q 632

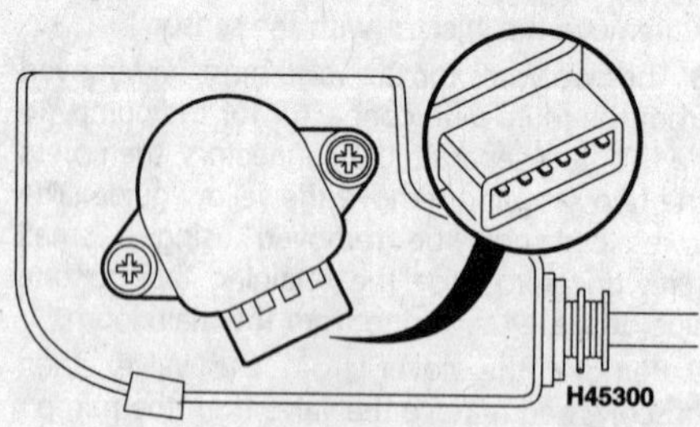

47 The accelerator potentiometer is located beneath the brake master cylinder. It is connected to the accelerator pedal by cable.

48 Remove the unit from under the master cylinder, then prise open the plastic cover.

49 Turn the potentiometer segment slightly and unhook the cable end fitting.

50 Release the cable ferrule from the potentiometer.

51 Refitting is a reversal of removal, but note that the engine management ECU should be checked by a fault code reader for faults residing in its memory.

ECU

Note: *The ECU is electronically-coded to match the engine immobiliser. If the ECU is being removed to enable a new unit to be fitted, the new unit must be programmed with the vehicle code as described.*

52 The ECU is located in the left-hand side of the engine compartment, behind the battery. First disconnect the battery negative lead.

53 Unclip the power steering hydraulic fluid reservoir from the bulkhead and position it to one side.

54 Where applicable, unbolt the bracket from the top of the ECU and release the strap *(see paragraph 67 in previous Section)*. Alternatively the bracket can remain on the ECU until the assembly is removed.

55 Undo the mounting screws and remove the ECU and mounting bracket *(see paragraph 68 in previous Section)*.

56 Disconnect the wiring connector and remove the ECU from the engine compartment.

57 Refitting is a reverse of the removal procedure ensuring that the wiring connector is securely reconnected. If a new ECU has been fitted, reprogramme it as follows. Turn the ignition on for a few seconds, then turn it off. Now remove the key to operate the immobiliser. After 10 seconds, the red immobiliser warning light should start to flash.

Renault Mégane & Scénic '99 – '02

22

Engine codes

1.4 litre models:	
SOHC	E7J
DOHC	K4J
1.6 litre models	K4M
2.0 litre models	F4R

System type

E7J engine	Siemens-Fenix 3 single-point injection with Bosch throttle body
K4J engine	Siemens-Sirius 32 sequential multi-point injection
K4M engine	Siemens-Sirius 32 sequential multi-point injection
F4R engine	Siemens-Sirius 32 sequential multi-point injection

Fuel system data

Fuel pressure regulator control pressure:	
E7J engine:	
Zero vacuum	3.0 ± 0.2 bars
500 mbars vacuum	2.5 ± 0.2 bars
K4J and K4M engines:	
With return circuit	3.0 ± 0.2 bars
Without return circuit	3.5 ± 0.2 bars
F4R engine	3.5 ± 0.2 bars

Fuel system data (continued)

Air temperature sensor resistance:	
E7J engine:	
At 0°C	7470 to 11 970 ohms
At 20°C	3060 to 4045 ohms
At 40°C	1315 to 1600 ohms
K4J and K4M engines:	
At 0°C	5290 to 6490 ohms
At 20°C	2400 to 2600 ohms
At 40°C	1070 to 1270 ohms
F4R engine:	
At -10°C	8525 to 10 450 ohms
At 25°C	1880 to 2120 ohms
At 50°C	760 to 860 ohms
Coolant temperature sensor resistance:	
E7J engine:	
At 20°C	2600 to 3000 ohms
At 40°C	1100 to 1300 ohms
At 80°C	270 to 300 ohms
At 90°C	200 to 215 ohms
K4J and K4M engines:	
At 20°C	3060 to 4045 ohms
At 40°C	1315 to 1600 ohms
At 80°C	300 to 370 ohms
At 90°C	210 to 270 ohms
F4R engine:	
At 25°C	2140 to 2360 ohms
At 50°C	770 to 850 ohms
At 80°C	275 to 290 ohms
At 110°C	112 to 117 ohms
Throttle potentiometer:	
E7J engine:	
Voltage	5.0 volts
Track resistance	4100 ± 800 ohms
Cursor resistance	1500 ± 150 ohms
K4J, K4M and F4R engines:	
Voltage	5.0 volts
Resistance:	
Track A-B	1250 ohms (no load), 1250 ohms (full load)
Track A-C	1245 ohms (no load), 2230 ohms (full load)
Track B-C	2230 ohms (no load), 1245 ohms (full load)
Stepper motor resistance:	
E7J engine	50.0 ohms
K4J, K4M and F4R engines	53.0 ± 5 ohms
Injector resistance:	
Single-point injection	2.0 ohms
Multi-point injection	14.5 ± 1.0 ohms
Crankshaft speed/position sensor resistance:	
E7J engine	220 ohms
K4J and K4M engines	200 to 270 ohms
F4R engine	200 to 270 ohms
Specified idle speed (non-adjustable)	750 ± 50 rpm

Torque wrench settings

	Nm	lbf ft
Fuel rail mounting bolts	9	7
Injector mounting shim	20	15
Knock sensor	20	15
Lambda (oxygen) sensor:		
K4J and K4M engines	45	33
F4R engine upstream	45	33
Throttle body:		
E7J, K4J and K4M engines	13	10
F4R engine	15	11

General information

Single-point injection system

1.4 litre models with the E7J engine are equipped with a Siemens-Fenix 3 single-point fuel injection/ignition system with a Bosch throttle body. The system incorporates a closed-loop catalytic converter and an evaporative emission control system, and complies with the latest emission control standards. The fuel injection side of the system operates as follows;

The fuel pump, immersed in the fuel tank, pumps fuel from the fuel tank to the fuel injector, via a filter mounted underneath the rear of the vehicle. Fuel supply pressure is controlled by the pressure regulator in the throttle body assembly. The regulator operates by allowing excess fuel to return to the tank.

The electrical control system consists of the ECU, along with the following sensors:

a) *Throttle potentiometer – informs the ECU of the throttle position, and the rate of throttle opening or closing.*
b) *Coolant temperature sensor – informs the ECU of engine temperature.*
c) *Inlet air temperature sensor – informs the ECU of the temperature of the air passing through the throttle body.*
d) *Lambda (oxygen) sensor – informs the ECU of the oxygen content of the exhaust gases.*
e) *Microswitch (built into idle speed stepper motor) – informs the ECU when the throttle is closed (ie, when the accelerator pedal is released).*
f) *Crankshaft speed/position (TDC) sensor – informs the ECU of engine speed and crankshaft position.*
g) *Power steering pressure switch – informs the ECU when the power steering pump is working so the engine idle speed can be increased to prevent stalling.*
h) *Knock sensor (where fitted) – informs the ECU when pre-ignition ('pinking') is occurring.*
i) *Manifold absolute pressure (MAP) sensor – informs the ECU of the engine load by monitoring the pressure in the inlet manifold.*
j) *Fuel vapour recirculation valve – operates the fuel evaporative control system.*

All the above information is analysed by the ECU and, based on this, the ECU determines the appropriate ignition and fuelling requirements for the engine. The ECU controls the fuel injector by varying its pulse width – the length of time the injector is held open – to provide a richer or weaker mixture, as appropriate. The mixture is constantly varied by the ECU, to provide the best setting for cranking, starting (with either a hot or cold engine), warm-up, idle, cruising, and acceleration.

The ECU also has full control over the engine idle speed, via a stepper motor which is fitted to the throttle body. The motor pushrod rests against a cam on the throttle spindle. When the throttle is closed (accelerator pedal released), the ECU uses the motor to vary the opening of the throttle valve and so control the idle speed.

The ECU also controls the exhaust and evaporative emission control systems, which are described in detail in Part C of this Chapter.

If there is an abnormality in any of the readings obtained from either the coolant temperature sensor, the inlet air temperature sensor or the lambda (oxygen) sensor, the ECU enters its back-up mode. In this event, the ECU ignores the abnormal sensor signal, and assumes a pre-programmed value which will allow the engine to continue running (albeit at reduced efficiency). If the ECU enters this back-up mode, the warning light on the instrument panel will come on, and the relevant fault code will be stored in the ECU memory.

A fuel cut-off inertia switch is incorporated into the fuel injection system. In the event of an impact the switch cuts off the electrical supply to the fuel pump and so prevents fuel being expelled should the fuel pipes/hoses be damaged in an accident.

Multi-point injection system

1.4 litre models with the K4J engine, and all 1.6 litre and 2.0 litre engines are equipped with a Siemens-Sirius sequential multi-point fuel injection/ignition system. The system is of closed-loop type incorporating two lambda (oxygen) sensors, one located upstream and the other downstream of the catalytic converter. An evaporative emission control system is fitted.

The multi-point injection system uses one injector and one ignition coil for each cylinder, and the injectors are operated individually and sequentially at the beginning of the inlet stroke. The electronic control unit (ECU) is able to determine which cylinder is on its inlet stroke without the use of a camshaft position sensor, however if the unit is renewed, the car must be taken for a road test lasting at least 25 minutes to enable the ECU to reprogram itself; the stepper motor must also be reset.

The system incorporates a closed-loop catalytic converter and an evaporative emission control system, and complies with the latest emission control standards. The fuel injection side of the system operates as follows;

The fuel pump, immersed in the fuel tank, pumps fuel from the fuel tank to the fuel rail, via a filter mounted beneath the rear of the vehicle. Fuel supply pressure is controlled by the pressure regulator in the fuel rail (except F4R engine) or fuel pump (F4R engine). The regulator operates by allowing excess fuel to return to the tank. There are four injectors (one per cylinder) located in the inlet manifold downstream of the throttle valve. All the injectors are fed from the fuel rail.

The electrical control system consists of the ECU, along with the following sensors:

a) *Throttle potentiometer – informs the ECU of the throttle position, and the rate of throttle opening or closing.*
b) *Coolant temperature sensor – informs the ECU of engine temperature.*
c) *Inlet air temperature sensor – informs the ECU of the temperature of the air passing through the throttle body.*
d) *Lambda (oxygen) sensor – informs the ECU of the oxygen content of the exhaust gases.*
e) *Idle speed regulation stepper motor – controls the idle speed.*
f) *Crankshaft speed/position (TDC) sensor – informs the ECU of engine speed and crankshaft position.*
g) *Power steering pressure switch – informs the ECU when the power steering pump is working so the engine idle speed can be increased to prevent stalling.*
h) *Knock sensor (where fitted) – informs the ECU when pre-ignition ('pinking') is occurring.*
i) *Manifold absolute pressure (MAP) sensor – informs the ECU of the engine load by monitoring the pressure in the inlet manifold.*
j) *Fuel vapour recirculation valve – operates the fuel evaporative control system.*

All the above information is analysed by the ECU and, based on this, the ECU determines the appropriate ignition and fuelling requirements for the engine. The ECU controls the fuel injector by varying its pulse width – the length of time the injector is held open – to provide a richer or weaker mixture, as appropriate. The mixture is constantly varied by the ECU, to provide the best setting for cranking, starting (with either a hot or cold engine), warm-up, idle, cruising, and acceleration. On automatic transmission models, information from sensors on the transmission is sent to the ECU for processing to determine the most efficient settings for the engine.

The ECU also has full control over the engine idle speed, via a stepper motor which is fitted to the throttle body. The motor pushrod rests against a cam on the throttle spindle. When the throttle is closed (accelerator pedal released), the ECU uses the motor to vary the opening of the throttle valve and so control the idle speed.

The ECU also controls the exhaust and evaporative emission control systems.

If there is an abnormality in any of the readings obtained from either the coolant temperature sensor, the inlet air temperature sensor or the lambda (oxygen) sensor, the ECU enters its back-up mode. In this event, the ECU ignores the abnormal sensor signal, and assumes a pre-programmed value which will allow the engine to continue running (albeit at reduced efficiency). If the ECU enters this back-up mode, the warning light on the instrument panel will come on, and the relevant fault code will be stored in the ECU memory.

A fuel cut-off inertia switch is incorporated into the fuel injection system. In the event of an impact the switch cuts off the electrical supply to the fuel pump and so prevents fuel being expelled should the fuel pipes/hoses be damaged in an accident. The switch is located in the left-hand rear corner of the engine compartment.

Fuel injection system – depressurisation

Note: *The fuel system referred to in this Section includes the tank-mounted fuel pump, the fuel filter, the fuel injector(s) and the pressure regulator in the throttle body/fuel rail, and the metal pipes and flexible hoses of the fuel lines between these components. All these contain fuel which will be under pressure while the engine is running, and/or while the ignition is switched on. The pressure will remain for some time after the ignition has been switched off, and it must be relieved when any of these components are disturbed for servicing work.*

Method 1

1 Disconnect the battery negative lead.

2 Place a suitable container beneath the connection or union to be disconnected, and have a large rag ready to soak up any escaping fuel not being caught by the container.

3 Slowly loosen the connection or union nut to avoid a sudden release of pressure, and position the rag around the connection, to catch any fuel spray which may be expelled. Once the pressure is released, disconnect the fuel line. Plug the pipe ends, to minimise fuel loss and prevent the entry of dirt into the fuel system.

Method 2

4 Remove the fuel pump relay located in the engine compartment fuse/relay box.

5 Start the engine and allow it to idle until it stops due to lack of fuel. Operate the starter motor a couple more times, to ensure that all fuel pressure has been relieved.

6 Switch off the ignition and refit the fuel pump relay.

Single-point fuel injection system components – removal and refitting

Fuel injector

Note: *If a faulty injector is suspected, before condemning the injector, it is worth trying the effect of one of the proprietary injector-cleaning treatments.*

1 Remove the inlet air temperature sensor as described later in this Section.

2 Note its fitted position, then lift out the injector and recover its sealing rings.

3 Refitting is a reversal of the removal procedure ensuring that the injector sealing rings and injector cap O-ring are in good condition. When refitting the injector cap ensure that the injector pins are correctly aligned with the cap terminals; the terminals are marked + and – for identification.

Fuel pressure regulator

4 The fuel pressure regulator is an integral part of the upper half of the throttle body. If it is defective, the complete throttle body assembly must be renewed.

Inlet air temperature sensor

5 Remove the air cleaner assembly.

6 Disconnect the sensor multi-plug.

7 Remove the single securing screw, and free the sensor cover.

8 Free the connector lugs, and withdraw the sensor together with the cover, wiring and connector.

9 Refit by reversing the removal operations.

Idle speed control stepper motor

10 Remove the air cleaner assembly.

11 Remove the throttle body securing screws. Carefully lift the throttle body to improve access to the motor.

12 Disconnect the motor multi-plug.

13 Remove the three mounting screws and withdraw the motor.

14 Refit by reversing the removal operations, then perform an initial setting operation as follows.

15 Place a shim or feeler gauges of approximately 5 mm thickness between the throttle quadrant and the motor plunger. Switch the ignition on for a few seconds, then switch it off again. Remove the shim or gauges, then switch on and off again.

16 Run the engine and check that the idle speed and quality are satisfactory.

Throttle potentiometer

17 The throttle potentiometer is an integral part of the lower half of the throttle body. If it is defective, the complete throttle body assembly must be renewed. The idle speed control motor can be transferred from the old half-body to the new one.

18 After renewal, the initial setting of the potentiometer must be verified using the Renault XR25 diagnostic tester.

Coolant temperature sensor

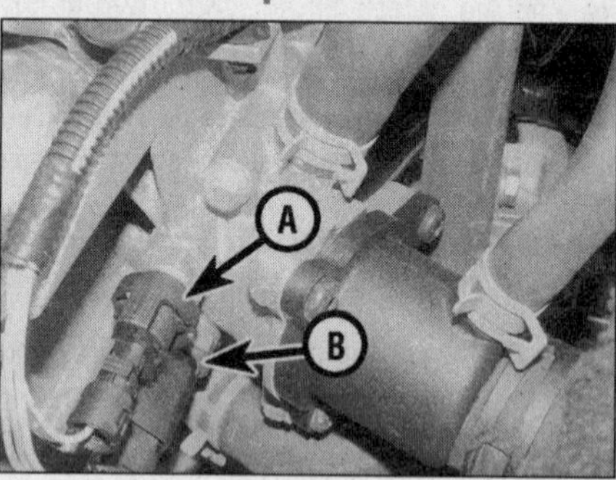

Temperature gauge sender (A) and injection system temperature sensor (B)

19 The coolant temperature sensor is located on the front of the thermostat housing. Drain the coolant system.

20 Disconnect the wiring plug, and unscrew the sensor.

21 Refitting is a reversal of removal. If a sealing ring was originally fitted, renew it. If no sealing ring was fitted, apply a little sealant to the sensor threads prior to refitting.

Electronic control unit (ECU)

22 The ECU is located in the front right-hand corner of the engine compartment. To remove the ECU first disconnect the battery negative lead.

23 Unbolt the bracket from the top of the ECU and release the strap. Alternatively the bracket can remain on the ECU until the assembly is removed.

24 Undo the mounting screws and remove the ECU and mounting bracket, then disconnect the wiring connector and remove the ECU from the engine compartment.

25 Refitting is a reverse of the removal procedure ensuring that the wiring connector is securely reconnected.

Manifold absolute pressure sensor

26 The sensor is mounted on the left-hand side of the engine compartment bulkhead.

27 Disconnect the wiring from the sensor.

28 Unscrew the mounting nuts and remove the sensor.

29 Refitting is a reversal of removal.

Relays

30 The fuel injection system and fuel pump relays are located in the engine compartment fuse/relay box.

31 Remove the cover from the box.

32 Remove the relevant relay from the fuse/relay box.

33 Refitting is the reverse of removal.

Crankshaft speed/position (TDC) sensor

34 The sensor is mounted on the top of the transmission housing at the left-hand end of the cylinder block.

35 To remove the sensor, disconnect the battery negative terminal then remove the air cleaner housing.

36 Trace the wiring back from the sensor to the wiring connector, and disconnect it from the main harness.

37 Unscrew the retaining bolts and remove the sensor.

38 Refitting is a reverse of the removal procedure. Ensure that the sensor retaining bolts are securely tightened – note that only the special shouldered bolts originally fitted must be used to secure the sensor; these bolts locate the sensor precisely to give the correct air gap between the sensor tip and the flywheel/driveplate.

Fuel cut-off inertia switch

39 The fuel cut-off inertia switch is located in the left-hand side of the engine compartment.

40 Unscrew and remove the switch retaining screws then disconnect its wiring connector and remove the switch from the engine compartment.

41 Refitting is the reverse of removal. On completion, reset the switch by depressing its button.

Power steering pressure switch

42 The pressure switch is screwed into the power steering feed pipe from the pump to the steering gear.

43 To remove the switch, disconnect the wiring connector.

44 Wipe clean the area around the switch then unscrew the switch and remove it from the pipe. Plug the pipe aperture to prevent excess fluid leakage and prevent dirt entry into the hydraulic system.

45 Refitting is the reverse of removal. On completion check the power steering fluid level.

Knock sensor (where fitted)

46 On the E7J engine, then knock sensor is located on the right-hand rear of the cylinder block. On K4J, K4M and F4R engines, it is located on the front of the cylinder block.

47 To remove the sensor, disconnect the wiring plug and unscrew it from the cylinder block.

48 Refitting is a reversal of removal, tightening the sensor to the specified torque.

Multi-point fuel injection system components – removal and refitting

Fuel rail and injectors

Note: *If a faulty injector is suspected, before condemning the injector, it is worth trying the effect of one of the proprietary injector-cleaning treatments.*

K4J and K4M engines

49 Depressurise the fuel system, then disconnect the battery negative lead.

50 Remove the cover for access to the fuel rail.

51 Disconnect the fuel feed and return hoses from the fuel rail.

52 Disconnect the vacuum hose connecting the pressure regulator (on the fuel rail) to the inlet manifold.

53 Disconnect the wiring from the injectors and move the loom to one side.

54 Unscrew and remove the mounting bolts and carefully ease the fuel rail together with the injectors from the inlet manifold.

55 Note the fitted positions of the injectors, then remove the clips and ease the injectors from the fuel rail.

56 Remove the sealing rings from the grooves at each end of the injectors and obtain new ones.

57 Refitting is a reversal of the removal procedure, noting the following points:

a) Renew all sealing rings, using a smear of engine oil to aid installation.

b) Refit the fuel rail assembly to the manifold, making sure the sealing rings remain correctly positioned, and tighten the retaining bolts to the specified torque.

c) On completion start the engine and check for fuel leaks.

F4R engine

58 Depressurise the fuel system, then disconnect the battery negative lead.

59 Remove the cover for access to the fuel rail. Disconnect the fuel feed hose and, where applicable, the return hose from the fuel rail. Also disconnect the vacuum hose connecting the pressure regulator (on the fuel rail) to the inlet manifold.

60 Remove the auxiliary drivebelt, then unbolt the power steering pump pulley, unscrew the pump mounting bolts and move the pump to one side without disconnecting the hydraulic fluid hoses.

61 Disconnect the wiring from the injectors and move the loom to one side.

62 Unscrew and remove the mounting bolts and carefully ease the fuel rail together with the injectors from the inlet manifold.

63 Note the fitted positions of the injectors, then remove the clips and ease the injectors from the fuel rail.

64 Remove the sealing rings from the grooves at each end of the injectors and obtain new ones.

65 To remove the lower inlet manifold, first remove the upper inlet manifold, then unbolt the lower manifold from the cylinder head and recover the gasket.

66 Refitting is a reversal of the removal procedure, noting the following points:

a) Fit a new gasket when refitting the lower inlet manifold, but use a straight-edge to align the right-hand end of the manifold with the cylinder head.

b) Renew all sealing rings, using a smear of engine oil to aid installation.

c) Refit the fuel rail assembly to the manifold, making sure the sealing rings remain correctly positioned, and tighten the retaining bolts to the specified torque.

d) On completion start the engine and check for fuel leaks.

Fuel pressure regulator

67 Disconnect the vacuum pipe from the regulator.

68 Place a wad of rag over the regulator to catch any spilled fuel, then extract the retaining spring and ease the regulator from the fuel rail.

69 Remove the sealing rings from the grooves in the pressure regulator and obtain new ones.

70 On refitting, fit new sealing rings to the regulator grooves and apply a smear of engine oil to them to ease installation. Ease the regulator back into the end of the fuel rail and refit the retaining spring and vacuum pipe.

Throttle potentiometer

71 Remove the throttle housing.

72 Undo the retaining screws and remove the potentiometer from the throttle housing.

73 Refitting is a reverse of the removal procedure ensuring that the potentiometer is correctly engaged with the throttle spindle. **Note:** *Renault recommend that the potentiometer operation should be checked, whenever it is disturbed, using the XR25 diagnostic tester.*

Inlet air temperature sensor

74 Unscrew and remove the inlet air temperature sensor from the air inlet duct or air cleaner body as applicable.

75 Refitting is a reversal of removal.

Coolant temperature sensor

76 The sensor is located on the thermostat housing at the left-hand end of the cylinder head. Refer to paragraphs19 to 21 for removal and refitting details.

Power steering pressure switch

77 The switch is screwed into the feed pipe from the power steering pump to the steering gear.

78 To remove the switch, disconnect the wiring connector.

79 Wipe clean the area around the switch then unscrew the switch and remove it from the pipe. Plug the pipe aperture to prevent excess fluid leakage and prevent dirt entry into the hydraulic system.

80 Refitting is the reverse of removal. On completion check the power steering fluid level.

Knock sensor (where fitted)

81 Refer to paragraphs 46 to 48.

Idle speed control stepper motor

82 The idle speed control stepper motor is mounted on the top of the throttle housing. To remove it first remove the throttle housing.

83 Undo the retaining screws and remove the stepper motor from the throttle housing. Recover the gasket and discard it; a new one should be used on refitting.

84 Refitting is a reversal of the removal procedure using a new gasket.

Manifold absolute pressure sensor

85 The pressure sensor is located on the left-hand end of the inlet manifold. Where necessary, remove the air cleaner assembly for access to the sensor.

86 Disconnect the wiring from the sensor.

87 Unscrew and remove the sensor.

88 Refitting is a reversal of removal.

Relays

89 The fuel injection system and fuel pump relays are located in the engine compartment fuse/relay box.

90 Remove the cover from the box.

91 Remove the relevant relay from the fuse/relay box.

92 Refitting is the reverse of removal.

Crankshaft speed/position (TDC) sensor

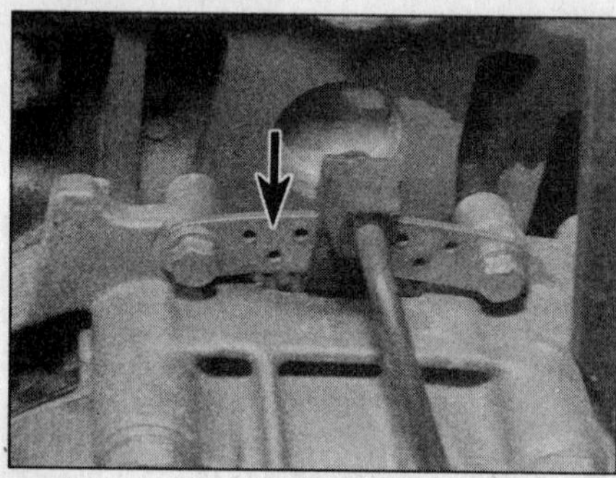

93 The sensor is mounted on the top of the transmission bellhousing at the left-hand end of the cylinder block.

94 To remove the sensor, disconnect the battery negative terminal then, remove the air cleaner housing.

95 Trace the wiring back from the sensor to the wiring connector, and disconnect it from the main harness.

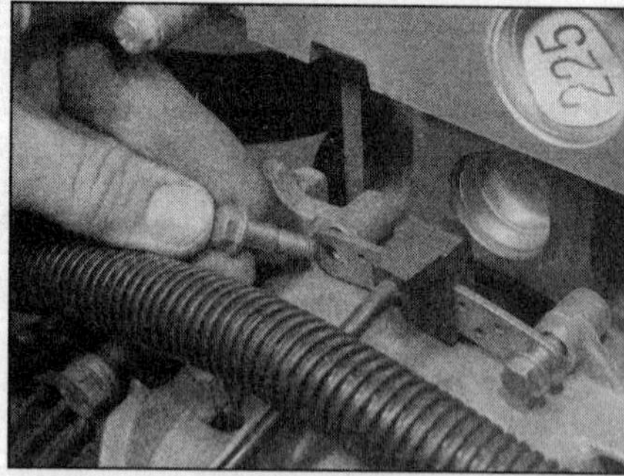

96 Unscrew the retaining bolts and remove the sensor.

97 Refitting is a reverse of the removal procedure. Ensure that the sensor retaining bolts are securely tightened – note that only the special shouldered bolts originally fitted must be used to secure the sensor; these bolts locate the sensor precisely to give the correct air gap between the sensor tip and the flywheel/driveplate.

Fuel cut-off inertia switch

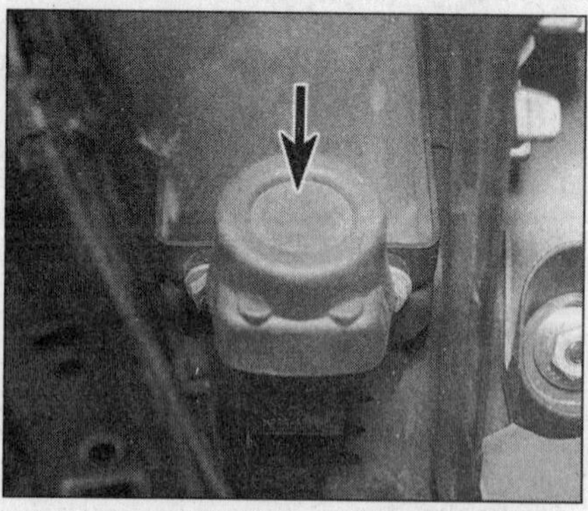

98 The switch is located in the left-hand side of the engine compartment.

99 Unscrew and remove the switch retaining screws then disconnect its wiring connector and remove the switch from the engine compartment.

100 Refitting is the reverse of removal. On completion, reset the switch by depressing its button.

Electronic control unit (ECU)

Note: *The ECU is electronically-coded to match the engine immobiliser. If the ECU is being removed to enable a new unit to be fitted, the new unit must be programmed with the vehicle code as described.*

101 The ECU is located in the left-hand side of the engine compartment. First disconnect the battery negative lead (refer to Chapter 5A).

102 Where applicable, unbolt the bracket from the top of the ECU and release the strap. Alternatively the bracket can remain on the ECU until the assembly is removed.

103 Undo the mounting screws and remove the ECU and mounting bracket.

104 Disconnect the wiring connector and remove the ECU from the engine compartment.

105 Refitting is a reverse of the removal procedure ensuring that the wiring connector is securely reconnected. If a new ECU has been fitted, reprogramme it as follows. Turn the ignition on for a few seconds, then turn it off. Now remove the key to operate the immobiliser. After 10 seconds, the red immobiliser warning light should start to flash.

Rover 25 & MG ZR '99 – '03

23

Petrol models

General

System type	Rover modular engine management system (MEMS 3) indirect multi-point fuel injection system integrated with ignition system

Fuel pump

Type	Electric roller vane pump in fuel tank
Maximum delivery pressure	4.1 bar
Regulated injection pressure	3.0 ± 0.2 bar

Idle settings (not adjustable – for reference only)

Idle speed	825 ± 50 rpm
Idle mixture (CO content)	Less than 0.5 %

Diesel models

General

System type	Electronic diesel control (EDC) system direct diesel injection system

Idle speed (not adjustable – for reference only)

Models without intercooler	850 ± 50 rpm
Models with intercooler	805 ± 50 rpm

Torque wrench settings

	Nm	lbf ft
Petrol models		
Crankshaft position sensor-to-flywheel housing bolt	6	4
Engine control module (ECM)-to-bracket nuts	4	3
Engine coolant temperature (ECT) sensor to housing	15	11
Fuel rail-to-manifold bolts	5	4
Heated oxygen sensor (HO_2S) to manifold/exhaust system	55	41
Idle air control valve (IACV)-to-manifold screws	6	4
Intake air temperature (IAT) sensor to manifold	7	5
Throttle position sensor securing screws	1.5	1.1
Diesel models		
Crankshaft position (CKP) sensor bolt	8	6
Engine control module (ECM) bolts	4	3
Engine coolant temperature (ECT) sensor	15	11
Fuel shut-off solenoid	20	15
Intake air temperature (IAT) sensor	12	9
Manifold absolute pressure (MAP) sensor bolt	9	7
Throttle position sensor nuts	4	3

Chapter 23

General information

Petrol models

The amount of fuel delivered, and the injection timing is controlled by the Modular Engine Management System (MEMS 3), which also controls the ignition system.

Fuel is supplied from the rear-mounted fuel tank by an electric pump mounted in the tank, via a fuel filter, to the fuel rail. A fuel pressure regulator mounted on the fuel rail maintains a constant fuel pressure to the fuel injectors. Excess fuel is returned from the regulator to the tank.

The fuel rail acts as a reservoir for the four fuel injectors, which inject fuel into the cylinder inlet tracts, upstream of the inlet valves. The injectors operate in pairs – the injectors for Nos 1 and 4 cylinders operate simultaneously, as do the injectors for Nos 2 and 3 cylinders.

The injectors are electronically controlled by the Engine Control Module (ECM). Each injector has a solenoid which, when energised by a pulse from the ECM, lifts the injector needle and fuel is sprayed into the cylinder.

The duration of the pulse determines the length of time the needle is lifted and, therefore, the amount of fuel injected. Pulse duration is calculated by the ECM, and is constantly varying to match engine operating and ambient conditions.

The fuelling, and ignition timing, calculation is based on information supplied to the ECM by the following sensors :

a) Throttle position sensor (TP) – mounted on the throttle body, the sensor provides a signal representing the amount and rate of throttle opening – ie, driver demand.

b) Crankshaft position sensor (CKP) – supplies a signal representing the speed and angular position of the crankshaft.

c) Camshaft position sensor (CMP) – attached to the camshaft cover, the sensor provides a signal representing the angular position of the camshaft, in relation to the angular position of the crankshaft.

d) Manifold absolute pressure (MAP) – located in the intake manifold, supplies a signal representing the amount of air entering the cylinders.

e) Engine coolant temperature (ECT) – located in the coolant outlet elbow from the cylinder head. The sensor provides a coolant temperature signal.

f) Engine oil temperature – located on the filter housing, the sensor supplies an oil temperature signal.

g) Intake air temperature (IAT) – located on the intake manifold, the sensor provides a signal

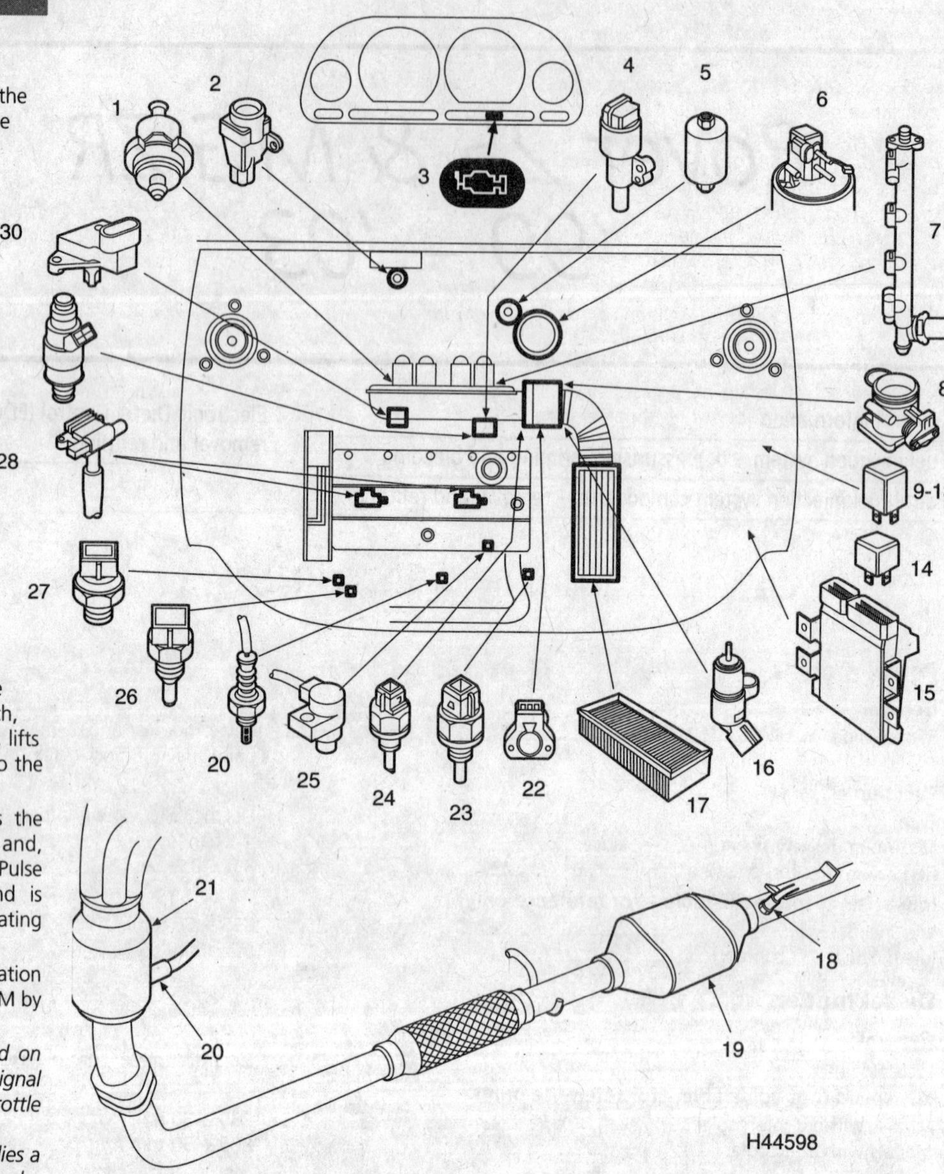

Engine management system components

1 Fuel pressure regulator
2 Inertia switch
3 Malfunction indicator lamp
4 Idle Air Control valve – IAC
5 Fuel filter
6 Purge valve and canister
7 Fuel rail
8 Throttle housing
9-13 Relays (in fusebox)
Main relay
Air conditioning relay
Cooling fan relay
14 Cooling fan relays – air conditioning only
15 Engine Control Module – ECM
16 Crankshaft position sensor – CKP
17 Air filter
*18 Downstream oxygen sensor – HO_2S**
19 Catalytic converter
20 Upstream oxygen sensor – HO_2S
21 Pre-catalyst
22 Throttle position sensor – TP
23 Engine Coolant Temperature sensor – ECT
24 Intake Air Temperature sensor – IAT
25 Camshaft Position sensor – CMP
26 Engine oil temperature sensor
27 Engine oil pressure switch
28 Ignition coils
29 Injectors x 4
30 Manifold Absolute Pressure sensor – MAP
** Certain models only*

representing the temperature of the air entering the engine.

h) Heated oxygen sensors (HO_2S) – located in the exhaust manifold, upstream of the catalytic converter. (Some models have a second sensor located in the exhaust pipe, downstream of the converter.) The sensors provide signals representing the amount of oxygen in the exhaust gas, and allow the ECM to precisely control the air/fuel ratio around the ideal combustion ratio of 14.7:1.

Idle speed is controlled by the ECM via an Idle Air Control Valve (IACV), which varies the amount of air bypassing the throttle valve.

Failure of certain sensors – either complete loss of signal, or sending an abnormal signal, will cause the ECM to default to preprogrammed values. In most cases this will be noticeable as poor or erratic engine performance.

Note that, if the crankshaft position (CKP) sensor signal is lost, the engine will not run.

Whenever a sensor fault occurs, a fault code is stored in a memory of the ECM. Using a fault code reader, the codes can be extracted.

All vehicle are fitted with a fuel inertia cut-off switch, which cuts off the fuel supply in the event of the vehicle being involved in an impact. The engine cannot be started if the switch is activated – the switch must be reset manually.

Diesel models

The amount of fuel delivered, and the injection timing is controlled by the Electronic Diesel Control system (EDC) which also controls the preheat system, Exhaust Gas Recirculation system (EGR) and fuel cut-off solenoid for engine stop.

The fuelling, and injection timing, calculation is based on information supplied to the ECM by the following sensors:

a) Crankshaft position sensor (CKP) sends signal representing the crankshaft speed and position.

b) Fuel injector needle lift sensor. Incorporated in number one injector, sends signal representing the start of No 1 cylinder injection sequence.

c) Engine coolant temperature (ECT).

d) Manifold absolute pressure (boost pressure) sensor (MAP), sends signal representing the pressure of the air entering the engine, The signal is used in conjunction with the intake air temperature sensor, to calculate the volume of oxygen in the air.

e) Road speed sensor, sends a signal representing the vehicle road speed, which operates the speedometer, and provides a signal to the ECM for idle stabilisation.

f) Throttle pedal position sensor (TP) The sensor is a potentiometer, which provides an electrical signal to the ECM representing the amount and rate of throttle opening – ie, driver demand.

g) Throttle control is 'drive by wire'. The sensor is connected to the pedal by a Bowden cable, and to the ECM by electrical cable.

h) Brake pedal switch, when closed by brake application, sends brake-applied signal.

i) Intake air temperature/Mass airflow sensor (IAT/MAF), supplies separate signals representing the mass air flow and intake air temperature.

Signals from these sensors are constantly updated and processed by the ECM. The resultant calculations are compared with preprogrammed 'maps' which provide datum points for the amount of fuel required and injection timing, for many combinations of engine operating conditions.

Failure of certain sensors – either complete loss of signal, or sending an abnormal signal, will cause the ECM to default to preprogrammed values. In most cases this will be noticeable as poor or erratic engine performance.

Whenever a sensor fault occurs, the engine malfunction warning lamp, on the instrument panel, is lit, and a fault code is stored in a memory of the ECM. Use a fault code reader to extract the codes.

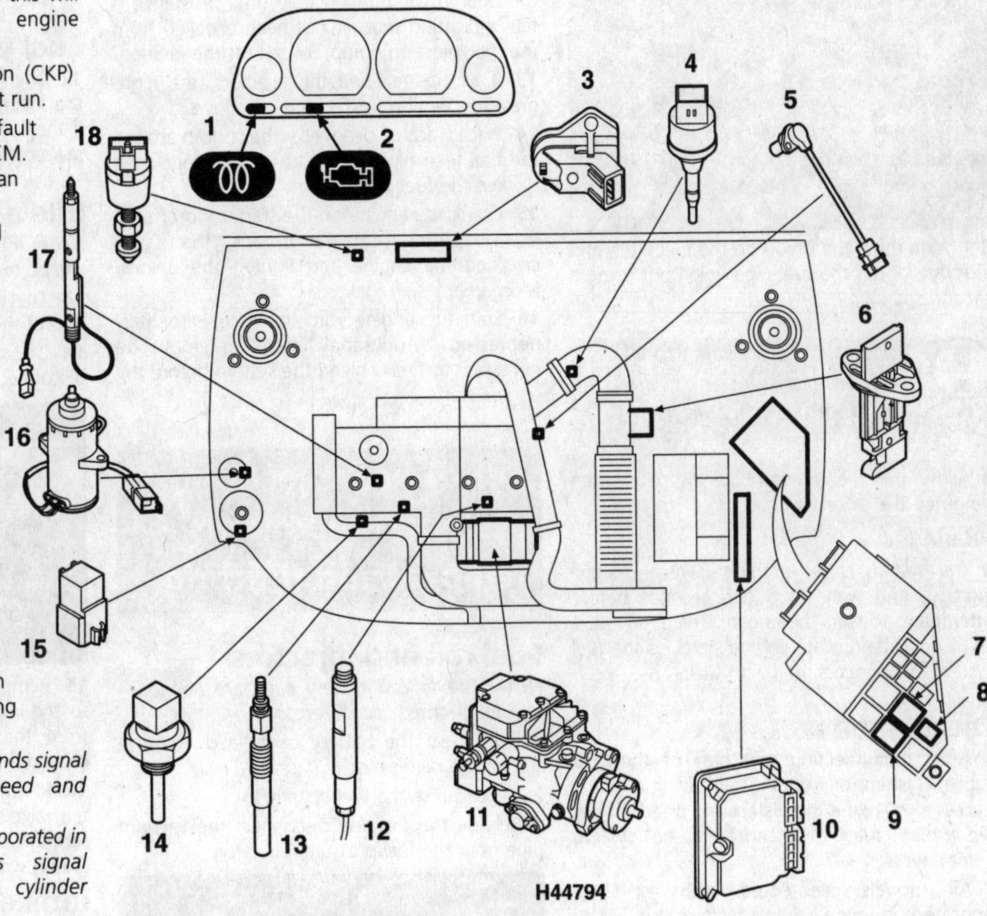

Electronic Diesel Control (EDC) components

1 Glow plug warning lamp
2 Engine malfunction lamp
3 Manifold Absolute Pressure sensor – MAP (turbo boost pressure)
4 Road speed sensor
5 Crankshaft Position sensor – CKP
6 Intake air temperature/Mass Air Flow sensor – IAT/MAF
7 Main relay
8 Air conditioning compressor clutch relay
9 Cooling fan relay
10 Engine Control Module – ECM
11 Fuel Injection Pump – FIP
12 Standard fuel injector x 3 (Nos. 2, 3 & 4)
13 Glow plug x 4
14 Engine Coolant Temperature sensor – ECT
15 Glow plug relay
16 Throttle Position sensor – TP
17 Fuel injector with needle lift sensor (No. 1)
18 Brake pedal switch

Fuel injection system – depressurising, priming and bleeding

Petrol models

Depressurising

1 Disconnect the battery negative lead.

2 Position a wad of absorbent cloth around the outlet pipe union on the top of the fuel filter.

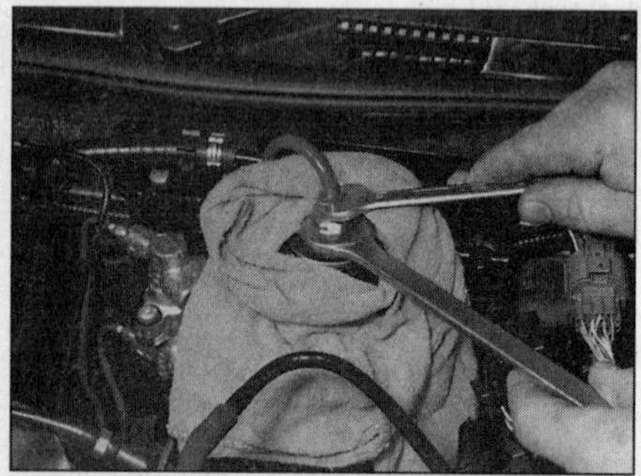

3 Counterhold the union on the fuel filter, and **slowly** slacken the fuel pipe union nut using a second spanner.

Warning: Be prepared for the escape of some fuel. Take adequate fire precautions.

4 Allow the fuel pressure to dissipate, then retighten the union nut.

Priming

5 The system is self-priming. Switch on the ignition, and wait for a few seconds before attempting to start the engine. There may be a short delay before the engine starts, as the fuel system components refill with fuel.

Diesel models

6 After disconnecting any part of the fuel supply system or running out of fuel, it is necessary to prime the system and bleed off any air which may have entered the system components.

7 All models are fitted with a hand-operated priming pump, consisting of a rubber bulb located at the rear of the engine compartment in the fuel supply line to the fuel filter, and a bleed screw on top of the fuel filter.

8 To prime the system, proceed as follows.

9 If none of the fuel system components downstream of the fuel filter (including the fuel filter itself) have been disturbed, and the vehicle has not run out of fuel, it should be possible to proceed directly to paragraph 7.

10 Place a wad of absorbent cloth around the bleed screw on top of the fuel filter, then slacken the bleed screw (1).

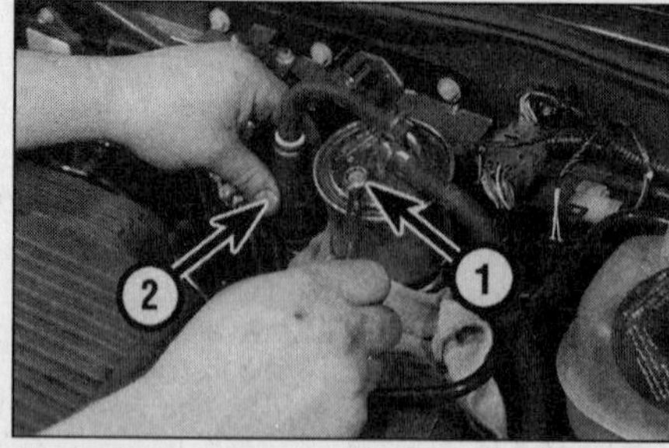

11 Switch on the ignition so that the stop solenoid is energised, then pump the rubber priming bulb (2) until fuel free from air bubbles emerges from the bleed screw. Tighten the bleed screw.

12 If not already done, switch on the ignition, then pump the priming bulb until resistance is felt, indicating that air has been expelled from the fuel injection pump. Switch off the ignition.

13 If air has reached the injectors, the high-pressure circuit must be bled as follows.

14 Place wads of clean, absorbent cloth around the fuel feed pipe unions at the injectors, then slacken the fuel pipe unions.

15 Crank the engine on the starter motor until fuel emerges from the unions, then stop cranking the engine and tighten the unions. Mop up any spilt fuel.

16 Start the engine with the accelerator fully depressed. Additional cranking may be necessary to finally bleed the system before the engine starts.

Petrol fuel injection system components – removal and refitting

Fuel rail and injectors

Note: *The following new parts are needed – injector O-rings, silicone grease.*

1 Disconnect the battery earth lead. Remove the engine oil dipstick,

2 Depressurise the fuel system.

3 Release the clip and disconnect the vacuum pipe from the fuel pressure regulator,

4 Pack absorbent cloth under the fuel rail. Disconnect the fuel supply hose from the fuel rail. Plug the pipe connections.

5 Remove two bolts securing the fuel rail to the inlet manifold.

6 Release the fuel rail and injectors from the inlet manifold, and remove the injector spacer.

7 Disconnect the multi-plugs from the injectors.

8 Remove the fuel rail complete with injectors.

9 Release the spring clips securing the injectors to the fuel rail, and remove the injectors.

10 Remove and discard two O-rings from each injector. Cover both ends of each injector with a protective cap.

11 Refitting is a reversal of the removal procedure, noting:

Lubricate new O-rings with silicone grease and fit them to the injectors.

Use a torque wrench to tighten the fuel rail bolts to the specified torque.

Fuel pressure regulator

12 The fuel pressure regulator is integral with the fuel rail, and cannot be removed separately. If the fuel regulator is faulty, the complete fuel rail/regulator assembly must be renewed.

Idle air control valve

Note: *A new O-ring and suitable silicone grease will be required on refitting.*

13 The idle air control valve is located at the rear of the engine.

14 Disconnect the battery negative lead, then disconnect the wiring plug from the valve.

15 Remove the two screws securing the valve to the inlet manifold, then withdraw the valve from the manifold. Recover the O-ring and discard it.

16 Refitting is a reversal of removal, tightening the valve securing screws to the specified torque setting. Use a new O-ring and lubricate the O-ring with a little silicone grease.

Intake air temperature sensor

17 The IAT sensor is located at the left-hand rear of the inlet manifold under the throttle cam assembly.

18 Disconnect the multi-plug from the sensor.

19 Unscrew the sensor from the inlet manifold.

20 Clean the sensor threads and mating faces.

21 Fit the sensor to the inlet manifold and tighten the specified torque.

22 Connect the multi-plug, and battery earth lead.

Engine coolant temperature sensor

Note: *Loctite 577 is needed to seal the sensor.*

23 The ECT sensor is screwed into the side of the coolant outlet elbow, and is connected to the engine harness via a multi-plug. **Note:** *The temperature gauge sensor is screwed into the underside of the coolant outlet elbow.*

24 Disconnect the battery earth lead.

25 Disconnect the multi-plug from the ECT sensor.

26 Locate a container under the sensor to collect coolant. Unscrew the sensor.

27 Refitting is a reversal of the removal procedure, noting the following points:

a) Clean and degrease the sensor threads, and apply a smear of Loctite 577.

b) Use a torque wrench to tighten the sender to the specified torque.

c) Top-up the coolant level.

Throttle position sensor

Note: *Whenever the sensor is removed or renewed, a 'throttle initialisation' procedure must be carried out. This can only be done by an MG Rover dealer or suitably equipped specialist, using special electronic equipment.*

Note: *The following new parts are needed – two Torx head screws and wave washers.*

28 Disconnect the battery earth lead.

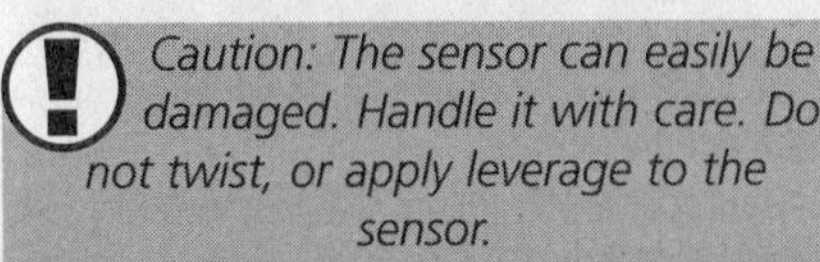

Caution: The sensor can easily be damaged. Handle it with care. Do not twist, or apply leverage to the sensor.

29 Disconnect the multi-plug from the sensor.

30 Remove and discard two Torx head screws and wave washers securing the sensor to the throttle housing.

31 Remove the sensor specification plate, and pull the sensor from the throttle spindle.

32 Clean the mating faces of the throttle housing and sensor.

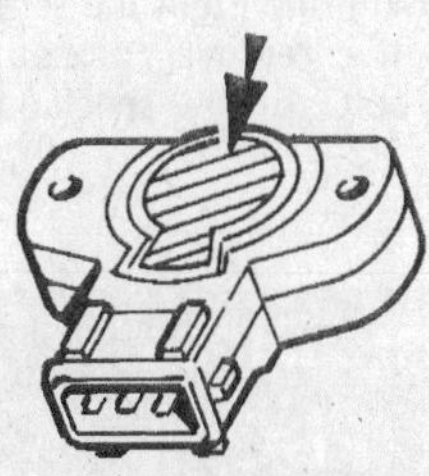

H28712

33 Align the flat on the sensor with the flat on the throttle spindle, and push the sensor onto the spindle.

Caution: When pushing the sensor onto the throttle spindle, use fingers only, and apply pressure the shaded area only.

Caution: Do not rotate the sensor in clockwise direction. Rotate only anti-clockwise, and do not rotate beyond the sensor internal stops.

34 Rotate the sensor in an anti-clockwise direction to align the screw holes.

35 Fit the specification plate, wave washers and screws.

36 Tighten the screws to the specified torque – DO NOT exceed the specified torque.

37 Connect the multi-plug to the sensor.

38 Check that throttle operation is correct.

39 Connect the battery earth lead.

40 Arrange to have throttle initialisation carried out by an MG Rover dealer or suitably equipped specialist.

Manifold Absolute Pressure sensor

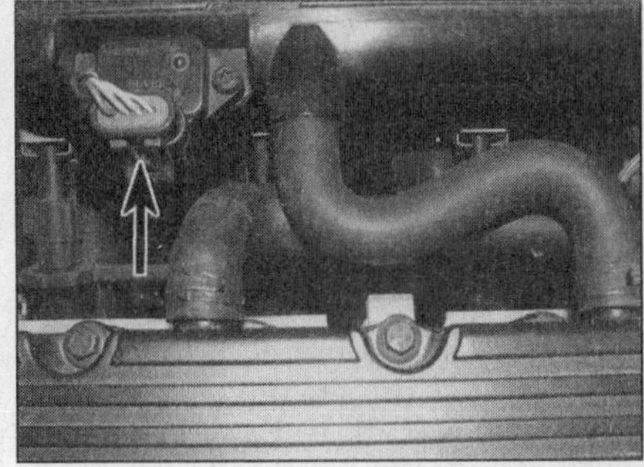

41 The sensor is located on the forward face of the inlet manifold.

42 Disconnect the multi-plug from the sensor.

43 Remove two Torx head screws, and remove the sensor from the inlet manifold.

44 Clean the sensor and manifold mating faces.

45 Fit the sensor to the manifold, and secure with two Torx head screws.

46 Connect the multi-plug to the sensor.

Heated oxygen sensor

Note: *The following new parts are needed – sensor sealing washer, exhaust manifold gasket.*

47 Remove the exhaust manifold.

48 Unscrew the sensor from the exhaust manifold, and discard the sealing washer.

49 Clean the sensor and manifold threads, and fit a new sealing washer to the sensor.

50 Fit the sensor to the manifold. Use a torque wrench to tighten the sensor to the specified torque.

51 Fit the exhaust manifold.

Camshaft position sensor

52 Disconnect the battery earth lead.

53 Remove three bolts securing the coil cover the camshaft cover, and remove the coil cover.

54 Release the sensor cable from the bracket, and disconnect the multi-plug.

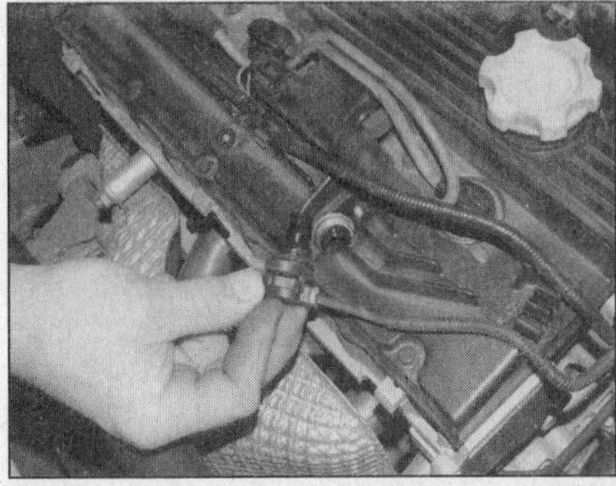

55 Remove the bolt securing the sensor to the camshaft cover. Release the cable from the clip, and remove the sensor.

56 Clean mating faces of sensor and camshaft cover.

57 Fit the sensor and bolt. Tighten the bolt securely.

58 Connect multi-plug, and secure in the bracket.

59 Fit the coil cover and bolts. Tighten the bolts securely.

60 Connect the battery earth lead.

61 Refitting is a reversal of the removal procedure.

Engine Control Module

Note: *If a new ECM is to be fitted it will need to be programmed with the code from the anti-theft security unit before the engine can be started. This can be done only by an MG Rover dealer or suitably equipped specialist.*

62 Disconnect the battery negative lead.

63 Remove two bolts securing the ECM to the batter carrier.

64 Release the ECM and disconnect the multi-plugs. Remove the ECM.

65 Refitting is a reversal of the removal procedure.

Fuel inertia cut-off switch

66 The switch is located at the right-hand side of the engine compartment bulkhead.

67 Disconnect the battery negative lead, then disconnect the wiring plug from the switch.

68 Remove the two securing screws and withdraw the switch from its mounting bracket.

69 Refitting is a reversal of removal, but ensure that the switch is reset by pressing the button on the top of the switch.

Crankshaft position sensor

70 Disconnect the battery negative lead.

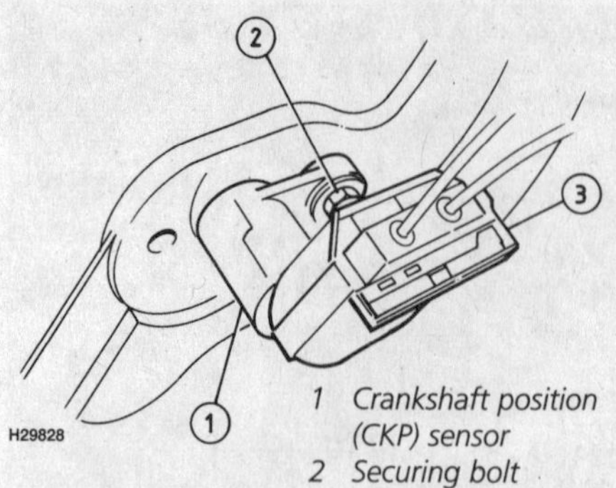

1 Crankshaft position (CKP) sensor
2 Securing bolt
3 Wiring plug

71 The sensor is mounted at the rear of the flywheel housing (at the left-hand end of the engine). To improve access to the sensor, apply the handbrake, then jack up the front of the vehicle and support securely on axle stands.

72 Disconnect the wiring plug from the sensor.

73 Unscrew the bolt securing the sensor to the flywheel housing, then remove the sensor.

74 Thoroughly clean the mating faces on the sensor and housing, then refit the sensor using a reversal of the removal procedure and tightening its bolt to the specified torque setting.

Electronic Diesel Control (EDC) system components – removal and refitting

Crankshaft position sensor

1 Disconnect the battery negative lead.

2 The sensor is mounted at the rear of the flywheel housing (at the left-hand end of the engine).

3 Disconnect the wiring plug from the sensor.

4 Unscrew the sensor securing bolt, then withdraw the sensor from its housing.

5 Refitting is a reversal of removal.

6 Tightening the sensor securing bolt to the specified torque setting.

Engine coolant temperature sensor

Note: Loctite 577 is needed to seal the sensor.

7 The ECM sensor is screwed into the coolant outlet elbow attached to the side of the cylinder head, adjacent to the dipstick. It is connected the engine harness via a multi-plug.

8 Disconnect the battery earth lead.

9 Disconnect the multi-plug from the sensor.

10 Locate a container under the sensor to collect coolant. Unscrew the sensor.

11 Refitting is a reversal of the removal procedure, noting:

a) Clean and degrease the sensor threads, and apply a smear of Loctite 577.
b Use a torque wrench to tighten the sender to the specified torque.
c) Top-up the coolant level.

Fuel temperature sensor

12 The sensor is integral with the fuel injection pump, and cannot be renewed separately.

Intake Air Temperature/Mass Air Flow sensor

Note: *The following new parts are needed – sensor O-ring.*

13 Disconnect the battery earth lead.

14 Release the clip and disconnect the multi-plug from the sensor.

15 Slacken the clip securing the air intake hose to the sensor assembly

16 Release two clips securing the sensor assembly to the air cleaner.

17 Remove the sensor assembly and discard the O-ring.

18 Refitting is a reversal of the removal procedure, ensuring that a new O-ring seal is fitted to the sensor.

Fuel injector needle lift sensor

19 The sensor is an integral part of the No 1 cylinder fuel injector, and cannot be renewed separately.

Vehicle speed sensor

20 The engine control module receives a vehicle speed signal from the same vehicle speed sensor used to provide a signal to the speedometer. The sensor is located on top of the transmission housing, next to the inboard end of the right-hand driveshaft. Jack up the front of the vehicle and support it securely on axle stands.

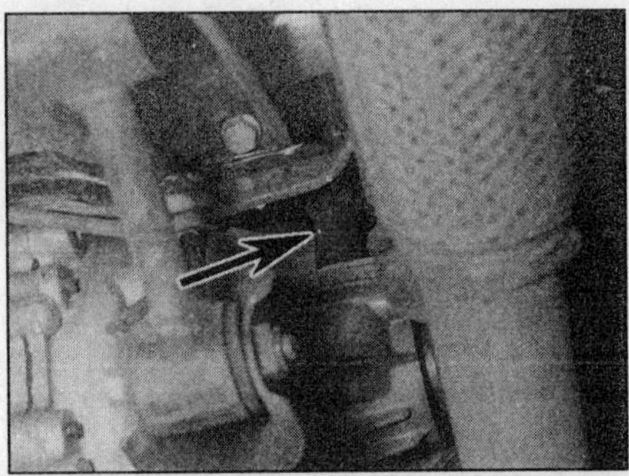

21 Disconnect the wiring plug, then unscrew the sensor retaining nut and remove the sensor from the top of the speedometer drive pinion.

22 Refitting is a reversal of removal, but smear the new O-ring seal with a little gearbox oil prior to refitting.

Fuel quantity servo position sensor

23 The sensor is integral with the fuel injection pump, and cannot be renewed separately.

Throttle position sensor

24 Disconnect the battery negative lead.

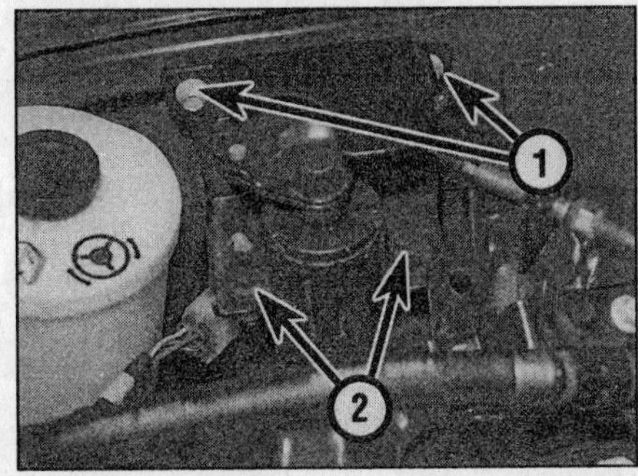

25 The sensor is located on the body panel at the right-hand side of the engine compartment. Unscrew the two bolts (1) securing the sensor mounting bracket to the body.

26 Slide the throttle cable adjuster from the cable bracket, then slide the cable end fitting from the groove in the throttle position sensor lever.

27 Disconnect the wiring plug from the sensor, then unclip the wiring from the sensor bracket.

28 Unscrew the two nuts and bolts securing the sensor to the mounting bracket (2), and remove the sensor.

29 Refitting is a reversal of removal, but reconnect and adjust the throttle cable.

Manifold absolute pressure sensor

30 Disconnect the battery negative lead.

31 The sensor is located on the right-hand side of the engine compartment bulkhead.

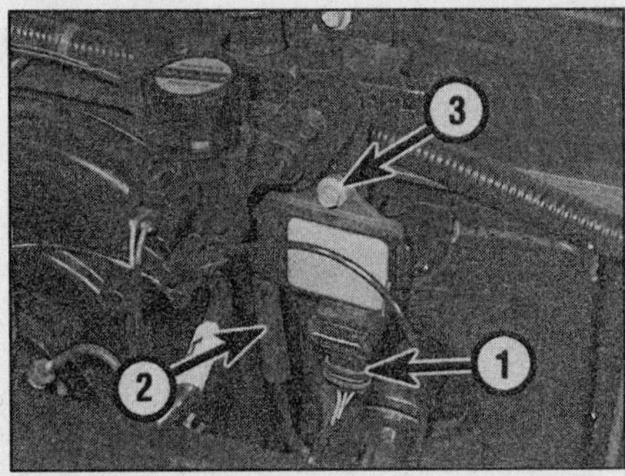

32 Disconnect the wiring plug (1) from the sensor, then release the hose clip and disconnect the vacuum hose (2) from the sensor.

33 Unscrew the bolt (3) securing the sensor to the mounting bracket, and remove the sensor.

34 Refitting is a reversal of removal, tightening the sensor securing bolt to the specified torque setting.

Brake pedal switch

35 The engine control module receives a signal from the brake light switch which indicates when the brake are being applied.

Engine Control Module

Note: *If a new ECM is to be fitted it will need to be programmed with the code from the anti-theft security unit before the engine can be started. This can be done only by an MG Rover dealer or authorised repairer.*

36 Disconnect the battery earth lead.

37 Remove two bolts and remove the ECM clamp from the battery carrier.

38 Release and disconnect two multi-plugs and remove the ECM.

39 Refitting is a reversal of the removal procedure, noting:

If necessary, arrange programming of the module at an MG Rover dealer or authorised repairer.

Notes

Vauxhall/Opel Corsa '00 – '03

24

Petrol models

General

System type:	
1.0 and 1.2 litre models	Bosch Motronic ME 1.5.5 sequential multi-point fuel injection
1.4 litre models	Multec S(F) sequential multi-point injection

Fuel pump

Type	Electric, mounted in fuel tank

Fuel injection system data

System pressure	3.8 bars
Idle speed (not adjustable – for reference only):	
1.0 litre engines	790 to 990 rpm
1.2 litre engines (manual and Easytronic)	750 to 970 rpm
1.4 litre engines	710 to 930 rpm
Idle mixture CO content (not adjustable – for reference only)	0.4% maximum

Diesel models

General

System type	ECD V5 direct injection system incorporating an electronically-controlled Nippon Denso V5 distributor fuel injection pump with integral transfer pump. Two stage injectors with pilot and post injection. Turbocharger fitted to all models and intercooler on Y17DT engined models

Adjustment data

Idle speed	825 to 875 rpm – controlled by ECU
Maximum speed	5060 to 5180 rpm – controlled by ECU

Injection pump

Direction of rotation	Clockwise, viewed from sprocket end
Pump timing (static):	
Checking	0.28 to 0.33 mm pump plunger travel @ TDC
Setting	0.33 mm pump plunger travel @ TDC
Transfer fuel pump pressure	8.0 ± 0.1 bars at 2200 rpm

Injectors

Opening pressure	175 bar (1st stage) to 335 bar (2nd stage)

Chapter 24

Torque wrench settings	Nm	lbf ft
Petrol models		
Camshaft position sensor to timing cover	6	4
Coolant temperature sensor	18	13
Crankshaft speed/position sensor to cylinder block baseplate	8	6
Engine control unit	8	6
Knock sensor	20	15
Manifold absolute pressure sensor	8	6
Oxygen sensor (catalytic converter control)	30	30
Oxygen sensor (mixture regulation)	30	30
Diesel models		
Charge pressure sensor	10	7
Coolant temperature sensor	22	16
Crankshaft position sensor	10	7
Electronic control unit	6	4

General information

Petrol models

1.0 and 1.2 litre models

1 The system is under the overall control of the Motronic ME 1.5.5 engine management system, which also controls the ignition system.

2 Fuel is supplied from the rear-mounted tank, via a fuel filter and a pressure regulator, to the fuel rail. Excess fuel is returned from the regulator to the tank. The fuel rail acts as a reservoir for the fuel injectors, which inject fuel into the cylinder inlet tracts, upstream of the inlet valves. The Motronic ME 1.5.5 system is a 'sequential' fuel injection system. This means that each of the three (1.0 litre engines), or four (1.2 litre engines) fuel injectors is triggered individually, just before the inlet valve on the relevant cylinder opens.

3 The duration of the electrical pulses to the fuel injectors determines the quantity of fuel injected. The pulse duration is computed by the Motronic electronic control unit (ECU) on the basis of information received from the following sensors:

a) Accelerator pedal position sensor – informs the ECU of accelerator pedal position, and the rate of throttle opening/ closing.

b) Air mass meter – informs the ECU of the load on the engine (expressed in terms of the mass of air passing from the air cleaner to the throttle body).

c) Crankshaft speed/position sensor – informs the ECU of the crankshaft speed and position.

d) Camshaft position sensor – informs the ECU when No 1 cylinder is at top dead centre (TDC) on the firing stroke (expressed in terms of the position of the inlet camshaft).

e) Coolant temperature sensor – informs the ECU of engine temperature.

f) Exhaust gas oxygen sensors (two) – inform the ECU of the oxygen content of the exhaust gases.

g) Knock sensor – informs the ECU when engine 'knock' (pre-ignition) occurs.

h) Vehicle speed sensor (where fitted) – informs the ECU of the vehicle speed.

4 The signals from the various sensors are processed by the ECU, and the optimum fuelling and ignition settings are selected for the prevailing engine operating conditions.

5 Idle speed is controlled by the idle speed control motor, which directly regulates the position of the throttle valve. The motor is controlled by the electronic control unit; there is no provision for direct adjustment of the idle speed.

6 Similarly, the fuel/air mixture is controlled within fine limits by the electronic control unit, via the fuel injectors. No manual adjustment of fuel/air mixture is possible.

7 A catalytic converter is incorporated in the exhaust manifold, to reduce harmful exhaust gas emissions.

8 If certain sensors fail, and send abnormal signals to the ECU, the ECU has a back-up programme. In this event, the abnormal signals are ignored, and a preprogrammed value is substituted for the sensor signal, allowing the engine to continue running, albeit at reduced efficiency. If the ECU enters its back-up mode, a warning light on the instrument panel will illuminate, and a fault code will be stored in the ECU memory. This fault code can be read using suitable specialist test equipment.

1.4 litre models

9 The multi-point injection system is under the overall control of the Multec S engine management system, which also controls the ignition system.

10 Fuel is supplied from the rear-mounted tank, via a fuel filter and a pressure regulator, to the fuel rail. Excess fuel is returned from the regulator to the tank. The fuel rail acts as a reservoir for the four fuel injectors, which inject fuel into the cylinder inlet tracts, upstream of the inlet valves. The Multec S system is a 'sequential' fuel injection system. This means that each of the four fuel injectors is triggered individually, just before the inlet valve on the relevant cylinder opens.

11 The duration of the electrical pulses to the fuel injectors determines the quantity of fuel injected. The pulse duration is computed by the Multec S electronic control unit (ECU) on the basis of information received from the following sensors:

a) Accelerator pedal position sensor – informs the ECU of accelerator pedal position, and the rate of throttle opening/closing.

b) Manifold absolute pressure (MAP) sensor – informs the ECU of the load on the engine (expressed in terms of inlet manifold vacuum).

c) Intake air temperature sensor – informs the ECU of the temperature of the air passing through the air intake ducting.

d) Crankshaft speed/position sensor – informs the ECU of the crankshaft speed and position.

e) Camshaft position sensor – informs the ECU when No 1 cylinder is at top dead centre (TDC) on the firing stroke (expressed in terms of the position of the exhaust camshaft).

f) Coolant temperature sensor – informs the ECU of engine temperature.

g) Exhaust gas oxygen sensors (two) – inform the ECU of the oxygen content of the exhaust gases.

h) Knock sensor – informs the ECU when engine 'knock' (pre-ignition) occurs.

12 The signals from the various sensors are processed by the ECU, and the optimum fuelling and ignition settings are selected for the prevailing engine operating conditions.

13 Idle speed is controlled by the idle speed control motor, which regulates the quantity of air bypassing the throttle valve. The motor is controlled by the electronic control unit; there is no provision for direct adjustment of the idle speed.

14 Similarly, the fuel/air mixture is controlled within fine limits by the electronic control unit, via the fuel injectors. No manual adjustment of fuel/air mixture is possible.

15 A catalytic converter is incorporated in the exhaust manifold to reduce harmful exhaust gas emissions.

16 If certain sensors fail, and send abnormal signals to the ECU, the ECU has a back-up programme. In this event, the abnormal signals are ignored, and a preprogrammed value is substituted for the sensor signal, allowing the engine to continue running, albeit at reduced efficiency. If the ECU enters its back-up mode, a warning light on the instrument panel will illuminate, and a fault code will be stored in the ECU memory. This fault code can be read using suitable specialist test equipment.

Diesel models

17 The fuel supply system consists of a fuel tank (which is mounted under the rear of the car), a fuel filter with integral water separator (in the engine compartment), an electronically-controlled distributor fuel injection pump with integral transfer pump, injectors and associated components.

18 Fuel is drawn from the fuel tank by the fuel injection pump. Before reaching the pump, the fuel passes through a fuel filter, where foreign matter and water are removed. Excess fuel lubricates the moving components of the pump, and is then returned to the tank.

19 The fuel injection pump is driven at half-crankshaft speed by the timing belt. The high pressure required to inject the fuel into the compressed air in the cylinder is achieved by radial plungers.

20 The injection pump is electronically-controlled to meet the latest emission standards. The system consists of the engine electronic control unit, the injection electronic control unit, and the following sensors:

a) Accelerator pedal position sensor – informs the ECUs of the accelerator pedal position.

b) Coolant temperature sensor – informs the ECUs of engine temperature.

c) Intake air temperature sensor – informs the ECUs of the temperature of the air passing through the intake duct.

d) Hot film mass airflow meter – informs the ECUs of the amount of air passing through the intake duct.

e) Crankshaft sensor – informs the ECUs of engine speed and crankshaft position.

f) Charge pressure sensor – informs ECUs of the pressure in the inlet manifold.

g) ABS control unit – informs the ECUs of the vehicle speed.

h) Atmospheric pressure sensor – informs the ECUs of the atmospheric pressure.

i) Fuel temperature sensor – informs the ECUs of the fuel temperature.

j) Injection pump shaft sensor – used by the ECUs to determine the exact injection timing.

k) Oil pressure switch – used by the engine ECU to control the instrument cluster oil warning lamp.

l) Clutch switch – used by the ECUs for cruise control functions.

m) Air conditioning system compressor switch (where fitted) – informs the ECUs when the air conditioning system is switched on.

21 All the above information is analysed by the ECUs and, based on this, the ECUs determine the appropriate injection requirements for the engine. The engine ECU controls the injection pump timing, via the pump control unit, to provide the best setting for cranking, starting (with either a hot or cold engine), warm-up, idle, cruising, and acceleration.

22 Basic injection timing is determined when the pump is fitted. When the engine is running, it is varied automatically to suit the prevailing engine speed by a mechanism which turns the cam plate or ring – controlled by the ECU.

23 The engine ECU also controls the exhaust gas recirculation (EGR) system and the pre-heating system.

23 The four fuel injectors produce a spray of fuel directly into the cylinders. The injectors are calibrated to open and close at critical pressures to provide efficient and even combustion. Each injector needle is lubricated by fuel which accumulates in the spring chamber and is channelled to the injection pump return hose by leak-off pipes.

24 The inlet manifold is a two-part assembly sealed by a metal gasket. The EGR (exhaust gas recirculation) valve and charge pressure sensor are mounted to the upper part of the manifold.

25 A turbocharger is fitted to increase engine efficiency by raising the pressure in the inlet manifold above atmospheric pressure. Instead of the air simply being sucked into the cylinders, it is forced in. Additional fuel is supplied by the injection pump in proportion to the increased air intake.

26 Energy for the operation of the turbocharger comes from the exhaust gas. The gas flows through a specially-shaped housing (the turbine housing) and in so doing, spins the turbine wheel. The turbine wheel is attached to a shaft, at the end of which is another vaned wheel known as the compressor wheel. The compressor wheel spins in its own housing, and compresses the inlet air on the way to the inlet manifold.

27 Between the turbocharger and the inlet manifold, the compressed air passes through an intercooler (except Y20DTL engine). This is an air-to-air heat exchanger mounted next to the radiator, and supplied with cooling air from the front of the vehicle. The purpose of the intercooler is to remove some of the heat gained in being compressed from the inlet air. Because cooler air is denser, removal of this heat further increases engine efficiency.

28 Charge pressure (the pressure in the inlet manifold) is limited by a wastegate, which diverts the exhaust gas away from the turbine wheel in response to a pressure-sensitive actuator. A pressure-operated switch operates a warning light on the instrument panel in the event of excessive charge pressure developing.

29 The turbo shaft is pressure-lubricated by an oil feed pipe from the engine main oil so that the shaft 'floats' on a cushion of oil. A drain pipe returns the oil to the sump.

30 The charge pressure wastegate is controlled by the ECU via a solenoid valve.

31 If there is an abnormality in any of the readings obtained from any sensor, the ECU enters its back-up mode. In this event, the ECU ignores the abnormal sensor signal, and assumes a preprogrammed value which will allow the engine to continue running (albeit at reduced efficiency). If the ECU enters this back-up mode, the warning light on the instrument panel will come on, and the relevant fault code will be stored in the ECU memory.

Fuel system – depressurising, priming and bleeding

Petrol models

Warning: The fuel system is pressurised all the time the ignition is switched on, and a high pressure will remain in the system even 20 minutes after switching off. It is therefore essential to depressurise the system before disconnecting fuel lines, or carrying out any work on the fuel system components. Failure to do this before carrying out work may result in a sudden release of pressure which may cause fuel spray – this constitutes a fire hazard, and a health risk. Note that, even when the system has been depressurised, fuel will still be present in the system fuel lines and components, and adequate precautions should still be taken when carrying out work.

1 The fuel system referred to in this Section is defined as the tank-mounted fuel pump and pressure regulator, the fuel filter, the fuel injectors, the fuel rail, and the metal pipes and flexible hoses of the fuel lines between these components. All these contain fuel which will be under pressure while the engine is running, and/or while the ignition is switched on. High pressure will remain for at least 20 minutes after the ignition has been switched off, and must be relieved in a controlled fashion when any of these components are disturbed for servicing work.

2 Where necessary, remove the engine top cover.

3 Place a container beneath the fuel pressure testing point on the fuel rail and have a cloth rag ready to soak up any escaping fuel not being caught by the container.

4 Slowly loosen the cap and allow the fuel to escape. With the cap fully unscrewed, all fuel pressure will have escaped, and the cap can then be refitted and tightened.

Diesel models

5 It is not necessary to manually prime and bleed the fuel system after any operation on the system components. Start the engine, noting that this may take longer than usual, especially if the fuel system has been allowed to run dry. Operate the starter in ten second bursts with 5 seconds rest in between each operation, for a maximum of 40 seconds. When the engine starts, run it at a fast idle speed for a minute or so to purge any trapped air from the fuel lines. After this time the engine should idle smoothly at a constant speed.

6 If the engine idles roughly, then there is still some air trapped in the fuel system. Increase the engine speed again for another minute or so then recheck the idle speed. Repeat this procedure as necessary until the engine is idling smoothly.

Petrol fuel injection system components (1.0 and 1.2 litre models) – removal and refitting

Throttle body

1 Disconnect the battery negative lead.

2 Remove the air cleaner cover, inlet trunking and airbox from the top of the throttle body.

3 Disconnect the wiring from the throttle body module.

4 Partially drain the cooling system (drain sufficient coolant to empty the coolant expansion tank).

5 Release the retaining clips and disconnect the two coolant hoses from the rear of the throttle body.

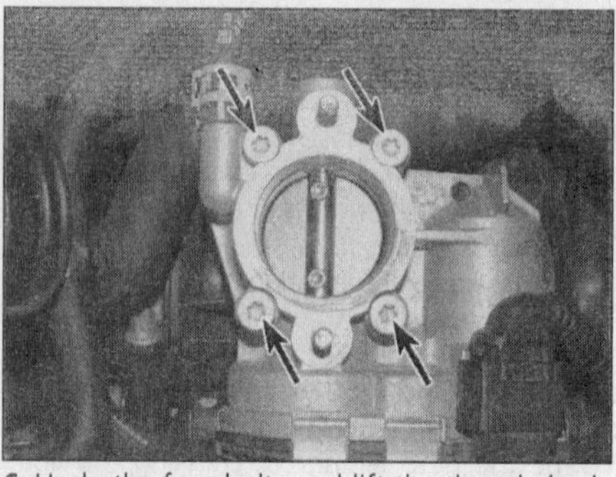

6 Undo the four bolts and lift the throttle body off the inlet manifold. Recover the gasket.

7 It is not possible to obtain the idle speed control motor or throttle valve position sensor separately, so if either is faulty, the complete throttle body must be renewed.

8 Refitting is a reversal of removal, but thoroughly clean the mating faces and use a new gasket. Tighten the bolts progressively and securely. Top-up the coolant level. Finally, switch on the ignition for 30 seconds without starting the engine.

Air mass meter

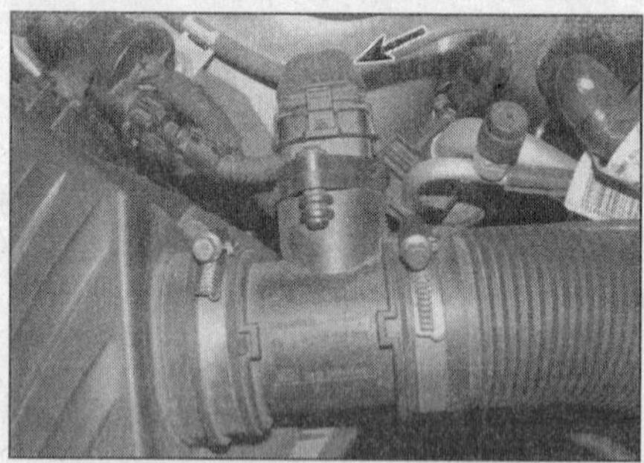

9 Disconnect the wiring from the air mass meter at the right-hand rear corner of the engine compartment.

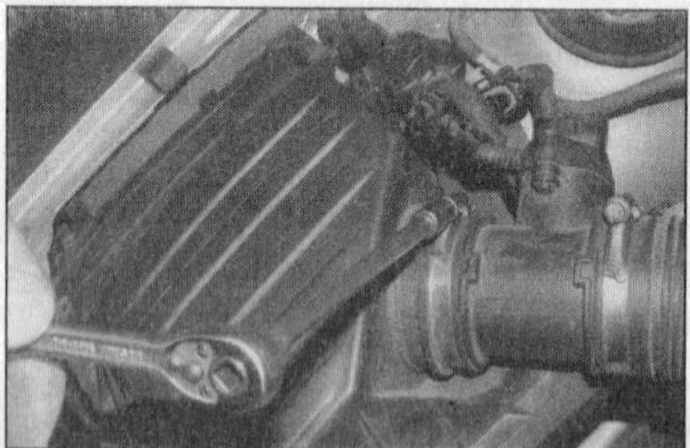

10 Release the retaining clips and remove the air mass meter from the air cleaner cover and inlet trunking.

11 Refitting is a reversal of removal, but ensure that the arrow on the air mass meter body points toward the throttle body when fitted.

Fuel injectors and fuel rail

Note: *The seals at both ends of the fuel injectors must be renewed on refitting.*

12 Disconnect the battery negative lead.

13 Depressurise the fuel system.

14 Remove the throttle body as described previously in this Section, however, it is not necessary to drain the cooling system, as the body can be placed to one side with the hoses still attached.

15 Loosen the clip and disconnect the engine breather hose from the camshaft cover.

16 Disconnect the wiring from the throttle body.

17 Disconnect the wiring from the injectors.

18 Disconnect the fuel feed hose quick-release connector at the fuel rail. Be prepared for some loss of fuel. A Vauxhall special tool is available to release the connector, but provided care is taken, it can be released using a pair of long-nosed pliers, or a similar tool, to depress the retaining tangs. Clamp or plug the open end of the hose, to prevent dirt ingress and further fuel spillage.

19 Unscrew the mounting bolts, then lift the fuel rail complete with the injectors off of the inlet manifold.

20 To remove an injector from the fuel rail, prise out the metal securing clip using a screwdriver or a pair of pliers, and pull the injector from the fuel rail. Remove and discard the injector sealing rings; new ones must be fitted on refitting.

21 Overhaul of the fuel injectors is not possible, as no spares are available. If faulty, an injector must be renewed.

22 Before refitting, clean thoroughly the mating surfaces of the throttle body and inlet manifold.

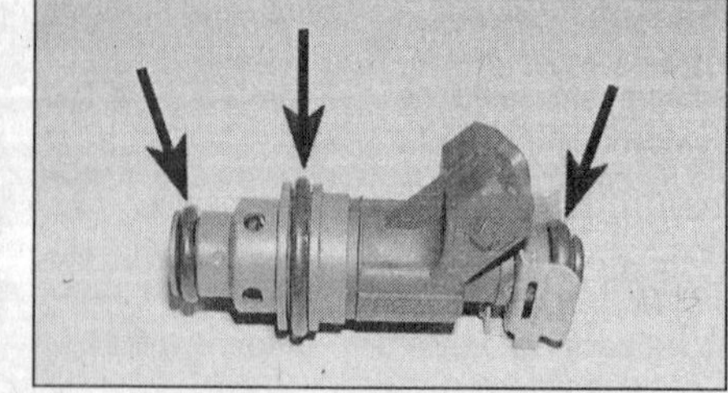

23 Commence refitting by fitting new O-ring seals to both ends of the fuel injectors. Coat the seals with a thin layer of petroleum jelly before fitting.

24 Refitting is a reversal of removal, bearing in mind the following points:

a) When refitting the injectors to the fuel rail, note that the groove in the metal securing clip must engage with the lug on the injector body.

b) Make sure that the quick-release connector audibly engages on the fuel rail.

c) Refit the throttle body as described previously in this Section.

d) Ensure that all wiring connectors are securely reconnected, and that the wiring is secured in the relevant clips and brackets.

e) Finally, it may be necessary to have a Vauxhall dealer or authorised repairer program all volatile memories.

Crankshaft speed/ position sensor

Note: *A new O-ring seal must be used on refitting.*

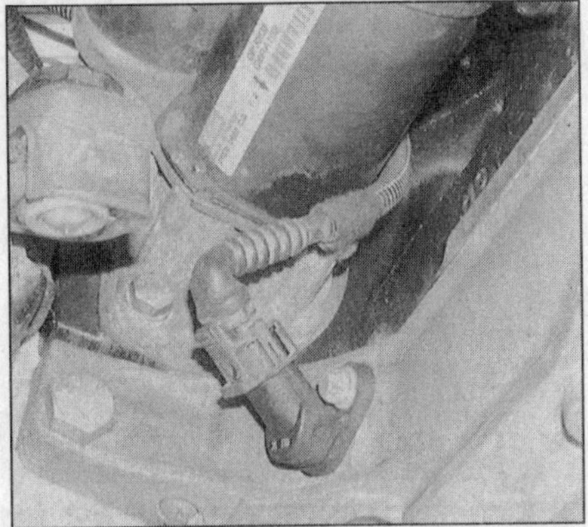

25 The crankshaft speed/position sensor is located at the rear left-hand end of the cylinder block baseplate, below the starter motor.

26 Apply the handbrake, then jack up the front of the vehicle and support it on axle stands. On ECO models, remove the engine undertray.

27 Disconnect the sensor wiring connector, then undo the retaining bolt and withdraw the sensor from the cylinder block baseplate.

28 Refitting is a reversal of removal, but ensure that the mating surfaces of the sensor and baseplate are clean and fit a new O-ring seal to the sensor before refitting. Tighten the bolt to the specified torque.

Camshaft position sensor

29 The camshaft position sensor is located on the timing cover, on the inlet camshaft side.

30 Disconnect the sensor wiring connector.

31 Undo the retaining bolt and withdraw the sensor from the timing cover.

32 Refitting is a reversal of removal, but ensure that the mating surfaces of the sensor and timing cover are clean before fitting.

Coolant temperature sensor

33 The coolant temperature sensor is located in the top of the coolant pump housing.

34 Partially drain the cooling system.

35 Disconnect the sensor wiring plug.

36 Unscrew the sensor, and withdraw it from the coolant pump. Recover the sealing ring.

37 Refitting is a reversal of removal, but use a new sealing ring, and tighten the sensor to the specified torque. On completion, top-up the cooling system.

Oxygen sensor (mixture regulation)

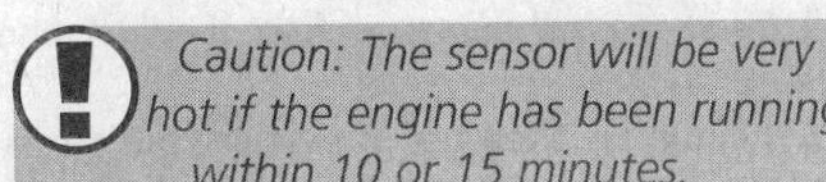

Caution: The sensor will be very hot if the engine has been running within 10 or 15 minutes.

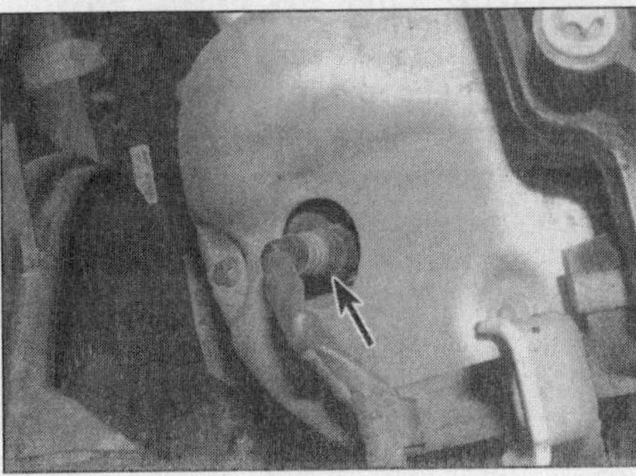

38 The mixture regulation oxygen sensor is located on the exhaust manifold/catalytic converter. First, trace the wiring back from the sensor to the connector next to the oil filter, and disconnect it. Release the wiring from the clip.

39 Unscrew the sensor from the exhaust manifold. Ideally, a special 'split' socket should be used, as this will locate over the sensor wiring.

40 Clean the threads of the sensor then coat them with Vauxhall special grease for oxygen sensors. If a new sensor is being fitted, it will be supplied with the threads already coated with the special grease to prevent it seizing in the manifold.

41 Screw the sensor into the exhaust manifold/catalytic converter and tighten to the specified torque.

42 Reconnect the wiring and clip it in place.

Oxygen sensor (catalytic converter control)

Caution: The sensor will be very hot if the engine has been running within 10 or 15 minutes.

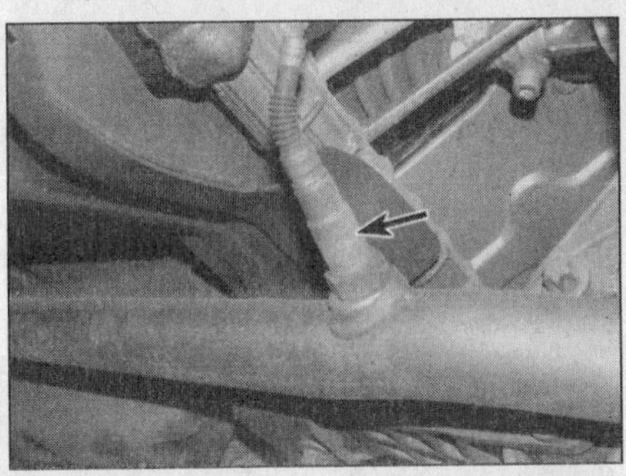

43 The catalytic converter control oxygen sensor is located in the exhaust front pipe, just behind the flexible section. First, apply the handbrake, then jack up the front of the vehicle and support it on axle stands.

44 Trace the wiring back from the sensor to the connector above the right-hand driveshaft, and disconnect it. Release the wiring from the clip.

45 Unscrew the sensor from the exhaust front pipe. Ideally, a special 'split' socket should be used, as this will locate over the sensor wiring.

46 Clean the threads of the sensor then coat them with Vauxhall special grease for oxygen sensors. If a new sensor is being fitted, it will be supplied with the threads already coated with the special grease to prevent it seizing in the pipe.

47 Screw the sensor into the exhaust front pipe and tighten to the specified torque.

48 Reconnect the wiring and clip it in place.

Knock sensor

49 The knock sensor is located on the rear of the cylinder block, just above the starter motor.

50 Apply the handbrake, then jack up the front of the vehicle and support it on axle stands. On ECO models, remove the engine undertray.

51 On the left-hand side of the engine, disconnect the wiring for the engine control unit by releasing it in the direction of the arrow.

52 Unscrew the bolt securing the earth wire to the cylinder head, then unbolt and remove the engine control unit.

53 Reach up behind the engine and disconnect the wiring from the knock sensor.

54 Note its position, then unscrew the bolt and remove the knock sensor from the block.

55 Clean the contact surfaces of the sensor and block. Also clean the threads of the sensor mounting bolt.

56 Locate the sensor on the block and insert the mounting bolt. Position the sensor as previously noted, then tighten the bolt to the specified torque. Note that the torque setting is critical for the sensor to function correctly.

57 Reconnect the wiring, then refit the engine control unit together with its wiring and earth wire.

58 Reconnect the battery negative lead.

59 Refit the engine undertray where applicable, then lower the vehicle to the ground. Finally, it may be necessary to have a Vauxhall dealer program all volatile memories.

Electronic control unit (ECU)

60 The engine management electronic control unit is located on the rear, left-hand side of the engine.

61 Disconnect the battery negative lead.

62 Disconnect the wiring for the engine control unit by releasing it in the direction of the arrow marked on the connector.

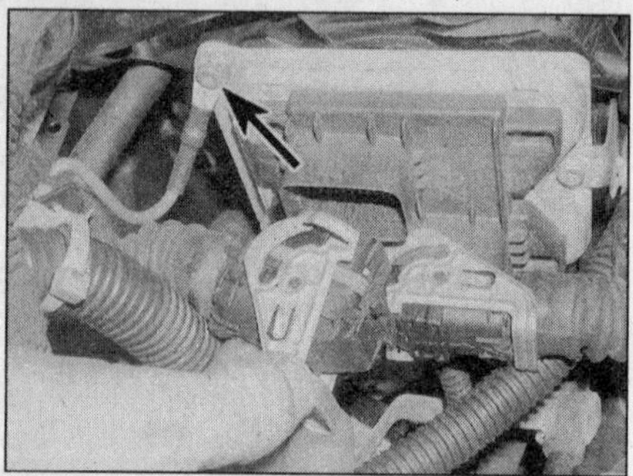

63 Unscrew the bolt securing the earth wire to the cylinder head, then unbolt and remove the engine control unit.

64 Refitting is a reversal of removal. Finally, it may be necessary to have a Vauxhall dealer program all volatile memories.

Petrol fuel injection system components (1.4 litre models) – removal and refitting

Throttle body

1 Disconnect the battery negative lead.

2 Remove the air cleaner cover, inlet trunking and airbox from the top of the throttle body.

3 Disconnect the wiring from the throttle body module.

4 Partially drain the cooling system (drain sufficient coolant to empty the coolant expansion tank).

5 Release the retaining clips and disconnect the two coolant hoses from the rear of the throttle body.

6 Disconnect the crankcase ventilation hose and carbon canister vent hose from the throttle body.

7 Undo the four bolts and lift the throttle body off the upper section of the inlet manifold. Recover the gasket.

8 It is not possible to obtain the idle speed control motor or throttle valve position sensor separately, so if either is faulty, the complete throttle body must be renewed.

9 Refitting is a reversal of removal, but thoroughly clean the mating faces and use a new gasket. Tighten the bolts progressively and securely. Top-up the coolant level. Finally. switch on the ignition for 30 seconds without starting the engine.

Manifold absolute pressure (MAP) sensor

10 The sensor is located on the inlet manifold on the front of the cylinder head.

11 Disconnect the wiring from the sensor.

12 Unscrew the clamp nut, remove the clamp plate, and withdraw the sensor from the cylinder head.

13 Refitting is a reversal of removal, but clean the contact faces of the sensor and manifold, and tighten the clamp nut to the specified torque.

Air temperature sensor

14 Disconnect the wiring from the air temperature sensor located on the rear of the air cleaner-to-throttle body duct.

15 Twist and pull the sensor from the duct, and recover the seal.

16 Refitting is a reversal of removal, but renew the seal.

Fuel injectors and fuel rail

Note: *The seals at both ends of the fuel injectors must be renewed on refitting.*

17 Disconnect the battery negative lead.

18 Depressurise the fuel system.

19 Unscrew the oil filler cap, then unscrew the engine top cover securing bolts and lift off the cover. Screw the cap back into the camshaft cover.

20 Disconnect the wiring from the air temperature sensor on the rear of the air cleaner air duct, then loosen the clip and disconnect the crankcase ventilation hose from the camshaft cover.

21 Loosen the clip and remove the airbox and duct from the throttle body. Place the duct to one side.

22 Disconnect the wiring from the oil pressure switch on the front of the cylinder block.

23 Unscrew the union bolt and disconnect the fuel feed line from the fuel rail. Tape over or plug the rail aperture and the end of the line.

24 Disconnect the wiring from the following components:

a) Engine management ECU.
b) Carbon canister purge valve located behind the air cleaner.
c) Coolant temperature sensor.
d) Camshaft position sensor.
e) Throttle body module.
f) Knock sensor.
g) Inlet manifold absolute pressure (MAP) sensor.

25 Remove the catalytic converter control oxygen sensor as described later in this Section.

26 Remove the EGR valve, then release the wiring harness from the cable ties.

27 Release the heater hose for the throttle body from the bracket on the camshaft cover.

28 Unscrew the two mounting bolts and remove the fuel rail, together with the injectors and wiring conduit, from the lower section of the inlet manifold.

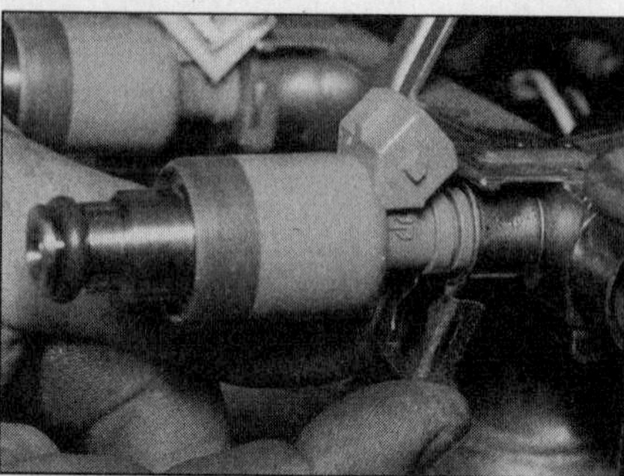

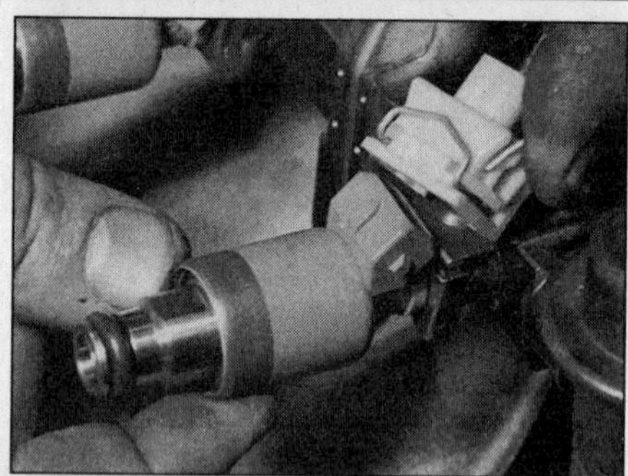

29 To remove an injector from the fuel rail and wiring conduit, first note the fitted position of the clips. Prise out the metal securing clip using a screwdriver or a pair of pliers, then release the wiring plug by carefully pulling the retaining spring clip downwards. Pull the injector from the fuel rail. Remove and discard the injector sealing rings; new ones must be fitted on refitting.

30 Overhaul of the fuel injectors is not possible, as no spares are available. If faulty, an injector must be renewed.

31 Commence refitting by fitting new O-ring seals to both ends of the fuel injectors. Coat the seals with a thin layer of petroleum jelly before fitting.

32 Refitting is a reversal of removal, bearing in mind the following points:

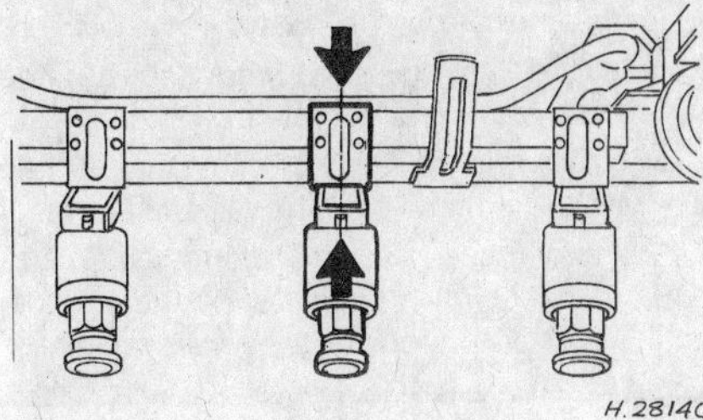

a) When refitting the injectors to the fuel rail, note that the groove in the metal securing clip must engage with the lug on the injector body.
b) Tighten the fuel rail mounting and fuel line union bolts securely.
c) Ensure that all wiring connectors are securely reconnected, and that the wiring is secured in the relevant clips and brackets.
d) Finally, it may be necessary to have a dealer or specialist program all volatile memories.

Crankshaft speed/position sensor

Note: *A new O-ring seal must be used on refitting.*

33 The crankshaft speed/position sensor is located at the front left-hand end of the cylinder block, below the oil filter. Access is most easily obtained from beneath the vehicle.

34 Apply the handbrake, then jack up the front of the vehicle and support it on axle stands. Where necessary, remove the engine undertray.

35 Disconnect the sensor wiring connector, then unscrew the retaining bolt and withdraw the sensor from the block.

36 Refitting is a reversal of removal, but ensure that the mating surfaces of the sensor and block are clean and fit a new O-ring seal to the sensor before refitting. Tighten the bolt to the specified torque.

Camshaft position sensor

37 The camshaft position sensor is located next to the right-hand side of the throttle body.

38 Disconnect the wiring, then unscrew the mounting bolt and withdraw the sensor.

39 Refitting is a reversal of removal.

Coolant temperature sensor

40 The coolant temperature sensor is located on the right-hand front end of the cylinder head.

41 Partially drain the cooling system. Alternatively, release any pressure in the cooling system by removing the expansion filler cap then refitting it, and have ready a means of plugging the sensor hole.

42 Disconnect the sensor wiring plug.

43 Unscrew the sensor, and withdraw it from the engine. Recover the sealing ring.

44 Refitting is a reversal of removal, but use a new sealing ring, and tighten the sensor to the specified torque. On completion, top-up the cooling system.

Exhaust gas oxygen sensor (mixture regulation)

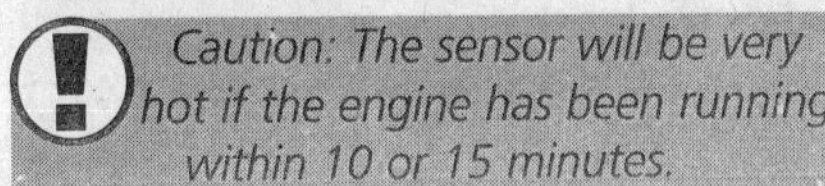

45 The mixture regulation oxygen sensor is located on the exhaust manifold/catalytic converter. With the engine top cover removed, trace the wiring back from the sensor to the connector located on the left-hand side of the cylinder head, and disconnect it.

46 Unscrew the sensor from the exhaust manifold/catalytic converter. Ideally, a special 'split' socket should be used, as this will locate over the sensor wiring.

47 Clean the threads of the sensor then coat them with Vauxhall special grease for oxygen sensors. If a new sensor is being fitted, it will be supplied with the threads already coated with the special grease to prevent it seizing in the manifold.

48 Screw the sensor into the exhaust manifold/catalytic converter and tighten to the specified torque.

49 Reconnect the wiring and clip it in place.

Exhaust gas oxygen sensor (catalytic converter control)

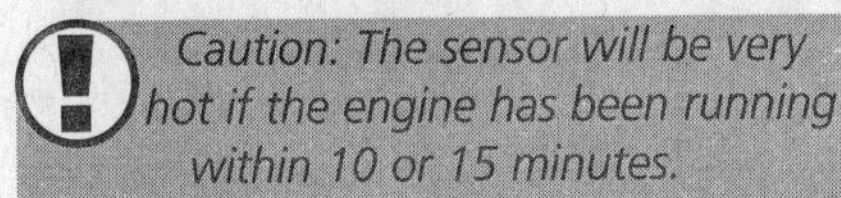

50 The catalytic converter control oxygen sensor is located in the exhaust front pipe, just behind the flexible section. First, apply the handbrake, then jack up the front of the vehicle and support it on axle stands.

51 Trace the wiring back from the sensor to the connector located on the bracket, and disconnect it. Unclip the wiring from the bracket.

52 Unscrew the sensor from the exhaust front pipe. Ideally, a special 'split' socket should be used, as this will locate over the sensor wiring.

53 Clean the threads of the sensor then coat them with Vauxhall special grease for oxygen sensors. If a new sensor is being fitted, it will be supplied with the threads already coated with the special grease to prevent it seizing in the pipe.

54 Screw the sensor into the exhaust front pipe and tighten to the specified torque.

55 Reconnect the wiring and clip it in place.

Knock sensor

56 The knock sensor is located on the rear, left-hand side of the cylinder block, just above the starter motor.

57 Apply the handbrake, then jack up the front of the vehicle and support it on axle stands.

58 Unscrew the oil filler cap, then unscrew the engine top cover securing bolts and lift off the cover. Screw the cap back into the camshaft cover.

59 Disconnect the wiring from the air temperature sensor located on the rear of the air cleaner-to-throttle body duct.

60 Loosen the clip and disconnect the airbox crankcase ventilation hose from the camshaft cover.

61 Loosen the clip and remove the airbox from the throttle body, then loosen the clip and separate the airbox and duct from the air cleaner cover.

62 Disconnect the knock sensor wiring at the connector on the top of the engine, then feed the wire between the fuel rail and wiring trough.

63 Note its position, then unscrew the mounting bolt and withdraw the sensor from the block.

64 Clean the contact surfaces of the sensor and block. Also clean the threads of the sensor mounting bolt.

65 Locate the sensor on the block and insert the mounting bolt. Position the sensor as previously noted, then tighten the bolt to the specified torque. Note that the torque setting is critical for the sensor to function correctly.

66 Reconnect the wiring, making sure that it is routed as previously noted.

67 Refit the airbox and duct and tighten the clips.

68 Reconnect the airbox crankcase ventilation hose, and reconnect the wiring to the air temperature sensor.

69 Refit the engine top cover and tighten the bolts.

70 Lower the vehicle to the ground.

Electronic control unit

71 The electronic control unit is located on the rear, left-hand side of the engine.

72 Disconnect the battery negative lead.

73 Unscrew the oil filler cap, then unscrew the engine top cover securing bolts and lift off the cover. Screw the cap back into the camshaft cover.

74 Disconnect the wiring from the following components:

a) Mixture regulating oxygen sensor.
b) Ignition module.
c) Combination plug.
d) Control unit.

75 Unscrew the bolt securing the earth wire to the control unit, then unbolt and remove the engine control unit.

76 Refitting is a reversal of removal. Finally, it may be necessary to have a Vauxhall dealer program all volatile memories.

Diesel injection system electronic components – removal and refitting

Hot film mass airflow meter

Note: *The intake air temperature sensor is built into the airflow meter.*

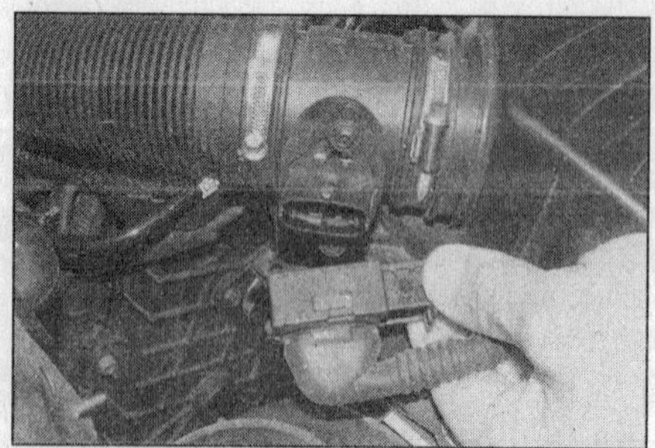

1 Ensure the ignition is switched off and disconnect the wiring plug from the hot film mass airflow meter. To do this, slide the lock towards the air cleaner.

2 Release the retaining clips then free the airflow meter from the air cleaner cover and intake duct. Remove it from the engine compartment.

3 Refitting is the reverse of removal, ensuring the airflow meter is correctly seated in the air cleaner and duct and the retaining clips are securely tightened.

Crankshaft position sensor

4 Apply the handbrake, then jack up the front of the vehicle and support it on axle stands. Remove the engine undertray.

5 The sensor is located on the rear of the cylinder block, beneath the starter motor. Working from under the vehicle, disconnect the wiring plug from the sensor.

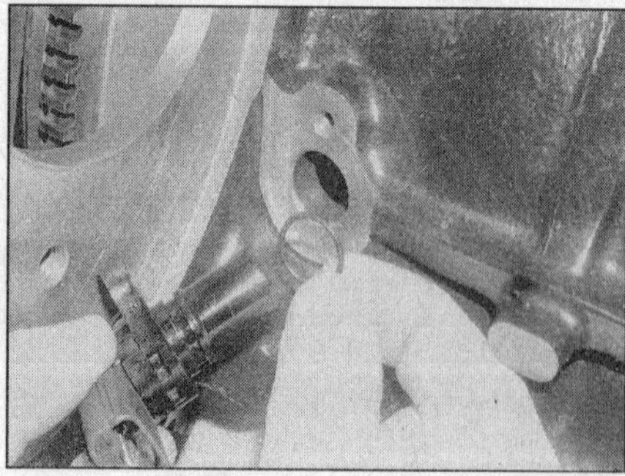

6 Undo the retaining bolt and remove the sensor. Recover the sealing ring.

7 Refitting is a reversal of removal. If necessary renew the sealing ring and tighten the sensor retaining bolt to the specified torque.

Coolant temperature sensor

8 Drain the coolant system, or be prepared for coolant spillage.

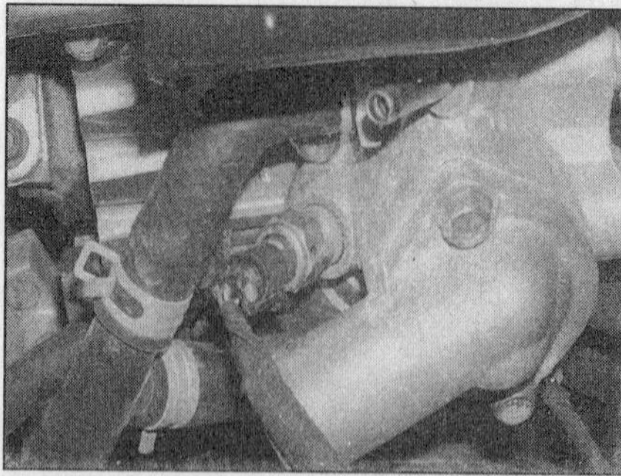

9 The coolant temperature sensor is located in the thermostat housing at the left-hand end of the cylinder head. Disconnect the wiring plug from the sensor.

10 Unscrew the sensor from the thermostat housing. Mop up any spilled coolant.

11 Apply a little locking compound to the threads of the sensor, and refit the sensor to the housing. Tighten the sensor to the specified torque.

12 Reconnect the wiring plug to the sensor.

13 Top-up the coolant system.

Charge pressure sensor

14 The charge pressure sensor is located on the right-hand rear end of the inlet manifold.

15 Loosen the clips and remove the air intake hose from above the inlet manifold.

16 Remove the exhaust gas recirculation (EGR) solenoid valve from the inlet manifold.

17 Disconnect the wiring plug from the sensor.

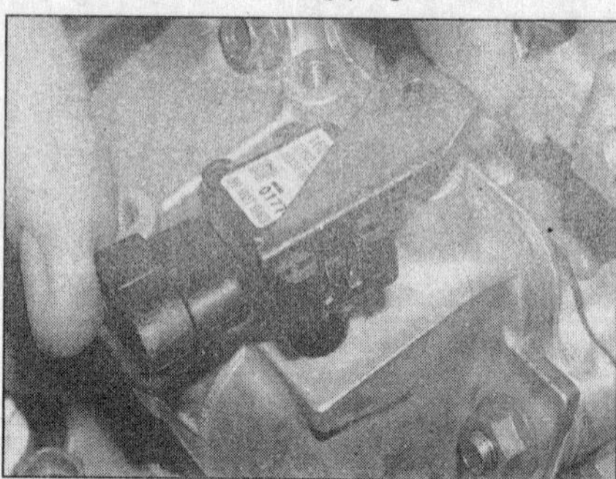

18 Unscrew the mounting bolt and remove the sensor.

19 Refitting is a reversal of removal. Tighten the sensor retaining bolt to the specified torque.

Oil pressure switch

20 The signal from the oil pressure switch is used by the injection control unit to determine whether or not to illuminate the oil pressure warning light.

21 The oil pressure switch is located at the left-hand, rear end of the cylinder block. For improved access, remove the battery and battery tray.

22 Unclip the coolant expansion tank hose and position to one side.

23 Disconnect the wiring plug from the switch.

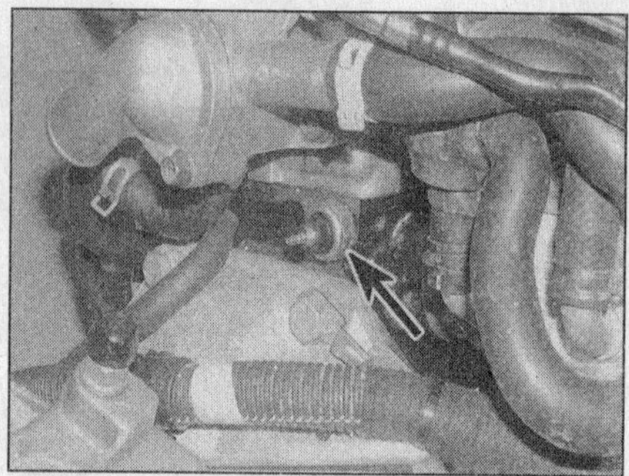

24 Unscrew the switch from the cylinder block. Be prepared for some oil spillage.

25 Apply some suitable silicone sealant (available from Vauxhall dealers) to the threads, and screw the switch into place. Tighten the switch to the specified torque.

26 Reconnect the wiring plug, locate the expansion tank hose in the clip, and refit the battery tray,and battery where removed.

27 Start the vehicle and check for oil leaks.

Engine electronic control unit (ECU)

28 The engine control unit is located on top of the engine above the exhaust gas recirculation valve.

29 Disconnect the battery negative lead.

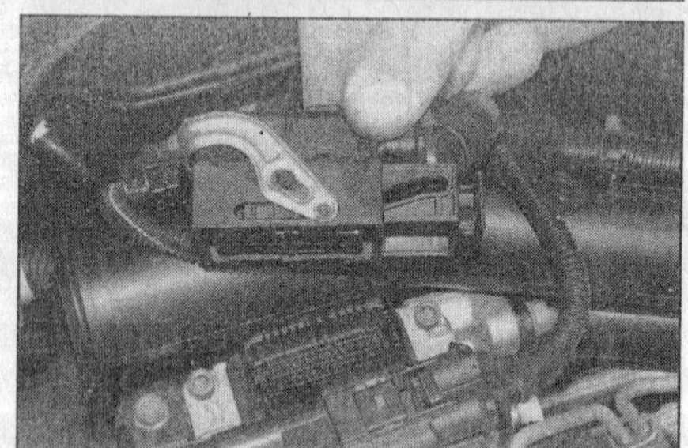

30 The two main plugs are unlocked by lifting up the metal levers, and this is best achieved using two screwdrivers. Disconnect the wiring plugs from the ECU.

31 Unbolt and remove the wiring harness bracket from the top of the ECU.

32 Unscrew the mounting bolts and nuts, and remove the ECU from the mounting bracket.

33 Release the glow plug wiring, then unscrew the two mounting bolts and single nut, and remove the ECU mounting bracket.

34 Refitting is a reversal of removal. Tighten the ECU retaining bolts/nuts to the specified torque. The ECU wiring plugs are shaped such that each plug will only fit its correct socket. The plug from the left-hand side of the engine compartment connects to the rearmost socket. Take great care when refitting the plugs, as the terminals are very delicate and easily damaged. Finally, it may be necessary to have a Vauxhall dealer or specialist program all volatile memories.

Injection electronic control unit (ECU)

35 The injection ECU is located on the rear of the engine, beneath the injection pump. Access is best from beneath the vehicle.

36 Disconnect the battery negative lead.

37 Apply the handbrake, then jack up the front of the vehicle and support it on axle stands. Remove the engine undertray.

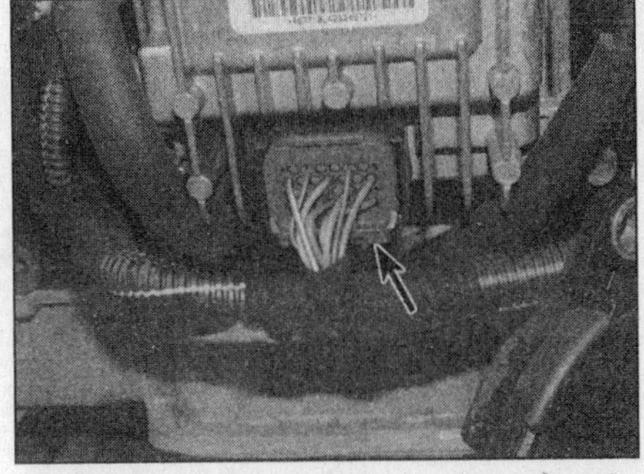

38 Working under the vehicle, disconnect the ECU wiring plug.

39 Undo the retaining bolts, and remove the ECU .

40 Refitting is a reversal of removal, but tighten the bolts to the specified torque. Finally, it may be necessary to have a Vauxhall dealer or specialist program all volatile memories.

Fuel temperature sensor

41 The fuel temperature sensor is fitted to the right-hand top of the injection pump.

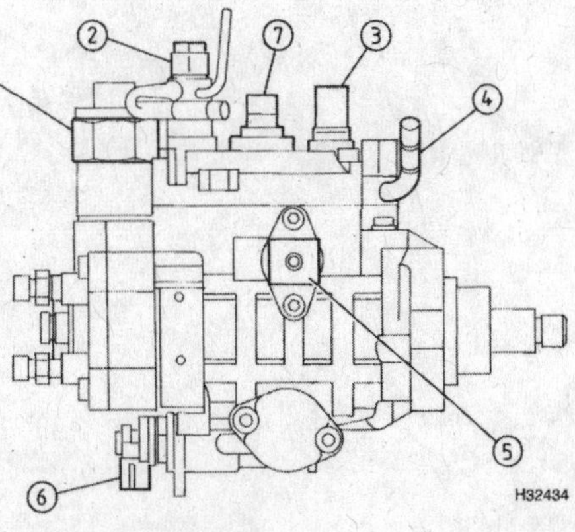

Fuel injection pump
1 Spill valve
2 Fuel return line
3 Fuel temperature sensor
4 Fuel supply line
5 Programmable read-only memory
6 Timing control solenoid
7 Injection pump shaft position sensor

42 To remove the sensor, disconnect the wiring plug and unscrew the sensor from the top of the injection pump.

43 With a new seal fitted, screw the sensor into the injection pump and tighten the nut securely.

44 Reconnect the wiring plug.

Injection pump shaft position sensor

45 The sensor is located on the side of the fuel injection pump.

46 The sensor may not be available as a separate item. Check with a Vauxhall dealer prior to any attempt to remove the sensor.

Injection commencement sensor

47 The sensor is located on the underside of the fuel injection pump at the flywheel end.

48 The sensor may not be available as a separate item. Check with a Vauxhall dealer prior to any attempt to remove the sensor.

Fuel cut-off solenoid

49 The fuel cut-off solenoid is located on top of the fuel injection pump, at the flywheel end.

50 The solenoid may not be available as a separate item. Check with a Vauxhall dealer prior to any attempt to remove the sensor.

Notes

Vauxhall/Opel Astra & Zafira '98 – '04

Petrol models

General

System type:

1.4 litre DOHC (X14XE and Z14XE) engines	Multec S multi-point injection
1.6 litre SOHC (X16SZR) engines	Multec single-point injection
1.6 litre DOHC (X16XEL and Z16XE) engines	Multec S multi-point injection
1.8 litre DOHC (X18XE1) engines	Simtec 70 multi-point injection
1.8 litre DOHC (Z18XE and Z18XEL) engines	Simtec 71 multi-point injection
2.0 litre DOHC (X20XEV) engines	Simtec 70 multi-point injection
2.2 litre DOHC (Z22SE) engines	GMPT – E15 multi-point injection

Fuel system data

Fuel pump type	Electric, immersed in tank
Fuel pump regulated constant pressure (approximate):	
Single-point injection system	0.8 bar
Multi-point injection system	3.0 to 3.8 bar
Specified idle speed	Not adjustable – controlled by ECU
Idle mixture CO content	Not adjustable – controlled by ECU

Live data/sensor values

The following values/data are included in the hope that they may be of assistance to anyone attempting fault diagnosis. The values were recorded using a fault code reader, from a vehicle running correctly with the engine at idle, and increased idle speed.

Engine type	1.6 litre DOHC	
	Idle	**Increased idle**
Battery	14.2V	14.1V
Throttle position sensor	0.64V	0.86V
MAP sensor	0.25 bar 1.27V	0.16 bar 0.86V
Intake air temperature	24°C	21°C
Injector pulse	3.2 ms	1.8 ms
Spark advance angle (BTDC)	5° to 8°	18°
Knock retard:		
Cylinder No 1	0.0°	1.1°
Cylinder No 2	0.0°	0.0°
Cylinder No 3	0.0°	0.4°
Cylinder No 4	0.0°	0.0°

Live data/sensor values (continued)

Engine speed	750 rpm	1980 rpm
Air/fuel ratio	14.6:1	14.6:1
Oxygen sensor	62 to 900 mV	58 to 940 mV
Desired idle	775 rpm	1900 rpm
Engine load	9%	14%
Coolant temperature	86°C 2.17V	89°C 2.05V

Diesel models

General

System type:	
1.7 litre SOHC models	EDC MSA 15M indirect injection system incorporating an electronically controlled VP29 distributor fuel injection pump. Low pressure turbocharger and intercooler fitted to all models
1.7 litre DOHC models	ECD V5 direct injection system incorporating an electronically-controlled Nippon Denso V5 distributor fuel injection pump with integral transfer pump. Two stage injectors with pilot and post injection. Turbocharger and intercooler fitted to all models
2.0 litre models	EDC 15M direct injection system incorporating an electronically-controlled VP 44 fuel injection pump. Turbocharger fitted to all models and intercooler fitted to high-pressure turbo models

Adjustment data

Idle speed:	
1.7 litre SOHC models	780 to 900 rpm – controlled by ECU
1.7 litre DOHC models	825 to 875 rpm – controlled by ECU
2.0 litre models	750 to 850 rpm – controlled by ECU
Maximum speed:	
1.7 litre SOHC models	5400 to 5500 rpm – controlled by ECU
1.7 litre DOHC models	5100 to 5300 rpm – controlled by ECU
2.0 litre models	4900 to 5100 rpm – controlled by ECU

Injectors

Opening pressure:	
1.7 litre SOHC models	130 to 143 bar
1.7 litre DOHC models	175 bar (1st stage) to 335 bar (2nd stage)
2.0 litre models:	
Low-pressure turbo models	180 to 365 bar
High-pressure turbo models	220 to 380 bar

Torque wrench settings

	Nm	lbf ft
Petrol models		
Camshaft sensor:		
1.4, 1.6, 1.8 and 2.2 litre DOHC engines	8	6
2.0 litre DOHC engines	6	4
Crankshaft sensor bolt:		
SOHC engines	10	7
DOHC engines	8	6
Fuel pressure regulator clamp – DOHC engines	5	4
Fuel rail bolts	8	6
Knock sensor bolt	20	15
Throttle body – SOHC engines:		
Retaining nuts	22	16
Upper body-to-lower body screws	6	4
Throttle housing bolts – 1.4, 1.6 and 1.8 litre DOHC engines	8	6
Throttle housing nuts – 2.0 and 2.2 litre DOHC engines	9	7
Throttle potentiometer bolts – SOHC engines	2	1.5
Diesel models		
Atmospheric pressure sensor (1.7 litre DOHC)	10	7
Charge pressure sensor	8	6
Coolant temperature sensor	20	15
Crankshaft sensor	8	6
Engine control unit	6	4
Sump drain plug (2.0 litre models)	18	13

General information

Petrol models

Multec single-point injection

1 The Multec engine management (fuel injection/ignition) system incorporates a closed-loop catalytic converter, an evaporative emission control system and an exhaust gas recirculation (EGR) system, and complies with the latest emission control standards. The fuel injection side of the system operates as follows;

2 The fuel pump, immersed in the fuel tank, pumps fuel from the fuel tank to the fuel injector, via a filter mounted underneath the rear of the vehicle. Fuel supply pressure is controlled by the pressure regulator in the throttle body assembly. The regulator operates by allowing excess fuel to return to the tank.

3 The electrical control system consists of the ECU, along with the following sensors.

a) Throttle potentiometer – informs the ECU of the throttle position, and the rate of throttle opening or closing.

b) Coolant temperature sensor – informs the ECU of engine coolant temperature.

c) Oxygen sensor – informs the ECU of the oxygen content of the exhaust gases.

d) Crankshaft sensor – informs the ECU of engine speed and crankshaft position.

e) Knock sensor – informs the ECU when pre-ignition ('pinking') is occurring.

f) Manifold absolute pressure (MAP) sensor – informs the ECU of the engine load by monitoring the pressure in the inlet manifold.

g) ABS control unit – informs the ECU of the vehicle speed.

h) Air conditioning system compressor switch (where fitted) – informs ECU when the air conditioning system is switched on.

4 All the above information is analysed by the ECU and, based on this, the ECU determines the appropriate ignition and fuelling requirements for the engine. The ECU controls the fuel injector by varying its pulse width – the length of time the injector is held open – to provide a richer or weaker mixture, as appropriate. The mixture is constantly varied by the ECU, to provide the best setting for cranking, starting (with either a hot or cold engine), warm-up, idle, cruising, and acceleration.

5 The ECU also has full control over the engine idle speed, via a stepper motor which is fitted to the throttle body. The motor pushrod controls the opening of an air passage which bypasses the throttle valve. When the throttle valve is closed (accelerator pedal released), the ECU uses the motor to vary the amount of air entering the engine and so controls the idle speed.

6 The ECU also controls the exhaust and evaporative emission control systems.

7 If there is an abnormality in any of the readings obtained from any sensor, the ECU enters its back-up mode. In this event, the ECU ignores the abnormal sensor signal, and assumes a pre-programmed value which will allow the engine to continue running (albeit at reduced efficiency). If the ECU enters this back-up mode, the warning light on the instrument panel will come on, and the relevant fault code will be stored in the ECU memory.

8 If the warning light comes on, a complete test of the engine management system can be carried out using a fault code reader which is simply plugged into the system's diagnostic connector. The connector is located under the trim beneath the handbrake lever handle; unclip the trim panel situated beneath the handbrake lever to gain access.

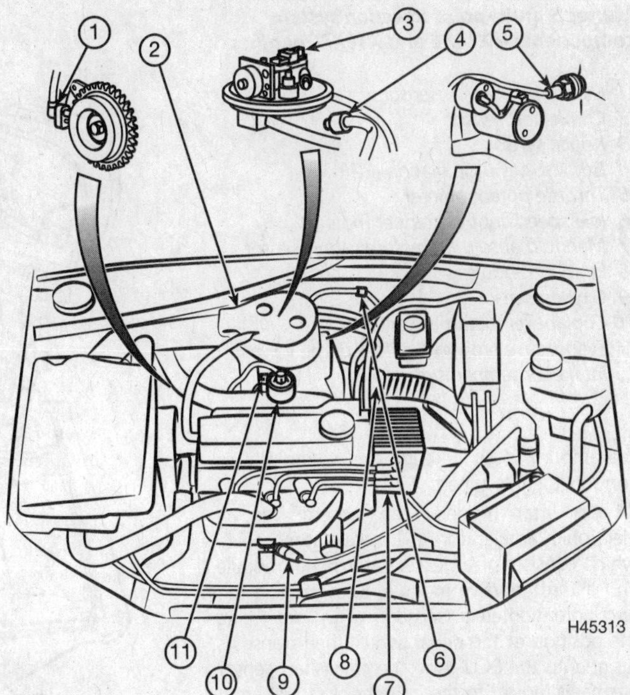

Multec single-point injection system components – X16SZR engines

1 Crankshaft sensor
2 Throttle body
3 Fuel injector
4 Throttle potentiometer
5 Knock sensor
6 Manifold absolute pressure (MAP) sensor
7 Ignition module
8 Evaporative emission system purge valve
9 Oxygen sensor
10 Exhaust gas recirculation (EGR) valve
11 Idle speed control stepper motor

Multec S multi-point injection

9 The Multec S engine management (fuel injection/ignition) system incorporates a closed-loop catalytic converter, an evaporative emission control system, and an exhaust gas recirculation (EGR) system, and complies with the latest emission control standards. The fuel injection side of the system operates as follows.

10 The fuel pump, immersed in the fuel tank, pumps fuel from the fuel tank to the fuel rail, via a filter mounted underneath the rear of the vehicle. Fuel supply pressure is controlled by the pressure regulator which allows excess fuel to be returned to the tank.

11 The electrical control system consists of the ECU, along with the following sensors.

a) Throttle potentiometer – informs the ECU of the throttle position, and the rate of throttle opening or closing.

b) Coolant temperature sensor – informs the ECU of engine temperature.

c) Intake air temperature sensor – informs the ECU of the temperature of the air entering the manifold.

d) Oxygen sensor – informs the ECU of the oxygen content of the exhaust gases.

e) Crankshaft sensor – informs the ECU of engine speed and crankshaft position.

f) Camshaft sensor – informs the ECU of speed and position of the exhaust camshaft.

g) Knock sensor – informs the ECU when pre-ignition (pinking) is occurring.

h) Manifold absolute pressure (MAP) sensor – informs the ECU of the engine load by monitoring the pressure in the inlet manifold.

i) ABS control unit – informs the ECU of the vehicle speed.

j) Air conditioning system compressor switch (where fitted) – informs ECU when the air conditioning system is switched on.

12 All the above information is analysed by the ECU and, based on this, the ECU determines the appropriate ignition and fuelling requirements for the engine. The ECU controls the fuel injectors by varying their pulse width – the length of time the injectors are held open – to provide a richer or weaker mixture, as appropriate. The mixture is constantly varied by the ECU, to provide the best setting for cranking, starting (with either a hot or cold engine), warm-up, idle, cruising, and acceleration. The Multec S system is a 'sequential' fuel injection system. This means that each of the four injectors is triggered individually just before the inlet valve of the relevant cylinder is about to open.

13 On early models, the ECU also has full control over the engine idle speed, via a stepper motor which is fitted to the throttle housing. The motor controls the opening of an air passage which bypasses the throttle valve. When the throttle valve is closed (accelerator pedal released), the ECU uses the motor to vary

Chapter 25

Multec S multi-point injection system components – X14XE and X16XEL engines

1 Camshaft position sensor
2 Crankshaft sensor
3 Knock sensor
4 Exhaust gas recirculation (EGR) valve
5 Throttle potentiometer
6 Idle speed control stepper motor
7 Manifold absolute pressure (MAP) sensor
8 Ignition module
9 Oxygen sensor
10 Coolant temperature sensor
11 Evaporative emission system purge valve
12 Intake air temperature sensor

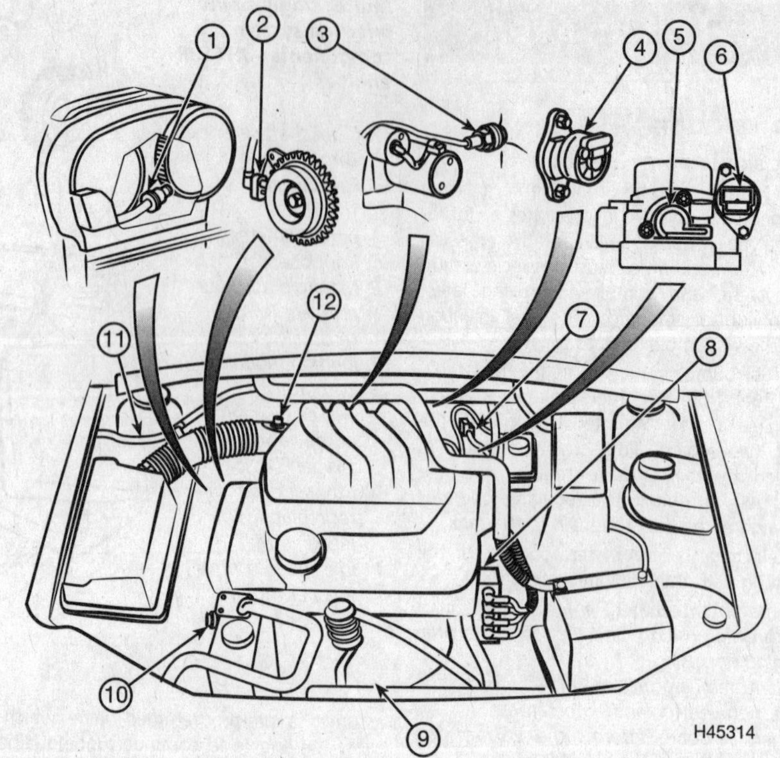

the amount of air entering the engine and so controls the idle speed.

14 On later models, the system has an electronic accelerator (no accelerator cable), which consists of an accelerator pedal module and a throttle valve module. Each module has two potentiometers (sensors) which determine the position of the pedal, which then transmits a signal to the ECU. Also the crankshaft sensor has been moved to the cylinder block, where it picks up the signal from an increment disc on number eight web on the crankshaft.

15 The ECU also controls the exhaust and evaporative emission control systems.

16 If there is an abnormality in any of the readings obtained from any sensor, the ECU may enter its back-up mode. In this event, the ECU ignores the abnormal sensor signal, and assumes a pre-programmed value which will allow the engine to continue running (albeit at reduced efficiency). If the ECU enters this back-up mode, the warning light on the instrument panel will come on, and the relevant fault code will be stored in the ECU memory.

17 If the warning light comes on, a complete test of the engine management system can be carried out using a fault code reader which is simply plugged into the system's diagnostic connector. The connector is located under the trim beneath the handbrake lever handle; unclip the trim panel situated just in front of the handbrake lever to gain access.

Multec S multi-point injection system components – Z14XE and Z16XE engines

1 Evaporative emission system purge valve
2 Oxygen sensor – catalytic converter control
3 Knock sensor
4 Throttle body
5 Exhaust gas recirculation (EGR) valve
6 Manifold absolute pressure (MAP) sensor
7 Engine electronic control unit (ECU)
8 Crankshaft sensor
9 Oxygen sensor – mixture regulation
10 Coolant temperature sensor
11 Coolant temperature sensor
12 Camshaft position sensor

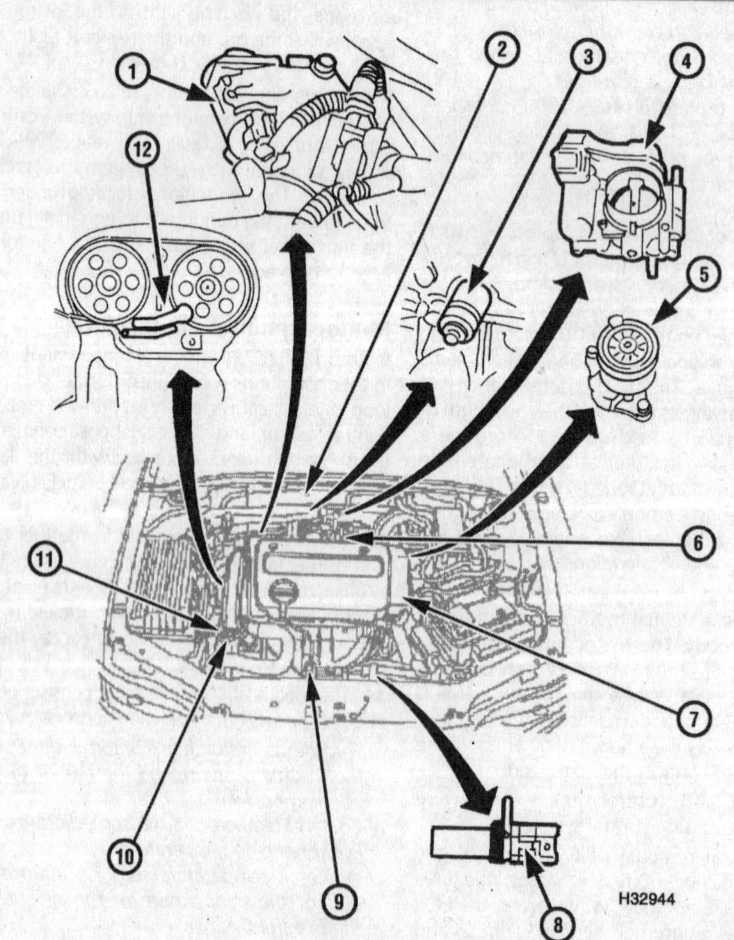

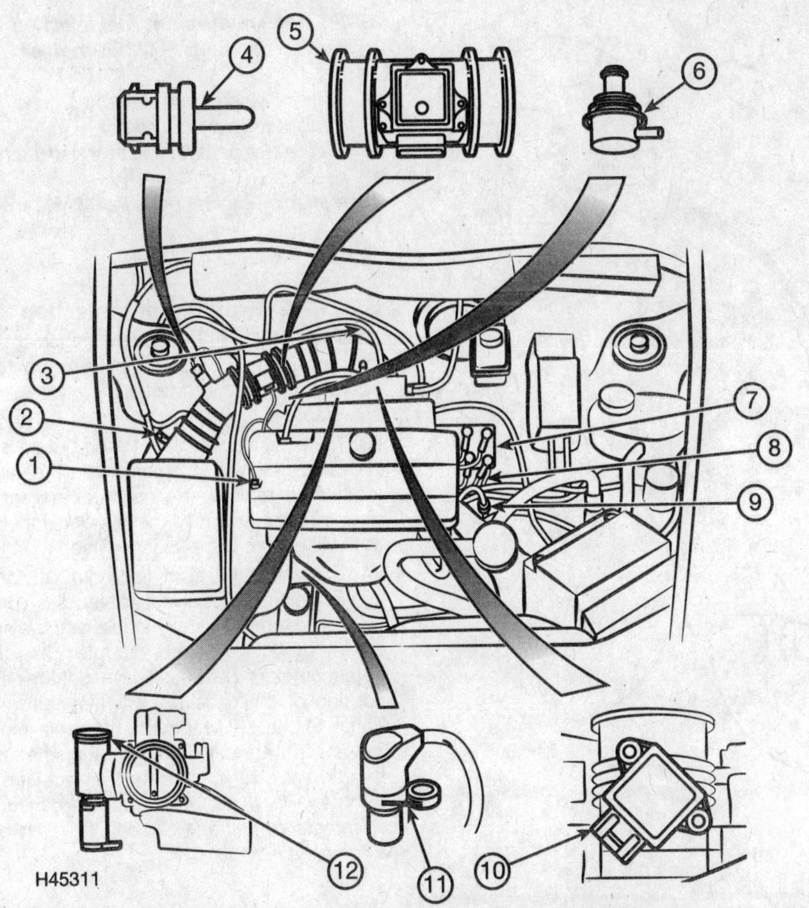

Simtec 70 multi-point fuel injection system components – X18XE1 and X20XEV engines

1 Camshaft position sensor
2 Evaporative emission system purge valve
3 Knock sensor (on rear of the cylinder block)
4 Intake air temperature sensor
5 Hot film mass airflow meter
6 Fuel pressure regulator
7 Ignition module
8 Coolant temperature sensor
9 Exhaust gas recirculation (EGR) valve
10 Throttle potentiometer
11 Crankshaft sensor
12 Idle speed control stepper motor

Simtec 70 and 71 multi-point injection

18 The Simtec 70 and 71 engine management (fuel injection/ignition) systems are almost identical in operation to the Multec S system (see paragraphs 9 to 17). The only major change to the system is that a hot film mass airflow meter is fitted in place of the manifold pressure (MAP) sensor. The airflow meter informs the ECU of the amount of air entering the inlet manifold.

19 Another additional feature of the Simtec system is that it incorporates a variable tract inlet manifold to help increase torque output at low engine speeds. Each inlet manifold tract is fitted with a valve. The valve is controlled by the ECU via a solenoid valve and vacuum diaphragm unit.

20 At low engine speeds (below approximately 3600 rpm) the valves remain closed. The air entering the engine is then forced to take the long inlet path through the manifold which leads to an increase in the engine torque output.

21 At higher engine speeds, the ECU switches the solenoid valve which then allows vacuum to act on the diaphragm unit. The diaphragm unit is linked to the valve assemblies and opens up each of the four valves allowing the air passing through the manifold to take the shorter inlet path which is more suited to higher engine speeds.

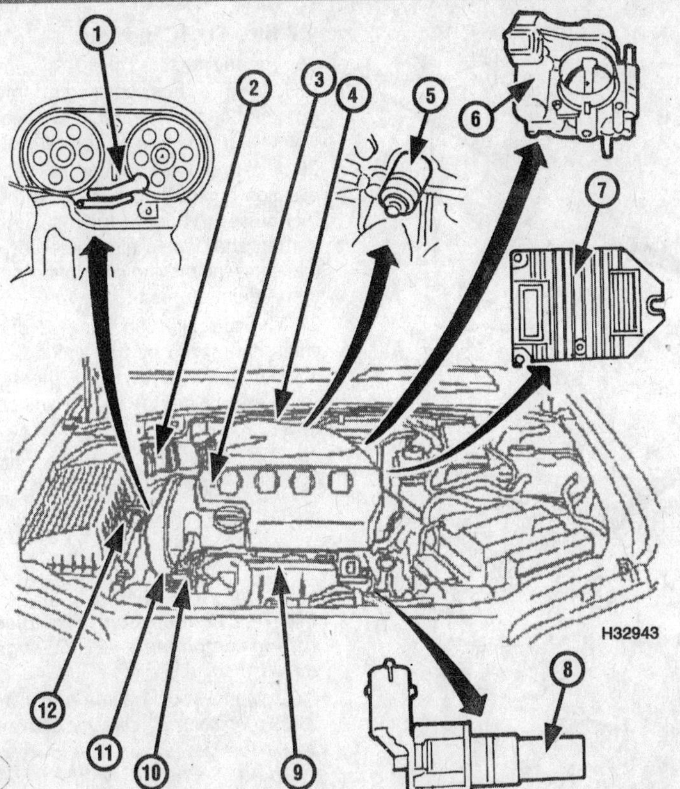

Simtec 71 multi-point fuel injection system components – Z18XE and Z18XEL engines

1 Camshaft position sensor
2 Hot film mass airflow meter
3 Ignition module
4 Oxygen sensor – catalytic converter control
5 Knock sensor
6 Throttle body
7 Engine electronic control unit (ECU)
8 Crankshaft sensor
9 Oxygen sensor – mixture regulation
10 Coolant temperature sensor
11 Coolant temperature sensor
12 Evaporative emission system purge valve

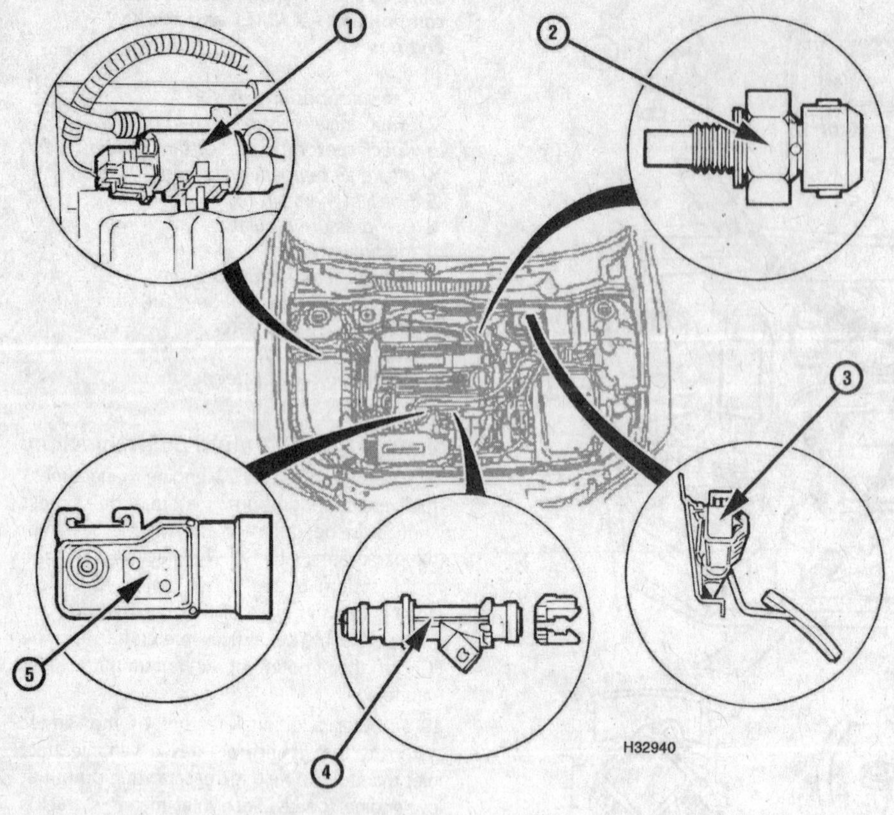

GMPT – E15 multi-point fuel injection system components – Z22SE engines

1 Evaporative emission system purge valve
2 Coolant temperature sensor
3 Accelerator pedal/accelerator pedal sensor
4 Injectors
5 Manifold absolute pressure (MAP) sensor

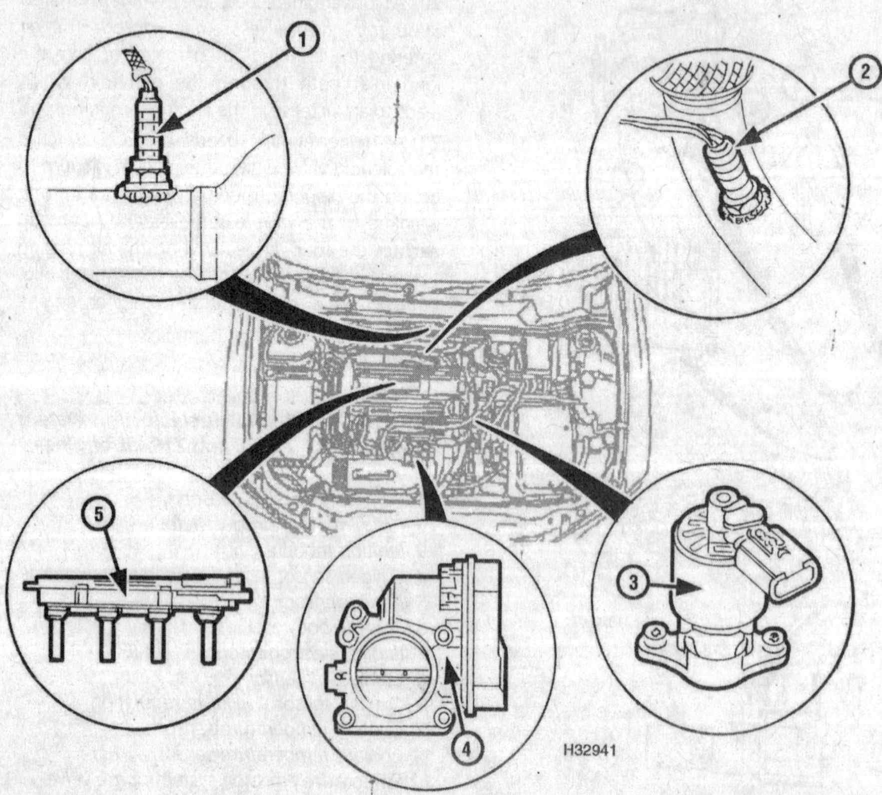

GMPT – E15 multi-point fuel injection system components – Z22SE engines

1 Oxygen sensor – mixture regulation
2 Oxygen sensor – catalytic converter control
3 Exhaust gas recirculation (EGR) valve
4 Throttle body
5 DIS ignition module

GMPT – E15 multi-point injection

22 The GMPT – E15 engine management (fuel injection/ignition) system is very similar in operation to the Multec S system (see paragraphs 9 to 17).

23 The only major change to the system is that special sensors are incorporated in the ignition module to remove the need for a camshaft sensor. The sensors read the injection impulse so as to determine the cylinder position.

24 This system also has an electronic accelerator (no accelerator cable), this consists of an accelerator pedal module and a throttle valve module. Each module has two potentiometers (sensors) which determine the position of the pedal, which then transmits a signal to the ECU. The throttle valve module consists of a throttle valve potentiometer and a throttle valve servo motor which actuates the throttle valve. **Note:** *The throttle valve module is an integrated unit and can only be renewed as a complete assembly.*

Diesel models

1.7 litre SOHC models

25 The fuel system consists of a rear-mounted fuel tank, a fuel filter with integral water separator and fuel heater, a fuel injection pump, injectors and associated components.

26 Fuel is drawn from the fuel tank by the fuel injection pump. Before reaching the pump, the fuel passes through a fuel filter, where foreign matter and water are removed. Excess fuel lubricates the moving components of the pump, and is then returned to the tank.

27 The fuel injection pump is driven at half-crankshaft speed by the timing belt. The high pressure required to inject the fuel into the compressed air in the swirl chamber is achieved by a radial piston pump.

28 The injection pump is electronically-controlled to meet the latest emission standards.

GMPT – E15 multi-point fuel injection system components – Z22SE engines

1 Engine electronic control unit (ECU)
2 Fuel pressure regulator
3 Crankshaft sensor
4 Knock sensor
5 Intake air temperature sensor

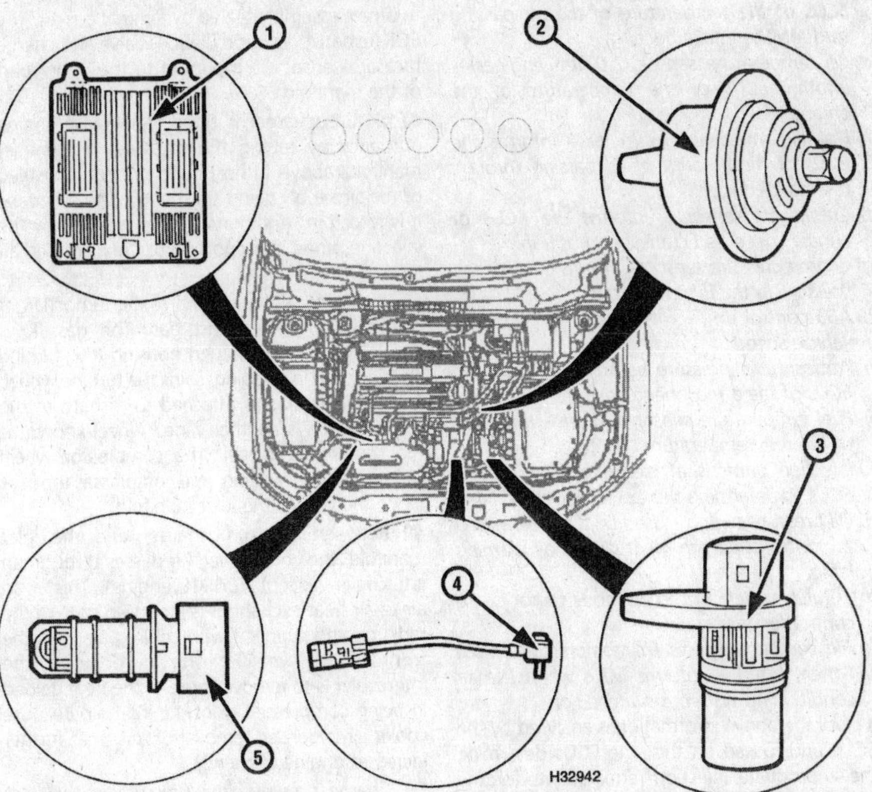

The system consists of the engine electronic control unit, injection pump electronic control unit (integral with the injection pump) and the following sensors:

a) Accelerator pedal position sensor – informs the ECUs of the accelerator pedal position.

b) Coolant temperature sensor – informs the ECUs of the engine temperature.

c) Hot film mass airflow meter – informs the ECUs of the mass and temperature of the air passing through the intake duct.

d) Crankshaft sensor – informs the ECUs of the engine speed and crankshaft position.

e) Charge pressure sensor – informs the ECUs of the pressure in the inlet manifold.

f) Injection pump sensors and regulators (integral with the injection pump) – used be the ECUs to determine and control injection timing and quantity.

g) Clutch switch – used by the ECUs for cruise control functions.

h) Air conditioning system compressor switch – informs the ECUs when the air conditioning system is switched on.

29 All the above information is analysed by the ECUs and, based on this, the ECUs determine the appropriate injection requirements for the engine. The engine ECU controls the injection pump timing, via the pump control unit, to provide the best setting for cranking, starting (with either a hot or cold engine), warm-up, idle, cruising, and acceleration.

30 Basic injection timing is determined when the pump is fitted. When the engine is running, it is varied automatically to suit the prevailing engine speed by a mechanism which turns the cam plate or ring – controlled by the ECU.

31 The engine ECU also controls the exhaust gas recirculation (EGR) system and the pre-heating system.

32 The four fuel injectors produce a spray of fuel indirectly into the cylinders via the swirl chambers. The injectors are calibrated to open and close at critical pressures to provide efficient and even combustion. Each injector needle is lubricated by fuel, which accumulates in the spring chamber and is channelled to the injection pump return hose by leak-off pipes.

33 A low-pressure turbocharger is fitted to increase engine efficiency by raising the pressure in the inlet manifold above atmospheric pressure. Instead of the air simply being sucked into the cylinders, it is forced in. Additional fuel is supplied by the injection pump in proportion to the increased air intake. Mounted between the exhaust manifold and front exhaust pipe, and driven by exhaust gases, the turbocharger takes its air supply from the filter housing and passes air under pressure to the inlet manifold via an intercooler (air cooler) which is mounted on the left-hand side of the radiator.

34 A dedicated oil supply pipe that runs from the cylinder block provides lubrication for the turbocharger. Oil is returned to the sump. The turbocharger has an integral wastegate valve and vacuum actuator diaphragm, which is used to control the charge pressure applied to the inlet manifold. The internal components of a turbocharger rotate at very high speed and are very sensitive to contamination. A great deal of damage can be caused by small particles of dirt, particularly if they strike the delicate turbine blades. To prevent the ingress of dirt during maintenance, thoroughly clean the area around all connections before disturbing them. Always stored dismantled components in a sealed container to prevent contamination. Cover the turbocharger air inlet ducts to prevent the ingress of debris and only use lint-free cloths when cleaning.

35 Never run the engine with the turbocharger air inlet hose disconnected. Depression at the inlet can build up very suddenly if the engine speed is raised, thereby increasing the risk of foreign objects being sucked in and ejected at high speed.

36 If there is an abnormality in any of the readings obtained from any sensor, the ECU enters its back-up mode. In this event, the ECU ignores the abnormal sensor signal, and assumes a pre-programmed value, which will allow the engine to continue running (albeit at reduced efficiency). If the ECU enters this back-up mode, the warning light on the instrument panel will come on, and the relevant fault code will be stored in the ECU memory.

1.7 litre DOHC and 2.0 litre models

37 The fuel system consists of a rear-mounted fuel tank, a fuel filter with integral water separator and fuel heater, a fuel injection pump, injectors and associated components.

38 Fuel is drawn from the fuel tank by the fuel injection pump. Before reaching the pump, the fuel passes through a fuel filter, where foreign matter and water are removed. Excess fuel lubricates the moving components of the pump, and is then returned to the tank.

39 On 1.7 litre engines the fuel injection pump is driven at half-crankshaft speed by the timing belt. On 2.0 litre engines, the fuel injection pump is driven at half-crankshaft speed by the timing chain. The high pressure required to inject the fuel into the compressed air in the cylinder is achieved by a radial piston pump.

40 The injection pump is electronically-controlled to meet the latest emission standards. The system consists of the engine electronic control unit, the injection electronic control unit, and the following sensors:

a) Accelerator pedal position sensor – informs the ECUs of the accelerator pedal position.

b) Coolant temperature sensor – informs the ECUs of engine temperature.

c) Intake air temperature sensor – informs the

ECUs of the temperature of the air passing through the intake duct.

d) Oil temperature sensor (2.0 litre engines) – informs ECUs of the temperature of the engine oil.

e) Hot film mass airflow meter – informs the ECUs of the amount of air passing through the intake duct.

f) Crankshaft sensor – informs the ECUs of engine speed and crankshaft position.

g) Charge pressure sensor – informs ECUs of the pressure in the inlet manifold.

h) ABS control unit – informs the ECUs of the vehicle speed.

i) Atmospheric pressure sensor – informs the ECUs of the atmospheric pressure.

j) Fuel temperature sensor – informs the ECUs of the fuel temperature.

k) Injection pump shaft sensor – used by the ECUs to determine the exact injection timing.

l) Oil pressure switch – used by the Engine ECU to control the instrument cluster oil warning lamp.

m) Clutch switch – used by the ECUs for cruise control functions.

n) Air conditioning system compressor switch (where fitted) – informs ECUs when the air conditioning system is switched on.

41 All the above information is analysed by the ECUs and, based on this, the ECUs determine the appropriate injection requirements for the engine. The engine ECU controls the injection pump timing, via the pump control unit, to provide the best setting for cranking, starting (with either a hot or cold engine), warm-up, idle, cruising, and acceleration.

42 Basic injection timing is determined when the pump is fitted. When the engine is running, it is varied automatically to suit the prevailing engine speed by a mechanism which turns the cam plate or ring – controlled by the ECU.

43 The engine ECU also controls the exhaust gas recirculation (EGR) system and the pre-heating system.

44 The four fuel injectors produce a spray of fuel directly into the cylinders. The injectors are calibrated to open and close at critical pressures to provide efficient and even combustion. Each injector needle is lubricated by fuel, which accumulates in the spring chamber and is channelled to the injection pump return hose by leak-off pipes.

45 On 2.0 litre engines, the inlet manifold is fitted with a butterfly valve arrangement to improve efficiency at low engine speeds. Each cylinder has two intake tracts in the manifold, one of which is fitted a valve; the operation of the valve is controlled by the ECU via a solenoid valve and vacuum diaphragm unit. At low engine speeds (below approximately 1500 rpm) the valves remain closed, meaning that air entering each cylinder is passing through only one of the two manifold tracts. At higher engine speeds, the ECU opens up each of the four valves allowing the air passing through the manifold to pass through both inlet tracts.

46 On 1.7 litre engines, the inlet manifold is a two-part assembly sealed by a metal gasket. The EGR (exhaust gas recirculation) valve and charge pressure sensor are mounted to the upper part of the manifold.

47 A turbocharger is fitted to increase engine efficiency by raising the pressure in the inlet manifold above atmospheric pressure. Instead of the air simply being sucked into the cylinders, it is forced in. Additional fuel is supplied by the injection pump in proportion to the increased air intake.

48 Energy for the operation of the turbocharger comes from the exhaust gas. The gas flows through a specially-shaped housing (the turbine housing) and in so doing, spins the turbine wheel. The turbine wheel is attached to a shaft, at the end of which is another vaned wheel known as the compressor wheel. The compressor wheel spins in its own housing, and compresses the inlet air on the way to the inlet manifold.

49 Between the turbocharger and the inlet manifold, the compressed air passes through an intercooler (except Y20DTL engine). This is an air-to-air heat exchanger is mounted next to the radiator, and supplied with cooling air from the front of the vehicle. The purpose of the intercooler is to remove some of the heat gained in being compressed from the inlet air. Because cooler air is denser, removal of this heat further increases engine efficiency.

50 Charge pressure (the pressure in the inlet manifold) is limited by a wastegate, which diverts the exhaust gas away from the turbine wheel in response to a pressure-sensitive actuator. A pressure-operated switch operates a warning light on the instrument panel in the event of excessive charge pressure developing.

51 The turbo shaft is pressure-lubricated by an oil feed pipe from the engine main oil so that the shaft 'floats' on a cushion of oil. A drain pipe returns the oil to the sump.

52 The charge pressure wastegate is controlled by the ECU via a solenoid valve.

53 If there is an abnormality in any of the readings obtained from any sensor, the ECU enters its back-up mode. In this event, the ECU ignores the abnormal sensor signal, and assumes a pre-programmed value which will allow the engine to continue running (albeit at reduced efficiency). If the ECU enters this back-up mode, the warning light on the instrument panel will come on, and the relevant fault code will be stored in the ECU memory.

Fuel injection system – depressurisation, priming and bleeding

Petrol models

1 The fuel system referred to in this Section is defined as the tank-mounted fuel pump, the fuel filter, the fuel injector(s) and the pressure regulator, and the metal pipes and flexible hoses of the fuel lines between these components. All these contain fuel which will be under pressure while the engine is running, and/or while the ignition is switched on. The pressure will remain for some time after the ignition has been switched off, and it must be relieved in a controlled fashion when any of these components are disturbed for servicing work.

SOHC engines

2 Ensure the ignition is switched off then remove the cover from the engine compartment relay box, which is situated next to the left-hand suspension turret. Remove the fuel pump relay (the relay should be coloured purple) from the box.

3 Start the engine and allow it to idle until the engine starts to run roughly then switch the ignition off.

4 Disconnect the battery negative terminal then refit the relay and cover to the relay box.

DOHC engines

5 On these engines, the fuel system can either be depressurised as described above in paragraphs 2 to 4, or as follows.

6 Locate the valve assembly which is fitted to the fuel rail on the inlet manifold. On 1.4 and 1.6 litre engines the valve is on the right-hand

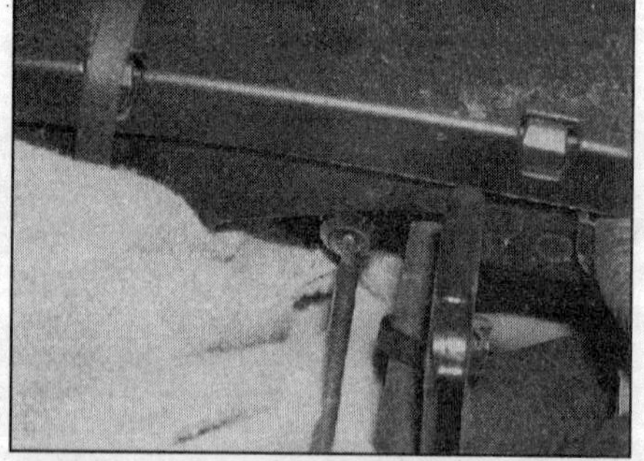

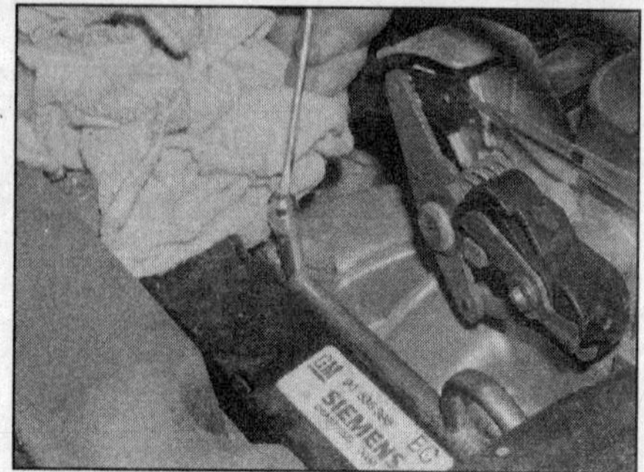

end of the rail and, on 1.8, 2.0 and 2.2 litre engines, it can be found on the top of the rail.

7 Unscrew the cap from the valve and position a container beneath the valve. Hold a wad of rag over the valve and relieve the pressure in the fuel system by depressing the valve core with a suitable screwdriver. Be prepared for the squirt of fuel as the valve core is depressed and catch it with the rag. Hold the valve core down until no more fuel is expelled from the valve.

8 Once all pressure is relieved, securely refit the valve cap.

Diesel models

9 It is not necessary to manually prime and bleed the fuel system after any operation on the system components. Start the engine (this may take longer than usual, especially if the fuel system has been allowed to run dry – operate the starter in ten second bursts with 5 seconds rest in between each operation) and run it a fast idle speed for a minute or so to purge any trapped air from the fuel lines. After this time the engine should idle smoothly at a constant speed.

10 If the engine idles roughly, then there is still some air trapped in the fuel system. Increase the engine speed again for another minute or so then recheck the idle speed. Repeat this procedure as necessary until the engine is idling smoothly.

Multec single-point injection system components – removal and refitting

Fuel injector

Note: *Before condemning an injector, it is worth trying the effect of one of the proprietary injector-cleaning treatments.*

1 Depressurise the fuel system, then disconnect the battery negative terminal.

2 Slacken the retaining clip and disconnect the air intake duct from the throttle body cover. Disconnect the vacuum and breather hoses from the cover then undo the retaining screws and remove the cover and sealing ring from the top of the throttle body.

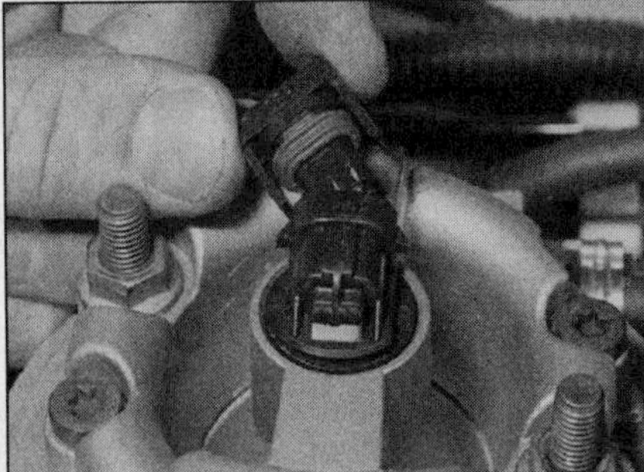

3 Release the retaining clips and disconnect the wiring connector from the injector.

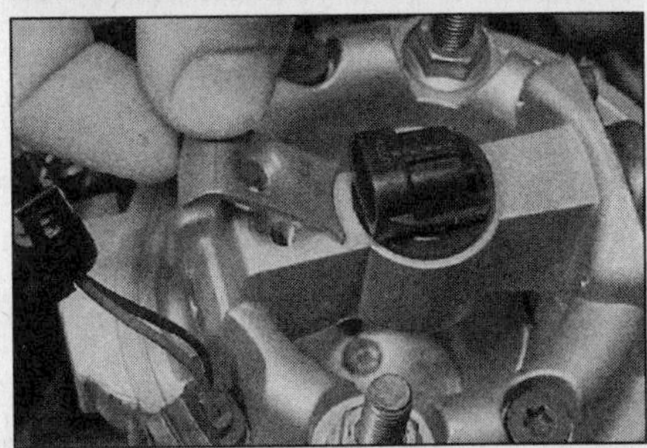

4 Undo the retaining screw and remove the injector retaining plate.

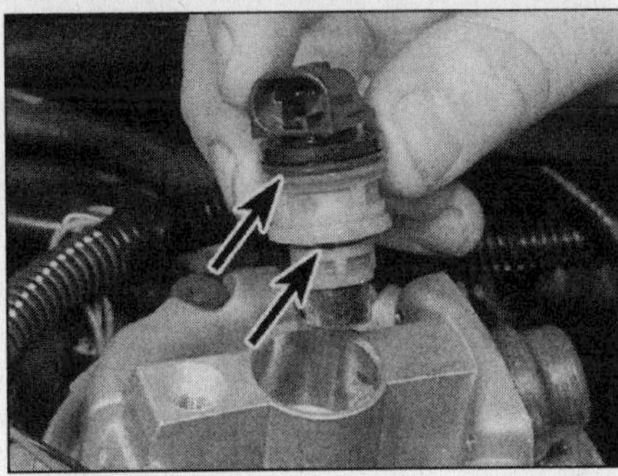

5 Ease the injector out from the throttle body along with its sealing rings. Discard the sealing rings, they must be renewed whenever the injector is disturbed.

6 Refitting is a reversal of the removal procedure using new injector sealing rings. When refitting the retaining clip, ensure it is correctly engaged with the injector and securely tighten its retaining screw.

Fuel pressure regulator

Note: *At the time of writing it appears that the regulator assembly is not available separately; if it is faulty the complete throttle body upper section must be renewed. Although the unit can be dismantled for cleaning, if required, it should not be disturbed unless absolutely necessary.*

7 Depressurise the fuel system, then disconnect the battery negative terminal.

8 Slacken the retaining clip and disconnect the air intake duct from the throttle body cover. Disconnect the vacuum and breather hoses from the cover then undo the retaining screws and remove the cover and sealing ring from the top of the throttle body.

9 Using a marker pen, make alignment marks between the regulator cover and throttle body, then slacken and remove the cover retaining screws.

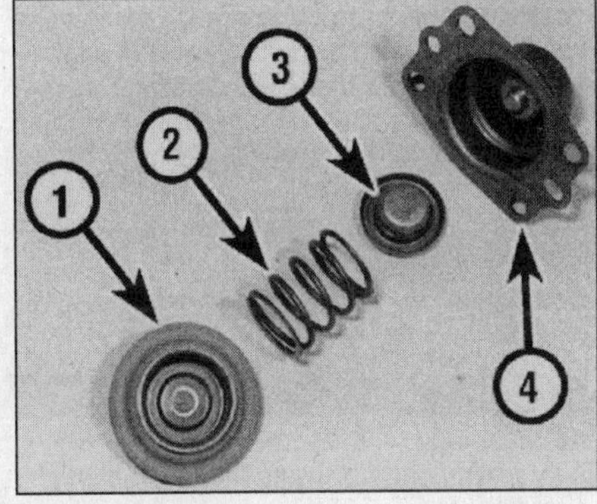

10 Lift off the cover (4), then remove the spring seat (3) and spring (2) then withdraw the diaphragm (1), noting its correct fitted orientation. Remove all traces of dirt, and examine the diaphragm for signs of splitting. If damage is found, it will probably be necessary to renew the throttle body assembly.

11 Refitting is a reverse of the removal procedure, ensuring that the diaphragm and cover are fitted the correct way round, and that the retaining screws are securely tightened.

Idle speed control stepper motor

12 Disconnect the battery negative terminal.

13 To improve access to the motor, slacken the retaining clip and disconnect the air intake duct from the throttle body cover. Disconnect the vacuum and breather hoses from the cover then undo the retaining screws and remove the cover and sealing ring from the top of the throttle body.

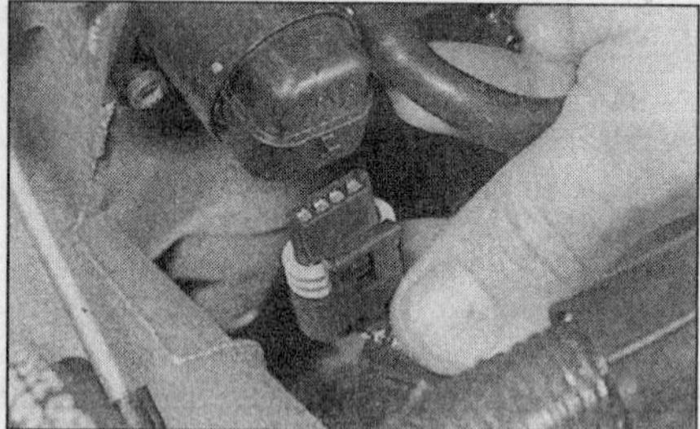

14 Release the retaining clip and disconnect the wiring connector from the motor which is fitted to the front of the throttle body assembly.

15 Undo the retaining screws and carefully manoeuvre the motor out of position, taking great care not to damage the motor plunger. Remove the sealing ring from the motor and discard it; a new one should be used on refitting.

16 Refitting is the reverse of removal, using a new sealing ring. To ensure the motor plunger is not damaged on refitting, prior to installation, check that the plunger tip does not extend more than 28 mm from the motor mating flange (A). If necessary, **gently** push the plunger into the body until it is correctly positioned.

Throttle potentiometer

17 Disconnect the battery negative terminal.

18 Release the retaining clip and disconnect the wiring connector from the throttle potentiometer which is fitted to the left-hand

side of the throttle body. Slacken and remove the retaining screws and remove the potentiometer.

19 Prior to refitting, clean the threads of the retaining bolts and apply a drop of fresh locking compound to each one. Ensure the potentiometer is correctly engaged with the throttle valve spindle then tighten its retaining bolts to the specified torque and reconnect the wiring connector.

Coolant temperature sensor

20 The coolant temperature sensor is screwed into the rear of the inlet manifold. Partially drain the cooling system.

21 Remove the air cleaner and air ducts.

22 Disconnect the wiring plug, then unscrew the sensor from the manifold. Recover the sealing ring.

23 Refitting is a reversal of removal, ensuring a new sealing ring if fitted.

Manifold absolute pressure sensor

24 The MAP sensor is mounted onto the engine compartment bulkhead, just to the left of the throttle body. Ensure the ignition is switched off then disconnect the wiring connector and vacuum hose from the sensor. The MAP sensor can then be unclipped and removed from its mounting.

25 Refitting is the reverse of removal.

Crankshaft sensor

26 The sensor is mounted on the rear of the cylinder block and is accessible from underneath the vehicle. Firmly apply the handbrake, then jack up the front of the car and support it securely on axle stands.

27 Trace the wiring back from the sensor, releasing it from all the relevant clips and ties whilst noting its correct routing. Disconnect the wiring connector so the wiring is free to be removed with the sensor.

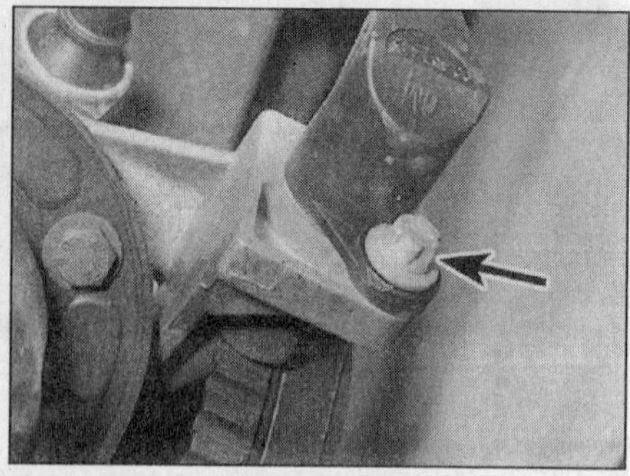

28 Unscrew the retaining bolt and remove the sensor from underneath the vehicle.

29 Refitting is the reverse of removal, tightening the retaining bolt to the specified torque. Ensure the wiring is correctly routed and retained by all the necessary clips and ties.

30 On completion, using feeler gauges, check that the clearance between the sensor tip and the crankshaft pulley teeth is 1.0 ± 0.7 mm. If the clearance is not within the specified range, renew the sensor mounting bracket.

Knock sensor

31 The knock sensor is mounted on the rear of the cylinder block and is accessible from underneath the vehicle. Firmly apply the handbrake, then jack up the front of the car and support it securely on axle stands.

32 Trace the wiring back from the sensor, noting its correct routing, and disconnect it at the connector.

33 Slacken and remove the retaining bolt and remove the sensor from the engine.

34 On refitting ensure the mating surfaces are clean and dry then fit the sensor and tighten its retaining bolt to the specified torque. Ensure the wiring is correctly routed and securely reconnected then lower the vehicle to the ground.

Electronic control unit

35 The ECU is located beside the battery on the left-hand side of the engine compartment. Disconnect the battery negative terminal.

36 Remove both windscreen wiper arms.

37 Remove the protective cover from the ECU, then pull the unit upwards out of the retaining bracket. Disconnect the ECU wiring plugs. **Note:** *Prior to renewing an ECU, the security code must be reset using dedicated diagnostic equipment. Refer to your local Vauxhall dealer or specialist.*

38 Refitting is a reversal of removal.

Fuel pump relay

39 The fuel pump relay is located in the engine compartment main relay box.

40 Unclip the cover and remove it from the relay box. Identify the fuel pump relay which is coloured purple. Ensure the ignition is switched off then pull out the relay.

41 Refitting is the reverse of removal.

Air conditioning system switch

42 The air conditioning system switch is screwed into one of the refrigerant pipes and cannot be removed without first discharging the refrigerant. Renewal of the switch should therefore be entrusted to a suitably-equipped garage.

Multec S multi-point injection system components – removal and refitting

Fuel rail and injectors

Note: *If a faulty injector is suspected, it is worth trying the effect of one of the proprietary injector-cleaning treatments.*

1 Depressurise the fuel system, then disconnect the battery negative terminal.

2 Remove the oil filler cap, undo the retaining screws, and remove the engine cover.

3 Disconnect the wiring plug from the intake air temperature sensor, and detach the intake trunking from the air cleaner housing and the throttle body. On Zafira models, it will be necessary to remove the engine compartment seal and water deflector cover from in front of the windscreen, then undo the retaining nuts and bolts, and remove the bulkhead cover plate to allow access to the air intake trunking.

4 Disconnect the engine breather hoses from the camshaft cover.

5 Detach all of the wiring connectors which are associated with the wiring harness plastic tray which runs above the fuel rail. Note the cable routing. The items to disconnect are:

Earth connections on manifold (two).
Evaporative canister purge valve.
Crankshaft position sensor.
Oil pressure switch.
Idle speed stepper motor.
Coolant temperature sensor.
Throttle position sensor.
Camshaft position sensor.
MAP sensor.
EGR valve.
Ignition module.
Oxygen sensor.
ECU (top connector).
Knock sensor.
Interference suppressor (on Ignition module).
The grey connector in front of the engine compartment relay box.

6 On X14XE and X16XEL engines, detach the vacuum hose from the fuel pressure regulator on the left-hand end of the fuel rail.

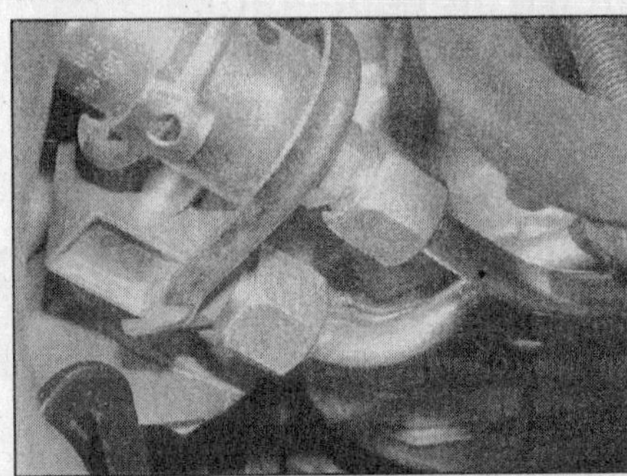

7 Disconnect the supply pipe and, on X14XE and X16XEL engines, the return pipe from the fuel rail or fuel pressure regulator. Note that the unions are different sizes.

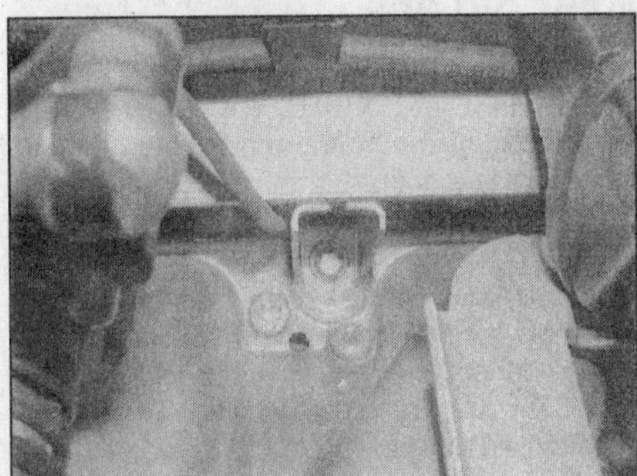

8 Unscrew the two retaining bolts and remove the fuel rail with the injectors complete with the plastic wiring tray.

9 If required, the injectors can be separated from the fuel rail by sliding off the relevant retaining clip and withdrawing the injector from the fuel rail. Remove the upper sealing ring from the injector and discard it; all disturbed sealing rings must be renewed.

10 Refitting is a reversal of the removal procedure, noting the following points.

a) Renew all disturbed sealing rings and apply a smear of engine oil to them to aid installation.
b) Ease the injector(s) into the fuel rail, ensuring that the sealing ring(s) remain correctly seated, and secure in position with the retaining clips.
c) On refitting the fuel rail, take care not to damage the injectors and ensure that all sealing rings remain in position. Once the fuel rail is correctly seated, tighten its retaining bolts to the specified torque.
d) On completion start the engine and check for fuel leaks.

Fuel pressure regulator

Note: *On 1.4 litre (Z14XE) and 1.6 litre (Z16XE) engines, fuel pressure regulation is a function of the fuel tank pump module. The regulator is located under the fuel pump access cover. The regulator does not appear to be available as a separate part – check with your Vauxhall dealer.*

11 Undo the two retaining screws and the oil filler cap, and remove the engine cover. Depressurise the fuel system, then disconnect the battery negative terminal.

12 Detach the vacuum hose from the regulator.

13 Undo the screw and remove the regulator retaining clamp.

14 Carefully pull the regulator from the fuel rail along with its sealing rings.

15 Refitting is the reverse of removal, using new sealing rings. On completion start the engine and check for fuel leaks.

Idle speed control stepper motor

Note: *On 1.4 litre (Z14XE) and 1.6 litre (Z16XE) engines, there is an electronic accelerator system fitted (no accelerator cable). The throttle valve potentiometer and servo motor are a combined part of the throttle body housing. Individual parts for the throttle body housing are not available separately and, if defective, it must be renewed as a complete unit.*

16 Undo the two retaining screws and the oil filler cap, and remove the engine cover.

17 Disconnect the intake air temperature sensor and remove the air intake trunking from the air cleaner housing and the throttle body. On Zafira models, it will be necessary to remove the engine compartment seal and water deflector cover from in front of the windscreen, then undo the retaining nuts and bolts, and remove the bulkhead cover plate to allow access to the air intake trunking.

18 Detach the wiring harness plug from the stepper motor.

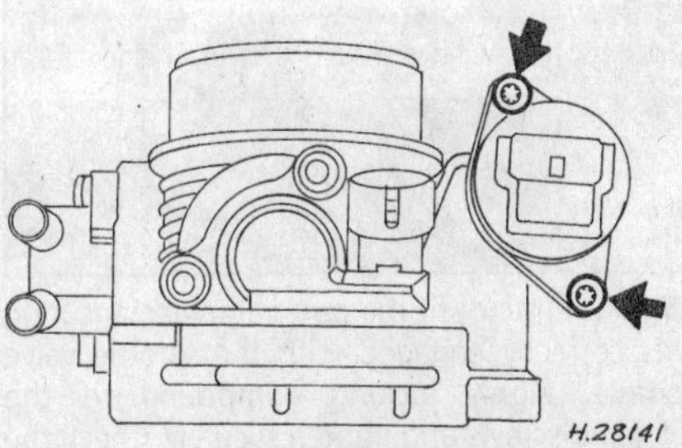

19 Undo the two screws and remove the motor from the throttle body.

20 Refitting is the reverse of removal, noting the following points.

a) Prior to installation, check that the plunger tip does not extend more than 33 mm from the motor mating flange (see paragraph 16 on page 25•10). If necessary, ***gently*** *push the plunger into the body until it is correctly positioned. Failure to do this could lead to the motor being damaged.*

b) Apply locking compound to the motor screws.

Throttle potentiometer

Note: *On 1.4 litre (Z14XE) and 1.6 litre (Z16XE) engines, there is an electronic accelerator system fitted (no accelerator cable). The throttle valve potentiometer and servo motor are a combined part of the throttle body housing. Individual parts for the throttle body housing are not available separately and, if defective, it must be renewed as a complete unit.*

21 Undo the two retaining screws and the oil filler cap, and remove the engine cover.

22 Disconnect the intake air temperature sensor and remove the air intake trunking from the air cleaner housing and the throttle body. On Zafira models, it will be necessary to remove the engine compartment seal and water deflector cover from in front of the windscreen, then undo the retaining nuts and bolts, and remove the bulkhead cover plate to allow access to the air intake trunking.

23 Disconnect the wiring harness plug from the potentiometer.

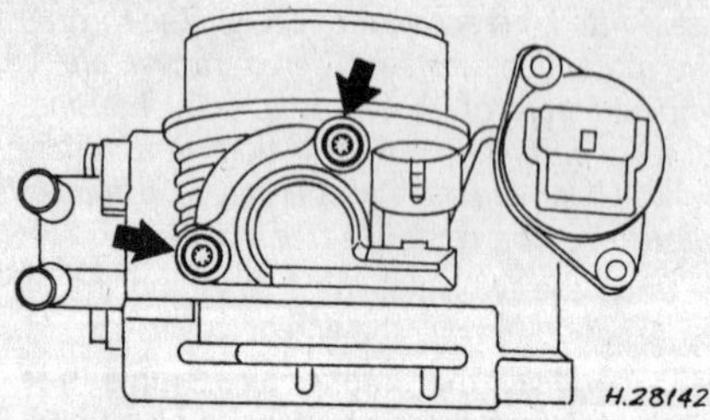

24 Undo the two retaining screws and remove the potentiometer from the throttle body.

25 On refitting, fit the potentiometer, ensuring it is correctly engaged with the throttle valve spindle. Apply locking compound to the retaining screws and tighten securely. Check the operation of the throttle valve then reconnect the wiring plug and refit the air intake trunking.

Coolant temperature sensor

26 The coolant temperature sensor is screwed into the thermostat housing. Partially drain the cooling system.

27 Remove the air cleaner and air ducts.

28 Disconnect the wiring plug, then unscrew the sensor from the housing.

29 Refitting is a reversal of removal. Apply a little sealant to the sensor threads prior to refitting.

Intake air temperature sensor

30 The intake air temperature sensor is mounted in the intake duct which connects the air cleaner housing to the inlet manifold. Ensure the ignition is switched off, then disconnect the wiring connector from the sensor.

31 Carefully ease the sensor out of position and remove its sealing grommet from the intake duct. If the sealing grommet shows signs of damage or deterioration it should be renewed.

32 Refitting is the reverse of removal, ensuring the sensor and grommet are correctly located in the duct.

Manifold absolute pressure sensor

33 On X14XE and X16XEL engines, the MAP sensor is mounted onto the engine compartment bulkhead, just to the left of the inlet manifold. On Z14XE and Z16XE engines, the MAP sensor is located on the left-hand side of the inlet manifold.

34 On X14XE and X16XEL engines, ensure the ignition is switched off then disconnect the wiring connector and vacuum hose from the sensor. The MAP sensor can then be unclipped and removed from its mounting. On Z14XE and Z16XE engines, undo the two retaining screws and the oil filler cap, and remove the engine cover. Ensure the ignition is switched off then disconnect the wiring connector from the sensor. The MAP sensor can then be unbolted and removed from the manifold.

35 Refitting is the reverse of removal.

Crankshaft sensor

36 On X14XE and X16XEL engines, the sensor is mounted on a bracket at the rear of the right-hand end of the cylinder block and is accessible from underneath the vehicle. Firmly apply the handbrake, then jack up the front of the car and support it securely on axle stands. On Z14XE and Z16XE engines, the sensor is mounted on the front of the cylinder block below the oil filter.

37 Trace the wiring back from the sensor, releasing it from all the relevant clips and ties whilst noting its correct routing. Disconnect the wiring connector so the wiring is free to be removed with the sensor.

38 Unscrew the retaining bolt and remove the sensor from underneath the vehicle.

39 Refitting is the reverse of removal, tightening the retaining bolt to the specified torque. Ensure the wiring is correctly routed and retained by all the necessary clips and ties.

40 On X14XE and X16XEL engines, using feeler gauges, check that the clearance between the sensor tip and the crankshaft pulley teeth is 1.0 ± 0.7 mm. If the clearance is not within the specified range, check that the bracket mounting surface is clean and flat with no corrosion.

Camshaft sensor

41 Remove the timing belt upper cover.

42 Trace the wiring back from the sensor, releasing it from all the relevant clips and ties whilst noting its correct routing. Disconnect the wiring connector so the wiring is free to be removed with the sensor.

43 Unscrew the retaining bolts and remove the sensor from the end of the cylinder head.

44 Refitting is the reverse of removal, apply a little locking compound, and tighten the retaining bolts to the specified torque. Ensure the wiring is correctly routed and retained by all the necessary clips and ties.

Knock sensor

45 The knock sensor is mounted onto the rear of the cylinder block, just to the right of the starter motor, and is accessible from underneath the vehicle. Firmly apply the handbrake, then jack up the front of the car and support it securely on axle stands.

46 Trace the wiring back from the sensor, noting its correct routing, and disconnect it at the connector.

47 Slacken and remove the retaining bolt and remove the sensor from the engine.

48 On refitting ensure the mating surfaces are clean and dry then fit the sensor and tighten its retaining bolt to the specified torque. Ensure the wiring is correctly routed and securely reconnected then lower the vehicle to the ground.

Electronic control unit

49 Disconnect the battery negative terminal.

50 Undo the two retaining screws and the oil filler cap, and remove the engine cover.

51 Release the retaining clips then disconnect

the wiring connectors from the ECU. Undo the three retaining bolts and remove the ECU from the vehicle.

52 Refitting is the reverse of removal, ensuring the wiring connectors are securely reconnected. **Note:** *Prior to renewing an ECU, the security code must be reset using dedicated diagnostic equipment. Refer to your local Vauxhall dealer or specialist.*

Fuel pump relay

53 The fuel pump relay is located on the left-hand side of the engine compartment in the main relay box *(see paragraph 2 on page 25•8)*. Unclip the cover and remove it from the relay box. Identify the fuel pump relay which is coloured purple. Ensure the ignitioh is switched off then pull out the relay.

54 Refitting is the reverse of removal.

Air conditioning system switch

55 The air conditioning system switch is screwed into one of the refrigerant pipes and cannot be removed without first discharging the refrigerant. Renewal of the switch should therefore be entrusted to a suitably-equipped garage.

Simtec 70 and 71 multi-point injection system components – removal and refitting

Fuel rail and injectors

Note: *Before condemning an injector, it is worth trying the effect of one of the proprietary injector-cleaning treatments.*

1 Undo the retaining screws, remove the oil filler cap, and remove the engine cover.

2 Depressurise the fuel system, then disconnect the battery negative terminal.

3 Disconnect the wiring connectors from the hot film mass airflow meter.

4 Release the retaining clips then disconnect the intake duct from the air cleaner and throttle housing and remove the duct assembly, complete with hot film mass airflow meter, from the engine compartment, freeing it from the wiring harness. On Zafira models, it will be necessary to remove the engine compartment seal and water deflector cover from in front of the windscreen, then undo the retaining nuts and bolts, and remove the bulkhead cover plate to allow access to the air intake trunking.

5 Release the retaining clips and disconnect the breather hoses from the rear of the camshaft cover and throttle body.

6 Detach all wiring connectors which are connected to the plug strip located on top of the injectors. Release all the plug strip wiring loom retaining clips.

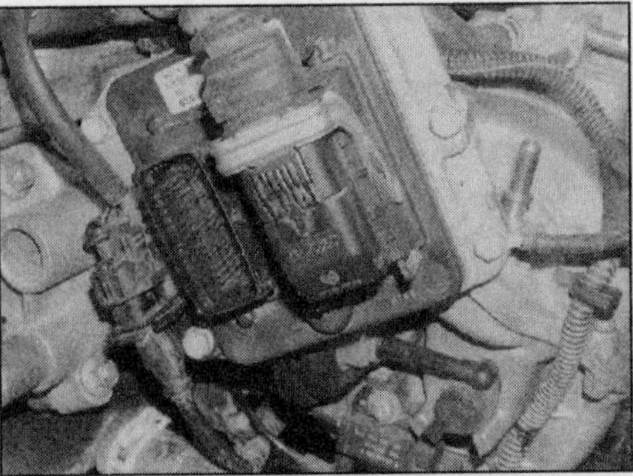

7 Undo the bolts securing the plug strip earth connections on the right-hand end of the inlet manifold, and the engine ECU at the left-hand end of the manifold. Undo the plug strip retaining bolt.

8 Unscrew the union nut(s) and disconnect the fuel hose(s) from the fuel rail. Whilst the unions are being slackened, retain the fuel rail adapters with an open-ended spanner.

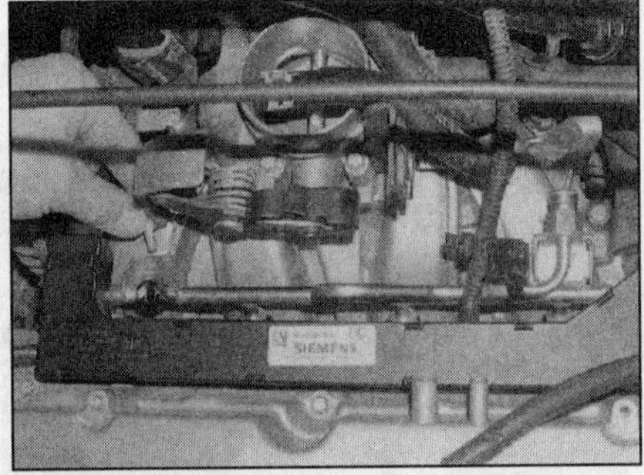

9 Slacken and remove the two retaining bolts then carefully ease the fuel rail and injector assembly, complete with plug strip and loom, out of position and remove it from the manifold. Remove the lower sealing rings from the injectors and discard them; they must be renewed whenever they are disturbed.

10 Carefully release the retaining clips and lift the plug strip assembly away from the top of the injectors; the wiring connectors are an integral part of the plug strip.

11 Slide off the relevant retaining clip and withdraw the injector from the fuel rail. Remove the upper sealing ring from the injector and discard it; all disturbed sealing rings must be renewed.

12 Refitting is a reversal of the removal procedure, noting the following points.

a) *Renew all disturbed sealing rings and apply a smear of engine oil to them to aid installation.*

b) *Ease the injector(s) into the fuel rail, ensuring that the sealing ring(s) remain correctly seated, and secure in position with the retaining clips.*

c) *On refitting the fuel rail, take care not to damage the injectors and ensure that all sealing rings remain in position. Once the fuel rail is correctly seated, tighten its retaining bolts to the specified torque.*

d) *Once the fuel rail has been refitted, the plug strip can be pressed onto the top of the injectors. Ensure that the plug strip connectors engage fully with the injectors.*

e) *On completion start the engine and check for fuel leaks.*

Fuel pressure regulator

***Note:** On all 1.8 litre engines, fuel pressure regulation is a function of the fuel tank pump module. The regulator is located under the fuel pump access cover. The regulator does not appear to be available as a separate part – check with your Vauxhall dealer.*

13 On 2.0 litre engines, remove the engine cover, depressurise the fuel system, then disconnect the battery negative terminal.

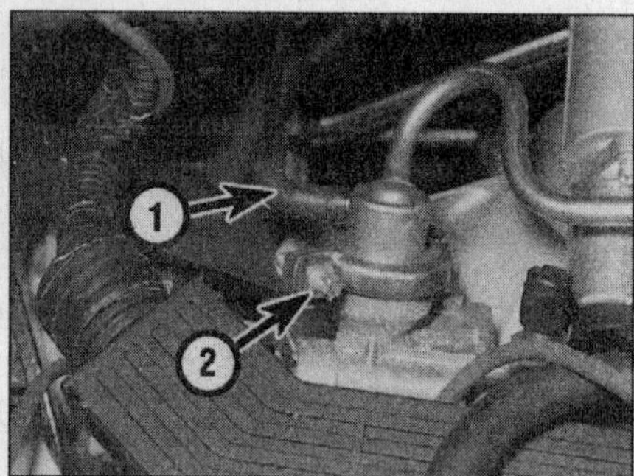

14 Detach the vacuum hose from the regulator (1). Undo the screw and remove the regulator retaining clamp (2).

15 Carefully pull the regulator from the fuel rail along with its sealing rings.

16 Refitting is the reverse of removal, using new sealing rings. On completion start the engine and check for fuel leaks.

Throttle valve adjuster

1.8 litre engines

17 The engine idle speed (X18XE1 engines) and throttle valve position (Z18XE and Z18XEL engines) is controlled by the ECU by means of a throttle valve adjuster. The throttle potentiometer is integrated into the adjuster, which is part of the throttle body housing. Individual parts for the throttle body housing are not available separately and, if defective, the throttle valve adjuster must be renewed as a complete assembly with the throttle body.

Idle speed control stepper motor

2.0 litre engines

18 Disconnect the wiring connectors from the hot film mass airflow meter. Release the retaining clip and disconnect the breather hose from the rear of the camshaft cover then slacken the retaining clips and remove the intake duct assembly from the engine compartment. On Zafira models, it will be necessary to remove the engine compartment seal and water deflector cover from in front of the windscreen, then undo the retaining nuts and bolts, and remove the bulkhead cover plate to allow access to the air intake trunking.

19 Disconnect the wiring connector then undo the two retaining screws and remove the motor

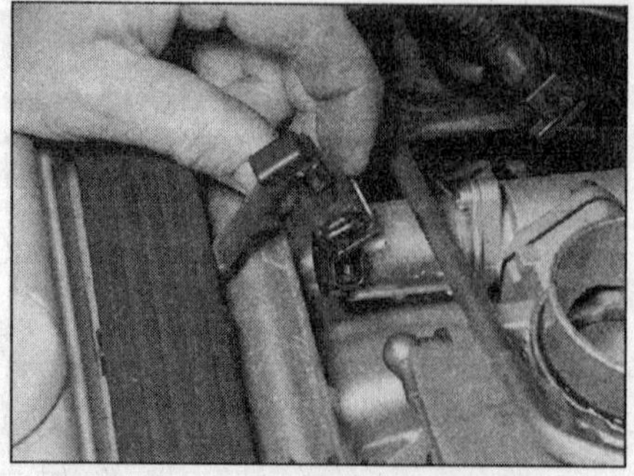

assembly from the side of the throttle housing. Remove the gasket and discard it; a new one should be used on refitting.

20 Refitting is the reverse of removal using a new gasket.

Throttle potentiometer

21 On 1.8 litre engines, the function of the throttle potentiometer is incorporated within the throttle valve adjuster. On 2.0 litre engines, remove the throttle body.

22 Slacken and remove the two retaining screws then remove the potentiometer from the housing.

23 Refitting is the reverse of removal, making sure the potentiometer is correctly engaged with the throttle valve spindle.

Coolant temperature sensor

24 The coolant temperature sensor is screwed into the thermostat housing on the left-hand end of the cylinder head. In order to unscrew the sensor the thermostat housing must be removed.

Intake air temperature sensor

25 The function of the intake air temperature sensor is incorporated within the hot film mass airflow meter. Therefore, no removal or refitting procedure is applicable.

Hot film mass airflow meter

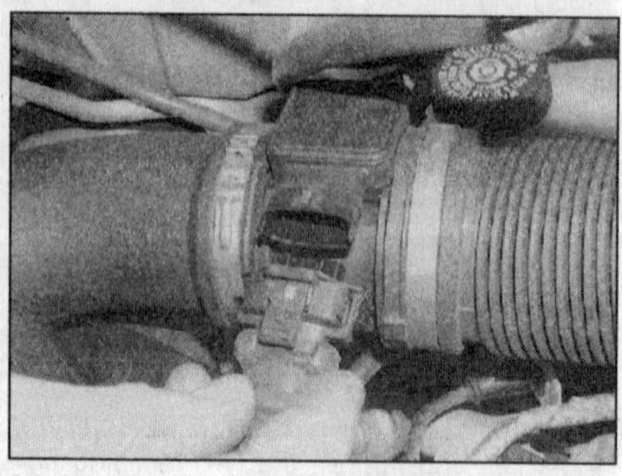

26 Ensure the ignition is switched off then disconnect the wiring connector from the hot mass airflow meter.

27 Slacken the retaining clips then disconnect the intake ducts and remove the airflow meter from the vehicle. On Zafira models, it will be necessary to remove the engine compartment seal and water deflector cover from in front of the windscreen, then undo the retaining nuts and bolts, and remove the bulkhead cover plate to allow access to the air intake trunking. Inspect the meter for signs of damage and renew if necessary.

28 Refitting is the reverse of removal, ensuring the intake ducts are correctly engaged with the meter recesses.

Crankshaft sensor

1.8 litre engines

29 On X18XE1 engines, the sensor is mounted on a bracket at the rear of the right-hand end of the cylinder block and is accessible from underneath the vehicle. Firmly apply the handbrake, then jack up the front of the car and support it securely on axle stands. On Z18XE and Z18XEL engines, the sensor is mounted on the front of the cylinder block below the oil filter.

30 Trace the wiring back from the sensor, releasing it from all the relevant clips and ties whilst noting its correct routing. Disconnect the wiring connector so the wiring is free to be removed with the sensor.

31 Unscrew the retaining bolt and remove the sensor from underneath the vehicle.

32 Refitting is the reverse of removal, tightening the retaining bolt to the specified torque. Ensure the wiring is correctly routed and retained by all the necessary clips and ties.

2.0 litre engines

33 Remove the air cleaner housing.

34 On models with air conditioning, undo the retaining bolt and remove the engine oil dipstick guide tube from the front of the cylinder block.

35 Trace the wiring back from the sensor, releasing it from all the relevant clips and ties whilst noting its correct routing. At the rear of the cylinder head, disconnect the wiring connector so the wiring is free to be removed with the sensor.

36 Undo the single retaining screw and remove the sensor from the block. Discard the seal ring.

37 Refitting is a reversal of removal. Fit a new seal ring to the sensor and smear a little clean engine oil on the ring before the sensor is refitted to the block. Tighten the sensor retaining screw to the specified torque.

Camshaft sensor

38 On 1.8 litre engines, remove the oil filler cap, undo the retaining screws, and remove the engine cover. On 2.0 litre engines, undo the retaining screws and remove the spark plug cover from the top of the camshaft cover.

39 Ensure the ignition is switched off then disconnect the wiring connector from the camshaft sensor.

40 Remove the timing belt upper cover.

41 Unscrew the retaining bolts and remove the camshaft sensor from the top of the cylinder head.

42 Refitting is the reverse of removal. Apply a little locking compound, tighten the sensor retaining bolts to the specified torque.

Knock sensor

43 The knock sensor is mounted on the rear of the cylinder block and is accessible from underneath the vehicle. Firmly apply the handbrake, then jack up the front of the car and support it securely on axle stands.

44 On 2.0 litre engines remove the starter motor.

45 Trace the wiring back from the sensor, releasing it from all the relevant clips and ties whilst noting its correct routing, and disconnect the wiring plug.

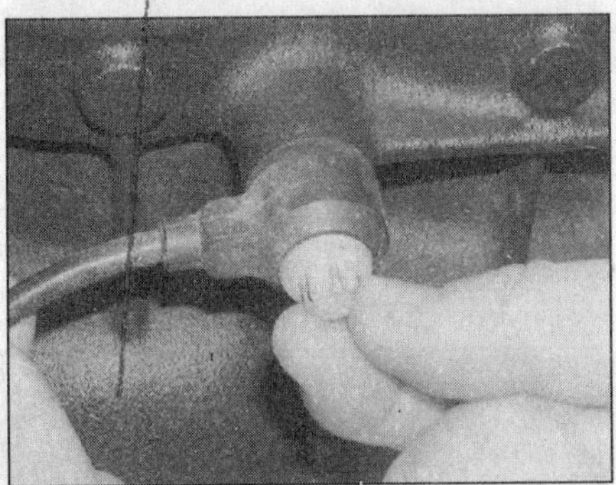

46 Slacken and remove the retaining bolt and remove the sensor from the engine.

47 On refitting, ensure the mating surfaces are clean and dry then fit the sensor and tighten its retaining bolt to the specified torque. Ensure the wiring is correctly routed and securely reconnected then clip the wiring cover assembly back into position. On 2.0 litre engines, refit the starter motor.

Electronic control unit

48 Disconnect the battery negative terminal.

49 On 1.8 litre engines, the ECU is bolted directly onto the left-hand side of the inlet manifold. Whilst on 2.0 litre engines, the ECU is located next to the battery.

50 Release the retaining clips then disconnect the wiring connectors from the ECU. Detach the earth connection from the ECU frame. On 1.8 litre engines, release the knock sensor wiring harness plug from the bracket on the ECU frame, but do not disconnect it.

51 Undo the three retaining bolts and remove the ECU from the vehicle. Note that the lower retaining bolt also secures a wiring plug support bracket.

52 Refitting is the reverse of removal, ensuring the wiring connectors are securely reconnected. **Note:** *Prior to renewing an ECU, the security code must be reset using dedicated diagnostic equipment. Refer to your local Vauxhall dealer or specialist.*

Fuel pump relay

53 Refer to the Multec S system Section, paragraphs 53 to 54.

Air conditioning system switch

54 The air conditioning system switch is screwed into one of the refrigerant pipes and cannot be removed without first discharging the refrigerant. Renewal of the switch should therefore be entrusted to a suitably-equipped garage.

Inlet manifold valve switching solenoid

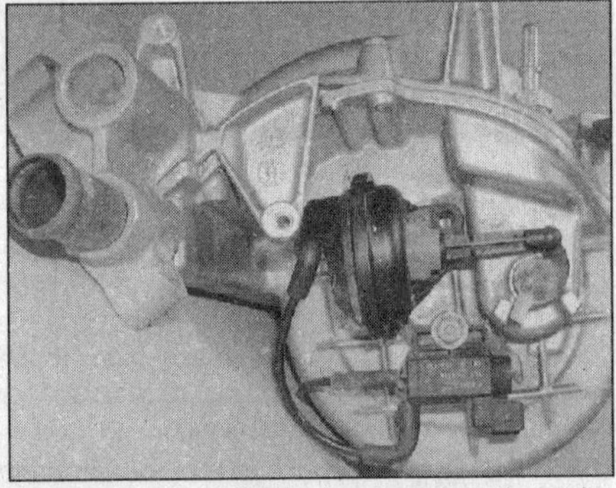

55 The solenoid which operates the inlet manifold valve assemblies is mounted on the left-hand end of the manifold.

56 Ensure the ignition is switched off then disconnect the wiring connector from the solenoid.

57 Disconnect both vacuum hoses, undo the retaining screw, then remove the solenoid from the manifold.

58 Refitting is the reverse of removal.

Inlet manifold valve vacuum diaphragm unit

59 The inlet manifold valve vacuum diaphragm unit is fitted to the left-hand end of the manifold *(see paragraph 55)*.

60 On 1.8 litre engines, remove the ECU as described in paragraphs 48 to 52. On 2.0 litre engines, remove the inlet manifold.

61 On 1.8 litre engines, disconnect the vacuum hose, unclip the diaphragm rod from its balljoint, undo the two retaining bolts and remove the unit from the manifold. On 2.0 litre engines, undo the nut that secures the diaphragm rod to the valve actuating arm, drift out the pin securing the diaphragm housing to the inlet manifold, and disconnect the vacuum hose.

62 Refitting is the reverse of removal.

GMPT – E15 multi-point injection system components – removal and refitting

Fuel rail and injectors

Note: *Before condemning an injector, it is worth trying the effect of one of the proprietary injector cleaning treatments.*

1 Unclip the engine cover, depressurise the fuel system, then disconnect the battery negative terminal.

2 Disconnect the wiring connector from the air temperature sensor in the air intake pipe.

3 Release the retaining clip and disconnect the breather hose from the camshaft cover.

4 Release the retaining clips then disconnect the intake pipe from the air cleaner assembly and throttle housing and remove the pipe assembly from the engine compartment, detach the vacuum pipe(s) from the intake pipe on removal.

5 Unscrew the union nuts and disconnect the fuel pipes from the fuel rail, release the fuel pipes from the retaining clips and move to one side.

6 Release the wiring harness plugs and disconnect them from the injectors, the ignition module and the EGR valve.

7 Unclip the coolant hose from the top of the wiring trough, undo the two retaining nuts and carefully unclip the wiring trough and move it to one side.

8 Undo the wiring trough bracket retaining nut and bolt and move the bracket and wiring harness to one side.

9 Slacken and remove the two retaining bolts then carefully ease the fuel rail and injector assembly out of position and remove it from the manifold. Remove the lower sealing rings from the injectors and discard them; they must be renewed whenever they are disturbed.

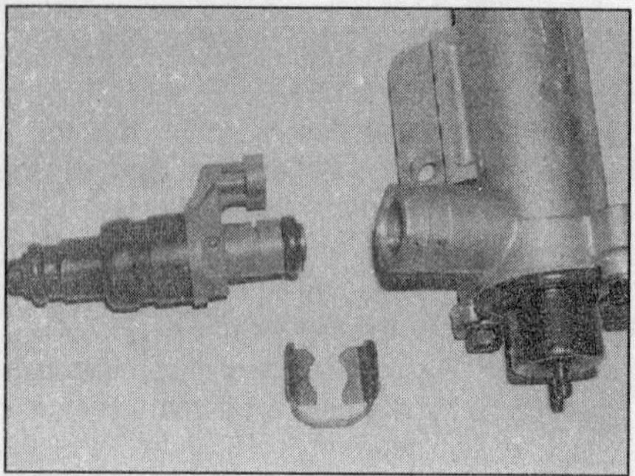

10 Slide off the relevant retaining clip and withdraw the injector from the fuel rail. Remove the upper sealing ring from the injector and discard it; all disturbed sealing rings must be renewed.

11 Refitting is a reversal of the removal procedure, noting the following points.

a) *Renew all disturbed sealing rings and apply a smear of engine oil to them to aid installation.*
b) *Ease the injector(s) into the fuel rail, ensuring that the sealing ring(s) remain correctly seated, and secure in position with the retaining clips.*
c) *On refitting the fuel rail, take care not to damage the injectors and ensure that all sealing rings remain in position. Once the fuel rail is correctly seated, tighten its retaining bolts to the specified torque.*
d) *Ensure that all the wiring connectors engage fully to make a good connection.*
e) *On completion start the engine and check for fuel leaks.*

Fuel pressure regulator

12 Unclip the engine cover, depressurise the fuel system, then disconnect the battery negative terminal.

13 Disconnect the vacuum line from the pressure regulator and from the inlet manifold.

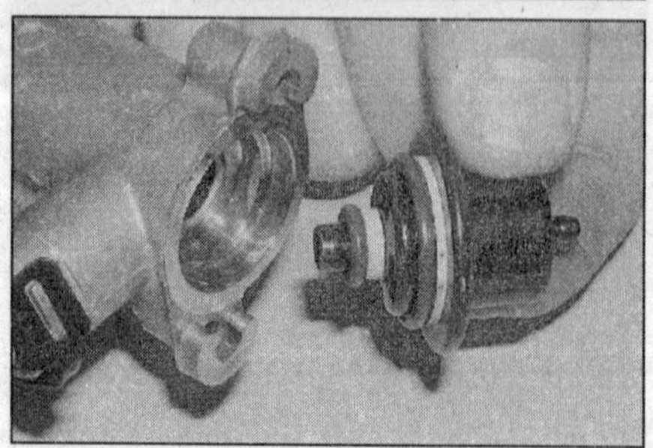

14 Slacken and remove the two retaining bolts, then carefully ease the pressure regulator from the end of the fuel rail along with its sealing rings.

15 Refitting is the reverse of removal, using new sealing rings. On completion start the engine and check for fuel leaks.

Accelerator pedal position sensor

16 From inside the vehicle, unscrew the fasteners and remove the lower trim panel from underneath the driver's side of the facia to gain access to the accelerator pedal.

17 Disconnect the wiring connector from the pedal position sensor by sliding it to the rear of the vehicle.

18 Undo the retaining bolts and remove the pedal position sensor from the pedal assembly.

Throttle valve servo motor

19 The servo motor is a combined part of the throttle body housing. Individual parts for the throttle body housing are not available separately and, if defective, it must be renewed as a complete unit.

Throttle valve potentiometer

20 The throttle valve potentiometer is a combined part of the throttle body housing. Individual parts for the throttle body housing are not available separately and, if defective, it must be renewed as a complete unit.

Coolant temperature sensor

21 The coolant temperature sensor is screwed into the thermostat housing at the left-hand end of the cylinder head. Remove the DIS ignition module.

22 Partially drain the cooling system.

23 Disconnect the wiring plug and unscrew the sensor.

24 Refitting is a reversal of removal. Apply a little sealant to the sensor threads prior to refitting.

Intake air temperature sensor

25 The intake air temperature sensor is mounted in the intake pipe which connects the air cleaner housing to the inlet manifold.

26 Ensure the ignition is switched off, then disconnect the wiring connector from the sensor.

27 Carefully ease the sensor out of position (note position for refitting) and remove its sealing grommet from the intake pipe. If the sealing grommet shows signs of damage or deterioration it should be renewed.

28 Refitting is the reverse of removal, ensuring the sensor and grommet are correctly located in the intake pipe.

Manifold absolute pressure sensor

29 Unclip the engine cover, depressurise the fuel system, then disconnect the battery negative terminal.

30 Disconnect the wiring connector from the air temperature sensor in the air intake pipe.

31 Release the retaining clip and disconnect the breather hose from the camshaft cover.

32 Release the retaining clips then disconnect the intake pipe from the air cleaner assembly and throttle housing and remove the pipe assembly from the engine compartment, detach the vacuum pipe(s) from the intake pipe on removal.

33 Disconnect the wiring connector from the throttle body, then undo the securing bolts/stud and remove the throttle body unit from the inlet manifold. Note the position of the cable retaining clip on the securing stud.

34 Disconnect the wiring connector from the MAP sensor, turn the sensor 1/4 of a turn clockwise and carefully pull out of the inlet manifold.

35 Refitting is the reverse of removal, using a new sealing ring.

Crankshaft sensor

36 The sensor is mounted on the front of the cylinder block above the starter motor. **Note:** *It may be necessary to remove the starter motor to gain better access to the sensor.*

37 Trace the wiring back from the sensor, releasing it from all the relevant clips and ties whilst noting its correct routing. Disconnect the wiring connector from the sensor.

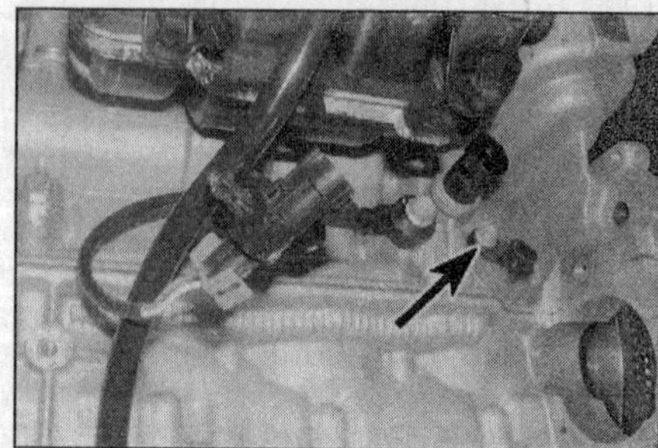

38 Unscrew the retaining bolt and remove the sensor from the cylinder block.

39 Refitting is the reverse of removal, tightening the retaining bolt to the specified torque. Ensure the wiring is correctly routed and retained by all the necessary clips and ties.

Knock sensor

40 The sensor is mounted on the front of the cylinder block above the starter motor. **Note:** *It may be necessary to remove the starter motor to gain better access to the sensor.*

41 Disconnect the wiring block connector at the bottom of the oil dipstick tube, so the wiring is free to be removed with the sensor.

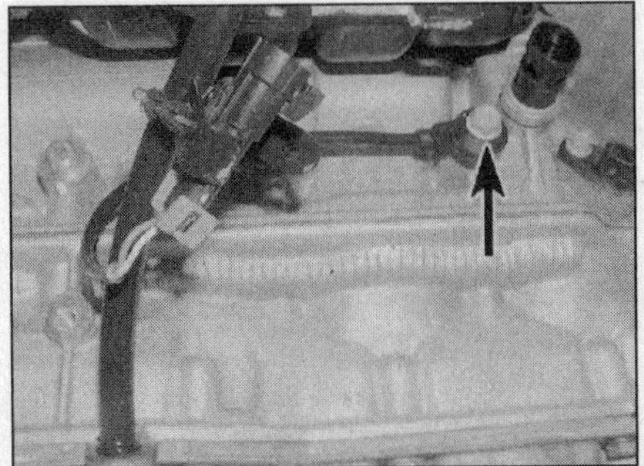

42 Unscrew the retaining bolt and remove the sensor from the cylinder block.

43 On refitting ensure the mating surfaces are clean and dry then fit the sensor and tighten its retaining bolt to the specified torque. Ensure the wiring is correctly routed and retained by all the necessary clips and ties.

Electronic control unit

44 Unclip the engine cover, depressurise the fuel system, then disconnect the battery negative terminal.

45 Disconnect the wiring connector from the air temperature sensor in the air intake pipe.

46 Release the retaining clip and disconnect the breather hose from the camshaft cover.

47 Release the retaining clips then disconnect the intake pipe from the air cleaner assembly and throttle housing and remove the pipe assembly from the engine compartment, detach the vacuum pipe(s) from the intake pipe on removal.

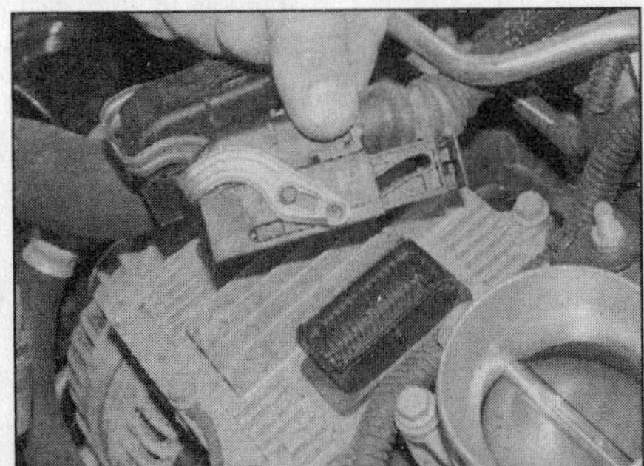

48 Release the two wiring harness plugs from the ECU, then undo the securing bolts and remove the unit from the inlet manifold. Note the position of the earth cable on one of the securing bolts.

49 Refitting is the reverse of removal, ensuring the wiring connectors are securely reconnected. **Note:** *Prior to renewing an ECU, the security code must be reset using dedicated diagnostic equipment. Refer to your local Vauxhall dealer or specialist.*

Fuel pump relay

50 Refer to the Multec S system Section, paragraphs 53 to 54.

Air conditioning system switch

51 The air conditioning system switch is screwed into one of the refrigerant pipes and cannot be removed without first discharging the refrigerant. Renewal of the switch should therefore be entrusted to a suitably-equipped garage.

Diesel injection system electronic components – removal and refitting

1.7 litre SOHC models

Hot film mass airflow meter

Note: *The intake air temperature sensor is built into the airflow meter.*

1 Ensure the ignition is switched off and disconnect the wiring plug from the hot film mass airflow meter.

2 Release the retaining clips then free the airflow meter from the intake ducts and remove it from the engine compartment.

3 Refitting is the reverse of removal, ensuring the intake ducts are correctly seated and their retaining clips are securely tightened.

Accelerator pedal position sensor

4 At the time of writing, the pedal position sensor was only available as a complete assembly with the accelerator pedal. If the sensor is defective, the complete assembly will have to be renewed.

5 From inside the vehicle, unscrew the fasteners and remove the lower trim panel from underneath the driver's side of the facia to gain access to the accelerator pedal.

6 Unscrew the retaining nuts and remove the pedal assembly from the bulkhead.

7 Disconnect the wiring connector from the accelerator pedal position sensor by sliding out the locking element of the wiring plug.

8 Reconnect the wiring connector.

9 Refit the pedal assembly and tighten its retaining nuts to the specified torque setting, then refit the trim panel to the facia.

Crankshaft sensor

10 The crankshaft sensor is located at the base of the cylinder block front face. Disconnect the wiring plug, and note the cable routing.

11 Undo the retaining bolt and remove the sensor.

12 Refitting is a reversal of removal. Tighten the sensor retaining bolt to the specified torque.

Coolant temperature sensor

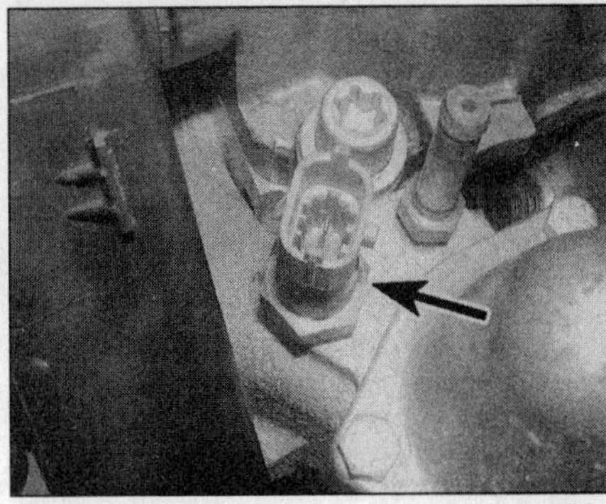

13 The coolant temperature sensor is located in the thermostat housing at the right-hand end of the cylinder head. Disconnect the sensor wiring plug.

14 Drain the cooling system, or be prepared for fluid spillage.

15 Unscrew the sensor from the thermostat housing.

16 Apply a little locking compound to the threads of the sensor, and refit the sensor to the housing. Tighten the sensor to the specified torque.

17 Reconnect the wiring plug to the sensor.

18 Top up the coolant system.

Charge pressure sensor

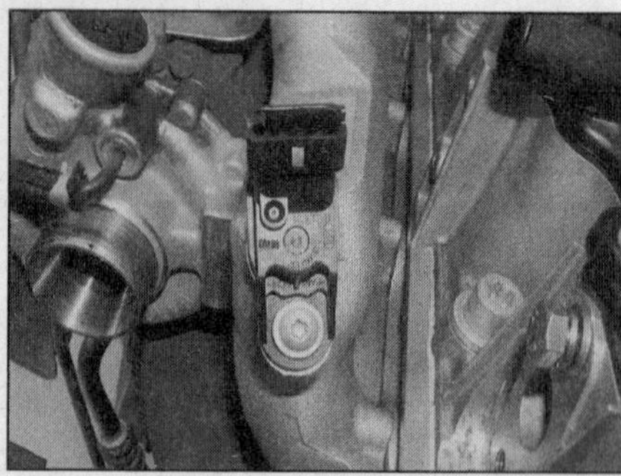

19 The charge pressure sensor is fitted to the right-hand end of the inlet manifold. Disconnect the wiring plug from the sensor.

20 Undo the screw and remove the sensor from the manifold.

21 Refitting is a reversal of removal. Tighten the sensor retaining bolt to the specified torque.

Engine electronic control unit

Caution: Before removing an ECU, the control unit security code must be reset using specialist test equipment. Refer to your Vauxhall dealer or specialist.

22 The engine ECU is located behind the left-hand front inner wing protective panelling. Firmly apply the handbrake, jack up the front of the vehicle and support it on axle stands. Remove the front-left roadwheel.

23 Release the panelling retaining screws and clips. Remove the panelling from under the wheel arch.

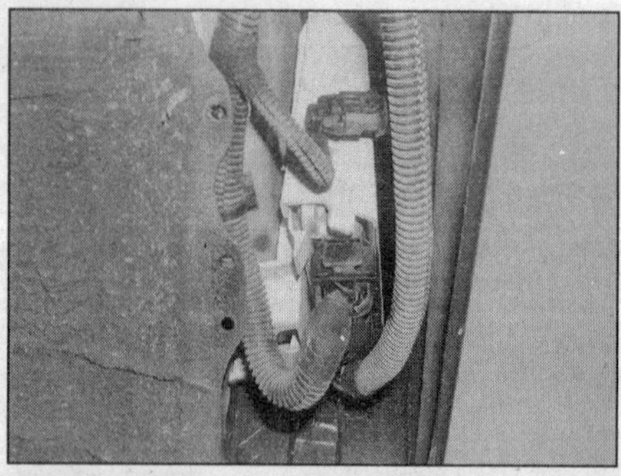

24 Disconnect the two ECU wiring plugs and remove the unit from the retaining bracket.

25 Refitting is a reversal of removal, noting that the ECU wiring plugs should be locked in place.

Injection electronic control unit (ECU)

26 The injection ECU is integral with the injection pump, and it is recommended that no attempt be made to separate the ECU from the pump.

Fuel temperature sensor

27 The fuel temperature sensor is integral with the fuel pump. It is recommended that no attempt be made to remove the sensor.

Injection pump shaft position sensor

28 The injection pump shaft position sensor is integral with the fuel pump. It is recommended that no attempt be made to remove the sensor.

Injection commencement sensor

29 The injection commencement sensor is integral with the fuel pump. It is recommended that no attempt be made to remove the sensor.

Fuel cut-off solenoid

30 The fuel cut-off solenoid is integral with the fuel pump. It is recommended that no attempt be made to remove the sensor.

1.7 litre DOHC models

Hot film mass airflow meter

Note: *The intake air temperature sensor is built into the airflow meter.*

31 Ensure the ignition is switched off and disconnect the wiring plug from the hot film mass airflow meter *(see paragraph 1 on page 25•17)*.

32 Release the retaining clips then free the airflow meter from the intake ducts and remove it from the engine compartment.

33 Refitting is the reverse of removal, ensuring the intake ducts are correctly seated and their retaining clips are securely tightened.

Accelerator pedal position sensor

34 At the time of writing, the pedal position sensor was only available as a complete assembly with the accelerator pedal. If the sensor is defective, the complete assembly will have to be renewed.

35 From inside the vehicle, unscrew the fasteners and remove the lower trim panel from underneath the driver's side of the facia to gain access to the accelerator pedal.

36 Unscrew the retaining nuts and remove the pedal assembly from the bulkhead *(see paragraph 6 opposite)*.

37 Disconnect the wiring connector from the accelerator pedal position sensor by sliding out the locking element of the wiring plug.

38 Reconnect the wiring connector.

39 Refit the pedal assembly and tighten its retaining nuts to the specified torque setting, then refit the trim panel to the facia.

Crankshaft sensor

40 Firmly apply the handbrake then jack up the front of the vehicle and support it on axle stands. Where necessary, undo the retaining bolts/clips and remove the undertray from beneath the engine/ transmission unit.

41 The sensor is located on the rear of the cylinder block, underneath the starter motor. Working from under the vehicle, disconnect the wiring plug from the sensor.

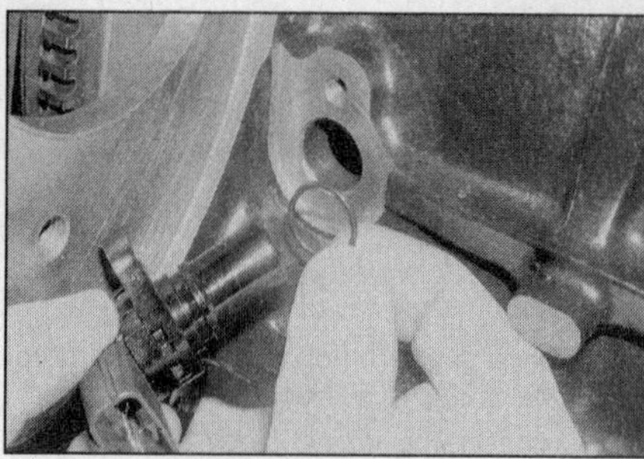

42 Undo the retaining bolt and remove the sensor. Recover the sealing ring.

43 Refitting is a reversal of removal. If necessary renew the sealing ring and tighten the sensor retaining bolt to the specified torque.

Coolant temperature sensor

44 Drain the coolant system, or be prepared for coolant spillage.

45 The coolant temperature sensor is located in the thermostat housing at the left-hand end of the cylinder head. Disconnect the wiring plug from the sensor.

46 Unscrew the sensor from the thermostat housing. Mop up any spilled coolant.

47 Apply a little locking compound to the threads of the sensor, and refit the sensor to the housing. Tighten the sensor to the specified torque.

48 Reconnect the wiring plug to the sensor.

49 Top up the coolant system.

Charge pressure sensor

50 Remove the air cleaner ducting at the rear of the engine compartment.

51 Remove the exhaust gas recirculation solenoid valve from the inlet manifold.

52 Disconnect the wiring plug from the sensor on the right-hand end of the inlet manifold.

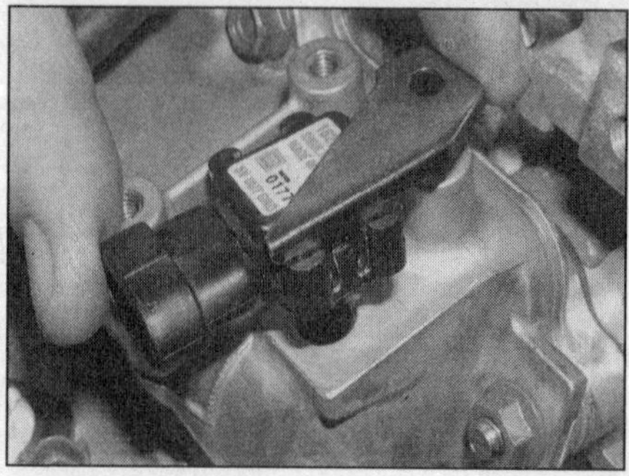

53 Undo the retaining bolts and remove the sensor.

54 Refitting is a reversal of removal. Tighten the sensor retaining bolts to the specified torque.

Oil pressure switch

55 The signal from the oil pressure switch is used by the injection control unit to determine whether or not to illuminate the oil pressure warning light.

56 To remove the oil pressure switch, remove the battery and battery tray.

57 The oil pressure switch is located at the left-hand end of the cylinder block. Disconnect the wiring plug from the switch.

58 Unscrew the switch from the cylinder block. Be prepared for oil spillage.

59 Apply some suitable silicone sealant (available from Vauxhall dealers) to the threads, and screw the switch into place. Tighten the switch to the specified torque.

60 Reconnect the wiring plug, and refit the battery tray, and battery.

61 Start the vehicle and check for oil leaks.

> *Caution: Before removing an ECU, the control unit security code must be reset using specialist test equipment. Refer to your Vauxhall dealer or specialist.*

Engine electronic control unit (ECU)

62 The engine control unit is located on top of the engine above the exhaust gas recirculation valve.

63 Disconnect the battery negative lead.

64 Disconnect the two wiring plugs from the ECU. The plugs are unlocked by lifting up the metal levers.

65 Undo the retaining bolts/nuts, and remove the ECU.

66 Refitting is a reversal of removal. Tighten the ECU retaining bolts/nuts to the specified torque. The ECU wiring plugs are shaped such that each plug will only fit its correct socket. The plug from the left-hand side of the engine compartment connects to the rearmost socket. Take great care when refitting the plugs, as the terminals are very delicate and easily damaged.

Injection electronic control unit (ECU)

67 The injection ECU is located at the rear of the injection pump.

68 Firmly apply the handbrake then jack up the front of the vehicle and support it on axle stands. Where necessary, undo the retaining bolts/clips and remove the undertray from beneath the engine/transmission unit.

69 Working under the vehicle, disconnect the ECU wiring plug.

70 Undo the retaining bolts, and remove the ECU.

71 Refitting is a reversal of removal.

Turbocharger wastegate control solenoid valve

72 Firmly apply the handbrake then jack up the front of the vehicle and support it on axle stands. Where necessary, undo the retaining bolts/clips and remove the undertray from beneath the engine/transmission unit.

73 Remove the injection ECU as described in paragraphs 67 to 70.

74 Disconnect the vacuum pipes, and wiring plug from the solenoid valve.

75 Undo the two retaining bolts and remove the solenoid valve.

76 Refitting is a reversal of removal. Tighten the solenoid valve retaining bolts securely. Note that the vacuum pipes are of two different diameters. The smaller pipe is fitted to the outer connection.

Atmospheric pressure sensor

77 The atmospheric pressure sensor is located at the rear of the cylinder block, between the injection pump and the starter motor.

78 Firmly apply the handbrake then jack up the front of the vehicle and support it on axle stands. Where necessary, undo the retaining bolts/clips and remove the undertray from beneath the engine/transmission unit.

79 Disconnect the wiring plug from the pressure sensor.

80 Lift the sensor up and off the retaining bracket.

81 Refitting is a reversal of removal.

Fuel temperature sensor

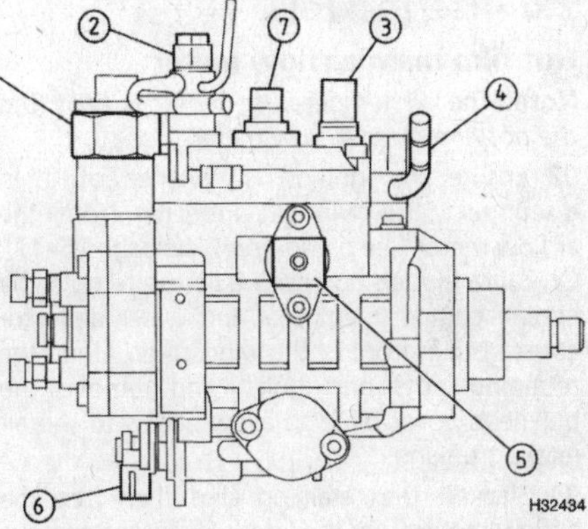

Fuel injection pump – 1.7 litre DOHC models
1 Spill valve
2 Fuel return line
3 Fuel temperature sensor
4 Fuel supply line
5 Programmable read-only memory
6 Timing control solenoid
7 Injection pump shaft position sensor

82 The fuel temperature sensor is fitted to the right-hand top of the injection pump.

83 To remove the sensor, disconnect the wiring plug and unscrew the sensor from the top of the injection pump.

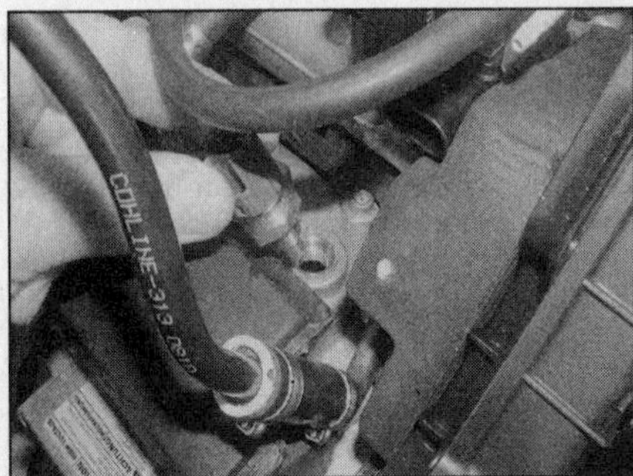

84 With a new seal fitted, screw the sensor into the injection pump and tighten the nut securely.

85 Reconnect the wiring plug.

Injection pump shaft position sensor

86 The sensor is located on the side of the fuel injection pump *(see paragraph 82)*.

87 The sensor may not be available as a separate item. Check with a Vauxhall dealer prior to any attempt to remove the sensor.

Injection commencement sensor

88 The sensor is located on the underside of the fuel injection pump at the flywheel end.

89 The sensor may not be available as a separate item. Check with a Vauxhall dealer prior to any attempt to remove the sensor.

Fuel cut-off solenoid

90 The fuel cut-off solenoid is located on top of the fuel injection pump, at the flywheel end.

91 The solenoid may not be available as a separate item. Check with a Vauxhall dealer prior to any attempt to remove the sensor.

2.0 litre models

Hot film mass airflow meter

Note: *The air temperature sensor is built into the hot film mass airflow meter.*

92 Ensure the ignition is switched off then disconnect the wiring connector from the airflow meter *(see paragraph 1 on page 25•17)*. On Zafira models, to improve access, remove the engine compartment seal and water deflector cover from in front of the windscreen. Undo the retaining nuts and bolts, and remove the bulkhead cover plate to allow access to the air intake trunking

93 Slacken the retaining clips then free the airflow meter from the intake ducts and remove it from the engine compartment.

94 Refitting is the reverse of removal, ensuring the intake ducts are correctly seated and their retaining clips are securely tightened.

Accelerator pedal position sensor

95 At the time of writing, the pedal position sensor was only available as a complete assembly with the accelerator pedal. If the sensor is defective, the complete assembly will have to be renewed.

96 From inside the vehicle, unscrew the fasteners and remove the lower trim panel from underneath the driver's side of the facia to gain access to the accelerator pedal.

97 Unscrew the retaining nuts and remove the pedal assembly from the bulkhead *(see paragraph 6 on page 25•17)*.

98 Disconnect the wiring connector from the accelerator pedal position sensor by sliding out the locking element of the wiring plug.

99 Reconnect the wiring connector.

100 Refit the pedal assembly and tighten its retaining nuts to the specified torque setting, then refit the trim panel to the facia.

Crankshaft sensor

101 Undo the retaining screws and remove the plastic cover from the top of the cylinder head.

102 To gain access to the sensor from below, firmly apply the handbrake then jack up the front of the vehicle and support it on axle stands. Where necessary, undo the retaining clips/bolts and remove the engine undertray.

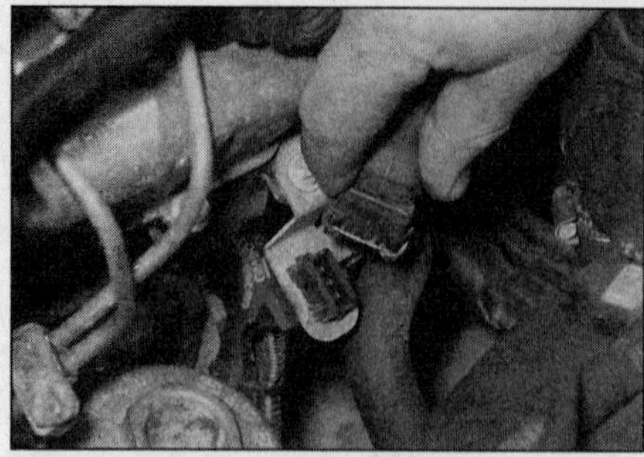

103 Trace the wiring back from the crankshaft sensor to its wiring connector then free the connector from its bracket and disconnect it from the main harness.

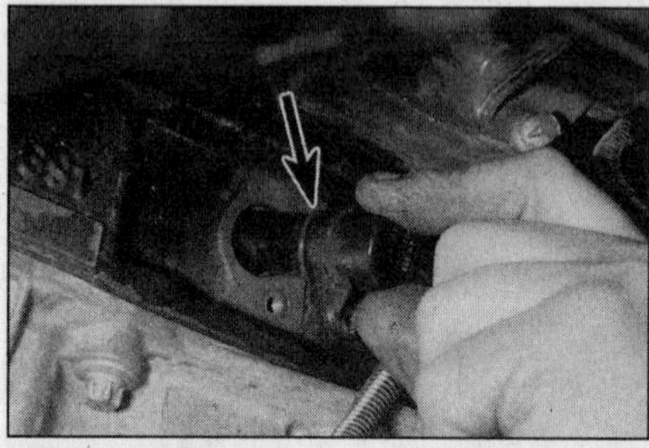

104 Wipe clean the area around the crankshaft sensor then slacken and remove the retaining bolt. Remove the sensor from the front of the cylinder block and recover the sealing ring.

105 Refitting is the reverse of removal, using a new sealing ring. Tighten the sensor retaining bolt to the specified torque.

Coolant temperature sensor

106 The coolant temperature sensor is screwed into top of the thermostat housing and the right-hand end of the cylinder head. Partially drain the cooling system.

107 Remove the engine top cover.

108 Disconnect the wiring plug and unscrew the sensor from the housing. Recover the sealing ring.

109 Refitting is a reversal of removal. Renew the sealing ring.

Charge pressure sensor

110 Undo the retaining screws and remove the plastic cover from the top of the cylinder head.

111 Slacken and remove the screws securing the wiring harness tray to the top of the inlet manifold and disconnect the wiring connector from the charge pressure sensor.

112 Undo the retaining bolt and remove the sensor from the top of the inlet manifold, noting the sealing ring fitted to the sensor shaft.

113 Refitting is the reverse of removal, using a new sealing ring and tightening the retaining bolt to the specified torque.

Oil temperature sensor

114 Firmly apply the handbrake then jack up the front of the vehicle and support it on axle stands. Where necessary, undo the retaining clips/bolts and remove the engine undertray.

115 Drain the engine oil. Once the oil has finished draining, fit a new sealing ring then refit the drain plug and tighten it to the specified torque.

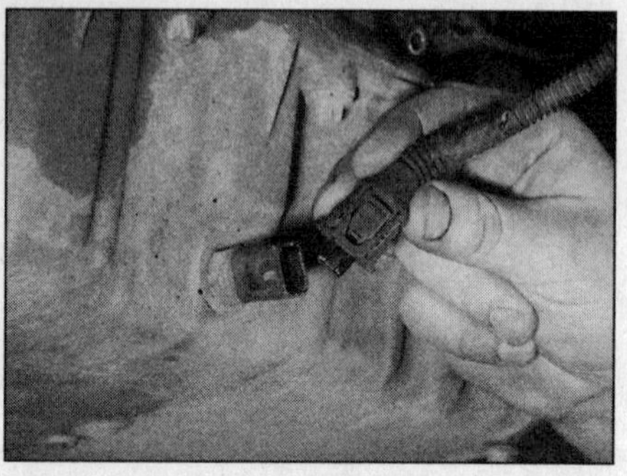

116 Disconnect the wiring connector then unscrew the sensor from the front of the sump.

117 On refitting is the reverse of removal, refilling the engine with oil.

Electronic control unit (ECU)

Caution: Before removing an ECU, the control unit security code must be reset using specialist test equipment. Refer to your Vauxhall dealer or specialist.

118 Jack up the front of the vehicle and support it on axle stands. Remove the front-left roadwheel.

119 Release the three retaining clips/screws and remove the inner wing protective panelling.

120 Detach the wiring harness plugs from the engine ECU.

121 Release the retaining clips and remove the ECU from the retaining bracket.

122 Refitting is the reverse of removal, ensuring the wiring connectors are securely reconnected.

Injection pump control unit

123 The control unit is an integral part of the injection pump and should not be disturbed. **Never** attempt to separate the control unit and pump.

Inlet manifold switchover solenoid valve

124 Note that there are two valves, the EGR system valve and the manifold switchover valve;

the manifold switchover valve is nearest the front of the vehicle.

125 Disconnect the wiring connector and vacuum hoses from the valve then undo the retaining screws and remove the valve from its mounting bracket.

126 Refitting is the reverse of removal.

Turbocharger wastegate solenoid valve

127 The solenoid valve is located in the right-hand rear corner of the engine compartment.

128 To gain access to the valve, remove the intake duct assembly (complete with hot film mass airflow meter) linking the air cleaner housing to the turbocharger.

129 Disconnect the wiring connector and vacuum hoses from the valve then undo the retaining screws and remove the valve from its mounting bracket.

130 Refitting is the reverse of removal.

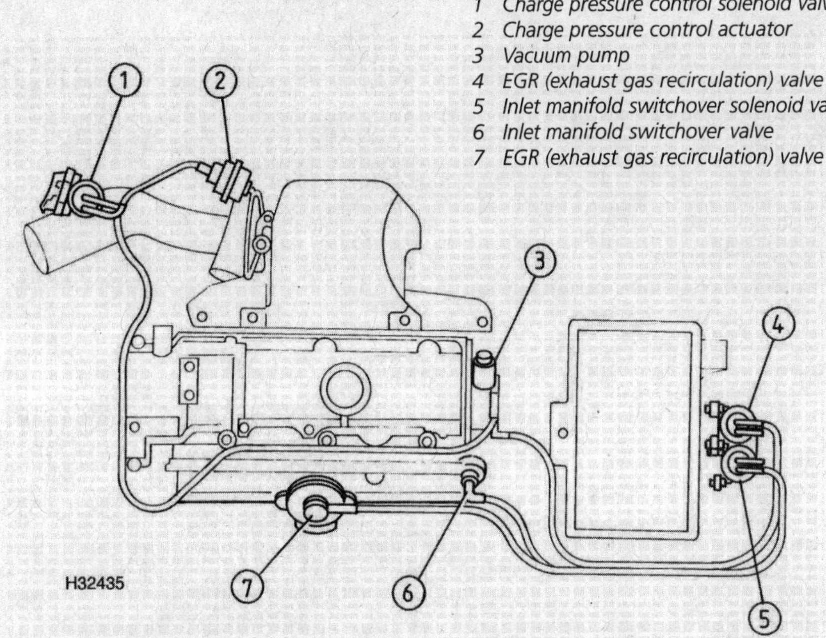

Vacuum pipe layout – 2.0 litre high-pressure models

1 Charge pressure control solenoid valve
2 Charge pressure control actuator
3 Vacuum pump
4 EGR (exhaust gas recirculation) valve
5 Inlet manifold switchover solenoid valve
6 Inlet manifold switchover valve
7 EGR (exhaust gas recirculation) valve

Notes

Vauxhall/Opel Vectra '99 – '02

Petrol models

General

System type:	
1.6 litre SOHC (X16SZR) models	Multec (single-point injection)
1.6 litre DOHC (X16XEL and Z16XE) models	Multec S (multi-point injection)
1.8 litre (X18XE1) models	Simtec 70 (multi-point injection)
1.8 litre (Z18XE) models	Simtec 71 (multi-point injection)
1.8 litre (X18XE) and 2.0 litre (X20XEV) models	Simtec 56.5 (multi-point injection)
2.2 litre (Z22SE) models	GMPT – E15 (multi-point injection)

Fuel system data

Fuel pump type	Electric, immersed in tank
Fuel pump regulated constant pressure (approximate):	
Single-point injection system	0.8 bar
Multi-point injection system*	3.0 to 3.5 bar
Specified idle speed	Not adjustable – controlled by ECU
Idle mixture CO content	Not adjustable – controlled by ECU

** Fuel pressure regulator vacuum hose disconnected and plugged (on X16XEL, X18XE and X20XEV models)*

Diesel models

General

System type	EDC 15M direct injection system incorporating an electronically-controlled VP44 fuel injection pump. Turbocharger fitted to all models and intercooler fitted to high-pressure turbo models
Manufacturer's code:	**Engine type**
X20DTH	2.0 litre high-pressure turbo diesel engine
X20DTL	2.0 litre low-pressure turbo diesel engine
Y20DTH	2.0 litre high-pressure turbo diesel engine
Y22DTR	2.2 litre high-pressure turbo diesel engine

Adjustment data

Idle speed:	
X20DTL and X20DTH	790 to 890 rpm – controlled by ECU
Y20DTH	750 to 850 rpm – controlled by ECU
Y22DTR	760 to 920 rpm – controlled by ECU
Maximum speed:	
X20DTL and X20DTH	4750 rpm – controlled by ECU
Y20DTH	4950 to 5050 rpm – controlled by ECU
Y22DTR	4930 to 5070 rpm – controlled by ECU

Injection pump		
Direction of rotation	Clockwise, viewed from sprocket end	
Pump timing (static)	Preset – controlled by ECU	
Injectors		
Opening pressure:		
X20DTL	180 to 365 bar	
X20DTH, Y20DTH and Y22DTR	220 to 380 bar	

Torque wrench settings	**Nm**	**lbf ft**
Petrol models		
Accelerator pedal position sensor nuts	20	15
Camshaft sensor*:		
All engines except X18XE and X20XEV	8	6
X18XE and X20XEV engines	6	4
Coolant temperature sensor:		
X16SZR, X16XEL, Z16XE, Z18XE	14	10
X18XE, X20XEV	10	7
Z22SE	20	15
Crankshaft sensor bolt:		
All engines except X18XE, X20XEV and Z22SE	8	6
X18XE and X20XEV engines	6	4
Z22SE engine	10	7
Fuel rail bolts	8	6
Knock sensor bolt:		
All engines except Z22SE	20	15
Z22SE engine	25	18
Throttle potentiometer bolts – X16SZR engine	2	1
** Use new fasteners*		
Diesel models		
Coolant temperature sensor:		
To thermostat housing	10	7
To cylinder head	18	13
Charge (boost) pressure sensor bolt	8	6
Crankshaft sensor bolt	8	6

General information

Petrol models

Multec single-point injection system

1 The Multec engine management (fuel injection/ignition) system incorporates a closed-loop catalytic converter, an evaporative emission control system and an exhaust gas recirculation (EGR) system, and complies with the latest emission control standards. The fuel injection side of the system operates as follows;

2 The fuel pump, immersed in the fuel tank, pumps fuel from the fuel tank to the fuel injector, via a filter mounted underneath the rear of the vehicle. Fuel supply pressure is controlled by the pressure regulator in the throttle body assembly. The regulator operates by allowing excess fuel to return to the tank.

3 The electrical control system consists of the ECU, along with the following sensors.

a) Throttle potentiometer – informs the ECU of the throttle position, and the rate of throttle opening or closing.

b) Coolant temperature sensor – informs the ECU of engine temperature.

c) Oxygen sensor – informs the ECU of the oxygen content of the exhaust gases.

d) Crankshaft sensor – informs the ECU of engine speed and crankshaft position.

e) Knock sensor – informs the ECU when pre-ignition ('pinking') is occurring.

f) Manifold absolute pressure (MAP) sensor – informs the ECU of the engine load by monitoring the pressure in the inlet manifold.

g) ABS control unit – informs the ECU of the vehicle speed.

h) Air conditioning system compressor switch (where fitted) – informs ECU when the air conditioning system is switched on.

4 All the above information is analysed by the ECU and, based on this, the ECU determines the appropriate ignition and fuelling requirements for the engine. The ECU controls the fuel injector by varying its pulse width – the length of time the injector is held open – to provide a richer or weaker mixture, as appropriate. The mixture is constantly varied by the ECU, to provide the best setting for cranking, starting (with either a hot or cold engine), warm-up, idle, cruising, and acceleration.

5 The ECU also has full control over the engine idle speed, via a stepper motor which is fitted to the throttle body. The motor pushrod controls the opening of an air passage which bypasses the throttle valve. When the throttle valve is closed (accelerator pedal released), the ECU uses the motor to vary the amount of air entering the engine and so controls the idle speed.

6 The ECU also controls the exhaust and evaporative emission control systems.

7 If there is an abnormality in any of the readings obtained from any sensor, the ECU enters its back-up mode. In this event, the ECU ignores the abnormal sensor signal, and assumes a pre-programmed value which will allow the engine to continue running (albeit at reduced efficiency). If the ECU enters this back-up mode, the warning light on the instrument panel will come on, and the relevant fault code will be stored in the ECU memory.

8 If the warning light comes on, a complete test of the engine management system can then be carried out, using a fault code reader which is simply plugged into the system's diagnostic connector. The connector is located behind the centre console; unclip the trim panel situated just in front of the handbrake lever to gain access.

Multec S multi-point injection system

9 The Multec S engine management (fuel injection/ignition) system incorporates a closed-loop catalytic converter, an evaporative emission control system, an exhaust gas recirculation (EGR) system and a secondary air system, and complies with the latest emission control standards. The fuel injection side of the system operates as follows;

10 The fuel pump, immersed in the fuel tank,

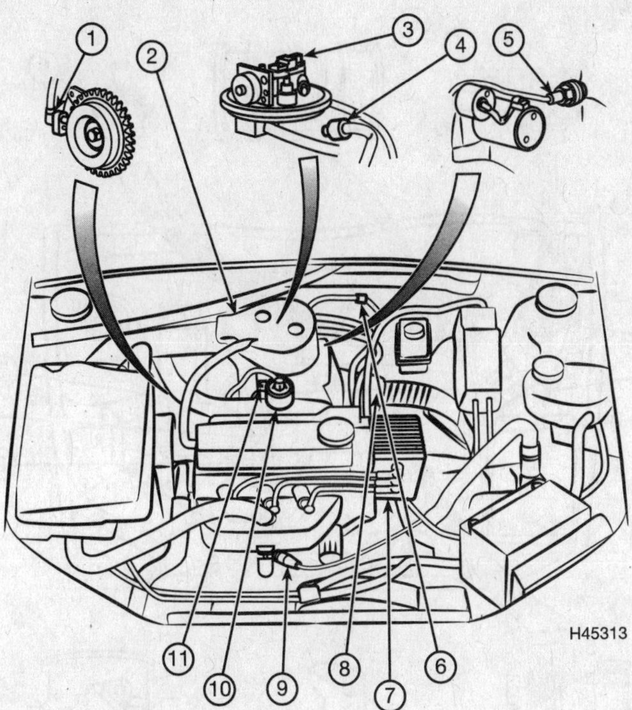

Multec single-point injection system components (X16SZR)

1 Crankshaft sensor
2 Throttle body
3 Fuel injector
4 Throttle potentiometer
5 Knock sensor
6 MAP sensor
7 Evaporative emission system purge valve
8 Ignition module
9 Exhaust gas recirculation (EGR) valve
10 Oxygen sensor
11 Idle speed control stepper motor

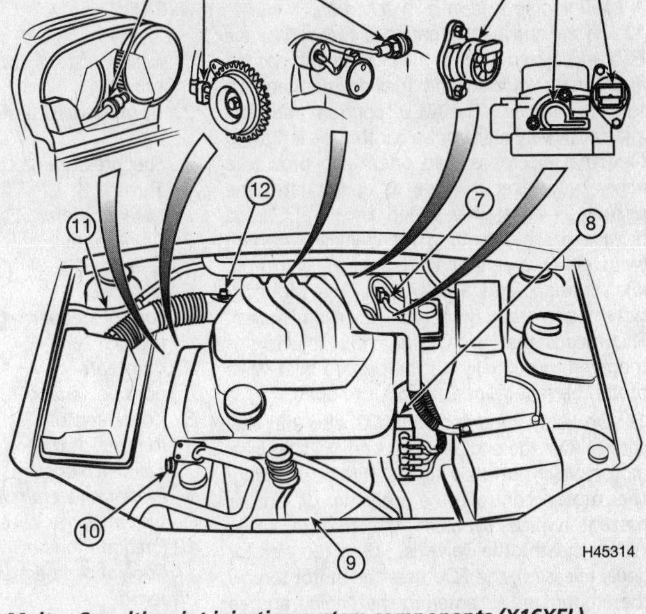

Multec S multi-point injection system components (X16XEL)

1 Crankshaft sensor
2 Knock sensor
3 Exhaust gas recirculation (EGR) valve
4 Throttle potentiometer
5 Idle speed control stepper motor
6 MAP sensor
7 Evaporative emission system purge valve
8 Ignition module
9 Oxygen sensor
10 Coolant temperature sensor
11 Intake air temperature sensor
12 Camshaft position sensor

pumps fuel from the fuel tank to the fuel rail, via a filter mounted underneath the rear of the vehicle. Fuel supply pressure is controlled by the pressure regulator which allows excess fuel to be returned to the tank. **Note:** *On some later models the pressure regulator is located in the fuel tank as part of the fuel tank sender unit/pump.*

11 The electrical control system consists of the ECU, along with the following sensors.

a) Throttle potentiometer – informs the ECU of the throttle position, and the rate of throttle opening or closing.
b) Coolant temperature sensor – informs the ECU of engine temperature.
c) Intake air temperature sensor – informs the ECU of the temperature of the air entering the manifold.
d) Oxygen sensor – informs the ECU of the oxygen content of the exhaust gases.
e) Crankshaft sensor – informs the ECU of engine speed and crankshaft position.
f) Camshaft sensor – informs the ECU of speed and position of the exhaust camshaft.
g) Knock sensor – informs the ECU when pre-ignition ('pinking') is occurring.
h) Manifold absolute pressure (MAP) sensor – informs the ECU of the engine load by monitoring the pressure in the inlet manifold.

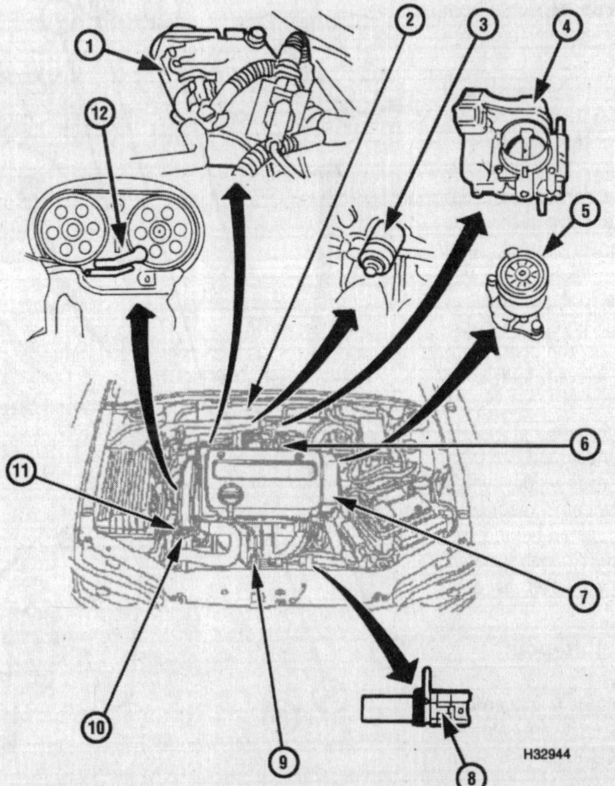

Multec S multi-point injection system components (Z16XE)

1 Evaporative emission system purge valve
2 Oxygen sensor – catalytic converter control
3 Knock sensor
4 Throttle body
5 Exhaust gas recirculation (EGR)
6 Intake manifold pressure (MAP) sensor
7 Engine control unit
8 Crankshaft sensor
9 Oxygen sensor – mixture regulation
10 Coolant temperature sensor
11 Coolant temperature sensor
12 Camshaft position sensor

i) *ABS control unit – informs the ECU of the vehicle speed.*

j) *Air conditioning system compressor switch (where fitted) – informs ECU when the air conditioning system is switched on.*

12 All the above information is analysed by the ECU and, based on this, the ECU determines the appropriate ignition and fuelling requirements for the engine. The ECU controls the fuel injector by varying its pulse width – the length of time the injector is held open – to provide a richer or weaker mixture, as appropriate. The mixture is constantly varied by the ECU, to provide the best setting for cranking, starting (with either a hot or cold engine), warm-up, idle, cruising, and acceleration. The Multec S system is a 'sequential' fuel injection system. This means that each of the four injectors is triggered individually just before the inlet valve of the relevant cylinder is about to open.

13 On early models, the ECU also has full control over the engine idle speed, via a stepper motor which is fitted to the throttle housing. The motor controls the opening of an air passage which bypasses the throttle valve. When the throttle valve is closed (accelerator pedal released), the ECU uses the motor to vary the amount of air entering the engine and so controls the idle speed.

14 On later models (Z16XE, Z18XE and Z22SE), the system has an electronic accelerator (no accelerator cable), this consists of an accelerator pedal module and a throttle valve module. Each module has two potentiometers (sensors) which determine the position of the pedal, which then transmits a signal to the ECU. Also the crankshaft sensor has been moved to the cylinder block, where it picks up the signal from an increment disc on number eight web on the crankshaft.

15 The ECU also controls the exhaust and evaporative emission control systems, which are described in detail in Part C of this Chapter.

16 If there is an abnormality in any of the readings obtained from any sensor, the ECU enters its back-up mode. In this event, the ECU ignores the abnormal sensor signal, and assumes a pre-programmed value which will allow the engine to continue running (albeit at reduced efficiency). If the ECU enters this back-up mode, the warning light on the instrument panel will come on, and the relevant fault code will be stored in the ECU memory.

17 If the warning light comes on, a complete test of the engine management system can then be carried out, using a fault code reader which is simply plugged into the system's diagnostic connector. The connector is located behind the centre console; unclip the trim panel situated just in front of the handbrake lever to gain access.

Simtec 56.5, 70 and 71 multi-point injection systems

18 The Simtec 56.5/70/71 engine management (fuel injection/ignition) systems are almost identical in operation to the Multec S system

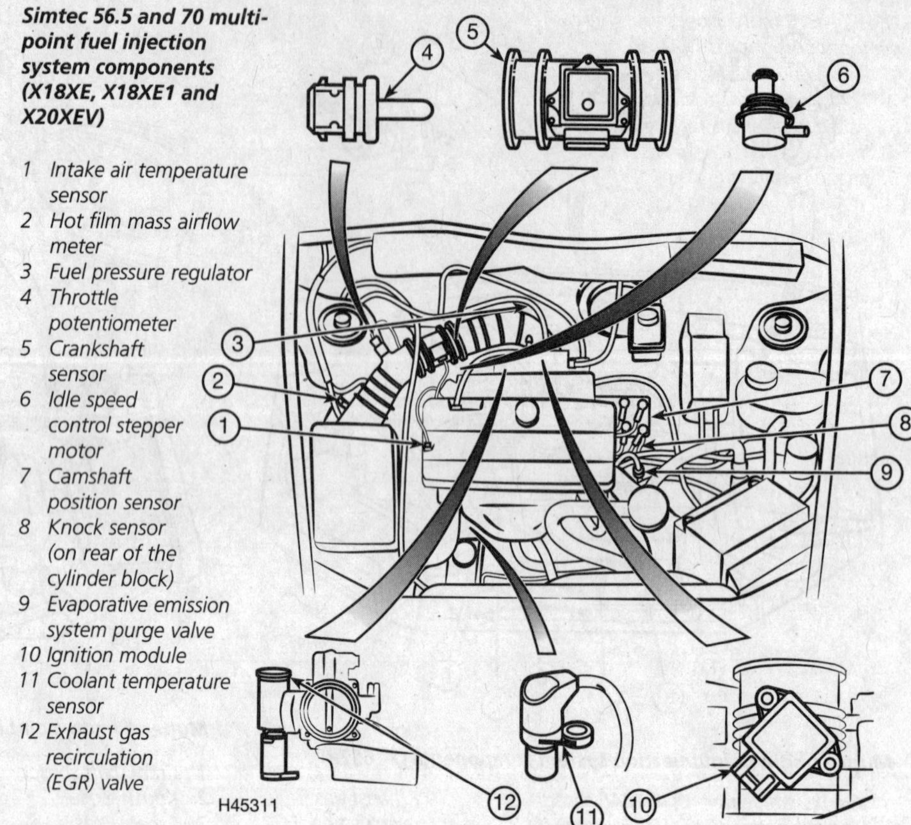

Simtec 56.5 and 70 multi-point fuel injection system components (X18XE, X18XE1 and X20XEV)

1 *Intake air temperature sensor*
2 *Hot film mass airflow meter*
3 *Fuel pressure regulator*
4 *Throttle potentiometer*
5 *Crankshaft sensor*
6 *Idle speed control stepper motor*
7 *Camshaft position sensor*
8 *Knock sensor (on rear of the cylinder block)*
9 *Evaporative emission system purge valve*
10 *Ignition module*
11 *Coolant temperature sensor*
12 *Exhaust gas recirculation (EGR) valve*

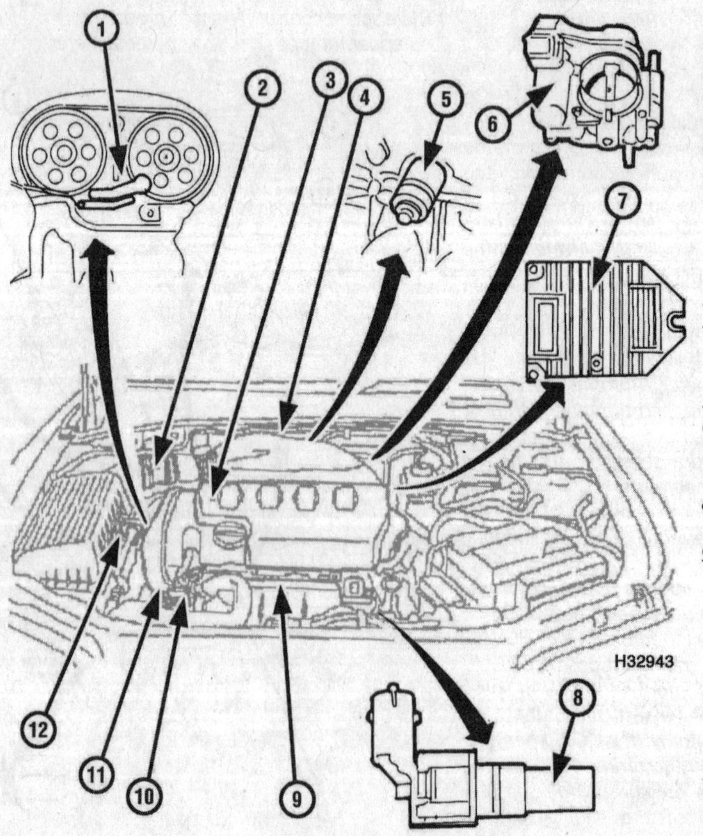

Simtec 71 multi-point fuel injection system components (Z18XE)

1 *Camshaft position sensor*
2 *Hot film mass airflow meter*
3 *Ignition module*
4 *Oxygen sensor – catalytic converter control*
5 *Knock sensor*
6 *Throttle body*
7 *Engine control unit*
8 *Crankshaft sensor*
9 *Oxygen sensor – mixture regulation*
10 *Coolant temperature sensor*
11 *Coolant temperature sensor*
12 *Evaporative emission system purge valve*

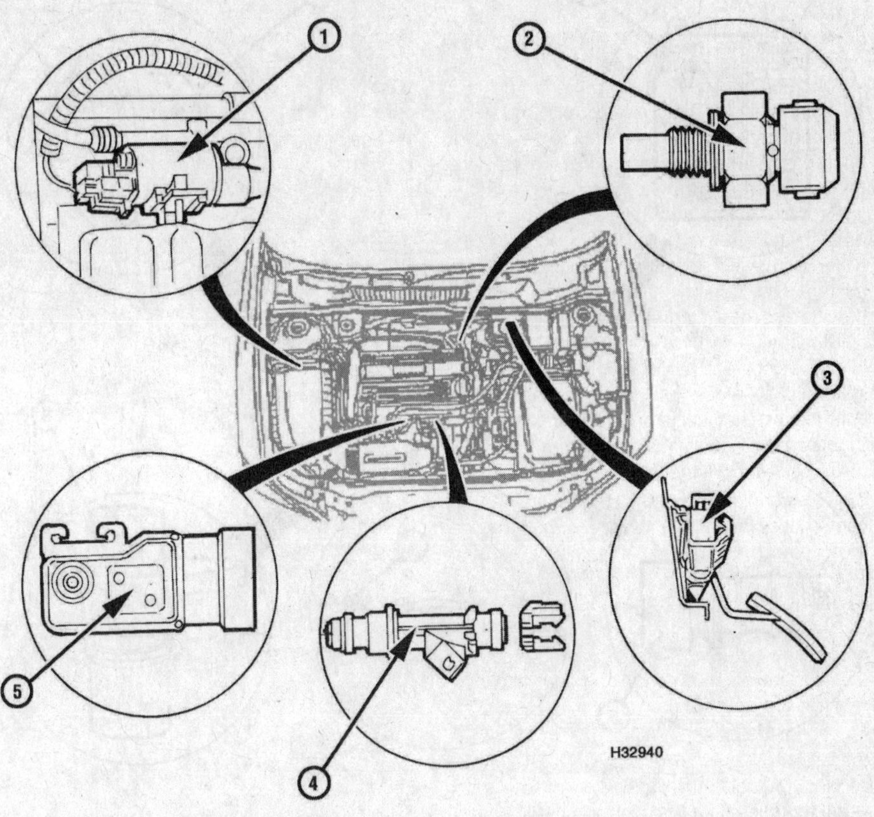

GMPT – E15 multi-point fuel injection system components (Z22SE)

1 Evaporative emission system purge valve
2 Coolant temperature sensor
3 Accelerator pedal/accelerator pedal sensor
4 Injectors
5 Intake manifold pressure (MAP) sensor

(see paragraphs 9 to 17). The only major change to the system is that a Hot Film Mass airflow sensor is fitted in place of the manifold pressure (MAP) sensor. The airflow meter informs the ECU of the amount of air entering the inlet manifold.

19 Another additional feature of the Simtec system is that it incorporates a variable tract inlet manifold to help increase torque output at low engine speeds. Each inlet manifold tract is fitted with a valve. The valve is controlled by the ECU via a solenoid valve and vacuum diaphragm unit.

20 At low engine speeds (below approximately 3600 rpm) the valves remain closed. The air entering the engine is then forced to take the long inlet path through the manifold which leads to an increase in the engine torque output.

21 At higher engine speeds, the ECU switches the solenoid valve which then allows vacuum to act on the diaphragm unit. The diaphragm unit is linked to the valve assemblies and opens up each of the four valves allowing the air passing through the manifold to take the shorter inlet path which is more suited to higher engine speeds.

GMPT – E15 multi-point injection system

22 The GMPT – E15 engine management (fuel injection/ignition) system is very similar in operation to the Multec S system (see paragraphs 9 to 17).

23 The only major change to the system is that special sensors are incorporated in the ignition module to replace the need for a camshaft sensor. The sensors read the injection impulse so as to determine the cylinder position.

24 This system also has an electronic accelerator (no accelerator cable), this consists of an accelerator pedal module and a throttle valve module. Each module has two potentiometers (sensors) which determine the

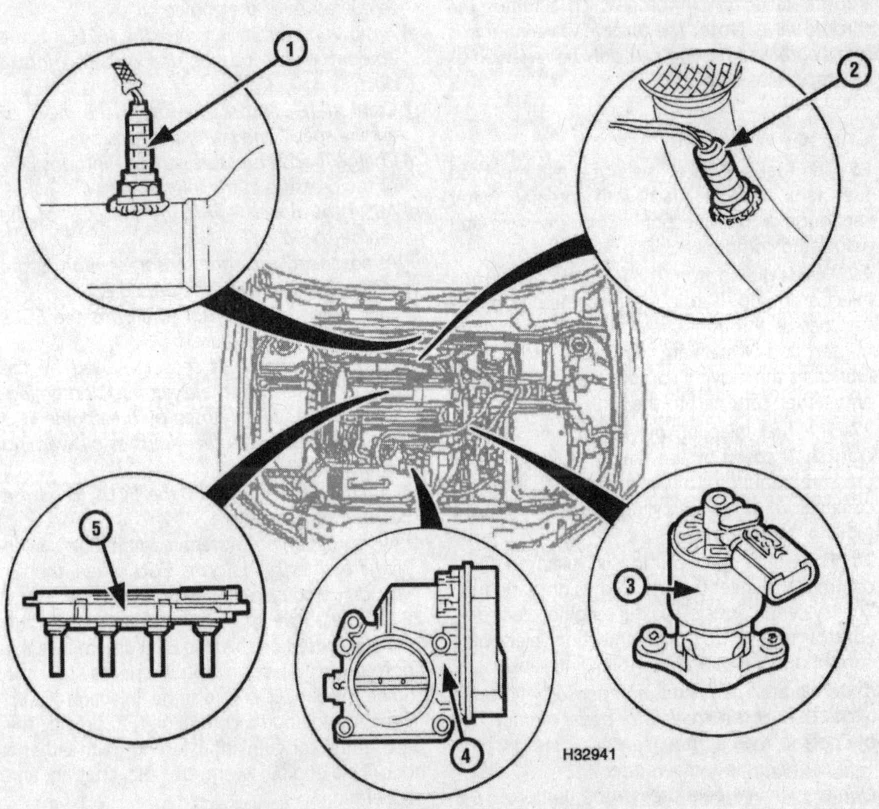

GMPT – E15 multi-point fuel injection system components (Z22SE)

1 Oxygen sensor – mixture regulation
2 Oxygen sensor – catalytic converter control
3 Exhaust gas recirculation (EGR) valve
4 Throttle body
5 DIS ignition module

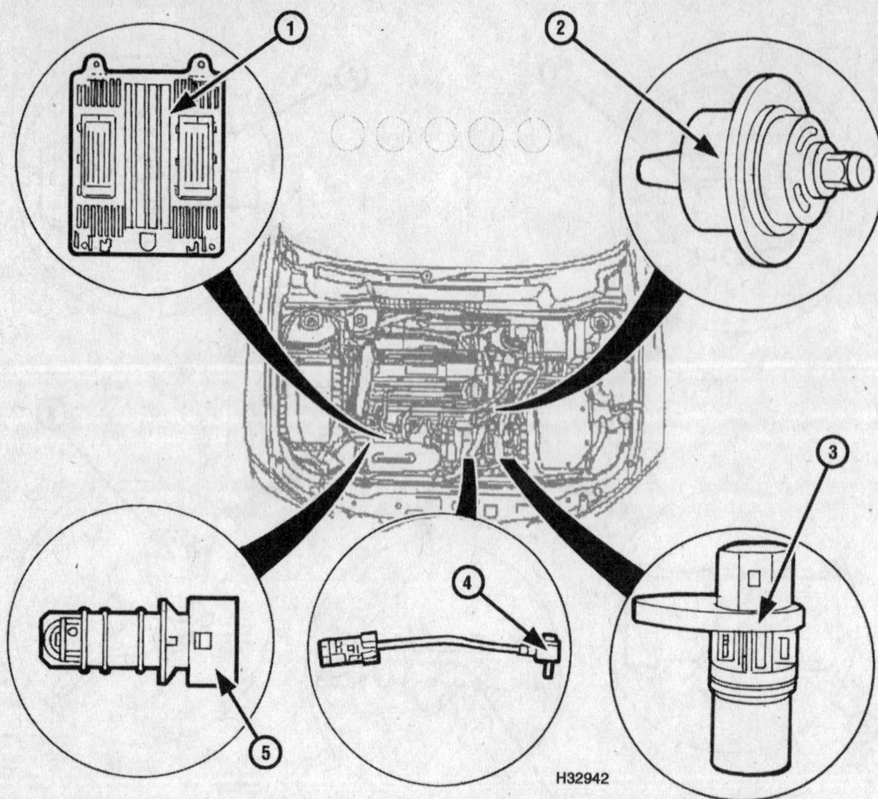

GMPT – E15 multi-point fuel injection system components (Z22SE)
1 Engine control unit
2 Fuel pressure regulator
3 Crankshaft sensor
4 Knock sensor
5 Intake air temperature sensor

position of the pedal, which then transmits a signal to the ECU. The throttle valve module consists of a throttle valve potentiometer and a throttle valve servo motor which actuates the throttle valve. **Note:** *The throttle valve module is an integrated unit and can only be renewed as a complete assembly.*

Diesel models

25 The fuel system consists of a rear-mounted fuel tank, a fuel filter with integral water separator, a fuel injection pump, injectors and associated components.

26 Fuel is drawn from the fuel tank by the fuel injection pump. Before reaching the pump, the fuel passes through a fuel filter, where foreign matter and water are removed. Excess fuel lubricates the moving components of the pump, and is then returned to the tank.

27 The fuel injection pump is driven at half-crankshaft speed by the timing chain. The high pressure required to inject the fuel into the compressed air in the cylinder is achieved by a radial piston pump.

28 The injection pump is electronically-controlled to meet the latest emission standards. The system consists of the engine electronic control unit (ECU), the injection electronic control unit (ECU) and the following sensors.

a) Accelerator pedal position sensor – informs the ECU of the accelerator pedal position.
b) Coolant temperature sensor – informs the ECU of engine temperature.
c) Intake air temperature sensor – informs the ECU of the temperature of the air passing through the intake duct.
d) Oil temperature sensor – informs ECU of the temperature of the engine oil.
e) Airflow meter – informs the ECU of the amount of air passing through the intake duct.
f) Crankshaft sensor – informs the ECU of engine speed and crankshaft position.
g) Charge (boost) pressure sensor – informs ECU of the pressure in the inlet manifold.
h) ABS control unit – informs the ECU of the vehicle speed.
i) Atmospheric pressure sensor – informs the ECUs of the atmospheric pressure.
j) Fuel temperature sensor – informs the ECUs of the fuel temperature.
k) Injection pump shaft sensor – used by the ECUs to determine the exact injection timing.
l) Oil pressure switch – used by the engine ECU to control the instrument cluster oil warning lamp.
m) Clutch switch – used by the ECUs for cruise control functions.
n) Air conditioning system compressor switch (where fitted) – informs ECU when the air conditioning system is switched on.

29 All the above information is analysed by the ECU and, based on this, the ECU determines the appropriate injection requirements for the engine. The ECU controls the injection pump timing, via the pump control unit, to provide the best setting for cranking, starting (with either a hot or cold engine), warm-up, idle, cruising, and acceleration.

30 Basic injection timing is determined when the pump is fitted. When the engine is running, it is varied automatically to suit the prevailing engine speed by a mechanism which turns the cam plate or ring – controlled by the ECU.

31 The ECU also controls the exhaust gas recirculation (EGR) system and the preheating system.

32 The four fuel injectors produce a spray of fuel directly into the cylinders. The injectors are calibrated to open and close at critical pressures to provide efficient and even combustion. Each injector needle is lubricated by fuel, which accumulates in the spring chamber and is channelled to the injection pump return hose by leak-off pipes.

33 The inlet manifold is fitted with a butterfly valve arrangement to improve efficiency at low engine speeds. Each cylinder has two intake tracts in the manifold, one of which is fitted a valve; the operation of the valve is controlled by the ECU via a solenoid valve and vacuum diaphragm unit. At low engine speeds (below approximately 1500 rpm) the valves remain closed, meaning that air entering each cylinder is passing through only one of the two manifold tracts. At higher engine speeds, the ECU opens up each of the four valves allowing the air passing through the manifold to pass through both inlet tracts.

34 A turbocharger is fitted to increase engine efficiency by raising the pressure in the inlet manifold above atmospheric pressure. Instead of the air simply being sucked into the cylinders, it is forced in. Additional fuel is supplied by the injection pump in proportion to the increased air intake.

35 Energy for the operation of the turbocharger comes from the exhaust gas. The gas flows through a specially-shaped housing (the turbine housing) and in so doing, spins the turbine wheel. The turbine wheel is attached to a shaft, at the end of which is another vaned wheel known as the compressor wheel. The compressor wheel spins in its own housing, and compresses the inlet air on the way to the inlet manifold.

36 Between the turbocharger and the inlet manifold, the compressed air passes through an intercooler. This is an air-to-air heat exchanger is mounted next to the radiator, and supplied with cooling air from the front of the vehicle. The purpose of the intercooler is to remove some of the heat gained in being compressed from the inlet air. Because cooler air is denser, removal of this heat further increases engine efficiency.

37 Charge (boost) pressure which is the pressure in the inlet manifold, is limited by a wastegate. This diverts the exhaust gas away from the turbine wheel in response to a pressure-sensitive actuator. A pressure-operated switch operates a warning light on the instrument panel in the event of excessive charge pressure developing.

38 The turbo shaft is pressure-lubricated by an oil feed pipe from the engine main oil so that the shaft 'floats' on a cushion of oil. A drain pipe returns the oil to the sump.

39 The charge (boost) pressure wastegate is controlled by the ECU via a solenoid valve. Note that an intercooler is only fitted to high-pressure turbo models.

40 If there is an abnormality in any of the readings obtained from any sensor, the ECU enters its back-up mode. In this event, the ECU ignores the abnormal sensor signal, and assumes a pre-programmed value which will allow the engine to continue running (albeit at reduced efficiency). If the ECU enters this back-up mode, the warning light on the instrument panel will come on, and the relevant fault code will be stored in the ECU memory.

41 If the warning light comes on, a complete test of the injection system can then be carried out using a fault code reader which is simply plugged into the system's diagnostic connector. The connector is located behind the centre console; unclip the trim panel situated just in front of the handbrake lever to gain access.

Fuel injection system – depressurisation, priming and bleeding

Petrol models

1 The fuel system referred to in this Section is defined as the tank-mounted fuel pump, the fuel filter, the fuel injector(s) and the pressure regulator, and the metal pipes and flexible hoses of the fuel lines between these components. All these contain fuel which will be under pressure while the engine is running, and/or while the ignition is switched on. The pressure will remain for some time after the ignition has been switched off, and it must be relieved in a controlled fashion when any of these components are disturbed for servicing work.

SOHC engines

2 Ensure the ignition is switched off then remove the cover from the engine compartment

relay box, which is situated next to the battery. Remove the fuel pump relay (the relay should be coloured purple) from the box.

3 Start the engine and allow it to idle until the engine starts to run roughly then switch the ignition off.

4 Disconnect the battery negative terminal then refit the relay and cover to the relay box.

DOHC engines

5 On these engines, the fuel system can either be depressurised as described above in paragraphs 2 to 4, or as follows.

6 Locate the valve assembly which is fitted to the fuel rail on the inlet manifold. On 1.6 litre engines the valve is on the right-hand end of the rail and on 1.8, 2.0 and 2.2 litre engines it can be found on the top of the rail.

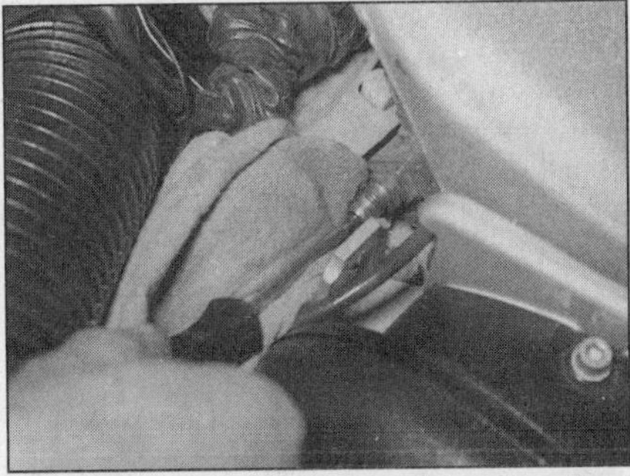

7 Unscrew the cap from the valve and position a container beneath the valve. Hold a wad of rag over the valve and relieve the pressure in the fuel system by depressing the valve core with a suitable screwdriver. Be prepared for the squirt of fuel as the valve core is depressed and catch it with the rag. Hold the valve core down until no more fuel is expelled from the valve.

8 Once all pressure is relieved, securely refit the valve cap.

Diesel models

9 It is not necessary to manually prime and bleed the fuel system after any operation on the system components. Start the engine (this may take longer than usual, especially if the fuel system has been allowed to run dry – operate the starter in ten second bursts with 5 seconds rest in between each operation) and run it a fast idle speed for a minute or so to purge any trapped air from the fuel lines. After this time the engine should idle smoothly at a constant speed.

10 If the engine idles roughly, then there is still some air trapped in the fuel system. Increase the engine speed again for another minute or so then recheck the idle speed. Repeat this procedure as necessary until the engine is idling smoothly.

Caution: Do not operate the starter for more than 10 seconds at a time. Allow 30 seconds between attempts for the starter to cool.

Multec single-point injection system components – removal and refitting

Fuel injector

Note: *Before condemning an injector, if a faulty injector is suspected, it is worth trying the effect of one of the proprietary injector-cleaning treatments.*

1 Depressurise the fuel system, then disconnect the battery negative terminal.

2 Slacken the retaining clip and disconnect the

air intake duct from the throttle body cover. Disconnect the vacuum and breather hoses from the cover then undo the retaining screws and remove the cover and sealing ring from the top of the throttle body.

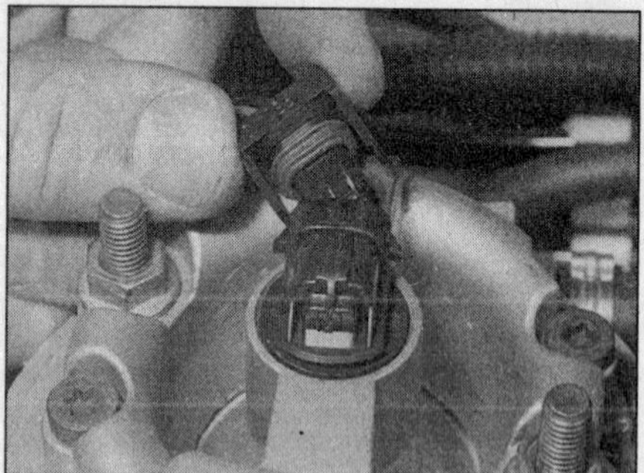

3 Release the retaining clips and disconnect the wiring connector from the injector.

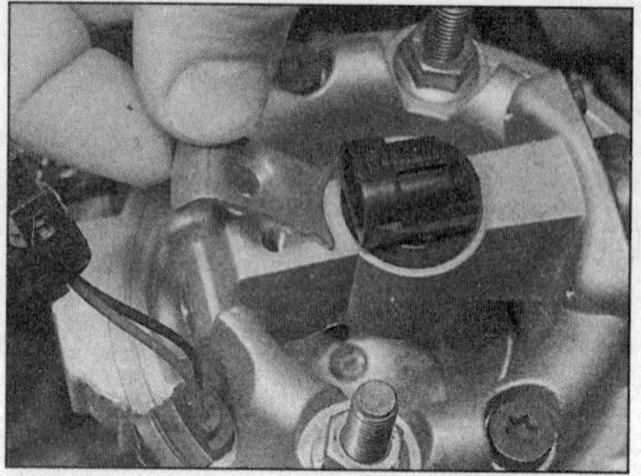

4 Undo the retaining screw and remove the injector retaining plate.

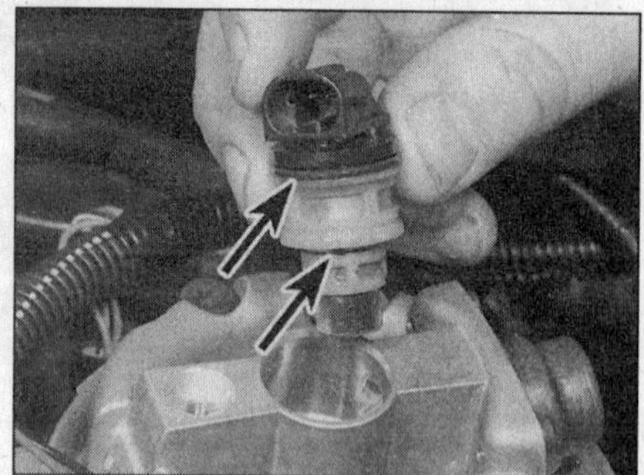

5 Ease the injector out from the throttle body along with its sealing rings. Discard the sealing rings, they must be renewed whenever the injector is disturbed.

6 Refitting is a reversal of the removal procedure using new injector sealing rings. When refitting the retaining clip, ensure it is correctly engaged with the injector and securely tighten its retaining screw.

Fuel pressure regulator

Note: *At the time of writing it appears that the regulator assembly is not available separately; if it is faulty the complete throttle body upper section must be renewed. Although the unit can be dismantled for cleaning, if required, it should not be disturbed unless absolutely necessary.*

7 Depressurise the fuel system, then disconnect the battery negative terminal.

8 Slacken the retaining clip and disconnect the air intake duct from the throttle body cover. Disconnect the vacuum and breather hoses from the cover then undo the retaining screws and remove the cover and sealing ring from the top of the throttle body.

9 Using a marker pen, make alignment marks between the regulator cover and throttle body, then slacken and remove the cover retaining screws.

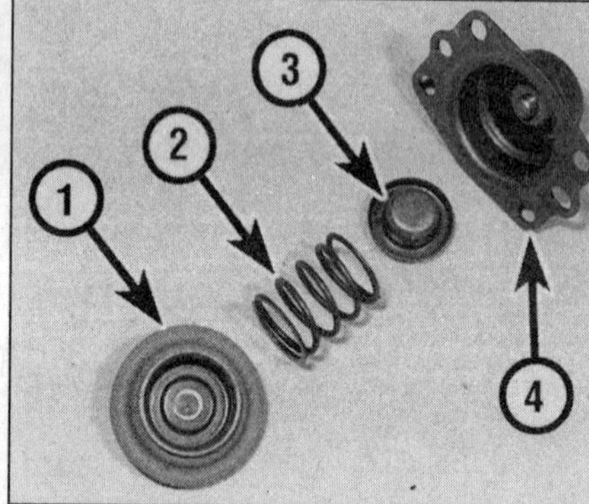

10 Lift off the cover (4), then remove the spring seat (3) and spring (2) then withdraw the diaphragm (1), noting its correct fitted orientation. Remove all traces of dirt, and examine the diaphragm for signs of splitting. If damage is found, it will probably be necessary to renew the throttle body assembly.

11 Refitting is a reverse of the removal procedure, ensuring that the diaphragm and cover are fitted the correct way round, and that the retaining screws are securely tightened.

Idle speed control stepper motor

12 Disconnect the battery negative terminal.

13 To improve access to the motor, slacken the retaining clip and disconnect the air intake duct from the throttle body cover. Disconnect the vacuum and breather hoses from the cover then undo the retaining screws and remove the cover and sealing ring from the top of the throttle body.

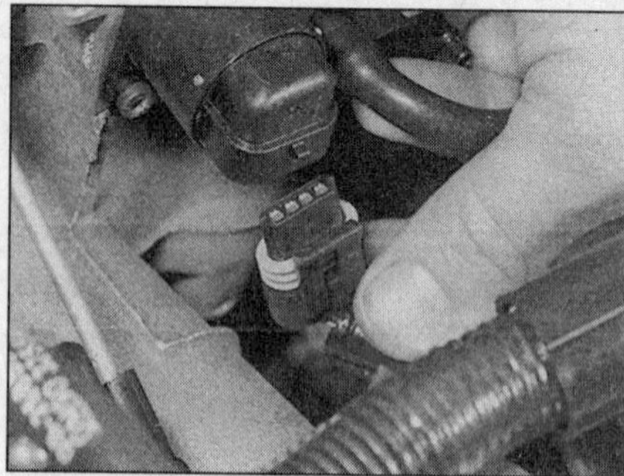

14 Release the retaining clip and disconnect the wiring connector from the motor which is fitted to the front of the throttle body assembly.

15 Undo the retaining screws and carefully manoeuvre the motor out of position, taking great care not to damage the motor plunger. Remove the sealing ring from the motor and discard it; a new one should be used on refitting.

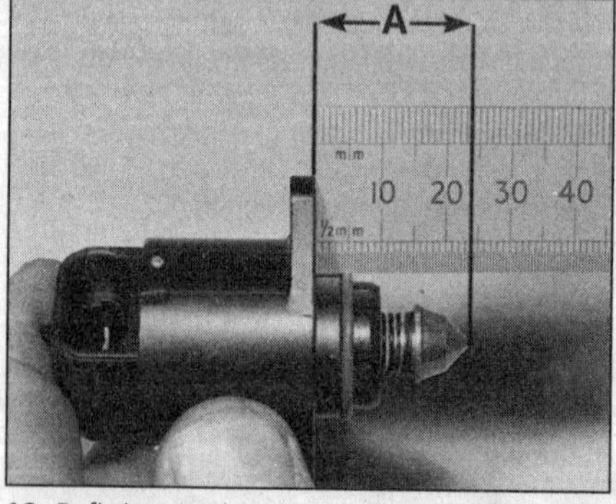

16 Refitting is the reverse of removal, using a new sealing ring. To ensure the motor plunger is not damaged on refitting, prior to installation, check that the plunger tip does not extend more than 28 mm from the motor mating flange (A). If necessary, **gently** push the plunger into the body until it is correctly positioned.

Throttle potentiometer

17 Disconnect the battery negative terminal then release the retaining clip and disconnect

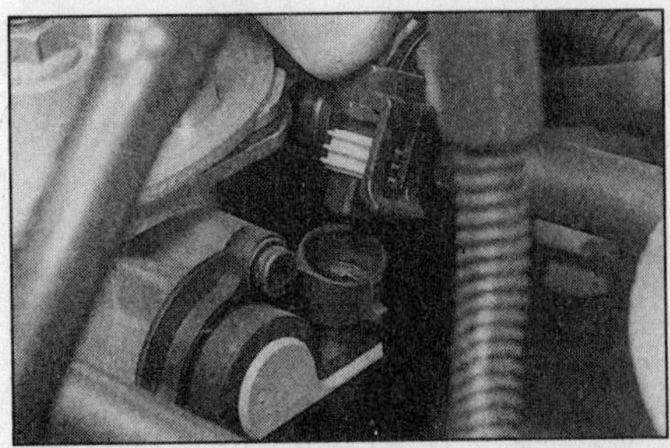

the wiring connector from the throttle potentiometer which is fitted to the left-hand side of the throttle body.

18 Slacken and remove the retaining screws and remove the potentiometer.

19 Prior to refitting, clean the threads of the retaining bolts and apply a drop of fresh locking compound to each one. Ensure the potentiometer is correctly engaged with the throttle valve spindle then tighten its retaining bolts to the specified torque and reconnect the wiring connector.

Coolant temperature sensor

20 The coolant temperature sensor is screwed into the rear of the inlet manifold. Partially drain the cooling system.

21 Disconnect the wiring plug and unscrew the sensor from the manifold.

22 Refitting is a reversal of removal. Apply a little sealant to the sensor threads prior to refitting and tighten it to the specified torque.

Manifold absolute pressure (MAP) sensor

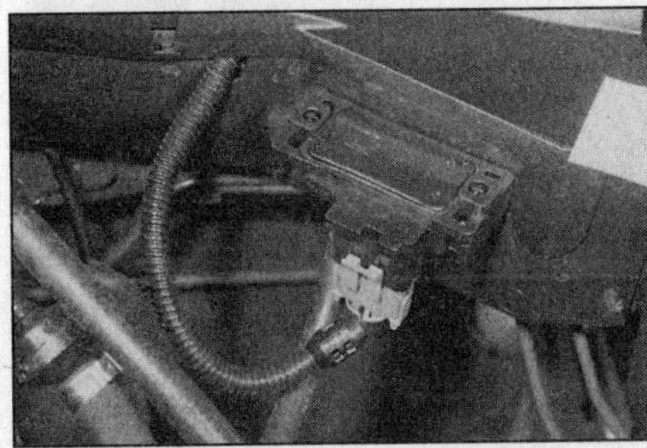

23 The MAP sensor is mounted onto the engine compartment bulkhead, just to the left of the throttle body. Ensure the ignition is switched off then disconnect the wiring connector and vacuum hose from the sensor. The MAP sensor can then be unclipped and removed from its mounting.

24 Refitting is the reverse of removal.

Crankshaft sensor

25 The sensor is mounted on the rear of the cylinder block and is accessible from underneath the vehicle. Firmly apply the handbrake then jack up the front of the vehicle and support it on axle stands.

26 Trace the wiring back from the sensor, releasing it from all the relevant clips and ties whilst noting its correct routing. Disconnect the wiring connector so the wiring is free to be removed with the sensor.

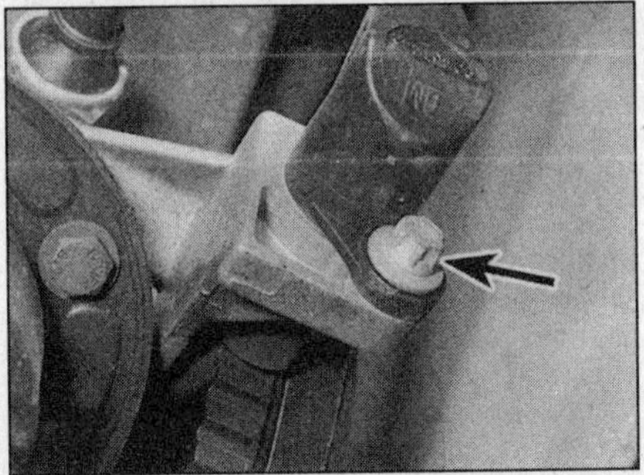

27 Unscrew the retaining bolt and remove the sensor from underneath the vehicle.

28 Refitting is the reverse of removal, tightening the retaining bolt to the specified torque. Ensure the wiring is correctly routed and retained by all the necessary clips and ties.

29 On completion, using feeler gauges, check that the clearance between the sensor tip and the crankshaft pulley teeth is 1.0 ± 0.7 mm. If the clearance is not within the specified range, renew the sensor mounting bracket.

Knock sensor

30 The knock sensor is mounted on the rear of the cylinder block and is accessible from underneath the vehicle. Firmly apply the handbrake then jack up the front of the vehicle and support it on axle stands.

31 Trace the wiring back from the sensor, noting its correct routing, and disconnect it at the connector.

32 Slacken and remove the retaining bolt and remove the sensor from the engine.

34 On refitting ensure the mating surfaces are clean and dry then fit the sensor and tighten its retaining bolt to the specified torque. Ensure the wiring is correctly routed and securely reconnected then lower the vehicle to the ground.

Electronic control unit (ECU)

35 The ECU is located behind the water deflector panel at the base of the windscreen. Disconnect the battery negative terminal then remove both windscreen wiper arms.

36 Peel off the rubber seal from the engine compartment bulkhead then unclip the water deflector panel from the base of the windscreen and remove it from the vehicle.

37 Undo the retaining nuts and remove the protective cover to gain access to the ECU.

38 Disconnect the wiring connectors then undo the retaining nuts and remove the ECU from the vehicle.

39 Refitting is the reverse of removal, ensuring the wiring connectors are securely reconnected and the ECU cover and water deflector are correctly seated.

Fuel pump relay

40 The fuel pump relay is located in the engine compartment main relay box. Unclip the cover and remove it from the relay box, the fuel pump relay is coloured purple. Ensure the ignition is switched off then pull out the relay.

41 Refitting is the reverse of removal.

Air conditioning system switch

42 The air conditioning system switch is screwed into one of the refrigerant pipes and cannot be removed without first discharging the refrigerant. Renewal of the switch should therefore be entrusted to a suitably-equipped garage.

Multec-S injection system components – removal and refitting

Fuel rail and injectors

Note: *If a faulty injector is suspected, before condemning the injector, it is worth trying the effect of one of the proprietary injector-cleaning treatments.*

1 Depressurise the fuel system, then disconnect the battery negative terminal.

2 Remove the upper section of the inlet manifold.

3 Disconnect the wiring connectors from the four injectors then free the wiring harness from the fuel rail.

4 Unscrew the union nuts and disconnect the fuel hoses from the fuel rail. Whilst the unions are being slackened, retain the fuel rail adapters with an open-ended spanner.

5 Slacken and remove the three retaining bolts then carefully ease the fuel rail and injector assembly out of position and remove it from the manifold. Remove the lower sealing rings from the injectors and discard them; they must be renewed whenever they are disturbed.

6 Slide off the relevant retaining clip and withdraw the injector from the fuel rail. Remove the upper sealing ring from the injector and discard it; all disturbed sealing rings must be renewed.

7 Refitting is a reversal of the removal procedure, noting the following points.

a) Renew all disturbed sealing rings and apply a smear of engine oil to them to aid installation.

b) Ease the injector(s) into the fuel rail, ensuring that the sealing ring(s) remain correctly seated, and secure in position with the retaining clips.

c) On refitting the fuel rail, take care not to damage the injectors and ensure that all sealing rings remain in position. Once the fuel rail is correctly seated, tighten its retaining bolts to the specified torque.

d) On completion start the engine and check for fuel leaks.

Fuel pressure regulator

Note: *On Z16XE models: Fuel pressure regulation is a function of the fuel tank pump module. The regulator is located under the fuel pump access cover. The regulator does not appear to be available as a separate part – check with your Vauxhall dealer. To remove the regulator, remove the fuel pump.*

8 On X16XEL models, depressurise the fuel system, then disconnect the battery negative terminal.

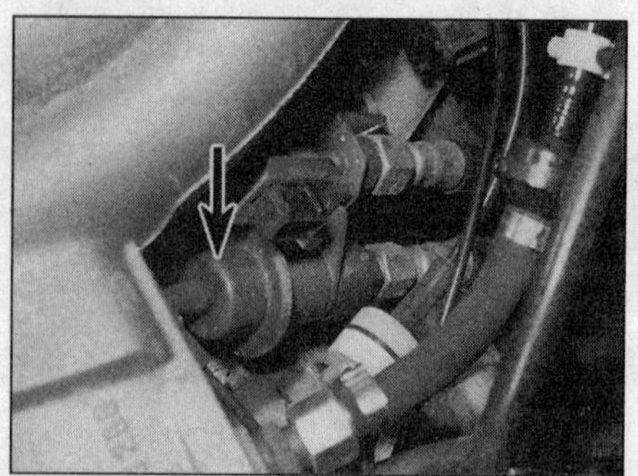

9 Access to the fuel regulator is poor and can be improved slightly by freeing the wiring harness plastic tray from the inlet manifold.

10 Retain the regulator with an open-ended spanner and unscrew the fuel hose union nut. Disconnect the fuel hose and the vacuum hose from the regulator.

11 Slacken and remove the retaining bolt and free the wiring bracket from the fuel rail.

12 Ease the regulator out from the end of the fuel rail and remove it along with its sealing ring.

13 Refitting is the reverse of removal, using a new sealing ring. On completion start the engine and check for fuel leaks.

Idle speed control stepper motor

Note: *On Z16XE models, there is an electronic accelerator system fitted (no accelerator cable). The throttle valve potentiometer and servo motor are a combined part of the throttle body housing. Individual parts for the throttle body housing are not available separately and, if defective, it must be renewed as a complete unit.*

X16XEL

14 Access to the idle control stepper motor is very poor, to improve it remove the throttle housing.

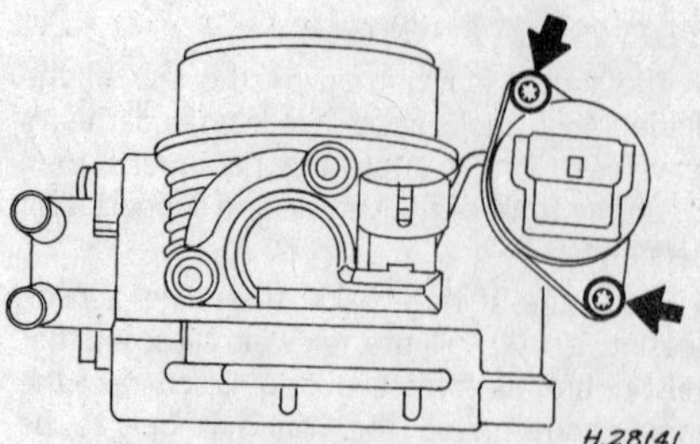

15 Undo the retaining screws and remove the motor from the housing along with its sealing ring.

16 Refitting is the reverse of removal, noting the following points.

*a) Prior to installation, check that the plunger tip does not extend more than 33 mm from the motor mating flange (*see paragraph 16 on page 26•8*). If necessary,* ***gently*** *push the plunger into the body until it is correctly positioned. Failure to do this could lead to the motor being damaged.*

b) Apply locking compound to the motor screws.

Throttle potentiometer

Note: *On Z16XE models, there is an electronic accelerator system fitted (no accelerator cable). The throttle valve potentiometer and servo motor are a combined part of the throttle body housing. Individual parts for the throttle body housing are not available separately and, if defective, it must be renewed as a complete unit.*

X16XEL

17 Remove the throttle housing.

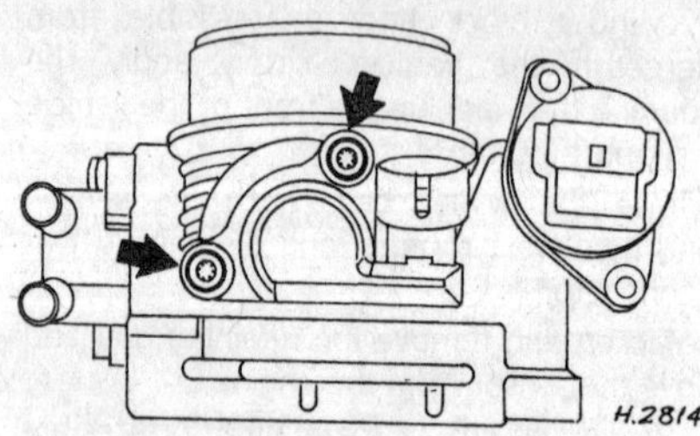

18 Undo the retaining screws and remove the potentiometer from the side of the housing.

19 On refitting, fit the potentiometer, ensuring it is correctly engaged with the throttle valve spindle. Apply locking compound to the retaining screws and tighten them securely. Check the operation of the throttle valve then refit the housing.

Coolant temperature sensor

20 The coolant temperature sensor is screwed into the thermostat housing on the right-hand end of the cylinder head. Partially drain the cooling system.

21 Disconnect the wiring plug then unscrew the sensor from the housing.

22 Refitting is a reversal of removal. Apply a little sealant to the threads, then tighten the sensor to the specified torque.

Intake air temperature sensor

23 The intake air temperature sensor is mounted in the intake duct which connects the air cleaner housing to the inlet manifold. Ensure the ignition is switched off, then disconnect the wiring connector from the sensor.

24 Carefully ease the sensor out of position and remove its sealing grommet from the intake duct. If the sealing grommet shows signs of damage or deterioration it should be renewed.

25 Refitting is the reverse of removal, ensuring the sensor and grommet are correctly located in the duct.

Manifold absolute pressure (MAP) sensor

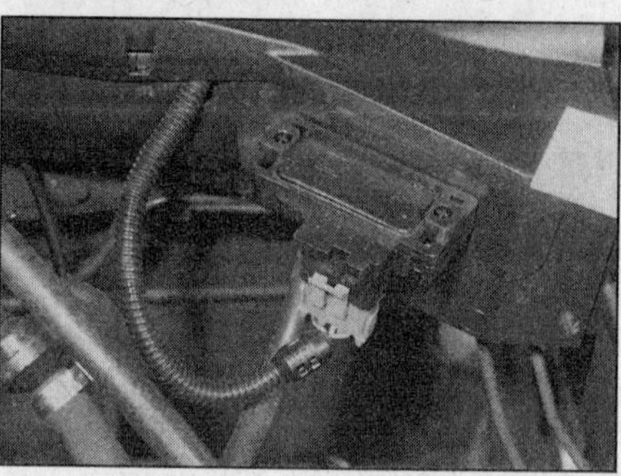

26 The MAP sensor is mounted onto the engine compartment bulkhead, just to the left of the inlet manifold. Ensure the ignition is switched off then disconnect the wiring connector and vacuum hose from the sensor. The MAP sensor can then be unclipped and removed from its mounting.

27 Refitting is the reverse of removal.

Crankshaft sensor

28 On X16XEL engines, the sensor is mounted on a bracket at the rear of the right-hand end of the cylinder block and is accessible from underneath the vehicle. Firmly apply the handbrake then jack up the front of the vehicle and support it on axle stands.

29 On Z16XE engines, the sensor is mounted on the front of the cylinder block below the oil filter.

30 Trace the wiring back from the sensor, releasing it from all the relevant clips and ties whilst noting its correct routing. Disconnect the wiring connector so the wiring is free to be removed with the sensor. Unscrew the retaining bolt and remove the sensor.

31 Refitting is the reverse of removal, tightening the retaining bolt to the specified torque. Ensure the wiring is correctly routed and retained by all the necessary clips and ties.

32 On X16XEL engines, using feeler gauges, check that the clearance between the sensor tip and the crankshaft pulley teeth is 1.0 ± 0.7 mm. If the clearance is not within the specified range, renew the sensor mounting bracket.

Camshaft sensor

33 Remove the timing belt upper cover.

34 Trace the wiring back from the sensor, releasing it from all the relevant clips and ties whilst noting its correct routing. Disconnect the wiring connector so the wiring is free to be removed with the sensor.

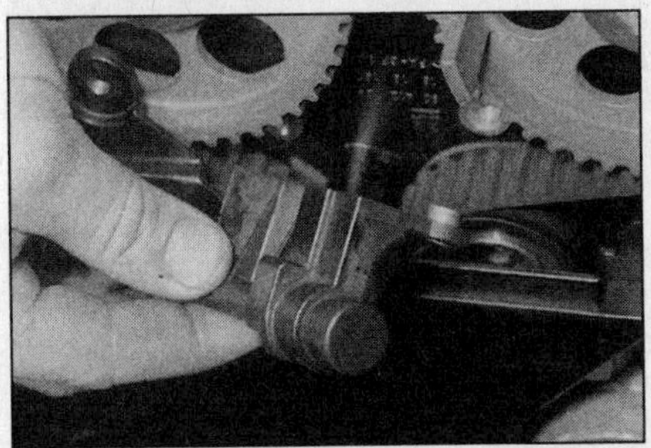

35 Unscrew the retaining bolts and remove the sensor from the side of the cylinder head.

36 Refitting is the reverse of removal, tightening the retaining bolts to the specified torque. Ensure the wiring is correctly routed and retained by all the necessary clips and ties.

Knock sensor

37 Firmly apply the handbrake then jack up the front of the vehicle and support it on axle stands. The knock sensor is mounted onto the rear of the cylinder block, just to the right of the starter motor, and is accessible from underneath the vehicle.

38 Trace the wiring back from the sensor, noting its correct routing, and disconnect it at the connector.

39 Slacken and remove the retaining bolt and remove the sensor from the engine.

40 On refitting ensure the mating surfaces are clean and dry then fit the sensor and tighten its retaining bolt to the specified torque. Ensure the wiring is correctly routed and securely reconnected then lower the vehicle to the ground.

Electronic control unit (ECU)

41 The ECU is located behind the water deflector panel at the base of the windscreen.

42 Disconnect the battery negative terminal then remove both windscreen wiper arms.

43 Peel off the rubber seal from the engine compartment bulkhead then unclip the water deflector panel from the base of the windscreen and remove it from the vehicle.

44 Undo the retaining nuts and remove the protective cover to gain access to the ECU.

45 Release the retaining clip(s) then disconnect the wiring connector(s) from the ECU. Undo the retaining nuts and remove the ECU from the vehicle.

46 Refitting is the reverse of removal, ensuring the wiring connectors are securely reconnected and the ECU cover and water deflector are correctly seated.

Fuel pump relay

47 The fuel pump relay is located in the engine compartment main relay box.

48 Unclip the cover and remove it from the relay box, the fuel pump relay is coloured purple. Ensure the ignition is switched off then pull out the relay.

49 Refitting is the reverse of removal.

Air conditioning system switch

50 The air conditioning system switch is screwed into one of the refrigerant pipes and cannot be removed without first discharging the refrigerant. Renewal of the switch should therefore be entrusted to a suitably-equipped garage.

Simtec injection system components – removal and refitting

Fuel rail and injectors

Note: *Before condemning an injector, if a faulty injector is suspected, it is worth trying the effect of one of the proprietary injector-cleaning treatments.*

1 Depressurise the fuel system, then disconnect the battery negative terminal.

2 Disconnect the wiring connectors from the intake air temperature sensor and the Hot Film Mass airflow meter.

3 Release the retaining clips and disconnect the breather hoses from the rear of the cylinder head cover.

4 Slacken the retaining clips then disconnect the intake duct from the air cleaner and throttle housing and remove the duct assembly from the engine compartment, freeing it from the wiring harness.

5 On X18XE, X18XE1 and X20XEV models, remove the retaining clip and detach the accelerator cable from the throttle cam balljoint and unclip the cable from its mounting bracket. On models with cruise control it will also be necessary to detach the cruise control cable.

6 Disconnect the wiring connector from the idle control stepper motor then undo the retaining bolts and free the cable mounting bracket from the throttle housing.

7 Carefully release the retaining clips and lift the injector wiring cover assembly squarely away from the top of the injectors; the wiring connectors are an integral part of the cover.

8 Position the injector wiring cover assembly clear of the fuel rail.

9 Unscrew the union nuts and disconnect the fuel hoses from the fuel rail. Whilst the unions are being slackened, retain the fuel rail adapters with an open-ended spanner.

10 Slacken and remove the three retaining bolts then carefully ease the fuel rail and injector assembly out of position and remove it from the manifold. Remove the lower sealing rings from the injectors and discard them; they must be renewed whenever they are disturbed.

11 Slide off the relevant retaining clip and withdraw the injector from the fuel rail. Remove the upper sealing ring from the injector and discard it; all disturbed sealing rings must be renewed.

12 Refitting is a reversal of the removal procedure, noting the following points.

a) *Renew all disturbed sealing rings and apply a smear of engine oil to them to aid installation.*

b) *Ease the injector(s) into the fuel rail, ensuring that the sealing ring(s) remain correctly seated, and secure in position with the retaining clips.*

c) *On refitting the fuel rail, take care not to damage the injectors and ensure that all sealing rings remain in position. Once the fuel rail is correctly seated, tighten its retaining bolts to the specified torque.*

d) *On completion start the engine and check for fuel leaks.*

Fuel pressure regulator

13 On X18XE and X20XEV models, depressurise the fuel system, then disconnect the battery negative terminal. **Note:** *On Z18XE and X18XE1 models, fuel pressure regulation is a function of the fuel tank pump module. The*

regulator is located under the fuel pump access cover. However, the regulator does not appear to be available as a separate part – check with your Vauxhall dealer. To remove the regulator, remove the pump.

14 To improve access to the regulator, release the retaining clips and disconnect the breather hose from the rear of the cylinder head cover.

15 Disconnect the vacuum hose (1) from the regulator.

16 Slacken the retaining clamp (2) then carefully ease the pressure regulator out from the top of the fuel rail. Remove the both sealing rings from the regulator and discard them; new ones must be used on refitting.

17 Refitting is the reverse of removal, using new sealing rings. Lubricate the sealing rings with a smear of engine oil to ease installation and tighten the retaining clamp screw to the specified torque. On completion check for signs of fuel leakage.

Idle speed control stepper motor

Note: *On Z18XE models, there is an electronic accelerator system fitted (no accelerator cable). The throttle valve potentiometer and servo motor are a combined part of the throttle body housing. Individual parts for the throttle body housing are not available separately and, if defective, it must be renewed as a complete unit.*

Except Z18XE

18 Disconnect the wiring connectors from the intake air temperature sensor and the airflow meter. Release the retaining clip and disconnect the breather hose from the rear of the cylinder head cover then slacken the retaining clips and remove the intake duct assembly from the engine compartment.

19 Disconnect the wiring connector then undo the retaining screws and remove the motor assembly from the side of the throttle housing.

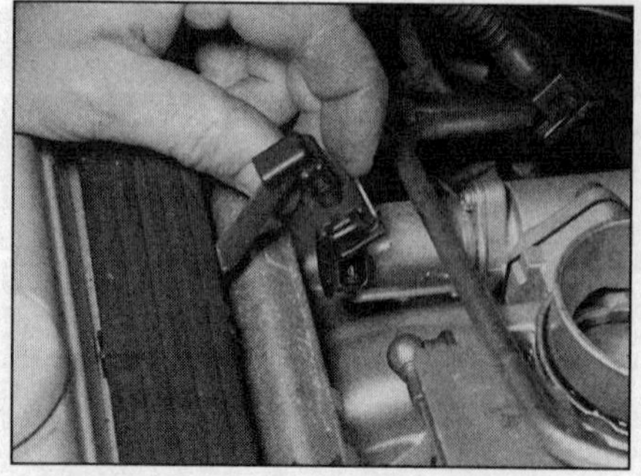

Remove the gasket and discard it; a new one should be used on refitting.

20 Refitting is the reverse of removal using a new gasket.

Throttle potentiometer

Note: *On Z18XE models, there is an electronic accelerator system fitted (no accelerator cable). The throttle valve potentiometer and servo motor are a combined part of the throttle body housing. Individual parts for the throttle body housing are not available separately and, if defective, it must be renewed as a complete unit.*

Except Z18XE

21 Ensure the ignition is switched off then disconnect the wiring connector from the throttle potentiometer which is fitted to the rear of the throttle housing.

22 Slacken and remove the retaining screws then remove the potentiometer from the housing.

23 Refitting is the reverse of removal, making sure the potentiometer is correctly engaged with the throttle valve spindle.

Coolant temperature sensor

24 The coolant temperature sensor is screwed into the housing on the left-hand end of the cylinder head. Partially drain the cooling system.

25 On X18XE engines, remove the DIS module.

26 On all engines, disconnect the wiring plug, then unscrew the sensor from the housing.

27 Refitting is a reversal of removal. Apply a little sealant to the sensor threads then tighten it to the specified torque.

Intake air temperature sensor

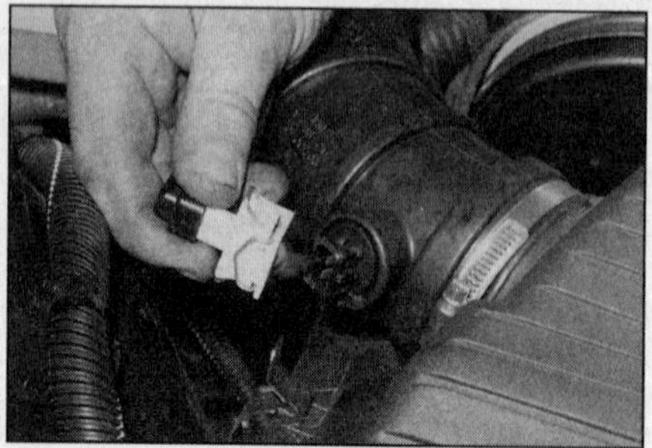

28 Ensure the ignition is switched off then disconnect the wiring connector from the sensor.

29 Slacken the retaining clips then detach the intake duct from the airflow meter and air cleaner housing and remove it from the vehicle.

30 Carefully ease the sensor out of position, taking great care not to damage the duct.

31 Refitting is the reverse of removal, noting that the sensor must be fitted so that its flat edge aligns with the flat on the intake duct.

Hot Film Mass airflow meter

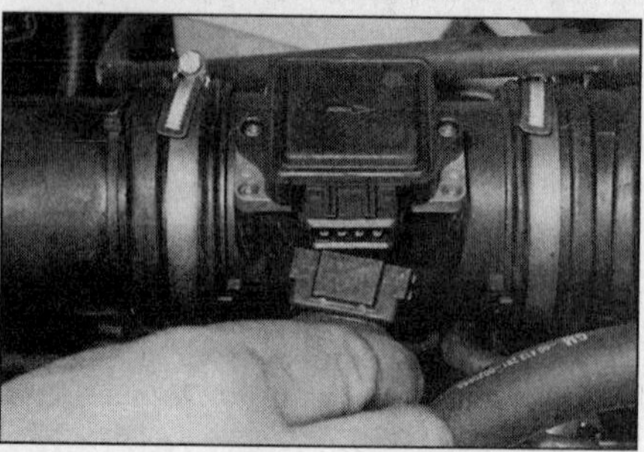

32 Ensure the ignition is switched off then disconnect the wiring connector from the airflow meter.

33 Slacken the retaining clips then disconnect the intake ducts and remove the airflow meter from the vehicle. Inspect the meter for signs of damage and renew if necessary.

34 Refitting is the reverse of removal, ensuring the intake ducts are correctly engaged with the meter recesses.

Crankshaft sensor

35 On Z18XE engines, the sensor is mounted on the front of the cylinder block below the oil filter.

36 On X18XE1 engines, the sensor is mounted on a bracket at the rear of the right-hand end of the cylinder block. Follow the procedures as described in the Multec-S system Section, paragraphs 28 to 32, for the X16XEL engine.

37 On X18XE and X20XEV engines, the sensor is mounted on the front of the cylinder block at the timing belt end. Remove the air cleaner housing, and unbolt the engine lifting shackle from the top of the engine. At the rear of the cylinder head, disconnect the wiring connector so the wiring is free to be removed with the sensor.

38 Trace the wiring back from the sensor, releasing it from all the relevant clips and ties whilst noting its correct routing.

39 Unscrew the retaining bolt and remove the sensor from the front of the cylinder block, along with its sealing ring. Discard the sealing ring, a new one should be used on refitting.

40 Refitting is the reverse of removal using a new sealing ring and tightening the sensor bolt to the specified torque. Ensure the wiring cover assembly is clipped securely in position.

Camshaft sensor

41 Undo the retaining screws and remove the spark plug cover from the top of the cylinder head cover.

42 Ensure the ignition is switched off then disconnect the wiring connector from the camshaft sensor.

43 Remove the timing belt front cover.

44 Unscrew the retaining bolt and remove the camshaft sensor from the cylinder head.

45 Refitting is the reverse of removal, tightening the sensor retaining bolt to the specified torque.

Knock sensor

46 The knock sensor is mounted on the rear of the cylinder block and is accessible from underneath the vehicle. Firmly apply the handbrake then jack up the front of the vehicle and support it on axle stands. Depending on model, it may be necessary to remove the starter motor to make access easier.

47 Trace the wiring back from the sensor, releasing it from all the relevant clips and ties whilst noting its correct routing.

48 Carefully release the retaining clips and lift the wiring cover assembly squarely away from rear of the cylinder head to gain access to the sensor wiring connector. Disconnect the connector so the wiring is free to be removed with the sensor.

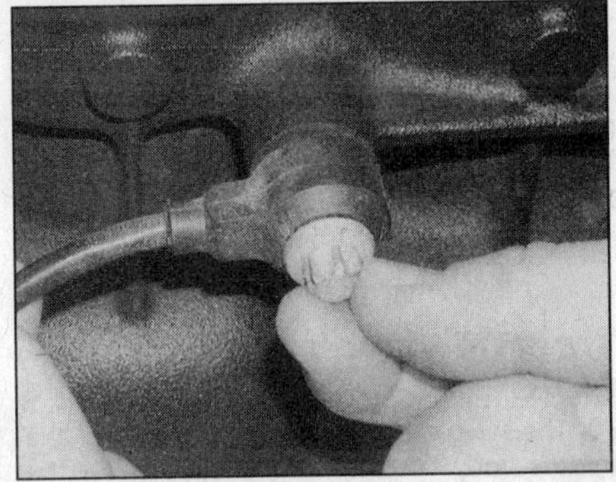

49 Slacken and remove the retaining bolt and remove the sensor from the engine.

50 On refitting, ensure the mating surfaces are clean and dry then fit the sensor and tighten its retaining bolt to the specified torque. Ensure the wiring is correctly routed and securely reconnected then clip the wiring cover assembly back into position.

Electronic control unit (ECU)

51 Refer to the Multec-S system Section, paragraphs 41 to 46.

Fuel pump relay

52 Refer to the Multec-S system Section, paragraphs 47 to 49.

Air conditioning system switch

53 The air conditioning system switch is screwed into one of the refrigerant pipes and cannot be removed without first discharging the refrigerant. Renewal of the switch should therefore be entrusted to a suitably-equipped garage.

Inlet manifold valve switching solenoid

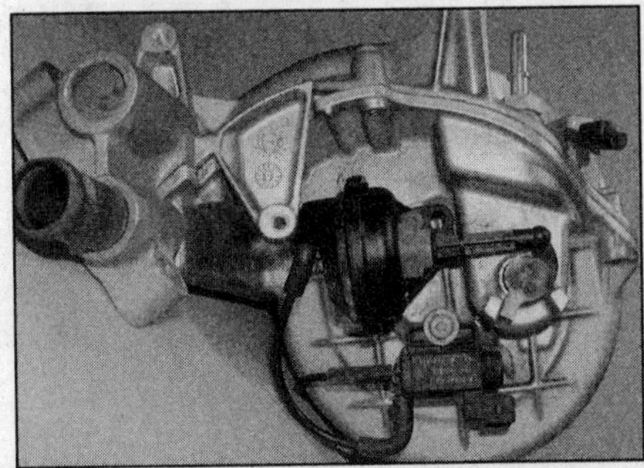

54 The solenoid which operates the inlet manifold valve assemblies is mounting on the left-hand end of the manifold.

55 Ensure the ignition is switched off then disconnect the wiring connector from the solenoid.

56 Disconnect both vacuum hoses then remove the solenoid from the manifold.

57 Refitting is the reverse of removal.

Inlet manifold valve vacuum diaphragm unit

58 The inlet manifold valve vacuum diaphragm unit is fitted to the left-hand end of the manifold *(see paragraph 54)*.

56 Disconnect the vacuum hose, unclip the diaphragm rod from the its balljoint and remove the unit from the manifold.

57 Refitting is the reverse of removal.

GMPT – E15 injection system components – removal and refitting

Fuel rail and injectors

Note: *Before condemning an injector, it is worth trying the effect of one of the proprietary injector-cleaning treatments.*

1 Unclip the engine cover and depressurise the fuel system, then disconnect the battery negative terminal.

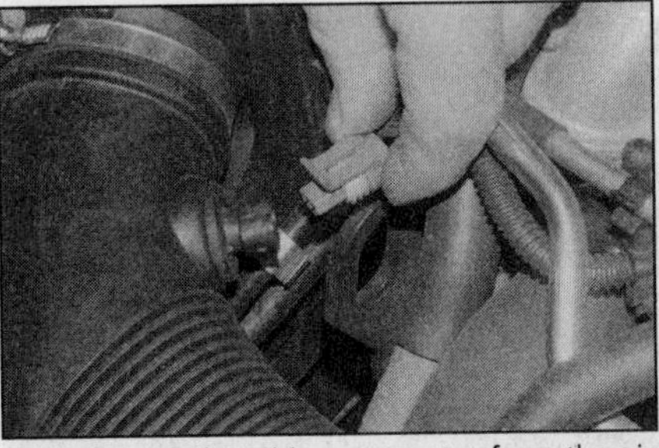

2 Disconnect the wiring connector from the air temperature sensor in the air intake pipe.

3 Release the retaining clip and disconnect the breather hose from the cylinder head cover.

4 Release the retaining clips then disconnect the intake pipe from the air cleaner and throttle housing and remove the pipe assembly from the engine compartment, detach the vacuum pipe(s) from the intake pipe on removal.

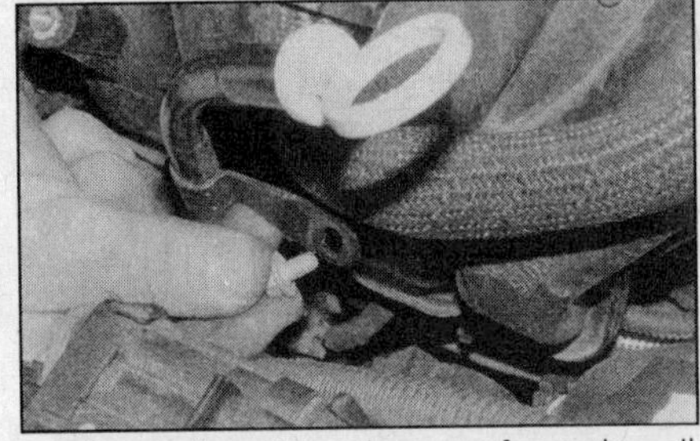

5 Release the wiring harness from the oil dipstick guide tube, then remove the retaining bolt from the guide tube and twist it away from the intake manifold.

6 Undo the two retaining bolts and release the earth cables on the cylinder block below the intake manifold.

7 Release the wiring harness from the lower part of the intake manifold and the knock sensor wiring block connector from its retaining clip.

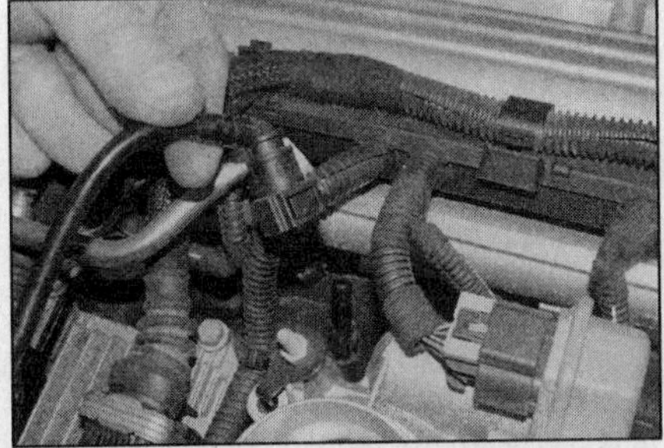

8 Disconnect the tank vent/purge valve (which is clipped onto the air filter housing) vacuum line from the intake manifold.

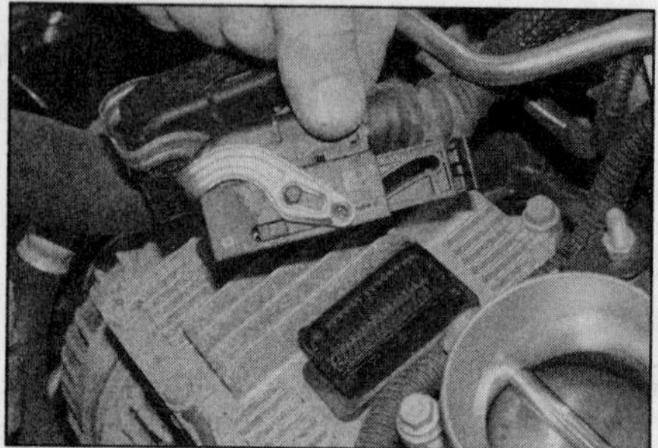
9 Release the two wiring harness plugs from the engine electronic control unit, then undo the securing bolts and remove the control unit from the intake manifold. Note the position of the earth cable on one of the securing bolts.

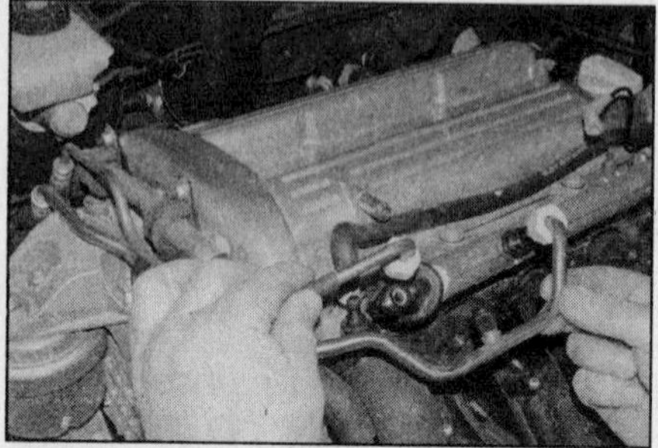
10 Unscrew the union nuts and disconnect the fuel pipes from the fuel rail, release the fuel pipes from the retaining clips and move to one side.

11 Disconnect the wiring connector from the throttle body, then undo the securing bolts and remove the throttle body unit from the intake manifold. Note the position of the cable retaining clip on one of the securing bolts.

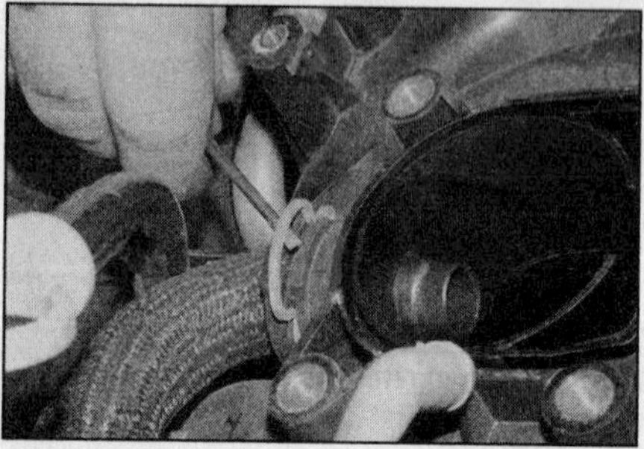
12 Release the retaining clip for the EGR (exhaust gas recirculation) pipe, and disconnect it from the intake manifold.

13 Unclip the brake vacuum line and remove the securing nuts from the bracket on the EGR pipe to cylinder head retaining studs. Undo the two studs and remove the EGR pipe from the engine.

14 Disconnect the wiring connector from the air intake manifold pressure sensor.

15 Release the wiring harness plugs and disconnect them from the injectors, the ignition coil and the EGR valve.

16 Disconnect the brake servo vacuum pipe and release it from its retaining clips.

17 Unclip the coolant hose from the top of the wiring trough, undo the two retaining nuts and carefully unclip the wiring trough and move it to one side.

18 Slacken and remove the two retaining studs/bolts then carefully ease the fuel rail and injector assembly out of position and remove it from the manifold. Remove the lower sealing rings from the injectors and discard them; they must be renewed whenever they are disturbed.

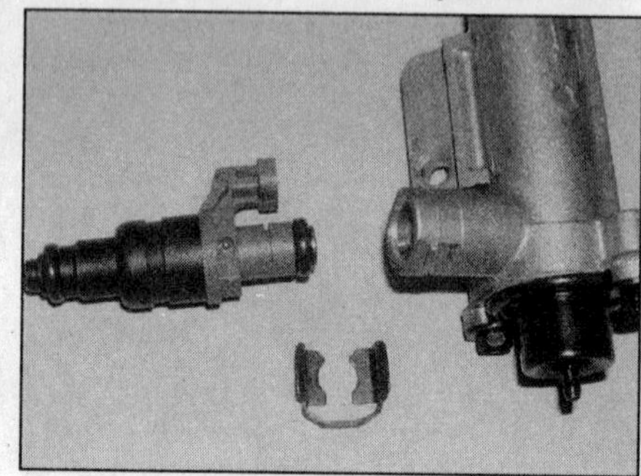
19 Slide off the relevant retaining clip and withdraw the injector from the fuel rail. Remove the upper sealing ring from the injector and discard it; all disturbed sealing rings must be renewed.

20 Refitting is a reversal of the removal procedure, noting the following points.

a) Renew all disturbed sealing rings and apply a smear of engine oil to them to aid installation.

b) Ease the injector(s) into the fuel rail, ensuring that the sealing ring(s) remain correctly seated, and secure in position with the retaining clips.

c) On refitting the fuel rail, take care not to damage the injectors and ensure that all sealing rings remain in position. Once the fuel rail is correctly seated, tighten its retaining bolts to the specified torque.

d) Ensure that all the wiring connectors engage fully to make a good connection.

e) On completion start the engine and check for fuel leaks.

Fuel pressure regulator

21 Unclip the engine cover and depressurise the fuel system, then disconnect the battery negative terminal.

22 Release the retaining clip and disconnect the breather hose from the cylinder head cover.

23 Unclip the coolant hose from the top of the wiring trough, undo the two retaining nuts and carefully unclip the wiring trough and move it to one side.

24 Disconnect the vacuum line from the pressure regulator and from the intake manifold.

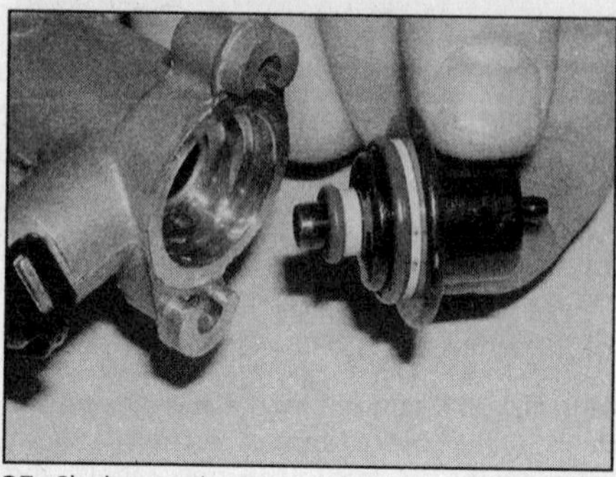
25 Slacken and remove the two retaining bolts, then carefully ease the pressure regulator from the end of the fuel rail along with its sealing rings.

26 Refitting is the reverse of removal, using new sealing rings. On completion start the engine and check for fuel leaks.

Accelerator pedal position sensor

27 From inside the vehicle, unscrew the fasteners and remove the lower trim panel from underneath the driver's side of the facia to gain access to the accelerator pedal.

28 Disconnect the wiring connector from the pedal position sensor by sliding it to the rear of the vehicle.

29 Undo the retaining bolts and remove the pedal position sensor from the pedal assembly.

Throttle valve servo motor

30 The servo motor is a combined part of the throttle body housing. Individual parts for the throttle body housing are not available separately and, if defective, it must be renewed as a complete unit.

Throttle valve potentiometer

31 The throttle valve potentiometer is a combined part of the throttle body housing. Individual parts for the throttle body housing are not available separately and, if defective, it must be renewed as a complete unit.

Coolant temperature sensor

32 The coolant temperature sensor is screwed into the thermostat housing at the left-hand end of the cylinder head. Partially drain the cooling system.

33 Disconnect the wiring plug, then unscrew the sensor from the housing.

34 Refitting is a reversal of removal. Apply a little sealant to the sensor threads, then tighten it to the specified torque.

Intake air temperature sensor

35 The intake air temperature sensor is mounted in the intake pipe which connects the air cleaner housing to the inlet manifold. Ensure the ignition is switched off, then disconnect the wiring connector from the sensor *(see paragraph 2 on page 26•19)*.

36 Carefully ease the sensor out of position (note position for refitting) and remove its sealing grommet from the intake pipe. If the sealing grommet shows signs of damage or deterioration it should be renewed.

37 Refitting is the reverse of removal, ensuring the sensor and grommet are correctly located in the intake pipe.

Manifold Absolute Pressure (MAP) sensor

38 Unclip the engine cover and depressurise the fuel system, then disconnect the battery negative terminal.

39 Disconnect the wiring connector from the air temperature sensor in the air intake pipe.

40 Release the retaining clip and disconnect the breather hose from the cylinder head cover.

41 Release the retaining clips then disconnect the intake pipe from the air cleaner and throttle housing and remove the pipe assembly from the engine compartment, detach the vacuum pipe(s) from the intake pipe on removal.

42 Disconnect the wiring connector from the throttle body, then undo the securing bolts/stud and remove the throttle body unit from the intake manifold. Note the position of the cable retaining clip on the securing stud.

43 Disconnect the wiring connector from the air intake manifold pressure sensor, turn the

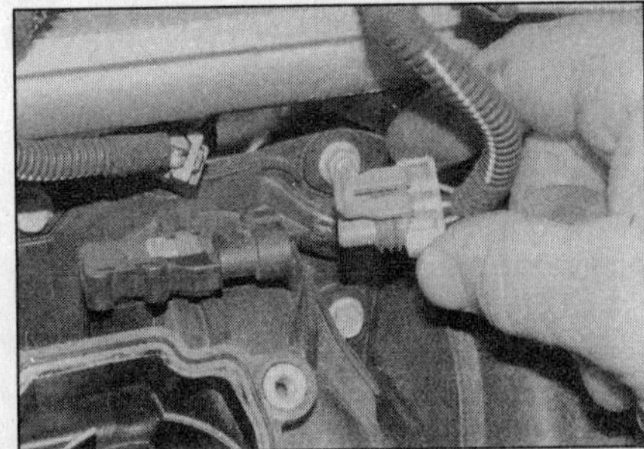

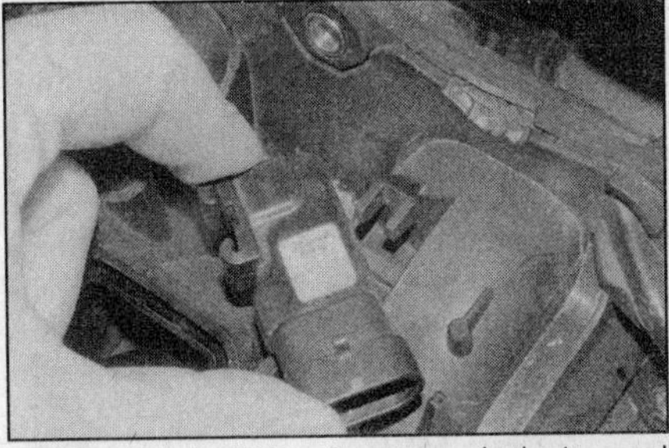

pressure sensor 1/4 of a turn clockwise and carefully pull out of the intake manifold.

44 Refitting is the reverse of removal, using a new sealing ring.

Crankshaft sensor

45 The sensor is mounted on the front of the cylinder block above the starter motor. **Note:** *It may be necessary to remove the starter motor to gain better access to the sensor.*

46 Trace the wiring back from the sensor, releasing it from all the relevant clips and ties whilst noting its correct routing. Disconnect the wiring connector from the sensor.

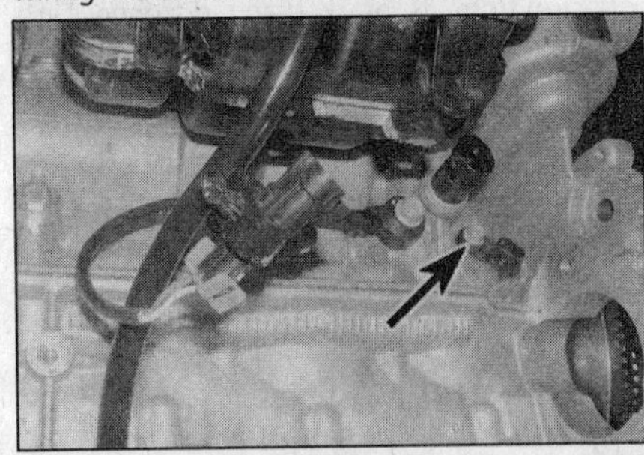

47 Unscrew the retaining bolt and remove the sensor from the cylinder block.

48 Refitting is the reverse of removal, tightening the retaining bolt to the specified torque. Ensure the wiring is correctly routed and retained by all the necessary clips and ties.

Knock sensor

49 The sensor is mounted on the front of the cylinder block above the starter motor. **Note:** *It may be necessary to remove the starter motor to gain better access to the sensor.*

50 Disconnect the wiring block connector at the bottom of the oil dipstick tube, so the wiring is free to be removed with the sensor.

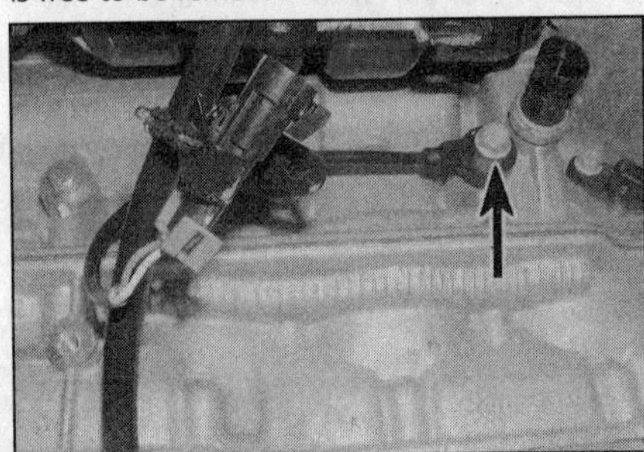

51 Unscrew the retaining bolt and remove the sensor from the cylinder block.

52 On refitting ensure the mating surfaces are clean and dry then fit the sensor and tighten its retaining bolt to the specified torque. Ensure the wiring is correctly routed and retained by all the necessary clips and ties.

Electronic control unit (ECU)

53 Unclip the engine cover and depressurise the fuel system, then disconnect the battery negative terminal.

54 Disconnect the wiring connector from the air temperature sensor in the air intake pipe.

55 Release the retaining clip and disconnect the breather hose from the cylinder head cover.

56 Release the retaining clips then disconnect the intake pipe from the air cleaner and throttle housing and remove the pipe assembly from the engine compartment, detach the vacuum pipe(s) from the intake pipe on removal.

57 Release the two wiring harness plugs from the engine control unit, then undo the securing bolts and remove the control unit from the intake manifold *(see paragraph 9 on page 26•12)*. Note the position of the earth cable on one of the securing bolts.

Fuel pump relay

58 Refer to the Multec-S system Section, paragraphs 47 to 49.

Air conditioning system switch

59 The air conditioning system switch is screwed into one of the refrigerant pipes and cannot be removed without first discharging the refrigerant. Renewal of the switch should therefore be entrusted to a suitably-equipped garage.

Tank vent/purge valve

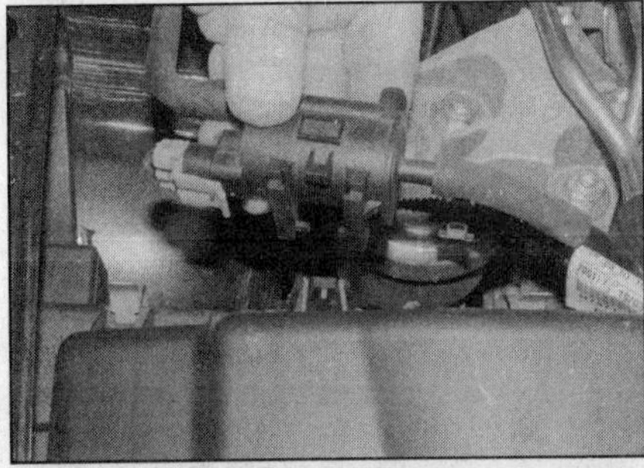

60 The tank vent/purge valve is clipped onto the side of the air filter housing.

61 Carefully detach the vacuum lines from the tank vent/purge valve, taking care not to damage the connections. If the valve is damaged in any way, then it must be renewed.

62 Disconnect the wiring connector from the tank vent/purge valve, then unclip the valve from the upper part of the air filter housing.

63 Refitting is the reverse of removal.

Diesel injection system electronic components – removal and refitting

Airflow meter

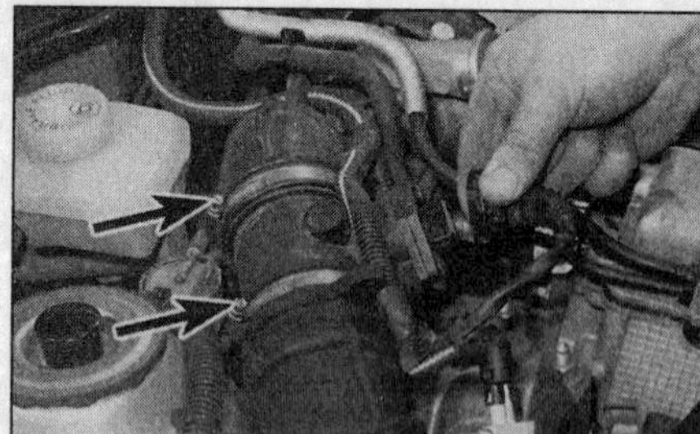

1 Ensure the ignition is switched off then disconnect the wiring connector from the airflow meter .

2 Slacken the retaining clips then free the airflow meter from the intake ducts and remove it from the engine compartment.

3 Refitting is the reverse of removal, ensuring the intake ducts are correctly seated and their retaining clips are securely tightened.

Intake air temperature sensor

Note: *On high-pressure turbo models, the air temperature sensor is built into the airflow meter.*

Low-pressure turbo models

4 Ensure the ignition is switched off, then disconnect the wiring connector and carefully ease the sensor out from the intake duct.

5 Refitting is the reverse of removal ensuring the sensor is correctly located in the duct.

Accelerator pedal position sensor

6 From inside the vehicle, unscrew the fasteners and remove the lower trim panel from underneath the driver's side of the facia to gain access to the accelerator pedal.

7 Disconnect the wiring connector from the sensor which is mounted onto the top of the accelerator pedal.

8 Remove the retaining clip from the link rod ball joint then unclip the rod from the sensor.

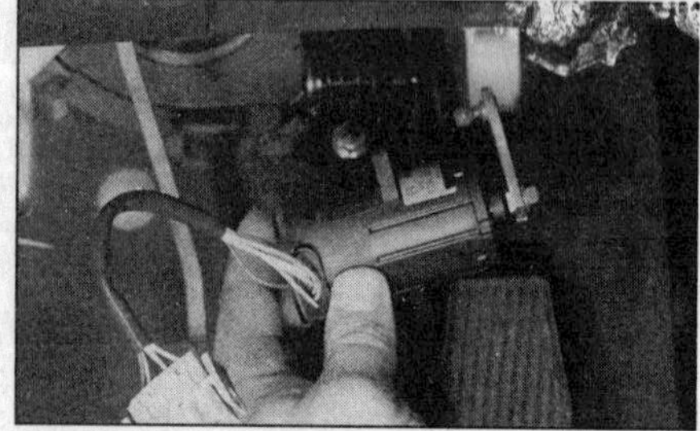

9 Undo the retaining screws and remove the sensor from the pedal bracket.

10 Refitting is the reverse of removal.

Crankshaft sensor

11 Undo the retaining screws and remove the plastic cover from the top of the cylinder head.

12 To gain access to the sensor from below, firmly apply the handbrake then jack up the front of the vehicle and support it on axle stands. Where necessary, undo the retaining bolts and remove the undercover from beneath the engine/transmission unit.

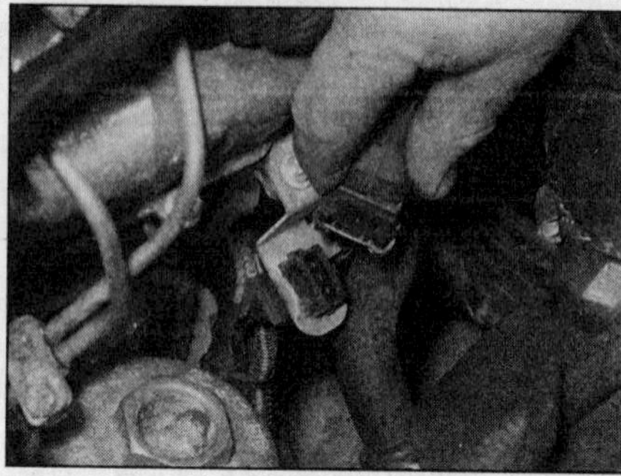

13 Trace the wiring back from the crankshaft sensor to its wiring connector then free the connector from its bracket and disconnect it from the main harness.

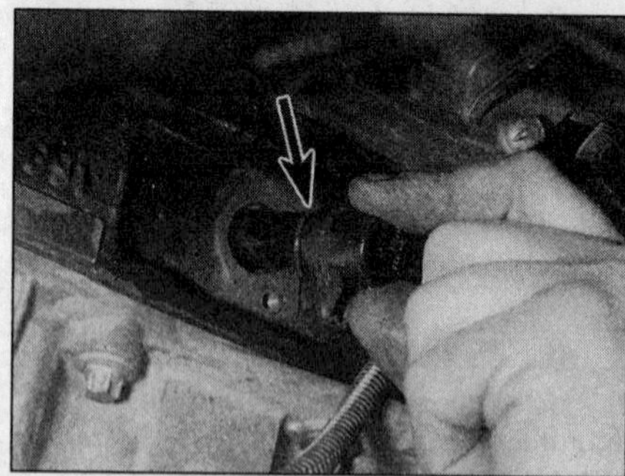

14 Wipe clean the area around the crankshaft sensor then slacken and remove the retaining bolt. Remove the sensor from the front of the cylinder block and recover the sealing ring.

15 Refitting is the reverse of removal, using a new sealing ring. Tighten the sensor retaining bolt to the specified torque.

Coolant temperature sensor

16 The coolant temperature sensor is screwed into the front of the cylinder head, at the right-hand end. Partially drain the cooling system.

17 Unbolt and remove the engine top cover.

18 Disconnect the wiring plug and unscrew the sensor.

19 Refitting is a reversal of removal. Apply a little sealant to the sensor threads and tighten it to the specified torque.

Charge (boost) pressure sensor

20 Undo the retaining screws and remove the plastic cover from the top of the cylinder head.

21 Slacken and remove the screws securing the wiring harness tray to the top of the inlet manifold and disconnect the wiring connector from the charge pressure sensor.

22 Undo the retaining bolt and remove the sensor from the top of the inlet manifold, noting the sealing ring fitted to the sensor shaft.

23 Refitting is the reverse of removal, using a new sealing ring and tightening the retaining bolt to the specified torque.

Oil temperature sensor

24 Firmly apply the handbrake then jack up the front of the vehicle and support it on axle stands. Where necessary, undo the retaining bolts and remove the undercover from beneath the engine/transmission unit.

25 Drain the engine oil. Once the oil has finished draining, fit a new sealing ring then refit the drain plug and tighten it to the specified torque.

26 Disconnect the wiring connector then unscrew the sensor from the front of the sump.

27 On refitting is the reverse of removal, refilling the engine with oil.

Electronic control unit (ECU)

28 Remove the windscreen wiper arms, then disconnect the battery negative terminal.

29 Unscrew the plastic nuts from the wiper arm spindles.

30 Peel off the rubber sealing strip from the top of the engine compartment bulkhead.

31 Carefully unclip the plastic vent panel from the base of the windscreen and remove it from the vehicle to gain access to the ECU.

32 Undo the retaining nuts and free the ECU protective cover from the body.

33 Release the retaining clip and disconnect the wiring connector from the ECU.

34 Undo the retaining nuts and remove the ECU from the vehicle.

35 Refitting is the reverse of removal, ensuring the wiring connector is securely reconnected.

Injection pump control unit

36 The control unit is an integral part of the injection pump and should not be disturbed. **Never** attempt to separate the control unit and pump.

Inlet manifold switchover solenoid

37 Remove the battery to gain access to the valve which is located in the front left-hand corner of the engine compartment. Note that there are two valves, the EGR system valve and the manifold switchover valve; the manifold switchover valve can be identified by its grey wiring connector.

38 Disconnect the wiring connector and vacuum hoses from the valve then undo the retaining screws and remove the valve from its mounting bracket.

39 Refitting is the reverse of removal.

Turbocharger wastegate solenoid

40 The wastegate (charge pressure) solenoid valve is located in the right-hand rear corner of the engine compartment *(see paragraph 37)*.

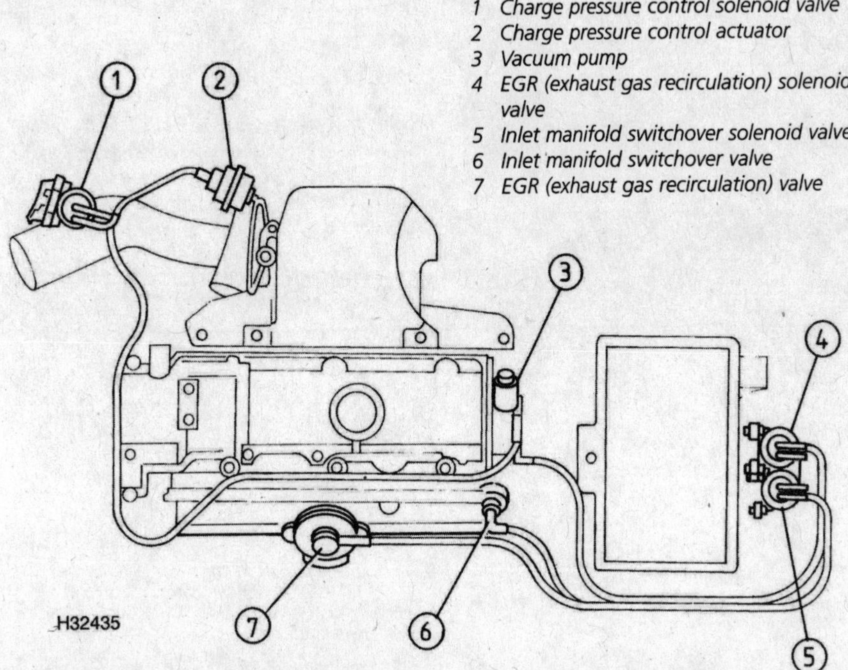

Vacuum pipe layout – high-pressure models

41 To gain access to the valve, remove the intake duct assembly linking the air cleaner housing to the turbocharger.

42 Disconnect the wiring connector and vacuum hoses from the valve then undo the retaining screws and remove the valve from its mounting bracket.

43 Refitting is the reverse of removal.

Notes

VW Golf & Bora '98 – '00

27

Petrol models

General

System type:	
1.4 litre engines:	
AHW engine code:	
Early (pre May 99) engines	Magneti-Marelli 4AV
Later (May 99 on) engines	Magneti-Marelli 4CV
Engine code AKQ	Magneti-Marelli 4AV
Engine codes APE and AXP	Bosch Motronic ME7.5.10
1.6 litre SOHC engines:	
Engine codes AEH and AKL:	
Vehicles with cruise control	Simos 2.2
Vehicles without cruise control	Simos 2.1
Engine code APF	Simos 3.3
1.6 litre DOHC engines	Magneti-Marelli 4LV
1.8 litre engines:	
Engine code AGN	Bosch Motronic M3.8.5
Engine code AGU	Bosch Motronic M3.8.3
Engine codes AQA, ARZ	Bosch Motronic ME7.5
2.0 litre engines:	
Engine code APK	Bosch Motronic M5.8.2
Engine code AQY	Bosch Motronic M5.9.2

Fuel system data

Fuel pump type	Electric, immersed in fuel tank
Fuel pump delivery rate	400 cm^3/ min (battery voltage of 12.5 V)
Regulated fuel pressure	2.5 bar
Engine idle speed (non-adjustable, electronically controlled):	
Engine codes:	
AHW and AKQ	700 to 800 rpm
APE, AXP	650 to 850 rpm
AEH, AKL	760 to 880 rpm
APF	640 to 900 rpm
ATN, AUS	630 to 730 rpm
AGN	760 to 880 rpm
AGU, AQA	800 to 920 rpm
ARZ:	
Manual gearbox	700 to 820 rpm
Automatic transmission	640 to 760 rpm
APK	760 to 880 rpm
AQY	740 to 820 rpm
Idle CO content (non-adjustable, electronically controlled)	0.5 % max
Injector electrical resistance (typical)	12 to 17 ohms

Diesel models

General

Firing order	1-3-4-2	
Maximum engine speed	N/A (ECU controlled)	
Engine idle speed	900 ± 30 rpm (ECU controlled)	
Engine fast idle speed	N/A (ECU controlled)	
Turbocharger		
Type	Garrett or KKK	

Torque wrench settings

	Nm	lbf ft
Petrol models		
All models:		
Knock sensor(s)	20	15
Oxygen sensor(s)	50	37
1.4 and 1.6 litre DOHC engines:		
Fuel rail mounting bolts	10	7
1.6 litre SOHC engine:		
Throttle housing mounting bolts	8	6
1.8 litre non-turbo engine:		
Camshaft position sensor inner element mounting bolt	25	18
Fuel rail mounting bolts	10	7
Inlet air temperature sensor mounting bolt	10	7
Throttle housing mounting bolts	10	7
1.8 litre turbo:		
Camshaft position sensor inner element mounting bolt	25	18
Fuel rail mounting bolts	10	7
Inlet air temperature sensor mounting bolt	10	7
Throttle housing mounting bolts	10	7
2.0 litre:		
Fuel rail mounting bolts	10	7
Throttle housing mounting bolts	10	7
Diesel models		
Fuel cut-off solenoid valve	40	30

General information

Petrol models

The systems described are all self-contained engine management systems, which control both the fuel injection and ignition.

The fuel injection system comprises a fuel tank, an electric fuel pump/level sender unit, a fuel filter, fuel supply and return lines, a throttle housing, a fuel rail, a fuel pressure regulator, four electronic fuel injectors, and an Electronic Control Unit (ECU) together with its associated sensors, actuators and wiring. The fuel systems used are essentially very similar, but there are significant detail differences, particularly in the sensors used and in the inlet manifold arrangements.

The fuel pump is immersed in the fuel inside the tank, and delivers a constant supply of fuel through a cartridge filter to the fuel rail, at a slightly higher pressure than required – the fuel pressure regulator maintains a constant fuel pressure to the fuel injectors, and returns excess fuel to the tank via the return line. This constant flow system also helps to reduce fuel temperature, and prevents vaporisation.

The fuel injectors are opened and closed by an Electronic Control Unit (ECU), which calculates the injection timing and duration according to engine speed, crankshaft/camshaft position, throttle position and rate of opening, inlet manifold depression, inlet air temperature, coolant temperature, road speed and exhaust gas oxygen content information, received from sensors mounted on and around the engine.

On 2.0 litre models, the fuel injectors are each enclosed in a shroud, into which a small supply of inlet air is drawn. At the injector tip, the air mixes with the fuel, to form a more finely-atomised fuel spray.

Some 1.6, 1.8 and 2.0 litre models may also be equipped with a secondary air injection system, which feeds air into the exhaust gases, to promote combustion of any unburnt fuel during engine warm-up; this process also helps to heat the catalytic converter more quickly to its effective operating temperature.

Inlet air is drawn into the engine through the air cleaner, which contains a renewable paper filter element. On some non-turbo models, the inlet air temperature is regulated by a valve mounted in the air cleaner inlet trunking, which blends air at ambient temperature with hot air, drawn from over the exhaust manifold.

The temperature and pressure of the air entering the throttle housing is measured either by a sensor mounted on the inlet manifold, or by the air mass meter attached to the air cleaner. This information is used by the ECU to fine-tune the fuelling requirements for different operating conditions. Turbocharged engines have an additional air temperature sensor mounted downstream of the throttle housing, which monitors the (compressed) air temperature after it has been through the turbocharger and intercooler.

On 1.6 litre SOHC engines, and on 1.8 litre non-turbo engines, a variable-length inlet manifold is fitted. A vacuum-controlled flap is fitted inside the manifold assembly, which is used to divert the inlet air into one of two paths through the manifold, the paths being of different lengths. Controlling the inlet air in this way has the effect of altering the engine's torque characteristics at different engine speeds and loads.

Idle speed control is achieved partly by an electronic throttle valve positioning module, which is part of the throttle housing, and partly by the ignition system, which gives fine control of the idle speed by altering the ignition timing. As a result, manual adjustment of the engine idle speed is not necessary or possible.

The exhaust gas oxygen content is constantly monitored by the ECU via an oxygen sensor (also known as a lambda sensor), which is mounted in the front section of the exhaust pipe. On all engines except the earliest 1.4 and 1.6 litre units, two oxygen sensors are fitted, one before the catalytic converter, and one after – this improves sensor response time and accuracy, and the ECU compares the signals from each sensor to confirm that the converter is working correctly. The ECU uses the information from the sensor(s) to modify the injection timing and duration to maintain the optimum air/fuel ratio – a result of this is that manual adjustment of the idle exhaust CO content is not necessary or possible. All models are fitted with a catalytic converter.

Where fitted, the ECU controls the operation of the activated charcoal filter evaporative loss system.

Diesel models

The fuel system comprises a fuel tank, a fuel injection pump, an engine-bay mounted fuel filter with an integral water separator, fuel supply and return lines, and four fuel injectors. Most engines are fitted with a turbocharger.

The injection pump is driven at half crankshaft speed by the camshaft timing belt. Fuel is drawn from the fuel tank, through the filter by the injection pump, which then distributes the fuel under very high pressure to the injectors via separate delivery pipes.

The direct-injection fuelling system is controlled electronically by a diesel engine management system, comprising an Electronic Control Unit (ECU) and its associated sensors, actuators and wiring.

Basic injection timing is set mechanically by the position of the pump on its mounting bracket. Dynamic timing and injection duration are controlled by the ECU, and are dependant on engine speed, throttle position and rate of opening, inlet air flow, inlet air temperature, coolant temperature, fuel temperature, ambient pressure (altitude) and manifold depression information, received from sensors mounted on and around the engine. Closed-loop control of the injection timing is achieved by means of an injector needle lift sender. Note that injector No 3 is fitted with the needle lift sender.

Two-stage injectors are used, which improve the engine's combustion characteristics, leading to quieter running and better exhaust emissions.

In addition, the ECU manages the operation of the Exhaust Gas Recirculation (EGR) emission control system, the turbocharger boost pressure control system and the glow plug control system. On non-turbo models, an electrically-operated flap valve is fitted to the inlet manifold to increase the vacuum when the engine speed is less than 2200 rpm; this is necessary to operate the EGR system efficiently.

On turbo models, the flap valve fitted to the inlet manifold is closed by the ECU for 3 seconds as the engine is switched off, to minimise the air intake as the engine shuts down. This

minimises the vibration felt as the pistons come up against the volume of highly compressed air present in the combustion chambers. A vacuum reservoir mounted on the front of the cylinder head provides the vacuum supply to a vacuum capsule which operates the flap.

Fuel injection system – depressurisation, priming and bleeding

Petrol models

Warning: The following procedure will merely relieve the pressure in the fuel system – remember that fuel will still be present in the system components and take precautions accordingly before disconnecting any of them.

1 The fuel system referred to in this Section is defined as the tank-mounted fuel pump, the fuel filter, the fuel injectors, the fuel pressure regulator, and the metal pipes and flexible hoses of the fuel lines between these components. All these contain fuel, which will be under pressure while the engine is running and/or while the ignition is switched on. The pressure will remain for some time after the ignition has been switched off, and must be relieved before any of these components are disturbed for servicing work. Ideally, the engine should be allowed to cool completely before work commences.

2 Locate and remove the fuel pump relay. Alternatively, identify and remove the fuel pump fuse from the fusebox.

3 With the fuel pump disabled, crank the engine for about ten seconds. The engine may fire and run for a while, but let it continue running until it stops. The fuel injectors should have opened enough times during cranking to considerably reduce the line fuel pressure, and reduce the risk of fuel spraying out when a fuel line is disturbed.

4 Disconnect the battery negative lead and position it away from the terminal.

5 Place a suitable container beneath the relevant connection/union to be disconnected, and have a large rag ready to soak up any escaping fuel not being caught by the container.

6 Slowly open the connection to avoid a sudden release of pressure, and position the rag around the connection to catch any fuel spray which may be expelled. Once the pressure has been released, disconnect the fuel line. Insert plugs to minimise fuel loss and prevent the entry of dirt into the fuel system.

Diesel models

7 Although no specific procedure is given for depressurising the fuel system, common sense dictates that the engine should be allowed to cool for while before commencing any repair work. During this time, the residual pressure in the system should have subsided sufficiently. Place a rag over any fuel connection as the union is slackened.

8 It is not necessary to manually prime and bleed the fuel system after any operation on the system components. Start the engine (this may take longer than usual, especially if the fuel system has been allowed to run dry – operate the starter in ten second bursts with 5 seconds rest in between each operation) and run it a fast idle speed for a minute or so to purge any trapped air from the fuel lines. After this time the engine should idle smoothly at a constant speed.

9 If the engine idles roughly, then there is still some air trapped in the fuel system. Increase the engine speed again for another minute or so then recheck the idle speed. Repeat this procedure as necessary until the engine is idling smoothly.

Caution: Do not operate the starter for more than 10 seconds at a time. Allow 30 seconds between attempts for the starter to cool.

Petrol fuel injection system components – removal and refitting

Throttle housing

1 On non-turbo 1.8 litre and 2.0 litre models, the throttle housing is coolant-heated, so removing it entails disconnecting two coolant pipes. Even if the cooling system is drained, it is likely that the throttle housing supply pipes will not be drained, and spillage will result. If the coolant is not due for renewal, it may be preferable not to drain the system, but be prepared to plug the pipes once they have been disconnected.

2 On non-turbo models, remove the air cleaner.

3 On turbo models, remove the engine top cover(s). Removal details vary according to model, but the cover retaining nuts are concealed under circular covers, which are prised out of the main cover. Where plastic screws or turn-fasteners are used, these can be removed using a wide-bladed screwdriver. Remove the nuts or screws, and lift the cover from the engine, releasing any wiring or hoses attached.

4 Detach the accelerator cable from the throttle valve lever.

5 Disconnect the battery negative lead and position it away from the terminal.

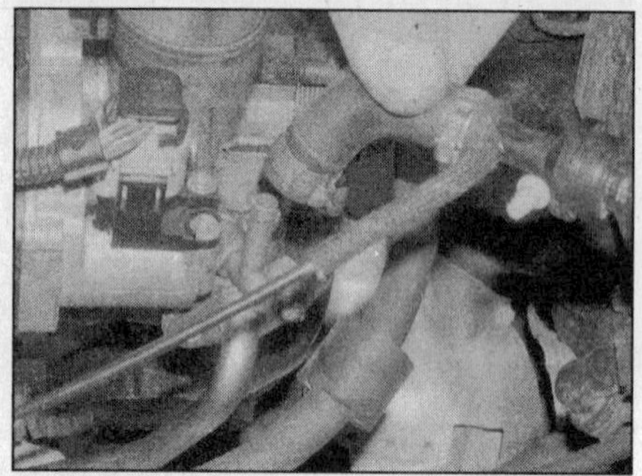

6 Disconnect the hose for the charcoal canister from the port on the throttle housing. Also disconnect the brake servo vacuum supply hose, where applicable.

7 Unplug the wiring connector from the throttle potentiometer.

8 Where applicable, disconnect the coolant pipes from the throttle housing, noting their positions carefully for refitting. Be prepared for coolant spillage, and plug the pipe ends to prevent too much coolant loss.

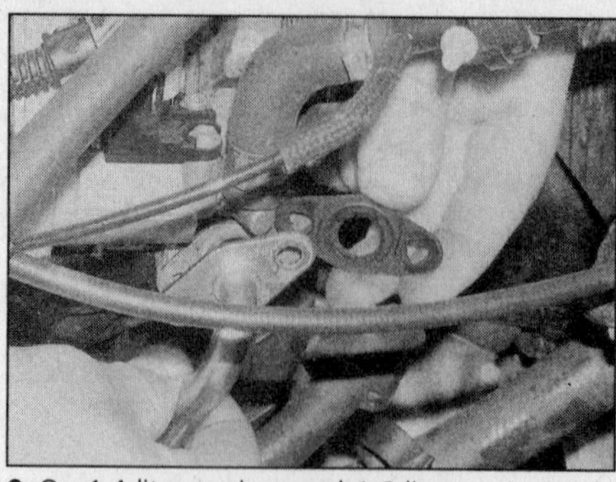

9 On 1.4 litre engines and 1.6 litre engine code APF, unscrew the two bolts securing the metal pipe for the EGR system. Separate the pipe flange from the throttle housing, and recover the gasket – a new gasket must be used when refitting.

10 On turbo models, loosen the charcoal canister hose clip (B) and remove the large-diameter air inlet trunking (A) from the throttle housing.

11 Slacken and withdraw the through-bolts, then lift the throttle housing away from the inlet manifold. On 1.4 litre engines, recover the mounting flange to which the EGR pipe is bolted, noting its orientation. Recover and discard the gasket(s). Note that, on some models, one of the bolts secures the throttle housing earth strap.

12 Refitting is a reversal of removal, noting the following:

a) Use a new throttle housing-to-inlet manifold gasket.

b) Tighten the throttle housing through-bolts evenly to the specified torque, to prevent air leaks.

c) Ensure that all hoses and electrical connectors are refitted securely.

d) Check and if necessary adjust the accelerator cable.

Fuel injectors and fuel rail

Note: *If a faulty injector is suspected, before removing the injectors, it is worth trying the effect of one of the proprietary injector-cleaning treatments. These can be added to the petrol in the tank, and are intended to clean the injectors as you drive.*

13 Disconnect the battery negative lead, and position it away from the terminal.

14 Where applicable, remove the engine top cover(s). Removal details vary according to model, but the cover retaining nuts are concealed under circular covers, which are prised

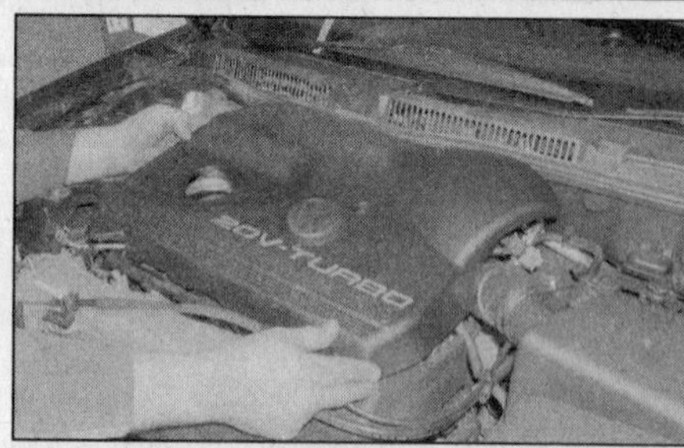

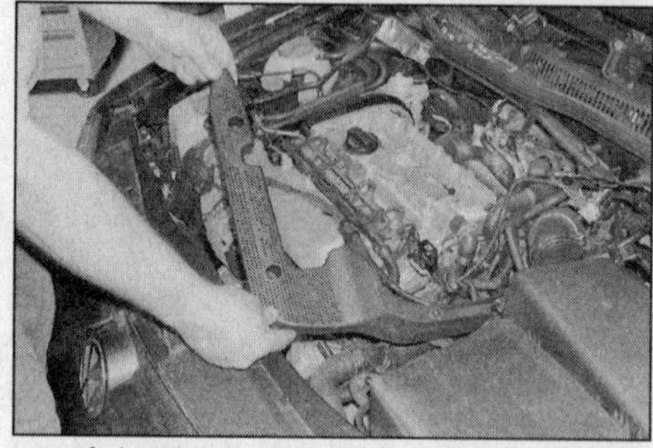

out of the main cover. Where plastic screws or turn-fasteners are used, these can be removed using a wide-bladed screwdriver. Remove the nuts or screws, and lift the cover from the engine, releasing any wiring or hoses attached.

15 On 1.6 and 2.0 litre SOHC engines, and on 1.8 litre non-turbo engines, remove the upper part of the inlet manifold.

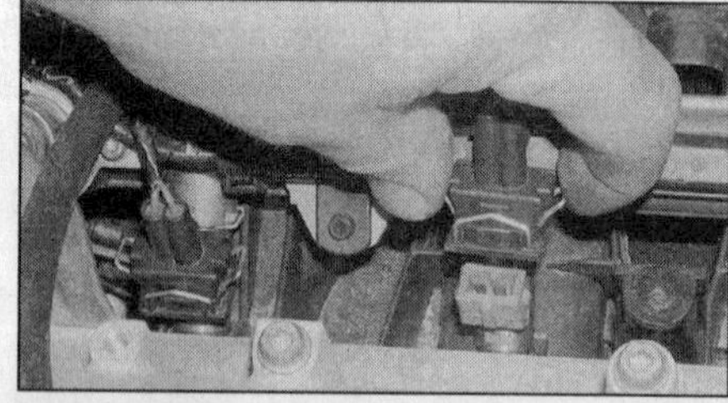

16 Unplug the injector harness connectors, labelling them to aid correct refitting later. Release the wiring harness clips from the top of the fuel rail, and lay the harness to one side.

17 Depressurise the fuel system.

18 Disconnect the vacuum hose from the port on the fuel pressure regulator.

19 On 2.0 litre models, disconnect the air supply hose for the injector air shrouds.

20 Squeeze the catches on the quick-release fittings, and disconnect the fuel supply and return hoses from the entry points at the bulkhead. Alternatively, release the spring clips securing the hoses to the fuel rail. *Carefully* note the fitted positions of the hoses – the supply hose is marked with a black or white arrow, and the return hose is marked with a blue arrow.

21 Slacken and withdraw the fuel rail mounting bolts, then carefully lift the rail away from the inlet manifold, together with the injectors. Recover the injector lower O-ring seals as they emerge from the manifold.

22 The injectors can be removed individually from the fuel rail by extracting the relevant metal clip and easing the injector out of the rail. Recover the injector upper O-ring seals.

23 If required, remove the fuel pressure regulator.

24 Check the electrical resistance of the injectors using a multimeter, and compare it with the Specifications.

25 Refit the injectors and fuel rail by following the removal procedure in reverse, noting the following points:

a) Renew the injector O-ring seals if they appear worn or damaged.
b) Ensure that the injector retaining clips are securely seated.
c) Check that the fuel supply and return hoses are reconnected correctly – refer to the colour coding described in Removal.
d) Check that all vacuum and electrical connections are remade correctly and securely.
e) On completion, check exhaustively for fuel leaks before bringing the vehicle back into service.

Fuel pressure regulator

26 Disconnect the battery negative lead, and position it away from the terminal.

27 Depressurise the fuel system.

28 Where applicable, remove the engine top cover(s). Removal details vary according to model, but the cover retaining nuts are concealed under circular covers, which are prised out of the main cover. Where plastic screws or turn-fasteners are used, these can be removed using a wide-bladed screwdriver. Remove the nuts or screws, and lift the cover from the engine, releasing any wiring or hoses attached.

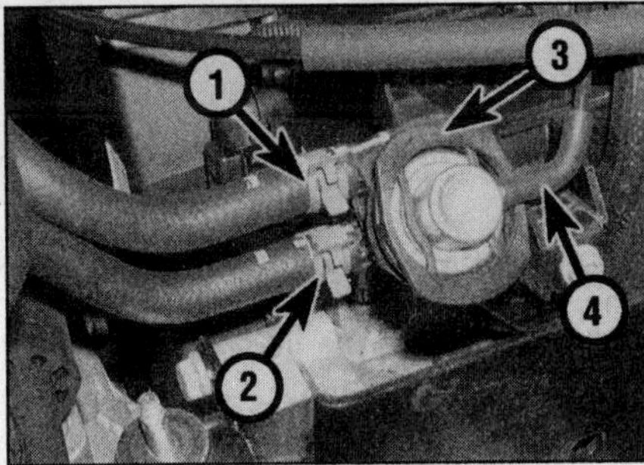

1 Fuel supply hose
2 Fuel return hose
3 Spring clip
4 Vacuum hose

29 Disconnect the vacuum hose from the port on the bottom (or side) of the fuel pressure regulator.

30 Release the spring clip and temporarily disconnect the fuel supply hose from the end of the fuel rail. This will allow the majority of fuel in the regulator to drain out. Be prepared for an amount of fuel loss – position a small container and some old rags underneath the fuel regulator housing. Reconnect the hose once the fuel has drained. **Note:** *The supply hose is marked with a black or white arrow.*

31 On 1.4 and 1.6 litre DOHC engines, remove the regulator collar securing screw, and lift off the collar. Lift out the regulator, and recover the O-ring seals.

32 On all other engines, extract the retaining spring clip from the top of the regulator housing and lift out the regulator body, recovering the O-ring seals.

33 Refit the fuel pressure regulator by following the removal procedure in reverse, noting the following points:

a) Renew the O-ring seals if they appear worn or damaged.
b) Ensure that the regulator retaining clip is securely seated, or that the collar is correctly fitted and the screw securely tightened.
c) Refit the regulator vacuum hose securely.

Throttle valve potentiometer/positioner

34 The potentiometer (or motor/positioner, on models with a 'fly-by-wire' throttle) is matched to the throttle housing during manufacture, and is not available separately – if defective, a complete throttle housing assembly will be required. Although the unit can be removed from the housing by removing the retaining screws, doing so will damage the seal between the two, and it does not appear that a new seal is available as a replacement part.

Throttle position sensor

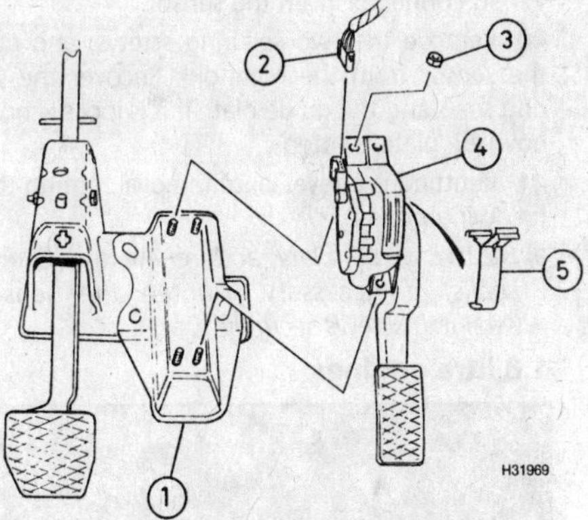

Accelerator pedal and throttle position sensor

1 Bracket
2 Position sensor wiring connector
3 Mounting nuts
4 Position sensor
5 Retainer for footwell cover

35 On models with a conventional accelerator cable, the throttle position sensor function is performed by the potentiometer attached to the throttle housing. As described above, the potentiometer is not available separately.

36 On models with the 'fly-by-wire' throttle, the position sensor is integral with the accelerator pedal. The pedal assembly can be removed (once access has been gained by removing the driver's lower facia panel) by disconnecting the sensor wiring plug and unscrewing the nuts securing the pedal to its mounting bracket.

Inlet air temperature/ pressure sensors

37 All models except those with the 1.4 and 1.6 litre DOHC engines have an air temperature sensor built into the air mass meter. This sensor is an integral part of the air mass meter, and cannot be renewed separately. On 1.8 litre engines, an additional air temperature sensor is fitted to the inlet manifold, and this can be renewed as described below.

1.4 and 1.6 litre DOHC engines

38 The sensor is attached to the right-hand side of the inlet manifold (right as seen from the driver's seat).

39 Disconnect the battery negative lead, and position it away from the terminal. Unplug the wiring connector from the sensor.

40 Remove the two securing screws, and pull the sensor from the manifold. Recover the O-ring seals and the guide plate if it is loose – note how the plate is fitted.

41 Refitting is a reversal of removal, noting the following points:

a) *Fit the guide plate and renew the O-ring seal(s) if necessary. Tighten the sensor retaining screws securely.*

1.8 litre engines

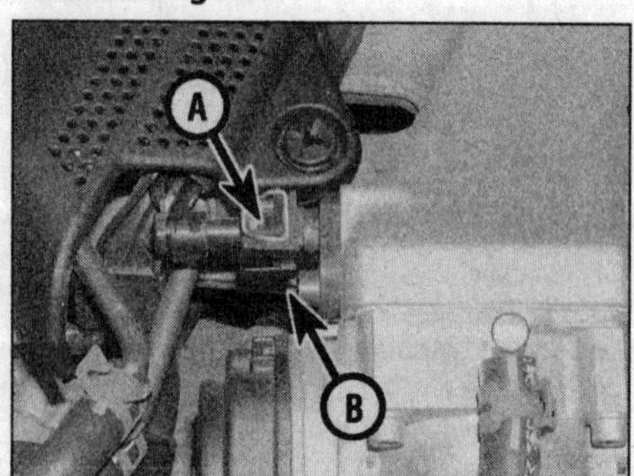

42 The sensor is located on the inlet manifold, next to the throttle housing. Disconnect the battery negative lead, and position it away from the terminal. Unplug the wiring connector (A) from the sensor.

43 Unscrew the sensor retaining bolt (B), then pull the sensor out of its location in the inlet manifold. Recover the O-ring seal.

44 Refitting is a reversal of removal, noting the following points:

b) *Renew the O-ring seal if it shows signs of damage. Tighten the sensor bolt to the specified torque.*

Road speed sensor

45 The road speed sensor is mounted on the top rear of the transmission casing. Locate the sensor and disconnect the wiring plug.

46 On 02J transmissions, the sensor is fitted directly on top of the drive pinion. If required, hold the housing with a spanner, then unscrew the sensor from the top of the pinion housing and recover the washer.

47 On 02K transmissions, the sensor and pinion unit is retained by a single Allen screw. Undo the screw and withdraw the unit from the transmission.

48 Refitting is a reversal of removal.

Coolant temperature sensor

1.4 litre engines

49 The sensor is clipped into the underside of the thermostat housing at the left-hand end of the cylinder head. Partially drain the cooling system and remove the engine top cover.

1.6 (except engine code APF), 1.8 and 2.0 litre engines

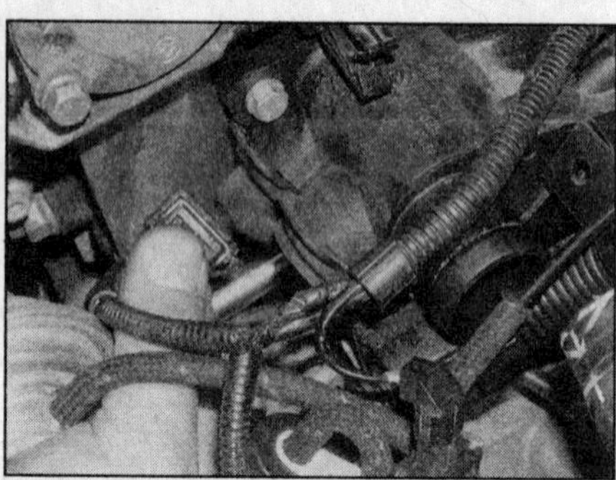

50 The sensor is clipped into the top of the hose connector, bolted to the left-hand end of the cylinder head. Partially drain the cooling system and remove the engine top cover.

1.6 litre engine (code APF)

51 As this engine has an 'electronically mapped' cooling system two sensors are fitted which supply information to the engine management system. One is clipped into the upper chamber of the coolant distribution housing bolted to the left-hand end of the cylinder head, whilst the other is screwed into the left-hand end of the radiator. Partially drain the cooling system, where necessary, remove the engine top cover.

All engines

52 Disconnect the wiring plug and unscrew/ unclip the sensor from place.

53 Refitting is a reversal of removal. Renew any seals where necessary.

Oxygen (lambda) sensor(s)

54 All models have one sensor, threaded into the exhaust downpipe or manifold, ahead of the catalytic converter. Most models have an additional oxygen sensor, mounted downstream of the converter. Disconnect the battery negative lead and position it away from the terminal.

> *Warning: Working on the sensor(s) is only advisable with the engine (and therefore the exhaust system) completely cold. The catalytic converter in particular will be very hot for some time after the engine has been switched off.*

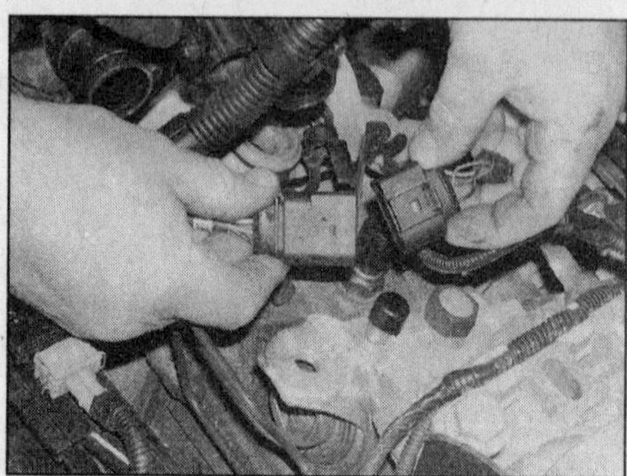

55 Working from the sensor, trace the wiring harness from the oxygen sensor back to the connector, and disconnect it. Typically, the wiring plug is coloured black for the front sensor, and brown for the rear sensor. Unclip the sensor wiring from any retaining clips, noting how it is routed.

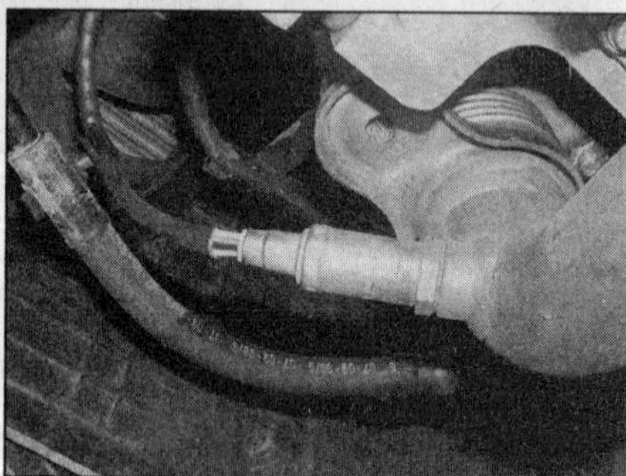

56 Access to the front sensor is possible on some models from above, while the rear sensor (where fitted) is only accessible from below. On some models, access to a front pipe-mounted sensor is improved by unbolting and removing the cover from the right-hand driveshaft inner CV joint.

57 Slacken and withdraw the sensor, taking care to avoid damaging the sensor probe as it is removed. **Note:** *As a flying lead remains connected to the sensor after it has been disconnected, if the correct-size spanner is not available, a slotted socket will be required to remove the sensor.*

58 Apply a little high-temperature anti-seize grease to the sensor threads – avoid contaminating the probe tip.

59 Refit the sensor, tightening it to the correct torque. Restore the harness connection.

Engine speed sensor

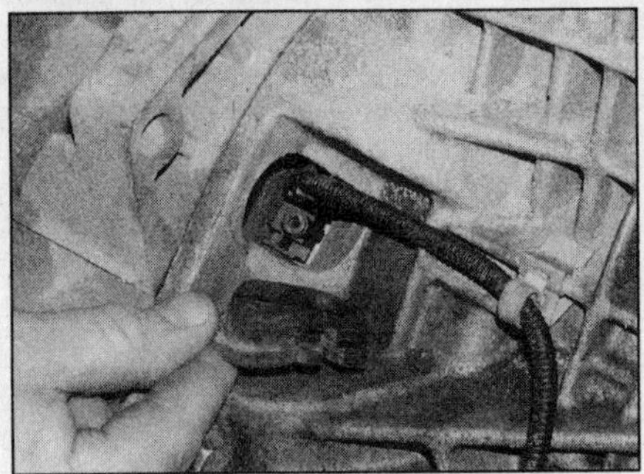

60 On 1.4 and 1.6 litre DOHC engines, the engine speed sensor is mounted at the left-hand rear of the cylinder block, next to the transmission bellhousing, and access is very difficult. Prise out the rubber bung for access to the sensor.

61 On all other engines, the engine speed sensor is mounted on the front, left-hand side of the cylinder block, adjacent to the mating surface of the block and transmission bellhousing, next to the oil filter. If necessary, drain the engine oil and remove the oil filter and cooler to improve access.

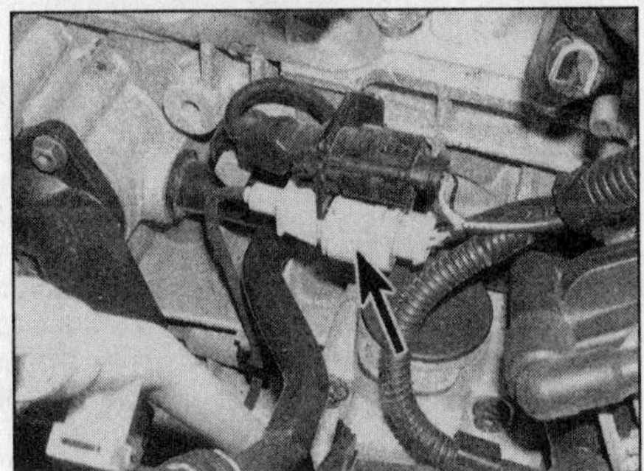

62 Trace the wiring back from the sensor, and unplug the harness connector.

63 Unscrew the retaining bolt and withdraw the sensor from the cylinder block.

64 Refitting is a reversal of removal.

Camshaft position sensor

1.4 and 1.6 litre DOHC engines

65 Where applicable, remove the engine top cover(s). Removal details vary according to model, but the cover retaining nuts are concealed under circular covers, which are prised out of the main cover. Where plastic screws or turn-fasteners are used, these can be removed using a wide-bladed screwdriver. Remove the nuts or screws, and lift the cover from the engine, releasing any wiring or hoses attached.

66 Remove the oil filler cap.

67 Remove the four screws securing the cover over the camshaft housings, and lift the cover off the engine.

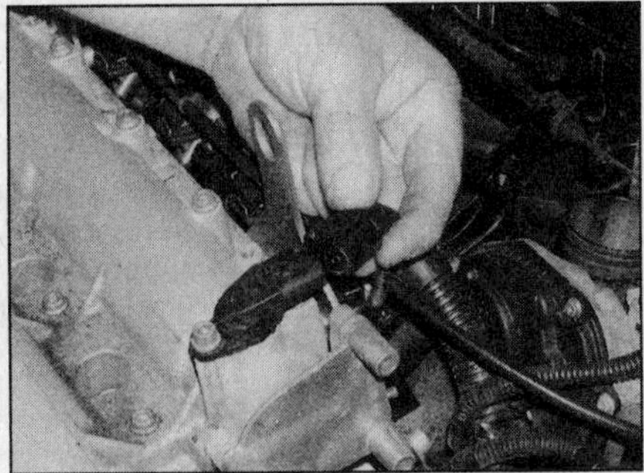

68 Pull the wiring connector upwards off the sensor, which is located next to the engine lifting eye.

69 Unscrew the sensor mounting bolt, and pull the sensor out of the camshaft housing.

70 Refitting is a reversal of removal.

1.8 litre engines

71 Remove the timing belt outer cover.

72 Release the clip and disconnect the wiring multiplug from the sensor.

73 Unscrew the mounting bolts and withdraw the sensor from the cylinder head. If required, unscrew the central bolt and remove the inner element and hood, noting how they are fitted.

74 Refitting is a reversal of removal; tighten the mounting bolts securely.

1.6 and 2.0 litre SOHC engines

75 Remove the camshaft sprocket.

76 Note the location of the sensor and if necessary mark it in relation to the cylinder head. Disconnect the wiring from the sensor.

77 Unbolt the timing belt inner cover from the cylinder head.

78 Unscrew the remaining bolts and remove the sensor from the cylinder head.

79 Refitting is a reversal of removal, but make sure that the sensor base plate is central before tightening the retaining bolts.

Clutch pedal switch

80 Fitted to 1.6, 1.8 and 2.0 litre engines, the clutch switch is mounted on the clutch pedal, and sends a signal to the ECU. The purpose of the switch is to disable the throttle closing damper during gearchanges, allowing the engine revs to die down more quickly than would otherwise happen when the accelerator is released. The switch also deactivates the cruise control system (where fitted) when the pedal is pressed.

81 To remove the switch, first remove the facia lower trim panel on the driver's side.

82 Locate the switch wiring plug in front of the clutch pedal, and disconnect it.

83 Release the switch retaining lugs, and withdraw it from the pedal.

84 When refitting the switch, first extend the switch plunger to its fullest extent, then hold the clutch pedal depressed when offering it into position. Once the switch has been clipped into place, release the pedal – this sets the switch adjustment. Further refitting is a reversal of removal.

Power steering pressure switch

85 When the steering is at or near full left or right lock, this places a greater load on the power steering pump. Since the pump is driven by the engine, this could result in the engine idle speed dropping, risking the engine stalling. The pressure switch fitted to the pump detects the rise in system fluid pressure, and signals the ECU, which raises the idle speed temporarily to compensate for the extra load.

86 The pressure switch is screwed into the top of the steering pump fluid supply union, and is most easily accessed from below. Disconnect the wiring plug from the top of the switch.

87 Hold the (slim) union nut against rotation with one spanner, and use another to unscrew the pressure switch from the union. Recover the sealing washer, where fitted. Anticipate some fluid spillage as the switch is unscrewed. Once the switch has been removed, cover the open connection to prevent dirt from entering.

88 Refitting is a reversal of removal, noting the following points:

a) *Use a new sealing washer, where applicable. Tighten the switch securely, holding the union nut as for removal.*

b) *Top-up the power steering system. If a large amount of fluid was lost, bleed the system.*

c) *On completion, start the engine and have an assistant turn the steering wheel from lock-to-lock, while you check around the switch for leaks.*

Electronic control unit (ECU)

> *Caution: Always wait at least 30 seconds after switching off the ignition before disconnecting the wiring from the ECU. When the wiring is disconnected, all the learned values are erased, although any contents of the fault memory are retained. Note also that if the ECU is renewed, the identification of the new ECU must be transferred to the immobiliser control unit by a VW dealer or specialist.*

89 The ECU is located centrally behind the engine compartment bulkhead, under one of the windscreen cowl panels. Remove the wiper arms and cowl panel.

90 Release the locking clip or lever on each ECU wiring connector, and disconnect the plug. On most models, there are two separate plugs to be disconnected.

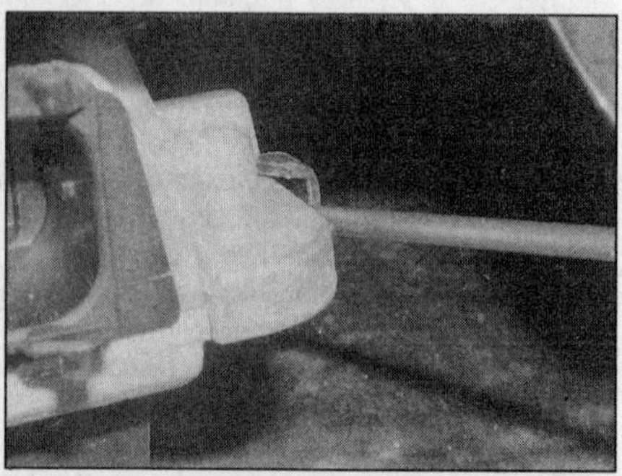

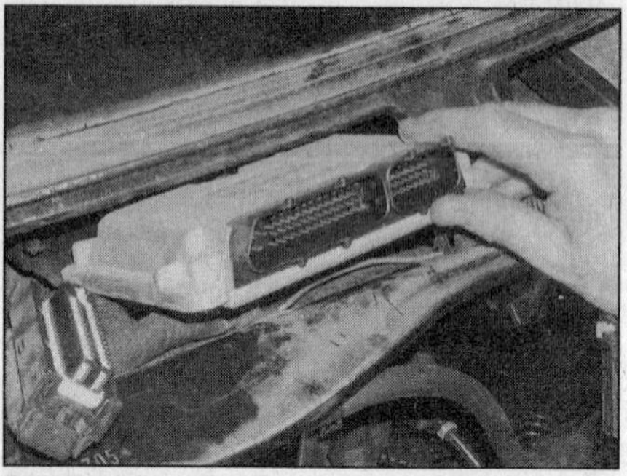

91 Press the control unit to the right, to release its retaining clip, and pull it from its location. Alternatively, the clip on the right-hand side of the unit can be released using a screwdriver.

92 Refitting is a reversal of removal. Press the unit to the left, once in position, to secure it. Bear in mind the comments made in the caution above – the ECU may not work correctly until it has been electronically coded.

Altitude sensor

1.8 litre engine code AGU

93 The altitude sensor is mounted on the engine compartment bulkhead, behind the air cleaner housing.

94 Disconnect the wiring connector from the sensor then undo the retaining screws and remove the sensor from the vehicle.

95 Refitting is the reverse of removal.

Camshaft chain adjuster solenoid valve

1.8 litre engine code AGN

96 The solenoid valve is an integral part of the inlet camshaft drive chain adjuster/ tensioner. At the time of writing it was unclear if the solenoid valve was available separately or whether it is an integral part of the adjuster/tensioner assembly. Refer to your VW dealer for solenoid valve availability. The chain adjuster/tensioner assembly is removed with the camshafts.

Diesel engine management system components – removal and refitting

Throttle position sensor

1 The position sensor is integral with the accelerator pedal. The pedal assembly can be removed (once access has been gained by removing the driver's lower facia panel) by disconnecting the sensor wiring plug and unscrewing the nuts securing the pedal to its mounting bracket. Depending on the transmission fitted, a separate sensor may be fitted above the pedal bracket, secured by two screws.

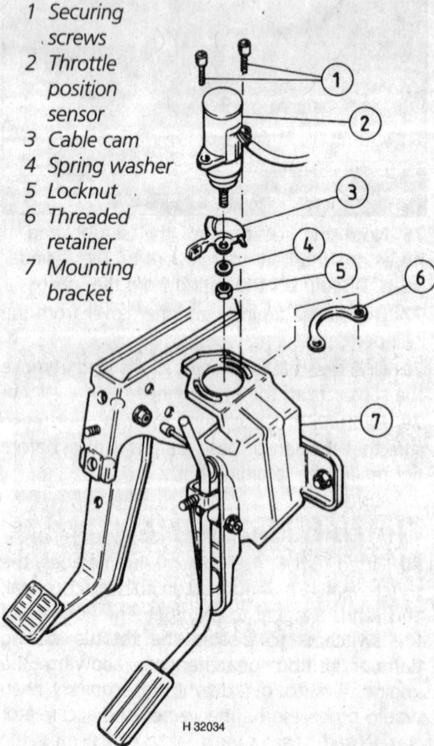

1 Securing screws
2 Throttle position sensor
3 Cable cam
4 Spring washer
5 Locknut
6 Threaded retainer
7 Mounting bracket

Coolant temperature sensor

2 Drain approximately one quarter of the coolant from the engine. Alternatively, be prepared for coolant spillage as the sensor is removed.

3 Where necessary for access, remove the engine top cover(s). Removal details vary according to model, but the cover retaining nuts are concealed under circular covers, which are prised out of the main cover. Remove the nuts, and lift the cover from the engine, releasing any wiring or hoses attached.

4 The sensor is at the top coolant outlet elbow, at the front of the cylinder head. Unplug the wiring from it at the connector.

5 Remove the securing clip, then extract the sensor from its housing and recover the O-ring seal.

6 Refit the coolant temperature sensor by reversing the removal procedure, using a new O-ring seal. Top-up the cooling system.

Fuel temperature sensor

7 The fuel temperature sensor is incorporated into the fuel quantity adjuster, which is fitted to the top of the injection pump. It appears from the information available at time of writing that the sensor is not available separately.

Inlet air temperature sensor

Non-turbo models

8 The sensor is mounted on the rear of the air cleaner top cover. Disconnect the sensor wiring plug, then remove the securing clip and extract the sensor. Recover the O-ring seal.

9 Refit the inlet air temperature sensor by reversing the removal procedure, using a new O-ring seal.

Turbo models

10 All models have an air temperature sensor built into the air mass meter. This sensor is an integral part of the air mass meter, and cannot be renewed separately. An additional air temperature/pressure sensor is fitted, either on top of the intercooler, or on the air hose from the intercooler to the inlet manifold, and this can be renewed as described below.

11 Trace the air hose back from the inlet manifold to the point where it passes through the inner wing.

12 If the sensor is mounted on the hose, disconnect the wiring plug, then remove the

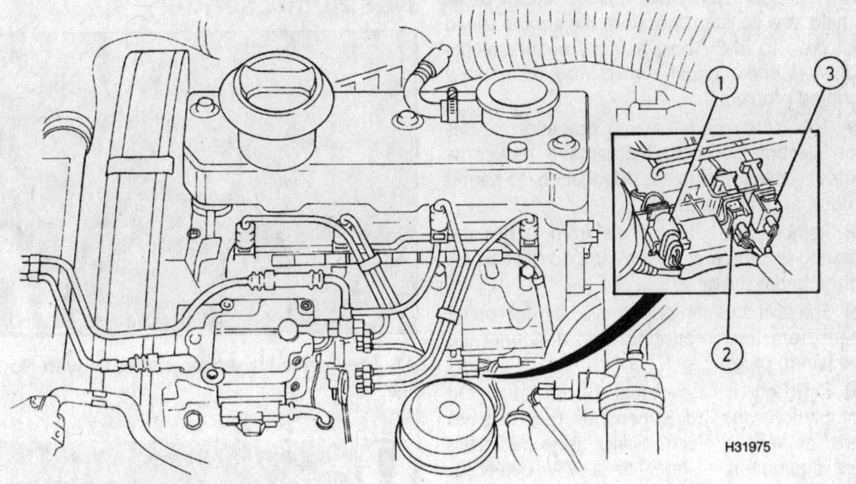

Wiring connectors behind oil filter housing

1 Fuel temperature sensor, quantity adjuster, shut-off valve and commencement of injection valve
2 Engine speed sensor
3 Needle lift sensor

two retaining screws and withdraw the sensor. Recover the O-ring seal.

13 If the sensor is mounted on the intercooler, remove the right-hand headlight. The sensor can then be removed in the same way as the pipe-mounted type.

14 Refit the inlet air temperature sensor by reversing the removal procedure, using a new O-ring seal.

Engine speed sensor

15 The engine speed sensor is mounted on the front cylinder block, adjacent to the mating surface of the block and transmission bellhousing.

16 Where necessary for access, remove the engine top cover(s). Removal details vary according to model, but the cover retaining nuts are concealed under circular covers, which are prised out of the main cover. Remove the nuts, and lift the cover from the engine, releasing any wiring or hoses attached.

17 Trace the wiring back from the sensor, and disconnect it at the plug behind the oil filter housing.

18 Remove the retaining screw and withdraw the sensor from the cylinder block.

19 Refit the sensor by reversing the removal procedure.

Air mass meter

Turbo models

20 Slacken the clips and disconnect the air ducting from the air mass meter, at the rear of the air cleaner housing.

21 Disconnect the wiring from the air mass meter, and the vacuum hose beneath the wiring connector.

22 Remove the retaining screws and extract the meter from the air cleaner housing. Recover the O-ring seal.

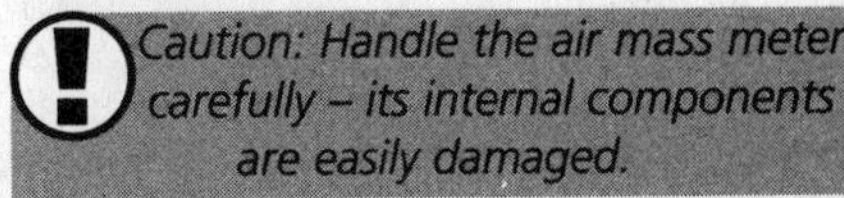
Caution: Handle the air mass meter carefully – its internal components are easily damaged.

23 Refitting is a reversal of removal. Renew the O-ring seal if it appears damaged.

Absolute pressure (altitude) sensor

24 The absolute pressure sensor is an integral part of the ECU, and hence cannot be renewed separately.

Inlet manifold flap housing

All models

25 Where necessary for access, remove the engine top cover(s). Removal details vary according to model, but the cover retaining nuts are concealed under circular covers, which are prised out of the main cover. Remove the nuts, and lift the cover from the engine, releasing any wiring or hoses attached.

26 Loosen the clip (or release the spring clip) and disconnect the air trunking from the flap housing. Disconnect the hose from the top of the housing.

Non-turbo models

27 Disconnect the flap control motor wiring plug from the front of the housing.

28 Unscrew the four housing retaining bolts, and withdraw the housing from the inlet manifold. Recover the O-ring seal.

29 Refitting is a reversal of removal. Renew the O-ring seal if it appears damaged.

Turbo models

30 Unscrew the two bolts securing the EGR pipe flange to the base of the housing. Separate the flange from the housing slightly, and recover the gasket – do not strain the EGR pipe.

31 Trace the hose back from the flap vacuum unit to the solenoid valve, and disconnect the pipe from the valve. The vacuum unit can be removed with the housing.

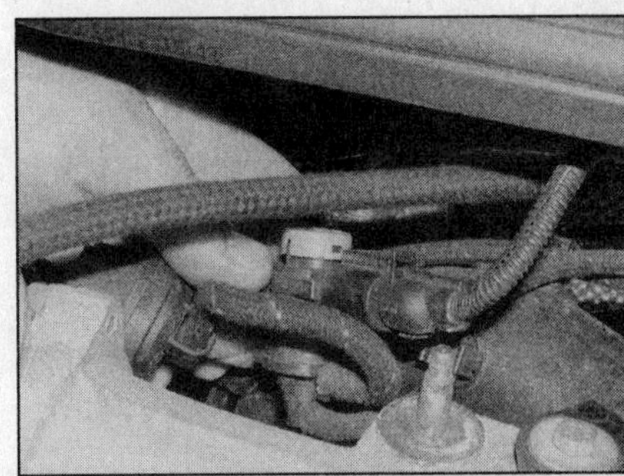

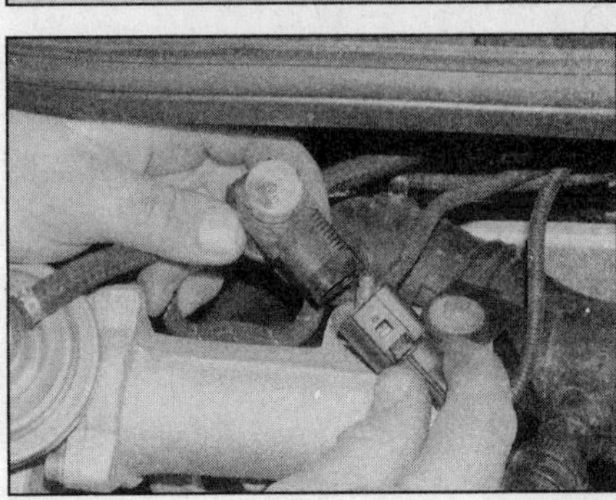

32 Unclip the solenoid valve for the inlet manifold flap operating vacuum capsule; disconnect the wiring plug and vacuum hose from the solenoid, and remove it.

33 Unscrew and remove the three housing retaining bolts, and withdraw the housing from the inlet manifold. Recover the O-ring seal.

34 Refitting is a reversal of removal. Renew the O-ring seal if it appears damaged.

Clutch and brake pedal switches

35 The clutch and brake pedal switches are clipped to mounting brackets directly above their respective pedals.

36 The brake pedal switch operates as a safety device, in the event of a problem with the accelerator position sensor. If the brake pedal switch is depressed while the accelerator pedal is held at a constant position, the engine speed will drop to idle. Thus, a faulty or incorrectly-adjusted brake pedal switch may result in a running problem.

37 The clutch pedal switch operation causes the injection pump to momentarily reduce its output while the clutch is disengaged, to permit smoother gear changing.

38 Remove the trim panels from under the steering column area of the facia, to gain access to the pedal cluster.

39 The switches can be removed by unclipping them from their mountings and disconnecting the wiring plugs.

40 Refitting is a reversal of removal. On completion, the adjustment of the switches must be verified electronically, using dedicated test equipment – refer to a VW dealer or specialist for advice.

Needle lift sender

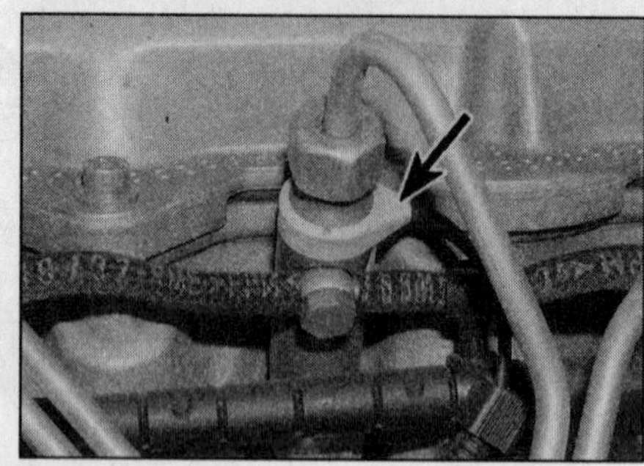

41 The needle lift sender is integral with No 3 injector.

Electronic control unit (ECU)

42 The ECU is located centrally behind the engine compartment bulkhead, under one of the windscreen cowl panels. Remove the wiper arms and cowl panel.

43 Release the locking clip or lever on the ECU wiring connector, and disconnect the plug. On most models, there are two separate plugs to be disconnected.

44 Press the control unit to the right, to release its retaining clip, and pull it from its location.

45 Refitting is a reversal of removal. Press the unit to the left, once in position, to secure it. Bear in mind the comments made in the caution above – the ECU will not work correctly until it has been electronically coded.

Caution: Always wait at least 30 seconds after switching off the ignition before disconnecting the wiring from the ECU. When the wiring is disconnected, all the learned values are erased, however any contents of the fault memory are retained. After reconnecting the wiring, the basic settings must be reinstated by a VW dealer or specialist using a special test instrument. Note also that if the ECU is renewed, the identification of the new ECU must be transferred to the immobiliser cohtrol unit by a VW dealer or specialist.

VW Polo '94 – '99

28

General information
Fuel injection system – depressurisation, priming and bleeding
Bosch Mono-Motronic system components – removal and refitting
Bosch Motronic and Magneti Marelli system components – removal and refitting
Diesel engine management system – component removal and refitting

Petrol models

General

Engine code AEV	Bosch Mono-Motronic 1.2.3
Engine codes ADX and AEA	Bosch Mono-Motronic 1.3
Engine codes AER, ALL, AEX, AKV, ANX and APQ	Bosch Motronic 9.0
Engine code AEE	Magneti-Marelli 1AV

Fuel system data

Fuel pump type	Electric, immersed in fuel tank
Bosch Mono-Motronic:	
Fuel pump delivery rate	1000 cm^3 / min (battery voltage of 12.5 V)
Regulated fuel pressure	0.8 to 1.2 bar
Engine idle speed	750 to 850 rpm (non-adjustable, electronically controlled)
Idle CO content	0.5% (non-adjustable)
Maximum engine speed	6300 rpm (governed electronically)
Injector electrical resistance	1.2 to 1.6 ohms at 15 to 30°C
Inlet manifold heater resistance	0.25 to 0.5 ohms (engine cold)
Multi-point petrol injection models:	
Fuel pump delivery rate	1100 cm^3 / min (battery voltage of 12.5 V)
Regulated fuel pressure	2.5 bar
Engine idle speed (non-adjustable, electronically controlled):	
Engine codes AER and ALL	660 to 740 rpm
Engine codes AEX, AKV, ANX and APQ:	
Manual transmission models:	
Control unit 030 906 027A	600 to 700 rpm
Control unit 030 906 027K	770 to 970 rpm
Automatic transmission models	750 to 850 rpm
Engine code AEE:	
Manual transmission models	830 to 930 rpm
Automatic transmission models:	
Control unit 032 906 030A	750 to 850 rpm
Control unit 032 906 030J	630 to 730 rpm
Idle CO content (non-adjustable, electronically controlled)	0.5 % max
Injector electricai resistance	14 to 17 ohms

Diesel models

General

Firing order	1-3-4-2	
Maximum engine speed (engine code AEF)	5050 ± 100 rpm	
Engine idle speed (engine code AEF)	940 ± 20 rpm	
Engine idle speed boost (engine code AEF)	1050 ± 50 rpm	

Torque wrench settings	**Nm**	**lbf ft**
Petrol models		
Bosch Mono-Motronic models:		
Injector cap/inlet air temperature sensor housing screw	5	4
Lambda sensor	50	37
Throttle body mounting flange-to-inlet manifold nuts	10	7
Throttle valve positioning module screws	6	4
Multi-point petrol injection models:		
ECU mounting bracket nuts	10	7
Fuel rail mounting bolts	10	7
Lambda sensor	55	41
Throttle body earth strap bolt	10	7
Diesel models		
Fuel cut-off solenoid (except engine code AEF)	40	30

General information

Petrol models

Bosch Mono-Motronic

The Bosch Mono-Motronic system is a self-contained engine management system, which controls both the fuel injection and ignition.

The fuel injection system comprises a fuel tank, an electric fuel pump, a fuel filter, fuel supply and return lines, a throttle body with an integral electronic fuel injector, and an Electronic Control Unit (ECU) together with its associated sensors, actuators and wiring.

The fuel pump delivers a constant supply of fuel through a cartridge filter to the throttle body, at a slightly higher pressure than required – the fuel pressure regulator (integral with the throttle body) maintains a constant fuel pressure at the fuel injector, and returns excess fuel to the tank via the return line. This constant flow system also helps to reduce fuel temperature and prevents vaporisation.

The fuel injector is opened and closed by an Electronic Control Unit (ECU), which calculates the injection timing and duration according to engine speed, throttle position and rate of opening, inlet air temperature, coolant temperature, road speed and exhaust gas oxygen content information, received from sensors mounted on the engine.

Inlet air is drawn into the engine through the air cleaner, which contains a renewable paper filter element. The inlet air temperature is regulated by a vacuum-operated valve mounted in the air cleaner inlet trunking, which blends air at ambient temperature with hot air, drawn from over the exhaust manifold. Vacuum supply to the valve is regulated by a temperature switch mounted in the air cleaner.

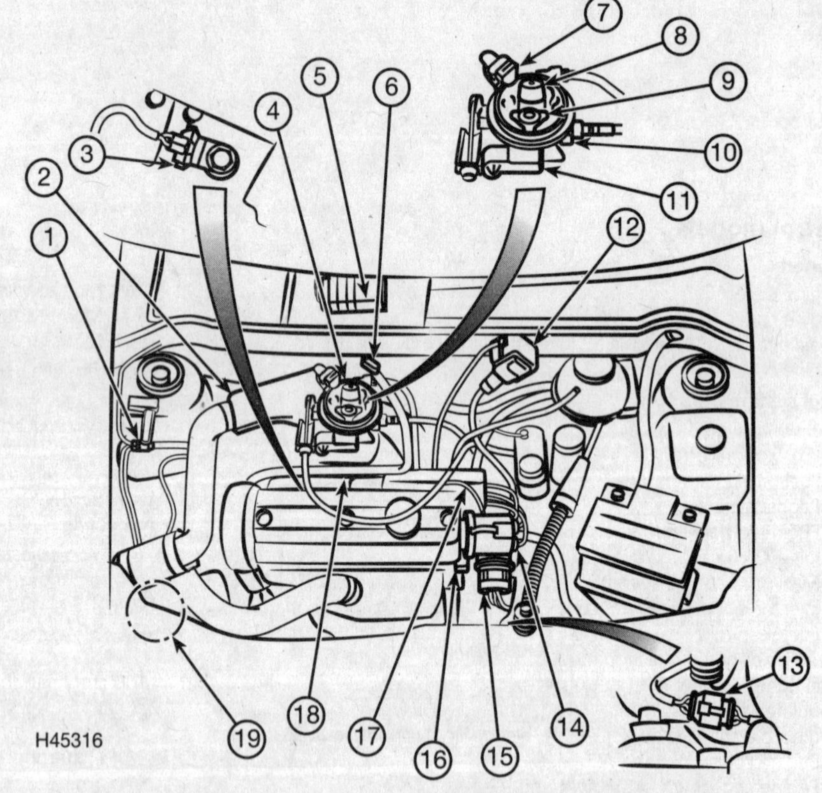

Bosch Mono-Motronic engine management system component locations

1 Charcoal filter solenoid valve
2 Air cleaner and air inlet trunking
3 Knock sensor – engine codes ADX and AEA only
4 Injection unit/throttle body
5 ECU
6 Inlet air temperature vacuum switch
7 Inlet air temperature sensor connector
8 Fuel injector and inlet air temperature sensor
9 Fuel pressure regulator
10 Throttle valve potentiometer
11 Throttle valve positioning module and idle switch
12 Ignition coil
13 Lambda sensor connector
14 Distributor
15 Main wiring harness connector
16 Coolant temperature sender
17 Earth connection
18 Spark plug
19 Charcoal canister

The temperature of the air entering the throttle body is measured by a sensor mounted directly above the injector. This information is used by the ECU to fine-tune the fuelling requirements for different operating temperatures.

Idle speed control is achieved partly by an electronic throttle positioning module, mounted on the side of the throttle body and partly by the ignition system, which gives fine control of the idle speed by altering the ignition timing. As a result, manual adjustment of the engine idle speed is not necessary, or possible.

To improve cold starting and idling (and fuel economy), an electric heating element is mounted on the underside of the inlet manifold; this prevents fuel vapour condensation when the engine is cold. Power is supplied to the heater by a relay, which is in turn controlled by the ECU. The heater is sometimes known as the 'hedgehog', due to its appearance.

The exhaust gas oxygen content is constantly monitored by the ECU via the Lambda sensor, which is mounted in the exhaust pipe. The ECU then uses this information to modify the injection timing and duration to maintain the optimum air/fuel ratio – a result of this is that manual adjustment of the idle exhaust CO content is not necessary. All models are fitted with a catalytic converter.

In addition, the ECU controls the operation of the activated charcoal filter evaporative loss system.

Multi-point petrol injection models

The Bosch Motronic and Magneti-Marelli 1AV systems are self-contained engine management systems, which control both the fuel injection and ignition.

The fuel injection system comprises a fuel tank, an electric fuel pump, a fuel filter, fuel supply and return lines, a throttle body, a fuel rail, a fuel pressure regulator, four electronic fuel injectors, and an Electronic Control Unit (ECU) together with its associated sensors, actuators and wiring. The two systems used are essentially very similar – the only significant differences lie within the ECUs.

The fuel pump delivers a constant supply of fuel through a cartridge filter to the fuel rail, at a slightly higher pressure than required – the fuel pressure regulator maintains a constant fuel pressure to the fuel injectors, and returns excess fuel to the tank via the return line. This constant flow system also helps to reduce fuel temperature, and prevents vaporisation.

The fuel injectors are opened and closed by an Electronic Control Unit (ECU), which calculates the injection timing and duration according to engine speed, crankshaft position, throttle position and rate of opening, inlet manifold depression, inlet air temperature, coolant temperature, road speed and exhaust gas oxygen content information, received from sensors mounted on and around the engine.

Inlet air is drawn into the engine through the air cleaner, which contains a renewable paper filter element. The inlet air temperature is regulated by a valve mounted in the air cleaner inlet trunking, which blends air at ambient temperature with hot air, drawn from over the exhaust manifold.

The temperature of the air entering the throttle body is measured by a sensor mounted on the right-hand side of the inlet manifold. This sensor also monitors the pressure in the inlet manifold. This information is used by the ECU to fine-tune the fuelling requirements for different operating conditions.

Idle speed control is achieved partly by an electronic throttle valve positioning module, on the rear of the throttle body and partly by the ignition system, which gives fine control of the idle speed by altering the ignition timing. As a result, manual adjustment of the engine idle speed is not necessary or possible.

The exhaust gas oxygen content is constantly monitored by the ECU via the Lambda sensor, which is mounted in the exhaust pipe. The ECU then uses this information to modify the injection timing and duration to maintain the optimum air/fuel ratio – a result of this is that manual adjustment of the idle exhaust CO content is not necessary or possible. All models are fitted with a catalytic converter.

Where fitted, the ECU controls the operation of the activated charcoal filter evaporative loss system.

Diesel models

Engine code AEF

The fuel system comprises a fuel tank, a fuel injection pump, an engine-bay mounted fuel filter with an integral water separator, fuel supply and return lines, and four fuel injectors.

The injection pump is driven at half crankshaft speed by the camshaft timing belt.

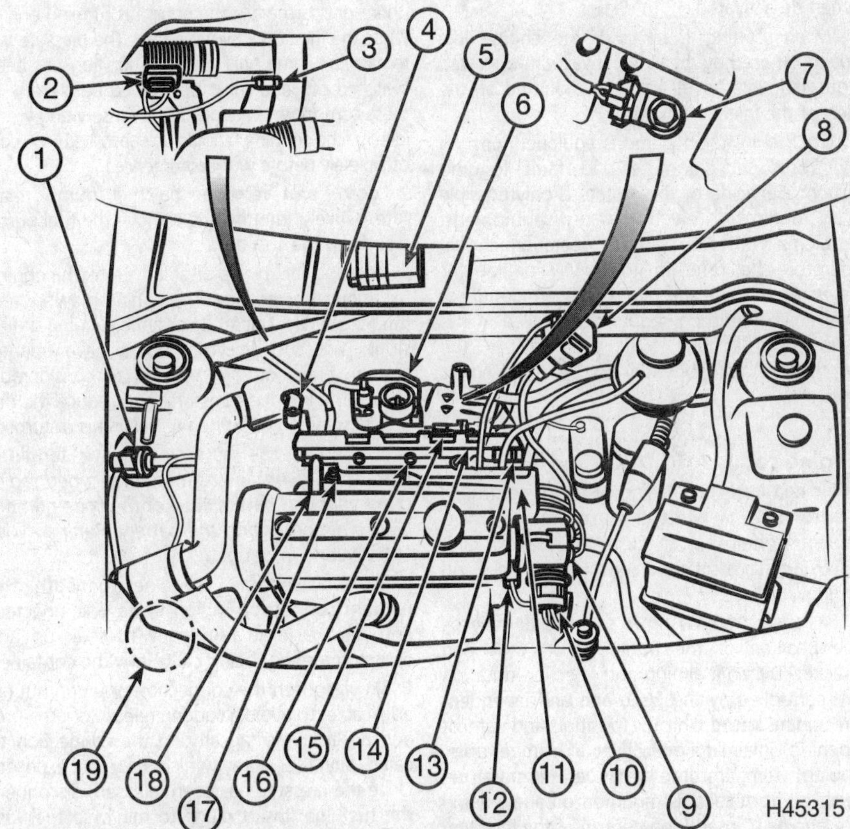

Bosch Motronic/Magneti-Marelli 1AV engine management system component locations

1 Charcoal filter solenoid valve
2 Lambda sensor connector
3 Lambda sensor
4 Inlet manifold pressure and inlet air temperature sensor
5 ECU
6 Throttle body
7 Knock sensor
8 Ignition coil
9 Distributor
10 Main wiring harness connector
11 Earth connection
12 Coolant temperature sender
13 Fuel pressure regulator
14 Fuel rail
15 Air cleaner housing
16 Fuel injector
17 Spark plug
18 Oil pressure switch
19 Charcoal canister

Fuel is drawn from the fuel tank, through the filter by the injection pump, which then distributes the fuel under very high pressure to the injectors via separate delivery pipes.

The injectors are spring-loaded mechanical valves, which open when the pressure of the fuel supplied to them exceeds a specific limit. Fuel is then sprayed from the injector nozzle into the cylinder via a swirl chamber (indirect injection). Two-stage injectors are fitted, which open in steps as the supplied fuel pressure rises; this improves the engines combustion characteristics.

The basic injection timing is set by the position of the injection pump on its mounting bracket. When the engine is running, the injection timing is advanced and retarded mechanically by the injection pump itself, and is influenced primarily by the accelerator position and engine speed.

The engine is stopped by means of a solenoid-operated fuel cut-off valve, which interrupts the flow of fuel to the injection pump when de-activated.

When starting from cold, the engine idle speed is raised by means of a vacuum-operated automatic idle boost actuator, mounted on the side of the injection pump.

The fuel injection pump is equipped with an electronic self-diagnosis and fault logging system. Servicing of this system is only possible with dedicated electronic test equipment. Problems with the system's operation should therefore be referred to a VW dealer for assessment. Once the fault has been identified, the removal/refitting sequences detailed in the following Sections will then allow the appropriate component(s) to be renewed as required.

Engine codes AHG, AKU and AGD

The direct-injection fuelling system is controlled electronically by a diesel engine management system, comprising an Electronic Control Unit (ECU) and its associated sensors, actuators and wiring.

Basic injection timing is set mechanically by the position of the pump on its mounting bracket. Dynamic timing and injection duration are controlled by the ECU, and are dependent on engine speed, throttle position and rate of opening, inlet air flow , inlet air temperature, coolant temperature, and fuel temperature, received from sensors mounted on and around the engine. Closed-loop control of the injection timing is achieved by means of an injector needle lift sensor, which is fitted to injector No 3.

Two-stage injectors are used, which improve the engine's combustion characteristics, leading to quieter running and better exhaust emissions.

In addition, the ECU manages the operation of the Exhaust Gas Recirculation (EGR) emission control system and the glow plug control system.

Fuel injection system – depressurisation, priming and bleeding

Petrol models

Warning: The following procedure will merely relieve the pressure in the fuel system – remember that fuel will still be present in the system components and take precautions accordingly before disconnecting any of them.

1 The fuel system referred to in this Section is defined as the tank-mounted fuel pump, the fuel filter, the fuel injector, the throttle body-mounted fuel pressure regulator, and the metal pipes and flexible hoses of the fuel lines between these components. All these contain fuel, which will be under pressure while the engine is running and/or while the ignition is switched on. The pressure will remain for some time after the ignition has been switched off, and must be relieved before any of these components are disturbed for servicing work. Ideally, the engine should be allowed to cool completely before work commences.

2 Locate and remove the fuel pump relay. Alternatively, identify and remove the fuel pump fuse from the fusebox.

3 With the fuel pump disabled, crank the engine for about ten seconds. The engine may fire and run for a while, but let it continue running until it stops. The fuel injector should have opened enough times during cranking to considerably reduce the line fuel pressure, and reduce the risk of fuel spraying out when a fuel line is disturbed.

4 Disconnect the battery negative terminal. **Note:** *If the vehicle has a security-coded radio, check that you have a copy of the code number before disconnecting the battery. Refer to your VW dealer if in doubt.*

5 Place a suitable container beneath the relevant connection /union to be disconnected, and have a large rag ready to soak up any escaping fuel not being caught by the container.

6 Slowly loosen the connection or union nut (as applicable) to avoid a sudden release of pressure, and position the rag around the connection to catch any fuel spray which may be expelled. Once the pressure has been released, disconnect the fuel line. Insert plugs to minimise fuel loss and prevent the entry of dirt into the fuel system.

Diesel models

7 Although no specific procedure is given for depressurising the fuel system, common sense dictates that the engine should be allowed to cool for while before commencing any repair work. During this time, the residual pressure in the system should have subsided sufficiently. Place a rag over any fuel connection as the union is slackened.

8 It is not necessary to manually prime and bleed the fuel system after any operation on the system components. Start the engine (this may take longer than usual, especially if the fuel system has been allowed to run dry – operate the starter in ten second bursts with 5 seconds rest in between each operation) and run it a fast idle speed for a minute or so to purge any trapped air from the fuel lines. After this time the engine should idle smoothly at a constant speed.

9 If the engine idles roughly, then there is still some air trapped in the fuel system. Increase the engine speed again for another minute or so then recheck the idle speed. Repeat this procedure as necessary until the engine is idling smoothly.

Caution: Do not operate the starter for more than 10 seconds at a time. Allow 30 seconds between attempts for the starter to cool.

Bosch Mono-Motronic system components – removal and refitting

Throttle body

1 Remove the air cleaner housing.

2 Depressurise the fuel system, then disconnect the battery negative lead and position it away from the terminal. **Note:** *If the vehicle has a security-coded radio, check that you have a copy of the code number before disconnecting the battery. Refer to your VW dealer if in doubt.*

3 Disconnect the fuel supply and return hoses from the ports on the side of the throttle body. Note the arrows that denote the direction of fuel flow, and mark the hoses accordingly. The fuel supply hose is colour-coded white, and the return hose has blue markings.

4 Unplug the wiring harness from the throttle body at the connectors, labelling them to aid correct refitting later.

5 Disconnect the accelerator cable from the throttle body.

6 Loosen and remove the four through-bolts which secure the throttle body to the inlet manifold. Lift the throttle body away from the inlet manifold, recovering the gasket (where applicable). Unless specifically required, it is not recommended that the upper and lower halves of the throttle body are separated – these are held together by two inner through-bolts. If the two halves are split, a new gasket must be used on reassembly.

7 Refitting is a reversal of removal; renew all gaskets where appropriate. Tighten the through-bolts securely. On completion, check and if necessary adjust the accelerator cable.

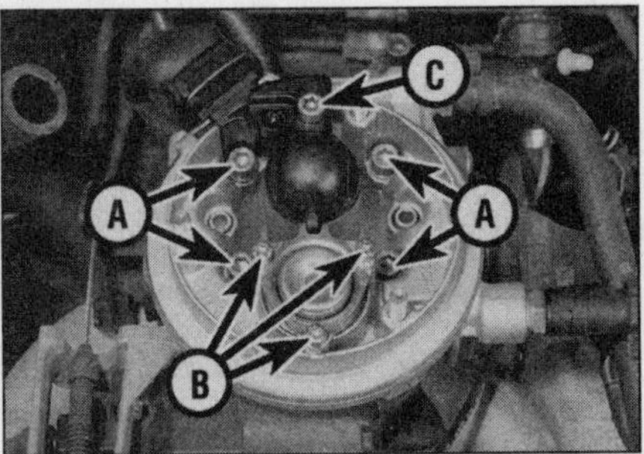

Throttle body through-bolts (A), fuel pressure regulator screws (B) and injector securing screw (C)

Fuel injector

8 Remove the air cleaner housing.

9 Depressurise the fuel system, then disconnect the battery negative lead and position it away from the terminal – refer to the note in paragraph 2.

10 Unplug the wiring harness from the injector at the connector(s), labelling them to aid correct refitting later.

11 Remove the screw and lift off the injector retaining cap/inlet air temperature sensor housing. Recover the gasket.

12 Release the securing washer, then lift the injector out of the throttle body, recovering the O-ring seals.

13 Check the injector electrical resistance using a multimeter and compare the result with the Specifications.

14 Refit the injector by following the removal procedure in reverse, renewing all O-ring seals and gaskets. Apply a suitable sealant to the screw threads, then insert and tighten the retaining screw to the specified torque.

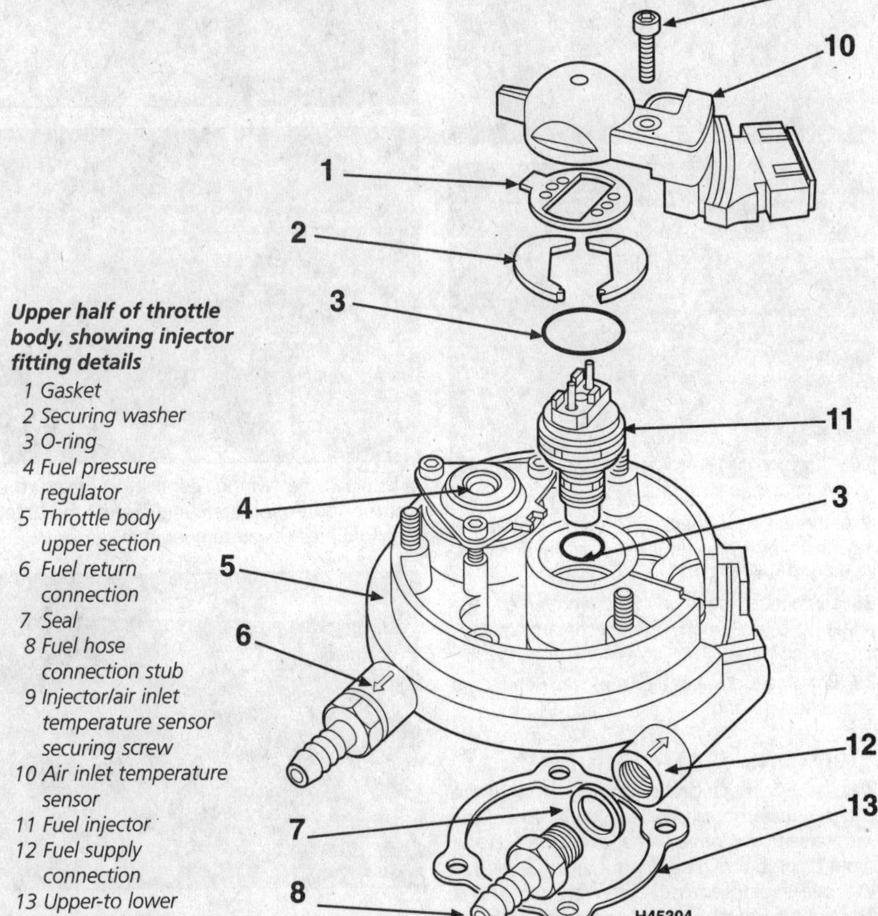

Upper half of throttle body, showing injector fitting details

1 Gasket
2 Securing washer
3 O-ring
4 Fuel pressure regulator
5 Throttle body upper section
6 Fuel return connection
7 Seal
8 Fuel hose connection stub
9 Injector/air inlet temperature sensor securing screw
10 Air inlet temperature sensor
11 Fuel injector
12 Fuel supply connection
13 Upper-to lower section gasket

Inlet air temperature sensor

15 The inlet air temperature sensor is an integral part of the injector retaining cap. Removal is as described in the previous sub-Section.

Fuel pressure regulator

16 If the operation of the fuel pressure regulator is in question, dismantle the unit as described below, then check the cleanliness and integrity of the internal components. **Note:** *The fuel pressure regulator components are matched to the upper part of the throttle body during manufacture. If the pressure regulator is defective, VW state that the upper part of the throttle body must be renewed complete. Consult your parts supplier for the latest advice.*

17 Remove the air cleaner housing.

18 Depressurise the fuel system, then disconnect the battery negative lead and position it away from the terminal – refer to the note in paragraph 2.

19 With reference to the relevant sub-Section, remove the screw and lift off the inlet air temperature sensor/injector cap.

20 Slacken and withdraw the three Torx retaining screws and lift off the fuel pressure regulator retaining frame.

21 Lift out the upper cover, spring and membrane.

22 Clean all the components thoroughly, then inspect the membrane for cracks or splits.

23 Reassemble the pressure regulator by following the removal procedure in reverse.

Throttle valve positioning module

24 Disconnect the battery negative lead and position it away from the terminal – refer to paragraph 2. Remove the air cleaner housing.

25 Disconnect the accelerator cable from the throttle body.

26 Unplug the connector from the side of the throttle valve positioning module.

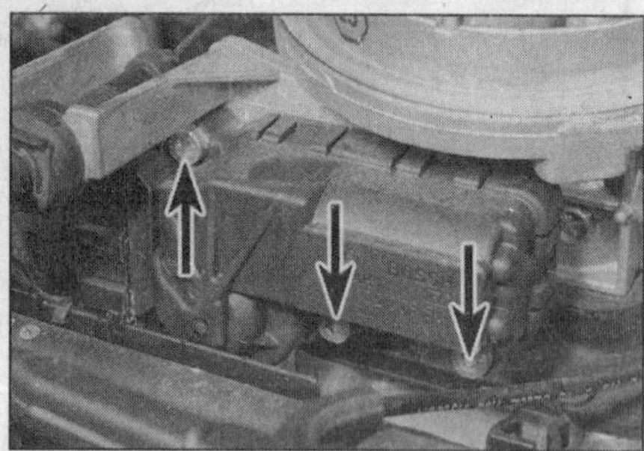

27 Remove the retaining screws and lift the module, together with the accelerator cable outer mounting bracket, away from the throttle body.

28 Refitting is a reversal of removal. Note that if a new module has been fitted, the adjustment of the idle switch will need to be checked – refer to a VW dealer or specialist for advice as this operation requires access to dedicated test equipment.

Throttle valve potentiometer

29 Refer to the relevant sub-Section and remove the throttle body. The throttle valve potentiometer is an integral part of the lower section of the throttle body, and cannot be renewed separately.

Idle switch

30 Refer to the relevant sub-Section and remove the throttle valve positioning module. The idle switch is an integral part of the module, and cannot be renewed separately.

31 Where a new throttle valve positioning module has been fitted, the adjustment of the idle switch will need to be checked – refer to a VW dealer or specialist for advice as this operation requires access to dedicated test equipment.

Lambda sensor

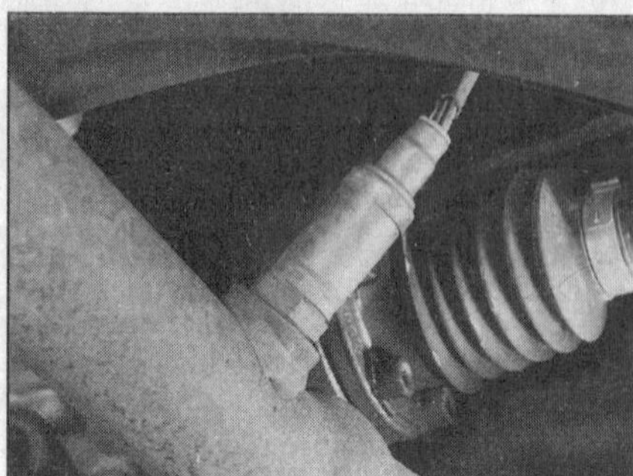

32 The lambda sensor is threaded into the exhaust pipe, ahead of the catalytic converter.

33 Disconnect the battery negative lead and position it away from the terminal – refer to the note in paragraph 2. Unplug the wiring harness from the lambda sensor at the connector, which is located in a bracket mounted on the transmission.

34 **Note:** *As a flying lead remains connected to the sensor after it has been disconnected, if the correct-size spanner is not available, a slotted socket will be required to remove the sensor.* Working under the vehicle, slacken and withdraw the sensor, taking care to avoid damaging the sensor probe as it is removed.

35 Apply a little anti-seize grease to the sensor threads only – keep the probe tip clean.

36 Refit the sensor to its housing, tightening it to the correct torque. Restore the harness connection. Note that the type of lambda sensor fitted depends on vehicle specification – the sensor may not be interchangeable with one obtained from another model.

Coolant temperature sensor

37 The sensor is combined with the coolant temperature gauge sensor. The sensor is clipped into the thermostat housing. Partially drain the cooling system.

38 Disconnect the wiring plug, and remove the retaining clip. Remove the sensor and recover the sealing ring.

39 Refitting is a reversal of removal. Renew the sensor sealing ring.

Electronic control unit (ECU)

40 The ECU is located centrally behind the engine compartment bulkhead, under one of the windscreen cowl panels. The unit is coded, and should not be removed without consulting a VW dealer or specialist, otherwise it may not function correctly when the multi-plug is reconnected.

Bosch Motronic and Magneti Marelli system components – removal and refitting

Throttle body

1 Remove the air cleaner housing.

2 Detach the accelerator cable from the throttle valve lever.

3 Disconnect the battery negative lead, and position it away from the terminal. **Note:** *If the vehicle has a security-coded radio, check that you have a copy of the code number before disconnecting the battery. Refer to your VW dealer if in doubt.*

4 Unplug the wiring connectors from the throttle potentiometer and from the inlet manifold pressure/air temperature sensor.

5 Disconnect the hose for the charcoal canister from the port on the throttle body.

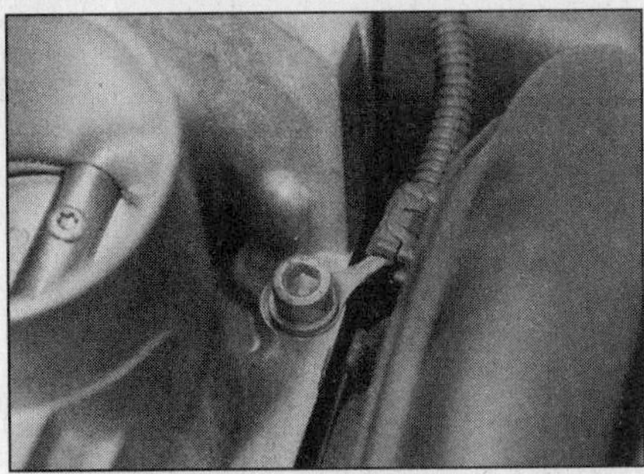

6 Note that one of the bolts secures the throttle body earth strap. Slacken and withdraw the through-bolts, then lift the throttle body away from the inlet manifold. Recover and discard the gasket.

7 If required, refer to the relevant sub-Section and remove the throttle potentiometer.

8 Refitting is a reversal of removal, noting the following:

a) Use a new throttle body-to-inlet manifold gasket.

b) Tighten the throttle body through-bolts securely, to prevent air leaks. A tightening torque for these bolts is not specified by the manufacturers.

c) Ensure that all hoses and electrical connectors are refitted securely.

d) Check and if necessary adjust the accelerator cable.

Fuel injectors and fuel rail

Note: *If a faulty injector is suspected, before removing the injectors, it is worth trying the effect of one of the proprietary injector-cleaning treatments. These can be added to the petrol in the tank, and are intended to clean the injectors as you drive.*

9 Disconnect the battery negative lead, and position it away from the terminal – refer to the note in paragraph 3.

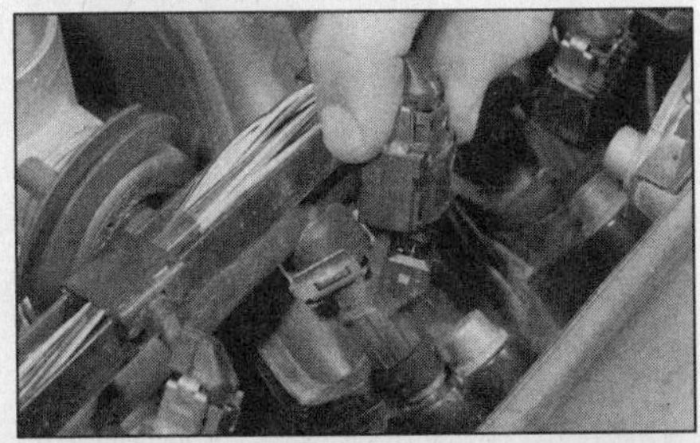

10 Unplug the injector harness connectors, labelling them to aid correct refitting later. Unclip the wiring harness clips from the top of the fuel rail, and lay the harness to one side.

11 Depressurise the fuel system.

12 Disconnect the vacuum hose from the port on the bottom of the fuel pressure regulator.

13 Slacken the clips and disconnect the fuel supply and return hoses from the end of the fuel rail. *Carefully* note the fitted positions of the hoses – the supply hose is marked with a black or white arrow, and the return hose is marked with a blue arrow.

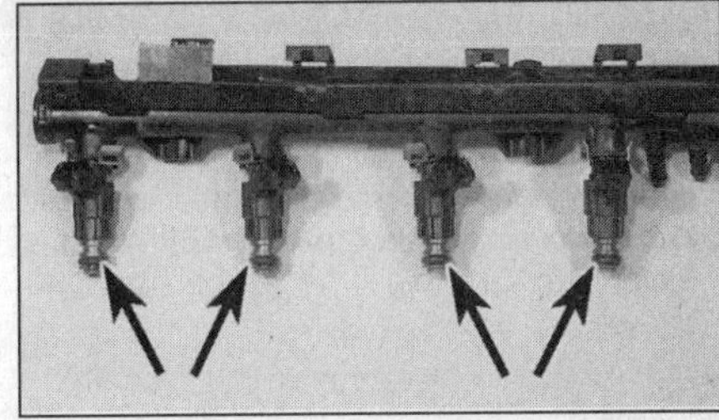

14 Slacken and withdraw the fuel rail mounting bolts, then carefully lift the rail away from the inlet manifold, together with the injectors. Recover the injector lower O-ring seals as they emerge from the manifold.

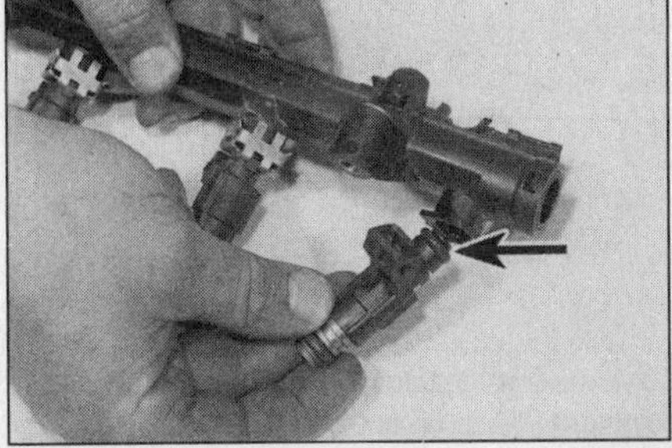

15 The injectors can be removed individually from the fuel rail by extracting the relevant metal clip and easing the injector out of the rail. Recover the injector upper O-ring seals.

16 If required, remove the fuel pressure regulator, referring to the relevant sub-Section for guidance.

17 Check the electrical resistance of the injector using a multimeter and compare it with the Specifications.

18 Refit the injectors and fuel rail by following the removal procedure in reverse, noting the following points:

a) *Renew the injector O-ring seals if they appear worn or damaged.*

b) *Ensure that the injector retaining clips are securely seated.*

c) *Check that the fuel supply and return hoses are reconnected correctly – refer to the colour coding described in Removal.*

d) *Check that all vacuum and electrical connections are remade correctly and securely.*

e) *On completion, check exhaustively for fuel leaks before bringing the vehicle back into service.*

Fuel pressure regulator

19 Disconnect the battery negative lead, and position it away from the terminal – refer to the note in paragraph 3.

20 Depressurise the fuel system.

21 Disconnect the vacuum hose from the port on the bottom of the fuel pressure regulator.

22 Slacken the clip and disconnect the fuel supply hose from the end of the fuel rail. This will allow the majority of fuel in the regulator to drain out. Be prepared for an amount of fuel loss – position a small container and some old rags underneath the fuel regulator housing. **Note:** *The supply hose is marked with a black or white arrow.*

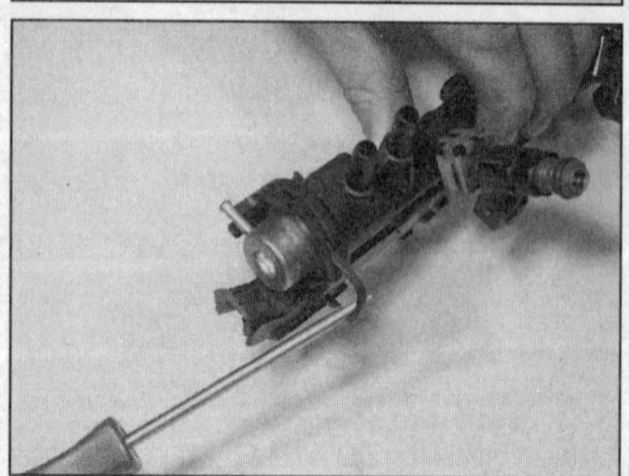

23 Extract the retaining clip from the top of the regulator housing and lift out the regulator body, recovering the O-ring seals.

24 Refit the fuel pressure regulator by following the removal procedure in reverse, noting the following points:

a) *Renew the O-ring seals if they appear worn or damaged.*

b) *Ensure that the regulator retaining clip is securely seated.*

c) *Refit the regulator vacuum hose securely.*

Throttle valve potentiometer

Note: *The potentiometer is matched to the throttle body during manufacture, and is not available separately – if defective, a complete throttle body assembly will be required.*

25 Disconnect the battery negative lead, and position it away from the terminal – refer to the note in paragraph 3.

26 Unplug the harness connector from the potentiometer.

27 Remove the retaining screws and lift the potentiometer away from the throttle body. Recover the gasket.

28 Refitting is a reversal of removal, noting the following:

a) *Renew the gasket if it is damaged.*

b) *Ensure that the potentiometer drive engages correctly with the throttle spindle extension.*

c) *On vehicles with automatic transmission, the potentiometer must be matched to the automatic transmission Electronic Control Unit (ECU) – this operation requires access to dedicated electronic test equipment, refer to a VW dealer for advice.*

Inlet manifold temperature sensor

29 The sensor is attached to the right-hand side of the inlet manifold.

30 Disconnect the battery negative lead, and position it away from the terminal – refer to the note in paragraph 3. Unplug the harness connector from the sensor.

31 Remove the two securing screws, and pull the sensor from the manifold. Recover the O-ring seal.

32 Refitting is a reversal of removal, renewing the O-ring seal if necessary, and tightening the securing screws securely.

Coolant temperature sensor

33 Refer to the Bosch Mono-Motronic System Section.

Lambda sensor

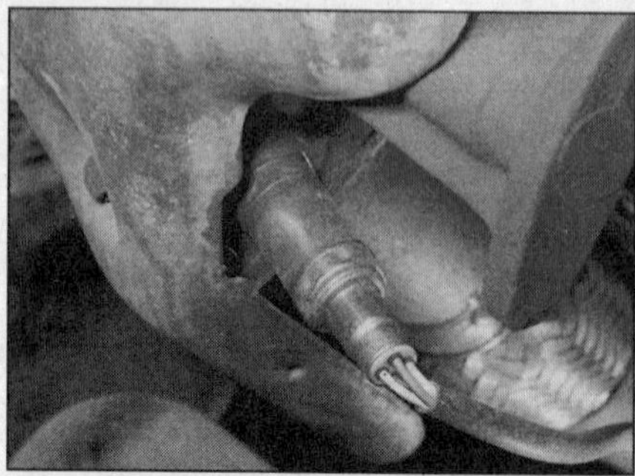

34 The lambda sensor is threaded into the exhaust downpipe, ahead of the catalytic converter.

35 Disconnect the battery negative lead and position it away from the terminal – refer to the note in paragraph 3. Unplug the wiring harness from the lambda sensor at the connector, located on the right-hand side of the engine rear mounting.

36 On early models, the sensor is only accessible from below – later models have the sensor situated higher up the downpipe, and it can be accessed from above. Slacken and withdraw the sensor, taking care to avoid damaging the sensor probe as it is removed. **Note:** *As a flying lead remains connected to the sensor after is has been disconnected, if the correct-size spanner is not available, a slotted socket will be required to remove the sensor.*

37 Apply a little anti-seize grease to the sensor threads – avoid contaminating the probe tip.

38 Refit the sensor to its housing, tightening it to the correct torque. Restore the harness connection.

Electronic control unit (ECU)

39 The ECU is located centrally behind the engine compartment bulkhead, under one of the windscreen cowl panels. Remove the cowl panel. The unit is coded, and should not be removed without consulting a VW dealer, otherwise it may not function correctly when the multi-plug is reconnected.

Diesel engine management system – component removal and refitting

Note: *The following procedures apply to vehicles with engine codes AHG, AKU and AGD only.*

Accelerator position sensor

1 Disconnect the battery negative lead, and position it away from the terminal.

2 Remove the trim panels from under the steering column area of the facia, to gain access to the pedal cluster.

3 Unplug the position sensor from the wiring harness at the connector.

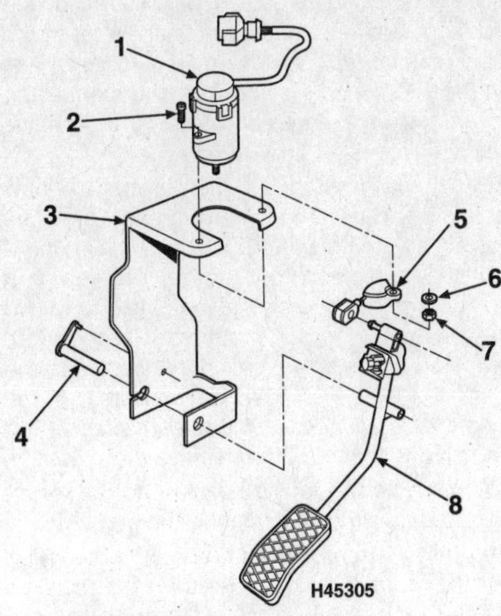

Accelerator position sensor details

1 Accelerator position sensor
2 Securing screw
3 Mounting bracket
4 Accelerator pedal spindle
5 Cable cam plate
6 Spring washer
7 Adjustment bolt
8 Accelerator pedal

4 Remove the two screws securing the position sensor mounting bracket, recover the nuts and washers, and lift the sensor out of the car.

5 If wished, the cable cam plate can be unhooked from the accelerator fitting, and removed.

6 Refitting is a reversal of removal, noting the following points:

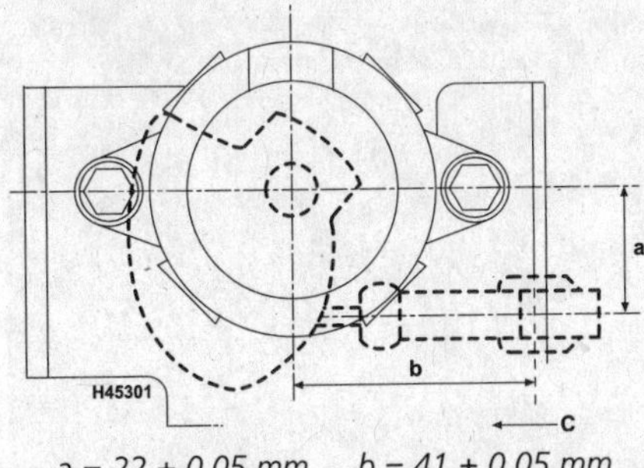

a = 22 ± 0.05 mm b = 41 ± 0.05 mm
C To front of car

a) The cable cam plate must be fitted to the position sensor spindle according to the dimensions shown (see above).
b) On completion, the adjustment of the position sensor must be verified electronically, using dedicated test equipment – refer to a VW dealer or specialist for advice.

Coolant temperature sensor

7 Refer to the Bosch Mono-Motronic System Section.

Fuel temperature sensor

8 Disconnect the battery negative lead and position it away from the terminal.

9 Slacken and withdraw the retaining screws and lift the top cover from the injection pump. Recover the gasket.

10 Remove the screws and lift out the fuel temperature sensor.

11 Refitting is a reversal of removal. Tighten the pump top cover screws to the specified torque.

Inlet air temperature sensor

12 The air temperature sensor is incorporated into the diesel engine management system ECU, and is not available separately.

Engine speed signal sensor

13 The engine speed sensor is mounted on the front cylinder block, adjacent to the mating surface of the block and transmission bellhousing.

14 Disconnect the battery negative lead and position it away from the terminal.

15 Unplug the harness connector from the sensor.

16 Remove the retaining screw and withdraw the sensor from the cylinder block.

17 Refit the sensor by reversing the removal procedure.

Throttle valve housing

18 Disconnect the battery negative cable and position it away from the terminal.

19 Slacken the clip and disconnect the air trunking from the throttle valve housing.

20 Pull the vacuum hose off the base of the throttle valve housing, and disconnect the wiring plug.

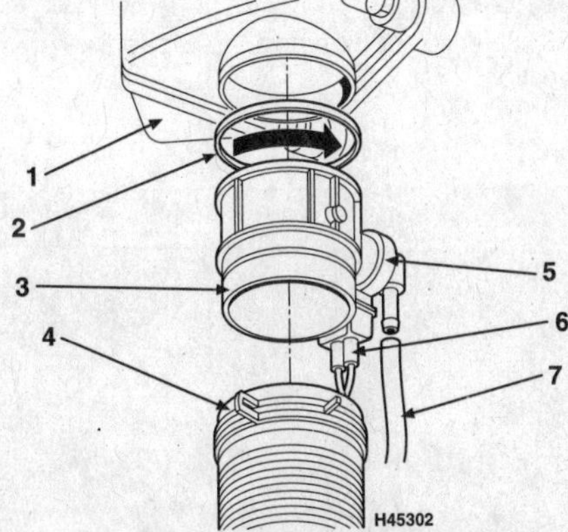

21 Twist the throttle valve housing towards the front of the car to release it from its bayonet fitting, and remove it from the engine compartment. Recover the O-ring seal from the inlet manifold cover.

22 Refitting is a reversal of removal. Renew the O-ring seal if it appears damaged.

Clutch and brake pedal switches

23 The clutch and brake pedal switches are clipped to mounting brackets directly above their respective pedals.

24 The brake pedal switch operates as a safety device, in the event of a problem with the accelerator position sensor. If the brake pedal switch is depressed while the accelerator is held at a constant position, the engine speed will drop to idle. Thus, a faulty or incorrectly-adjusted brake pedal switch may result in a running problem.

25 The clutch pedal switch operation causes the injection pump to momentarily reduce its output while the clutch is disengaged, to permit smoother gear changing.

26 To remove either switch, remove the trim panels from under the steering column area of the facia, to gain access to the pedal cluster.

27 The switches can be removed by unclipping them from their mountings and disconnecting the wiring plugs.

28 Refitting is a reversal of removal. On completion, the adjustment of the switches must be verified electronically, using dedicated test equipment – refer to a VW dealer or specialist for advice.

Electronic control unit (ECU)

29 The ECU is located centrally behind the engine compartment bulkhead, under one of the windscreen cowl panels. The unit is coded, and should not be removed without consulting a VW dealer or specialist, otherwise it may not function correctly when the multi-plug is reconnected.

Throttle valve housing details

1 Inlet manifold cover
2 O-ring seal
3 Throttle valve housing
4 Inlet air trunking
5 Vacuum unit
6 Wiring plug
7 Vacuum hose

Notes

Abbreviations and glossary of technical terms

A

A/C
Air Conditioning.

AATS
Ambient Air Temperature Sensor.

AAV
Auxiliary Air valve.

AC
Alternating Current.

An electric current that first flows in one direction and then the opposite. AC voltage is produced by an alternator or by a pulse generator. AC voltage must be rectified to DC before it can be used in the vehicle charging system. AC voltage from a pulse generator is converted to DC by an analogue to Digital converter.

ACAV
Variable Acoustic Characteristic Induction – Citroën/Peugeot term.

ACC
Air Conditioning compressor Clutch – Ford term.

ACS
Air Conditioning Switch – Ford term.

ACT
Air Charge Temperature – Ford term – see ATS.

Usually refers to a totally separate sensor, as distinct to one that is integrated into the AFS.

Actuator
A device controlled by the ECM. Examples would be injectors, ISCV etc.

Actuator driver
Refer also to driver (actuator), relay driver, control signal and final stage.

ACW
Anti Clock Wise.

Direction of rotation.

Adaptive system
An EMS that is able to learn or relearn the best setting for each application is said to be adaptive.

ADC
Analogue to Digital Converter.

Advance curve
Progressive increase in ignition timing advance as rpm increases. The curve is determined by the manufacturer, and may be mechanical or electronic in operation. It can be checked with a stroboscopic light over the rpm range. An accurate advance curve, matched to the engine, will give economical and powerful running.

AEI
See Renix.

AFM
Air Flow Meter – see AFS.

AFR
Air Fuel Ratio.

Ratio of air to fuel by weight in a vaporised charge: ie 14.7 kilos of air to 1.0 kilo of fuel. The ideal AFR for complete combustion is 14.7: 1, and the EMS (via the Lambda sensor) will be set up to maintain this ratio as closely as possible (the Lambda 'window') – see AFR / Lambda chart. The AFR in carburettor systems varies according to intake air temperature and density. This makes accurate control of the correct mixture strength virtually impossible. In contrast, the electronic fuel injection system uses a number of sensors to monitor all conditions that will affect the AFR. This means that very accurate control of the mixture is possible.

AFR/Lambda Chart
AFR = 14.7 : 1 by weight

AFR	Lambda
0.7	*10.29*
0.8	*11.76*
0.9	*13.23*
0.97	*14.26) Lambda.*
1.0	*14.70) 'window'*
1.03	*15.14)*
1.1	*16.17*
1.2	*17.64*
1.3	*19.11*

AFS
Air Flow Sensor.

A sensor that measures the volume of air being drawn into the engine and passes this information to the ECM as an electrical signal. The AFS signal determines the load factor from which the ECM looks up the appropriate injection duration. Common types of AFS would include the hot-wire or hot-film types and the mechanical vane type.

Glossary

ALDL

Assembly Line Diagnostic Link.

The name given to the serial data port used mainly on GM vehicles.

Air

Air is a mixture of nitrogen (79%), oxygen (20%), carbon dioxide (0.04%) and inert gases (0.06%).

ALT

Alternator – Ford term.

Alternator

A current generating device used in a vehicle charging system.

Ammeter

An instrument for measuring current in amperes.

Amp

Abbreviation for ampere.

A unit measurement of current flow.

Amp

Abbreviation for amplifier.

Amplifier operation

In a conventional ignition, the CB switches the negative side of the coil on and off to produce a spark. The electronic system is very similar in operation but uses (typically) a pulse generator and amplifier to achieve the same result.

A pulse generator provides the amplifier with correctly timed electrical pulses to trigger the ignition. The generated pulse is too weak to operate the switching transistor and must be amplified. The amplifier senses the trigger pulse and amplifies the voltage to the correct level to operate the switching transistor. The coil negative (-) terminal is thus switched on by the amplifier circuitry to build the magnetic field, and switched off by the switching transistor to collapse the magnetic field and induce the secondary spark. In an EMS, the amplifier may be an integral part of the ECM, or may be a separate amplifier that is switched by the ECM.

When located within the ECM, no separate wiring is required. However, if the amplifier fails, the only course of action is to renew the ECM.

A safety circuit is used to prevent coil overheating. The primary current is switched off after (typically) one second if the ignition is switched on and the engine not started. The amplifier also contains the constant energy limiting circuitry.

Amplitude

Square waveform: Difference between the maximum and minimum voltage.

AC waveform: Difference between zero and either the maximum or minimum peak. The positive amplitude is likely to be slightly greater than the negative amplitude in CAS waveforms.

Analogue signal

An analogue signal is defined as a continuous signal that can change by an infinitely small amount. Any sensor that meets these conditions can also be called an analogue sensor. Typically, an analogue signal is measured by an instrument that uses a needle to progressively sweep across a fixed scale. Any change in the signal will cause the needle to move by a similar amount. One example would be the throttle pot. As the throttle is opened and closed, the voltage output signal from the throttle pot increases and decreases and an analogue signal is passed to the ECM.

Annular coil

A type of signal generator that utilises a coiled wire magnet attached to a stator plate. The plate contains a number of magnetised upright arms equal to the number of cylinders and also equal to the number of arms on the reluctor.

API

American Petroleum Institute.

Refers to body who sets a worldwide standard for motor lubrication.

APS

Absolute Pressure Sensor – see MAP sensor.

Arcing

Unwanted electrical bridging in a circuit.

ASV

Air Switching Valve.

A vacuum switching valve – often found on Japanese vehicles.

Asynchronous

Refers usually to an injection system that is not synchronised to the ignition. Asynchronous pulses may occur at a set time interval or be intermittent in operation.

AT

Automatic Transmission.

ATA

Automatic Transmission Actuator – Ford term.

ATDC

After Top Dead Centre.

After the piston has reached TDC and is descending. Refers to retarded ignition timing OR measurement of valve timing.

ATF

Automatic Transmission Fluid.

Atmospheric pressure

The weight of atmosphere per unit area. At sea level the atmospheric pressure is 14.7 psi absolute or 102 kPa.

See Pressure conversion table under Pressure.

ATR

Automatic Transmission Relay – Ford term.

ATS

Air Temperature Sensor.

A thermistor that changes in resistance according to temperature. Most ATS usually work on the NTC principle. However, ATS fitted to Renix equipped EMS may be NTC or PTC. As the temperature changes, the thermistor resistance changes, and thus the ECM is able to calculate the air temperature from the level of voltage (or current) that is registered on the sensor signal wire.

Available HT voltage

The maximum secondary voltage that an ignition system is capable of producing.

Backprobe

A method of obtaining a voltage measurement from the multiplug pin of an electronic component or sensor. The multiplug must remain connected to the component. The multiplug insulating boot should be peeled back and the voltmeter positive probe attached to the relevant pin – ignition key on. Note: In this book, the multiplug diagram shows the terminals of the harness connector. When back-probing the multiplug (or viewing the sensor connector terminals), the terminal positions will be reversed.

Backcycling

Reversal of crankcase fumes due to a blocked PCV system.

Backfire

Noise of exploding air/ fuel charge in the intake or exhaust system. Usually occurs during cranking or deceleration. Often due to leaking exhaust, weak air/ fuel mixture, retarded timing, other ignition fault or valve related compression fault.

Back pressure

Unwanted additional pressure in the exhaust system of a running engine. Caused by a plugged exhaust and results in loss of power and efficiency.

Ballast resistor

A current compensating device that alters current flow in direct proportion to the temperature of the resistor. When used in the primary ignition circuit It serves two purposes.

1) *By providing the proper current level to a low primary resistance coil, it promotes cool coil running under all operating conditions.*
2) *When a full 12 volt by-pass supply is introduced to the coil under cranking, the coil output will be greater and starting will therefore be aided.*

The ballast resistor was mainly use in conventional CB ignition systems to compensate in part for some of the deficiencies of that system. A number of the early electronic ignition systems, that were not of the constant energy type, also utilised a ballast resistor for current control purposes. The ballast resistor can also be found in other circuits where current compensation is necessary. An example would be the fuel pump circuit on the Lucas LH system fitted to some Rover fuel injected systems and in the voltage supply to the fuel injector solenoid on some early systems.

Banked or simultaneous injection

Injectors connected in parallel circuits. In some four cylinder engines the four injectors may be connected together so that they all open at the same moment. In other four cylinder systems, the four injectors are connected in two banks of two cylinders. However, all injectors may still be actuated simultaneously. In a six cylinder engine the six injectors may be connected in two banks of three cylinders, In an eight cylinder engine the eight injectors may be connected in two banks of four cylinders, In a twelve cylinder engine the twelve injectors may be connected in four banks of three cylinders.

Bar

A unit of pressure. One bar is almost equal to atmospheric pressure. See pressure conversion table under Pressure.

Barometric pressure

Equal to atmospheric pressure. At sea-level, atmospheric pressure is 100 kPa. See pressure conversion table under MAP.

Battery

A storage device for electrical energy in chemical form. The primary function of the battery is to provide ignition current during the starting period and power to operate the starter motor. This must be accomplished irrespective of adverse temperature conditions. The battery also serves, for a limited time, as a current source to satisfy the electrical demands of the vehicle which are in excess of the generator output. A battery is now rated in terms of its CSP or CCA. There are three Internationally known standards and virtually all batteries in Europe, the Far East and the Americas are marked in accordance with these standards.

The three standards are:

BS / IEC:

The current in amperes that can be drawn from the battery for one minute at -18°C before cell voltage drops below 1.4 volts per cell. (BS3911 1982).

SAE / BCI:

The current in amperes that can be drawn from the battery for thirty seconds at -18°C before cell voltage drops below 1.2 volts per cell. (SAE J537 June 1982).

DIN:

The current in amperes that can be drawn from the battery for thirty seconds at -18°C before cell voltage drops below 1.5 volts per cell. (DIN 425-39 PT2 1983).

BBDC

Before Bottom Dead Centre.

BCI

See Battery.

An SAE standard for measuring battery performance.

BDC

Bottom Dead Centre.

Piston at exact bottom of stroke.

BHP

Brake Horse Power.

Measurement of engine power.

Bimetallic

Two strips of different metals bonded together. When heated or cooled the different expansion rates of the metals cause the strips to bend. Often used to operate a carburettor automatic choke.

Blowby

Combustion fumes that are blown past the piston rings during the power stroke. Worn or gummed pistons and rings will admit excessive blowby.

BOB

Break Out Box.

BOO

Brake On / Off switch – Ford term.

BPC

Boost Pressure Control solenoid (turbo) – Ford term.

Breather

As the pistons ascend and descend the crankcase air is displaced and proper ventilation via a breather is required if a pressurised crankcase is to be avoided.

BS / IEC

See Battery.

A standard for measuring battery performance.

BTDC

Before Top Dead Centre.

Ascending piston before top of stroke. Refers to advanced ignition timing OR measurement of valve timing.

Burn time

See spark duration.

C

°C

Celsius or Centigrade.

Measurement of temperature. Temperature conversion scale.

Cable

Heavy electrical wire used to conduct high voltage or high current. ie spark plug cables or battery cables.

Calibrate

The setting of an instrument to a base zero for greatest accuracy.

CANP

CANister Purge solenoid valve.

Capacitor

A device that absorbs electricity by offering an alternative path.

Carbon brush

A brush centred in the distributor cap. It contacts the rotor and passes secondary voltage for distribution to the correct cylinder.

Carburettor

A device that atomises air and fuel in the correct proportion over the engine operating range.

CAS

Crank Angle Sensor.

The CAS works on the same principle as the inductive permanent magnet pick-up. A number of steel pegs or pins are set at regular intervals around the circumference of the flywheel or crankshaft. Typically, a peg may be set at every 10° – 36 in all. One or more of the pegs may be removed at TDC, BDC or a known distance from these points. The flywheel thus becomes a reluctor.

A permanent magnet inductive signal generator is mounted in close proximity to the flywheel. As the flywheel spins, an alternating (AC) waveform is produced which signals RPM and flywheel position to the ECM. Although most modern systems utilise a single CAS, some of the older systems use two CAS – one for RPM and one for position. The waveform produced by a each type of CAS will be slightly different.

Cat

Catalytic converter.

Since January 1993, all new UK vehicles are required to be fitted with a catalytic converter. A catalyst is something which promotes a reaction, but itself remains unaffected by the reaction. The catalytic converter consists of a stainless steel housing containing a ceramic monolith with a honeycomb of passages called cells.

CB

Contact Breaker points.

A switch which turns the flow of primary current on and off to induce ignition. Used in vehicles that utilise conventional ignition systems.

CCA

Cold Cranking Amps – see battery.

A standard for measuring battery performance.

CCMC

The EC standard for lubricating oil, it is the European equivalent to API.

CCO

Clutch Converter lock-up solenoid – Ford term.

CD

Constant Depression.

A type of carburettor that uses a piston to vary the choke area according to speed and load. This results in a constant air velocity and depression over the jet at all engine speeds and does away with the need for separate compensating and idle jets.

Cell

A single battery unit consisting of positive and negative plates covered by Electrolyte. A twelve volt battery consists of six cells connected in series and measuring approximately 2.1 volts each.

Glossary

CEL
Check Engine Light – diagnostic warning lamp.

Celsius
See °C.

Centrifugal advance
See Mechanical advance.

Ceramic block
Insulating block used in construction of a certain type of ballast resistor.

CFI
Central Fuel Injection – Ford/Vauxhall term – see Spi.

CFCV
Carbon Filter Control Valve.

Mechanically operated valve used in the evaporation control system to control recycling of vapours from the carbon canister.

CFSV
Carbon Filter Solenoid Valve.

Electrically operated solenoid valve used in the evaporation control system to control recycling of vapours from the carbon canister. Sometimes termed a purge valve.

CFCOSV
Carbon Filter Cut-off Solenoid Valve.

This valve is often fitted to Peugeot and Citroën vehicles. The CFCOSV is actuated by the ignition key and is used in conjunction with the CFSV. When the ignition is switched off, the valve closes and fumes are retained in the system.

CID
Cylinder IDentification camshaft sensor – Ford term.

CIH
Cam In Head.

A camshaft located in the cylinder head. Differs from the OHC design in that the camshaft uses short pushrods to open the valves.

Circuit
An electrical path through which current can flow and that begins and ends at the current source. A circuit is NOT complete unless the current can return to the source. In modern systems the current flows from the positive terminal of the battery, through wires or cables and switches to the load (ie a starter motor). The return is through earth to the negative terminal of the battery.

CIS
Continuous Injection System.

A Bosch fuel injection system where the injectors spray fuel continuously during running. Another name for Bosch K-Jetronic.

Clear flood
Usually, if a fuel injected vehicle fails to start, continued cranking will result in excessive fuel being injected into the cylinders. Where the fuel injection system has a 'clear flood' facility, fully opening the throttle will result in a reduced volume of fuel being injected whilst the engine is cranked.

Closed loop control
An engine with a Lambda sensor is known as a closed loop engine. This is because of the measure of tight engine control about the stoichiometric point.

CLT
Cylinder Leakage Tester.

A gauge for measuring compression loss. Compressed air is pumped into the cylinder under test with the piston at TDC on the compression stroke. The rate of loss is recorded upon the CLT gauge. The higher the rate of loss, the less efficient is the cylinder. By listening at the carburettor intake, exhaust pipe, oil filler cap and radiator cap – the reason for the high loss can usually be determined.

CMH
Cold Mixture Heater Plate.

See Manifold heater.

CO
Carbon Monoxide.

Carbon Monoxide is formed by the partial burning of the fuel due to a lack of oxygen. A low proportion of CO indicates how well the air/fuel ratio is maintained. A high proportion of CO in the exhaust indicates a rich fuel mixture, choked air filter, choked PCV valve or low idle speed. Low CO would indicate a lean fuel mixture or a vacuum leak or even an exhaust leak. CO (and HC) emissions decrease as load (heat and temperature) rises to give a more efficient engine.

The CO content in the exhaust is an indicator of the AFR, but only when the engine is functioning normally. Any engine condition that causes a misfire will reduce the amount of CO that is burnt. CO is produced from the partial burning of fuel and if you don't burn the fuel you wont produce the CO.

It is therefore obvious that an engine with a burnt valve or a dead spark plug would produce less CO. In this instance, any attempt to adjust the fuel mixture would result in an over-rich mixture even though the Gas Analyser indicated a lean mixture.

Only a Gas Analysers that could calculate a 'corrected' CO reading would provide the complete picture. It is essential therefore that all mechanical and ignition faults are corrected before making fuel adjustments.

CO is a very poisonous, tasteless, colourless and odourless gas. It is a serious health hazard in dense traffic in cities and semi-enclosed areas. A concentration of 0.3% can be fatal if breathed in continually for 30 minutes. CO combines with red blood cells at the expense of oxygen and causes suffocation. By weight CO accounts for about 47% of air pollution but is thought to have little effect on the environment.

One molecule of CO contains one atom of carbon and one of oxygen and it is measured in % volume. CO is inversely proportional to the AFR, the less fuel the lower the CO.

CO_2
Carbon Dioxide.

CO_2 is the product of an efficient engine. With low CO and HC levels, the percentage of CO_2 in the exhaust is likely to be 13 to 15%. Less than 8% CO_2 indicates an incorrect AFR, or a misfire or a leaky exhaust. CO_2 is directly proportional to the AFR, but inversely proportional to CO. The less fuel, the higher the CO_2. At speeds over 2000 rpm, the level will be 1-2% higher than at idle due to an increase in engine efficiency.

One molecule of CO_2 contains one atom of carbon and two of oxygen. CO_2 is chemically stable and does not easily react with other substances. Not poisonous it is produced by all breathing animals including fish. Oxygen is inhaled and CO_2 exhaled at a concentration of about 5%. CO_2 is absorbed by all green plants by a process called 'photo-synthesis', which only happens in daylight and which also releases oxygen (O_2) into the atmosphere.

Any burning process produces CO_2 and the contribution from automotive sources is said to be less than half that of industrial and domestic sources. The contribution from people and animals is insignificant. A heavy concentration of CO_2 is like placing a blanket over the atmosphere and this prevents heat loss by radiation. At present, more CO_2 is being produced than is being consumed and the disappearance of the rain forests is another significant factor. As the forests fade away, less CO_2 is absorbed and the increase in atmospheric concentration is said to contribute towards global warming and the so-called 'greenhouse effect'.

In the automobile, the only way to produce less CO_2 is to burn less petrol or even none at all. This means an efficient engine with good economy (a lean-burn engine), or a diesel engine with high economy or even a car with no engine – ie an electric motor. But electric cars need electricity, and electricity is produced by power stations, and power stations also produce CO_2 . . .

Coil
A device that transforms low (battery) voltage into the high voltage required to bridge the rotor and spark plug gaps.

Cold start device
A choke or a starting device to enrich the air/fuel ratio during cold starting and during the warm-up period.

Combustion
During the combustion process, oxygen combines with carbon to form carbon dioxide (CO_2), and with hydrogen to form water (H_2O). If the air and fuel were homogenised prior to combustion and all the petrol was completely burnt then the ideal engine would inhale a perfect mixture of fuel and air and exhale Carbon Dioxide (CO_2) and water (H_2O). For every gallon of petrol that an engine consumes, a gallon of water is produced. When the engine is at normal operating temperature, this water is exhausted as steam.

Unfortunately, this ideal engine does not exist and for a number of reasons incomplete combustion occurs in all engines to a degree in even the most efficient engine. In addition to CO_2 and H_2O the exhaust thus contains varying quantities of CO, HC, O_2 and NOx. Some of these gases are harmless such as CO_2, H_2O and O_2, whilst others ie HC, CO and NOx are atmospheric pollutants. A less efficient engine will exhaust a greater volume of the more harmful pollutants. A regular tune-up and gas analysis should reduce pollution to acceptable levels. However, the more efficient the engine, the more CO_2 will be exhausted.

Common rail

Type of diesel fuel system where fuel at high pressure is delivered to the injectors from a high-pressure accumulator, known as a common rail.

Compression

The charging of a maximum volume of air and fuel into a minimum volume.

Compression tester

A gauge to measure the compression pressure of an engine and usually graduated in bars or psi.

Conductor

A material that will pass electrical current efficiently. A good conductor depends on material used, length, cross sectional area and temperature.

Constant energy

The use of high primary current limited, to a pre-set figure, for efficient electronic ignition operation.

Constant energy operation (electronic ignition)

Electronic ignition, used with a coil of lower primary resistance, produces increased primary current resulting in increased coil output. Coupled with higher coil energy, this will produce an arc of longer duration at the spark plug, and enable a weaker mixture to be ignited with a corresponding improvement in economy and exhaust emissions. Improved reliability, better control of timing and longer periods between 'tune-ups' are other benefits over the conventional CB system. Virtually all modern types of electronic ignition use the variable dwell current limiting constant energy system.

Control signal

See also Relay driver, Driver, Final stage.

Conventional ignition system

A system that uses CB and a condenser to induce ignition. Replaced generally by electronic ignition in recent years.

Conversion tables

See Pressure conversion table, Vacuum conversion table, Temperature conversion table.

Coolant

A mixture of water, anti-freeze and corrosion inhibitor to allow efficient cooling system operation.

Cooling system

The energy produced by combustion generates tremendous amounts of heat. About 25% of this heat energy is turned into the power that drives the road wheels. A further 50% is expelled with the exhaust gases which leaves approximately 25%. It is the function of the cooling system to dissipate this excess heat.

All water cooled engines should use a mixture of anti-freeze and water known as a coolant. The coolant mixture is usually 40-50% by volume and this will give protection down to about -40°C. Water alone will affect the temperature rating of all thermal engine temperature sensors and can cause incorrect metering of fuel in electronic carburettor and electronic fuel injection systems.

Corrected CO reading

A calculation that takes improper combustion into consideration. If the corrected CO and CO readings are dramatically different, then the engine has a combustion problem.

Corrosion

Deterioration and crumbling of a component by chemical action. Sensor terminals and multiplugs are particularly susceptible to this complaint.

Corrosion inhibitor

A preparation to prevent corrosion. Often used to prevent corrosion of the radiator internal channels by water action.

CPS

Crankshaft Position Sensor.

Ford term corresponding to CAS.

See CAS.

CPU

Central processing unit.

Cranking

Rotating the engine by use of the starter motor.

CSP

Cold Start Performance.

A standard for measuring battery performance.

CSI

Cold start injector.

See Cold Start Valve.

CSV

Cold Start Valve.

An injector mounted in the inlet manifold that is only actuated during cold starts when the engine temperature is below a certain threshold. Usually earthed through a TTS that completes the CSV earth path. The CSV is not fitted to modern EMSs.

CTS

Coolant Temperature Sensor.

A thermistor that changes in resistance according to temperature. Most CTS usually work on the NTC principle. However, CTS fitted to Renix equipped EMS may be NTC or PTC. As the temperature changes, the thermistor resistance changes, and thus the ECM is able to calculate the engine coolant temperature from the level of voltage (or current) that is registered on the sensor signal wire.

Current

The flow of electrons through a conductor and measured in amps.

Current controlled or pulse modulation injection

See EFi systems.

CVH

Compound Valve angle Head – Ford term.

Cylinder head with valves arranged in two planes in a V configuration.

CW

Clockwise.

Direction of rotation.

Cylinder balance

See Power balance.

Cylinder contribution

A method of comparing the relative power output of each cylinder without removing the spark as in conventional power balance operation. The acceleration time between each pair of ignition sparks is compared. A cylinder that is less efficient will have a lower acceleration time compared with the other cylinders. This method is much safer on catalyst equipped vehicles.

D

Dashpot

A device that enables the throttle to close slowly rather than suddenly, thus preventing the removal of droplets of fuel from the inlet manifold walls due to the high vacuum present during deceleration. These droplets are emitted as excess HC during this operation.

Datastream

Once the FCR has decoded a fault, a Datastream enquiry (some systems only) is a quick method of determining where the fault might lie. This data may take various forms but is essentially electrical data on voltage, frequency, dwell or pulse duration, temperature etc, provided by the various sensors and actuators. Unfortunately, such data is not available from all vehicle systems.

Since the data is in real time, various tests can be made and the response of the sensor or actuator evaluated.

Actuating the idle control valve, relays and injectors through the ECM is an excellent method of testing effectiveness of the actuator and associated wiring circuit.

DC

Direct Current.

An electrical current source which flows in only one direction.

Glossary

DC – ISC
Throttle plate control motor – Ford term.

Decarbonisation
Removing the cylinder head and scraping away the accumulated levels of carbon build-up from the head, valves and tops of pistons.

Deceleration
Closing the throttle and allowing the engine speed to reduce to idle.

Decoke
Abbreviated term for Decarbonisation.

Degree
1/360 part of a circle.

DEI
De-ice switch – Ford term.

Detonation
Refer to Knock.

Diaphragm
A thin sheet of rubber that is moved by vacuum to actuate a mechanical device.

Digital signal
A digital signal is represented by a code that has two states, on and off. In simple terms, the signal consists of a series of digital pulses when the frequency, pulse width or number of pulses is used to indicate a specific value.

Because the ECM works in a digital fashion, all analogue signals must pass through an analogue to digital converter when the signal will be stored by the ECM in digital format. A digital signal from a digital sensor does not need converting, and processing by the ECM is therefore much faster.

DIN
International standard used in the automotive industry.

Diesel
A fuel injected engine that uses the high temperature generated in compression to ignite the charge.

Dieseling
A fault condition where a petrol engine continues running after the ignition has been switched off. Often caused by cylinder hot spots or carbon deposits that continue to glow and which are hot enough to explode the air/fuel charge. May be cured by a decoke.

Differential pressure
The method by which air is drawn through a carburettor and into an engine. By the rules of physics, air will flow from high (atmospheric) pressure to low pressure (depression caused by fall of piston).

Diode
A transistor that allows current flow in one direction alone.

DIS
Direct Ignition System sometimes termed DIStributorless.

An ignition system where a distributor is not used.

DIS
Alternate Digital Idle Stabiliser.

Distributor
A component that distributes the secondary voltage to the correct spark plug in firing order. It is also used to house and operate the CB mechanism in conventional ignition and the reluctor and stator in some electronic ignition systems. The distributor turns at the same speed as the camshaft and at half the speed of the flywheel.

Distributor cam
Located in the distributor, it is mounted upon the distributor shaft and contains a number of peaks (equivalent to the number of cylinders). The distributor can often be adjusted to time the ignition and distributes the HT spark to the correct spark plug in firing order.

Distributor cap
An insulated cap with a centre tower and a circular series of terminals, one for each cylinder. The secondary HT pulses travel from the coil to the centre tower and are delivered, in firing order, to each terminal by the rotor.

DME Digital Motor Electronics
Generic term often used to describe the Bosch Motronic EMS. The term is used in particular by BMW.

DMM
Digital Multi-Meter.

An instrument designed for automotive use that can measure voltage, current, resistance and sometimes covers other functions such as dwell, duty cycle, frequency, rpm and amperage etc.

DOHC
Double Over Head Camshaft.

A set of two camshafts mounted in the cylinder head. Operation is similar to the SOHC type except that one of the camshafts opens the inlet valves, and the other one opens the exhaust valves. This leads to more efficient valve operation and improved engine efficiency.

DPFE
Delta Pressure Feedback Electronic system – Ford term.

An ECM controlled valve that regulates the flow of exhaust gas to the EGR valve.

Driver (actuator)
Refer also to Relay driver, Control signal and Final stage.

The system actuators are supplied with a voltage feed from either the ignition switch or from one of the EFi system relays. The earth connection is then connected to an ECM earth pin. When the ECM actuates the component, it drives the appropriate ECM pin to earth by completing the circuit internally for as long as the actuation is required. In general, the earth connection will only be completed after the ECM has received one or more signals from relevant sensors and either looked up the associated maps or calculated the correct actuator 'on' time.

This signal could be termed a 'driver' or a 'final stage' or a 'control' signal. In the this book we have settled for the 'driver' term. Examples of an actuator driver are: injector, relay, ISCV, CFSV, EGR solenoid etc.

DTR
Distributor – Ford term.

Duty cycle
The period of time in % or ms, during which a component is switched on or energised. By connecting the dwell meter between the pulse terminal and earth on actuators such as the coil, ignition module, hall-effect switch, injector, ISCV or in fact any other switchable device, a duty cycle may be obtained. By comparing this dwell with known operating parameters, correct operation of the device can be determined. Refer to dwell below for more information.

DVM
Digital VoltMeter.

Dwell
Traditionally, a dwell angle is defined as being the number of degrees through which a distributor rotates when the CB points are closed. However, in modern times we should consider the wider context of the meaning of 'dwell'. A good definition of a dwell angle would be the time or rotational period through which a device passes when it is energised.

Dwell could thus be measured in terms of degrees of rotation, time 'on' – (or off) in percentage (%) compared with the total time for one occurrence, or time on or off in milliseconds (ms). All we need is the appropriate meter. Usually, dwell is measured in degrees, but if we use either (%) or (ms), it is more common to refer to duty cycle.

To convert dwell degrees to dwell percent and vice versa, use the following formulae:

Dwell° x (CYLS/360) x 100 = Dwell%
ie 45° x (4/360) x 100 = 50%
(Dwell% / 100) x (360/CYLS) = Dwell°
ie (50% /100) x (360/6) = 30°

Dwell angle
Number of rotational degrees during which a device is switched 'on'. Normally used in reference to CB points. and refers to the degrees of rotation of the distributor cam whilst the points are closed. See also Duty Cycle.

Dwell meter
An instrument used to measure dwell angle.

Dwell variation
The difference in dwell reading taken at any two different engine speeds. Normally refers to CB equipped distributors.

Dynamic testing
The testing of a device whilst it is running under load as opposed to static testing.

Dynamic volt drop
In vehicles with electronic ignition, the dynamic volt drop refers to the volt drop over the primary circuit from the coil negative terminal to earth

through the amplifier final driver stage. Equivalent to the so-called distributor resistance or distributor volt drop in CB systems. This measurement is only available with the engine cranking or running, because current must be flowing in the circuit before a measurement can be taken. Not all DMMs are capable of measuring this circuit.

E

Earth
A path for current to return to the power source.

Earthing probe
A tool used to test for current leaks. Often used to test for secondary voltage insulation faults.

EACV
Electronic Air Control Valve Honda, Rover Term.

EAI
Electronic Advance Ignition GM term.

EBCV
Electronic Air Bleed Control Valve.

EC
European Community.

ECM
Electronic control module.

A computer control unit that assimilates information from various sensors and computes an output. Can be used to control the engine ignition timing, injection duration, opening of the ISCV, ABS brakes, air bag etc etc.

ECR
Electric Choke Resistor – Ford term.

ECT
Engine Coolant Temperature.
Refer to CTS. Ford term corresponding to CTS.

ECOTEC
Emission Consumption Optimised TECnology.
GM term Applied to the late series of engines.

ECU
Electronic Control Unit.
Refer to ECM.

EDIS
Electronic Distributorless Ignition System – Ford term.

EDIS-4
Applied to EDIS – 4 cylinder engines.

EDF
Electro Drive Fan – Ford term.

EDM
EDIS Diagnostics Monitor Signal – Ford term.

EEC
Electronic Engine Control – Ford term.

EEC IV
Electronic Engine Control 4th generation – Ford term.

EFi
Electronic Fuel injection.

A fuel injection system where the injectors are opened (pulsed) by an ECM. There are a several different kinds of injection system in current use and a description of each type now follows.

EFi systems

Current controlled or pulse modulation injection
Some systems rely on the principle that more current is required to open the injector than to actually keep it open. The injector solenoid circuit is also earthed – but for only about one millisecond, which is just long enough to open the injector. The opening circuit is then switched off, and another circuit rapidly closed and opened, to apply a small holding current to the injector so that it remains open for the correct time duration. The pulsing is so blindingly fast that the injector does not have time to close, and current flow is therefore much reduced. This type of system is known as current controlled or pulse modulated. A variation of this method is termed 'Peak and hold'. After the initial pulse, the second circuit holds the injector open without rapid pulse modulation.

Standard injection
In the standard EFi system, the injector is simply earthed by the ECM for a calculated period of time. During this time, known as the pulse duration or the injector 'on' time, the injector solenoid operates and fuel is injected.

EFi pulse duration
The period of time that the injector is held open. Can be measured in ms or by a dwell meter as a duty cycle.

EGOS
Exhaust Gas Oxygen Sensor.
See Oxygen Sensor.

EGR
Exhaust Gas Re-circulation.

A method of recycling a small amount of exhaust gas into the intake system. This leads to lower peak combustion temperature with a reduction in NOx emissions.

EHPR
Electro-Hydraulic Pressure Regulator – Ford term.

Electrode
An electrical conductor.

Electrolyte
A sulphuric acid and water solution used in a lead / acid batteries. Chemical reaction between the acid and battery plates produce voltage and current.

Electro-magnet
A magnet that requires electrical energy to create an electrical field.

Electronic 'MAP'
See Mapped timing / injection advance.

EI
Electronic Ignition.

An ignition system that uses a magnetic sensor and transistors to switch the coil negative terminal on and off.

Emissions
Pollution of the atmosphere by fumes from the exhaust, breather vent or fuel tank.

Emission standards

__US 79:__ This standard was set in the USA in 1979 and has been superseded by the US83 standard. The vehicle must be equipped with a regulated three way catalyst with OS.

__US 83:__ This is the most stringent of the current European emission levels and the standard was set in the USA in 1983. The vehicle must be equipped with a regulated three way catalyst with OS and evaporative emission control.

__US 88:__ LDT (Light Duty Truck): This standard sets the same requirements as the US83 standard. However, commercial vehicles over a certain weight will fall into this category.

NEEC 5th amendment: This is an European standard for emission control and vehicles equipped with at least one of the following systems will meet the standard.

__15.04:__ This is not a standard and is a category applied to vehicles that do not meet a particular emission standard. Vehicles without a catalytic converter, EGR, pulse air system or evaporative emission control will fall into this category.

Emission control
Devices used to control and minimise poisonous fume emissions.

EMR
Engine Management Relay – Ford term.

EMS
Engine Management System.

An EMS is essentially an electronic system whereby the engine ignition and fuelling functions are controlled by one or more ECM. The distributor, when used, is provided purely to distribute the HT spark to the correct cylinder in firing order. When separate ECMs are provided for ignition and fuelling, the two units do not operate independently and connections are made so that they can communicate with each other. An Ignition ECM would signal the Injection ECM to initiate injection. In some EMSs, a separate amplifier is pulsed by the ECM, which in turn switches the coil negative to initiate ignition.

Energised
The period during which an electrical device is switched on.

Engine sensor
See Sensors.

ENR
ENgine run Relay – Ford term.

Glossary

EOBD
European On-Board Diagnostics.

EPT
Electronic Pressure Transducer – Ford term.

ERIC
Electronically Regulated Ignition & Carburettor – Rover term.

ESA
Electronic Spark Advance – Toyota term.

ESC
Electronic Spark Control – Ford term.

ESC II
Electronic Spark Control 2nd generation module – Ford term.

EVAP
EVAPorative emission control systems – Ford term.

EVR
Electronic Vacuum Regulator – Ford term.

Exhaust gas
Burned and unburned gases that are exhausted after combustion.

External influences
An influence that is not directly attributable to a particular component but could affect the operation of that component.

F

Fahrenheit
Temperature scale.

Fast codes
Digital fault codes emitted by an EMS that are too fast to be displayed on an LED lamp or on a dash mounted warning lamp. A digital FCR instrument is required for capturing fast codes.

Fault codes
Electronics are now extensively used throughout the modern vehicle and may control functions such as the transmission, suspension, automatic gearbox, air conditioning and myriad others.

Most modern vehicle EMS or ECM have the facility of making self-diagnostic checks upon the sensors and actuators that interface with the vehicle computer or computers. A fault in one of the component circuits causes a flag or code to be set in the ECM memory.

If a suitable code reading device is attached to the serial port on the vehicle harness, these faults can then be read out from the vehicle computer and displayed in the form of a two or three digit output code.

FCR
Fault Code Reader.

A device that can be connected to the vehicle serial (diagnostic port) to interrogate the vehicle ECM. Fault codes and Datastream information can then be read from the ECM. In some instances, vehicle actuators can be actuated from the controls on the FCR. A FCR is sometimes termed a scanner.

Where adjustments to the ignition timing or fuel system are possible, for example on some Ford or Rover systems, then these adjustments must be made through an FCR.

The codes may be described as slow or fast and some ECMs are capable of emitting both types. Slow codes can be captured by an LED tool, whereas fast codes must be captured by a digital FCR. Future EMSs are more likely to utilise fast codes.

FI
Fuel injection.

Final stage
See Driver, Relay driver and Control signal.

FIR
Fuel Injection Relay – Ford term.

Firing line
The actual firing voltage as represented on an oscilloscope.

Firing order
The order in which the cylinders are fired.

Firing voltage
The secondary voltage required to overcome the rotor and spark plug gaps.

Fixed jet
A type of carburettor that uses fixed and calibrated fuel, air and compensating jets to ensure correct air / fuel mixture over the engine operating range.

Flash codes
Fault codes of the slow variety that are output on a dash mounted warning lamp or via an LED lamp.

Flashshield
A cover used in the distributor to prevent secondary arcing interfering with primary operation.

Flat spots
Hesitation of the engine under acceleration.

Flow rate
Describes the volume of fuel pumped during a pre-determined period of time in order to test fuel system output.

Flywheel sensor
See CAS.

FLW
Fuse Link Wire – Ford term.

FO
Fuel Octane – Ford term.

Fouled spark plug
Formation of deposits on the electrodes. This often cause a short to earth of the secondary voltage and the charge is not correctly fired.

FP
Fuel Pump.

FPR
Fuel Pump Relay – Ford term.

Frequency
Pulse Frequency. Usually measured in Hz.

FRS
Fuel Restrictor Solenoid – Rover term.

FRTS
Fuel Rail Temperature Sensor.

FSOR
Fuel Shut-Off Relay – Ford term.

A sensor to measure the temperature of fuel in the fuel rail.

FSOS
Fuel shut off solenoid – Ford term.

FT
Fuel Temperature sensor.

FTS
Fuel Temperature Switch.

FTVV
Fuel Tank Vent Valve.

A solenoid valve used to control evaporation emissions in GM vehicles.

Fuel atomisation
Proper mixing of air and fuel to ensure good combustion.

Fuel injector (EFi systems)
The injector is a solenoid operated valve that delivers an exact amount of fuel according to an opening duration signal from the ECM. A fine filter is used to prevent debris damaging the precision action. However, gums and lacquers can build-up on this filter and on the injection pintle eventually reducing fuel flow. Injection fouling is a serious problem on many injection systems.

Fuse
A small component containing a sliver of metal that is inserted into a circuit. The fuse will blow at a specified current rating, in order to protect the circuit from voltage overload.

Fuselink
Also known as fusible link.

A heavy duty circuit protection component that can burnout if the circuit becomes overloaded.

G

Gas analyser
A device used to sample gases at the exhaust pipe, so that an analysis may be made of the exhaust constituents.

Generator
An alternator or dynamo that produces voltage and current. See also alternator.

GM
General Motors.

Manufacturer of Opel and Vauxhall in Europe. The parent company is based in the USA.

GND

Ground.

USA term for earth.

See also Earth.

Hall-Effect generator

A type of pulse generator which returns a small digital voltage to trigger the coil negative.

Hall-Effect Switch operation

A constant 12 volt supply is passed through the Hall Effect switch in the distributor. Opposite the Hall switch is a magnet whose field causes the Hall switch to return a small voltage back to the amplifier. Attached to the distributor shaft is a rotor vane with the same number of cut-outs as cylinders. Passing the rotor between the switch and the magnet will cause the switch to be turned off and on. As the cut-out space proceeds past the switch then a voltage is returned to the amplifier. When the solid portion comes between the switch and magnet then the voltage is turned off as the magnetic field is diverted. The number of voltages returned per four-stroke engine cycle will equal the number of cut-outs.

Hard faults

Generally refers to faults logged by an ECM self-diagnosis routine. The faults are usually present at the moment of testing,.

HC

High compression engine.

HC

Hydrocarbons.

Typically 15% Hydrogen and 85% carbon (petrol is almost pure hydrocarbons). HC is a generic term and refers to unburnt fuel and partially burnt fuel; it is measured in PPM – parts per million).

There are many different kinds of HC in the exhaust, and HC is generally capable of serious damage to eyes, nose and lungs. When mixed with NOx and in the presence of bright sunshine, photochemical smog is formed. HC emissions are also said to be a reason for the death of the rainforests.

During combustion, the Hydrogen atoms combine with the O_2 atoms to produce H_2O. The Carbon atoms combine with O_2 atoms to produce CO_2. High levels of HC in the exhaust signifies ignition problems such as defective plugs or HT leads, incorrect timing, vacuum leaks, incorrect air/fuel ratio or engine mechanical faults. In fact, anything that causes inefficient engine operation will increase the level of unburnt HC in the exhaust.

As the AFR weakens, the HC emissions increase due to a lean misfire. This is why a black exhaust is often the result of a too-lean idle mixture. Careful design of the combustion chamber can overcome this problem.

HCS

High Compression Swirl – Ford term.

Refers to a Ford design of piston/combustion chamber, intended to promote efficient combustion.

Heat range

With reference to a spark plug, the operating range in which the plug will safely and effectively operate.

Heat sink

A component to dissipate high operating temperatures.

HEDF

High speed Electro Drive Fan – Ford term.

HEGOG

see HEGOS Heated Exhaust Gas Oxygen sensor Ground – Ford term.

HEGOS

HEGO Heated Exhaust Gas Oxygen Sensor.

The oxygen sensor will not operate under 250° C, and so it may be equipped with a heater to provide a faster warm-up. Heated oxygen sensors are often called HEGOS. The heater element is a resistance within the oxygen sensor, with a battery voltage supply from the fuel system relay and a connection to earth. See also oxygen sensor.

HES

Hall Effect Switch (or Sensor).

HG

Chemical symbol for measurement of mercury.

HLG

Hall effect generator.

Hot-film AFS

Very similar in operation to the hot-wire sensor.

Hot-wire AFS operation

The hot-wire AFS is mounted in the air flow trunking between the air filter and the engine. A box containing the electronics for hot-wire operation sits over the AFS body. A voltage of either 5 or 12 volts, according to system, is applied to the AFS unit.

Hot-wire sensor

A type of AFS in which the resistance of an electrically heated wire is measured. The hot-wire AFS is becoming increasingly popular as an alternative to the Vane and MAP sensor types. This is because the volume, temperature and density of air, at all altitudes can be more accurately measured than by the Vane or MAP sensor systems.

Air passes through the AFS body into the engine. A small quantity of air is drawn into a by-pass channel containing two wires. These wires are known as the sensing wire and the compensating wire. A small current is applied to the compensating wire which remains unheated. As air passes over the wire, its resistance and current change and the AFS is able to determine the temperature of the incoming air. The sensing wire is heated to a temperature of 100° C above that of the compensating wire. Air passing over the sensing wire causes it to become cooler and its current and resistance value change. More current is passed through the sensing wire so that it remains 100° C above that of the compensating wire. An output (signal) voltage, proportional to the current applied to the sensing wire, is returned to the ECM.

The value of this voltage is directly related to the volume, temperature and density of air introduced into the engine. The hot-wire system thus allows automatic compensation for altitudes from sea-level to mountain top, and the ECM will accurately calculate the AFR under virtually all conditions.

HT High tension

High voltage induced in the secondary windings of the ignition coil.

HT lead

High Tension lead.

Cable used to distribute secondary ignition to the distributor cap and to the spark plugs.

Hydrogen

Odourless highly-explosive gas. Forms two-thirds of the chemical make-up of water.

Hz

Hertz.

Frequency in cycles per second.

I

IA

Idle Adjust – Ford term.

IBR

Injector Ballast Resistor – Ford term.

ID

Identification.

Idle speed adjustment (EFi systems)

Most modern vehicles have fully automatic idle control with no means of adjustment. Where adjustment is possible, this is usually effected by a by-pass idle speed air screw. Turning the screw one way will reduce the air flow and therefore the idle speed. Turning the screw the other way will increase the air flow and therefore the idle speed.

Although most later systems use an ECM controlled ISCV or stepper motor to maintain idle speed under engine load, some versions of the ISCV or stepper motor may be adjustable. Generally, this only possible with early units.

Idle speed control

Idle speed control devices are actuated by the ECM and are therefore known as actuators – as distinct to sensors. On most modern engines, speed at idle is maintained at a constant speed irrespective of engine load or temperature. As idle conditions alter, or a temperature or an electrical load condition occurs, the ECM actuates either a solenoid controlled ISCV or a stepper motor to maintain the correct idle position – no matter the speed or load. This prevents poor idle and stalling with heavy electrical loads and a lean mixture. Some versions of the ISCV or stepper motor may be adjustable, although generally this only possible with early units.

Idle-up

Far Eastern term.

Any mechanical or electronic system that is used to increase the idle speed according to temperature or engine load could be termed an idle-up system.

IDM

Ignition Diagnostics Monitor signal – Ford term.

IGC

Ignition coil Ford term.

IGf

Ignition confirmation signal – Toyota term.

IGN

Ignition switch – Ford term.

Igniter

Ignition module amplifier.

Term used by Far Eastern vehicle manufacturers to describe the ignition amplifier.

Ignition module

Term used to describe the ignition amplifier.

Ignition switch

An on-off switch that provides current to the primary ignition circuit. When the switch is closed and the module is switched on (or the CB is closed), then current will flow through the primary circuit and return to the battery via the engine and body earths.

Ignition timing

The correct moment to ignite the compressed air/fuel charge for maximum downforce to be exerted upon the piston.

IGt

Ignition trigger signal from the ECM – Toyota term.

IIA

Integrated Ignition Assembly – Toyota term.

Ignition module integral with the distributor.

IMA

Idle Mixture adjuster Honda, Rover term.

Impedance

Resistance to the flow of current and often used to describe the resistance of a voltmeter. A minimum 10 megohm impedance is recommended for instruments used to measure values in electronic circuits.

IMPH

Inlet Manifold Pre-Heater.

See Manifold heater.

Intake system

The components responsible for the intake of the air/fuel mixture. ie air filter, carburettor (where used), manifold and inlet valve.

Inductive permanent magnet pick-up

The pick-up is a permanent magnet and inductive coil wound around a pole piece. It is usually fixed securely in the distributor and radiates a magnetic field. The two most common types in current service are the pick-up limb or the annular coil.

The reluctor or trigger wheel is mounted on the rotating distributor shaft with a number of triggering lugs equal to the number of engine cylinders. As the rotating lug passes the pick-up a small alternating voltage is generated that is strongest as the pick-up and lug are exactly opposite. This voltage is the ignition trigger and is sent to the amplifier for amplification.

Injection system

See MPi & SPi systems.

Injector fouling

Build-up of deposits on the injector internal filter, or on the injector head, so that flow is reduced or disturbed – resulting in improper injector operation.

Insulator

A material that will not pass current readily, and therefore used to prevent electrical leakage.

Inst. panel

Instrument panel on the vehicle dashboard.

Intercooler

A device for cooling the air charge supplied to the engine from the TurboCharger. Cooler air is denser than hot air and so a greater volume of air is inducted into the engine. The greater the volume of air inducted, the greater will be the horsepower produced by the engine.

IS

Inertia switch.

ISC

Idle speed control.

See ISCV – Ford term.

ISC – BPA

Idle Speed Control – By-Pass Air solenoid – Ford term.

ISCV

Idle Speed Control Valve.

A gate or rotary valve that is actuated by the ECM to maintain the correct idle speed, no matter the load or temperature. The ISCV is also used to provide a higher idle during engine warm-up. Early versions of engines fitted with an ISCV may be adjustable, although generally this is not the case with later units.

The ISCV contains an electro-magnet to open a by-pass port, thus allowing a small volume of air to by-pass the throttle plate. This air may pass through a hose, or through a port in the inlet manifold. The ISCV is mounted in situ, thus allowing the by-pass air to pass through the body of the valve. As the temperature decreases, or the load increases, the ECM pulses the valve for longer intervals (the pulse width increases) and the valve is further opened to allow more air to by-pass the throttle.

This results in a higher idle speed when the engine is cold, or no drop in idle speed when the engine is hot.

Early Bosch systems used an ISCV attached to an electric motor that can be rotated clockwise or anti-clockwise by virtue of two opposing circuits. The motor is supplied with a voltage supply and two earth paths that are made through the ECM. When one path is earthed, the motor will attempt to rotate in one direction. When the other path is earthed, the motor will attempt to rotate in the opposite direction. By varying the time that each circuit is energised, the ECM will place the ISCV in the exact position required. A duty cycle for the time energised can be obtained on each of the ISCV earth terminals, or at the corresponding ECM terminal. The waveform viewed at each terminal on an oscilloscope is that of a square waveform. This type is characterised by three wires at the electrical multi-plug (a battery voltage supply and the two earth wires).

Later Bosch systems use a solenoid that is opposed by a strong spring. The solenoid is supplied with a voltage supply and one earth path, made through the ECM. When the ISCV is earthed by the ECM, the solenoid will overcome the spring force and open the ISCV. If the solenoid fails, it will normally fail-safe closed. However, even when closed, a small amount of air will travel through the valve to give a basic (but low) idle speed. The waveform viewed on an oscilloscope is that of a square waveform.

The longer the time that the ECM pulses (holds open) the ISCV, the further open it will become. Pulsing occurs many times a second (the frequency is about 110), and by varying the time that the circuit is energised, the ECM will place the ISCV in the exact position required. A duty cycle for the time energised can be obtained on the ISCV earth terminal, or at the corresponding ECM terminal.

Ford use an ISCV which is very similar in operation to the later Bosch type. However, the waveform viewed on an oscilloscope is that of a sawtooth waveform.

ISO

International Standards Organisation.

ITS

Idle Tracking Switch – Ford term.

IV PWR

Ignition voltage power – Ford term.

J

Jumper lead

A small electrical cable that is used to bridge a component or wire on a temporary basis.

J1930

SAE standard for acronyms describing electrical and electronic components. Adopted by Ford and Mazda in 1994, other VMs may soon follow.

K

KA PWR
Keep Alive PoWeR – Ford term.

KAM
Keep Alive Memory – Ford term.

Ford term for a dynamic memory in the EEC IV ECM. This memory retains soft faults and also the vehicle idle speed settings.

KCM
Knock Control Module – Ford term.

KDS
Kick-Down Switch – Ford term.

KEM
KE Module – Ford term.

KEMKE
Fuelling module – Ford term.

King HT lead
The cable that carries secondary voltage from the coil to the distributor cap. 'King lead' is the term in common use.

KNK
Knock signal from knock sensor – Ford term.

Knock
The spontaneous explosion of the remaining air/fuel charge in the combustion chamber when only a portion of it has burnt progressively. A direct result of excessive combustion chamber temperature. Known also as detonation.

Knock threshold
The moment during engine operation when the onset of knocking is imminent.

KOHMS
Kilohms.

A resistance measurement equal to 1000 ohms. Many DMMs and engine analysers give a measurement in k.Ohms.

kPa
KiloPascals.

International standard for the measurement of pressure and vacuum.

See Pressure conversion table and Vacuum conversion table.

KS
Knock Sensor.

A sensor that outputs a small electrical signal on detecting 'engine knock'. On receiving a knock signal, the ECM will temporarily adjust (retard) the ignition timing to prevent the condition. Some engine systems with KS can detect engine knock in an individual cylinder. Timing for that cylinder alone will be retarded by the ECM until knock ceases.

KV
KiloVolt.

A unit of secondary voltage measurement equal to 1000 volts.

L

LAF
Linear Air Flow sensor.

Honda
Digital OS.

Lambda
Greek word for the 'stoichiometric symbol'.

As the engine operates, fuel and air are mixed together and drawn into each cylinder. The AFR at which fuel burns most efficiently is called the Stoichiometric point, and this is where HC and CO are lowest and CO_2 is highest. This ratio is 14.7:1 by weight, and it is also called Lambda = 1 which is the Greek word for correct.

A catalyst equipped engine will attempt to maintain the AFR between a Lambda factor of 0.97 and 1.03.

Although Lambda = 1 is not the best point for best fuel consumption, we have already established that it is the best compromise ratio for using a catalytic converter to oxidise CO, HC and NOx. Therefore, if the engine's AFR can be contained within the 'window' of 0.97 to 1.03, the resultant engine emissions will raise the efficiency of the catalytic converter to about 95 %. The lower the emissions from the engine, the less work the cat has to do, and the more efficient it will become. Moreover, by reducing the engine emissions, the cat will also last much longer.

Lambda sensor
See Oxygen Sensor.

A sensor that monitors the amount of oxygen in the exhaust stream, and passes a voltage signal back to the ECM. The ECM then alters the amount of fuel passed to the engine. An attempt is therefore made to keep the AFR at the most suitable ratio for perfect combustion.

Lb/in²
Pounds per Square Inch.

An Imperial measurement of pressure.

See Pressure conversion table under Pressure.

LDT
Light Duty Truck.

See Emission standards.

Refers to the US88 LDT emission standards for commercial vehicles.

Lead
A substance (tetra-ethyl or TEL) that is added to petrol to assist the fuel's ability to resist knocking and pre-ignition. Lead also lubricates the valves and seats in the engine. Lead levels in petrol have gradually been reduced in recent years and even leaded petrol contains a far smaller concentration than at one time.

Lead is a poisonous substance, and progressively and irreversibly reduces the efficiency of the blood to transport oxygen. It functions as cellular poison for blood, bone marrow and nerve cells. Lead also poisons the catalytic converter and clogs the cells thus quickly reducing efficiency.

LED
Light Emitting Diode.

LHS
Left Hand Side.

Viewed from the driver's seat.

Limp home
See LOS.

LOS
Limited Operating Strategy.

Often called LIMP HOME, this is a safety system that allows the vehicle to be driven to a service area if a fault occurs. Some LOS systems are so smart that the driver may be unaware from the way that the vehicle operates, that a fault has occurred.

When the system perceives that a sensor is operating outside of its design parameters, a substitute value is used which allows the engine to operate. However, this value is usually that for a hot or warm engine, and this means that the engine may be difficult to start, and run badly, when it is cold.

The instrument panel warning light (where fitted) may illuminate to indicate that a fault has occurred.

Some systems (for example Ford) may also lock the timing to a set figure (with no timing advance) and allow the fuel pump to run continuously.

LT
Low tension.

Primary ignition circuit.

LUS
Lock-up Solenoid (clutch of automatic transmission) – Ford term.

M

mA
Milli-amperes.

MAF
Mass Air Flow sensor – term for hot-wire AFS.

Magnet
A substance that has the ability to attract iron.

Magnetic field
The space around a magnet that is filled by invisible lines of magnetic force.

Manifold heater
The inlet manifold heater is sometimes termed a 'hedgehog' because of its distinctive shape. The MH may be found in many single point injection engines or some carburettor equipped engines and usually functions on the PTC principle.

The voltage supply to the MH is often made through a thermal switch or a relay when the engine coolant is cold. As the engine coolant rises above a pre-determined temperature, the thermal switch or relay opens and the voltage supply is to the MH is cut. The switch is usually placed in a coolant hose, or located in the coolant passage of the inlet manifold.

MAP table

Atmospheric condition	Pressure	Vacuum	MAP
Engine off/ign on	*1.0 ± 0.1*	*Zero*	*1.0 ± 0.1*
Idle speed	*1.0 ± 0.1*	*0.72 to 0.45*	*0.28 to 0.55*
High load (wide-open throttle)	*1.0 ± 0.1*	*Zero*	*1.0 ± 0.1.*
Deceleration	*1.0 ± 0.1*	*0.80 to 0.75*	*0.20 to 0.25.*

All units are in bars and typical rather than definitive.

Refer to the Vacuum conversion table and the Pressure conversion table for conversion to/from other units.

Note: Atmospheric pressure – vacuum = MAP.

MAP

Manifold Absolute Pressure sensor.

This is an inexpensive and less accurate alternative to the AFS. The MAP sensor measures the manifold vacuum or pressure, and uses a transducer to pass an electrical signal back to the ECM. The unit may be located in the engine compartment or in the ECM. Used in both simultaneous MPi and SPi systems, the MAP sensor is particularly popular in SPi systems. MAP is calculated from the formula: Atmospheric pressure – vacuum = MAP. Refer to the MAP table.

Where the manifold is of the 'wet' type (ie SPi), the changing pressures in the manifold will cause fuel to enter the vacuum hose where it may eventually reach the MAP sensor. Installation of a fuel trap and careful routing of the vacuum hose may slow down the ingress of fuel. However, once fuel reaches the MAP sensor, its diaphragm may be adversely affected. If the MAP sensor is a separate unit, renewal is comparatively inexpensive.

MAP sensors may take one of two forms. Older generation vehicle use an analogue sensor where the voltage signal output is proportional to the load. A newer system that is fast gaining popularity is the digital type. Digital MAP sensors send a square waveform in the form of a frequency. As the load increases, the frequency also increases and the time in ms between pulses becomes shorter. An ECM will respond much faster to a digital signal, because an analogue to digital converter is unnecessary in this system.

Mapped ignition timing or injection

Electronic timing advance or injection pulse that is controlled by the ECM from a 'map' located within the ECM. A two dimensional map contains settings for a number of engine load and speed variations. A three dimensional map contains settings for a number of engine load, speed and temperature variations. The timing and injection settings are usually contained in separate maps within the ECM.

Max.

Abbreviation for maximum.

Mechanical advance

A unit that uses centrifugal action to advance and retard the ignition timing. Usually mounted in a distributor in conventional engines.

MEMS

Modular Engine Management System.

A type of EMS manufactured by Rover.

MH

Manifold Heater.

See Manifold heater.

MHR

Manifold Heater Relay.

Mixture adjustment

This is a trimming device and only a small change in CO at the idle condition is generally possible. Once the throttle is opened from its idle stop, the quantity of injected fuel depends upon the pulse duration. Where CO adjustment is possible, there are two methods of CO adjustment in current use. Catalyst equipped vehicles often have no form of CO adjustment.

1) *An air screw that varies the air flow through an idle passage in the AFS. As the screw position is varied, the airflow acting upon the metering flap varies, causing the flap to change its idle position. The changed position results in an altered voltage signal to the ECM, and the idle mixture will either change appropriately. This type is usually fitted to older type vehicles.*
2) *A potentiometer with a variable resistance. As the adjustment screw is turned, the resistance varies, causing a change in the voltage signal being returned to the ECM. This sensor may be mounted upon the ECM, on the airflow meter, or even upon the inner engine compartment wing.*

Mixture screw

See Volume screw.

Molecule

The smallest particle into which a chemical compound may be divided.

Motronic

A type of EMS manufactured by Bosch.

MPi

Multi-Point Injection.

One injector per cylinder. May be triggered in banks (simultaneous) or sequentially.

MPi

Multi-Point Injection – simultaneous.

This is the most common type of EFi system in current use. A number of injectors are looped together in a parallel 'bank' with a single connection to the ECM. Where an engine has more than one bank, each bank has its own ECM connection.

In a 4-cylinder engine, one bank connects all of the injectors. In a 6-cylinder engine the injectors are placed in two groups of three, and in an 8-cylinder engine the injectors are placed in two groups of four termed left and right bank. In a 12-cylinder engine the injectors are placed into four groups of three cylinders. Two power resistors control two groups each.

The injectors are triggered from a reference signal which may originate from the ignition system, or from a timing pulse at the CAS. Normally, the injectors operate twice per complete engine cycle. Half of the fuel required is injected onto the back of a closed inlet valve, waiting for it to open, and the other half is injected as the valve opens for the intake stroke. Once the valve has opened, the fuel enters the cylinder in the normal way.

This system is fairly effective and usually works quite well. It is also cheaper to develop than a full-blown sequential system, which makes it very popular amongst vehicle manufacturers.

MPi

Multi-Point Injection – sequential.

Eventually, both SPi and simultaneous MPi systems are likely to succumb to the sequential MPi system, where the injectors open in cylinder sequence. Emissions can be significantly reduced with this type – particularly if the engine suffers a mechanical or ignition problem. The sequential MPi system uses the same sensors as other injection systems. However, an additional sensor pinpoints the correct cylinder for the sequential system. In some instances this is a Hall-effect trigger located in the distributor.

ms

Millisecond.

1/1000 second (0.001 s).

MSTS-h

Microprocessor Spark Timing System – HES ignition – GM term.

MSTS-i

Microprocessor Spark Timing System – inductive ignition – GM term.

MT

Manual Transmission.

Multimeter

See DMM.

Multiplug

A connecting plug in the wiring harness. Often used to connect the harness to a sensor or actuator. In this book, the multiplug diagram shows the terminals of the harness connector. When back-probing the multiplug (or viewing the sensor connector terminals), the terminal positions will be reversed.

mV

Millivolt.

One millivolt = 1/1000 of a volt (0.001 V).

MY Model Year
Most VMs start manufacturing their models for the new year in the months leading up to the end of the preceding year. The actual date when manufacturing commences is usually termed the 'model year' date, and the year used is usually that of the following year. In the UK, the new model year starts soon after the registration letter changes on August 1st; many VMs would for example call a vehicle manufactured in August or September 1996 a '1997 model'.

N

nbv
Nominal battery voltage.

Nominally 12 volt, the voltage will vary under engine operating conditions:

Engine stopped: 12 – 13 volts.
Engine cranking: 9.0 to 12.0 volts.
Engine running: 13.8 to 14.8 volts.

NDS
Neutral Drive Switch – Ford term.

Ne
RPM signal from the PU coil Toyota term.

Nearside
Side nearest to the kerb on any vehicle – irrespective of whether LH or RH drive.

NEEC
New European Economic Community.

Newton (N)
An international unit of force that is independent of gravity. This unit was introduced because gravity varies in different parts of the world. The Newton is defined as the force required to accelerate a mass of 1kg at 1 metre per second per second. Newton units of force are measured as N/m^2 and called Pascal units. This unit is very small, and is measured in MPa (1,000,000 Pascals) or kPa (1,000 Pascals).

See also Pascal.

Nitrogen
An atmospheric gas.

Non-cat
Non catalyst. Vehicles without a catalytic converter.

Non-sinusoidal
Waveforms such as sawtooth (ie Ford ISCV), square, ripple etc.

Non-volatile memory
ECM memory that is able to retain information – even when the vehicle battery is disconnected.

NOx
Oxides of Nitrogen.

NOx is a poisonous gas formed due to high temperatures (exceeding 1300° C – 2500° F). and high compression. There are many different kinds of NOx (ie NO, NO_2, NO_3 etc) and they are all lumped together under the term 'NOx'. The N representing one Nitrogen atom, and Ox representing any number of Oxygen atoms.

The Nitrogen (N_2 content of air (an inert gas) passes through the combustion process unchanged until high temperature (over 1370° C) and pressures are reached. Under these conditions, Nitrogen and oxygen react to form Nitrogen Monoxide – sometimes called Nitric oxide – (NO). The breeding conditions for NOx are wide-open throttle, acceleration and high-speed cruising. When NO is combined with HC in the presence of strong sunshine, NO_2 (Nitrogen Dioxide), Ozone (O_3) and NO_3 (nitrogen nitrate) are the result. NO_2 is a light brown gas commonly called 'SMOG'. Unfortunately, NOx emissions reach their peak at Lambda = 1, the so-called perfect combustion point.

The Diesel engine, whilst producing low levels of CO and HC, has a poor record where NOx is concerned. NOx is a particular problem due to the high temperature and high pressures present in a Diesel engine combustion chamber. NOx causes irritation of eyes and respiratory organs and symptoms of poisoning. Inhalation over a long period leads to destruction of lung tissues.

One method of controlling NOx is to recycle a small amount of exhaust gas into the combustion chamber (EGR, or Exhaust Gas Recirculation.) This reduces combustion temperature (and power) by recycling the inert exhaust gas.

NTC
Negative Temperature Co-efficient.

A thermistor in which the resistance falls as the temperature rises. An NTC resistor decreases in resistance as the temperature (ie coolant temperature) rises.

O_2
Oxygen.

A harmless gas that is present in about 21% of air and is necessary for proper combustion.

O_2 consists of two oxygen atoms and is measured in % volume. A small proportion of oxygen (1 – 2%) will be left after proper combustion. Too much or too little would indicate an incorrect air / fuel ratio, ignition or mechanical problems or an exhaust leak.

The amount of O_2 that is expelled into the exhaust is that which is left over after combustion, and is a good indicator of the AFR – so long as the engine is operating correctly.

OA
Octane Adjuster.

A device to finely tune the engine timing for fuels of differing octane levels.

OAI
Octane Adjust Input – Ford term.

Octane level
The level of fuel resistance to knock. The higher the octane level the more resistance to knock.

OHC
Over Head Camshaft.

Ohm
A unit of resistance that opposes the flow of current in a circuit.

Ohmmeter
An instrument that is used to measure resistance in ohms.

Ohms Law
Volts = Amps x Ohms (V = I x R)

Amps = Volts/Ohms (I = V / R)

Ohms = Volts/Amps (R = V / I)

Also:

Power (Watts) = Volts x Amps

Open circuit
A break in an electrical circuit which prevents the flow of current.

Open loop control
When an engine with Lambda control is operating outside of that control, it is in 'open loop'. This may occur during acceleration, wide-open throttle, or during the warm-up period, and when in LOS. some system may go into open loop at idle. When the system is under 'open loop' control, a richer mixture is allowed to prevent hesitation or poor driveability.

Optical distributor
Alternative CAS that utilises LEDs. Mainly used on some Japanese and other Far Eastern vehicles. Refer to Nissan for a detailed description of operation.

OS
Oxygen Sensor.

Also refer to Lambda.

An oxygen Sensor is a ceramic device placed in the exhaust manifold on the engine side of the catalytic converter for measuring the amount of oxygen that is left in the exhaust after combustion.

Essentially, the OS contains two porous platinum electrodes. The outer surface electrode is exposed to exhaust air and coated in porous ceramic. The inner surface electrode is exposed to ambient atmospheric air.

The difference in oxygen at the two electrodes generates a voltage signal which is transmitted to the ECM. This voltage is inversely proportional to the amount of oxygen The quantity of oxygen remaining after combustion is an excellent indicator of a deficit or surplus of air (rich or weak mixture). The oxygen sensor measures the surplus or deficit of air and sends a signal back to the ECM, which almost instantaneously adjusts the injection duration (within 50 milliseconds). By controlling the engine electronically so that the air/ fuel ratio is always in a small window around the Lambda point (ie Lambda = 0.97 to 1.03), during most operating conditions, almost perfect combustion could be achieved. Thus the catalyst has less work to do and it will last longer with fewer emissions at the tail pipe.

The oxygen sensor closed loop voltage is quite low and switches between 100 mVolts (weak) to 1.0 volt (rich). The signal actually takes the form of a switch and switches from weak to rich at the rate of approximately 1 Hz.

Various names have been given to this sensor, and it could equally be called a lambda sensor, oxygen sensor or even an EGOS.

OS heater
Because the sensor does not operate efficiently below about 300° C, many Oxygen sensors incorporate a heater element for rapid warm-up. Such sensors may also be termed HEGOS.

Oscilloscope
A high speed voltmeter that visually displays a change in voltage against time. Used to display ignition, alternator and engine sensor or actuator waveforms.

OTS
Oil Temperature Sensor.

Overrun
See deceleration.

OVP
Over Voltage Protection – Ford term.

Oxidation
A chemical change in a lubricating oil caused by combustion, heat and oxygen.

Oxides of nitrogen
Refer to NOx.

P

Parade
An oscilloscope pattern where all cylinders are displayed in line.

Pascal
International standard for the measurement of pressure and vacuum.
See Pressure conversion table and Vacuum conversion table.
Refer also to Newton.

PA
Pressure atmospheric Honda, Rover term.

PAS
Power Assisted Steering.

PCS
Pressure Control Switch – Ford term.

PCV
Positive Crankcase Ventilation.
A control system to recycle crankcase fumes into the intake system for burning in the combustion chamber.

Percent
Parts of a hundred.

Permanent magnet
A magnet that has a magnetic field at all times. Compare 'Electro-magnet'.

Petrol
A hydrocarbon based fuel composed of a mixture of hydrogen and carbons.

Pick-up
See also Inductive.
Used as a trigger in an electronic system. The pick-up generates a small voltage which signals the amplifier or ECM to switch off and thus instigate ignition. The pick-up is usually some form of permanent magnet fixed in the distributor or on the flywheel. When a reluctor is rotated in the magnetic field, the signal to switch occurs when the signal is at its strongest.

Pick-up air gap
A clearance between the reluctor and pick-up, which is often adjustable.

PIM
MAP sensor signal – Honda/Toyota term.

Pinging
The audible sound produced by pre-ignition/detonation.

Pinking
A commonly used aberration of pinging.

P / N
Park Neutral switch.
A switch to cut the electrical supply to the starter motor and so prevent the engine from being started in gear, most often found on vehicles with automatic transmission. Sometimes called a 'Starter Inhibitor' switch.

PIP
Profile Ignition Pick-up – Ford term.
Ford term for the basic timing signal.

Plugged exhaust
An exhaust blockage causing back pressure and lack of performance. Can occur in catalyst equipped vehicles when the catalyst exceeds its normal operating temperature and melts thus obstructing the exhaust system.

Polarity
A positive or negative state with reference to two electrical poles.

Pollutants
See Emissions.

Ported vacuum
A vacuum source located in front of the throttle valve. The valve must be opened before a vacuum signal is produced.

Pot
Potentiometer .
A variable resistance.

Power balance
For an engine to give maximum power then each cylinder must contribute an equal share to the workload. By cutting the spark to each cylinder in turn, and noting the rpm drop, it is possible to measure the amount of work, or balance, that each cylinder is contributing to the overall power. A weak cylinder will drop less in rpm than a strong one.

PPM
Parts Per Million.
A measurement value of unburned HC.

Pre-emission
Engines that do not have emission control devices.

Pre-ignition
The premature explosion of the compressed air/fuel charge before proper ignition by the spark plug. Usually caused by excessive combustion temperature.

Pressure conversion table

bar	*lb/in²*	*kPa*
0.1	1.45	10
0.2	2.90	20
0.3	4.35	30
0.4	5.80	40
0.5	7.25	50
1.0	14.50	100
1.02	14.75	102 *
1.1	15.95	110
1.2	17.40	120
1.3	18.85	130
1.4	20.30	140
1.5	21.75	150
1.6	23.20	160
1.7	24.65	170
1.8	26.10	180
1.9	27.55	190
2.0	29.00	200
3.0	43.50	300
4.0	58.00	400
5.0	72.50	500

* *Approximate atmospheric pressure at sea level.*

Pressure regulator
The fuel pump supplies fuel at a pressure that exceeds the required system pressure. A spring-loaded diaphragm relieves this pressure by allowing excess fuel to flow back to the tank via the fuel return line.

Primary ignition circuit
The low-voltage circuit required to begin the ignition process. Components are the ignition switch, ballast resistor (if fitted), ignition coil, distributor (if fitted), amplifier (if fitted), CB points & condenser (if fitted), distributor cap & rotor (if fitted) and the wiring cables between these components.

Primary switching
Primary windings. The outer windings of relatively heavy wire in an ignition coil in which the primary current flows.

Probe
A method of obtaining voltage from the multiplug pin of an electronic component or sensor. The multiplug should be disconnected from the component and the voltmeter positive test lead used to probe the relevant pin.

Primary trigger
This is the speed and position sensor that signals the ECM to begin actuating ignition, injection and relay control. The ECM will not function without the primary trigger. Examples of primary triggers include the CAS, inductive trigger or the Hall Effect sensor.

PROM
Programmable read only memory.

PS
Phase Sensor – Ford term.

PSA
Citroën and Peugeot group.

PSI
Pounds per Square Inch.
An Imperial measurement of pressure.
See Pressure conversion table under Pressure.

PSPS
Power Steering Pressure Switch – Ford term.

PU
Inductive pick-up coil.

PUA
Pulse Air solenoid – Ford term.

PTC
Positive Temperature Co-efficient.
A thermistor in which the resistance rises as the temperature rises. A PTC resistor increases (positively) in resistance as the temperature (ie coolant temperature) rises.

Pulse
A digital signal actuated by the ECM.

Pulse generator
The pulse generator is a trigger used to initiate ignition. It sends a correctly timed signal to the amplifier, which then amplifies the signal to switch the coil negative terminal. Examples of pulse generators are:

1) *An inductive permanent magnet pick-up located inside the distributor.*
2) *An inductive permanent magnet located adjacent to the flywheel (CAS).*
3) *Hall Effect trigger located inside the distributor.*

Pulse width
The time period during which electronic components (especially fuel injectors) are energised. It is usually measured in milliseconds.

Purge valve
Refer to CFSV.

PVS
Ported Vacuum Switch (valve).

R

RAM
Random Access Memory – computer term.

Raster
Display of all cylinders on an oscilloscope, one below the other in firing order beginning with number one. The order may be from the top down or bottom up, depending on the 'scope.

Reference voltage
During normal engine operation, battery voltage could vary between 9.5 (cranking) and 14.5 (running). To minimise the effect on engine sensors (for which the ECM would need to compensate), many ECM voltage supplies to the sensors are made at a constant value (known as a reference voltage) of 5.0 volts.

REG
Regulator.

Relay
An electro-magnetic switching solenoid controlled by a fine shunt coil. A small current activates the shunt winding, which then exerts magnetic force to close the relay switching contacts. The relay is often used when a low current circuit is required to connect one or more circuits that operate at high current levels. The relay terminal numbers are usually annotated to the DIN standard, to which most (but not all) European VMs subscribe.

Relays – EFi system
One system relay may be used to control the whole fuel injection system. In that instance the relay will have double contacts. Alternatively, two or more relays may be used to control the system.

Relay typical annotation to DIN standard

30 *Supply voltage direct from the battery positive terminal.*
31 *Earth return direct to battery.*
85 *Relay earth for energising system. May be connected direct to earth, or 'driven' to earth through the ECM.*
85b *Relay earth for output. May be connected direct to earth, or 'driven' to earth through the ECM.*
86 *Energising system supply. May arrive from battery positive or through the ignition switch.*
87 *Output from first relay or first relay winding. This terminal will often provide power to the second relay terminal 86 and provide voltage to the ECM, injectors, ISCV etc.*
87b *Output from second relay or second relay winding. Often provides power to the fuel pump and OS.*

Relay control/relay driver
The system relays are supplied with a voltage feed from either the battery, ignition switch or from another of the EFI system relays. The earth connection is then connected to an ECM earth pin. When the ECM actuates the relay, it drives the appropriate ECM pin to earth by completing the circuit internally for as long as the actuation is required. In general, the relay earth connection will only be completed once the ECM receives a pre-determined sensor input signal.

Depending upon the relay, the input signal may be instigated by switching on the ignition or cranking the engine (ie CAS signal). Once the ECM has received the signal, the ECM will 'drive' the relay to earth by completing the circuit internally. The signal could be termed a 'driver' or a 'final stage' or a 'control' signal. In the this book we have settled for the 'driver' term. Examples of other actuator drivers are: injector, ISCV, CFSV etc.

Reluctor
A metal rotor with a series of tips equal to the number of cylinders.

REMCO
Remote adjustment for CO pot – Ford term.

Remote start
A device to operate the starter solenoid directly from under the bonnet and which enables more efficient turning or inching over of the engine.

Renix
A type of EMS used mainly on Renault and Volvo vehicles.

Required voltage
The minimum amount of secondary voltage that must be produced to bridge the rotor and spark plug gaps.

Res.
Abbreviation for resistance.

Resistance
Opposition to the flow of current.

Retarded timing
Opposite to advance. When the ignition timing fires AFTER the correct moment. Can also be used to describe ignition timing that occurs ATDC.

Return
Term used to describe the earth return path to an ECM or module of typically a sensor, CAS or relay when the return is not directly connected to earth. The ECM or module will internally connect the return to one of its own earth connections. By this method the number of earth connections is much reduced.

RFI
Radio Frequency Interference.
The EMS is susceptible to outside interference. Radiated RFI can be a problem if the levels are high enough and this can emanate from items such as a faulty secondary HT circuit or a faulty alternator. Excess RFI can disrupt and affect ECM and EMS operation – particularly where both ignition and fuelling are located in the same ECM.

RHS
Right Hand Side.
Viewed from the driver's seat.

RMS
Root mean square.
AC equivalent to DC voltage.
Can be calculated from AC amplitude by the formula:
AC amplitude x 0.707.

ROM
Read Only Memory – computer term.

Rotor
Rotating part of a component such as a rotor arm or an electro-magnet used in an alternator.

Rotor air gap
The space between the rotor tip and the distributor cap terminal.

Rotor arm
The rotor is an electrical contact, keyed to the distributor shaft so that it points directly at the correct distributor cap terminal when a plug is fired.

Rotor register
The alignment of the rotor tip to the distributor cap terminal. Where the register is misaligned, the resulting large air gap will cause high firing voltages.

RPM
Revolutions per minute.
A measure of engine speed.

RSS
Remote starter switch.

Rubbing block
The part of the CB that rubs against the distributor cam. When the cam peak touches the rubbing block the points will open to instigate ignition.

S

SAE
Society of Automotive Engineers.
The Society sets standards for automotive engineering. See also BCI and J1930.

SAW
Spark Advance Word.
A Ford term for the modified timing signal passed from the EEC IV ECM to the EDIS module.

Scanner
See FCR US term for a FCR.

Scope
See Oscilloscope.
Abbreviation for an oscilloscope.

SD
Self Diagnostics.

Secondary ignition circuit
The high voltage circuit used to distribute secondary voltage to the spark plug.

Secondary voltage
Output from ignition coil.

Secondary windings
Ignition coil HT windings.

SEFI
Sequential Electronic Fuel Injection – Ford term.

Self-diagnosis of serial data
See Fault codes.

Sensor
A device that can measure one or more of the following parameters: temperature, position, airflow, pressure etc. and returns this information to the ECM, in the form of a voltage or current signal, for processing by the ECM.

Sequential injection
See MPi systems.

Serial data port
The serial port is an output terminal from the ECM. Signals have therefore been processed and faults or values are output to the terminal as a coded digital signal.

SG
Signal Generator.
Distributor pick-up coil.

Short
Short to earth, or short circuit. When electricity goes to earth and takes a shorter path back to the power source. Because extremely high current values are present, the condition can cause an electrical fire.

Signal generator
See Pulse generator.

Signal voltage
A varying voltage returned to the ECM by a sensor so that the ECM can detect load, or temperature.

Simultaneous injection
Refer to MPi systems. An injection system in which all the injectors are pulsed simultaneously (all fire at the same time).

Sinusoidal
A sine wave, ie a CAS or inductive pick-up waveform where the amplitude of the positive part of the waveform is roughly equal to the amplitude of the negative part of the waveform.

Slow codes
Fault codes emitted by an EMS that are slow enough to be displayed on an LED lamp or on a dash mounted warning lamp.

Smog
So-called 'photo-chemical smog'. Formed by combining HC and NOx in the presence of strong sunshine. A particular problem in car-dense, sunny climates such as California in the USA.

Soft faults
Generally refers to intermittent faults logged by an ECM self-diagnosis routine. The faults are often not present at the moment of testing, but have been logged at some period in the past.

SOHC
Single Over Head Camshaft.
A single rotating camshaft that controls the opening and closing of both inlet and exhaust valves. The camshaft is mounted above the valves in the cylinder head and acts directly upon them.

Solenoid
An electrical device that produces a mechanical effort when energised.

SP
Sensor phase – Ford term.

Spark advance
See Timing advance.

Spark control
Spark advance control by electronic or thermostatic means.

Spark duration
The time taken for a spark to bridge the spark plug electrodes. Shown as a spark line on an oscilloscope.

Spark line
See Spark duration.
Also known as burn time.

Spark plug
A device screwed into the cylinder head for igniting the compressed air / fuel charge.

Spark plug electrodes
1) The centre rod passing through the spark plug insulator.
2) The 'earth' rod welded to the outer shell.

Spark plug gap
The spark plug electrodes can be adjusted to an exact gap for the spark to jump across.

SPi
Single Point injection.
Sometimes known as throttle body injection (TBI), the SPi system has gained much popularity over recent years. Essentially less costly, SPi uses the same sensors as the MPi systems. A single injector (normally of the current controlled type) injects fuel into a distributing manifold in much the same fashion as a carburettor.

Although the injection of fuel is much more precise, the problems of manifold heating becomes critical and the warm-up period must be carefully controlled if driveability is not to be impaired. Furthermore, the manifold is said to be of the 'wet' type. This term means that fuel is present in the manifold. An MPi system is said to be of the 'dry' type because fuel is injected into the inlet stub to the valve, and thus only air is present in the manifold.

SPOUT
Spark Out. A Ford term for the modified timing signal passed from the EEC IV ECM to the TFI module.

Square waveform
A waveform that illustrates the switching on and off of a particular circuit. The higher voltage line at supply voltage & the lower voltage at earth potential. The transitions is straight and the distance between the transitions defines the time of 'switch on' and 'switch off'.

STA
Starter motor signal – Toyota term.

STAR
Self Test Automatic Readout (electronic FCR test) – Ford term.

Ford scanner or FCR.

Starter motor
An electrical motor that rotates the engine to starting speed.

Static ignition
This term is often used by European VMs to describe a Direct Ignition System.

Stator
Used in electronic ignition or an alternator. As the rotating reluctor and stationery stator become opposite then AC voltage is induced.

STC
Self-Test ConnTector – Ford term

Refer to Self Diagnosis.

Stepper motor
The stepper motor may take several forms and here are described the two most common systems.

1) *A motor is used to drive a valve which opens or closes an air by-pass passage in the inlet manifold.*
2) *A motor is used to increment the throttle plate by so many steps, thereby allowing more or less air through the opening.*

Usually the motor is supplied with nbv from the fuel system relay. The windings of the motor are then connected to four earth paths. By pulsing the motor using a combination of the earth paths, the ECM can step the motor to the correct position.

STI
Self-test Input – Ford term.

STO
Self-test Output – Ford term.

Stoichiometric ratio
The point at which fuel burns most efficiently is called the Stoichiometric ratio and this is where HC and CO are lowest and CO´is highest. The air / fuel ratio by weight at this point is approximately 14.7:1. ie 14.7 kilograms of air to 1 kilogram of fuel.

Strobe
Abbreviation for stroboscope.

Stroboscopic
A light that flashes in unison with the number one spark plug, giving the impression of a 'frozen' rotating timing mark.

Sulphation
Slow formation of a hard insoluble compound upon the lead battery plates during discharge. Can be reduced by long slow charge. A heavily sulphated battery will reject a charge.

Superimposed
An oscilloscope display pattern where all cylinder traces are placed on top of each other. Differences between the various cylinders will tend to 'stick out'.

Suppression
Reduction of radio or television interference generated by the high voltage ignition system. Typical means used are radio capacitors or resistive components in the secondary ignition circuit.

Suppresser
Used to prevent radio interference. See capacitor.

SVC Service Connector
octane/ idle adjustment – Ford term.

Synchronised
Usually refers to an injection pulse that is synchronised with the ignition system. The injector will be pulsed at a pre-determined time interval before ignition occurs.

System overview
A term to describe the technical description of how the system operates.

T

Tacho
Tachometer.

A device used to indicate engine speed in RPM.

Tachometric relay
A relay that requires a speed signal from the ignition to function.

TAD
Thermactor Air Diverter (vacuum solenoid valve) – Ford term.

TBH
Throttle Body Heater.

A PTC device that quickly warms-up the throttle area; thereby preventing ice from forming during engine operation at low and moist temperatures.

TBI
Throttle Body Injection.

See SPi.

TBV
Turbo Boost Valve.

TCATS
Turbo Charge Air Temperature Sensor.

TDC
Top Dead Centre.

Position of the piston at the top of its stroke.

TDCL
Total Diagnostic Communication Link.

Used for extraction of fault codes on some Toyota vehicles.

Temp
Abbreviation for temperature.

Temperature conversion table

°C	value	°F
-17.8	*zero*	*32*
-17.2	*1*	*33.8*
-15	*5*	*41.0*
-12.2	*10*	*50.0*
-9.4	*15*	*59.0*
-6.7	*20*	*68.0*
-3.9	*25*	*77.0*
-1.1	*30*	*86.0*
zero	*32*	*89.6*
4.4	*40*	*104.4*
7.2	*45*	*113.0*
10.0	*50*	*122.0*
12.8	*55*	*131.0*
15.6	*60*	*140.0*
18.3	*65*	*149.0*
21.1	*70*	*158.0.*
23.8	*75*	*166.6*
26.7	*80*	*176.0*
29.4	*85*	*185.0*
32.2	*90*	*194.0*
35.0	*95*	*203.0*
37.8	*100*	*212.0*
40	*105*	*221*
43	*110*	*230*
46	*115*	*239*
49	*120*	*248*
52	*125*	*257*
54	*130*	*266*
57	*135*	*275*
60	*140*	*284*
63	*145*	*293*
66	*150*	*302*
68	*155*	*311*
71	*160*	*320*
74	*165*	*329*
77	*170*	*338*
79	*175*	*347*
82	*180*	*356*
85	*185*	*365*
88	*190*	*374*
91	*195*	*383*
93	*200*	*392*
96	*205*	*401*
99	*210*	*410*
102	*215*	*419*
149	*300*	*572*
204	*400*	*752*
260	*500*	*932*
316	*600*	*1112*
371	*700*	*1292*
427	*800*	*1472*
482	*900*	*1652*
538	*1000*	*1832*
743	*1370*	*2500*
1206	*2202*	*4000*

Conversion formula

°C x 1.8 + 32 = °F

°F – 32 x 0.56 = °C

Glossary

Terminal
An electrical connecting point.

TFI

Thick Film Ignition
Ford term for ignition module.

THA
Air Temperature Sensor – Toyota term, see ATS.

Thermistor
A potentiometer controlled by temperature.

Three wire sensor
The three wire sensor has a voltage supply of 5 volts, an earth connection (often made through the ECM) and an output (signal) wire. The output wire returns a variable voltage signal to the ECM. The two most common forms of output are by resistance track and wiper arm, or via a transducer. Examples would include the AFS and TPS (wiper arm) and MAP (transducer).

Throttle valve
A valve that controls the volume of airflow into the engine. Sometimes known as throttle plate or throttle disc.

Throttle valve positioner
VAG term.
See Stepper motor.

THS 3/4
Transmission Hydraulic Switch (3rd/4th gear solenoid) – Ford term.

THW
Coolant Temperature Sensor – Toyota term.
See CTS.

Timing advance
As engine speed increase, combustion must occur earlier so that a correctly timed maximum force is exerted upon the piston.

Timing belt or chain
A belt or chain that connects crankshaft and camshaft.

Timing light
A stroboscopic light used to check and set ignition timing.

Timing marks
Two marks, or a scale and a mark, to indicate TDC or the timing point when aligned. These marks may be on the timing case and front pulley, or on the flywheel and viewed through an inspection hatch.

TP
Throttle plate.

TPS
Throttle Potentiometer Sensor.
The throttle pot is a potentiometer that sends a variable voltage signal to the ECM to indicate (depending on system), throttle position, idle mode, WOT and rate of throttle opening. It is adjustable on some models. Additionally, the TPS may be used in conjunction with a TS. If so, the TS will indicate idle position and a non-adjustable Throttle pot will only be used to indicate rate of opening.

Transducer
A device that converts pressure or vacuum etc into an electrical signal. ie Manifold vacuum may be taken to a transducer which turns it into an electrical load signal.

Transistor
An electronic switching component.

Trigger
See Pulse generator.

Trigger wheel
See Reluctor.

Trouble codes
US term for fault codes.

TS
Throttle Switch.
The throttle switch informs the ECM when the engine is in idle mode. An additional contact may indicate wide-open throttle. Additional enrichment may be provided at idle and during full throttle running. The TPS is adjustable on some models. Some systems may support both a TS and a TPS, although most systems support only one of either type.

TSS
Throttle stop screw.

TSS
Turbo Speed Sensor.

TTS
Thermo Time Switch.
A switch dependant on time and temperature.

Turbocharger
An exhaust gas driven compressor that compresses the air inducted by the engine to increase the horsepower for any given engine capacity.

TVS
Thermal Vacuum Switch.
Used to control vacuum according to engine temperature. Mainly used in carburettor systems.

TVSV
Thermostatic Vacuum Switching Valve.
See VSV.

Two wire sensor
The two wire sensor utilise an earth wire and a 5 volt supply wire in a circuit that begins and ends at the ECM. The supply wire also doubles as the output wire in the following manner. Once the supply and earth wires are connected to the sensor, the resistance value of the sensor causes the voltage value of the supply to vary. Thus, if we take an example of a two wire CTS, the supply value of 5 volts will reduce (typically) to between two and three volts if the engine is cold (20° C), and to between 0.6 and 0.8 volts once the engine has become warm (80° C). Examples of two wire sensors would include the ATS and CTS.

U

UCL
Upper Cylinder Lubricant.

UESC
Universal Electronic Spark Control module – Ford term.

Unleaded petrol
A hydrocarbon fuel that is blended without the addition of lead. Even unleaded petrol contains a very small amount of natural lead that is not usually removed during refining. This amount is insignificant from an emissions viewpoint, and has no adverse effect upon the catalytic converter.

Unported vacuum
A vacuum source located on the manifold side of the throttle plate. A vacuum signal is produced irrespective of throttle valve position.

V

Vacuum
A negative pressure or a pressure less than atmospheric. Measured in millibars or inches of mercury. A perfect vacuum exists in a space which is totally empty. It contains no atoms or molecules and therefore has no pressure. In practice a perfect vacuum cannot be achieved.

A vacuum occurs in the inlet manifold of a petrol engine because the fall of the piston continually attempts to draw air into each cylinder at a greater speed than the partially closed throttle disc will allow air to flow past it. The level of vacuum will depend on engine speed and throttle opening. The lowest reading will occur with the engine on full load (wide open throttle), and the highest reading when the throttle is closed with the engine running at high rpms ie. when the engine is descending a hill.

Vacuum conversion table

in.Hg	mm.Hg	kPa	millibar	
0.5	*12.75*	*1.7*	*17*	
1.0	*25.395*	*3.386*	*33.86*	
1.003	*25.50*	*3.4*	*34*	
2.0	*51.00*	*6.8*	*68*	
3.0	*76.50*	*10.2*	*102*	
4.0	*102.00*	*13.6*	*136*	
5.0	*127.50*	*17.0*	*170*	
6.0	*153.00*	*20.4*	*204*	
7.0	*178.50*	*23.8*	*238*	
8.0	*204.00*	*27.2*	*272*	
9.0	*229.50*	*30.5*	*305*	
10.0	*255.00*	*34.0*	*340*	
11.0	*280.50*	*37.3*	*370*	
12.0	*306.00*	*40.8*	*408*	
13.0	*331.50*	*44.2*	*442*	
14.0	*357.00*	*47.6*	*476*	
15.0	*382.50*	*51.0*	*510*	
16.0	*408.00*	*54.0*	*544*	
17.0	*433.50*	*57.8*	*578*	*normal*
18.0	*459.00*	*61.2*	*612*	*engine*
19.0	*484.50*	*64.6*	*646*	*operating*
20.0	*510.00*	*68.0*	*680*	*range*
21.0	*535.50*	*71.4*	*714*	*at idle*
22.0	*561.00*	*74.8*	*748*	
23.0	*586.50*	*78.2*	*782*	
24.0	*612.00*	*81.6*	*816*	
25.0	*637.50*	*85.0*	*850*	
26.0	*663.00*	*88.4*	*884*	
27.0	*688.50*	*91.8*	*918*	
28.0	*714.00*	*95.2*	*952*	
29.0	*739.50*	*98.6*	*986*	
29.53	*750.00*	*100.0*	*1000*	
30.0	*765.00*	*102.0*	*1020*	

Note: in.Hg figures rounded to nearest whole number.

Vacuum gauge

A gauge used to measure the amount of vacuum in the engine intake system.

Valve timing

The timing of valve opening and closing in relation to the piston and crankshaft position.

VAF

Vane Air Flow – Ford term.

Refers to a particular type of AFS.

See Vane type AFS.

VAT

Vane air temperature sensor – Ford term.

Vane type AFS

As air is drawn through the sensor, the Vane or AFS door is pushed open. The vane is attached to a potentiometer that varies as the door position varies. A voltage signal that varies according to the position of the sensor door is thus returned to the ECM. The ECM is therefore able to compute an injection duration that relates to the actual volume of air being drawn into the engine. No recognition is made of density and this AFS is less accurate than the hot-wire / hot-film types. This sensor is an example of a three wire sensor.

Vb batt +

Voltage supplied from the ECM – Toyota term.

Vc

AFM reference voltage – Toyota term.

Vcc

PIM (MAP sensor reference voltage) – Toyota term.

Venturi

A restriction in the carburettor throat which results in a speeding up of the air flow.

Vf

Feedback voltage.

VIN

Vehicle Identification Number.

A serial number to identify the vehicle, The number often contains coded letters to identify model and year.

VM

Vehicle Manufacturer.

Volt

A unit of electrical pressure.

Voltage

Electrical pressure.

Voltage drop

Voltage drop is voltage expended when a current flows through a resistance. The greater the resistance then the higher the voltage drop. The total voltage drop in any automotive circuit should be no more than 10%.

Voltage regulator

A device use to limit the voltage output of a generator such as an alternator or dynamo.

Voltage reserve

The ignition system must provide sufficient secondary HT voltage to bridge the rotor and sparkplug gaps under normal operating conditions. In addition, an adequate reserve of coil voltage must be maintained to meet the greater demands made by the ignition system during conditions such as hard acceleration or high engine rpms. If at some point during engine operation the coil reserve becomes lower than the voltage demanded by the ignition, misfiring and loss of power will be the result. A low voltage reserve can be caused by poor ignition components (ie plugs, HT leads etc) or poor primary ignition connections.

Voltmeter

An instrument used to measure voltage in a circuit in volts.

Volume screw

A screw that regulates fuel in the idle circuit of a carburettor by gradually restricting the idle mixture channel.

VRS

Variable Reluctance Sensor – Ford term.

Vs

Variable signal from the AFM to the ECM – Toyota term.

VSS

Vehicle Speed Sensor.

A sensor to measure the road speed of the vehicle.

VSTP

Vacuum Solenoid Throttle Plate – Ford term.

VSV

Vacuum Switching Valve – Japanese vehicle manufacturer term.

VTEC

Variable Valve Timing and Electronic Control – Honda Term.

W

WAC

Wide-open throttle A/C cut-off.

WCS

Wastegate Control Solenoid – Ford term.

Watt

A unit of electrical power. 746 watts are equal to one mechanical horsepower.

Wiggle test

With the engine running, a suspect connection is wiggled, or gently tapped, or gentle heated or cooled. If the engine misfires or otherwise misbehaves, that connection may be suspect.

WOT

Wide Open Throttle.

Throttle position when fully open. Many EFi systems provide more fuel when this condition is met.

Warnings: Precautions to be taken with automotive electronic circuits

1 The electronic ignition high tension (HT) system generates a high secondary voltage. Care must be taken that no parts of the body contact HT components. Shock or injury may be caused by allowing the HT to pass to earth through the human body. DO NOT work on electronic vehicle systems if you have a heart condition or any form of heart pacemaker. Pacemaker operation can also be disrupted by radiated RFI (eg the alternator).

2 The ECM and other electronic components can easily be damaged by an open HT circuit. When HT is faced with an impossible gap to jump, it will look for an alternative path. This path may be via the ECM, and sensitive components such as transistors may be damaged. In addition, spurious electrical signals from the HT circuit or from other sources of RFI (eg the alternator) may disrupt ECM operation.

3 VERY IMPORTANT: Avoid severe damage to the ECM, or amplifier by switching the ignition OFF before disconnecting the multiplug to these components. It is generally safe to disconnect the multiplug to other sensors, actuators and components with the ignition switched on, or even with the engine running.

4 Many modern radios are coded as a security measure, and the radio will lose its coding and its pre-selected stations when the battery is disconnected. The code should be obtained from the vehicle owner before disconnecting the battery for renewal or to make other repairs.

5 When taking voltage readings at a multiplug or terminal block, the use of meter leads with thin probes is strongly recommended. However, it is useful to attach a paper clip or split pin to the terminal and attach the voltmeter to the clip. Be very careful not to short out these clips. A number of ECMs employ gold plated pins at the ECM multiplug. Particular care should be taken that this plating is not removed by insensitive probing.

6 DO NOT use an analogue voltmeter, or a digital voltmeter with an electrical impedance of less than 10 megohms, to take voltage readings at an ECM or AFS with the ECM in circuit.

7 To prevent damage to a DMM or to the vehicle electronic system, the appropriate measuring range should be selected before the instrument probes are connected to the vehicle.

8 During resistance tests with an ohmmeter, always ensure that the ignition is OFF and that the circuit is isolated from a voltage supply. Resistance tests should NOT be made at the ECM pins. Damage could be caused to sensitive components, and in any case results would be meaningless.

9 When removing battery cables, good electrical procedure dictates that the earth (negative) cable is disconnected before the live (positive) cable. This will prevent spurious voltage spikes that can cause damage to electronic components.

10 Use protected jumper cables when jump starting a vehicle equipped with an ECM. If unprotected cables are used, and the vehicle earth cables are in poor condition, a voltage spike may destroy the ECM.

11 When a battery is discharged, by far the best course of action is to recharge the battery (or renew the battery if it is faulty), before attempting to start the vehicle. The ECM is put at risk from defective components such as battery, starter, battery cables and earth cables.

12 Do not use a boost charger, nor allow a voltage higher than 16.0 volts, when attempting to start an engine. The battery leads should be disconnected before a boost charger is used to quick charge the battery.

13 All fuel injection systems operate at high pressure. Keep a fire extinguisher handy, and observe all safety precautions. Before loosening fuel banjos or fuel hoses, it is a good idea to de-pressurise the system.

14 A number of diagnostic procedures – such as engine cranking and power balance – may result in unburnt fuel passing into the exhaust, and this is potentially harmful to catalyst-equipped vehicles. Each test must be completed quickly, and back-to-back tests must not be attempted if damage to the catalytic converter is to be avoided. Do not therefore, make repeated cranking or power balance tests with catalyst equipped vehicles. Always run the engine at a fast idle for at least 30 seconds between such tests to clear the exhaust of the fuel residue. Where the engine is a non-runner, the catalyst may need to be disconnected to allow cranking to continue. If this advice is not followed the petrol in the catalyst may explode once the temperature in the exhaust reaches a certain temperature.

15 Catalyst damage can be caused when the catalyst temperature exceeds 900° C. When unburnt fuel passes into the catalyst, as the result of any engine malfunction or misfire, the catalyst temperature could easily pass the 900° C mark causing the catalyst to melt. Apart from catalyst destruction, the melted catalyst will often cause an exhaust blockage with loss of engine power.

16 Disconnect all ECMs when welding repairs are to be made upon a vehicle.

17 The ECM must not be exposed to a temperature exceeding 80° C. If the vehicle is to be placed in a vehicle spray booth, or if welding is to be carried out near the ECM, the ECM must be disconnected and removed from the car to a place of safety.

18 Compression test: Where possible, disable both ignition and injection systems before attempting a compression test. The above advice about avoiding catalyst damage should also be heeded.

19 The following precautions must be taken with vehicles that utilise Hall-Effect electronic ignition:

a) Do not connect a suppresser or condenser to the negative coil terminal.
b) If the electronic ignition is suspect, the HES connection to the distributor and to the amplifier should be disconnected before the vehicle is towed.
c) During engine cranking tests – compression test or otherwise – remove the HES connection to the distributor.
d) All other precautions, as detailed above, should also be taken.

20 Do not run the fuel pump or by-pass the relay when the fuel tank is empty; the pump or pumps will overheat and may be damaged.

21 All modern vehicles are now equipped with SRS (Supplemental Restraint System) which is an airbag assembly installed in the steering column or passenger compartment. Extreme caution must be exercised when repairing components situated close to the wiring or components of the SRS. In some vehicles, the SRS wiring runs under the dash, and related SRS components are situated in the steering wheel, in and around the under-dash area, and adjacent to some components used in the vehicle EMS. Any damage to the SRS wiring must be repaired by renewing the whole harness. Improper removal or disturbance of SRS components or wiring could lead to SRS failure or accidental deployment. Failure to observe these precautions can lead to unexpected deployment of the SRS and severe personal injury. In addition, the SRS must be repaired and serviced according to the procedures laid down by the manufacturer. Any impairment of the SRS could lead to its failure to deploy in an emergency, leaving the vehicle occupants